AF843766

PRINCIPLES AND APPLICATIONS OF SOIL MICROBIOLOGY

PRINCIPLES AND APPLICATIONS OF SOIL MICROBIOLOGY

THIRD EDITION

Edited by

TERRY J. GENTRY

Department of Soil and Crop Sciences, Texas A&M University, College Station, TX, United States

JEFFRY J. FUHRMANN

Department of Plant and Soil Sciences, Department of Biological Sciences, University of Delaware, Newark, DE, United States

DAVID A. ZUBERER

Department of Soil and Crop Sciences, Texas A&M University, College Station, TX, United States

ELSEVIER

Elsevier
Radarweg 29, PO Box 211, 1000 AE Amsterdam, Netherlands
The Boulevard, Langford Lane, Kidlington, Oxford OX5 1GB, United Kingdom
50 Hampshire Street, 5th Floor, Cambridge, MA 02139, United States

Copyright © 2021 Elsevier Inc. All rights reserved.

No part of this publication may be reproduced or transmitted in any form or by any means, electronic or mechanical, including photocopying, recording, or any information storage and retrieval system, without permission in writing from the publisher. Details on how to seek permission, further information about the Publisher's permissions policies and our arrangements with organizations such as the Copyright Clearance Center and the Copyright Licensing Agency, can be found at our website: www.elsevier.com/permissions.

This book and the individual contributions contained in it are protected under copyright by the Publisher (other than as may be noted herein).

Notices

Knowledge and best practice in this field are constantly changing. As new research and experience broaden our understanding, changes in research methods, professional practices, or medical treatment may become necessary.

Practitioners and researchers must always rely on their own experience and knowledge in evaluating and using any information, methods, compounds, or experiments described herein. In using such information or methods they should be mindful of their own safety and the safety of others, including parties for whom they have a professional responsibility.

To the fullest extent of the law, neither the Publisher nor the authors, contributors, or editors, assume any liability for any injury and/or damage to persons or property as a matter of products liability, negligence or otherwise, or from any use or operation of any methods, products, instructions, or ideas contained in the material herein.

Library of Congress Cataloging-in-Publication Data
A catalog record for this book is available from the Library of Congress

British Library Cataloguing-in-Publication Data
A catalogue record for this book is available from the British Library

ISBN: 978-0-12-820202-9

For information on all Elsevier publications
visit our website at https://www.elsevier.com/books-and-journals

Publisher: Candice Janco
Editorial Project Manager: Leticia M. Lima
Production Project Manager: Kumar Anbazhagan
Designer: Miles Hitchen

Typeset by Thomson Digital

ELSEVIER Book Aid International
Working together to grow libraries in developing countries
www.elsevier.com • www.bookaid.org

Contents

5. Bacteria and archaea

Mary Ann Bruns

6. Fungi

Joseph B. Morton

7. Cyanobacteria and algae

Ferran Garcia-Pichel, Jayne Belnap

8. The soil fauna

Josef H. Görres, Jose A. Amador

9. Viruses

Kurt E. Williamson

10. Microbial ecology

Stephanie A. Yarwood

11. Rhizosphere

Lindsey Slaughter

12. Mycorrhizal symbioses

Rhae A. Drijber, Morgan R. McPherson

13. Carbon transformations and soil organic matter formation

Jeffry J. Fuhrmann, David A. Zuberer

14. Applied aspects of soil carbon

Alan Joseph Franzluebbers

15. Transformations of nitrogen

David D. Myrold

16. Biological dinitrogen (N_2) fixation: introduction and nonsymbiotic

David A. Zuberer

17. Biological dinitrogen fixation: symbiotic

Janice E. Thies

18. Microbial transformations of sulfur in soil

V.V.S.R. Gupta, James J. Germida

19. Phosphorus and selected metals and metalloids

Shiping Deng

20. Global gases

Wendy H. Yang, Steven J. Hall, Gavin McNicol

21. Biodegradation of environmental pollutants

William J. Hickey

22. Bioremediation of contaminated soils

Robyn A. Barbato, C. Mike Reynolds

23. Biological control of soilborne plant pathogens and nematodes

James H. Graham, Sarah L. Strauss

24. Composting: the microbiological processing of organic wastes

David A. Zuberer, Larry M. Zibilske

25. Soil microbial influences on "One Health"

Ian L. Pepper, John P. Brooks

Contributors

Jose A. Amador Department of Natural Resources Science, University of Rhode Island, Kingston, RI, United States

Robyn A. Barbato US Army Cold Regions Research and Engineering Laboratory, Hanover, NH, United States

Jayne Belnap US Geological Survey, Southwest Biological Center, Moab, UT, United States

John P. Brooks USDA-ARS, Genetics and Sustainable Agriculture Unit, Mississippi State, MS, United States

Mary Ann Bruns The Pennsylvania State University, University Park, PA, United States

Shiping Deng Department of Plant and Soil Science, Oklahoma State University, Stillwater, OK, United States

Rhae A. Drijber Department of Agronomy and Horticulture, University of Nebraska-Lincoln, Lincoln, NE, United States

Alan Joseph Franzluebbers USDA—Agricultural Research Service, Raleigh, NC, United States

Jeffry J. Fuhrmann Department of Plant and Soil Sciences, University of Delaware, Newark, DE, United States

Josef H. Görres Plant and Soil Science Department, University of Vermont, Burlington, VT, United States

Ferran Garcia-Pichel Center for Fundamental and Applied Microbiomics, Biodesign Institute, and School of Life Sciences, Arizona State University, Tempe, AZ, United States

Terry J. Gentry Department of Soil and Crop Sciences, Texas A&M University, College Station, TX, United States

James J. Germida Department of Soil Science, University of Saskatchewan, Saskatoon, SK, Canada

James H. Graham University of Florida, Citrus Research and Education Center, Lake Alfred, FL, United States

V.V.S.R. Gupta CSIRO Agriculture and Food, Locked Bag 2, Glen Osmond, SA, Australia

Steven J. Hall Department of Ecology and Evolutionary Biology, Iowa State University, Ames, IA, United States

William J. Hickey Department of Soil Science, University of Wisconsin, Madison, WI, United States

Julie A. Howe Department of Soil and Crop Sciences, Texas A&M University, College Station, TX, United States

Gavin McNicol Department of Earth and Environmental Sciences, University of Illinois at Chicago, Chicago, IL, United States

Morgan R. McPherson Department of Agronomy and Horticulture, University of Nebraska-Lincoln, Lincoln, NE, United States

C. Mike Reynolds US Army Cold Regions Research and Engineering Laboratory, Hanover, NH, United States

Joseph B. Morton West Virginia University, Morgantown, WV, United States

David D. Myrold Department of Crop and Soil Science, Oregon State University, Corvallis, OR, United States

Cindy H. Nakatsu Department of Agronomy, Purdue University, West Lafayette, IN, United States

Ian L. Pepper Water and Energy Sustainable Technology Center (WEST), University of Arizona, Tucson, AZ, United States

Lindsey Slaughter Department of Plant and Soil Science, Texas Tech University, Lubbock, TX, United States

A. Peyton Smith Department of Soil and Crop Sciences, Texas A&M University, College Station, TX, United States

Sarah L. Strauss University of Florida, Southwest Florida Research and Education Center, Immokalee, FL, United States

Janice E. Thies Cornell University, Ithaca, NY, United States

Kurt E. Williamson Biology Department, The College of William & Mary, Williamsburg, VA, United States

Wendy H. Yang Departments of Plant Biology and Geology, University of Illinois at Urbana-Champaign, Urbana, IL, United States

Stephanie A. Yarwood Department of Environmental Science and Technology, University of Maryland, College Park, MD, United States

Larry M. Zibilske Vice-President for Research, Texas Plant and Soil Laboratory, Edinburg, TX, United States

David A. Zuberer Department of Soil and Crop Sciences, Texas A&M University, College Station, TX, United States

About the Editors

Terry J. Gentry

Dr. Terry J. Gentry is a Professor of Soil and Aquatic Microbiology in the Department of Soil and Crop Sciences at Texas A&M University. He teaches an undergraduate course in soil and water microbiology and a graduate course in environmental microbiology. His research focuses on the use of molecular technologies to detect and identify microbial pathogens from animal, human, and natural sources along with the characterization of microbial populations and communities contributing to applied processes such as the bioremediation of organic and metal contaminants and sustainability of intensive cropping systems.

Affiliations and Expertise

Professor of Soil and Aquatic Microbiology, Department of Soil and Crop Sciences, Texas A&M University, College Station, TX, United States

Jeffry J. Fuhrmann

Jeffry J. Fuhrmann is a Professor of Soil Microbiology in the Department of Plant and Soil Sciences, with a joint appointment in Biological Sciences, at the University of Delaware. He teaches a graduate course in soil microbiology and undergraduate courses in introductory soil science and soils and sustainability. His research has focused primarily on the ecology and diversity of soybean bradyrhizobia using combined phenotypic and genotypic approaches. More recently, this emphasis has extended into analogous studies of associated bacteriophages. His research has additionally examined broader aspects of the microbial ecology of soils and the rhizosphere. He has served as an associate editor for the Soil Science Society of America Journal.

Affiliations and Expertise

Professor of Soil Microbiology, Department of Plant and Soil Sciences, Department of Biological Sciences, University of Delaware, Newark, DE, United States

David A. Zuberer

Dr. David A. Zuberer is a Professor Emeritus of Soil Microbiology in the Department of Soil and Crop Sciences at Texas A&M University. He taught soil microbiology continuously for >30 years in addition to maintaining an active research program in nitrogen fixation and rhizosphere microbiology. He served as a Technical Editor of the Soil Biology and Biochemistry Division (S3) of the Soil Science Society of America Journal. He is a Fellow of both the American Society of Agronomy and the Soil Science Society of America. He was the recipient of the 2003 Soil Science Education Award and the 2012 Agronomic Resident Education Award of the American Society of Agronomy. He is also a Fellow of the American Association for the Advancement of Science.

Affiliations and Expertise

Professor Emeritus of Soil Microbiology, Department of Soil and Crop Sciences, Texas A&M University, College Station, TX, United States

Preface

Perhaps no area of soil science has seen greater transformation over the past few decades than soil microbiology. Thus, an updated, third edition of *Principles and Applications of Soil Microbiology* became necessary to reflect the new discoveries which have occurred over the ~15 years since the second edition was published in 2005.

The first edition of this book initially grew out of discussions in the 1990s among scientists in Regional Research Project S-226 (later S-262 and S-297) who saw the need for an updated, comprehensive textbook for teaching soil microbiology. The field had become so broad that it was almost impossible for any one individual to stay current in all aspects of the science. Thus, adopting King Solomon's adage that "as iron sharpens iron, so one man sharpens another," the editors took the nontraditional approach of having multiple authors write the textbook. Many of the original authors were members of the Regional Project, while others were invited to write chapters related to their specific area of research expertise. In this way, the editors hoped to better capture the rapid advances in both fundamental knowledge and potential applications of soil microbiology. They also believed that a multi-authored approach would serve to bridge students from a traditional, introductory, single-authored text to a typical multi-authored scientific treatise. Then and now, students need to learn the concept of consulting the works of individual scientists to keep abreast of current developments. Nevertheless, care was taken in the editing process to ensure a uniform style and consistent usage of terminology. This overall process was continued in the second edition, published in 2005, and again in the current, third, edition.

Principles and Applications of Soil Microbiology is designed for use by advanced undergraduate and beginning graduate students who require a comprehensive treatment of the field of soil microbiology. Professionals in agricultural, environmental, and industrial fields should also find this book a valuable reference.

This third edition of the book is divided into four major sections: (1) Fundamentals, (2) Organisms and Interactions, (3) Cycles in Soil, and (4) Environmental and Agricultural Perspectives. The book has been reorganized and updated throughout with many citations to the literature published since the second edition was printed. Significant new material has been added, and certain sections have been expanded. A partial list includes new chapters on (1) Applied Aspects of Soil Carbon and (2) Soil Microbial Influences on One Health, along with more comprehensive treatment of Archaea, biological soil crusts, bioremediation technologies, carbon sequestration, microbial interactions, mycorrhizal specificity, nucleic-acid techniques for soil microbiome analysis, rhizosphere, soil health, soil organic matter formation, symbiotic nitrogen fixation, and virus diversity and survival in soils.

Instructors may find that this book contains more information than they cover in a typical, one-semester course in soil microbiology. The fundamentals chapters are provided as background information for students who do not have prior courses in soil science or biochemistry. The chapters on the important groups of microorganisms found in soil may include more detail than is required of beginning students, but they will serve as an important reference on the taxonomy, physiology, and ecology of microorganisms for advanced students and professionals. Our short-term goal was to provide chapters on topics where active research shows promise of practical application. Our long-term goal is to provide timely updates of this book so students can keep abreast of the most recent advances in soil microbiology.

Although many methods are referred to in individual chapters in connection with discussions of specific principles and applications, the reader should be aware that this is not a book of methods. This is the reason why we avoided a specific chapter on methods. For a detailed treatise on methods in soil microbiology, the reader should consult other works (e.g., *Methods of Soil Analysis, Part 2. Microbiological and Biochemical Properties* published by the Soil Science Society of America; *Manual of Environmental Microbiology*, fourth ed. published by the American Society for Microbiology) and the current scientific literature.

ACKNOWLEDGMENTS

We thank the editors and authors of the first and second editions of the textbook for their previous contributions, many of which form the foundation for the current edition. Special thanks to our co-editors of the first and second editions, Drs. David Sylvia and Peter Hartel for their original contributions to the text and for their support and encouragement for the third edition. We thank Kim Luoma for her many excellent drawings and many internal and external reviewers of the individual chapters. We also thank other members of the S-226, S-262, and S-297 Regional Research Projects for their stimulating discussions. Finally, we thank all of our family and friends who provided the support and encouragement that was necessary to bring this project to completion.

CHAPTER

1

Introduction and historical perspective

Terry J. Gentry, David A. Zuberer

Department of Soil and Crop Sciences, Texas A&M University, College Station, TX, United States

"Society has its roots in the soil"—***Charles Kellogg.***

Soil microbiology is a branch of soil science concerned with microbes found in the soil and their relations to soil management, agricultural production, and environmental quality. Hence, soil microbiologists study the numbers and kinds of microbes found in soil and the effects these and introduced microbes have on soil-ecological processes (e.g., nutrient cycling). The applications of these studies have important consequences for crop production, environmental quality, and the restoration of disturbed environments. In fact, they are paramount to the sustained health and productivity of the world's soils and the populations that depend on them for sustenance.

Soil habitat

The soil is a complex habitat for microbial growth (Fig. 1.1; see also Chapter 2). It differs markedly from the environment that microorganisms encounter when grown on traditional laboratory culture media in two crucial ways. First, in its natural state, the soil is a heterogeneous medium of solid, liquid, and gaseous phases, varying in its properties, both across the landscape, with depth in a profile, and over time. Microbes generally exist as isolated microcolonies or biofilms on mineral particles, organic matter, roots, etc. Here they are dependent on the movement of nutrients to them by mass flow of soil water or by diffusion. Similarly, they must depend on passive means for the removal of toxins from their locale unless they are motile and can move from site to site. Second, in soil, **competition** exists among an enormous variety of organisms for nutrients, space, and moisture. Competition occurs among bacteria, actinomycetes (actinobacteria), and fungi, as well as with other life forms in soil, including animals and plant roots. If we are to understand soil microbes, then developing a working knowledge of the habitat in which they grow is of utmost importance.

Principles and Applications of Soil Microbiology. http://dx.doi.org/10.1016/B978-0-12-820202-9.00001-0
Copyright © 2021 Elsevier Inc. All rights reserved.

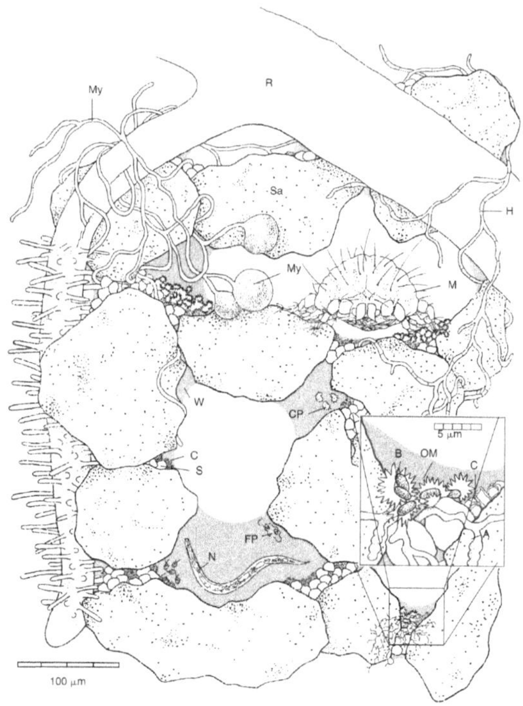

FIGURE 1.1 **A soil habitat containing mineral soil particles (sand (Sa), silt (Si), and clay (C)), organic matter (OM), water (W), plant root with root hairs (R), and soil organisms (bacteria (B), actinomycetes (A), mycorrhizal spores and hyphae (My), hyphae of a saprophytic fungus (H), a nematode (N), ciliate protozoa (CP), flagellate protozoa (FP), and a mite (M).** This soil can be a habitat of enormous complexity and diversity even over small distances. For example, the actual size of the soil in this drawing is <1 mm in both directions yet may contain habitats that are acid to alkaline, wet to dry, aerobic to anaerobic, reduced to oxidized, and nutrient-poor to nutrient-rich. Realizing this complexity and diversity is a key to understanding soil microbiology. *Source: Original drawing by Kim Luoma.*

Nature of cellular organisms

The basis of living matter is the cell (Box 1.1). Each cell is a unique entity made up of a complex mixture of chemical materials and subcellular components. The cell is bounded by the **cytoplasmic membrane**, separating the interior of the cell, known as the **cytoplasm**, from the external environment.

Characteristics of living cells

Two fundamental types of living cells are recognized: **prokaryotes** (from *pro*, meaning "before," and *karyon*, meaning "nucleus") and **eukaryotes** (from *eu* meaning "true"). The term "prokaryote" has taken on some controversy among modern scientists, with some even advocating for the disuse of the term. However, here we will continue to the use the term in its classical sense as it is useful in differentiating two types of organisms, that is, it describes cells whose nuclear material (often called the nucleoid) is not bounded by a distinct nuclear

BOX 1.1

Characteristics of living cells

Key characteristics that separate living cells from nonliving chemical systems include:

All cells:

- *Metabolism*: The capacity to take up and use chemicals from the environment and transform these chemicals into usable products, including energy to grow or survive, and to excrete waste products.
- *Growth*: The capacity to self-direct synthesis using nutrients from the environment to produce new cell materials, growing by division, forming two cells from one.
- *Evolution*: The capacity to change genetically, which may affect the overall fitness of the cell to survive in a particular environment.

Some cells:

- *Differentiation*: The capacity to undergo change in form or function (e.g., endospore formation), often in response to environmental changes or normal growth processes.
- *Communication*: The capacity to interact with other cells through chemical signaling.
- *Genetic exchange*: Cells exchange genes by several mechanisms.
- *Motility*: Some cells are capable of propelling themselves using "organelles" such as flagella or cilia or other means of movement.

Adapted from Madigan et al. (2019).

membrane (bacteria and archaea) in contrast to cells having a true nucleus. Major structural differences exist between the two types of cells. In eukaryotes, the nucleus is in the cytoplasm; it is bounded by a distinct nuclear membrane and contains several DNA molecules (**chromosomes**); membrane-bound organelles are also present. Eukaryotes undergo division by the well-known process of mitosis. The prokaryote does not have a well-defined nucleus (bounded by a nuclear membrane). Rather, it consists of a single, circular DNA molecule (the "chromosome"): the nucleoid. Cell division in the prokaryote is usually by binary fission (i.e., a simple nonmitotic splitting). Additional differences between prokaryotic and eukaryotic cells are presented in Table 1.1. **Bacteria** (including cyanobacteria and actinomycetes) and **Archaea** are prokaryotes, while all other cellular organisms are eukaryotes; viruses (Chapter 9) are unique in being acellular as described later.

Classification of organisms

The study and use of microorganisms are based on our ability to recognize and establish the identity of individuals. Most classification schemes are organized to show relationships among organisms. This orderly arrangement allows us to communicate descriptive

TABLE 1.1 Comparison of structural features of prokaryotic and eukaryotic cells.

	Prokaryotes	Eukaryotes		
Organelle	Bacteria, archaea, cyanobacteria	Fungi	Algae	Protozoa
Cytoplasmic membrane	+	+	+	+
Nuclear division	+	+	+	+
Nuclear membrane	–	+	+	+
Ribosomes	70S	80S	80S	80S[a]
Endoplasmic reticulum	–	+	+	+
Golgi complexes	–	+	+	+
Mitochondria	–	+	+	+
Cytoskeleton	–	+	+	+
Chloroplasts	–	–	+	–, +[b]
Vacuole	–	+	+	+
Cell wall	+	+	+	+

[a]*S is the Svedberg unit.*
[b]*Some phytoflagellates have chlorophyll.*

information about the organism to others. These data can also be entered into various microbial databases, allowing retrieval of information about related organisms.

Like other biologists, microbiologists use the Linnaean system of **binomial nomenclature** to name microbes. An organism's name consists of a genus and species (sort of a "first" name and "last" name). In the more evolutionarily advanced organisms, species are defined as groups of interbreeding or potentially interbreeding natural populations. However, many microbes do not reproduce sexually so this definition is not very useful. Microbiologists define a **species** as a group of similar individuals that are sufficiently different from other individuals to be considered a recognized taxonomic group. At the next higher taxonomic level, a **genus** (plural, genera) can be defined as a grouping of species that share a major property (or properties). For example, the Latin binomial name, *Acidithiobacillus thiooxidans* (abbreviated *A. thiooxidans* once it has been introduced by full name in the text) is representative of a group of individuals (species: *thiooxidans*) that have the capacity to oxidize elemental sulfur and share some common characteristics with other organisms in the genus *Acidithiobacillus*. Often microorganisms are named for an outstanding feature they possess (e.g., *A. thiooxidans* is a rod-shaped bacterium capable of oxidizing reduced sulfur for the generation of energy). In other cases, organisms are named to commemorate the contributions of an outstanding scientist in the field (e.g., *Nitrobacter winogradskyi*, named in honor of the Russian soil microbiologist Sergei Winogradsky).

Historically, microbes were classified on the basis of taxonomic features that were relatively easy to observe and measure. These characters include cell structure, morphology, staining reactions, and physiological parameters (e.g., ability to use a particular carbohydrate, nitrogen fixation, etc.). These features are **phenotypic** (based on physical characteristics, appearance, etc.) rather than **genotypic** (based on genetic relationships) and may obscure important genetic relations among related groups of organisms.

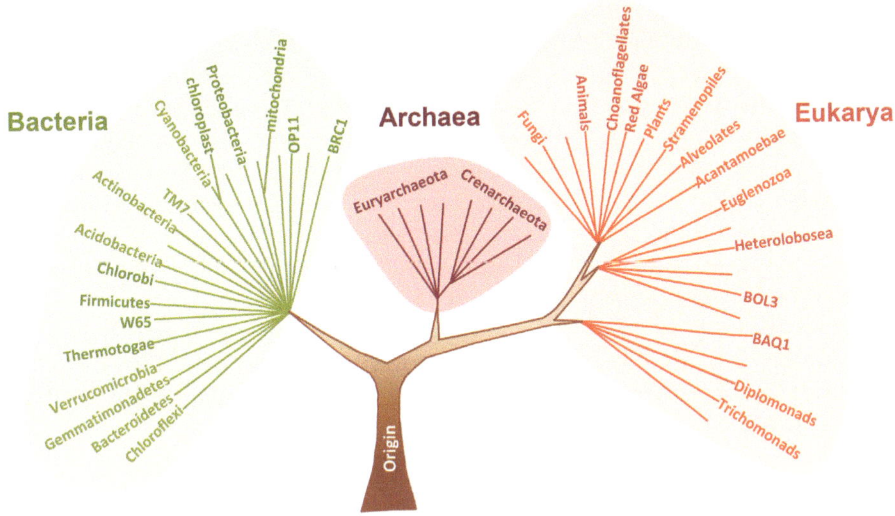

FIGURE 1.2 **Universal phylogenetic tree for living organisms, based on comparative sequencing of 16S or 18S ribosomal RNA genes.** Note that microorganisms comprise most of the biological diversity found on earth. *Source: Adapted from Pace (2009).*

The progressive development of DNA-based analysis tools over the last few decades introduced new ways of determining relations among organisms (Chapter 4). Currently, the methods most commonly used to identify and determine relatedness of organisms are based on comparison of ribosomal RNA (rRNA) gene sequences, although whole genome analysis is becoming more frequent as sequencing costs decrease. Insights from these analyses are revolutionizing microbial taxonomy. The traditional scheme of classification recognized five kingdoms of organisms: Bacteria, Fungi, Protista (including algae and protozoa), Animals, and Plants. More recent findings based on molecular phylogeny suggest there are three **domains** of living organisms: **Bacteria, Archaea,** and **Eukarya** (Fig. 1.2; Pace, 2009). Although placement of organisms into three domains was defined by differences within the rRNA gene, subsequent studies reveal that organisms in these domains also differ in cell-wall properties, lipid composition, and mechanisms of protein synthesis. Below the three domains, six or more kingdoms have been recognized.

Our new understanding of the tree of life contradicts several long-held beliefs (Pace, 2009). For example, the deep divergence between Archaea and Bacteria shatters the notion of evolutionary unity among prokaryotes, and it appears that the Eukarya line is as old as the prokaryotic lines. As promising as the early claims for molecular sequencing have been for determining relationships among organisms, the applications have not led to universal acceptance of the exact structure of the phylogenetic tree depicted in Fig. 1.2, except perhaps, at its most remote branches. Difficulties arise toward the root (base) of the tree. Woese (1998) stated it thusly; "Organismal lineages, and so organisms as we know them, did not exist at these early stages. The universal phylogenetic tree, therefore, is not an organismal tree at its base but gradually becomes one as its peripheral branchings emerge. The universal ancestor is not a discrete entity. It is, rather, a diverse community of cells that survives and evolves as a biological unit. This communal ancestor has a physical history but not a genealogical one. Over time, this ancestor refined into a smaller number of increasingly complex cell types with the ancestors of the three primary groupings of organisms arising as a result."

These organisms form an incredibly diverse community (**microbiome**) within soil; for example, soil microbiologists currently estimate that there may be up to 4,000–13,000 species of bacteria, not counting other organisms, in a single gram of soil. These estimates are based on the diversity of DNA extracted from soil samples, since the vast majority of soil bacteria (and other microorganisms) have not yet been cultured or characterized in the lab. The soil microbiome remains a largely undiscovered scientific frontier with much yet to be learned about these novel microorganisms.

In these classification schemes, there is no place for **viruses** (Chapter 9). Viruses are not cells because they lack a cytoplasmic membrane with internal cytoplasm; it is only when viruses are associated with another organism (e.g., bacterium, plant, animal) that they are able to fulfill the basic life processes as stated in Box 1.1.

Organisms in the soil

Soil organisms are both numerous and highly diverse. Many soil organisms are small and cannot be seen without the aid of magnification (Table 1.2). The smallest organisms, bacteria, actinomycetes, fungi, and algae, are referred to collectively as the **microflora**. Soil animals range in size from microscopic (**microfauna**) to mites, earthworms, and small mammals (**mesofauna** and **macrofauna**). With the exception of some soil animals and fungi, most soil organisms are single cells. Chapters 5–9 of this book describe the microbes present in soil.

Excluding viruses, bacteria are the most abundant microbes in soil (Table 1.2), attaining populations in excess of 100 million (10^8) individuals per gram (g^{-1}) of soil and representing perhaps as many as 10^3–10^4 different species. The actinomycetes and fungi are the next most numerous microbes in soil, numbering from 10^7 to 10^8 and 10^5 to 10^6 g^{-1} soil, respectively,

TABLE 1.2 Representative sizes, numbers[a], and biomass[a] of major microbial groups found in soil.

Microbial group	Example	Size (μm)	Numbers (# g^{-1} soil)	Biomass (kg wet mass ha^{-1} soil)
Viruses	Tobacco mosaic	0.02 × 0.3	10^{10}–10^{11}	—
Bacteria	*Pseudomonas*	0.5 × 1.5	10^8–10^9	300–3,000
Actinomycetes	*Streptomyces*	0.5–2.0 [b]	10^7–10^8	300–3,000
Fungi	*Mucor*	8.0 [b]	10^5–10^6	500–5,000
Algae	*Chlorella*	5 × 13	10^3–10^6	10–1,500
Protozoa	*Euglena*	15 × 50	10^3–10^5	5–200
Nematodes	*Pratylenchus*	1,000 [c]	10^1–10^2	1–100
Earthworms	*Lumbricus*	100,000 [c]		10–1,000

[a]*Data from Metting (1993)*
[b]*Diameter of hyphae*
[c]*Length*

although the filamentous nature of fungi (Chapter 6) makes enumeration problematic. Numbers of soil animals vary widely in the soil, ranging from just a few to as many as 10^6 g^{-1} soil.

It is important to note, however, that we must consider more than just the numbers of individuals in a gram of soil if we have to understand microbial functions in soil. Microbes have a wide range of sizes and morphologies; thus, numbers alone may not provide a very good indication of the importance of a microbial group. For example, even though bacterial numbers are usually several orders of magnitude greater than those of fungi, the fungi generally have a greater total biomass in the soil.

The importance of measurements in soil microbiology cannot be emphasized too strongly. Scientists are interested in numbers or activities of microbes in the soil system. Since microbes frequently range to as many as 10^8 organisms per gram of soil, it is necessary to dilute samples to be able to enumerate colonies on various media (Chapter 5). Because the numbers are large, frequently the values are expressed as a logarithmic transformation; hence 10^8 (100 million) organisms per gram of soil could also be expressed as $\log_{10} 8$ organisms per gram. The desire to express microbial activities as a rate function also requires a familiarity with logarithms either to the base 10 ($\log_{10}$) or base e (ln—natural logarithm) (Box 1.2).

The relative position and size of soil microbes within the soil habitat is illustrated in Fig. 1.1. The physical, chemical, and even biological properties of the soil habitat and their interactions with the resident community of soil microbes have a significant impact on their growth and activity. As our understanding of these complex relationships develops, we should be better able to manage the soil and its microbes for the maintenance and improvement of soil without damaging the soil as a resource.

Microbes exhibit an enormous diversity of functions in the soil. For example, they decompose organic compounds releasing inorganic elements, a process called mineralization (Chapters 13–15, 18, and 19), oxidize reduced forms of elements (e.g., elemental sulfur (S°) → sulfate (SO_4^{2-}; Chapters 13, 15, 18, and 19), and reduce oxidized forms of elements [e.g., nitrate (NO_3^-) → dinitrogen gas (N_2)]; Chapters 13, 15, 18, and 19). Additionally, the reduction of dinitrogen to a biologically utilizable form (ammonia, NH_3; Chapters 16 and 17) and degradation of organic wastes and pollutants to carbon dioxide and water (Chapters 21, 22, and 24) are important functions of soil microbes. Interactions among different organisms, including plant roots, can lead to benefits for some participating organisms, while other interactions may have a detrimental effect on growth or development (Chapters 10, 11, and 23). Still other

BOX 1.2

Scientific notation

For convenience, soil microbiologists express the huge numbers of microbes in soil using exponential numbers, a convention known as **scientific notation**. For example, "two million organisms per gram of soil" is written, 2×10^6 g^{-1} of soil. The same number can also be expressed as a logarithmic number, $\log_{10} 6.3$ g^{-1} of soil. Note that the superscripted "−1" translates as "per."

microbes may alter levels of global gases (Chapter 20). All of these functions are explored in the succeeding chapters of this book.

Historical context of soil microbiology

The development of soil microbiology is inextricably linked with the development of microbiology as a science and cannot be understood in isolation. Here we highlight significant accomplishments before 1950 that led to the development of both microbiology and soil microbiology (Table 1.3). After 1950, the list of those making significant contributions becomes too large to mention individual accomplishments; these contributions are considered more generally. However, during the 1980s and 1990s, notable advances in microbial genetics led to the development of the "genomics" era some of which are discussed in a later section.

Origins of soil microbiology

The first historical mention of the presence of microscopic organisms in soil is attributed to a Roman writer in about 60 B.C. (Waksman, 1927). Writing about marshes, Columella noted they gave up "noxious and poisonous steams," breeding "animals armed with poisonous stings," from which "hidden diseases are often contracted, the causes of which even physicians cannot properly understand." Even prior to that time, there are reports in the Old Testament of people who practiced strict isolation and cleanliness codes, particularly to those afflicted

TABLE 1.3 Outstanding historical figures who made major contributions to the development of soil microbiology.

Name	Dates	Country	Area of contribution
Leeuwenhoek	1632–1723	Netherlands	Inventor of the microscope
Pasteur	1822–1895	France	Repudiation of spontaneous generation, biological nature of nitrification
Tyndall	1820–1893	England	Repudiation of spontaneous generation, understanding of the sterilization processes
Cohn	1828–1898	Germany	Understanding of sterilization processes and taxonomy of *Bacillus*
Koch	1843–1910	Germany	Koch's postulates, gelatin plates for studying soil microbes
Winogradsky	1856–1953	Russia	Isolation and taxonomy of chemoautotrophic bacteria, especially nitrifiers and sulfur oxidizers
Beijerinck	1851–1931	Netherlands	Isolation of legume root-nodulating bacteria, Director of The Delft School of Microbiology
Gram	1853–1938	Denmark	Differential staining procedures
Lipman	1874–1939	United States	Concept of soil as a complex, living entity
Russell	1872–1965	England	Development of importance of soil-plant-microorganism interactions
Starkey	1899–1991	United States	Microbiology of the sulfur bacteria
Waksman	1888–1973	United States	Discovery of antibiotics from actinomycetes

with leprosy, suggesting they understood that disease had some relationship to an unseen cause. Likewise, there is good evidence that the Romans recognized that leguminous plants enhanced soil productivity and practiced crop rotations with leguminous plants. However, no one really understood the involvement of microscopic organisms in any of these phenomena.

Although Robert Hooke reported on the fruiting structures of molds in 1664, we most often think of the Dutch amateur microscope builder Antoni van Leeuwenhoek, with his newly constructed microscope, as the first individual to describe microbes (Madigan et al., 2019). In his paper to the Royal Society of London in 1684, van Leeuwenhoek reported the presence of "wee animalcules" in pond water. Others confirmed these observations, but little was known about these organisms and their relation to the environments in which they grew.

Practically coincidental with these scientific findings was the rise in the philosophical basis for science. Francis Bacon is often credited with being the founder of the "scientific method" for research. Bacon saw himself as the inventor of a method of investigation, which would kindle a light in nature—"a light that would eventually disclose and bring into sight all that is most hidden and secret in the universe." This method involved the collection of data, their judicious interpretation, and the execution of experiments, to learn the secrets of nature by organized observation of its regularities. Bacon argued that the only knowledge of importance to man was empirically rooted in the natural world and that a clear system of scientific inquiry would assure man's mastery over the world. He is said to be the originator of the expression, "Knowledge is power" (*ipsa scientia potestas est* ('knowledge itself is power')).

The 19th century

The scientific achievements of the 19th century were unprecedented for the recognition of soil microbiology as a legitimate branch of science. Less well known is the fact that certain political and social changes in the United States also provided the foundation for an explosion of science, particularly soil science and soil microbiology, in the 19th and 20th centuries.

Roots of American agricultural colleges

During the first half of the 19th century, two types of colleges and universities were emerging side-by-side, one publicly controlled and the other privately controlled. Both were greatly influenced by their European-educated professors, who thought that a university education was for the male leisure classes, government leaders, and members of the professions. Although science was gaining respectability, a scientific education was not widely available. By mid-century, the general and scientific media were making widespread demands for more agricultural and technical education. Agricultural societies in many states were insisting that colleges be made available where students could study agriculture.

In 1857, Vermont Representative Justin Smith Morrill introduced his first **land-grant** bill in Congress. The act was to give to every state a grant of 30,000 acres of public land for every member of its congressional delegation. Thus, even the smallest state received 90,000 acres since every state had at least two senators and one representative. Under the provisions of the legislation, the states were to sell the land and use the proceeds to establish colleges in engineering, agriculture, and military science. After a year of legislative maneuvering, Congress passed the **Morrill Act**; however, President Buchanan vetoed it, on the grounds that it violated the traditional policy of the federal government, which until then had left control of education to the states.

TABLE 1.4 Important legislation in the establishment of the Land-Grant College System and Associated Agricultural Experiment Stations and Extension Service.

Year	Act: provisions
1862	**Homestead Act**: 160 acres of public land, with purchase after 5 years of living on the land.
1862	**First Morrill Act (Land Grant College Act)**: Provisions set aside a minimum of 90,000 acres of federal land in each state for the support of institutions to teach, but not be limited to, agriculture and mechanical arts. These lands could be used by the states to generate income in support of higher education in the states.
1887	**Hatch Act**: Established the Agricultural Experiment Stations in the respective states.
1890	**Second Morrill Act**: Set out to extend the land grants to the southern states and a group of predominantly black institutions (colloquially known as the 1890 institutions).
1914	**Smith/Lever Act**: Completed the founding concept of the mission of the land grant institution by establishing the Cooperative States Extension Service.

The year 1862 proved to be monumental, as two acts were passed and signed into law that had a profound impact on education and agriculture (and by implication soil microbiology), perhaps more than any other legislation before or since. The first was the **Homestead Act**, which permitted any citizen, or any person who intended to become a citizen, to receive 160 acres of public land, and then to purchase it at a nominal fee after living on the land for 5 years. This act provided the most generous terms of any land act in American history and enabled people to settle and own their own farms, creating a large demand for agricultural information. A revised version of the act originally introduced by Morrill, now a US Senator, was passed and then signed into law by President Lincoln on July 2, 1862. This reconfigured Morrill Act gave to every state that had remained in the Union a grant of public land. Over 70 "land grant" colleges, as they came to be known, were established under the provisions of the original Morrill Act. A second act in 1890 extended the land grant provisions to the 16 southern states. The legislative history is summarized in Table 1.4.

The enormous impact of the land-grant concept on science and our democratic society is difficult to overstate, as we continue to reap the benefits of this visionary legislation over 100 years later. Although originally started as agricultural and technical schools, additional state aid allowed many of them to grow into large public universities that over the years have educated millions who otherwise might not have been able to afford higher education.

Scientific developments

During the 19th century, two important questions were answered that would lay the scientific foundations for both microbiology and soil microbiology:

- Does spontaneous generation occur?
- What is the origin of contagious disease?

The eminent French scientist Louis Pasteur, in a simple and elegant experiment (Box 1.3), demonstrated that it was microbes present in the air that were capable of initiating growth in an exposed sterile culture medium. Microbes simply did not arise spontaneously in a suitable medium. Pasteur's experiment effectively settled the controversy surrounding the prominently held theory of **spontaneous generation**, vigorously debated at the time. Similar

BOX 1.3

Pasteur's experiment disproving spontaneous generation

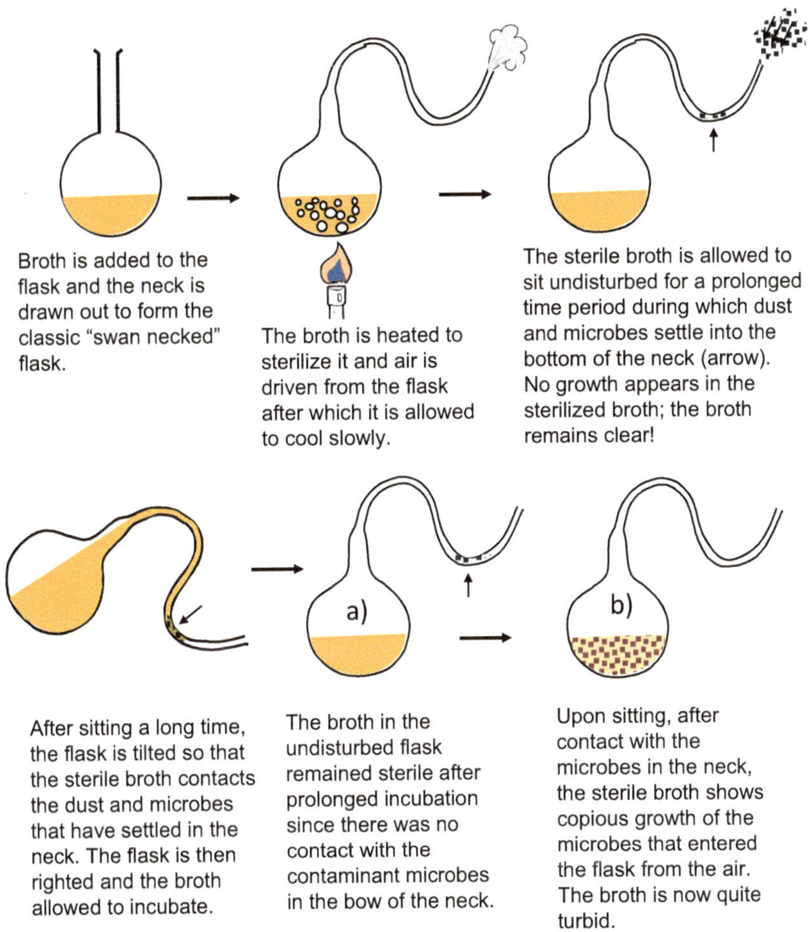

Pasteur introduced a nonsterile broth into each of two flasks, drew out the flasks to a swan-necked shape to provide a dust trap, and sterilized the contents of each by heating. The broth in one flask (a) was not allowed to contact the dust that settled in the trap and no growth occurred even after a long incubation. After the broth in the second flask (b) had sat for some time it was brought into contact with dust collecting at the low point near the open mouth of the swan neck. After a short period of incubation, the broth became turbid, indicating that growth had occurred in this flask. A translation of Pasteur's original report, *"On the organized bodies which exist in the atmosphere; examination of the doctrine of spontaneous generation"* can be found in Milestones in Microbiology (ed. Brock, T.D.) pp. 43–48 (Prentice-Hall, Englewood Cliffs, New Jersey, 1996).

experiments were also conducted by Spallanzani, the Italian scientist, and the Englishman John Tyndall. Tyndall's original flasks can still be viewed at the Royal Institute near Piccadilly Square in London. The rejection of the theory of spontaneous generation provided the foundations for **aseptic (sterile) technique** (the practice of working so as to avoid contamination of workers and cultures). While conducting his experiments, Tyndall also noted difficulties in trying to sterilize preparations from hay infusions. Further investigations by Tyndall, and also by Ferdinand Cohn of Germany, led to the discovery of organisms that produced **endospores** (the extremely resistant structures produced by bacteria like *Bacillus anthracis*, the anthrax bacillus), in these difficult-to-sterilize preparations.

During the same period, there was a widely held belief that disease was caused by something called "contagion." After microbes were discovered, they were generally accepted as the responsible agents, but rigorous proof was lacking. Although Ignaz Semmelweis and Joseph Lister (after whom the product Listerine was named) provided evidence that microbes caused human disease, it was not until the seminal work of Robert Koch that the **germ theory of disease** had a solid scientific basis. Koch reasoned that to prove that a microorganism was the causative agent of a disease, the following should apply:

- The organism should be consistently present in the subject exhibiting the disease symptoms, but not in the healthy subject
- The organism should be grown in pure culture away from the subject
- When the organism is used to inoculate a healthy susceptible subject, the disease symptoms should appear
- The organism should be re-isolated from the exposed subject and re-cultured in the laboratory to confirm its similarity with the original organism

These steps have been immortalized as **Koch's Postulates**, which are important for several reasons. First, by following these steps, it is possible to demonstrate that a specific organism is responsible for a specific disease or for some other microbiological process (e.g., sulfur oxidation or N_2 fixation). Second, they suggest the importance of the laboratory culture of organisms, and third, they recognize that specific organisms have specific functions. Acceptance of Koch's Postulates brought the science of microbiology to the point that others could now make specific contributions in more applied fields, such as soil microbiology. Another of Koch's significant developments, reported in 1881, was the use of a gelatin culture medium to study microbes *in vitro*. Five years later, the Danish physician Christian Gram described a staining procedure based on differential properties of cell walls, leading to important taxonomic distinctions. Thus, soil microbes could be studied in a systematic fashion (Waksman, 1927).

In the remaining few years until the turn of the century, significant developments in soil microbiology came in rapid succession. One notable advance was the discovery of **biological dinitrogen (N_2) fixation** (Chapters 16 and 17). In 1886, Hellriegel and Wilfarth showed that microorganisms in nodules on leguminous plants were able to convert atmospheric nitrogen to a form plants could use. Subsequently, the Dutch scientist Martinus W. Beijerinck (pronounced: "buy-a-rink," Fig. 1.3A) isolated the bacterium responsible for nodulation of leguminous plants and named it *Bacillus radicicola* (Chung and Ferris, 1996). In 1889, Frank renamed the bacterium *Rhizobium*, replacing the genus name used by Beijerinck. These actions set in motion a wave of activity in the study of biological N_2 fixation, which continues to this day.

A second important advance during this period was the discovery of **chemoautotrophy** (Chapter 3) through the study of **nitrification** (Chapter 15). Building on the foundational work of Koch (i.e., that specific organisms have specific functions), Sergei Winogradsky

FIGURE 1.3 **Several important individuals in the development of soil microbiology.** (A) Martinus Beijerinck, (B) Sergei Winogradsky, (C) Jacob Lipman, and (D) Selman Waksman. *Source: Figures courtesy of Waksman Institute at Rutgers University.*

(Fig. 1.3B) began studying the nitrifying bacteria. He was able to demonstrate that nitrification was really a two-step process, the first being the oxidation of ammonium (NH_4^+) to nitrite (NO_2^-) and the second being the oxidation of nitrite to nitrate (NO_3^-). Two distinctly different groups of bacteria mediated the two steps. Winogradsky isolated representatives of each group and named them *Nitrosomonas* for the NH_4^+ oxidizers and *Nitrobacter* for the NO_2^- oxidizers. He further demonstrated that these bacteria obtained their cellular carbon from CO_2. Given the meager understanding of bacterial metabolism at that time, this was a truly remarkable discovery. For these and many other groundbreaking accomplishments, Winogradsky is considered by many to be the founder of soil microbiology.

By 1895, developments in microbiology had reached such a point that the Delft School of Microbiology in the Netherlands was established. First headed by Martinus Beijerinck, considered by some to be the founder of microbial ecology, and later by A. J. Kluyver and C. B. van Niel, the Delft School had a lasting influence on the study of soil microbiology through discoveries in microbial biochemistry, biodiversity, and biotechnology. These individuals posed the following basic questions about microbial physiology that drew the study of microbiology and the environment closer together:

- How does the living organism interact with its abiotic and biotic environment?
- How can the fundamental principles of microbiology be brought to bear on applied problems?
- What is the place of microbes in the natural world?

These questions are as valid now as they were in the 19th century perhaps even more so. Certainly, the ubiquity and diversity of physiological functions of microbes in the environment is underscored by **Beijerinck's Rule**: "Everything is everywhere and the milieu selects."

The 20th century

If Winogradsky is considered the founder of soil microbiology, then Jacob G. Lipman (Fig. 1.3C) might be considered the founder of American soil microbiology. He considered soil as a complex and living entity that needed to be understood and studied from the standpoint of soil fertility and crop production. This revolutionary view stands as a milestone in soil microbiology (Clark, 1977) and has been carried across the United States and throughout the world, as evidenced by the concept of soil health (Chapter 25).

Two seminal works on soil microbiology were published in the early 1900s: *Handbuch der landwirtschaftlichen Bakteriologie* (Handbook of Agricultural Bacteriology), a comprehensive treatise on soil bacteriology by F. Löhnis (1910), and *Bacteria in Relation to Country Life* by Lipman (1908). Although these works emphasized the role of bacteria in soil fertility, they also directed attention to other organisms residing in soil, including fungi, actinomycetes (a specialized group of bacteria), algae, protozoa, nematodes, and insect larvae. Sir John Russell, Director of the Rothamsted Experiment Station in Great Britain (one of the oldest and continuing long-term agricultural research centers), was a particularly strong advocate of the importance of protozoa in soil fertility, suggesting that soil fertility improved as the number of soil protozoa present increased.

Two prominent individuals who shared Lipman's revolutionary concept of soil were Selman A. Waksman (Fig. 1.3D) and Robert L. Starkey. Besides promoting the "Lipman Philosophy," Waksman's book, *Principles of Soil Microbiology* (1927), and a later work by Waksman and Starkey (1931), *The Soil and the Microbe: An Introduction to the Study of the Microscopic Population of the Soil and Its Role in Soil Processes and Plant Growth*, became the standard soil microbiology texts for much of the period between 1925 and 1950. However, in many circles Waksman is remembered less for his contributions to soil microbiology than for his discovery of the antibiotic streptomycin, for which he was awarded the 1952 Nobel Prize in physiology and medicine. It should be noted that the native environment of the streptomycin-producing organism (an actinomycete) was the soil. Thus, the inscription on Waksman's headstone in a small cemetery in Woods Hole, Massachusetts, "The earth will open and bring forth salvation" (based on a biblical phrase and often quoted as "Out of the earth shall come thy salvation") is prophetic.

During the late 20th century, major advances in our understanding of soil microbes and their ecology were contributed by Francis Clark (USDA-ARS and Colorado State University), Martin Alexander (Cornell University), Eldor Paul (Colorado State University), and James Tiedje (Michigan State University). Perhaps the greatest contributions of these modern-era individuals (and those leading scientists mentioned earlier) were not so much their scientific papers, published books, inventions, or even patents, but the students they trained. The contributions of these individuals, now too numerous to mention, enhanced crop production and fostered the sound use of environmental resources worldwide and ushered in a new age of soil microbiology and microbial ecology.

Genomics era

A few pivotal developments in the 1980s ushered soil microbiology into the genomics era. One of the most important was invention of the polymerase chain reaction (PCR) in 1983 by Kary Mullis. Around this same time, Vigdis Torsvik and others demonstrated the ability to isolate microbial DNA directly from soil and its use for microbial diversity analysis. Along with the subsequent development of DNA sequencing technologies, direct DNA extraction from soil and PCR form the basis for many technologies still used in soil microbiology (Chapter 4). Scientists such as James Tiedje recognized the potential of these technologies for microbial community analysis resulting in major advances in microbial ecology in the 1990s. In 1998, Jo Handelsman and colleagues introduced the concept of the **metagenome**, defined as the "collective genomes of soil microflora," along with a cloning-based approach for characterizing or mining the functional capabilities (e.g., antibiotic production) of microorganisms without the need to first isolate and culture them. Subsequent work by Handelsman and others later broadened metagenomic approaches to include direct sequencing of extracted DNA. As discussed in the microbial ecology section later, recent advances and ongoing development of "omics-based" methods continue to provide exciting insights into soil microbiology (Chapter 10).

Current and future directions

Any look to the future should first be preceded by looking to see where we have been. If we neglect this first step, then we risk repeating mistakes of the past. Despite, or perhaps even because of, the urgency of some of the societal issues we face and the increasing need to better manage the resources at our disposal, it is prudent to step back and look at where we have been before deciding on the next meaningful step. In the previous section, we discussed the historical roots of soil microbiology. Here we summarize important topics in soil microbiology and suggest future research priorities.

Carbon sequestration

For most of the 20th century, studies were dominated by attempts to characterize the chemical composition of the **soil organic matter** (referred to by many as **humus**). Such research is still underway and has been complemented by newer studies designed to provide a clearer understanding of soil organic matter as a group of functional pools (i.e., which pools turn over most rapidly and what is in them). These ideas have brought a fresh approach to this complex topic (Chapter 13). This research was driven initially by the importance of soil organic matter in agriculture and soil fertility. More recently, a strong interest in global-scale transformations of carbon with respect to climate change has fueled an intensified interest in soil organic matter transformations and a desire to quantify the magnitude of carbon sequestration (storage) into various soil organic matter pools (Chapters 14 and 20). In years to come, such information should help us to understand the potential for nutrient cycling from within the soil organic pools and their relationship to sustainable agricultural practices. Additionally, this information contributes to our understanding of the magnitude of changes occurring in atmospheric carbon and is highly relevant to modeling global climate change, etc.

As we move closer to a better understanding of microbial processes involving soil organic carbon, we may gain new insights for further studies. A challenge for the future will be to manage our organic resources while maintaining clean air and water. One emphasis must be on efficiently and safely using resources such as plant nutrients from manures and human wastes. For example, some municipalities capture methane from landfills and wastewater treatment facilities for supplemental energy generation. Increased utilization of this technology could have wide-ranging applications and significant environmental impacts. Another topic deserving attention is the recycling of organic waste by converting it to high-grade, single-cell protein.

Other opportunities exist at the frontiers of space. If long-duration space travel is ever to become a reality, food production and waste-product recycling must be mastered in closed life-support systems. For the soil microbiologist, who already understands organic matter decomposition, the application of that knowledge to a closed spaceship would not require newly developed technologies (Alexander et al., 1989). The knowledge base transfer to this problem is only a matter of scale and control in a closed environment.

Microbial biogeochemistry (nutrient transformations)

From the moment Winogradsky first characterized the nitrifying bacteria, soil microbiologists have studied organisms involved in nutrient transformations (especially those of nitrogen) and tried to understand the factors affecting the various processes. The poor nutrient-use efficiencies of common crops and the expense of fertilizers for standard cropping systems drove soil microbiologists and agronomists to characterize the magnitude of nutrient losses, the conditions under which poor nutrient-use efficiencies might occur and, to a lesser extent, fertilizer management to reduce losses.

Emphasis on nutrient transformations has now shifted to new concerns. Environmental issues such as how much nitrous oxide, a "greenhouse" gas, is produced and released during denitrification and nitrification (Chapters 15 and 20) have become important topics that are currently studied by soil microbiologists. Other environmental issues include the microbial transformations of sulfur (Chapter 18) and of phosphorus (Chapter 19). Understanding the rates of nutrient release from organic residues (mineralization rates) as more people adopt agricultural approaches using crop or other organic residues as nutrient sources (organic farming and sustainable agriculture) has become increasingly important, especially by managers seeking to synchronize nutrient release to periods of greatest need by the crop. More recently, the biotransformations of the metallic cations (Chapter 19) such as arsenic, copper, mercury, iron, and selenium have become a priority research topic. Knowing the rate and amount of release of metallic cations grows more urgent as soils are increasingly used to recycle or dispose of a variety of waste products (Chapter 24).

Plant-microbe interactions

The track record of soil microbiologists is perhaps strongest in the area of discovering and manipulating important plant-microbe associations. The agricultural benefits of the development and exploitation of high-yielding leguminous crops, such as soybeans, alfalfa, and cowpeas, in combination with highly effective strains of rhizobia are enormous (Chapter 17). Through decades of research and development, efficient combinations of plant cultivars and inoculant bacteria have been brought to the market to facilitate the production of high-quality foods and forages for human and livestock consumption. Today, these fascinating associations are the subjects of intensive investigations of the molecular biology of the host plants and associated symbiotic bacteria and, perhaps more importantly, the elaborate communication between the two. These are model systems for understanding how plants and microbes "communicate" to form associations. Perhaps out of such studies will come the development of other important nonleguminous crops capable of meeting some of their nitrogen needs through microbial associations not unlike those of the legumes. Still more exciting are discoveries that bacteria involved in animal and plant disease have genomic similarities, for example, *Brucella suis* (an animal pathogen) and several bacteria from the *Agrobacterium/Rhizobium* group (Hoyle, 2003). Such findings may allow for novel methods to control animal, human, and plant diseases.

Soil microbiologists have been instrumental in the discovery of a wide range of symbiotic plant-fungus associations collectively referred to as "**mycorrhizae**" (Chapter 12). These associations occur with most land plants and exhibit a variety of forms. Most notable perhaps are the so-called ectomycorrhizae that one might observe on roots of trees in the forest visible as coatings of fungal mycelium covering the roots. Agricultural plants like corn and wheat are colonized in a manner not nearly so observable. Here the fungi colonize the internal tissues, actually penetrating the cells of the host root (hence they are called endomycorrhizae), and are generally only visible after special staining and microscopic observation. Soil microbiologists have had a significant role in unraveling the biology of these interesting and important associations. Today they continue to study these associations looking for ways to manipulate them for greater enhancement of plant growth and fertilizer-use efficiency. On a more basic level, there is intense interest in the fungi that participate in the endomycorrhizae because we have not yet discovered how to mass-produce them in pure cultures. They behave as obligate symbionts and do not grow in culture media. Thus, our ability to produce high-quality, inexpensive inoculants remains an elusive goal. Researchers believe that the endomycorrhizae may also play an important role in soil aggregation by producing a glycoprotein called glomalin; this is part of a mixed pool of biomolecules, currently referred to as glomalin-related soil proteins, which is important in cementing soil particles into larger aggregates (Chapter 2). There is still much to be learned about these very intriguing, ecologically important associations.

Soil microbiologists are also investigating other plant-microbe associations in efforts to enhance crop production or to reduce the use of chemical inputs in agricultural systems. These organisms have been referred to as "plant-growth-promoting rhizobacteria" or PGPR (Chapters 11 and 23). Still others are investigating plant-microbe associations for the cleanup of contaminated soils (phytoremediation, Chapter 22). The range of plant-bacteria associations appears to be very large, and we are at a rudimentary understanding of how to exploit these beneficial bacteria at a practical level. There is great interest in the development of microbial inoculants; however, this is a difficult area of research. There is a strong need for scientists to unravel the mysteries of how and why certain plants and microbes form associations with each other and why some microbes do well as root colonizers and some do not. These very basic questions of plant-microbe biology and many others await future researchers. The continued study of microbial interactions in the **rhizosphere** (soil around the root; Chapter 11) offers great promise.

Microbial ecology

The early phases of soil microbiology were often dominated by attempts to characterize soil populations using **culture techniques** through laboratory growth on selective (favors a particular group of microbes over others) or non-selective media. While these attempts produced valuable information, little more could be said other than that there was some number of **colony-forming-units per gram** (CFU g^{-1}) of soil. We are becoming increasingly aware that our knowledge of the soil microbial community is far from complete. We have learned that many microbes in the soil exist in a "viable (alive) but nonculturable" state (Chapter 10). It has long been observed that as many as 99% of microorganisms in a given soil sample, as quantified using microscopy or another direct technique, are not detected using traditional culture-based methods. This has led many to surmise that the vast majority (90%–99%) of soil microorganisms are not culturable. While it is true that lab culture remains elusive for many soil organisms, creative modifications in isolation methods, media, and incubation conditions have demonstrated that isolation and culture of many of these previously uncultured organisms may be possible if the appropriate conditions are provided. In fact, research using new culture-based methods has experienced somewhat of a renaissance in recent years due to the realization above and the need for cultured type-strains to characterize **phenotypes** that can be connected with the wealth of sequence data currently being produced through DNA-based approaches.

The explosion in the number and breadth of DNA-based (e.g., **genomics**) and other "omic" methods over the past few decades has given us unprecedented ability to detect and characterize soil microorganisms without having to culture them. This has led to the discovery of previously unknown organisms and changes in our understanding of organisms controlling major soil processes (e.g., the importance of archaea in ammonia oxidation (Chapter 15)). **Polymerase-chain-reaction** (PCR) procedures and high-throughput genomic sequencing techniques, especially, have transformed our understanding of the taxonomy and diversity of the soil microbial community (Woese, 1990; Chapters 4 and 10). Other methods, such as **transcriptomics** and **proteomics**, allow us to measure the activity and interactions of specific organisms from the level of a single cell to a complex microbial community. Undoubtedly, these new methodologies will continue to greatly advance our understanding of soil microorganisms and their complex environmental interactions.

Biological control

Compared to the chemical control of pests, the advantages of biological control are numerous. Biological control is a natural mechanism that does not employ pesticides and therefore leaves no **xenobiotic** residues after treatment (Chapters 21 and 23). Early attempts at biological control using soil microbes followed Waksman's discovery of antibiotic-producing microbes in the soil. Unfortunately, most if not all of those attempts failed outright.

Since that time, soil microbiologists have learned that it is not sufficient to just increase the number of biological control agents in the soil to achieve success; a basic understanding of the ecology of the biological control agent is also needed. Amelioration of the factor(s) limiting the growth or survival of the control agents is necessary for biological control to work. While successful biological control experiments are few in number in comparison to failures, biocontrol research continues to be an area where soil microbiologists make significant contributions. As we learn more about biological control agents and their pest-suppression mechanisms, ecology, and genetics, this area of soil microbiology research should continue to increase in importance, especially with increasing interest in organic agriculture and restrictions on chemical pesticides.

Bioremediation

During the last 100 years, an increasing variety of chemicals has been purposefully or accidentally applied to soils. While most compounds tend to decay to their elemental components, some compounds are more **recalcitrant** and remain in the soil for prolonged periods. Since many of the compounds are considered hazardous pollutants, it is desirable to devise methodologies to remove them from the environment. A process relying on indigenous or introduced organisms as a way to accelerate or encourage the degradation of pollutants is known as **bioremediation** (Chapter 22). In certain instances, bioremediation may involve a process other than degradation. For example, sometimes it is adequate to simply biologically extract pollutants such as many metals and other harmful elements from soils or to transform them into forms suitable for later processing. Algae and plants are particularly good at accumulating nutrients and metals from the environment—one of the varied options for treatment via the process of "**phytoremediation**." For other metals, such as uranium, it may be preferable under certain conditions to convert the pollutant to a form that is less toxic and/or less mobile and maintain it in place. Thus, technologies involved in bioremediation need to be very pollutant- and site-specific and range from the additions of organisms with superior capacities to degrade pollutants or accumulate metal ions to strategies for altering site conditions. Often the utilization of specific organisms is coupled with strategies that seek to remove the site-limiting factors, such as adding essential nutrients or sparging with oxygen (O_2) to enhance aerobic reactions.

If bioremediation is to be successful, scientists must find ways to make it cost effective, predictable, and timely. Strategies such as selecting or genetically engineering organisms with superior remediation properties or altering inherent site conditions to allow indigenous or introduced organisms to function efficiently must be evaluated. In the future, soil microbiologists need to work together with other scientists, engineers, and people from the humanities to achieve the maximum progress in the use and acceptance of bioremediation. Soil microbiologists are uniquely positioned to take a leadership role in these endeavors.

Genetically modified microbes and engineered communities

Soil microbiologists are interested in discovering microbes with superior traits, including better N_2 fixers, superior biocontrol agents, and agents for bioremediation processes. In the past, we relied on natural variation and selection from natural populations to obtain organisms with better environmental fitness; however, with various molecular techniques now available, it has become possible to make specific modifications in an organism's genome to obtain desired phenotypic properties. In fact, we are entering the era of synthetic biology where scientists have even demonstrated the ability to completely replace the genome of a bacterium with chemically synthesized DNA resulting in a viable bacterium with very different properties. The possibilities for genetically engineering microorganisms for various processes, from arsenic methylation to nitrogen fixation to xenobiotic degradation, appear almost limitless.

Although genetically modified microorganisms can be generated in the laboratory quite easily and have found tremendous use in some industries (e.g., commercial production of insulin), they have not been widely used in soil. One reason for this is the inability of most introduced organisms to survive within the challenging soil environment (Chapter 2), especially in competition with the large and diverse native community already established and specifically adapted to that environment. Scientists have developed a variety of techniques that can increase the potential for success of **bioaugmentation** (involving introduction of organisms), but the organisms often fail to survive (Chapter 22). To get around this limitation, some researchers are beginning to investigate use of CRISPR-Cas-based technologies, which have been widely used for gene-editing in a variety of organisms, to directly introduce desired traits into indigenous soil microorganisms. If successful, this may open a new area of soil microbiome engineering.

As these potential opportunities move to reality, questions on the fate of genetically engineered or modified microbes deliberately released into the environment assume primary importance. How do these genetically modified organisms grow, compete, move, and survive in the soil environment? Do introduced genes jump from one microbe to another? Conversely, can bacterial genes introduced into crop plants escape to other plants or microbes? Can the traditional models of growth that describe the carrying capacities of organisms help us understand what might happen upon release of genetically modified organisms? Perhaps of critical importance in this discussion is whether or not the current public fear of genetically modified organisms (GMOs) can be overcome so that progress can continue unimpeded by distrust among the general population. Questions and opportunities abound and the likelihood that soil microbiology will remain an important field of investigation is very strong in this arena.

Summary

Many microbes have their origin in the soil or are closely allied with the soil environment. Throughout history these microbes have had a substantial impact on humankind. In some instances, the impacts have been beneficial while others have been detrimental (Doyle and Lee, 1986). On the beneficial side, microbes are absolutely essential in nutrient cycling and the sustainability of life on earth. On the detrimental side, microbes contribute to environmental problems, such as production of "greenhouse gases" and groundwater contamination with nitrate.

As we proceed deeper into the 21st century, soil microbiologists are faced with at least two choices. Will we continue to emphasize those areas of studies that have brought us to our present position, or will we be willing to take advantage of opportunities that will permit us to step aggressively into future decades? Traditionally, soil microbiologists have seized upon opportunities for the betterment of society through their work. In light of the environmental problems associated with increasing amounts of anthropogenic wastes such as sewage and biosolids, animal manures, food processing wastes, and industrial wastes, the efforts of soil microbiologists will be needed to process these substrates into less objectionable forms and to find practical ways for their safe return to the soil. Solutions to these and other emerging problems will have to be found. New analytical procedures and perhaps genetic manipulations will be needed to tame future environmental problems likely to occur regardless of their source. In the future, we must advance our understanding of microbial ecology, enhance the opportunities for bioremediation, and learn how to harness the diversity of soil populations. If the past is any indication, the future of soil microbiology in both basic and applied research is very bright. Perhaps the solution of a pressing food-production or environmental problem will propel some soil microbiologist (perhaps yourself) into the ranks of the Nobel Prize winners, just as Selman Waksman was honored for his work on antibiotics.

Acknowledgments

Portions of this chapter are derived from the work of Dr. Arthur G. Wollum II (North Carolina State University) who authored a similar chapter for previous editions of the textbook. We thank him for use of this material in the Third Edition.

Study questions

1. In what ways is soil a unique environment/habitat for the growth of microbes?
2. What are the basic characteristics of a living cell?
3. Identify some individuals or events that had the greatest influence on the development of soil microbiology?
4. What modern issues draw soil microbiology and the environmental sciences closer together?
5. Why is Sergei Winogradsky often considered the founder of soil microbiology?
6. What did Jacob G. Lipman contribute to the advancement of soil microbiology?
7. What are some of the challenges/opportunities for soil microbiology in the 21st century?

References

Alexander, D.G., Zuberer, D.A., Hubbell, D.H., 1989. Microbiological considerations for lunar-derived soils. In: Ming, D.W., Henninger, D.L. (Eds.), Lunar Base Agriculture: Soils for Plant Growth. American Society of Agronomy, Madison, WI, pp. 245–255.

Chung, K.T., Ferris, D.H., 1996. Martinus Willem Beijerinck (1851-1931). Pioneer of general microbiology. ASM News 62, 539–543.

Clark, F.E., 1977. Soil microbiology—It's a small world. Soil Sci. Soc. Am. J. 41, 238–241.

Doyle, R.J., Lee, N.C., 1986. Microbes, warfare, religion and human institutions. Can. J. Microbiol. 32, 193–200.

Hoyle, B., 2003. *Brucella suis* genome like those of several plant, animal pathogens. ASM News 69, 9–10.

Lipman, J.G., 1908. Bacteria in Relation to Country Life. Macmillan, New York.

Löhnis, F., 1910. Handbuch der Landwirtschaftlichen Bacteriologie. Borntrager, Berlin.

Madigan, M.T., Bender, K.S., Buckley, D.H., Sattley, W.M., Stahl, D.A., 2019. Brock Biology of Microorganisms, fifteenth ed. Pearson, New York, NY.

Metting, F.B., 1993. Structure and physiological ecology of soil microbial communities. In: Metting, F.B. (Ed.), Soil Microbial Ecology: Applications in Agricultural and Environmental Management. Marcel Dekker, Inc, New York, pp. 3–25.

Pace, N.R., 2009. Mapping the tree of life: Progress and prospects. Microbiol. Molec. Biol. Rev. 73 (4), 565–576.

Waksman, S.A., 1927. Principles of Soil Microbiology. Williams and Wilkins, Baltimore, MD.

Waksman, S.A., Starkey, R.L., 1931. The Soil and the Microbe: An Introduction to the Study of the Microscopic Population of the Soil and Its Role in Soil Processes and Plant Growth. John Wiley and Sons, Inc, New York.

Woese, C.R., 1990. Towards a natural system of organisms. Proposal for the domains Archaea, Bacteria, and Eucarya. Proc. Natl. Acad. Sci. USA 87, 4576–4579.
Woese, C.R., 1998. The universal ancestor. Proc. Natl Acad. Sci. USA 95, 6854–6859.

Further reading

Alexander, M., 1986. Highlights of research in Division S—3, soil microbiology and biochemistry since 1961. Soil Sci. Soc. Am. J. 50, 839–840.
Allison, F.E., 1961. Twenty-five years of soil microbiology and a look to the future. Soil Sci. Soc. Am. Proc. 25, 432–439.
Lowdermilk, W.C., 1975. Conquest of the land through seven thousand years. USDA Soil Conservation Service. Agricultural Information Bulletin No. 99.
Lynch, J.M., Hobbie, J.E., 1988. Microbes in Action: Concepts and Applications Microbial Ecology, second ed. Blackwell Scientific Publications, Boston.

CHAPTER

2

The soil habitat

Julie A. Howe, A. Peyton Smith

Department of Soil and Crop Sciences, Texas A&M University, College Station, TX, United States

There can be no life without soil and no soil without life; they have evolved together.—**Charles E. Kellogg, 1938.**

Soil description

Soil is defined as a dynamic natural body because of its living component and ability to change over time and with management. The majority of soils on Earth are primarily composed of inorganic or mineral particles (i.e., they are **mineral soils**). **Organic soils**, which contain more than 20% organic materials, make up <1% of the world's soils (Gardiner and Miller, 2004). Most mineral soils contain <10% organic matter and often <2%. Organic materials provide the dark brown and black colors found in surface soils. Other soil colors are derived from either the bare mineral grains associated with the original geologic material from which the soil developed (e.g., rocks, minerals, sediments) or weathering products such as iron oxides or other secondary minerals that commonly coat mineral grain surfaces. For example, under oxic conditions, iron released by mineral weathering becomes oxidized to produce the red-, orange-, and yellow-hued minerals that are reflected as the color of many soils. However, under anoxic conditions (e.g., soils subject to long-term water saturation), iron can form compounds imparting gray, green, or blue coloration.

Horizons

A soil is composed of layers, each with distinct characteristics, called **horizons**. The layers exposed in a vertical slice of soil are called a **profile** (Fig. 2.1). Soils can have various combinations or subsets of what are referred to as master horizons, the most common are (top to bottom) O, A, E, B, and C. Sometimes rock is designated as R below the soil master horizons. The **O horizon** is a non-mineral layer formed from organic residues and is most commonly found in forest soils (e.g., leaf litter), organic soils (e.g., peat moss), and frozen soils. The **A horizon** is the topmost mineral horizon and usually is darkened by humified organic matter coating soil particles. Theoretically, this horizon is present in all soils, but there are some cases where it is extremely thin, blends into other horizons, or is eroded away. As soils form, soil materials

Principles and Applications of Soil Microbiology. http://dx.doi.org/10.1016/B978-0-12-820202-9.00002-2
Copyright © 2021 Elsevier Inc. All rights reserved.

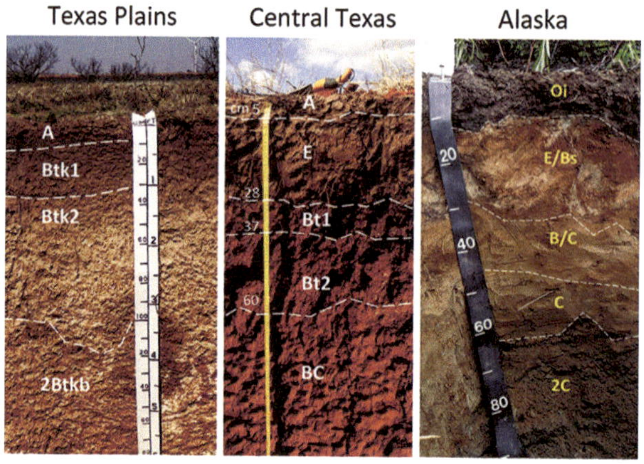

FIGURE 2.1 **Example soil profiles illustrating the diversity of soil types.** The Texas Plains soil profile developed from wind-deposited material in the panhandle of Texas. The A horizon is darkened from organic matter inputs from vegetation. The subsoil exhibits accumulations of clay and carbonates (i.e., Btk1 and Btk2) that formed over another soil (2Btkb). The Central Texas soil profile developed from alluvial deposits under savanna vegetation. The profile contains an eroded A horizon, a weakly-developed E horizon, and accumulations of 20%–30% clay in the underlying B horizons (i.e., Bt1, Bt2) that transition into the C horizon (BC). The Alaska soil profile developed in volcanic ash and loess (i.e., windblown silt) deposits overlying stratified sandy glacial outwash material in southern Alaska. The profile has an organic Oi horizon (the "i" indicates "fibric" or relatively non-decomposed material) and several horizons mixed from bioturbation (E/Bs and B/C) due to tree falls following fires. In addition, there are two types of parent materials: loess and glacial deposits (C and 2C). *Source: Photograph of Texas Plains soil courtesy of Craig Byrd, NRCS, and Central Texas and Alaska soils courtesy of J.A. Howe, Texas A&M University.*

in the A horizon move down in the profile with water in a process called **eluviation**. These materials can be mineral or organic matter. Common minerals that eluviate include silicate clay minerals, iron and aluminum oxides, and carbonates. Although the A horizon is continuously darkened by organic matter additions (e.g., leaf litter, root death), the depth to which darkening occurs is limited, and this depth may be less than the zone of eluviation. Therefore, a depleted horizon may eventually form below the A horizon. This depleted layer is called the **E horizon**, and it is typically lighter in color than the A horizon. The soil materials that are lost from the A and E horizons are deposited into the **B horizon**. This is a zone of **illuviation** (accumulation). In the early stages of B horizon development, only changes in structure and/or color may be evident. As the soil ages and the horizons develop further, the illuviation of materials, such as silicate clays, becomes more apparent. In semi-arid to arid environments, lack of water percolation causes the chemical precipitation and deposition of sparingly soluble minerals, such as salts and carbonates. In humid environments, water contributes to weathering and formation of silicate clays and oxides of iron and aluminum. The **C horizon** is located under the B horizon and represents the parent material the soil was derived from. Not every soil has all horizons, and horizons can be further subdivided based on their properties. Furthermore, sequences of horizons can sometimes repeat within a profile, especially in areas where soils are buried by deposition of sediments from flooding or windblown particles.

There are many different subdivisions (also called **subordinate distinctions**) of the master horizons denoted by a lowercase letter following the master horizon letter. The A horizon is often disturbed by human activity, primarily through plowing, and is then designated as an

Ap horizon. Plowing usually affects the top 6–8 inches (15–20 cm) of soil. Because plowing was historically associated with crop production, nutrient recommendations are still based on an **acre furrow slice** (AFS) or **hectare furrow slice** (HFS), which assumes an approximate plowing depth of 6 inches (15 cm) and bulk density of 1.5 g cm^{-3}. Roughly 2,000,000 lb or 2,250,000 kg of soil are in an AFS or HFS, respectively. In some soils, this is a mixture of A and E or A and B horizons. The most common subdivisions in the B horizon are Bw, Bk, Bh, Bt, and Bo, but there are many others. Bw ("w" for "weak") indicates a soil with some change in color or structure relative to the A and C horizons. It is often the initial B horizon formed and relatively coarse textured. As soils develop, weathering results in the dissolution of minerals and re-precipitation of new minerals. In arid and semi-arid regions, moisture may not be sufficient to leach some soluble ions from the profile (e.g., calcium) causing them to precipitate as secondary minerals. Calcium carbonate is a common secondary mineral in arid and semi-arid regions, and its accumulation in the soil profile is indicated by Bk. In more humid environments, stabilized organic materials, or humus, may move through the profile and be deposited in a Bh horizon. Similarly, silicate clay deposition is indicated by Bt ("t" for the German word "ton" for clay). Under conditions of intense weathering, silica in silicate clays dissolves and leaches from the soil profile causing iron and aluminum oxide minerals to form. These highly weathered horizons are designated with Bo ("o" for oxide) and are typically only found on stable landscapes in the tropics.

Soil nomenclature

Soil classification is hierarchically organized by order (broadest level), suborder, great group, subgroup, family, and series (most specific level). The taxonomic names, with the exception of those for series, are constructed from descriptive word roots (called formative elements) and can therefore provide considerable information about a soil. The soil order is indicated in the last 2–3 letters of the taxonomic name with the key letters of the order. There are 12 soil orders (key letters indicated by bold): **Alf**isol, **And**isol, Ar**id**isol, **Ent**isol, **Gel**isol, **Hist**osol, Incep**t**isol, **Moll**isol, **Ox**isol, Sp**od**osol, **Ult**isol, and **Vert**isol. Some orders indicate specific features, materials, or conditions required for formation (Table 2.1). The other parts of the taxonomic name represent factors such as soil moisture regimes (Table 2.2), soil temperature regime (Table 2.3), and special conditions of formation. For example, Houston Black clay is a soil series found in Texas, which has a taxonomic class of Fine, smectitic, thermic Udic Haplustert. The last three letters "ert" indicates that it is a Vertisol and thus has high clay content with shrink and swell properties. The "ust" indicates it is in the ustic moisture regime. The "hapl" prefix means minimum horizon development. The "udic" at the beginning of the taxonomic name is in the place of the subgroup and indicates that the soil moisture regime borders on udic. Lastly, "fine, smectitic, thermic" is the family, which represents specific characteristics that can be used to predict management concerns. In this example, the soil will have considerable amounts of smectitic clay (i.e., fine, smectitic) and be found in a warm temperature regime (i.e., thermic). Smectitic clays expand when water enters their structure during wet conditions and shrink upon drying. In dry conditions, soil cracking is common (Box 2.1).

Information about the specific characteristics of soil at a particular site can be investigated using the online tools created by the Natural Resource Conservation Service (NRCS) such as Web Soil Survey (http://websoilsurvey.sc.egov.usda.gov) and Soil Web (https://casoilresource.lawr.ucdavis.edu/gmap/), which is also available as a mobile phone app. In addition

TABLE 2.1 Soil order descriptions.

Soil order	General features	Implications for microbial habitat
Alfisols[a]	Moderately weathered soil. Silicate clay accumulation in the B horizon.	Clay accumulation in B horizon may be compacted. Typically form in relatively warm semi-arid to humid environments.
Andisols	Formed from volcanic materials.	Volcanic ash is extremely light and maintains good aeration. Volcanic materials are known to adsorb and bind P tightly.
Aridisols	Limited moisture conditions.	Limited moisture usually corresponds to sparse vegetation and low soil organic matter. Soil pH is usually > 7. B horizon present.
Entisols	Very young soil, typically only A horizon present.	This soil order is highly variable.
Gelisols	Permafrost near soil surface.	Temperature usually limits microbial activity. Soil organic matter is usually high. Often soils are seasonally wet due to freeze/thaw cycles.
Histosols	Organic soils.	Organic residue decomposition is reduced when oxygen is limited, which occurs when drainage is limited.
Inceptisols	Young soil, typically with A and newly formed B horizons; "Inception" of a B horizon.	This soil order is highly variable.
Mollisols	Grassland ecosystem leads to a relatively deep and dark-colored soil from high organic inputs from grass roots.	Mollisols often develop in semi-arid climates. The parent materials are often calcareous resulting in pH > 7. Soil organic matter can be relatively high for a mineral soil.
Oxisols	Highly weathered soil found in tropical areas on stable landforms. Fe and Al oxides accumulate in B horizon.	Fertility is typically low unless there is significant organic matter addition. Oxides do not retain nutrients, except for P, which is bound tightly and largely unavailable.
Spodosols	Relatively moist, often cool, coniferous forest with sandy parent materials. Must have eluviation of organic materials and deposition of Fe and Al oxides in subsoil.	Soil is typically sandy and acidic. Generally unproductive for row-crop agriculture.
Ultisols	Highly weathered soil in humid temperate to subtropical climates. Silicate clay and Fe and Al oxides accumulate in B horizon.	Soils tend to be acidic due to weathering. Fertility is typically low unless managed. Clay can be compacted in the B horizon.
Vertisols	High clay content, shrink/swell properties, and evidence of special features such as slickensides (planes of weakness).	Generally moderately high organic matter for a mineral soil. Soils are typically drained, but drainage is slow when cracks are not apparent.

[a] *Underlined portion of order name is used to indicate the soil order at the end of the taxonomic name.*

TABLE 2.2 Soil moisture regime names and descriptions.

Regime	Description
Aquic (aqu) [a]	Saturated with water for sufficient time to deplete oxygen
Udic (ud)	Humid climate, usually moist soil
Ustic (ust)	Semiarid climate, soil is moist ½–¾ of the time
Aridic (arid)	Arid climate, usually dry, soil not moist for >90 consecutive days
Xeric (xer)	Cool, moist winters and warm, dry summers (Mediterranean climate)

[a] *Typical abbreviation used in taxonomic classification in parenthesis.*

TABLE 2.3 Soil temperature regimes and corresponding mean annual soil temperatures (MAT).

Regime[a]	Mean annual temperature (MAT) at a soil depth of 50 cm
Gelic	< 1°C, permafrost
Cryic	< 8°C, but permafrost not present
Frigid	< 8°C (warmer in summer than cryic)
Mesic	8°C–15°C
Thermic	15°C–22 °C
Hyperthermic	> 22°C

[a] *If difference between mean summer and mean winter temperature is not > 6°C at a depth of 50 cm, then add "iso" prefix to temperature regime name.*

to information regarding soil classification, the website provides important information on the chemical and physical characteristics of each soil, erosion potentials, and land use characteristics that are helpful to researchers and land use planners.

Soil components

Soil contains both particles and pore space (i.e., voids). Particles are inorganic (e.g., mineral) or organic. Pore space is occupied by either air or water (Fig. 2.2). The soil microbial community, as well as soil meso- and macro-organisms, occupies the pore space. For many soil processes, pore space is ideally filled approximately half with water and half with air, but this varies due to factors affecting soil water content (e.g., precipitation, evaporation, plant transpiration). Thus, the properties of the particles and pores have a major influence on the microbial habitat.

Soil particles

The inorganic particles in soil are minerals (e.g., feldspars, mica, hornblende, and olivine) that form through the weathering and disintegration of rocks or are synthesized from degraded constituents of other minerals (e.g., iron and aluminum oxides and silicate clays

BOX 2.1

Cracking soils

Soils with high smectitic clay content tend to form cracks when dry (A, B). Smectites, particularly montmorillonite, swell when wet and contract when dry. The cracks allow rapid initial infiltration in these clayey soils when it starts raining. As the clay swells, the voids close reducing infiltration dramatically. Soils with a high tendency to shrink and swell are either placed into the Vertisol soil order or the term "vertic" is added to their taxonomic name if they cannot be classified as Vertisols. Cracking is problematic for roads (C), foundations, and other structures (D) built upon shrink and swell soils. Photographs from the Blackland Prairie of Texas.

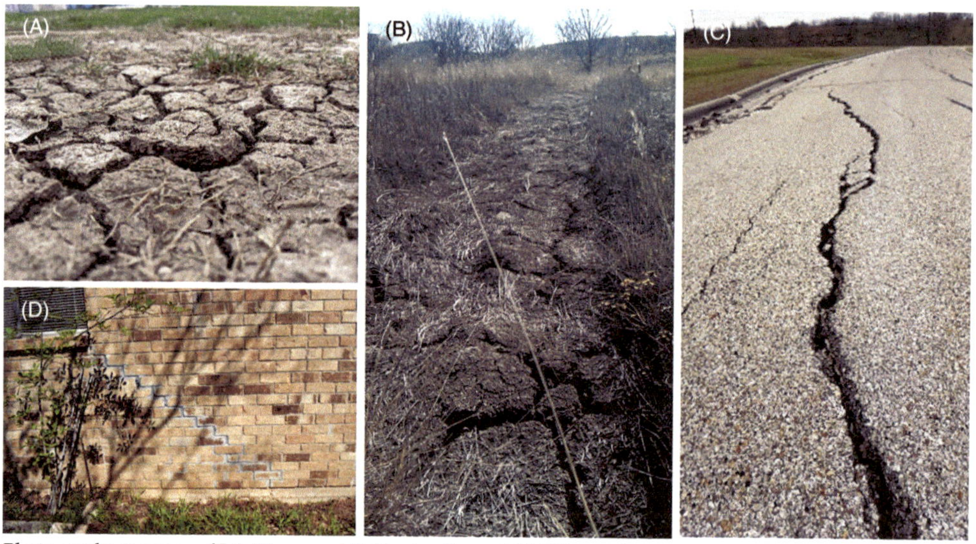

Photographs courtesy of J.A. Howe and T. Gentry, Texas A&M University.

such as smectite, vermiculite, and kaolinite). Soil mineral particles are usually further classified by size. The United States Department of Agriculture (USDA) defines soil inorganic particles as **clay** (< 0.002 mm or 2 μm), **silt** (between 0.002 and 0.05 mm), **sand** (between 0.05 and 2.0 mm), and gravel or stones (> 2.0 mm). In comparison, bacterial cells are typically 0.0005–0.005 mm (0.5–5 μm) in length.

The percentage particle size distribution of clay, silt, and sand in a soil on a mass basis represents the **soil texture** (Box 2.2). Organic matter as well as stones and gravel > 2 mm are removed prior to textural analysis. The percentages are used to define 12 textural classes according to the textural triangle (Box 2.2). Fine-textured soils are dominated by clay and silt, while coarse-textured soils are dominated by sands. Of the three particle sizes, clay has a

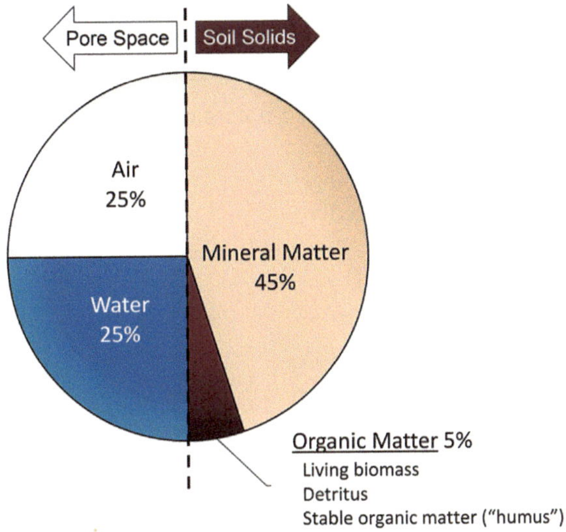

FIGURE 2.2 **Soil has four major components: mineral matter, organic matter, water, and air.** Ideally, 50% of the volume is made up of pore space for air and water, while the remaining 50% is solid mineral or organic materials. Organic matter content and composition varies considerably from soil to soil and depends largely on the climate and biota. It is composed of living biomass, detritus (i.e., non-living biomass), and stable organic matter.

disproportionate influence on soil properties. Even at 40%, clay dominates the physical properties of the soil (Box 2.2). The loam textural class indicates that the influence of sand, silt, and clay are relatively equal. Note that the loam textural class has approximately half as much clay as either silt or sand. Texture is considered an inherent property of soils because it is not affected by management or land use on a time scale that is relevant to humans.

Clay and soil organic matter are considered the most reactive components of the soil. Their small size translates to high surface area, which has a disproportionate influence on soil properties. For example, sand, silt, and clay are commonly reported to have average surface areas of approximately 50, 500, and 8,000,000 $cm^2\ g^{-1}$, respectively. Clay and soil organic matter particles have the ability to hold water and nutrients, join soil particles together into aggregates, moderate changes in temperature and pH, influence water movement, and provide a reservoir of nutrients for plants and microbes.

Soil structure

The arrangement of sand, silt, clay, and organic matter into groups of particles, called **aggregates**, defines **soil structure** (Fig. 2.3). Soil structure is classified by shape, size, and stability. Shapes include spheroidal (e.g., crumb, granular), blocky (e.g., angular blocky, subangular blocky), prism-like (e.g., columnar, prismatic), and platy (Fig. 2.4). Soils are structureless when no aggregates are formed and particles act as single grains ("single-grained" structureless state) or when they are compressed into continuous blocks (massive). Soil texture and structure largely determine overall porosity and the relative abundance of small and large pores, the latter being particularly important in providing aeration and water drainage. Spheroidal particles are usually smaller and have more surface area and generally more

BOX 2.2

Soil texture

Soil textures are based on the percentages of sand, silt, and clay in a soil. Together they add up to 100%.

Soil texture analysis methods

Laboratory analysis of soil particle size is performed on oven-dry soils that have been ground and sieved to remove organic materials and particles > 2 mm. There are several acceptable methods, but the Bouyoucos (1927) method illustrates the principles involved.

The Bouyoucos method (1927) utilizes Stoke's Law, which states that particles will fall in a column of water at a rate proportional to their size, that is, larger (heavier) particles settle faster than smaller particles. The soil sample is chemically and mechanically dispersed, typically in 1 L of water, and then allowed to settle. The density of suspended particles is measured by a hydrometer, which is a bulb-like floating scale that measures the density of a liquid at a defined temperature. The scale of a hydrometer used for soil texture has been adjusted to report the concentration of suspended particles in grams per liter. Hydrometer readings are taken at specified times, typically when sand will have settled, but silt and clay are suspended, then again after sand and silt have settled, but clay remains suspended. Using these two measurements and difference, the percent sand, silt, and clay can be determined.

Soil texture triangle

The soil textural triangle (next page) is used to determine the textural class of the soil. The three sides of the triangle have scales for sand, silt, and clay, respectively. To determine the textural class, follow the arrows corresponding to the particle size at the appropriate percentage. The point at which the percentages intersect indicates the textural class. For example, if a soil is 30% clay, 40% silt, and 30% sand, the lines intersect in the clay loam textural class.

(Continued)

BOX 2.2 (Cont.)

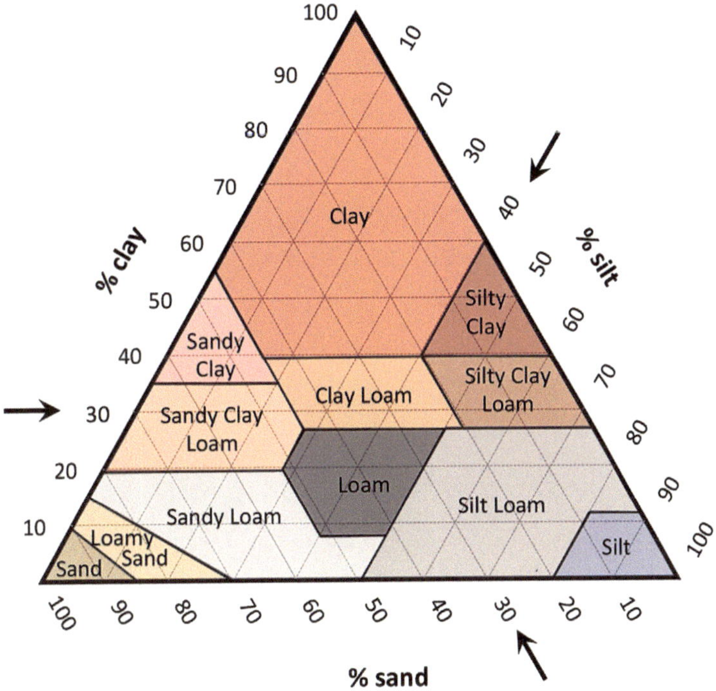

Textural triangle adapted from Soil Survey Staff (1999).

rapid permeability than other aggregate types. These structural types are commonly found in surface horizons.

A prevalence of water-stable aggregates is often a good indicator of soil health and the ability of a soil to resist erosion and sequester carbon (Chapter 14). Aggregate stability integrates the influences of many factors including clay content, type of clay, extractable cations, organic matter, soil organisms, and management. Management practices, such as tillage and intensive cropping, typically reduce aggregate size and stability through physical destruction and subsequent reduction in soil organic matter. Other physical processes, such as freeze-thaw and wet-dry cycles, can either disrupt or enhance soil aggregation.

The type and amount of clay in a soil plays a large role in aggregation. Some silicate clays, such as smectitic clays, have a strong tendency to shrink and swell, which causes natural fractures and fissures within the soil matrix that promote aggregation. Compared to sand and silt, clay particles have a higher tendency to flocculate and aggregate naturally, so as soil clay content increases, aggregation potential increases. Flocculation involving clay minerals is one of the main drivers of soil aggregation, especially in the formation of very small aggregates (~ 10 μm diameter). Aggregates of this size are predominantly composed of silt, clay, and

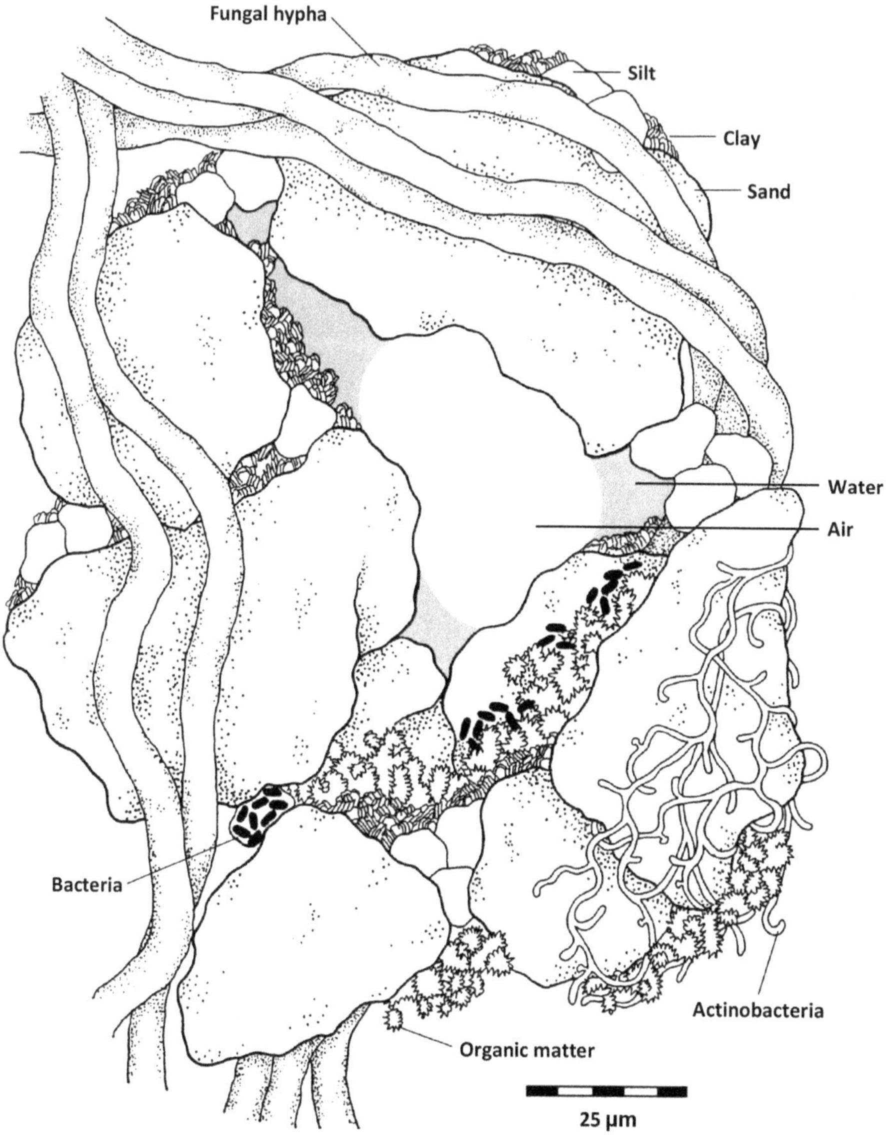

FIGURE 2.3 **A typical soil aggregate.** Here sand, silt, and clay particles, joined by organic matter, precipitated inorganic materials, and microorganisms, bind the soil particles together to form an aggregate. Note how the water *(light gray areas)* form a meniscus surrounding the air space (center). Bacteria (rods in organic matter, rods in a polysaccharide "plug," and actinobacterial filaments/hyphae) and fungus (hyphae only), as well as the sand, silt, and clay particles are all to scale. Also note how an aggregate can offer a diverse set of microsites for microbial habitation over very small distances. *Source: Original drawing by K. Luoma.*

FIGURE 2.4 **Soil structural shapes include spheroidal (e.g., crumb, granular), blocky (e.g., angular blocky, subangular blocky), prism-like (e.g., columnar, prismatic), and platy.** In addition, there are two structureless categories: single grain and massive. Within the blocky and prism-like categories, subtypes are commonly differentiated by the roundness of their edges with subangular blocky having rounder edges than angular blocky and columnar having rounder edges than prismatic. Image for platy structure is shown from two different aspects: top and side. *Source: Photograph courtesy of J.A.Howe, Texas A&M University.*

organic matter (Fig. 2.5). The type of exchangeable (extractable) cations associated with negatively charged clay particles also affect this tendency. Divalent cations, such as Ca^{2+} and Mg^{2+}, have high charge densities and generally promote flocculation by effectively countering the negative charges on clays, while monovalent cations are less effective. Sodium ions, in particular, have a very low charge density and therefore actually promote clay dispersion. Organic matter and other materials from soil organisms contribute to aggregation by grouping, binding, and/or cementing particles together, thereby increasing aggregate stability (Fig. 2.5).

Role of microorganisms in aggregation

Biotic factors, such as burrowing organisms, plant roots, and soil microorganisms, are central to forming and stabilizing soil aggregates both physically and chemically. Burrowing by soil animals (e.g., worms, insects) and the expansion of plant roots can disrupt soil particles but can also create stable aggregates by physically compressing particles together (Chapter 11). At the same time, plant roots secrete organic compounds, such as mucilages, that chemically and physically bind soil particles. Soil microorganisms also produce "glue-like" compounds, for example, extracellular polymeric substances (EPS), which cement soil particles, organic matter, and clay minerals together, forming water-stable aggregates. Soil fungi and filamentous bacteria (actinobacteria), in particular, produce a sticky network of EPS

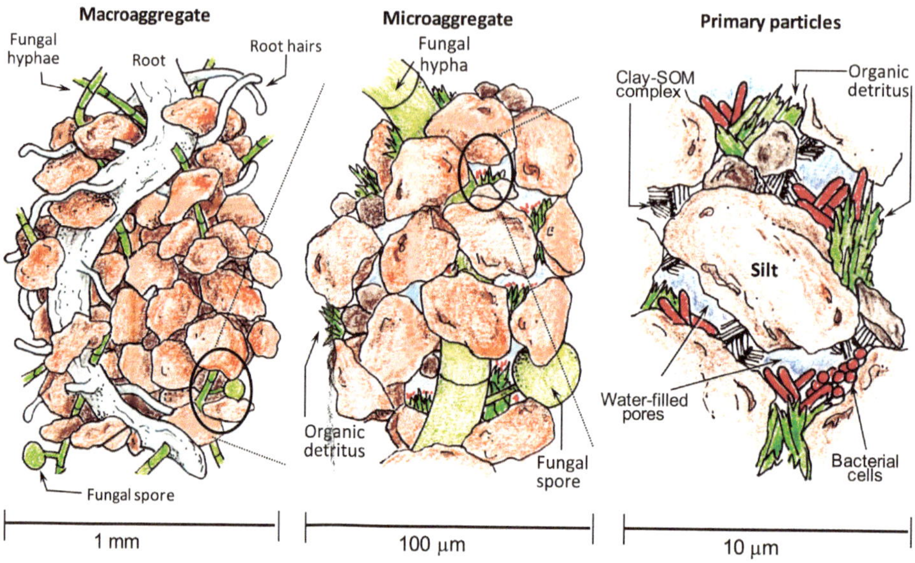

FIGURE 2.5 **Soil aggregates depicted at different scales.** Macroaggregates (left) consist of groups of particles and microaggregates that are bound together by plant roots and fungal hyphae. Microaggregates (center) are often found within macroaggregates and composed of clusters of clay, fine sand, and silt particles enmeshed by root hairs, fungal hyphae, and sticky microbial exudates. The interior of microaggregates (right) shows the attractive forces between particles as well as the intermingling and binding of organic materials that create fine clusters. *Source: Original drawing by K. Luoma.*

and hyphae that enmesh microaggregates together to form macroaggregates (Fig. 2.5). Similarly, glomalin-related soil proteins (GRSP) are reported to enhance soil aggregate stability by providing a chemically recalcitrant "glue" that coats and binds soil aggregates (Chapter 12). Historically described as being a glycoprotein derived from arbuscular mycorrhizal fungi, GRSP is now considered an extraction-defined substance that represents a complex mixture of fatty acids, lipids, and proteins that may or may not be entirely fungal in origin (Gillespie et al., 2011). Overall, microorganisms and their byproducts play a role in forming and stabilizing aggregates of every size (Fig. 2.5). At the same time, aggregates of different sizes create different habitats for soil microorganisms (Box 2.3).

Soil pores

The size and distribution of pore space in soil plays a major role in air and water movement and defines the confines of microbial habitats (microsites). Pores vary considerably in diameter and connectivity. In some soils with shrink-swell properties, such as Vertisols, voids can be around a centimeter in diameter or larger, while voids between some particles (e.g., compacted soil) may only be a nanometer across. The size and connectivity of the pores affect the flow and retention of water. Larger pores allow water to move more quickly and often remain air-filled when soils are drained. Smaller pores hold water tightly due to a greater influence of adhesive and cohesive forces, thereby slowing water flow and causing these pores to retain water against the force of gravity (Fig. 2.6).

The pore-size classification system most commonly used for soils recognizes five classes: macropores, mesopores, micropores, ultramicropores, and cryptopores (Soil Science Society of America, 1997). In this system, macropores are >75 μm, mesopores are between 30 and

BOX 2.3

Aggregates as microbial habitats

The arrangement of macroaggregates, microaggregates, and silt- and clay-sized particles generates a multitude of microsites that differ in the availability of nutrients and energy resources (O_2, H_2O, pH, organic matter, etc.). As such, aggregates of different sizes provide different habitats for soil microorganisms. In general, macroaggregates contain relatively more fresh litter substrates, compared to the silt and clay-sized fraction that contains highly-decomposed plant and microbial debris; thus, they may select for different microbial taxa. Aggregate stability and size distribution are dynamic soil properties that influence the potential for soils to sequester carbon. In an experiment investigating the effects of tropical land-cover change on soil aggregate stability and microbial communities, different soil aggregate sizes were separated using a wet-aggregate stability method (Smith et al., 2014). Aggregates were separated into the following size classes: macroaggregates (250 μm–2 mm), microaggregates (53–250 μm), and silt- and clay-sized fractions. Microbial community composition (phospholipid-fatty acid analysis) and microbial activity (potential reaction rates of select extracellular enzymes) were measured for each aggregate fraction. Results showed that the fungal-to-bacterial ratio and microbial enzyme activity potential was greatest in macroaggregates compared to the silt-and-clay fraction, which paralleled results showing that the silt-and-clay fraction contained more highly decomposed, and therefore less metabolically useful, organic matter than did the macroaggregates. These results suggest that the decreased fungal abundance associated with silt- and clay-sized fractions is due to reduced substrate bioavailability, differing microbial preference for substrate type, and restricted physical access of fungi to colonize the fine pore spaces associated with fine soil fractions. As illustrated in figure on the next page:

(A) Across the tropical land covers examined, β-glucosidase enzyme activity was higher in macroaggregates and microaggregates compared to silt and clay-sized soil fractions. β-glucosidase is a cellulose-degrading enzyme important in organic matter and plant decomposition. Lowercase letters denote significant differences.

(B) The fungal-to-bacterial ratio (F:B) differed among aggregate fractions and land cover, with higher F:B in macroaggregates compared to silt and clay-sized fractions in the pastures and early successional forests.

Used with permission by A. Vilches, E. Marin-Spiotta, and A.P. Smith.

(Continued)

BOX 2.3 (Cont.)

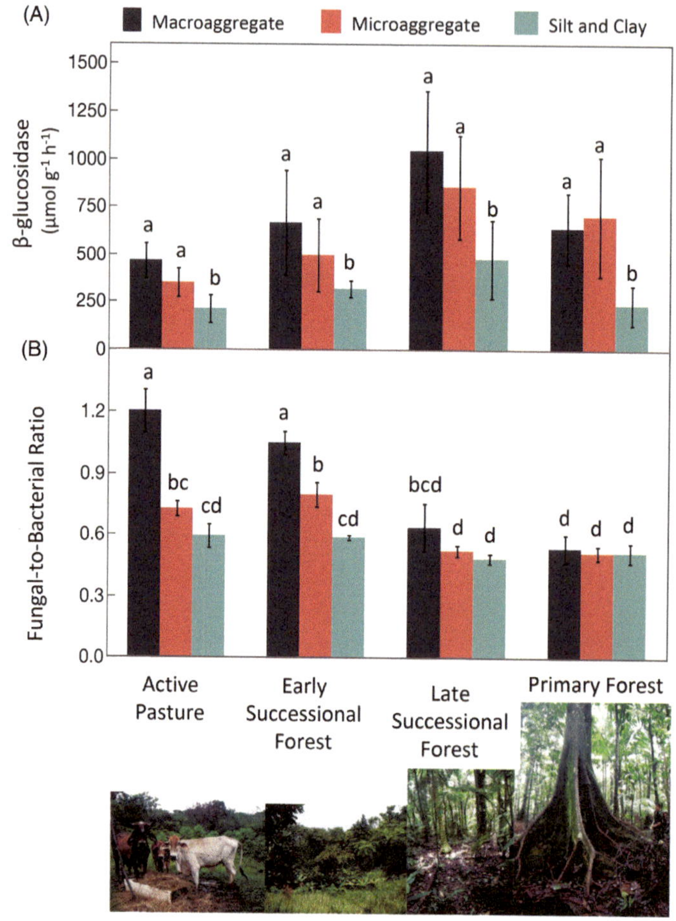

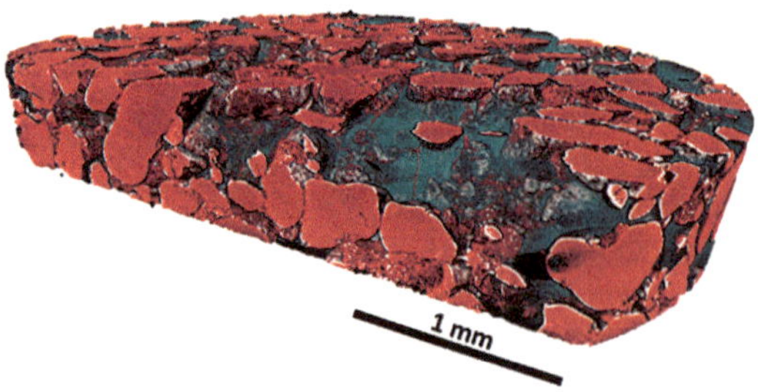

FIGURE 2.6 **Segment from a water-saturated sandy soil core with 3.3 mm diameter and 0.65 mm height imaged in 3D using µXCT at 1.3 µm voxel resolution.** Soil particles are shown in *red*, and the water-filled pore space is *dark teal*. Air-filled pore space, mostly absent from this image, is shown in *black*. *Source: Image courtesy of T. Varga, A.P. Smith, and V. Bailey.*

TABLE 2.4 Pore-size classifications from Soil Science Society of America (SSSA), International Union of Pure and Applied Chemistry (IUPAC), and International System of Units (SI).

Size class name	SSSA (Soil Science Society of America, 1997)	IUPAC (Rouquerol et al., 1994)	SI (Mays, 2007)
	μm		
Cryptopores	< 0.1	—	—
Nanopore	—	—	0.0001–0.1
Ultramicropores	0.1–5	—	—
Micropores	5–30	< 0.002	0.1–100
Mesopores	30–75	0.002–0.050	—
Macropores	> 75	> 0.050	—
Millipore	—	—	100–100,000

75 μm, micropores are between 5 and 30 μm, ultramicropores are between 0.1 and 5 μm, and cryptopores are <0.1 μm. Other pore size classification systems, such as International Union of Pure and Applied Chemistry (IUPAC) and International System of Units (SI), are used less commonly for soils and may differ in both the class name and size definition (Table 2.4). The characterization of soil pores is an active research area with a recent push to better define pore sizes based on their functional attributes (e.g., hydraulic functions such as water-holding capacity and drainage), inter- or intra-particle location, or pore connectivity (Kutílek and Jendele, 2008; Luxmoore, 1981; Mengistu et al., 2019).

Coarse-textured soils will generally have pores that do not retain water and thus drain well. Fine-textured soils are usually better at holding water and much slower to drain. Aggregation of clay particles promotes macroporosity, which facilitates water movement, drainage, and aeration in fine-textured soils. Organic matter similarly contributes to aggregation and drainage. Conversely, the micropores associated with organic matter and fine-textured aggregates retain water. Thus, aggregation creates pores of variable sizes that collectively provide both drainage and storage of water.

Bulk density

The relationship between soil solids and pore space can be represented by several different terms. **Bulk density** is one of the more commonly measured soil properties and is inversely related to porosity. Bulk density represents the mass of dry solid particles per unit volume of soil including the pore space. Because pore space has no mass in dry soil (i.e., all pores are filled with air), bulk density decreases with increasing pore space (Fig. 2.7). Bulk density of mineral soils ranges from approximately 1–2 g cm^{-3}; however, this represents a wide range of soil conditions and can be lower if a soil has significant organic matter content. For optimal plant growth, bulk density should be <1.6, 1.4, or 1.1 g cm^{-3} for sandy, silty, and clayey soils, respectively (USDA, 2008).

Bulk density should not be confused with **particle density**, which represents the mass of dry solid particles per unit volume of the particles alone (i.e., pore space is excluded; Fig. 2.7). Most inorganic soil particles have a particle density of 2.60–2.75 g cm^{-3}, but often 2.65 g cm^{-3} is used as a standard value. Particle densities of organic particles are more variable and

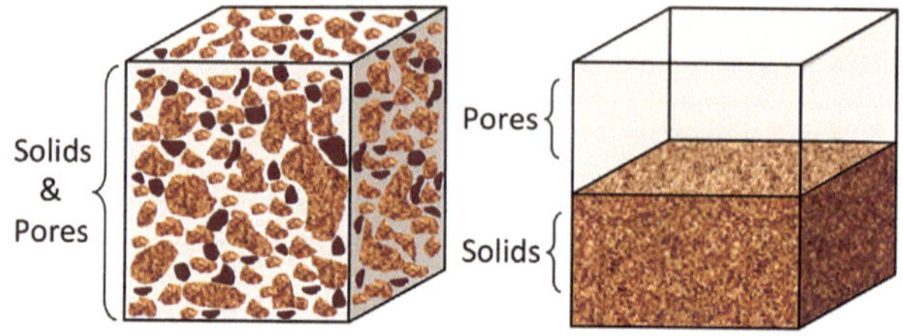

FIGURE 2.7 **Bulk density represents the mass of dry solid particles per unit volume of soil (image on left including solids and pores), while particle density represents the mass of dry solid particles per volume of particles (image on right where only the solid volume is included).** Bulk density decreases with increasing pore space as well as increasing soil organic matter content.

considerably lower than inorganic particles; therefore, soils with significant (>20%) organic material content may have bulk densities <1 g cm^{-3}.

Porosity can also be quantified more explicitly as total porosity, air-filled porosity, and water-filled porosity based on the bulk density and water content of the soil. Because the ratio of the bulk density (ρ_b) to particle density (ρ_p) represents the volume fraction of particles in the bulk soil, the **total porosity**, as a fraction, can be determined by subtracting this ratio from 1 (Eq. (2.1)).

$$\text{Total Porosity} = 1 - \frac{\rho_b}{\rho_p} \tag{2.1}$$

The fraction of the pore space that is air-filled and water-filled can be determined with the **gravimetric water content** (w), which is the mass of water per mass of dry soil. **Water-filled porosity**, which is also the **volumetric water content**, can be determined by multiplying the gravimetric water content by the bulk density of the soil and dividing by the density of water (ρ_w), which is usually assumed as 1 g cm^{-3} or 1 Mg m^{-3} (Eq. (2.2)).

$$\text{Water-filled Porosity or Volumetric Water Content} = w\frac{\rho_b}{\rho_w} \tag{2.2}$$

Air-filled porosity can be calculated by difference between the total porosity and the water-filled porosity (Eq. (2.3)).

$$\text{Air-filled Porosity} = \text{Total Porosity} - \text{Water-filled porosity} \tag{2.3}$$

Each of these fractional values can be expressed as a percentage of the soil volume (total porosity) or total pore space (water-filled and air-filled porosity) by multiplying by 100 (Box 2.4).

Impact of pore space on microorganisms

The architecture of the soil pore network, that is, the shape, size, and connectivity of pores within the soil matrix, has a well-established influence on the abundance, diversity, and activity of microorganisms. Differences in soil pore architecture create a multitude of diverse habitats for microorganisms, due in part to the movement and location of water, air, and other resources (e.g., organic matter, inorganic nutrients) needed for microbial growth. Soil can be

BOX 2.4

Example calculations of porosity

Total porosity

If the soil bulk density is 1.35 g cm^{-3} and the particle density is 2.65 g cm^{-3}, the pore space is calculated as shown. The answer (i.e., 0.49), which is the fraction of the total pore space, can be multiplied by 100 to get the percent porosity, which is 49% in this example. Note that the ratio of bulk density to particle density is the fraction of solids in a soil, which is 51% in this example.

$$\text{Total Porosity} = 1 - \frac{\rho_b}{\rho_p} = 1 - \frac{1.35}{2.65} = 0.49$$

Water-filled porosity

If the soil above has a gravimetric water content of 0.25 g g^{-1}, the fraction of the porosity filled with water is 0.34 or 34%.

$$\text{Water-filled Porosity or Volumetric Water Content} = w\frac{\rho_b}{\rho_w} = 0.25\frac{1.35}{1.0} = 0.34$$

Air-filled porosity

The air-filled porosity can be calculated by subtracting the water-filled porosity (0.34) from the total porosity (0.49), which is 0.15 or 15%.

$$\text{Air-filled Porosity} = \text{Total Porosity} - \text{Water-filled Porosity} = 0.49 - 0.34 = 0.15$$

viewed as an immense heterogeneous collection of microsites throughout which microorganisms are unevenly distributed. The effect of pore size on microbial community composition is often a function of microbial niche preference for the different microclimates, carbon sources, and other available resources (e.g., oxygen or other preferred electron acceptors; Chapter 3) associated with pores of different sizes and connectivities (Ruamps et al., 2011). As such, the diversity and distribution of soil microorganisms is not random. Recent technological advances in imaging soils (e.g., X-ray computed tomography, nanoscale secondary ion mass spectrometry), genomic sequencing, stable-isotope probing, and machine learning have expanded our understanding of how soil pore spaces influence microorganisms. For example, by combining μX-ray imaging with genomic sequencing, Negassa et al. (2015) observed that microbial groups associated with cellulose decomposition prefer soils dominated by macropores, while soils dominated by micropores harbor microbial groups able to thrive in nutrient- and oxygen-poor environments. Bacterial density is often greatest in proximity to well-connected, air-filled macropores with thin to moderate water films lining pore surfaces. Many soil microorganisms rely on water for diffusion of soluble substrates to their cell surfaces and across their cellular membranes, as well diffusional removal of waste products. Water is also essential for motility of many microorganisms. Therefore, water is an important driver of microbial

BOX 2.5

Nanopores and biodegradation

There are numerous examples of the long-term persistence of intrinsically biodegradable compounds in soil even when environmental conditions are not limiting for microbial growth. One possible mechanism for this reduced bioavailability is the slow (weeks to months) diffusion of organic materials or chemicals into soil pores so small (i.e., nanopores) that any organic compounds within them are beyond microbial or enzymatic attack. This concept may have important ramifications for carbon sequestration, biodegradation, and bioremediation (Chapters 14, 21, and 22).

organization in soils. One of the more notable differences in microbial community composition between macropores and micropores is the relative abundance of fungi and bacteria. The fungal-to-bacterial ratio is lower in micropores compared to macropores because fungi, which have an average diameter 4–6 μm, are physically excluded from micropores; the smaller size of bacteria (0.1–5.0 μm) allows them to colonize a wider range of pore sizes (Box 2.5).

Soil environment

Soil water

Water is essential for all organisms, including soil microorganisms. The size and spatial distribution of the pore network and the types of particles in a soil exert a major influence on water movement and water-holding capacity. In addition, water competes with air in the pore space and thus plays a major role in gas exchange, which affects many soil chemical and biological reactions (e.g., hydrolysis and redox reactions).

The polar nature of the water molecule is responsible for many of its properties including water movement and water retention in soils. Water is attracted to itself (i.e., **cohesion**) and to surfaces (i.e., **adhesion**) including soil particles. Cohesion and adhesion of water to soil particles in smaller pore spaces result in **matric force**, which allows water to move in any direction, even opposing the force of gravity. In larger pores, particles are further apart causing water to be held less tightly and become subject to **gravitational force**. Attraction of water to solutes, particularly soluble salts and ions, can exert an **osmotic force** in the soil. When a semi-permeable membrane is present, water moves from areas of low solutes to high solutes. When no membrane is present, the effect appears more as solute movement toward a more uniform concentration in the water (i.e., diffusion). For this reason, osmotic potential, in general, has little effect on the movement of water in soil.

Water potential reflects the potential (i.e., tendency) of water to move in soil. Technically, it is the measure of the potential energy (per unit mass or volume) of water at a point in a system relative to the potential energy of pure, free water. Soil water potential (represented by ψ) can be further divided into potentials associated with the various forces that cause water movement. Total soil water potential (ψ_t) is equal to the sum of these forces: **matric (or capillary) potential**

BOX 2.6

Units for expressing water potential

In practice, soil water potentials, whether gravitational, matric, or osmotic, are united in equivalents of pressure expressed as kilopascals (kPa) or megapascals (MPa). Note: 1 Pascal = 1 Newton m^{-2}; a Newton is measure of force required to accelerate 1 kg of mass at the rate of 1 m s^{-1}. Previously, soil water potentials were measured in bars. To convert bars to kPa or MPa, multiply bars by 100 or 0.1, respectively (e.g., −1 bar = − 100 kPa or −0.1 MPa).

(ψ_m), **gravimetric potential** (ψ_g), and **osmotic potential** (ψ_o) (Eq. (2.4)). The terms osmotic potential and **solute potential** (ψ_s) are sometimes used interchangeably. One principle remains inviolate: water always moves from areas of high potential to low potential (Box 2.6).

$$\psi_t = \psi_m + \psi_g + \psi_o \tag{2.4}$$

Movement of water due to matric or capillary forces, that is, **unsaturated flow,** occurs primarily when soil water is at or below **field capacity**, that is, the moisture content of soil after excess water has been allowed to drain completely by gravity, which is typically between −10 and −33 kPa (–1/10 to 1/3 bar) (Fig. 2.8). In this state, water can move in all directions due to differences in matric potential, but always from high water content (i.e., higher potential, wetter) to low water content (i.e., lower potential, drier). Water moves within films on particle surfaces or in smaller water-filled pores. **Mass flow** refers to the movement of nutrients with water via unsaturated flow in response to water deficits in the root zone caused by plant water uptake.

In addition to field capacity and saturation, permanent wilting point (i.e., wilting coefficient) and hygroscopic water are two additional, commonly defined moisture states in soil

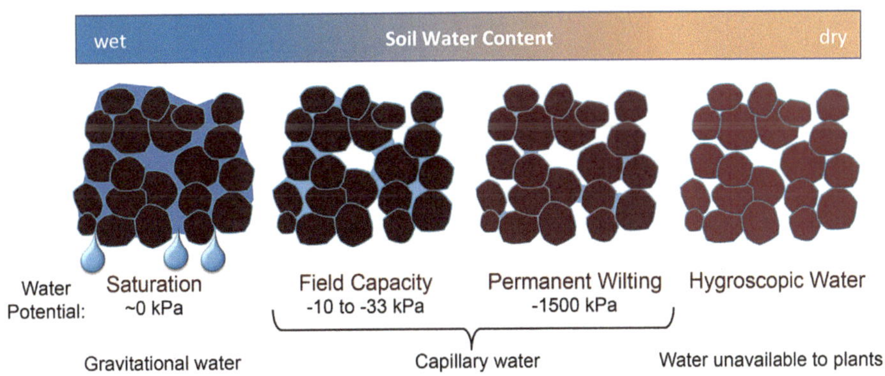

FIGURE 2.8 **Common points of soil moisture include saturation, when all pores are filled with water; field capacity, when free drainage has ceased (macropores are air-filled and micropores water-filled); permanent wilting point, when plants can no longer absorb water from the soil (only smallest micropores filled with water); and hygroscopic water.** As soil dries, water is lost from larger pores first then smaller pores. As soils get progressively drier, water clings to soil particles and aggregates in thin films to the point that no natural forces can remove it.

(Fig. 2.8). The **permanent wilting point** is the soil moisture content at which plants can no longer extract water from the soil, which is generally around −1500 kPa (−15 bar). **Plant available water** is considered water held in a soil between field capacity and the permanent wilting point. While the water above field capacity is technically available to the plant, its transient nature and variable effect on plant growth makes it an unreliable source of plant-available water. Water bound tightly in thin films on air-dry particle surfaces is referred to as **hygroscopic water**, which is unavailable to plants and cannot move due to natural forces.

Movement of water under unsaturated flow is relatively slow compared to **saturated flow** when all the pores are completely filled with water (i.e., **saturated**) or nearly so. During saturated flow, the adhesive and cohesive forces are not strong enough to span the distance across larger pores. The water is poorly held (i.e., water potential is high), allowing it to move due to gravitational forces relatively quickly and in a downward direction. Thus, water movement due to gravitational potential dominates when a soil is near saturation. **Preferential flow** (also called by-pass flow) is a special type of flow driven by gravitational forces that occurs where there are large cracks and connected macropores in the soil due to root channels (Fig. 2.9), shrinking and swelling clay (Box 2.1), and insect or worm pathways. Water movement is very rapid and can extend to considerable depths transporting suspended particles, microorganisms, nutrients, pesticides, and other chemicals with it.

Soil texture markedly affects soil water relations and has a dominant effect on pore size distribution. Water in fine-textured clayey soils is held more tightly due to greater adhesive forces in smaller pores that result in lower water potentials than would be found in the larger pores in coarser-textured sandy soils. Therefore, at any given water potential, a soil high in clay retains more water than a soil high in silt, which in turn retains more water than a soil high in sand. Changes in soil water content over a range of water potentials are expressed as **soil water characteristic curves** or **water release curves** (Box 2.7). The water release curves of most agricultural soils lie between the curves for clay and sandy soils. Greater amounts of organic matter typically increase water retention at any given water potential.

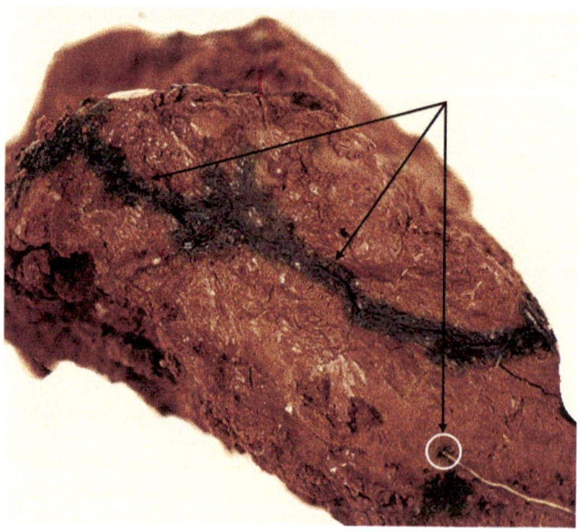

FIGURE 2.9 **A water-based blue dye was allowed to infiltrate into this soil and highlights a root channel that is acting as a preferential flow path.** At the bottom, the *circle* highlights a root surrounded by dye. *Source: Photo courtesy of Kevin McInnes, Texas A&M University.*

BOX 2.7

Capillary forces

To observe capillary forces on water, place a fine diameter glass tube in water. Because of the surface tension of water and the attraction of water molecules to the sides of the tube, the water rises in the tube according to the tube diameter; the smaller the tube, the greater the rise of water. This same effect translates to the pore sizes associated with soil texture. Clay soils hold more water and exhibit greater capillary rise due to their smaller particle size and correspondingly smaller pore size, while the larger particles in silt and sand result in larger pores, reduced water holding capacities, and less capillary rise. The graph depicts soil water characteristic curves for clay, silt loam, and sandy soils.

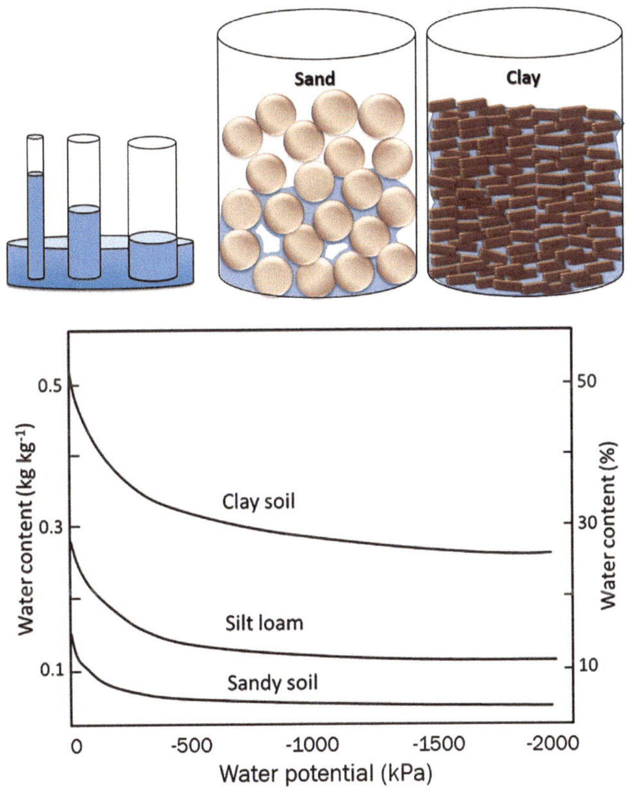

TABLE 2.5 Microbial tolerance to matric-controlled (Ψm) water stress.

Water potential kPa	Water film thickness	Microbial activity limited (example of species)
–33	4.0 μm	Movement of protozoa, zoospores, and bacteria
–100	1.5 μm	
–500	0.5 μm	
–1500	3.0 nm	Nitrification, sulfur oxidation
–4000	<3.0 nm	Bacterial growth (*Bacillus*)
–10,000	<1.5 nm	Fungal growth (*Fusarium*)
–40,000	<0.9 nm[a]	Fungal growth (*Penicillium*)

[a] *Less than 3 molecules of water thick.*
Adapted from Harris (1981).

The water potential in soil also has a profound effect on soil microorganisms and soil microbial processes. Some microorganisms and processes are tolerant of moisture stress, while others are not. In general, fungi, archaea, and some Gram-positive bacteria are more tolerant of low soil water content due to their thicker and stronger cell walls compared to Gram-negative bacteria. Earlier in the drying sequence, as soil water becomes limited, microbial movement starts to be restricted (Table 2.5). Microbial activities, such as carbon and nitrogen mineralization, are reduced at the permanent wilting point (–1500 kPa), whereas microbial growth ceases in drier soil at water potentials beyond –4000 kPa. As a soil dries, pores become hydrologically disconnected, which restricts the movement of nutrients to microbes. Furthermore, in order to access water, microorganisms must adjust their internal osmotic potentials relative to the surrounding environment; this is often done through the production of compatible solutes, an energy-consuming process that diverts metabolic resources which could otherwise be used for microbial growth and activity.

Thus, we see that multiple forces (ψ_m, ψ_g, ψ_o) exert control on the availability and movement of water in soil which in turn influence the growth and activities of soil microbes. Of these various forces, matric potential is the primary determinant of water availability in soil.

Soil pH

Soil pH (Box 2.8) is considered one of the master variables in soil due to its effect on numerous processes. Soil pH influences the availability and toxicity of nutrients and other elements present in the soil. For example, iron (Fe), manganese (Mn), and copper (Cu) become increasingly insoluble as pH increases, which reduces plant and microbial uptake. At low pH, these elements are highly soluble and thus can reach toxic levels. The availability of nutrients is generally optimal across the pH range of 6–7. For instance, phosphorus (P) is most soluble and available to plants and microorganisms in this pH range. Soil pH also affects the charge and binding properties of the soil, which affects the ability of soils to retain ions (e.g., nutrients). Plant and microbial habitats are often constrained by pH for a variety of reasons, including nutrient availability and ion toxicity as described previously. Plants and some microorganisms utilize pH gradients (i.e., protonmotive force) for nutrient uptake (Blaut and Gottschalk, 1997; Palmgren, 2001; Wang et al., 2014). When soil pH is outside of

BOX 2.8

Understanding soil pH

The concept of pH is often simplified as the negative logarithm of the hydrogen ion concentration or pH = $-\log_{10}$ [H^+], where brackets indicate concentration. The pH scale (0–14) is based on the dissociation of water into H^+ and OH^-, which is defined by the water dissociation constant: $K_w = 1 \times 10^{-14}$. The constant establishes a scale from 0 to 14 where pH and pOH ($-\log_{10}$ [OH^-]) must always total 14. Thus, when pH is high, pOH is low and vice versa. High concentrations of H^+ will have a low pH and indicate **acid** conditions, while low concentrations of H^+ will have **alkaline** or **basic** conditions. When H^+ and OH^- concentrations are equal, the pH is 7 and the solution is considered neutral. It is important to remember that pH is a log scale; if two solutions differ by 1 pH unit, then one solution has 10 times more H^+ ions than the other.

$$
\begin{aligned}
H_2O &\Leftrightarrow H^+ + OH^- \\
K_w &= [H^+][OH^-] = 1 \times 10^{-14} \\
-\log K_w &= -\log[H^+] + -\log[OH^-] \\
&= -\log(1 \times 10^{-14}) \\
pK_w &= pH + pOH = 14
\end{aligned}
$$

Pure water has pH of 7, but rarely water is pure. Rainwater, for example, has dissolved CO_2 or carbonic acid that lowers pH to ~5.6. Because the pH of water is easily influenced by dissolved ions, it is considered to be poorly buffered. Carbonate (CO_3^{2-}) and bicarbonate ions (HCO_3^-) can act as buffering agents in water to control water pH. **Alkalinity** is commonly assessed in water as a measure of these ions and the ability of water to resist changes in pH.

Soil pH is buffered by clay and organic matter, as well as carbonates and bicarbonates, in the soil. Between approximately pH 5.5 and 7, pH buffering often positively correlates with cation exchange capacity of the soil. Above pH 7, carbonates are usually the major cause of soil pH buffering. Calcium and magnesium carbonates, as well as calcium and magnesium oxides and hydroxides, can also be added as **liming materials** to increase the pH of acid soils. However, it is less common for additives (e.g., elemental sulfur) to be used to try to lower soil pH as this is usually more difficult to do.

In the soil, the simplified definition of pH does not reflect the true nature of H^+ and its influence on soil processes. A more accurate definition of soil pH is the negative logarithm of the hydrogen ion **activity**, where activity is the effective concentration, which accounts for the influence of other ions in the soil solution on hydrogen ions. The activity of H^+ is reduced when the concentration of other ions increases. Except for speciation calculations and geochemical modeling, it is generally sufficient to consider pH as a reflection of hydrogen ion concentration.

the organism's adapted pH range, considerable energy is required to maintain a gradient sufficient for nutrient uptake. Limited uptake of some ions, such as K^+ that have important roles in osmotic balance can further complicate water and nutrient imbalances creating additional stress.

Many studies over the past few decades have shown that soil pH is a strong determinant of soil microbial community structure. Some microbial groups may dominate at higher or lower pH conditions where they may have a competitive advantage. For example, actinobacteria prefer neutral to alkaline conditions and do not tolerate acidic conditions well. In contrast, fungi tend to out-compete other organisms under acidic conditions. Both plants and microorganisms have the ability to modify soil pH in their local environment through secretion of H^+ or other compounds such as organic acids.

Soil pH generally ranges from 4 to 10. Acidic soils are typically found in climates with higher rainfall. Weathering and development of soils is strongly enhanced by high rainfall that erodes, dissolves, and leaches basic cations (e.g., Ca^{2+}, Mg^{2+}, Na^+, and K^+) from the soil leaving acidic cations (Al^{3+}, H^+). Arid and semi-arid climates typically have high levels of basic cations and neutral to alkaline pH.

Colloid charge and nutrient retention (anion and cation exchange capacity)

The smaller clay-sized particles (<1 μm) in soil behave as colloids, which are microscopic particles that can be dispersed and remain suspended in solution. Soil colloids can be either inorganic or organic and exhibit positive and negative charges. These charges attract oppositely charged ions from the soil solution (e.g., K^+, Ca^{2+}, Mg^{2+}, Na^+, NO_3^-, Cl^-, and Al^{3+}). This property is called ion exchange (Fig. 2.10). The ability of a soil to hold cations represents the **cation exchange capacity** (CEC) of the soil, and the ability of a soil to hold anions is its **anion exchange capacity** (AEC). Although most soils will contain both positive and negative charges, typically one dominates. Surface soils are usually dominated by negative charges, and CEC is reported and discussed more commonly than AEC. Exchangeable cations and

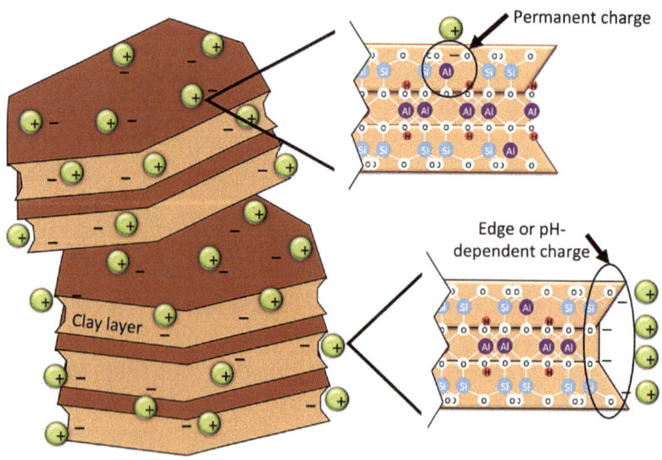

FIGURE 2.10 **Ion exchange.** Clay colloids are typically negatively charged. These charges attract cations (e.g., K^+, Ca^{2+}, Mg^{2+}, Na^+, and when pH is low H^+ and Al^{3+}). Cations are hydrated, weakly held, and can exchange with other cations rapidly, reversibly, and stoichiometrically with respect to charge. Some colloids (e.g., smectites and vermiculites) expand and allow exchangeable cations in between their layers. All colloids have charges at the edges of the minerals, which interrupt the crystal structure and are unbalanced. These edge charges are dependent on pH. Some minerals also have imperfections in the crystal structure, which results in permanent negative charge. Image depicts a structure with both types of charge.

anions are loosely bound to the soil through electrostatic attraction and thus can be replaced by other exchangeable ions. Ion-exchange reactions are rapid, reversible, and stoichiometric with respect to charge (e.g., a divalent cation will require one divalent or two monovalent cations for exchange). The AEC and CEC of a soil are expressed as centimoles of charge per kg of soil (i.e., $cmol_c\ kg^{-1}$ soil). The units are based on charge due to the stoichiometric property of the ion exchange reaction (Box 2.9). The exchange capacity of a soil is determined primarily by the type and amount of clay and organic matter in the soil (Table 2.6) and can be estimated through mineral composition or through analysis of exchangeable cations (Box 2.9).

BOX 2.9

CEC

Determination of cation exchange capacity (CEC) in the laboratory is generally a fairly tedious process and is often estimated by mineral composition of the soil or by sum of exchangeable cations (see example calculations). CEC and anion exchange capacity (AEC) are typically reported as $cmol_c\ kg^{-1}$ soil; however, milliequivalents (meq) 100 g^{-1} of soil is also used. Fortunately, the values are numerically equal. An equivalent and a mole of charge are also equal (1 eq = 1 mol_c). A mole of charge is equal to 6.022×10^{23} charges.

Example estimation of CEC by mineral composition

Given a soil that is 1% organic matter, 5% smectitic clay, 20% kaolinitic clay, and 5% Fe/Al oxides, and the remaining 69% is sand and silt, the CEC can be estimated using approximations of CEC for each component (Table 2.6).

Example estimation of CEC by sum of exchangeable cations

Assume the exchangeable cations of a soil are measured (in $cmol_c\ kg^{-1}$) as 1.1, 0.8, 0.1, 0, and 0.3 for Ca, Mg, Na, K, and Al, respectively. The estimated cation exchange capacity is the sum of molar charges. Since all the units are in $cmol_c\ kg^{-1}$, the estimated CEC in this example is simply 2.3 $cmol_c\ kg^{-1}$. Exchangeable cations can also be expressed in other units, such as ppm, mg kg^{-1}, or even cmol kg^{-1}, but it is not correct to sum units that are not reported on a molar charge basis.

Component	Approximate CEC for component		Fraction of component in soil		CEC contribution
	$cmol_c\ kg^{-1}$				$cmol_c\ kg^{-1}$
Organic matter	200	×	0.01	=	2
Smectitic clay	100	×	0.05	=	5
Kaolinitic clay	10	×	0.20	=	2
Fe/Al oxides	2	×	0.05	=	0.1
Sand/silt	0	×	0.69	=	0
Estimated CEC (sum of all contributions to CEC)					9.1

TABLE 2.6 Approximate ranges of cation exchange capacity (CEC) for selected soil minerals and texture classes. Anion exchange capacity (AEC) is generally not significant, except with organic matter, kaolinite, and Fe/Al oxides[a].

Soil material	CEC
	$cmol_c\ kg^{-1}$
Sand	<10
Silt	10–30
Clay	
Fe/Al oxides	<5
Kaolinite	3–15
Smectite/Montmorillonite	60–120
Vermiculite	80–150
Chlorite	10–40
Organic matter	100–300

[a] *Organic matter will yield both CEC and AEC, but CEC is usually greater. Kaolinite and Fe/Al oxides transition from CEC to AEC as pH decreases. The pH at which CEC transitions to AEC depends on the specific mineral properties (e.g., kaolinite pH ~4–5; goethite (Fe oxide) pH ~7–8).*

Colloid charge on inorganic mineral colloids is due to two sources: pH-dependent charge and permanent charge. The pH-dependent charge arises at the edges of colloid surfaces (small clay particles), where the crystalline structure, and thus factors balancing charges, are interrupted. Because the edges of minerals are usually dominated by oxygen, they are influenced by the H^+ activity (or concentration) in the soil solution. When pH is low (i.e., high H^+ concentration), exposed oxygen is often protonated, which results in a positive or neutral charge depending on the chemical environment (i.e., functional group). As pH increases, there are fewer and fewer H^+ that protonate the sites, so the electrons on oxygen are not participating in bonds and have a negative charge. Organic colloids also have similar pH-dependent charges, but these arise at organic functional groups, for example, carboxyl, phenolic, and alcoholic groups, that are protonated (neutral) at low pH or non-protonated (negative) at high pH. Organic matter has a CEC 2–5 times greater than montmorillonite (i.e., smectite) and up to 30 times greater than kaolinite. Permanent charge is found only on some silicate mineral structures that have "mistakes" in the crystalline lattice from substitution of similarly sized, but differentially charged, cations for cations that would normally be present. Examples of **isomorphic substitution** include substitution of Al^{3+} for Si^{4+} and Mg^{2+} for Al^{3+}. In both cases, a cation with less positive charge replaces a cation with greater positive charge, which is the most common type of substitution. When this occurs, the crystalline structure is deficient in positive charge to balance the negative charge from the surrounding oxygens and hydroxyls, resulting in a net negative charge. Some minerals, such as fine micas (e.g., illite), vermiculite, chlorites, and smectites, have considerable permanent charge. However, the interlayer (i.e., the space between the silicate layers) of mica and chlorite help balance this out with K^+ ions and a positively charged Mg-oxide layer, respectively.

All mineral and organic colloids have pH-dependent charge, while only some have considerable permanent charge. As pH decreases, minerals with permanent charge and organic colloids will still exhibit CEC, while minerals with only pH-dependent charge (e.g., kaolinite, Fe and Al oxides) can exhibit appreciable amounts of AEC. The CEC and AEC of soils help determine the nutrient holding capacities of soils. In addition, when the pH of a soil is between 5.5 and 7, the CEC is often an indicator of the soil's ability to buffer pH (resist changes in response to acidifying or alkalinizing inputs). This has important implications for management of acid soils as it affects the lime required to increase soil pH.

Some ions exhibit poor ion exchange properties. The most notable is phosphate. In most soils, phosphate exists as $H_2PO_4^-$ or HPO_4^{2-}. These ionic forms can bind tightly to soil colloids, particularly Fe and Al oxides, kaolinite, and amorphous volcanic materials (e.g., allophane, imogolite). Phosphate can also form highly insoluble precipitates with Ca at high pH and with Fe, Al, and Mn at low pH to produce phosphate minerals (e.g., strengite, variscite). Thus, phosphate solubility and sorption to soil colloids drastically reduces phosphate availability to organisms.

Soil temperature

Soil temperature influences the rates of biological, physical, and chemical processes in the soil. Within a limited range, the rates of chemical reactions and biological processes double for every 10°C increase in temperature (the so-called Q_{10} value, i.e., $Q_{10} = 2$). Soil temperature is intimately linked to soil water and pore space. Water has a greater specific heat (i.e., the energy required to heat a mass by 1°C) than soil minerals (e.g., water is 1 calorie g^{-1} and quartz is 0.19 calorie g^{-1}). Thus, wet soil requires more energy to heat than dry soil. This influences seasonal changes in soil temperatures. Wet soils will take more energy to warm in the spring and stay warmer in the winter than dry soils. As depth increases, soil temperature increasingly lags behind changes in aboveground temperature, both daily and seasonally (Fig. 2.11). This lag differentiates soil microbiology from aquatic microbiology as the mass of soil moderates the rapid fluctuation of environmental parameters that is more commonly found in aquatic systems (Alexander, 1977).

Soil temperature is also affected by ground cover. Vegetation and organic residues on the soil surface can moderate extremes in temperature by keeping soil cool in hot weather and insulating against heat loss in cold weather. They also play a role in moisture retention which in turn affects soil temperature. Soil color also has an impact on soil temperature. Bare, dark-colored soils warm more quickly than light-colored soils, which can have an impact on planting dates, etc.

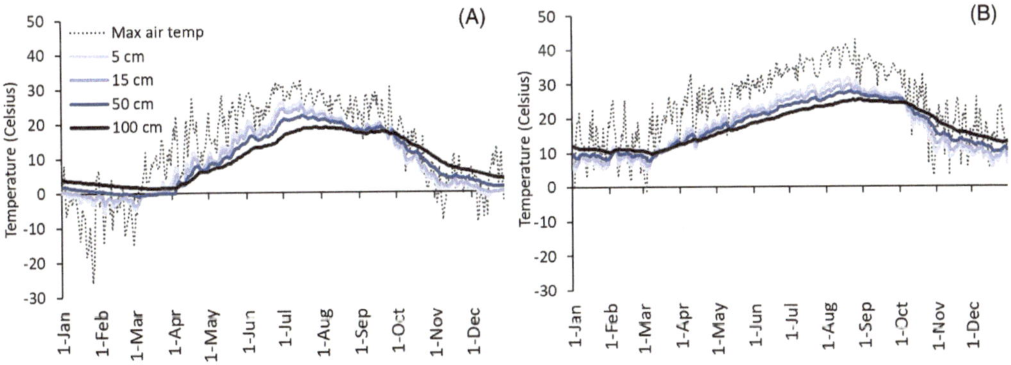

FIGURE 2.11 Daily soil temperatures at 5-, 15-, 50-, and 100-cm depths and maximum air temperature in 2019 in (A) Platteville, WI, and (B) Vernon, TX. Note the greater disparity in temperatures with increasing depth associated with the cooler climate in Wisconsin compared to Texas.

Soil air

The soil air differs from the overlying atmospheric air in that it has a higher concentration of CO_2 and lower concentration of O_2. In addition, unless very dry, the soil generally maintains nearly 100% humidity. However, the concentrations of CO_2 and O_2 can vary significantly based on water-filled porosity and biological demand for O_2 by plant roots and other macro- and microorganisms. When exchange of air between the atmosphere and the soil is low (e.g., wet or compacted soils), concentrations of O_2 typically decrease, while CO_2 increases in the soil. Aeration is the process of encouraging gas exchange to enhance O_2 in the soil. Ideally, a well-aerated soil will have sufficient O_2 and moisture to support respiration of aerobic organisms (i.e., **aerobes**). Generally, 50–60% water-filled pore space (40–50% air-filled) is considered optimal for microbial growth in soil. Changes in soil aeration affect the microbial community. **Facultative anaerobes** can grow in the presence or absence of oxygen, **microaerophiles** prefer or require low O_2, and **obligate anaerobes** grow only in the absence of O_2 (Chapter 3). Thus, as the aeration status of a soil changes, considerable changes can occur in the make-up of the microbial community.

Soil aeration is highly dependent on soil porosity, moisture, and depth. Soil texture and structure both affect diffusion. If a soil has a high percentage of clay (fine texture), then it will have a high percentage of small pores that will slow diffusion. Aggregation will add large pores resulting in a better mix of small and large pores to enhance gas diffusion. Sandy (coarse texture) soils typically have good aeration because diffusion is fast in the relatively large pores surrounding the larger sand particles. Soil water also affects diffusion of O_2. The diffusion coefficient of O_2 in air is 0.198 $cm^2\ sec^{-1}$, but in water it is only $1.9 \times 10^{-5}\ cm^2\ sec^{-1}$ (Hillel, 1998). In other words, O_2 diffuses through water 10,000 times more slowly than through air. Thus, when pores contain only water, it greatly reduces gas exchange, thereby causing CO_2 levels to rise and O_2 to decrease provided conditions conducive to respiration are present. Poor aeration is more common in clay-rich than in sandy soils due to the greater abundance of small pores which are more likely to be water-filled, especially when poorly aggregated. Soil depth also affects soil aeration. Generally, soil O_2 declines with depth, especially in soils with high clay contents.

Soil aeration is commonly measured in terms of its redox (*red*uction–*ox*idation) potential. The **redox potential** is the measure of the tendency of a compound to accept or donate electrons. As electrons are transferred, an electrical potential difference is created, and this difference is measured in millivolts (mV); the potential itself is abbreviated E_h. As a substance loses electrons (e.g., $Fe^{2+} \rightarrow Fe^{3+}$), it becomes more positive (more oxidized); as it gains electrons (e.g., $Fe^{3+} \rightarrow Fe^{2+}$), it becomes more negative (more reduced). The more oxidized a soil becomes, the more positive is its E_h. Waterlogged soils, especially those with an available carbon source to support microbial activity, generally have low redox potentials, and these environments are conducive to anaerobic processes like methane production and sulfate reduction (Box 2.10). So, in general, oxic soils are "oxidizing" environments where aerobic metabolism is active and anoxic soils are "reducing" environments where anaerobic metabolism is active (Chapter 3). Exchange of gases is largely mediated via diffusion. Because the complexity and heterogeneity of soils produces many diverse microenvironments, O_2 concentrations can vary from fully oxic to low (hypoxic) to completely anoxic over distances of < 1 mm. A wide range of microenvironments having oxygen concentrations spanning these extremes is likely the normal state in most soils.

BOX 2.10

Soil aeration, redox potential, and soil color

There is an interesting relationship between soil aeration (a chemical property) and soil color (a physical property). When soil is well aerated, Fe is oxidized (Fe^{3+}) and typically imparts colors of red, orange, yellow, and reddish-brown (Chapter 19). When soil is poorly aerated, Fe is reduced (Fe^{2+}), and soil colors of gray, green, and blue predominate, which are described as **gley** (indicated by the bracket in **A**). When soils have mixed zones of good and poor aeration, soils will have a mottled appearance (**B**). Soil scientists refer to these characteristics as "redoximorphic features." As a soil transitions from oxic to anoxic, microbial communities similarly transition from aerobic to anaerobic organisms, the latter utilizing terminal electron acceptors other than oxygen (Chapter 3). These alternate electron acceptors include nitrate, oxidized manganese (Mn^{3+} and Mn^{4+}) and iron (Fe^{3+}), sulfate, CO_2, and organic carbon substrates (e.g., fermentation). Each of these electron acceptors is reduced in the process.

(A) (B)

Photo courtesy of J.A. Howe, Texas A&M University.

Integrating soil properties and plants with microorganisms: the rhizosphere

The most complex of all soil environments is the rhizosphere, the zone of soil under the influence of plant roots (Fig. 2.12) (see Chapter 11). The rhizosphere represents a tripartite relationship among plant, microbes, and soil, and as such, each part affects the others. Soil microbial cells are mostly carbon, and most soils are low in readily available carbon (i.e., they

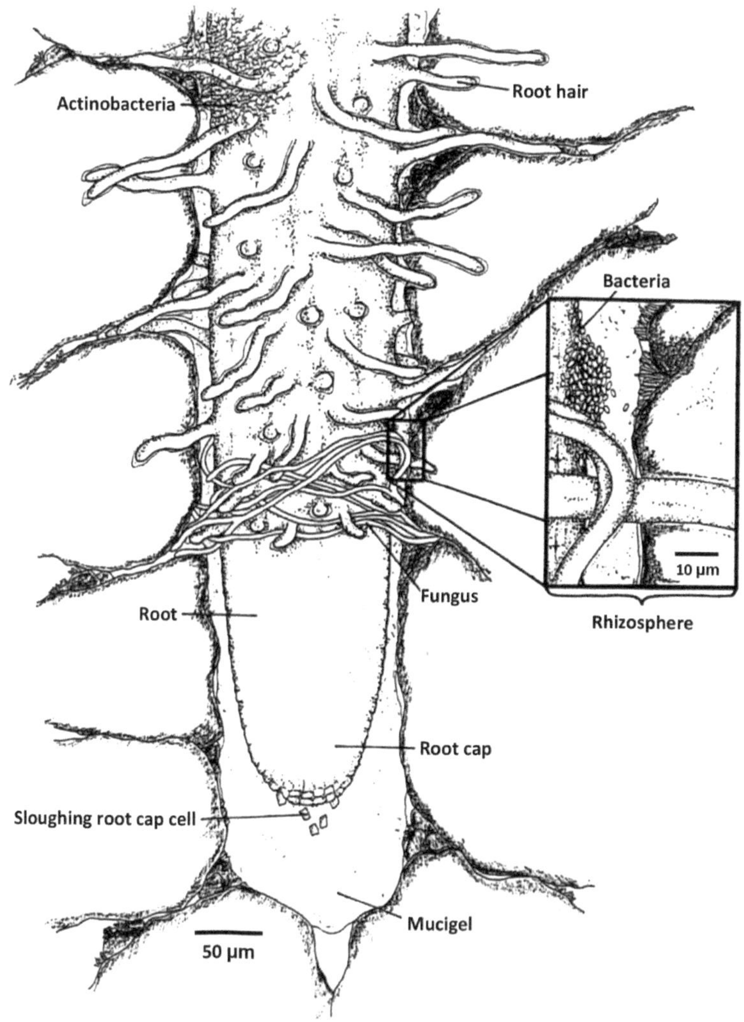

FIGURE 2.12 **The rhizosphere.** The mucigel is a mixture of gelatinous and other plant products, bacterial cells and their metabolic products, colloidal minerals, and organic matter, which together with other plant exudates, lysates, and secretions, form a gradient of carbon that is most concentrated near the root surface and decreases with increasing distance away from the root. Because of this abundance of organic substrates, the rhizosphere is a zone of intense microbial activity. *Source: Original drawing by K. Luoma.*

are "substrate limited"). Therefore, the natural state for most microbes in soil is starvation. Plant roots give off a rich variety of organic substances into the rhizosphere soil, which serve as nutrients for microbial growth. In return for these nutrients, microbes provide various benefits to the plant: production of plant growth hormones, solubilization of inorganic minerals, protection from plant pathogens, and other benefits.

Degradation of root and microbial tissues results in stable organic matter (Chapter 13) that alters soil physical and chemical characteristics (e.g., aggregation, nutrient retention, water holding capacity, CEC), which generally benefits plant growth and microbial activity. Soil pH is influenced by roots releasing H^+, HCO_3^-, or organic acids into the rhizosphere. The pH changes due to these excretions affect nutrient availability, weathering of soil minerals, and influences competitiveness of both plant and microbial communities.

Similarly, plants affect abiotic soil environmental factors like soil water, aeration, and temperature. Because plants transpire, they move water, at least until they reach their permanent wilting point. This movement of water not only affects soil microbes (e.g., by decreasing their motility) but also affects the diffusion of nutrients to the roots. Both effects result in the thinning and interruption of water films. As more water is removed, there is less water in the soil pores, and soil aeration increases. The increased microbial activity in the rhizosphere also changes soil aeration through increased O_2 uptake and CO_2 production. Additionally, decayed roots provide channels in the soil for preferential or bypass flow of water. Finally, plants provide a canopy over the soil, shading the soil and moderating soil temperatures. These dynamic interactions are covered in more detail in Chapter 11.

Summary

Soil is one of the most complex microbial habitats. Understanding the soil microbial habitat often requires reliance on soil description and classification, such as understanding differences between mineral and organic soils, horizons, series names, and textural classes. A full understanding of the soil habitat requires gathering and interpreting data regarding the effects of soil physical (e.g., bulk density, particle density, and pore spaces) and chemical properties (e.g., pH, ion exchange capacity) on the soil microbial environment (e.g., soil water, soil aeration, and soil temperature). The soil habitat is also influenced by the microbial community through altering nutrient and carbon cycling and affecting biological oxygen demand (soil aeration). Plants and human activity simultaneously affect and are affected by the soil microbiota. Knowledge of these interdependencies helps to explain the enormous heterogeneity in soil, which explains how, within the space of a few millimeters, some microorganisms will prosper and others will perish, and how some microbial processes will begin and others will stop. Knowing this interdependency is the key to understanding "soil" and, indeed, understanding soil microbiology.

Acknowledgments

Portions of this chapter, with permission, are derived from the work of Dr. Peter G. Hartel who authored a similar chapter for the Second Edition of the textbook. The current authors and editors thank him for permitting the use of this material and for graciously supporting production of the Third Edition.

Study questions

1. A sample from the A horizon of soil in the Houston Black series is 50% clay, 35% silt, and 15% sand. Using the soil textural triangle in Box 2.2, determine the textural class of this soil.
2. The Orangeburg soil series consists of a very deep, well-drained soil in the Southern Coastal Plain that is dominated by kaolinitic clay and Fe/Al oxides but has low soil organic matter (i.e., <1%). The surface is a sandy loam. The Houston Black series is also a very deep soil. However, it developed from residuum of calcareous mudstone and is dominated by smectitic clays having shrink/swell properties (for soil texture, see question 1). Soil organic matter is approximately 1.5%. Based on their respective textures and other properties, predict which of the two soils is likely to have the greatest infiltration and percolation rates and indicate how this may affect soil microorganisms.
3. Within 24 h after a good rainfall, the water potential of a soil is −30 kPa. After 3 days of dry weather, the water potential is −100 kPa. Which condition is drier and likely to have greater air-filled macropores?
4. Explain how changes in bulk density and porosity with soil depth influence the habitat for microorganisms. How will microbial abundance, composition, and activity change with soil depth?
5. Describe how microbial activity may differ between an Ultisol and a Mollisol (Hint: use soil properties described for each order in Table 2.1)?
6. Alfisols, Ultisols, and Oxisols (Table 2.1) are often considered a weathering sequence of soil orders in sub-humid to tropical environments. In this sequence, the clay minerals become progressively more weathered changing from silicate clay minerals to Fe/Al oxide minerals. How does weathering affect the fertility of the soil and thus the limitations to the growth of plants and microbial communities?
7. A 10-cm layer of coarse sand is on top of a 10-cm layer of clay. Water is poured on the top of the sand. (1) Draw a diagram to depict how the water moves through the sand into the clay. (2) Draw a diagram to depict how the water moves if the clay layer was above the sand instead. (3) Which layer has the greater water potential and why?
8. Explain why microbial activity is typically greater in the rhizosphere than in bulk soil?

References

Alexander, M., 1977. Introduction to Soil Microbiology. Wiley and Sons, New York.

Blaut, M., Gottschalk, G., 1997. Energetics of aerobic and anaerobic bacteria. Gräber, P., Milazzo, G., Walz, D. (Eds.), Bioenergetics. Bioelectrochemistry: Principles and Practice, vol. 4, Birkhäuser Basel, Switzerland.

Bouyoucos, G.J., 1927. The hydrometer as a new method for the mechanical analysis of soils. Soil Sci. 23, 343–354.

Gardiner, D.T., Miller, R.W., 2004. Soils in Our Environment. Pearson Prentice Hall, New Jersey.

Gillespie, A.W., Farrell, R.E., Walley, F.L., Ross, A.R., Leinweber, P., Eckhardt, K.-U., Regier, T.Z., Blyth, R.I.R., 2011. Glomalin-related soil protein contains non-mycorrhizal-related heat-stable proteins, lipids and humic materials. Soil Biol. Biochem. 43, 766–777.

Harris, R., 1981. Effect of water potential on microbial growth and activity. Parr, J.F., Gardner, W.G., Elliot, L. (Eds.), Water Potential Relations in Soil Microbiology, vol. 9, Soil Science Society of America, Wisconsin, pp. 23–95.

Hillel, D., 1998. Environmental Soil Physics: Fundamentals, Applications, and Environmental Considerations. Academic Press, Elsevier, California.

Kutílek, M., Jendele, L., 2008. The structural porosity in soil hydraulic functions-a review. Soil Water Res. 3, S7–S20.
Luxmoore, R., 1981. Micro-, meso-, and macroporosity of soil. Soil Sci. Soc. Am. J. 45, 671–672.
Mays, T., 2007. A new classification of pore sizes. Stud. Surf. Sci. Catal. 160, 57–62.
Mengistu, A.G., Mavimbela, S.S., Van Rensburg, L.D., 2019. Characterisation of the soil pore system in relation to its hydraulic functions in two South African aeolian soil groups. S. Afr. J. Plant Soil 36, 107–116.
Negassa, W.C., Guber, A.K., Kravchenko, A.N., Marsh, T.L., Hildebrandt, B., Rivers, M.L., 2015. Properties of soil pore space regulate pathways of plant residue decomposition and community structure of associated bacteria. PLoS One 10, e0123999.
Palmgren, M.G., 2001. Plant plasma membrane H^+-ATPases: powerhouses for nutrient uptake. Annu. Rev. Plant Bio. 52, 817–845.
Rouquerol, J., Avnir, D., Fairbridge, C.W., Everett, D.H., Haynes, J.M., Pernicone, N., Ramsay, J.D.F., Sing, K.S.W., Unger, K.K., 1994. Recommendations for the characterization of porous solids (technical report). Pure Appl. Chem. 66, 1739–1758.
Ruamps, L.S., Nunan, N., Chenu, C., 2011. Microbial biogeography at the soil pore scale. Soil Biol. Biochem. 43, 280–286.
Smith, A., Marín-Spiotta, E., De Graaff, M., Balser, T., 2014. Microbial community structure varies across soil organic matter aggregate pools during tropical land cover change. Soil Biol. Biochem. 77, 292–303.
Soil Science Society of America, 1997. Glossary of Soil Science Terms. Soil Science Society of America, Wisconsin.
Soil Survey Staff, 1999. Soil Taxonomy: A Basic System of Soil Classification for Making and Interpreting Soil Surveys, second ed. Natural Resources Conservation Service. U.S. Department of Agriculture Handbook, p. 436.
United States Department of Agriculture, 2008. Soil Quality Indicators—Bulk Density. USDA Natural Resources Conservation Service.
Wang, E., Yu, N., Bano, S.A., Liu, C., Miller, A.J., Cousins, D., Zhang, X., Ratet, P., Tadege, M., Mysore, K.S., Downie, J.A., Murray, J.D., Oldroyd, G.E.D., Schultze, M., 2014. A H^+-ATPase that energizes nutrient uptake during mycorrhizal symbioses in rice and *Medicago truncatula*. Plant Cell 26, 1818–1830.

Further reading

Brady, N.C., Weil, R.R., 2018. Elements of the Nature and Properties of Soils, fourth ed. Pearson, United Kingdom.
Paul, E. (Ed.), 2014. Soil Microbiology, Ecology and Biochemistry, fourth ed. Academic Press.

CHAPTER

3

Microbial metabolism

Jeffry J. Fuhrmann

Department of Plant and Soil Sciences, University of Delaware, Newark, DE, United States

Organic chemistry is the chemistry of carbon compounds.
Biochemistry is the study of carbon compounds that crawl.—**Anonymous**

The succeeding chapters will describe the high degree of morphological, structural, and genetic diversity within and among populations of soil microorganisms. This chapter, however, will consider microbial diversity from another perspective—metabolic diversity among microorganisms. The term metabolism refers to the total sum of the biochemical reactions that occur within living cells.

Although many of the topics and concepts discussed in this chapter are familiar, they warrant repetition because of the important role that metabolism plays in defining populations of soil microorganisms and in understanding their interactions with each other, plants, and the environment. It is simply impossible to appreciate fully the discipline of soil microbiology without a fundamental understanding of the underlying biochemical principles. In discussing metabolic diversity, a central concept is that the overall goal of all forms of metabolism is essentially the same—to obtain energy and carbon needed for the production of cellular constituents necessary for growth, survival, and reproduction. Thus, the metabolic diversity among microorganisms described in this chapter comes down to differences in strategies rather than the goals of these strategies.

An overview of microbial metabolism

Biochemical reactions in metabolism either yield energy (exergonic) or consume energy (endergonic). Exergonic reactions result in the production of energy needed to support cellular processes. They are often referred to as "spontaneous" because, given the proper conditions, they can be sustained without the addition of external energy. Endergonic activities result in

Principles and Applications of Soil Microbiology. http://dx.doi.org/10.1016/B978-0-12-820202-9.00003-4
Copyright © 2021 Elsevier Inc. All rights reserved.

the biosynthesis of cellular components and microbial biomass. These reactions are not spontaneous and require a supply of energy from exergonic reactions to proceed. Although such a separation of exergonic and endergonic reactions is helpful for purposes of discussion, metabolism involves both types of reactions and is a highly integrated and interdependent system.

Microbes obtain energy from a myriad of sources, including organic compounds, inorganic substances, and light. Furthermore, microorganisms as a group, and sometimes as individuals, can use these sources under both **oxic** (aerobic) and **anoxic** (anaerobic) conditions. Thus, to a large extent, the ubiquity of microorganisms reflects their diverse strategies for obtaining energy.

Two general types of energy-rich compounds are derived from the exergonic reactions of metabolism:

- high-energy phosphate compounds, and
- stored electrons associated with specialized carrier molecules (coenzymes).

Adenosine triphosphate (ATP) is the dominant high-energy phosphate compound in cells (Fig. 3.1). The potential energy stored in the "high-energy phosphate bonds" of ATP is useful in a wide variety of biosynthetic reactions. Electrons released during exergonic metabolism are generally captured by oxidized electron carriers such as **nicotinamide adenine dinucleotide** (NAD^+), NAD phosphate ($NADP^+$), or related compounds (Fig. 3.2). The chemically reduced forms of these carriers will be referred to as NADH and NADPH, respectively. An additional but less energy-rich electron carrier is flavin adenine dinucleotide ($FAD/FADH_2$). The electrons stored in these compounds are referred to as **reducing equivalents** or reducing power.

Most organisms obtain cellular energy from the biodegradation of energy-rich organic compounds, such as carbohydrates, proteins, and lipids; this "dismantling" of organic substances

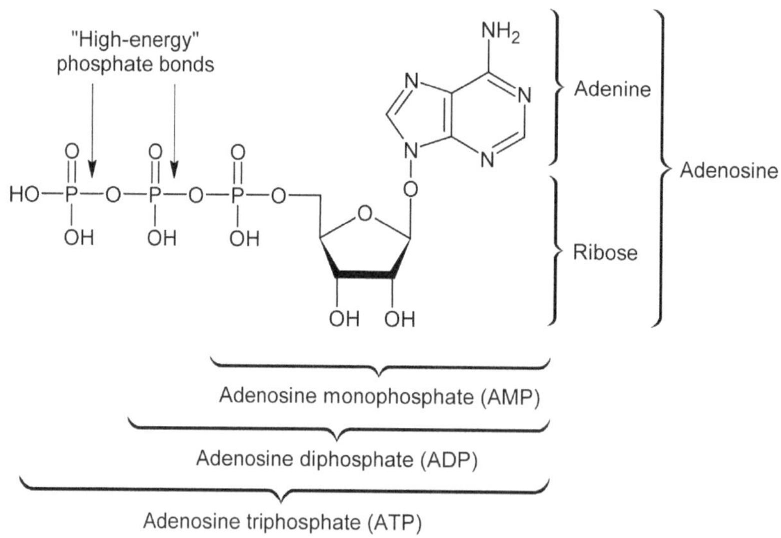

FIGURE 3.1 **Structure of adenosine triphosphate (ATP), adenosine diphosphate (ADP), and adenosine monophosphate (AMP).** *Source: Figure by J. Fuhrmann.*

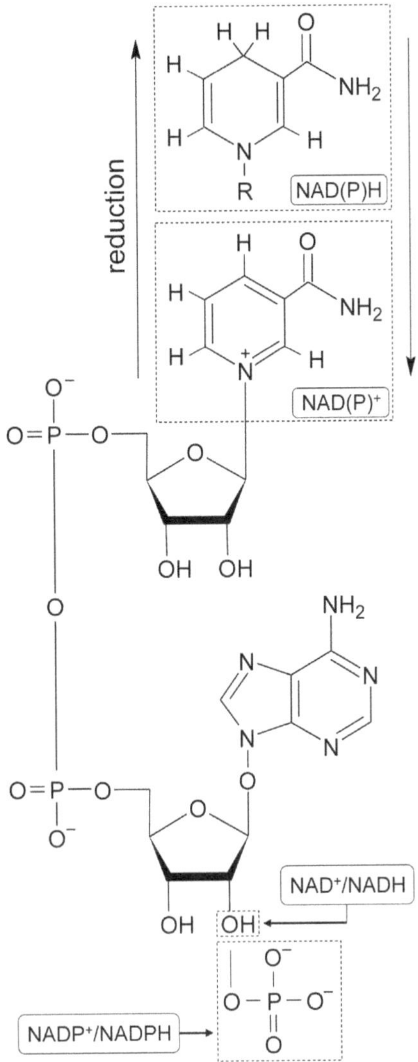

FIGURE 3.2 **Structure of nicotinamide adenine dinucleotide and nicotinamide adenine dinucleotide phosphate.** The oxidized (NAD^+ and $NADP^+$) and reduced (NADH and NADPH) forms are shown in the two upper *dotted boxes*. For $NADP^+$/NADPH, the phosphate group enclosed by the lower *dotted box* replaces the boxed hydroxyl group of NAD^+/NADH. *Source: Figure by J. Fuhrmann.*

is **catabolism**. Alternatively, some organisms obtain energy from transformations of various inorganic compounds, whereas others use light as an energy source. Organisms that derive energy from organic or inorganic substances are called **chemotrophs,** whereas **phototrophs** convert light energy into chemical energy to support metabolic processes. Organisms that obtain reducing equivalents from organic compounds are referred to as **organotrophs**, and those that use inorganic substances are called **lithotrophs**.

A portion of the energy derived from exergonic reactions is subsequently used for the endergonic biosynthesis of cellular components needed for cell growth, biomass production, and reproduction. This biosynthetic process is called **anabolism**, which is the functional opposite of catabolism.

One fundamental source of diversity among microorganisms relates to carbon nutrition. Most microorganisms obtain the carbon intermediates, or "building blocks", needed for biosynthesis from the degradation of organic compounds; these organisms are **heterotrophs**. The remaining organisms (algae and certain bacteria) can convert inorganic carbon from carbon dioxide (or carbonates) into organic compounds by the endergonic process known as carbon fixation; these organisms are called **autotrophs** (auto—self, troph—feeding).

A note on terminology

As indicated earlier, microorganisms can be categorized with respect to three important metabolic requirements (Fig. 3.3):

- source of energy (chemotrophs or phototrophs),
- source of reducing equivalents (organotrophs or lithotrophs), and
- source of carbon for anabolism (heterotrophs or autotrophs).

According to this system (summarized in Table 3.1), a filamentous fungus might be called a "chemoorganoheterotroph" or perhaps a "heterotrophic chemoorganotroph", whereas a photosynthetically active alga could be referred to as a "photolithoautotroph". In practice, however, many soil microbiologists have informally adopted an abbreviated system for referring to these metabolic categories, which recognizes that certain combinations of these traits nearly always occur together. For example, the description of the fungus given previously is typically reduced to simply "heterotroph" because the vast majority of heterotrophic organisms are also chemoorganotrophs.

The algal cell presents a somewhat more complicated case in that autotrophy commonly occurs with both phototrophy (e.g., eukaryotic algae and cyanobacteria) and chemotrophy (e.g., nitrifying bacteria). Given this ambiguity, the label "photoautotroph" would probably be used by most soil microbiologists. Lithotrophic organisms are usually also autotrophic, and this association can be assumed unless otherwise noted. Used alone, the term "lithotroph" generally refers to chemolithoautotrophs (commonly referred to simply as chemoautotrophs), such as the nitrifying bacteria, although a much less ambiguous label would also indicate whether the energy source is light or chemical compounds, as in photoautotroph or chemoautotroph.

In summary, the vast majority of microorganisms encountered by soil microbiologists fit into the three metabolic groups: chemoheterotrophs, photoautotrophs, and chemoautotrophs (see also Table 5.4).

Chemical foundations of metabolism

Discussion of metabolism requires a basic understanding of certain chemical principles. This section is provided for those readers wishing to review the more important concepts.

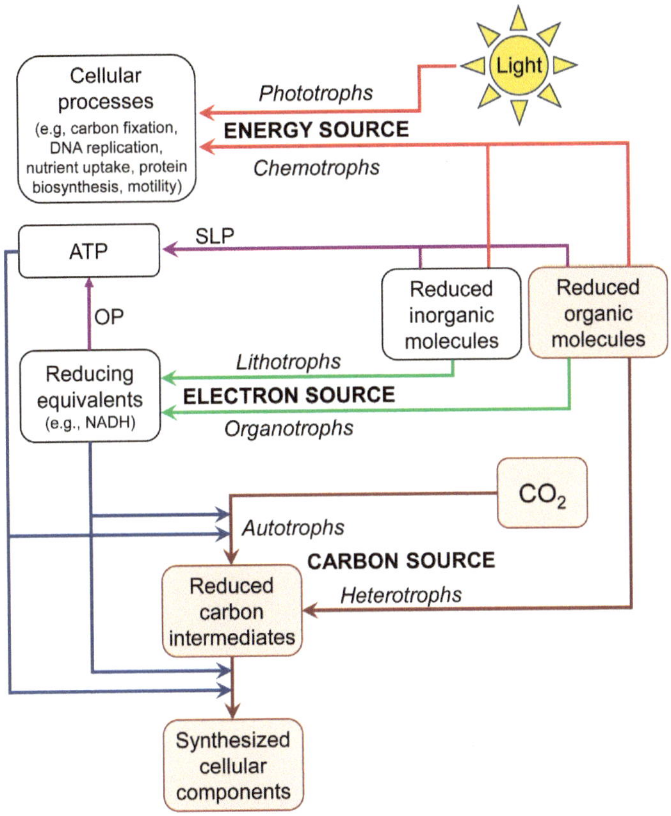

FIGURE 3.3 **Overview of different metabolic strategies used by soil microorganisms for acquiring cellular energy, reducing equivalents, and carbon for the synthesis of cellular components.** *Red arrows* indicate the ultimate source of cellular energy (light or chemical oxidations). *Green arrows* indicate the source of reducing equivalents (reduced inorganic or organic molecules). *Brown arrows* and *boxes* indicate the source of carbon (CO_2 or reduced organic molecules) and carbon flow. The labels next to these *arrows* indicate the associated type of microbial metabolism represented. Additionally, *purple arrows* indicate pathways by which ATP is produced (see the discussion on the production of ATP elsewhere in this chapter): SLP, substrate level phosphorylation; OP, oxidative phosphorylation. *Blue arrows* show the expenditure of ATP and reducing equivalents for carbon fixation and biosynthesis of cellular components. As indicated in the *boxes* for energy acquisition and carbon processing, the metabolic energy acquired (ATP and reducing equivalents) can be used for many purposes in cells, such as motility, nutrient uptake from the environment, carbon and N_2 fixation, and the synthesis of DNA, enzymes (proteins), and structural polymers. *Source: Figure by J. Fuhrmann.*

Oxidation/reduction reactions

Metabolism involves oxidation-reduction (redox) reactions. Oxidation refers to the loss of electrons by a substance. Oxygen may or may not be involved in the reaction. The substance oxidized is transformed into a second substance with a lower level of potential energy. Conversely, reduction refers to the acceptance of electrons by a substance. This represents a net gain in energy by the substance being reduced. Hydrogen is often transferred to the reduced compound simultaneously with the electrons. A critical point here is that oxidation

TABLE 3.1 Summary of metabolic classifications of microorganisms based on their source of energy, reducing equivalents, and carbon.

Metabolic requirement	Alternative metabolic strategies	Definition and comments	Representative microorganisms
Energy	Phototrophy	Light is used as the primary source of energy for metabolism and growth.	Algae, cyanobacteria
	Chemotrophy	Light-independent chemical reactions are the source of metabolic energy.	All non-photosynthetic cellular organisms
Reducing equivalents (electrons)	Lithotrophy	Inorganic substances used as electron donors; lithotrophic organisms may be either phototrophs or chemotrophs.	Algae and cyanobacteria (photolithotrophs); nitrifying and sulfur-oxidizing bacteria (chemolithotrophs)
	Organotrophy	Organic substances are used as electron donors; normally associated with chemotrophy, although also exhibited by certain photosynthetic bacteria (phototrophs).	Fungi, most bacteria, and protozoa
Carbon	Autotrophy	Cellular carbon is derived mostly or entirely from carbon dioxide; associated primarily with lithotrophy.	Algae, cyanobacteria, nitrifying and sulfur-oxidizing bacteria
	Heterotrophy	Cellular carbon is derived from preformed organic compounds; generally associated with chemoorganotrophy.	Fungi, most bacteria, and protozoa

Summarized from Singleton and Sainsbury (2006).

and reduction are inseparable; the oxidation of one compound necessarily implies the reduction of a second compound.

Oxidation state

Redox reactions result in changes in the oxidation state, or valence, of elements. When considering a reaction, it is often helpful to determine which elements are being reduced and oxidized. In many cases, the oxidation state of an element can be determined by applying the following general rules:

- Any element in the free state is assigned an oxidation number of zero; for example, elemental sulfur (S) has an oxidation state of zero; each oxygen atom in O_2 has an oxidation state of zero.
- Elements present as monoatomic ions have an oxidation state equal to the ionic charge; for example, magnesium as Mg^{2+} has an oxidation state of +2.
- When combined with other elements, hydrogen and oxygen typically have oxidation states of +1 and −2, respectively. (The exceptions are hydrides for hydrogen and peroxides for oxygen; in both cases, the oxidation state is −1.)

BOX 3.1

Example redox reactions

The numbers above the compounds indicate the oxidation state of the constituent elements.

$$2\overset{0}{H_2}+\overset{0}{O_2}\rightarrow 2\overset{+1}{H_2}\overset{-2}{O}$$

Each hydrogen loses one electron and is oxidized from 0 to +1 oxidation state; each oxygen gains two electrons and is reduced from 0 to -2 oxidation state.

$$2\overset{0}{S}+3\overset{0}{O_2}+2\overset{+1}{H_2}\overset{-2}{O}\rightarrow 2\overset{+1}{H_2}\overset{+6}{S}\overset{-2}{O_4}$$

Sulfur is oxidized from 0 to +6; each oxygen from O_2 is reduced from 0 to -2.

$$\overset{-3}{N}\overset{+1}{H_3}+1\tfrac{1}{2}\overset{0}{O_2}\rightarrow \overset{+3}{N}\overset{-2}{O_2^-}+\overset{+1}{H^+}+\overset{+1}{H_2}\overset{-2}{O}$$

Nitrogen is oxidized from -3 to +3; each oxygen from O_2 is reduced from 0 to -2.

$$\overset{+4}{C}\overset{-2}{O_2}+4\overset{0}{H_2}\rightarrow \overset{-4}{C}\overset{+1}{H_4}+2\overset{+1}{H_2}\overset{-2}{O}$$

Carbon is reduced from +4 to -4; hydrogen is oxidized from 0 to +1; oxygen is unchanged.

- The sum of the oxidation states of the atoms comprising a compound or polyatomic ion must equal the overall net charge of the substance; for example, the N in NO_3^- has an oxidation state of +5; the C in CH_4 has an oxidation state of –4; the S in H_2SO_4 has an oxidation state of +6.

In general, a relatively high ratio of hydrogen to oxygen atoms in a given substance indicates that the associated elements are in a comparatively high state of reduction (Box 3.1).

Reduction potentials

Although oxidation states are useful for monitoring changes to the specific elements comprising a redox reaction, they do not indicate the direction in which the reaction will tend to proceed. Nearly all metabolic reactions are reversible, but a given reaction will proceed spontaneously only in the direction yielding a net release of energy. For redox reactions, this will reflect the relative concentrations of the substances comprising a reaction and their tendencies to donate or accept electrons.

For example, the overall reaction:

$$H_2 + \tfrac{1}{2}O_2 \rightarrow H_2O \tag{3.1}$$

has two half-reactions:

an oxidation: $H_2 \rightarrow 2H^+ + 2e^-$ (3.2)

and a reduction: $\tfrac{1}{2}O_2 + 2H^+ + 2e^- \rightarrow H_2O$ (3.3)

Therefore, assuming equal concentrations of the reaction components, the tendency for reaction (3.1) to result in the formation of H_2O versus H_2 and O_2 will be determined by the relative tendencies of reactions (3.2) and (3.3) to proceed in the direction indicated.

By convention, the tendency for a half reaction to occur is determined by measuring the electrical potential it generates relative to a reference substance (H_2) under standardized conditions of temperature, acidity, and pressure (25°C, pH = 0, and 1 atm). The resulting potentials are then corrected to pH = 7 (and designated E_0') to reflect that metabolic reactions generally occur under near neutral conditions. Finally, to facilitate comparison among reactions, the half reactions are written as reductions and the sign of the corresponding electrical potential is adjusted accordingly. These standardized values are referred to as reduction potentials.

Some reduction potentials are given in Fig. 3.4. A relatively low (more negative) reduction potential indicates a relatively low tendency for the reduction to occur. Conversely, it indicates a relatively high tendency for the opposite oxidation reaction to occur. Thus, for our example, the half reactions (both written here as reductions):

$$2H^+ + 2e^- \rightarrow H_2 \qquad E_0' = -0.42\ \text{V} \tag{3.4}$$

$$\tfrac{1}{2}O_2 + 2H^+ + 2e^- \rightarrow H_2O \qquad E_0' = +0.82\ \text{V} \tag{3.5}$$

the E_0' for reaction (3.4) is more negative than that for reaction (3.5), indicating that reaction (3.5) has the greater tendency to proceed as written. In other words, oxygen at an oxidation state of 0 accepts electrons much more readily than does hydrogen at an oxidation state of +1. Thus, the combined reaction (3.3) can proceed spontaneously to form H_2O. The reverse reaction can also occur but only with the input of energy from an external source.

Free energy change

Perhaps the most useful measure of a metabolic reaction is its corresponding change in **free energy** (also called Gibb's free energy). Free energy ($G^{0\prime}$) refers to the intrinsic energy contained in a given substance. For the reaction $X \rightarrow Y$, the change in free energy ($\Delta G^{0\prime}$; also called free energy of formation) refers to the difference between the free energy of product Y and the reactant X, that is, $G^{0\prime}Y - G^{0\prime}X$. The $\Delta G^{0\prime}$ value is important because it indicates the amount of energy that is released by or required for a particular reaction. A negative or positive $\Delta G^{0\prime}$ indicates that the corresponding reaction is exergonic or endergonic, respectively. It is important to realize, however, that prevailing environmental conditions (e.g., concentrations of reactants and products) can have a large effect on the validity of this prediction (see Madigan et al. 2019). The value of $\Delta G^{0\prime}$ for a redox reaction can be calculated by using the following equation:

$$\Delta G^{0\prime} = -n\,F\,\Delta E_0' \tag{3.6}$$

where n = the number of electrons transferred, F = Faraday's constant (96.5 kJ V^{-1} mol^{-1} [= 23.1 kcal V^{-1} mol^{-1}]), and $\Delta E_0'$ = the difference, or numerical distance, between the reduction potentials for the two half reactions comprising the combined reaction when both are oriented so as to reflect the direction of the combined reaction.

For the reaction: $H_2 + \tfrac{1}{2}O_2 \rightarrow H_2O$,

Redox couple	Electrons transferred	E_0' (V)
SO_4^{2-}/HSO_3^-	2	−0.52
CO_2/glucose	24	−0.43
$2H^+/H_2$	2	−0.42
NAD^+/NADH	2	−0.32
S^0/H_2S	2	−0.28
FAD/$FADH_2$	2	−0.22
SO_3^{2-}/H_2S	2	−0.12
Fe^{3+}/Fe^{2+} (pH 7)	1	+0.20
NO_2^-/NO	2	+0.36
NO_3^-/NO_2^-	2	+0.42
NO_3^-/½N_2	5	+0.74
Fe^{3+}/Fe^{2+} (pH 2)	1	+0.76
Mn^{4+}/Mn^{2+}	2	+0.80
½O_2/H_2O	2	+0.82

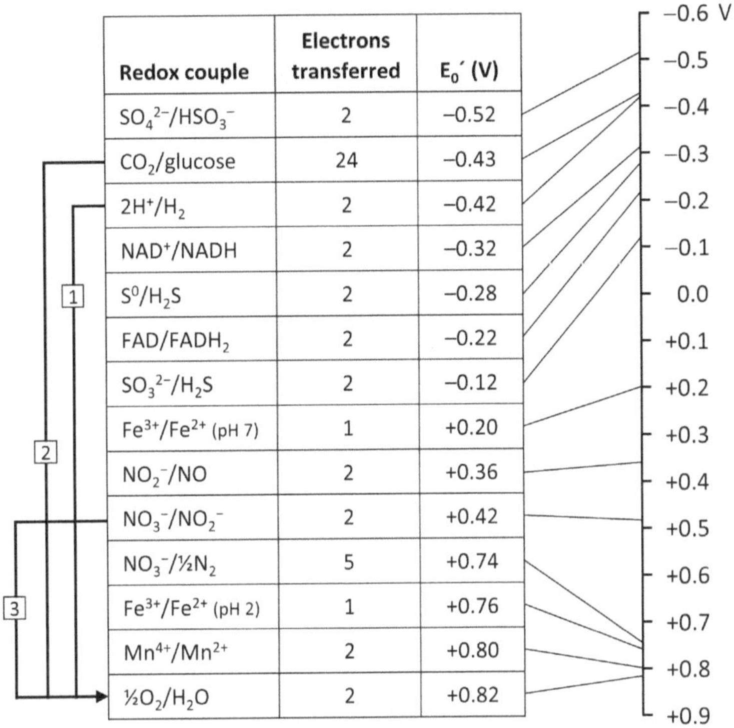

No.	Overall chemical reaction	$\Delta E_0'$ (volts)	$\Delta G^{0\prime}$ (kJ mol^{-1} donor)
1	H_2 + ½O_2 → H_2O	+1.24	−239
2	$C_6H_{12}O_6 + 6O_2$ → $6CO_2 + 6H_2O$	+1.25	−2895*
3	NO_2^- + ½O_2 → NO_3^-	+0.40	-77

FIGURE 3.4 **Upper table: Redox tower showing standard reduction potentials (E_0') in volts (V) and number of electrons transferred for selected redox couples that have metabolic importance to soil microorganisms.** The couples are arranged with the strongest electron donors at the top of the list and the strongest acceptors at the bottom. Lower table: Shown for each redox combination labeled 1 through 3 on the left side of the upper table, all of which use O_2 as an electron acceptor, are the overall chemical reaction, the corresponding difference in redox potentials ($\Delta E_0'$), and the free energy change ($\Delta G^{0\prime}$) calculated using Eq. (3.6) found in the text. Despite having similar values for $\Delta E_0'$, reaction 2 has a much greater $\Delta G^{0\prime}$ than does reaction 1 due to the greater number of electrons transferred (24 and 2, respectively). Comparing reactions 1 and 3, both reactions involve the transfer of 2 electrons; therefore, the lower $\Delta G^{0\prime}$ for 3 is a direct reflection of its lower $\Delta E_0'$. (* The magnitude of $\Delta G^{0\prime}$ shown for reaction 2 is slightly greater than the commonly accepted value of −2870 kJ mol^{-1} glucose; the latter more accurately accounts for the multistep nature of glucose oxidation to CO_2.) *Source: Figure by J. Fuhrmann; Redox potentials obtained from (Madigan et al., 2019).*

$$\Delta E_0' = +0.82 - (-0.42) = +1.24 \text{ V and} \quad (3.7)$$

$$\Delta G^{0\prime} = -(2)(96.5)(1.24) = -239 \text{ kJ mol}^{-1} \text{ of } H_2 \text{ consumed.} \quad (3.8)$$

This highly negative $\Delta G^{0\prime}$ value indicates that a large amount of free energy is released by the reaction. According to our definition of $\Delta G^{0\prime}$, this value is also the amount of energy that must be expended to convert one mole of water into its component elements.

Essential concepts

Metabolic processes in cells involve oxidation-reduction (redox) reactions. All redox reactions consist of two parts, an oxidation half reaction and a reduction half reaction. When a substance is oxidized, it donates electrons to a second substance, which thereby becomes reduced (more energy-rich). The tendency to be reduced or oxidized differs among substances and is commonly expressed as a standard reduction potential (E_0'). A combined redox reaction can proceed spontaneously in the direction written provided the numeric difference of the reduction potentials of the corresponding half reactions is positive. The free energy change ($\Delta G^{0\prime}$) associated with a reaction gives a measure of the energy released by ($\Delta G^{0\prime} < 0$) or required for ($\Delta G^{0\prime} > 0$) the reaction. The value of $\Delta G^{0\prime}$ is a function of the number of electrons transferred in a reaction and the corresponding difference in reduction potentials for the half reactions.

Role of enzymes in metabolism

Although redox chemistry and free energy principles are useful in predicting energy relationships in metabolic processes, they provide no information concerning the rate at which reactions proceed. Certain reactions may release energy too rapidly for cellular processes and are incompatible with living cells, generally because of heat production. Conversely, many chemical reactions do not occur at appreciable rates even though the corresponding free energy change is highly favorable. This results from the need to satisfy the **activation energy** for the reaction—the energy input necessary to break the existing chemical bonds within each reactant. For a reaction to proceed, its activation energy must be either met by an external energy source or reduced in magnitude by the presence of a **catalyst** (Fig. 3.5).

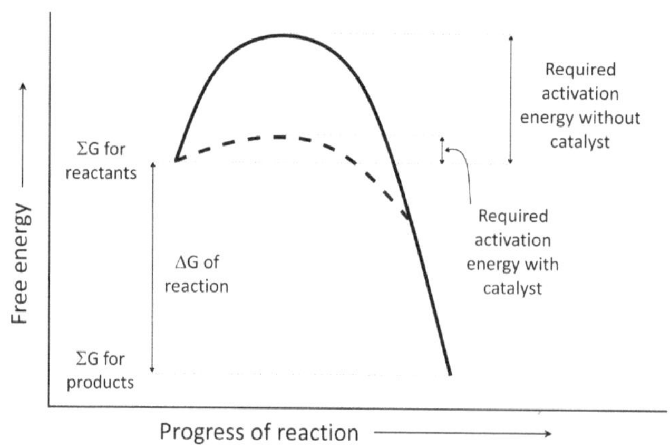

FIGURE 3.5 **Effect of a catalyst on the activation energy required for initiating an exergonic chemical reaction.** The catalyst lowers the amount of energy needed to initiate the reaction. *Source: Figure by J. Fuhrmann.*

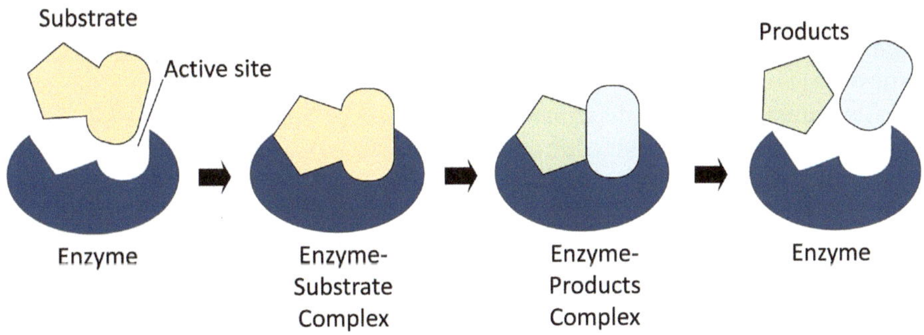

FIGURE 3.6 **The major steps in a degradative enzymatic reaction.** The substrate binds to the enzyme at the active site which is usually "shape specific". The bound substrate is cleaved while in the enzyme-substrate complex, and the products diffuse away from the enzyme leaving it ready to repeat the cycle. *Source: Figure by J. Fuhrmann.*

A catalyst is a substance that promotes a reaction by reducing the required energy of activation without itself being altered by the ensuing reaction. This catalytic role in living cells is provided by specialized proteins called **enzymes**. Unlike most nonbiological catalysts, enzymes are highly specific in that they generally promote only one particular reaction or class of reactions. Enzymes promote efficient metabolism not only by facilitating the initiation of biochemical reactions but also by controlling the rate and extent of these reactions. Degradative (catabolic) enzymes promote the cleavage of bonds between the substrate's components (Fig. 3.6). Synthetic (anabolic) enzymes do this by temporarily binding the substrates (reactants) at the enzyme's active site and thereby physically orienting them so as to both promote and control their interaction (essentially the reverse of Fig. 3.6). In this way, enzymes allow reactions to proceed in a step-by-step fashion with minimum energy input.

Enzymes often require the presence of various nonprotein substances called coenzymes to carry out their catalytic functions. The noncomplexed protein component is referred to as the **apoenzyme,** whereas the apoenzyme-coenzyme complex is called the **holoenzyme**. Coenzymes may bind either covalently or noncovalently to the apoenzyme. Covalently bound coenzymes are called prosthetic groups. For example, the **heme** group in many enzymes is usually present as a prosthetic group (Fig. 3.7). Noncovalently bound coenzymes (e.g.,

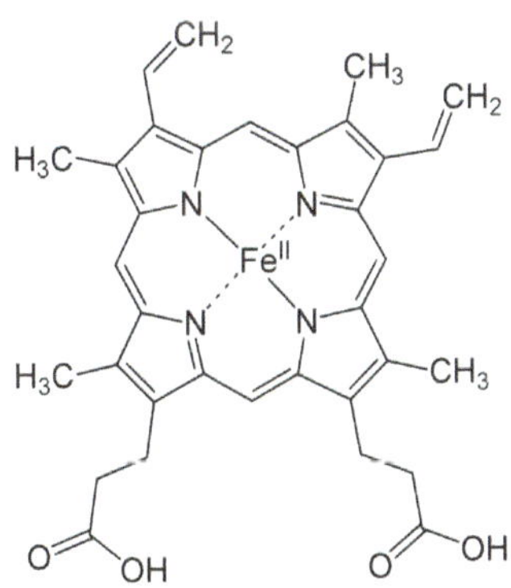

FIGURE 3.7 **Structure of heme.** The coordination iron at the center is the active site for oxidation and reduction. *Source: Adapted from public domain.*

NADH) are only transiently associated with the enzyme and may be thought of as cosubstrates. Many substances serve as coenzymes, including metal ions (e.g., iron, molybdenum, and magnesium) and carrier molecules (e.g., NADH, $FADH_2$, and various vitamins). The carrier molecules function to supply substances required by the reaction, such as electrons and hydrogen, or to aid in the transfer of chemical groups, such as phosphate, between molecules.

Microorganisms need control of enzyme production both to promote balanced cell growth and to minimize unnecessary energy expenditure for enzyme production. Therefore, various strategies have evolved to regulate the production and activity of cellular enzymes. Enzymes needed for fundamental cellular processes are generally maintained at constant levels in the cell, and their production is unaffected by concentrations of substrates or products; such enzymes are called **constitutive enzymes**. Alternatively, enzyme production may be inducible, when production occurs in the presence of an inducer molecule, or derepressible, when production occurs in the absence of a repressor molecule. Many catabolic enzymes are inducible in that the proper degradative enzymes are produced only in response to the presence of the corresponding substrate, which functions as an inducer; for example, lactose (milk sugar) acts as an inducer for the production of its degradative enzymes. Conversely, some anabolic enzymes are derepressible so as to limit enzyme production to periods when a specific enzymatic product has dropped below some critical concentration in the cell; that product functions as a repressor. The enzymes responsible for the conversion of dinitrogen gas to ammonia (N_2 fixation) are examples of derepressible enzymes in that their production is inhibited by elevated concentrations of ammonium and other forms of combined nitrogen (Chapter 16).

Enzyme activity can be controlled not only at the level of enzyme production but also by modifying the activity of existing enzymes in the cell. This general type of altered activity is known as **allosteric control**, meaning that a molecule affects a metabolic reaction by binding to the corresponding enzyme at an allosteric site distinct from the active site (Fig. 3.8). The most common form of allosteric control is allosteric inhibition, also referred to as **feedback inhibition** or end product inhibition. Allosteric inhibitors are generally metabolic end products which attenuate the metabolic pathway that produced them by inhibiting an early step

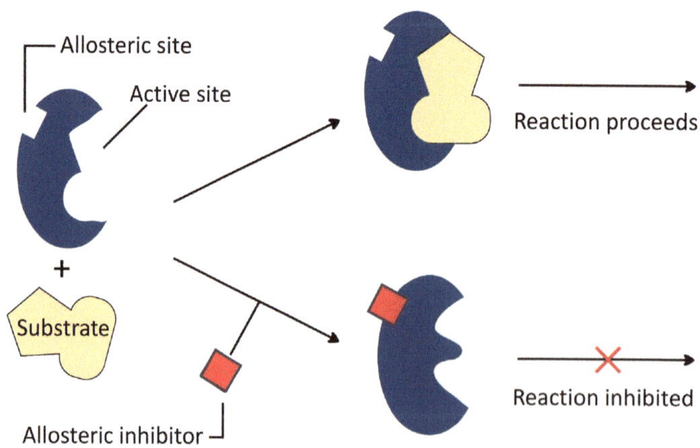

FIGURE 3.8 **Negative allosteric control of enzyme activity.** Allosteric inhibitors decrease the rate of enzymatic reactions by binding to a separate allosteric site on the enzyme and thereby indirectly altering the configuration (shape) of the active site so as to prevent substrate binding. *Source: Figure by J. Fuhrmann.*

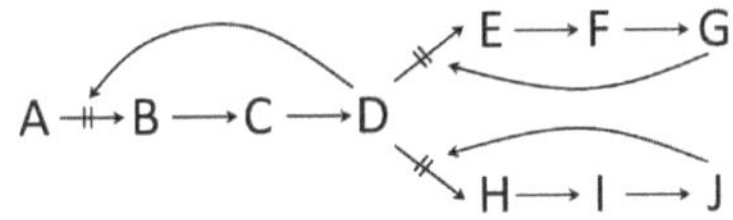

FIGURE 3.9 **Schematic of feedback inhibition.** The enzymatic products D, G, and J inhibit early steps in their respective metabolic pathways through feedback inhibition. In branched pathways, the inhibitor often works at the first enzyme of the branch. *Source: Figure by J. Fuhrmann.*

in the pathway (Fig. 3.9). Attachment of the inhibitor at the allosteric site reduces enzymatic activity by altering the conformation (shape) of the active site, thereby preventing attachment of the substrate(s). In other cases, the attachment of a substance at the allosteric site may result in increased enzyme activity.

Enzymes promote many types of reactions within living cells and are generally named for the particular reaction catalyzed. Enzyme names are further designated by the suffix "-ase". Table 3.2 summarizes the kinds of reactions carried out by different classes of enzymes. In many respects, the metabolic diversity discussed in the remainder of this chapter is a direct reflection of the enzymatic diversity that exists among soil microorganisms. Specific examples of enzymes important in soil microbiology include:

- cellulases, which degrade the polymer cellulose into smaller constituents
- "nitrogenase", which converts dinitrogen gas into biologically available ammonia (the term nitrogenase actually refers to an enzyme complex)
- sulfatases, which release sulfate from proteins and certain other organic compounds
- proteases, which degrade proteins into smaller peptides and amino acids
- lysozyme, which degrades peptidoglycan in the cell walls of bacteria

TABLE 3.2 Major classes and subclasses of enzymes and the corresponding types of reactions catalyzed.

Class	Representative subclasses	Types of reactions catalyzed by enzyme class
Oxidoreductases	Dehydrogenases Oxidases Reductases Oxygenases Peroxidases Catalases	Catalyze oxidation-reduction reactions. Important in fermentation and respiration pathways.
Transferases	Aminotransferases Kinases	Catalyze the transfer of molecular substituents among molecules.
Hydrolases	Glycosidases Peptidases Phosphatases Ribonucleases	Catalyze the hydrolytic cleavage of chemical bonds.
Lyases	Decarboxylases Synthases Lyases	Catalyze the addition or removal of chemical compounds such as carbon dioxide, ammonia, and water.
Isomerases	Racemases Isomerases	Catalyze inversions at asymmetric carbon atoms and the intramolecular transfer of molecular substituents.
Ligases	Synthetases Carboxylases	Catalyze the binding of two molecules with the expenditure of ATP. Important in anabolic pathways.

BOX 3.2

Common enzyme groupings

Soil microbiologists often find it useful to place enzymes into broad groupings based on certain functional characteristics. One common grouping refers to the intended location of enzymatic activity:

- *Intracellular enzyme*—an enzyme that catalyzes biochemical reactions occurring within cells; many intracellular enzymes can be found in soil apart from cells and are thought to be released to the environment by cell lysis.
- *Extracellular enzyme*—an enzyme purposefully released exterior to cells, generally to catalyze the degradation of a polymeric substance too large to cross the cellular membrane.

Another common grouping relates to the position within a polymer at which an enzyme is active:

- *Exoenzyme*—an enzyme, often extracellular, that catalyzes the removal of terminal monomers of a polymeric substance.
- *Endoenzyme*—an enzyme, often extracellular, that degrades polymers at positions other than the termini, thereby producing oligomers, chains containing more than two monomers, which can be further degraded by exoenzymes.

- chitinases, which degrade chitin in fungal cell walls and arthropod exoskeletons
- phosphatases, which remove phosphate groups from nucleic acids and other organic compounds

Not all enzymes are intracellular (Box 3.2), meaning that they are produced and active within the cell. Enzymes that catalyze the degradation of polymeric substances added to soil, such as cellulose from crop residues, are necessarily extracellular because the polymer is too large to be transported across the cellular membrane. However, once the polymer has been reduced to its smaller subunits, subsequent catabolism may proceed intracellularly. Because many enzymes are important in nutrient cycling and other soil-related processes, standardized assays have been developed for measuring their activities. For instance, an assay for phosphatase activity in soil is described in Chapter 19.

Production of ATP

Adenosine triphosphate (ATP) is synthesized in cells by one of two mechanisms—**substrate-level phosphorylation** or **oxidative phosphorylation**. Substrate-level phosphorylation is the direct production of ATP during the enzymatic oxidation of a substance and involves the transfer of a phosphate group from a substrate to an ADP molecule. This type of phosphorylation is reaction-specific and closely linked to the corresponding oxidative reaction. In contrast, oxidative phosphorylation, sometimes called electron-transport-chain phosphorylation, involves membrane-localized reactions that are physically removed from the initial energy-yielding reaction. Oxidative phosphorylation is implied to be active in respiration, which is discussed later in this chapter. In the case of oxidative phosphorylation, electrons released during oxidation are transferred to electron carriers such as NAD^+ and FAD and transported to an electron-transport chain rather than being directly used for ATP production. These electron-transport chains consist of membrane-bound electron carriers that function as a unit to accept electrons from NADH and $FADH_2$, thereby releasing NAD^+ and FAD back to the cell (Fig. 3.10). Within the electron transport chain, the electrons

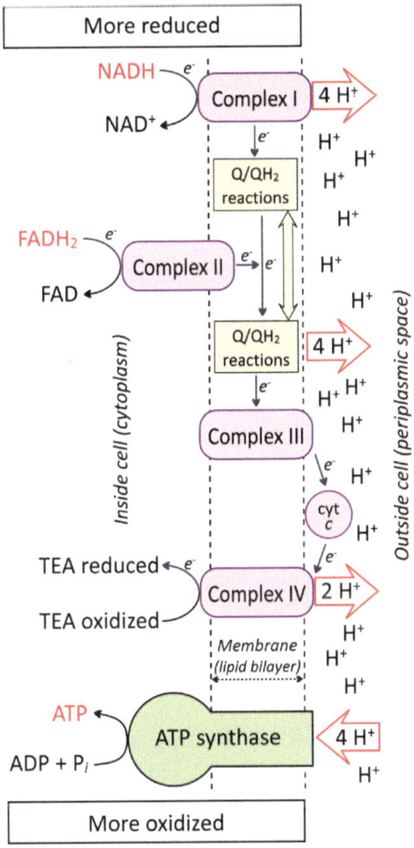

FIGURE 3.10 **Generalized depiction of oxidative phosphorylation via an electron transport chain (based on Madigan et al., 2019).** Reduced electron carriers are oxidized by a membrane-bound electron transport system which results in the eventual reduction of a terminal electron acceptor (TEA) (e.g., O_2 which is reduced to H_2O). These oxidations produce a proton gradient (proton motive force) which, in turn, is used to produce ATP. Note that NADH enters the transport chain at a higher energy (more reduced) level than does $FADH_2$ and, therefore, generates more hydrogen ions. As a result, NADH and $FADH_2$ yield three and two ATP molecules, respectively. The membrane-bound enzyme ATP synthase uses the proton gradient to make ATP; note that ATP synthase is not part of the transport chain per se. Electron (e^-) flow within the transport chain is indicated by *blue lines* and *arrows*. The *boxes* and *arrow shaded* in *yellow*, labeled "Q/QH_2 reactions"(also called "Q-cycle reactions"), refer to what is essentially a membrane-bound quinone shuttle system for protons and electrons as they travel down the transport chain and/or across the membrane (Q, ubiquinone; QH_2, reduced ubiquinone or ubiquinol). Unlike Complexes I–IV (called complexes because each has multiple components not described here), cytochrome *c* (cyt *c*) is not membrane-bound and functions to transfer electrons from Complex III to Complex IV. P*i*, inorganic phosphate. *Source: Figure by J. Fuhrmann.*

participate in a carefully controlled series of redox reactions that capture a portion of the energy released as ATP.

In comparison with substrate-level phosphorylation, oxidative phosphorylation is a more generalized means of ATP production because reducing equivalents derived from diverse oxidative reactions may be processed by the same electron transport chain. It is also a much more efficient means of producing ATP. However, oxidative phosphorylation requires that an external oxidant should be present to accept the electrons as they exit from the electron-transport chain. This **terminal electron acceptor** (TEA) (also called the final electron acceptor) is generally oxygen, but some microorganisms can use alternative compounds. Substrate-level phosphorylation can also take advantage of its ability to occur under more diverse conditions than can oxidative phosphorylation because it does not require specific external electron acceptors.

A discussion of quantifying ATP in soil is presented in Box 3.3.

BOX 3.3

Measurement of ATP in soils

Soil microbiologists who study general soil processes such as nutrient cycling often wish to know the biomass of the resident microbial population. Although several different approaches for the measurement of microbial biomass and metabolic status in soils have been proposed, many share the principle of measuring some component of biomass that is common to the entire microbial community. Because ATP is inherent to all forms of metabolism and persists only in living cells, it has received considerable attention as an index of these measures (Blagodatskaya and Kuzyakov, 2013).

Briefly, the general procedure for the measurement of soil ATP is to treat a soil sample so as to rupture, or lyse, the cells of the indigenous microorganisms and to then measure the ATP released by means of the *luciferin-luciferase reaction*. Luciferin is an aromatic compound found naturally in fireflies that reacts, via the enzyme luciferase, to produce light in the presence of ATP:

$$\text{reduced luciferin} + \text{ATP} + O_2 \xrightarrow{\text{luciferase}} \text{oxidized luciferin} + CO_2 + \text{AMP} + 2\,\text{Pi} + \text{light}$$

Light production is typically measured in a luminometer. Given the presence of excess luciferin, the amount of light produced will be proportional to the amount of ATP present in the sample of soil extract analyzed. Proposed refinements include comparing the relative concentrations of ATP, ADP, and AMP as an index of the energetic state or metabolic health of the microbial community (Chapter 25). Significant concerns have been raised about using ATP measurements for assessing soil microbial communities (Blagodatskaya and Kuzyakov, 2013). Nevertheless, there is increasing interest in assessing the overall health or metabolic activity of soil microbial communities, and measurement of ATP and related compounds is one approach that has been proposed.

Glycolysis

The glycolytic, or Embden-Meyerhof-Parnas, pathway is the most common pathway for the catabolism of glucose and other simple sugars to pyruvate (pyruvic acid) (Fig. 3.11). The products of **glycolysis** are pyruvate, ATP, and NADH. The initial six-carbon portion of glycolysis involves the utilization of two molecules of ATP to produce fructose-1,6-bisphosphate. This activated sugar is then split into two three-carbon molecules, dihydroxyacetone phosphate and glyceraldehyde-3-phosphate, which are interconvertible by enzymatic action. Therefore, subsequent metabolic steps occur twice for each molecule of glucose catabolized. During several energy-yielding steps; each molecule of glyceraldehyde-3-phosphate is converted to pyruvate with the coproduction of one molecule of NADH and two molecules of ATP; the latter by substrate-level phosphorylation. This results in a total net gain of two NADH and two ATP molecules per molecule of glucose.

Two additional points should be noted. The end product of glycolysis, pyruvate, is incompletely oxidized and constitutes a rich potential source of metabolic energy. Second, the NAD^+ needed to accept electrons from glyceraldehyde-3-phosphate has not been regenerated, a situation that must be corrected if energy production is to continue. As the following discussions illustrate, the alternative metabolic pathways subsequently employed by

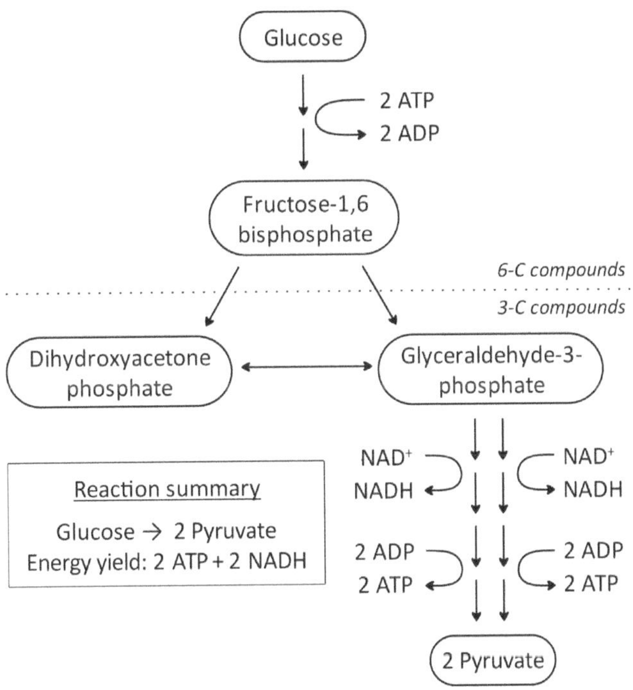

FIGURE 3.11 **Simplified overview of the Embden-Meyerhof-Parnas glycolytic pathway, that is, glycolysis.** Note that the end product, pyruvate, remains a rich source of energy and carbon. *Source: Figure by J. Fuhrmann.*

microorganisms differ primarily in the degree to which the energy contained in pyruvate is captured for cellular growth and the strategy employed to recover NAD^+ for metabolic processes.

Fermentation

Following glycolysis, some microorganisms recover NAD^+ by using NADH to reduce pyruvate or one of its derivatives to various organic compounds via any of several anaerobic pathways. The overall reaction of glucose to organic end products, including glycolysis, is called **fermentation** and is characterized by its use of internally produced (endogenous) organic electron donors and acceptors. Depending on the particular fermentative pathway used by an organism, compounds produced may include various alcohols (e.g., ethanol, butanol, and isopropanol), organic acids (e.g., lactic, acetic, propionic, and butyric acid), and, in some cases, carbon dioxide and hydrogen (Fig. 3.12).

Fermentative processes are often associated primarily with the catabolism of glucose and other sugars, but other compounds, such as amino acids, can serve as fermentation substrates for certain microorganisms. Fermentative processes are important commercially in the production of chemicals and in the baking and brewing of some foods and beverages. They are the basis for the extensive biofuels industry that has emerged in the last few decades. They

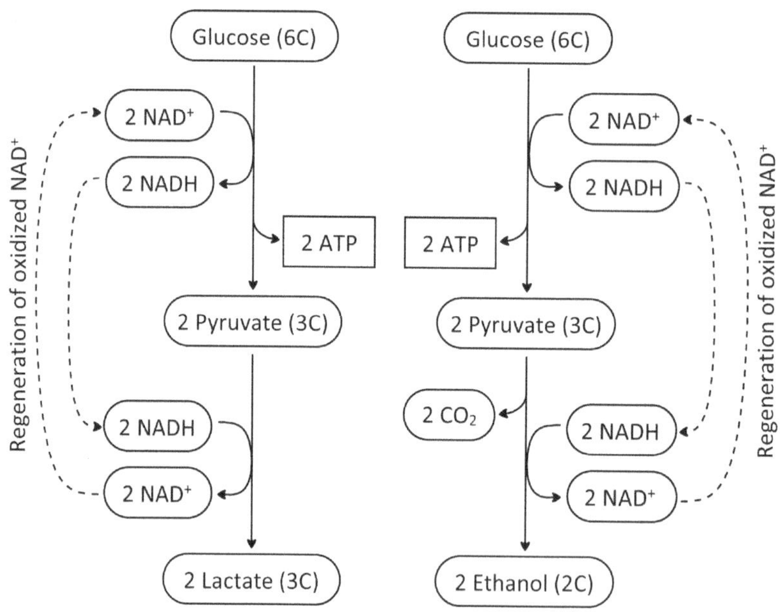

FIGURE 3.12 **Generalized pathways for the fermentation of glucose to lactic acid (left) and ethanol (right).** Note that the reactions proceeding from pyruvate serve primarily to regenerate NAD^+. As in the case of glycolysis, the end products are still rich sources of energy and carbon. *Source: Figure by J. Fuhrmann.*

are also significant ecologically as sources of carbon and energy for other microorganisms, metal chelators, and agents of mineral weathering in soils.

Respiration

Besides fermentation, the second possible catabolic fate of pyruvate is **respiration**. The defining characteristic of respiration is its use of an external TEA (rather than the internally produced organic acceptors found in fermentation). The overriding benefit derived from this change in metabolic strategy is that respiratory organisms have the potential to maximize their recovery of energy contained in reduced compounds such as pyruvate; these compounds are no longer needed as terminal electron acceptors themselves. The primary disadvantage is that the proper external electron acceptor must be present in the environment and available to the organism for respiration to proceed.

Respiratory catabolism of pyruvate is perhaps best thought of as a two-step process:

- Pyruvate is oxidized to carbon dioxide in a series of reactions called the tricarboxylic acid (TCA) cycle, also known as the Krebs cycle and the citric acid cycle (Fig. 3.13). As a result of these oxidative steps, the TCA cycle produces NADH, $FADH_2$, and a small amount of GTP, a high-energy compound functionally equivalent to ATP. GTP is produced by substrate-level phosphorylation. Although a detailed discussion of the TCA cycle is beyond the scope of this chapter, it is important to note that the only other products of the cycle are carbon dioxide and water. Thus, the carbon source is fully oxidized to yield maximum energy.
- The reduced electron carriers are further processed by electron-transport chains and oxidative phosphorylation to produce ATP and simultaneously regenerate NAD^+ and FAD for use during glycolysis and the TCA cycle (Fig. 3.14).

Acceptable external electron acceptors vary among organisms and may be either oxygen (**aerobic respiration**) or another highly oxidized compound like nitrate (NO_3^-) (**anaerobic respiration**). Furthermore, although most aerobic microorganisms use the respiratory catabolism of organic compounds such as pyruvate to obtain ATP, certain microorganisms are capable of capturing the energy released by the respiration of reduced inorganic compounds.

Aerobic respiration of organic compounds

The most widely studied soil microorganisms are those that rely on organic compounds as their source of energy and reducing equivalents. These microorganisms are classified metabolically as chemoorganotrophs. As noted previously, chemoorganotrophs are also nearly always heterotrophic (i.e., they obtain carbon from organic compounds). These organisms are essential to soil productivity and environmental quality because they have the primary responsibility for the degradation of organic substances added to soil, including plant residues, dead soil organisms, and natural and synthetic compounds from agricultural and industrial sources. Here we discuss the general metabolic characteristics of those chemoorganotrophs that carry out respiration in the presence of oxygen.

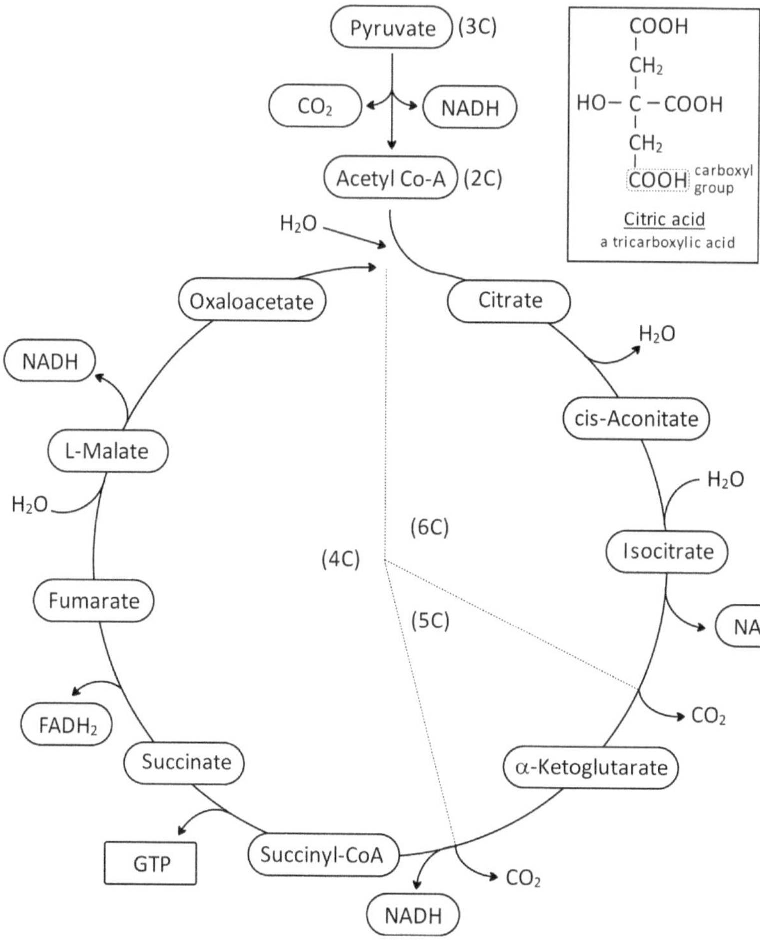

FIGURE 3.13 **The tricarboxylic acid (TCA) cycle, also called the Krebs cycle and citric acid cycle.** Inset shows the structure of citric acid, or citrate, which possesses three carboxyl groups. Unlike fermentation, the original glucose molecule is completely oxidized to CO_2 by the TCA cycle, with an accompanying generation of a large quantity of useable cellular energy. *Source: Figure by J. Fuhrmann.*

With glucose as an example substrate, the overall reaction for its oxidation to carbon dioxide under aerobic conditions via glycolysis and the TCA cycle can be given as:

$$C_6H_{12}O_6 + 6O_2 \rightarrow 6CO_2 + 6H_2O \quad \left(G^{0\prime} = -2870\,\text{kJ}\left[-686\,\text{kcal}\right]\,\text{mol}^{-1}\ \text{glucose}\right). \qquad (3.9)$$

This highly negative change in free energy for this reaction indicates a high energy yield from the aerobic respiration of glucose (see also Fig. 3.4.). This high potential energy yield is clearly evident in Fig. 3.14, which emphasizes the energy yielding reactions shown in

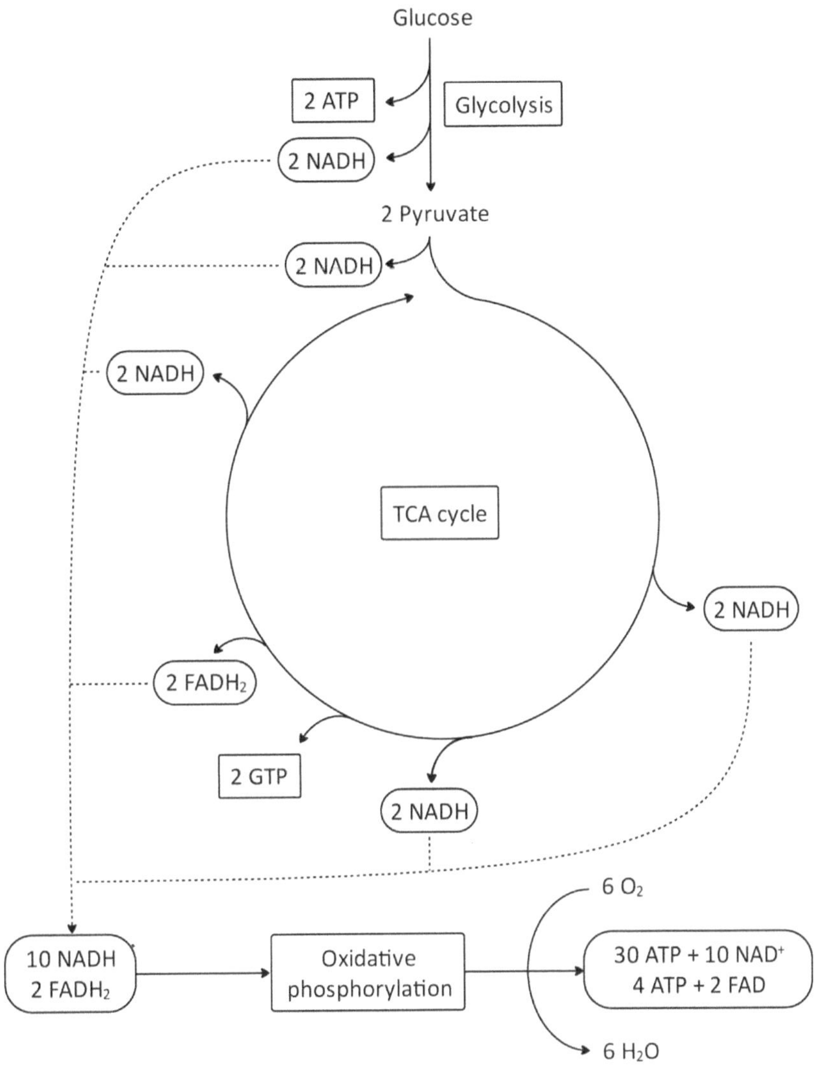

FIGURE 3.14 **Catabolism of glucose via aerobic respiration.** Note that each NADH and FADH$_2$ produces three and two ATP molecules, respectively. Aerobic catabolism of glucose yields 38 ATP—34 through the oxidation of NADH and FADH$_2$ via oxidative phosphorylation, 2 through substrate level phosphorylation (SLP) during glycolysis, and 2 (GTP) through SLP in the TCA cycle. This yield is in stark contrast to the 2 ATP produced by SLP during fermentation. *Source: Figure by J. Fuhrmann.*

Figs. 3.10, 3.11, and 3.13. Assuming that all of the energy contained in a molecule of glucose is allocated to ATP production, the effective energy yield from aerobic respiration can be summarized as shown in Table 3.3. Clearly, the theoretical production maximum of 38 moles of ATP per mole of glucose oxidized is far in excess of the 2 moles of ATP produced by the major fermentative pathways discussed earlier.

TABLE 3.3 Energy yield (expressed as ATP) for the aerobic respiration of glucose.

	Reducing equivalents	ATP yield
Glycolysis		
• Substrate-level phosphorylation	not applicable	2
• Oxidative phosphorylation	2 NADH	6
Tricarboxylic acid cycle		
• Substrate-level phosphorylation (GTP)	not applicable	2
• Oxidative phosphorylation	8 NADH	24
	2 $FADH_2$	4
Total energy yield		38

Anaerobic respiration of organic compounds

Some microorganisms, almost all of them bacteria or archaea, can substitute certain other oxidized compounds for oxygen as a terminal electron acceptor in a process known as anaerobic respiration. The most common inorganic alternative electron acceptors in soils are nitrate, manganic (+4 oxidation state) manganese, ferric (+3) iron, sulfate, and carbon dioxide; in some environments, oxidized anions of elements such as selenium, arsenic, and chromium may be used by certain microorganisms (Chapter 19). Some organisms can use organic compounds such as fumarate as acceptors. All of these substances accept electrons less readily than oxygen as reflected in their less positive reduction potentials (Fig. 3.4). Thus, the resulting energy yields are correspondingly reduced relative to aerobic respiration. Nevertheless, anaerobic respiration is more efficient than fermentation as a mode of metabolism and may confer a significant ecological advantage to those microorganisms possessing this ability, provided the proper electron acceptors are available.

Note that substances used as TEAs in respiration are not incorporated (assimilated) into microbial biomass but rather are used solely for energy production. The transfer of electrons to these external TEAs is called **dissimilatory reduction** because the reduced product is released back into the environment. An example of **assimilatory reduction** is the use of previously acquired metabolic energy to reduce N in nitrate (oxidation state = +5) to the −3 oxidation state of NH_4^+, followed by its incorporation into amino acids and proteins (i.e., the reduced N is used as a nutrient for biomass production).

Bacteria that can use nitrate as a TEA are generally **facultative anaerobes**, meaning that they carry out normal aerobic respiration in the presence of oxygen but can switch to nitrate in its absence. Oxygen is the preferred acceptor for these organisms, and its presence strongly inhibits nitrate reduction. Three types of dissimilatory nitrate reduction are common among soil microorganisms (Chapter 15):

- Denitrification is the reduction of nitrate to gaseous products, primarily nitrous oxide and dinitrogen. This process converts biologically available or "fixed" nitrogen to unavailable forms and is effectively the opposite of dinitrogen fixation (Chapter 16). Denitrification is extremely important because, depending on one's perspective, it alternatively represents a loss of soil fertility for plant growth, a means of reducing nitrate contamination of ground and surface waters, and a source of gaseous pollutants that contribute to global warming and the destruction of protective stratospheric ozone.

- Dissimilatory reduction of nitrate to ammonium (DNRA) is carried out under anoxic conditions by certain microbes and may be linked to energy production via either oxidative phosphorylation or fermentative processes.
- Various other bacteria carry out a less complete reduction to produce nitrite in a process called nitrate respiration. This process is less efficient than denitrification and DNRA, in that fewer electrons are accepted per nitrate molecule. It may also lead to the accumulation of nitrite in the soil environment, which can be toxic to plants and microorganisms.

A rather diverse group of bacteria can use sulfate and other oxidized forms of sulfur as terminal electron acceptors in respiration (Chapter 18). Unlike the nitrate reducers, the sulfate reducers are generally obligate anaerobes, also called strict anaerobes, which are only active in the absence of oxygen. The final product of microbial sulfate reduction is typically hydrogen sulfide (H_2S), which is partly responsible for the disagreeable "rotten egg" or "sulfur" smell associated with water-logged soils.

The utilization of carbon dioxide as a terminal electron acceptor to produce methane (CH_4) is called **methanogenesis**. Methanogenesis is of environmental concern because methane is one of several gases contributing to global warming. On the positive side, **methanogens** are useful in the production of "biogas" (for cooking and heating) from manures and other carbon sources. The methanogenic bacteria belong to the Archaea, which are noted for their unique cell wall composition, distinctive ribosomal characteristics, and certain other attributes (Chapter 5).

A unique and remarkable group of bacteria, **exoelectrogens**, carry out oxidative phosphorylation using TEAs located external to their cellular membranes in a process called **extracellular electron transfer** (EET) (Box 3.4).

Respiration of inorganic compounds

A restricted but extremely important group of bacteria and archaea can obtain energy and reducing equivalents necessary for oxidative phosphorylation from reduced inorganic compounds or elements. These organisms are called **chemolithotrophs** according to our previous discussion of metabolic groupings of organisms (Table 3.1). However, because these organisms generally also obtain carbon from carbon dioxide, they are often called chemoautotrophs. A number of reduced inorganic substances can be used as energy sources by this group, including hydrogen gas, methane, ferrous (divalent) iron, and certain reduced nitrogen and sulfur compounds as well as elemental sulfur. Most chemolithotrophs are aerobes, but others can use alternative electron acceptors such as nitrate and sulfate.

A given microorganism generally can oxidize only a single compound, and its common name often reflects this ability. For example, particularly important in normal agricultural soils are the aerobic nitrifying bacteria and archaea, which derive energy from the oxidation of ammonium to nitrate—a process called nitrification (Chapter 15). More precisely, as described in Chapter 15, it is ammonia that is the initial enzymatic substrate for the nitrification process (an equilibrium exists between ammonia and ammonium in aqueous systems, with ammonium being the dominant form). Nitrification involves two groups of organisms: the ammonia oxidizers, which convert ammonia to nitrite, and the nitrite oxidizers, which complete the conversion to nitrate.

BOX 3.4

Electric bacteria

The electron transport systems of bacteria described elsewhere in this chapter employ terminal electron acceptors present within the cell cytoplasm (e.g., O_2, NO_3^-). However, some bacteria use acceptors residing on the exterior of the cellular membrane in a process termed **extracellular electron transfer** (EET) (Light et al., 2018). As a result, these bacteria generate an electrical current, a phenomenon that has unique and potentially useful implications. These bacteria are referred to by various names including exoelectrogens, anode respiring bacteria, and electricigens. Both Gram-negative and Gram-positive bacteria (Chapter 5) have been found to possess this capability.

In an ecological or environmental context, exoelectrogens are thought to be involved in the cycling of metallic elements by chemically reducing the electron acceptor, thereby altering its bioavailability, for example, reducing the ferric iron (Fe^{3+}) in iron oxides to the more soluble ferrous state (Fe^{2+}). Additionally, some of these bacteria produce electrically conductive appendages called **nanowires** which function in distributing electrons within thick **biofilms**, thereby giving bacteria in anaerobic portions of the biofilm access to O_2, as one example.

However, exoelectrogens have received the greatest attention with respect to their use in microbial fuel cells (MFCs) (see figure). An MFC harnesses microorganisms to convert chemical energy from any of a wide range of organic compounds (e.g., sugars, wastewater constituents) to electrical energy. Although the quantity of electrical power produced is often small, even relatively weak MFCs can have useful applications, for example, wireless biosensors used for remote environmental monitoring. In soils, the large bacterial communities combined with often rich organic matter contents have be shown to yield MFCs that generate significant electrical currents. Moreover, technological advances offer the promise of using MFCs on a larger scale, at a level that can contribute to electrical grids powering homes and businesses.

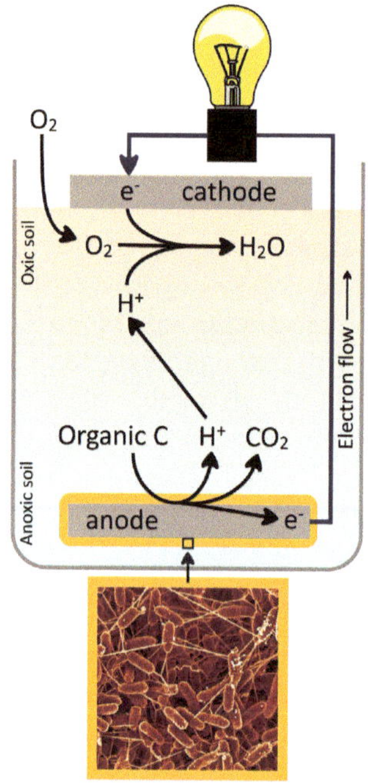

Schematic of a microbial fuel cell (MFC) constructed using soil. Bacteria called exoelectrogens (lower image) form a biofilm (orange) on the anode and oxidize organic carbon compounds, thereby releasing electrons. These electrons cause an electrical current that travels through wires (blue lines) to the cathode where O_2 is reduced to H_2O. Note that the soil in the lower portion of the MFC must be anoxic (e.g., resulting from water saturation; indicated by the blue shading), so that the electrons from the organic compounds are not diverted to O_2 at that location and instead are donated to the anode to generate the electrical current. Nanowires produced by the exoelectrogens are readily apparent in the lower image. *(Photo credit: R. Bencheikh and B. Arey. Image provided by EMSL, the Environmental Molecular Sciences Laboratory, a DOE Office of Science user facility. Used with permission.) Diagram of MFC by J. Fuhrmann.*

TABLE 3.4 Standard free energy of reactions utilized by chemolithotrophic microorganisms.

Reaction	$\Delta G^{o\prime}$ (kJ mol^{-1})
$HS^- + H^+ + \frac{1}{2} O_2 \rightarrow S + H_2O$	−212
$S + 1\frac{1}{2} O_2 + H_2O \rightarrow SO_4^{2-} + 2 H^+$	−587
$NH_3 + 1\frac{1}{2} O_2 \rightarrow NO_2^- + H^+ + H_2O$	−275
$NO_2^- + \frac{1}{2} O_2 \rightarrow NO_3^-$	−77
$H_2 + \frac{1}{2} O_2 \rightarrow H_2O$	−239
$2 Fe^{2+} + H^+ + \frac{1}{2} O_2 \rightarrow 2 Fe^{3+} + H_2O$	−31

Values for $\Delta G^{o\prime}$ are at pH = 7 except for Fe^{2+}/Fe^{3+}, which is at pH = 2 to 3. For reference, the aerobic oxidation of glucose to CO_2 yields −2870 kJ (−686 kcal) mol^{-1}, and the hydrolysis of ATP to ADP yields approximately −30 kJ (−7 kcal) mol^{-1}.
From various sources.

$$NH_3 + 1\frac{1}{2}O_2 \rightarrow NO_2^- + H^+ + H_2O \qquad \left(\Delta G^{0\prime} = -275\,\text{kJ mol}^{-1}\right) \tag{3.10}$$

$$NO_2^- + \frac{1}{2}O_2 \rightarrow NO_3^- \qquad \left(\Delta G^{0\prime} = -77\,\text{kJ mol}^{-1}\right) \tag{3.11}$$

These bacteria and archaea are important both agriculturally and environmentally. The first step in the process of nitrification is ammonia oxidation, which acidifies soils and is carried out by both bacteria and archaea; the second step that oxidizes nitrite (NO_2^-) to nitrate (NO_3^-) appears to be restricted to bacteria (Chapter 15). Importantly, by converting nitrogen from a cationic (NH_4^+) to an anionic form, nitrifiers promote nitrogen contamination of ground and surface waters because the anion exchange capacity of soil is usually low, allowing the highly mobile nitrate anion to leach through the soil profile with percolating water. Additionally, the transformation of ammonium to nitrate completes the nutrient cycling link between the opposing processes of nitrogen fixation and denitrification.

Other common groups of chemolithotrophic soil microorganisms are the sulfur-oxidizing bacteria, iron-oxidizing bacteria, and hydrogen-oxidizing bacteria. Additionally, some bacteria can obtain energy from the oxidation of reduced forms of arsenic and selenium (Chapter 19) and are important in the cycling and environmental impact of these elements. Values for the change in free energy associated with some of the reactions carried out by chemolithotrophs are given in Table 3.4. Note that these reactions yield significantly less energy for cellular processes than the aerobic respiration of glucose, for example. Balanced against these low energy yields, however, is the ability of chemolithotrophs to use energy sources that are unavailable to other microorganisms, thereby reducing competitive constraints on their growth.

Phototrophy

The metabolic process of obtaining chemical energy from light is known as **phototrophy** (Fig. 3.3, Table 3.1, and Chapter 7). Nearly all phototrophs are autotrophic and use energy captured from light to fix carbon from CO_2 by means of photosynthesis. Similarly, most phototrophs are lithotrophic because they obtain reducing equivalents from water or another inorganic substance. Photosynthesis is conveniently divided into its energy-gathering light reactions and the subsequent dark reactions that are involved in carbon fixation. Only the light reactions will be discussed here; the dark reactions are covered later in this chapter in the section on carbon nutrition.

The light reactions of photosynthesis (Fig. 3.15) require specialized light-gathering pigments called chlorophylls. Among microorganisms, chlorophylls are found in eukaryotic algae, cyanobacteria, and certain other bacteria. Several different types of chlorophylls exist but all are planer molecules that contain a magnesium-bearing core Fig. 3.16. A large

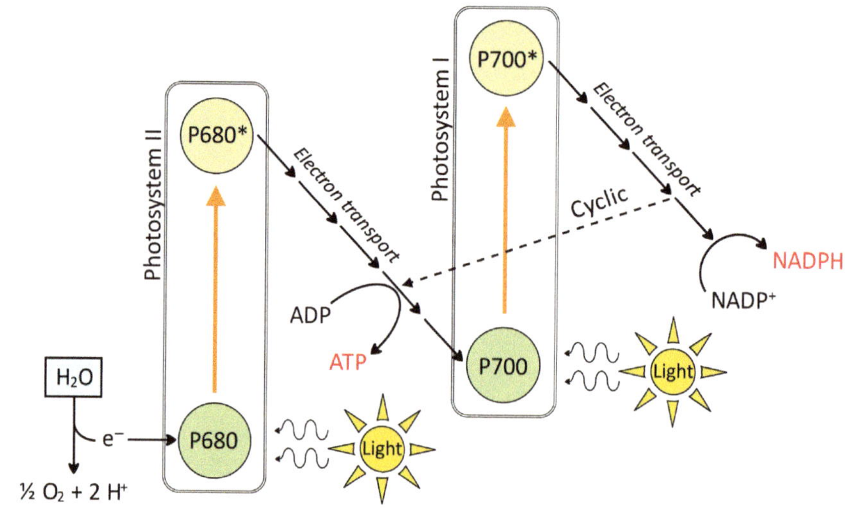

FIGURE 3.15 The light reactions of photosynthesis; often referred to as the "Z-scheme" because of its general appearance. Asterisks indicate the high-energy forms of the chlorophyll molecules caused by the absorption of light energy. Photosystem I can be active in the absence of Photosystem II; in this case, electron flow follows the path indicated by the *dashed arrow* labeled "Cyclic". When Photosystem II is active (non-cyclic photophosphorylation), the electrons that reduce P700 in Photosystem I to its original state are produced by the oxidation of H_2O to produce O_2. *Source: Figure by J. Fuhrmann.*

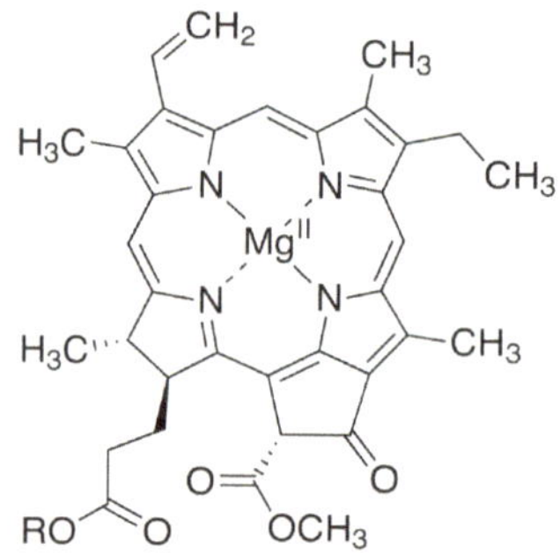

FIGURE 3.16 Structure of the planar portion of chlorophyll *a*, showing the central ring structure containing magnesium and surrounding side chains. Not shown and attached at the position labeled "R" is a hydrocarbon chain approximately 15 carbons in length. *Source: Adapted from public domain.*

number of chlorophyll molecules functions together to absorb light energy over their combined surface area and then funnel the absorbed energy to a reaction-center chlorophyll. Within this latter chlorophyll molecule, the captured light energy causes electrons to be elevated in energy and ejected from the pigment in a process called photooxidation. These ejected electrons then reduce specific acceptors to form high energy compounds. The high-energy electrons within these acceptor compounds then return to a lower energy state by traveling through an electron-transport system, thereby producing ATP by **photophosphorylation** in a manner analogous to that for oxidative phosphorylation during respiration.

Depending on the organism involved, electrons involved in photophosphorylation may simply lead to the reduction of the original chlorophyll molecule (i.e., return it to its original state) and simultaneously produce ATP in a process known as cyclic photophosphorylation. This set of reactions involves a reaction-center chlorophyll molecule designated P700 to indicate that its maximum light absorption occurs at a wavelength of 700 nm; P700 and its associated chlorophylls and electron-transport components are referred to as Photosystem I. Alternatively, many phototrophs carry out noncyclic photophosphorylation in which electrons from the initial photooxidation do not cycle back to P700 but rather reduce NADP to NADPH. In this case, the electrons needed to reduce P700 arise from the light-mediated oxidation of water (photolysis) to oxygen occurring at a second reaction-center chlorophyll designated P680, which has an absorption maximum at 680 nm (Photosystem II). As in the case of cyclic photophosphorylation, ATP is also produced. Thus, compared with cyclic photophosphorylation, noncyclic photophosphorylation additionally produces NADPH and molecular oxygen. Noncyclic photophosphorylation is associated with green plants, algae, and cyanobacteria and is sometimes referred to as oxygenic photophosphorylation. Conversely, because cyclic photophosphorylation does not produce oxygen, it is often called anoxygenic photophosphorylation; it is limited primarily to the green and purple bacteria (Chapter 5).

Carbon nutrition

One remaining major source of diversity among microorganisms is how they obtain carbon. As mentioned previously, organisms called autotrophs have the capacity to transform carbon dioxide into organic molecules in a process known as carbon, or CO_2, fixation. Autotrophic organisms are ultimately responsible for fixing essentially all of the carbon used by heterotrophs, those organisms which require preformed organic compounds for growth. Carbon nutrition is a fundamental attribute of microbial metabolism because it determines the degree to which a microorganism relies on other organisms to fix carbon needed for biosynthetic processes. Carbon is the major element comprising organisms on a dry mass basis—for example, bacteria are approximately 50% carbon—and its acquisition is correspondingly critical to cellular processes.

Autotrophs are defined as organisms that obtain essentially all of their cellular carbon from carbon dioxide. Implicit in this definition is the realization that, under certain circumstances, otherwise autotrophic organisms may obtain some carbon from organic sources. Autotrophy occurs primarily in green plants, algae, and certain bacteria such as cyanobacteria, nitrifying bacteria, and sulfur-oxidizing bacteria. Carbon fixation is an endergonic process and, therefore, requires energy captured by the cell by means of exergonic reactions. Autotrophic organisms typically obtain this energy by either phototrophy or chemolithotrophy.

The pathway utilized by nearly all autotrophs to fix carbon is the Calvin cycle (also known as the Calvin-Benson-Bassham cycle or the reductive pentose phosphate pathway) in which carbon from CO_2 is fixed into organic form with the expenditure of 9 ATP and 6 NADPH; the final product of the cycle is glyceraldehyde-3-phosphate (Fig. 3.17). The Calvin cycle is also sometimes referred to as the "dark reactions" of photosynthesis because it can proceed independent of light as long as sufficient captured light energy is available in the form of ATP and NADPH. Central to the cycle is the C-fixing enzyme ribulose-1,5-bisphosphate carboxylase/oxygenase (RuBisCO) which, using ribulose-1,5-bisphosphate and CO_2 as substrates, produces two molecules of 3-phospho-glycerate (glycerate-3-phosphate); RuBisCO is commonly thought to be perhaps the most abundant enzyme on Earth given its key role in carbon fixation. As you may recall, the final product of the Calvin cycle,

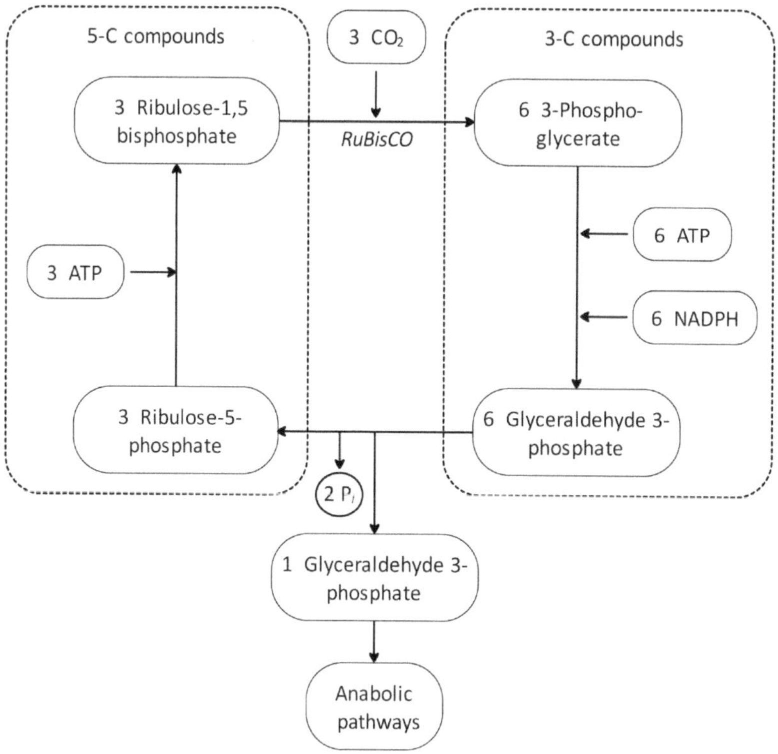

FIGURE 3.17 **Simplified Calvin cycle showing key reactants and products (alternatively called the Calvin-Benson-Bassham cycle, the reductive pentose phosphate pathway, or the "dark reactions" of photosynthesis).** Note that the product glyceraldehyde 3-phosphate is also an intermediate compound in glycolysis and can enter into both anabolic and catabolic pathways. (RuBisCO, ribulose-1,5-bisphosphate carboxylase/oxygenase; P_i, inorganic phosphate). *Source: Figure by J. Fuhrmann.*

glyceraldehyde-3-phosphate, is one of the 3-carbon products of glycolysis. This newly synthesized glyceraldehyde-3-phosphate is a potential precursor in the biosynthesis of needed cellular constituents.

Assuming carbon fixation is occurring in concert with noncyclic photophosphorylation, the production of glucose from carbon dioxide can be summarized as:

$$6CO_2 + 6H_2O \rightarrow C_6H_{12}O_6 + 6O_2 \qquad \left(\Delta G^{0\prime} = +478\,\text{kJ mol}^{-1}\ CO_2\right) \tag{3.12}$$

This reaction is the opposite of Eq. (3.9).

Heterotrophs, which include animals, protozoa, fungi, and the majority of bacteria, obtain essentially all of their carbon from preformed organic compounds. Although nearly all heterotrophs are chemoorganotrophs, some bacteria combine heterotrophy with either phototrophy or chemolithotrophy. Whereas autotrophs obtain biosynthetic intermediates by means of anabolic reactions, heterotrophs generally obtain these compounds via catabolic processes. Similarly, while autotrophs must produce and maintain the enzymes necessary for carbon fixation, heterotrophs often have evolved the capacity to enzymatically degrade large or complex organic compounds that are inaccessible to autotrophs. Thus, the two groups of organisms may be thought of as having the same metabolic goal (biosynthetic intermediates) but different means of obtaining this goal.

Integration of metabolic pathways

Although it is useful for purposes of discussion to introduce the various metabolic processes carried out by microorganisms as separate entities, metabolic pathways are actually highly integrated and interdependent. In particular, many of the compounds comprising glycolysis or the TCA cycle play an important role in both catabolic and anabolic reactions in cells. Central pathways such as these that serve in both energy-yielding and biosynthetic roles are called amphibolic pathways. Fig. 3.18 summarizes some of the general pathways by which glycolysis and the TCA cycle are linked to various other pathways, both catabolic and anabolic, in many microorganisms. Such integration serves to increase metabolic efficiency of cells by allowing them to allocate most of their enzymatic resources to a limited number of central pathways.

Summary

The different strategies by which soil microorganisms obtain carbon and energy are impressively diverse and innovative. For many soil microbiologists (this author included), this range of metabolic types and abilities, and the corresponding complexity of the interactions among the populations, is in large part the fascination of the discipline. It is hardly an overstatement to say that essentially all forms of metabolism and life strategies can be found among the microorganisms that inhabit soil environments. It should be evident that underlying this diversity is the unifying principle that life processes require that organisms acquire energy and nutrients to meet their needs of growth, survival, and reproduction. This

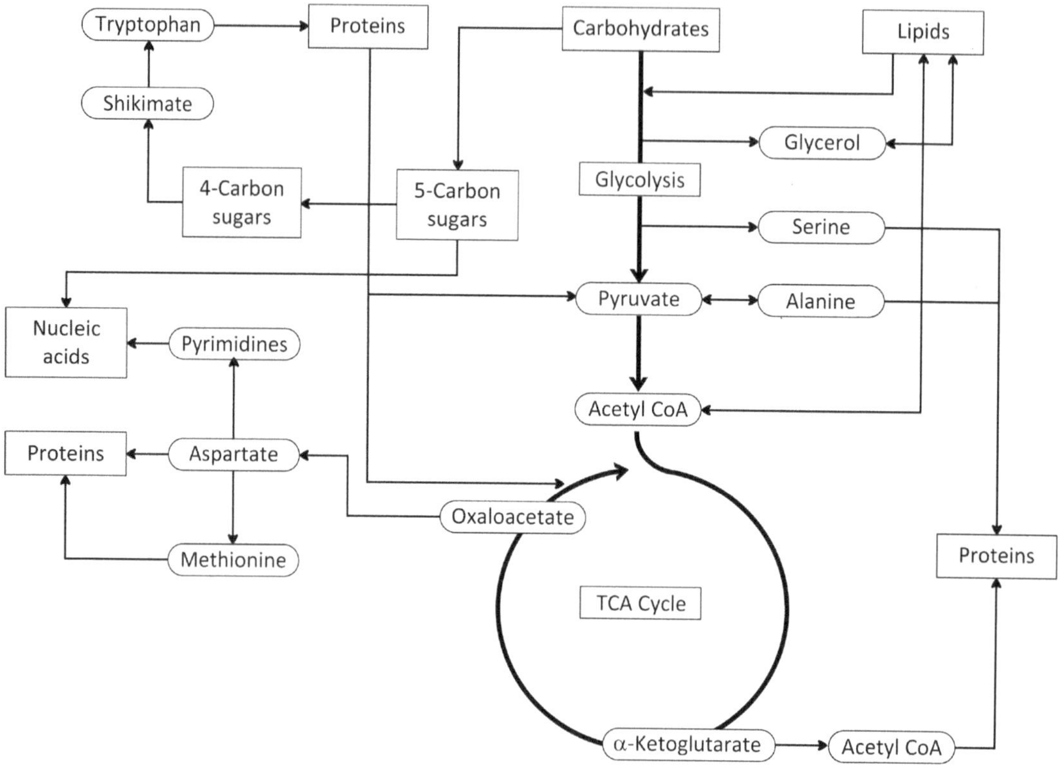

FIGURE 3.18 **Some of the major relationships of glycolysis and the TCA cycle to biosynthesis and catabolism of various cellular constituents.** Reactions that replenish the supply of TCA intermediates that have been "siphoned off" for biosynthetic pathways are referred to as anaplerotic reactions, that is, anaplerosis is the act of replenishing TCA cycle intermediates. *Source: Figure by J. Fuhrmann.*

marriage of metabolic diversity and uniformity of purpose makes the study of soil microorganisms a formidable but stimulating challenge.

Study questions

1. When moist, the exterior of an aggregate may be oxic (O_2 present), while the interior may be simultaneously anoxic. What changes in bacterial metabolism might one expect between the interior and exterior of the aggregate?
2. For each of the following types of microorganisms, specify which kind of culture medium and environmental conditions would be needed to support their growth: (1) nitrifying bacteria, (2) cyanobacteria, (3) fungi, (4) methanogens.
3. Determine the oxidation state of sulfur in each of the following compounds. Which among those listed is the richest potential source of energy for chemolithotrophic bacteria: S (elemental), SO_4^{2-}, H_2S, or SO_3^{2-}?

4. Many aquifer materials are aerobic despite being colonized by large numbers of heterotrophic bacteria. Some also contain high concentrations of nitrate, which may be of environmental and health concern. One means suggested to remediate such aquifers is to encourage nitrate removal by denitrification. What conditions must exist for denitrification to occur? How might these requirements be met in the field? Briefly describe the sequence of events that would lead to denitrification.
5. Explain why it is necessary that certain degradative enzymes be produced extracellularly (e.g., those involved in starch and cellulose degradation).
6. Control of enzyme activity and production is essential for balanced microbial growth. List the various means by which microorganisms exert this control.

References

Blagodatskaya, E., Kuzyakov, Y., 2013. Active microorganisms in soil: Critical review of estimation criteria and approaches. Soil Biol. Biochem. 67, 192–211, http://dx.doi.org/10.1016/j.soilbio.2013.08.024.

Light, S.H., Su, L., Rivera-Lugo, R., Cornejo, J.A., Louie, A., Iavarone, A.T., Ajo-Franklin, C.M., Portnoy, D.A., 2018. A flavin-based extracellular electron transfer mechanism in diverse Gram-positive bacteria. Nature 562, 140–144, https://doi.org/10.1038/s41586-018-0498-z.

Madigan, M.T., Bender, K.S., Buckley, D.H., Sattley, W.M., Stahl, D.A., 2019. Brock Biology of Microorganisms, fifteenth ed. Pearson, New York.

Singleton, P., Sainsbury, D., 2006. Dictionary of Microbiology and Molecular Biology, third ed. Wiley, New York, https://doi.org/10.1002/9780470056981.

Further reading

Cammack, R., Attwood, T.K., Campbell, P.N., Parish, J.H., Smith, A.D., Stirling, J.L., Vella, F., 2006. Oxford Dictionary of Biochemistry and Molecular Biology, second ed. Oxford University Press.

Cohen, G.N., 2014. Microbial Biochemistry, third ed. Springer, New Yorkdoi: 10.1007/978-94-017-8908-0.

Nelson, D.L., Cox, M.M., 2017. Lehninger Principles of Biochemistry, seventh ed. W.H. Freeman and Company, New York.

CHAPTER

4

Microbial genetics

Cindy H. Nakatsu

Department of Agronomy, Purdue University, West Lafayette, IN, United States

"Nor does the extreme minuteness of the gemmules, which can hardly differ much in nature from the lowest and simplest organism, render it improbable that they should grow and multiply" —***Charles Darwin***

Our knowledge of the diversity and physiology of soil microorganisms has improved vastly since we have gained a better understanding of microbial genetics. Classically defined, genetics is the study of the heredity of individuals. Basically, genetics is the study of why and how offspring are very similar to their parents but not identical. Therefore, the driving forces of genetics are the processes that conserve genetic information and those that contribute to changes. The beginning of the field of genetics is attributed by most to the work of Gregor Mendel and Charles Darwin in the mid-1800s (Brock, 1990). The field did not really explode until the latter half of the 20th century after DNA structure and replication was discovered and molecular genetic techniques were implemented. The 21st century began in the era of **genomics** (the determination of the entire genome of organisms), which has made us realize that we still have much more to learn and understand about the diversity of organisms and the functional roles they play. There are now many scientists, including soil scientists, who use these genetic principles and methods to study microorganisms from the level of single cells to whole communities. Recent advances in molecular genetics allow scientists to look at microbial species and communities in soil (including those that are presently unculturable) from a wide range of habitats and geographic locations in far more depth than was previously possible. In this chapter, we cover the fundamentals of microbial genetics and how these fundamentals can help us better understand soil microorganisms.

Nucleic acids

Basic chemistry

Nucleic acids are the basis of inheritance and are found in all living organisms. The genetic information is encoded in the sequence of nucleic acids, the **genotype**, which gives rise to

Principles and Applications of Soil Microbiology. http://dx.doi.org/10.1016/B978-0-12-820202-9.00004-6

Copyright © 2021 Elsevier Inc. All rights reserved.

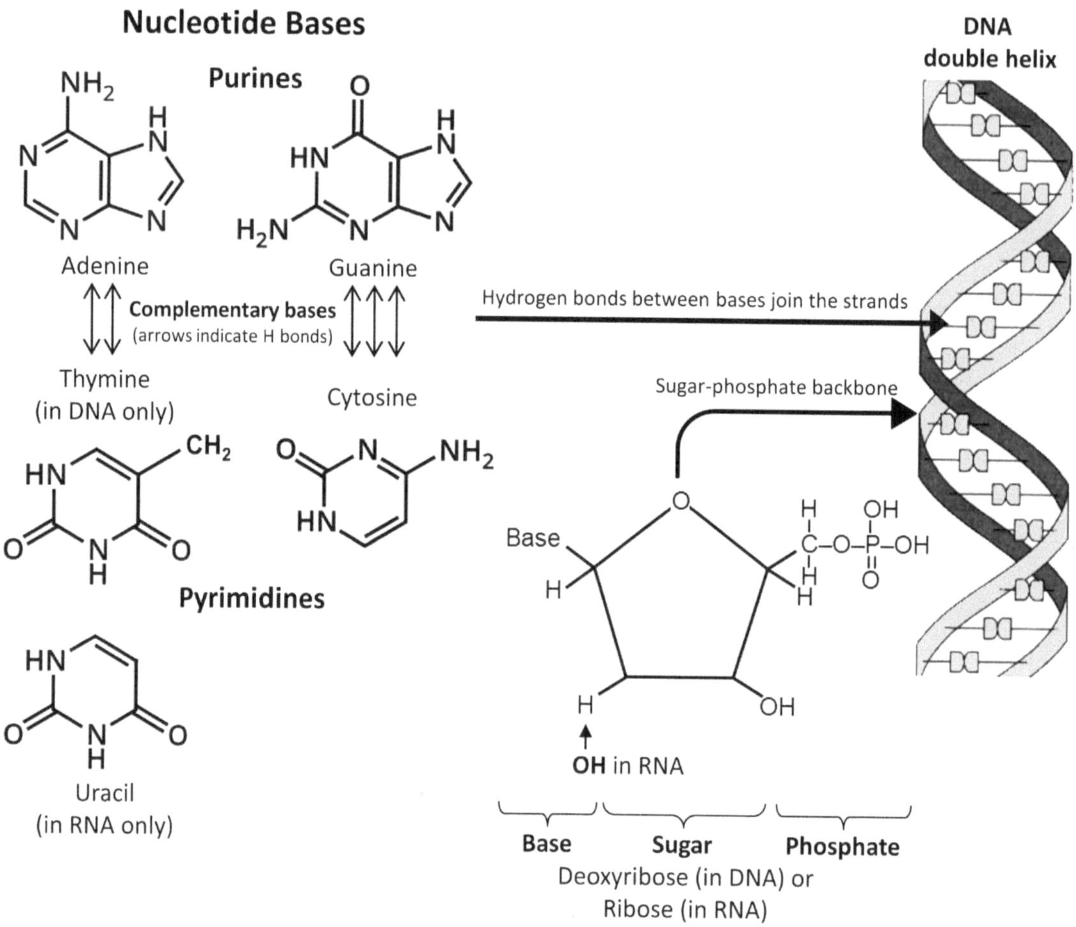

FIGURE 4.1 **Basic chemical components of nucleic acids and their arrangement in the DNA double helix.**

the traits expressed by the organism, the **phenotype**. The basic unit of information coded is the **gene**. There are two main categories of **nucleic acids, deoxyribonucleic acid (DNA)** and **ribonucleic acid (RNA)**. The basic chemical structure of all nucleic acids is similar (Fig. 4.1). They are composed of **nucleotides** that, in turn, are composed of a sugar, a phosphate, and a purine or pyrimidine base. There are two purine bases, adenine (A) and guanine (G); and three pyrimidine bases, cytosine (C), thymine (T, only in DNA), and uracil (U, only in RNA). The subsequent products derived from the genetic information are determined by the order or sequence of different nucleotide bases joined together by phosphodiester bonds (a type of covalent bond). Laboratory methods can now determine the sequence of nucleotides and even decipher the entire **genome** (all the DNA) of an organism.

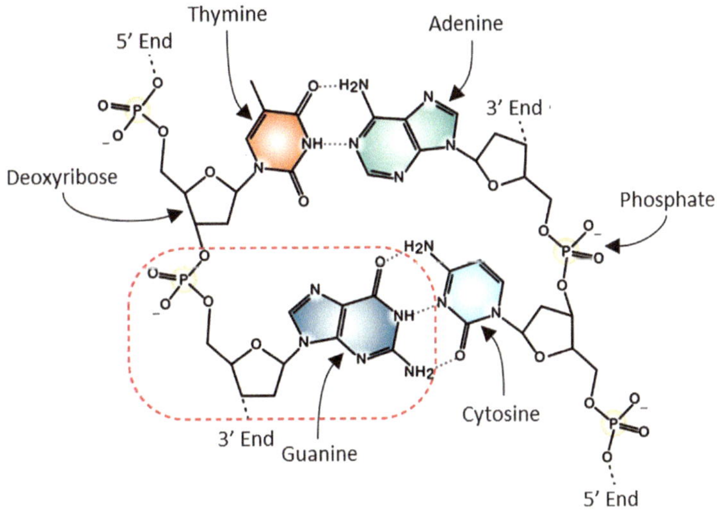

FIGURE 4.2 Portion of DNA showing two pairs of nucleotides and the hydrogen bonding between the bases in each pair. Note that the adenine-thymine (A-T) and guanine-cytosine (G-C) pairings have two and three hydrogen bonds, respectively, indicated by *dotted black lines*. Each nucleotide consists of a phosphate, sugar, and nitrogenous base joined by covalent bonds. The phosphates and sugars of adjacent nucleotides are joined to form the "backbone" of a DNA strand. The *dashed red oval* outlines one nucleotide.

DNA

DNA is the genetic material, which codes information that is the basis of life for all organisms with the exception of some viruses (Chapter 9). DNA is typically composed of two or double strands (ds) of nucleotides with a deoxyribose as the sugar component and the bases A, C, G, or T. The sequence or order of nucleotides codes for genes that provide information for making proteins. The two strands are joined by hydrogen bonds that form between complementary nucleotide bases (Fig. 4.2). The joining of the nucleotides and bonds between the two strands cause the strands to twist together, forming a double helix. The cytosine always complements (or pairs with) guanine due to the formation of three hydrogen bonds, and the adenine complements thymine due to the formation of two hydrogen bonds. This specific bonding results in the maintenance of the nucleotide sequence during DNA **replication**. **Denaturation** of DNA occurs when the weak hydrogen bonds are disrupted resulting in the separation of the two strands. More energy is required to disrupt DNA with high GC content because of greater number of hydrogen bonds. Commonly used ways of disrupting these bonds in the laboratory include applying heat or exposure to alkaline pHs. **Renaturation** or **reannealing** is the process in which the complementary strands are reformed by the formation of hydrogen bonds. For example, DNA that has been denatured by heating can be renatured by lowering the temperature. The weak nature of hydrogen bonds and nucleotide complementation are also important because many techniques used for molecular genetic analysis are possible because of these properties (Box 4.1).

BOX 4.1

Nucleic acid hybridization

Hybridization methods take advantage of the ability for DNA to denature and renature. Accurate and relatively easy methods to determine species diversity in a community are based on nucleic acid hybridization (Nakatsu and Forney, 1996). Hybridization is commonly accomplished by annealing (or hybridizing), a previously characterized nucleic acid sequence (called a probe) to its counterpart (called a target) within a mixture of DNA or RNA. Probe sequences can differentiate community members at phylogenetic levels ranging from narrow (species or strain) to broad (all organisms, domain, or classes). The intensity of the hybridization signals gives an indication of the evenness (proportion of the total) of groups in the community. In addition to species evenness, these methods can assess physical distribution by examining intact cells with fluorescent in situ hybridization (FISH; Amann et al., 2001). Even more informative methods, like microarrays can tell us about species richness (number of different species) and functional potential of the community. Microarrays are basically large-scale hybridization experiments conducted on a solid support, about the size of a microscope slide, in which thousands of targets are probed simultaneously.

RNA

RNA is a single strand (ss) of nucleotides with a ribose as the sugar component and the bases A, C, G, or U. There are three main types of RNA in the cell: **messenger RNA** (mRNA), **transfer RNA** (tRNA), and **ribosomal RNA** (rRNA). Messenger RNA is made from **transcription** of DNA and directs the order in which amino acids are assembled in the **ribosome** during **translation** and synthesis of the protein (described later). Transfer RNA is the adapter molecule responsible for matching the 3-base code on the mRNA to its specific amino acid. Ribosomal RNA (rRNA) is involved in the structure and catalytic activity of the ribosome components. The ribosome is composed of two subunits, often referred to as small and large, or by their size (determined by their sedimentation rate, S). In prokaryotes, the small subunit contains 16S rRNA; in eukaryotes, the small subunit is slightly larger and contains 18S rRNA. The large subunit in prokaryotes contains both 5S and 23S rRNA; in eukaryotes, the large subunit contains 5S, 5.8S, and 28S rRNA. In addition to their importance and necessity for cellular function, rRNA genes are important to our understanding of the diversity of microorganisms (Box 4.2).

Cellular organization

In prokaryotes, DNA is organized into genetic elements either as **chromosomes** or **plasmids**. Chromosomes are found in all cells, whereas plasmids typically do not encode genes that are essential for basic cell survival and therefore are not found in all cells. Both typically

BOX 4.2

ribosomal RNA (rRNA)

Many direct and indirect analyses of communities and individual species in the soil use rRNA sequences. These sequences are phylogenetically informative because some regions are highly conserved (have not changed over time because their function is critical), while other regions of sequence are not highly conserved (have changed over time as the organism evolved; Woese et al., 1983). The variability in these regions increases with evolutionary distance between two organisms. The 16S rRNA gene sequence is extensively used to study prokaryotic microorganisms. The intergenic region, or sequences between the genes, is often included in analyses of eukaryotes and, to some extent, prokaryotes; because these regions have less selective pressure, they are more prone to mutations thus allowing greater discrimination. The rRNA gene encoded in DNA or the rRNA itself can be isolated from soil and examined directly or after amplification using the polymerase chain reaction (PCR; described later).

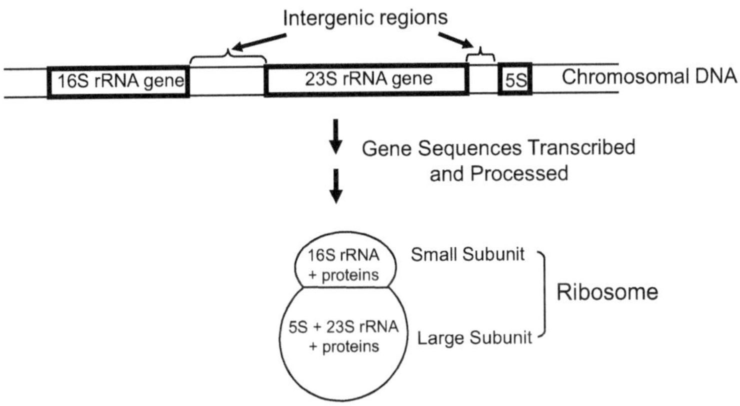

Example of ribosomal gene organization and cartoon of ribosome structure.

are circular, supercoiled in order to fit into the cell, and self-replicating. The chromosome forms a compact unit called a **nucleoid** and carries all the genetic information required for the survival and maintenance of the cell. Plasmids are smaller than chromosomes but have been studied extensively because they often carry valuable genetic information, such as genes for antibiotic resistance, catabolic functions (e.g., degradation of pesticides), and heavy metal tolerance. Plasmids are an important tool in molecular genetic methods, where their small size and ability to self-replicate have made it possible to use them as **vectors**, a means of transferring genetic information from one host to another.

In eukaryotes, DNA is contained in a membrane-bound structure called the **nucleus** and is also found in two cell organelles, mitochondria and chloroplasts. The DNA in the nucleus is organized as linear chromosomes that are held in place by proteins. The DNA in mitochondria and chloroplasts are similar to those in prokaryotes both in organization and in their encoded housekeeping genes.

In prokaryotes, the mRNA is produced in the cytoplasm. The close association between mRNA and ribosomes makes it possible to couple transcription and translation. In contrast, the mRNA in eukaryotes is made in the nucleus and then transported out to the cytoplasm before translation occurs.

Cellular nucleic acid processes

Replication

A new strand of DNA, which is the exact duplicate of the parent molecule, is synthesized in a process called **replication** (Fig. 4.3). The exact sequence is maintained because of the base pair complementarity, A with T and C with G. There are a number of different enzymes required for the entire process, but only the most fundamental will be covered here. Each of the two strands of DNA serves as a template, and they both replicate at the same time. There are various enzymes that are involved in unwinding the DNA strands (e.g., helicases, topoisomerases, gyrases) for replication to occur. The initiation of DNA replication requires the presence of a **primer** (short strands of RNA) as a starting place for nucleotides to be added. The **nucleotides** are added to the growing strand with an enzyme called **DNA polymerase**.

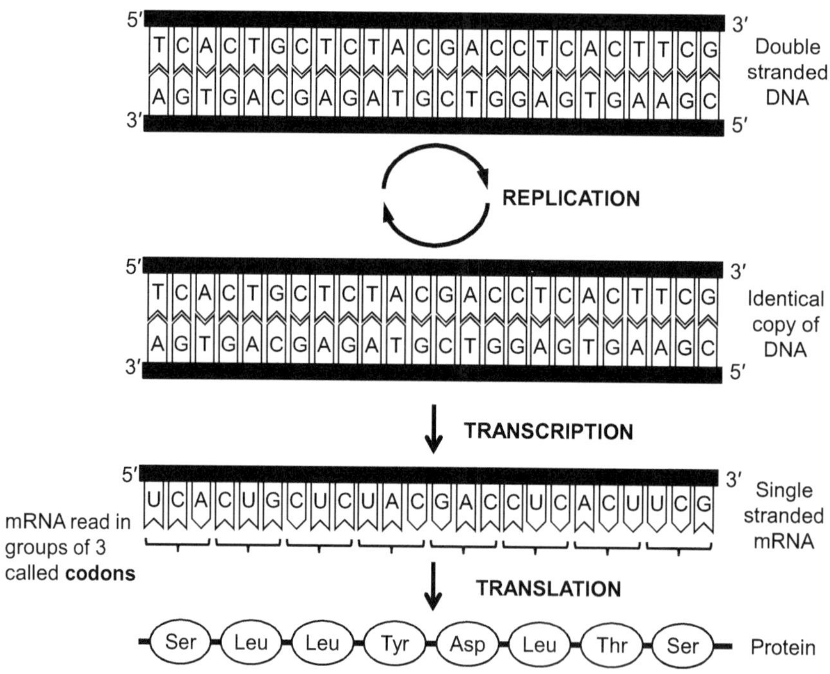

FIGURE 4.3 **Schematic representation of DNA replication, transcription of mRNA, and translation into protein.** Note only RNA transcription of the bottom DNA strand is illustrated.

DNA synthesis occurs only in one direction because the polymerase only forms bonds between the 5́ (five prime) triphosphates and the free 3́-OH group in the sugar (Fig. 4.2). Because replication is unidirectional, the opposite strand must replicate in a discontinuous fashion and is referred to as the lagging strand.

Transcription

RNA is synthesized by using DNA as a template in a process called **transcription**. The process differs from replication because the entire DNA strand is not transcribed into RNA and is instead limited to regions coding for a single **gene** or a group of associated genes under common regulatory control called an **operon**. A gene is defined as the basic unit of inheritance, a sequence of nucleotides that codes for a single polypeptide (protein). The nucleotides are added to the strand by the enzyme **RNA polymerase** starting at a specific sequence before a gene called the promoter region and ending at a termination sequence or protein. Some genes are always being transcribed whereas others are regulated by proteins that signal if a gene needs to be transcribed (Chapter 3). This is the main step at which gene expression is regulated.

Translation

The mRNA is the template for protein synthesis in a process called **translation** (Fig. 4.3). Translation occurs at the **ribosomes**, structures composed of proteins and rRNA responsible for protein synthesis. The mRNA binds to the ribosome to initiate translation. The tRNA reads the nucleotide sequence of the mRNA in sequential groups of three called **codons** and brings in the correct amino acid to the ribosome, elongating the protein chain. Of the 64 possible permutations of the 4 nucleotides in groups of 3, 61 codons correspond to 20 amino acids and 3 to stop codons. The redundancy in the genetic code means that one amino acid can be represented by more than one codon. The amino acid typically indicating the start of a coding region is methionine and, on occasion, valine. Translation is terminated at one of the three possible stop codons. Thus, the process occurs in three steps, that is, initiation, elongation, and termination.

Genetic mutations

Mutations are the driving force of evolution. They can result in a change in genotype compared to the wild-type organism. The wild-type strain is arbitrarily chosen and is usually the strain that is first isolated and expressed a phenotype of interest. Mutations are random events (i.e., the site of mutation in a gene cannot be predicted). Any change in the nucleotide sequence of DNA is considered a mutation, but not all mutations result in a change of phenotype, partly due to the redundancy of the genetic code. Genetic changes underlying mutations can be point mutations (changes in a single nucleotide) or larger genetic rearrangements caused by duplications, deletions, and inversions of a length of sequence. Even a single nucleotide change can alter a single amino acid that may be responsible for the folding or functional features of a protein and make the protein nonfunctional. Other mutations will cause the protein to be only partially functional or have no detectable effect on the gene product. The maintenance (retention) of a genetic mutation is typically dependent on natural selection. Natural selection (i.e., survival of the fittest) occurs for traits that give the organism a competitive advantage over other organisms that interact in a given habitat. Thus, mutations are the source of new genetic traits and the basis of evolution.

Gene transfer in the environment

Initially scientists did not think that genetic information was exchanged among prokaryotes, but now we know this is not true. However, prokaryotic genetic exchange mechanisms differ from sexual reproduction by eukaryotes. There are three basic mechanisms for gene transfer of microorganisms in the environment: **transconjugation, transformation,** and **transduction** (Fig. 4.4). These mechanisms, especially transconjugation, are the closest prokaryote equivalents to the sexual reproduction that occurs between eukaryotic organisms. However, unlike eukaryotes, gene transfer between bacteria is often not limited to occur only between

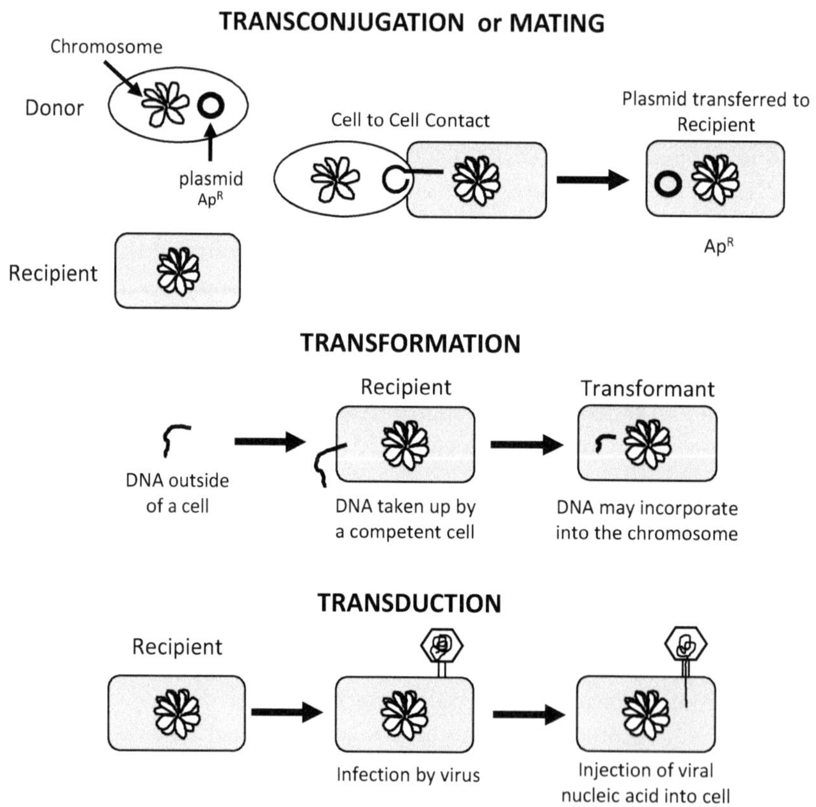

FIGURE 4.4 **Mechanisms of gene transfer among bacteria.** Transconjugation illustrated by transfer of a plasmid encoding ampicillin resistance (Ap^R) from a donor to recipient bacterium; transformation illustrated by bacterium taking up extracellular DNA; transduction illustrated by virus injecting nucleic acid into a bacterium which may ultimately incorporate into the bacterium's genome.

organisms belonging to the same species. Genes can be transferred between phylogenetically different bacteria and even from bacteria to plants and animals. The success of transfer of genetic information among diverse hosts is limited by its incorporation into the genome or the ability of a plasmid to replicate in alternate hosts. All of these mechanisms occur naturally, but they have also been used extensively in the laboratory to study basic and applied aspects of microbiology.

Transconjugation, also called conjugation or mating, is the transfer of genetic material between bacteria via cell-to-cell contact. The cell giving the genetic material is called the donor and the one receiving is the recipient. The genetic element typically transferred is a plasmid. Before transfer, in some cases, a quantity of chromosomal information can be incorporated into the plasmid. After transfer, the plasmid may replicate independently, or the whole plasmid or a portion of it may integrate into the chromosome of the new host. The frequency of transfer and the compatibility between individuals is donor and recipient dependent.

Transformation is the uptake of extracellular (naked) DNA into the cell without contact with another living cell. Extracellular DNA is found in the environment as a result of cell death and lysis. Unless the DNA carries the genetic information for self-replication, once the DNA is in the cell, it must be incorporated into the genome. Some, but not all, species of bacteria are naturally transformable. In the laboratory, cells can be chemically or electrically modified to be competent (transformable). In this case, the transferred information is usually carried on a plasmid.

Transduction is the virus-mediated transfer of bacterial DNA between two hosts (see Chapter 9). If the DNA of a temperate (integrative) virus is incorporated into the bacterial DNA, then some of the host genome may be included during excision. In this manner, a virus infects a new host with not only its own DNA, but also with some bacterial DNA from the previous host.

For gene maintenance after transfer into a new host, the DNA must become integrated into the genome, especially if it is not introduced on a plasmid or if the plasmid cannot self-replicate in the new host. Integration can occur by a temperate phage inserting itself into the host chromosome to form a **prophage** or through **recombination** or **transposition**. Integration by homologous recombination occurs by crossover between identical sequences of DNA in the newly acquired DNA and in the host genome. Elements that move by transposition are called **transposons**. There are many families of transposons and transposon-like elements. In general, transposons carry genes that allow themselves and any associated genes to be inserted into random sites in the genome. Recombination and transposition are not limited to the integration of exogenous DNA—they can also occur within the existing genome and cause rearrangements of the genes resulting in mutations that can affect their subsequent expression. Transposons are an important source of genetic variation.

In vitro nucleic acid processes

Many of the basic processes carried out by the cell have been adapted by scientists for use in the laboratory. Two processes used extensively to study soil microorganisms are polymerase chain reaction (PCR) and nucleotide sequencing. Most researchers conducting genetic studies use these two methods.

Polymerase chain reaction (PCR)

The PCR method amplifies a specific nucleotide sequence about a million times by mechanizing the steps required for replication. This method is extremely useful when: (1) the sequence of interest has a low copy number, (2) the presence of a sequence is unknown, (3) the location of the sequence in the genome is unknown, or (4) the sequence is present in an organism that cannot be cultivated. This method has been used extensively in soil microbiology to characterize microbial communities by amplifying rRNA gene sequences prior to DNA sequencing. Similar to all other in vitro procedures based on nucleic acid replication, the major components of a PCR include: template (DNA or RNA), all four nucleotides, **primers**, polymerase (additionally reverse transcriptase if the template is RNA), and proper solution conditions (e.g., pH, buffer, co-factors, and salts). A specific or nonspecific sequence of nucleic acid can be amplified by PCR depending on the primers that are chosen. Primers for PCR are **oligonucleotides** (a short sequence of nucleotides) that have been synthesized in a laboratory. The primers must be able to anneal to complementary regions of opposite strands of the DNA molecule. The sequence can be based on a known target, typically a gene of interest. However, sequences for unknown target areas can be used if one wishes to amplify random regions of the genome for genetic profiling or fingerprinting (Nakatsu and Marsh, 2007), but this approach is used less frequently now because of the availability of high throughput sequencing technology.

The sequence of steps in PCR is controlled by changes to the temperature (thermocycling) of the reaction mix that determines the nucleic acid form (either double or single stranded) and polymerase enzyme activity (Fig. 4.5). There are three basic steps in the process:

- Denaturation: The temperature of the solution is raised (typically 94°C or 95°C) in order to denature the DNA (convert from double stranded to single stranded).
- Annealing: The temperature of the solution is then lowered (typically 50°C–65°C) to cause the primer to anneal to the single-stranded DNA. There is an excess amount of primer in the solution and they are short; therefore, they preferentially anneal to the template DNA strands.
- Extension: The temperature is then raised for optimal activity of the polymerase (typically 72°C–76°C). Even though optimal activity occurs in this temperature range, most polymerases used for PCR are heat-stable thus enabling them to withstand the high temperatures used in the denaturation step (Box 4.3). During this stage, the polymerase adds nucleotides to each strand starting from the primers.

This process is repeated for ~25–40 cycles, increasing copy numbers exponentially by doubling the target sequence at each cycle (first cycle, two copies; second cycle, four copies; third cycle, eight copies, and so on). In the first cycle, random lengths of the template are amplified starting from each primer. The majority of strands being synthesized are the length between the two primers. As the number of cycles increases, the desired product preferentially amplifies and accumulates. The number of cycles used depends on the amount of PCR product desired and the life of the polymerase under the experimental conditions.

Although useful for detecting organisms and amplifying segments of DNA, conventional PCR is not a quantitative method and cannot be used to enumerate the number of targets in a sample. To do this, a different method, termed quantitative PCR (qPCR) or real time PCR, is

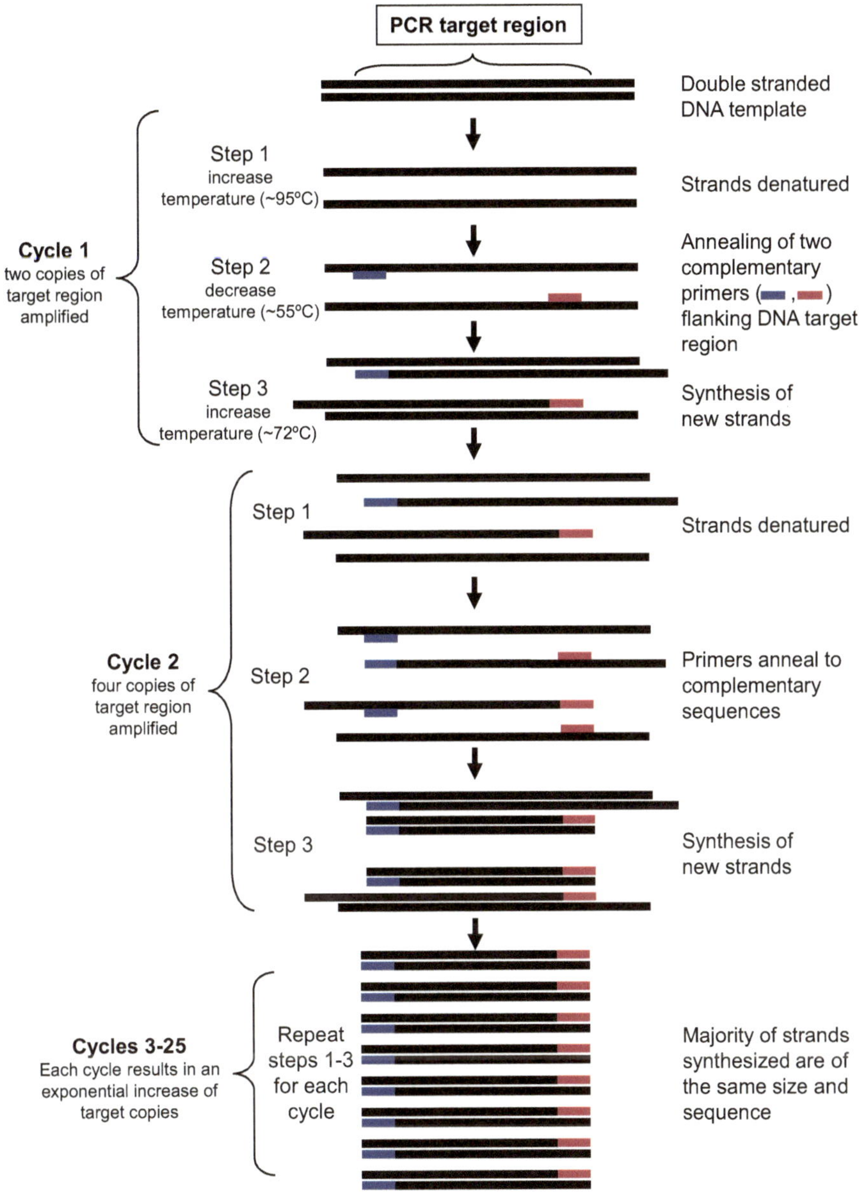

FIGURE 4.5 **Steps in the polymerase chain reaction (PCR) used to amplify specific sequences.**

used where copies are determined at the end of each PCR cycle. This method includes a fluorescent dye in the PCR solution that fluoresces when bound to or released from (two different approaches possible) the targeted double stranded DNA. This fluorescence can be detected by a specialized PCR instrument at the end of each PCR cycle, and the signal increases with each cycle. This allows quantification to be made during the exponential increase in copy

BOX 4.3

Taq DNA polymerase

As previously mentioned, DNA polymerase is the enzyme used to replicate DNA. Polymerase activity is typically optimized for the temperature of the host. For example, in human enzymes, activity is optimal at approximately 37°C. Enzymes from humans would not work well for PCR because it is dependent on using thermotolerant polymerases that are able to be active and survive at higher temperatures. PCR would not be possible without the discovery of thermotolerant polymerases. The first thermotolerant polymerase (*Taq* DNA polymerase) came from a bacterium called *Thermus aquaticus*, which was isolated from a hot spring sample collected in Yellowstone National Park. Since then, a number of different thermotolerant DNA polymerases have been obtained from thermophiles, microorganisms (mainly Archaea) living in extremely hot environments, such as deep-sea thermal vents. This shows the importance of curiosity in scientific endeavors. This original work was done to see organisms that could thrive in such hot environments but the polymerase from it has led to billions of dollars in profit to the biotechnology industry.

numbers. Template concentrations are calculated using standards of known quantities or relative quantities by comparing fluorescence to a reference gene. Relative quantities are typically used for quantifying gene expression in which RNA, usually mRNA, is the template. In this case, reverse transcriptase qPCR is used to first convert RNA to cDNA that serves as the template for PCR. All subsequent steps are the same as qPCR for DNA targets.

There are several cautions to keep in mind when using any PCR-based method (Box 4.4). The primers chosen for PCR are based on available sequence information; therefore, there is a bias toward these known sequences. Other artifacts include the formation of heteroduplexes (two strands from different origins), chimeric molecules (two sequences joined), and sequence heterogeneities due to inaccurate polymerization. When performing PCR, care must be taken for primer selection and conditions need to be optimized (e.g., concentration of DNA template, primer concentration, Mg co-factor concentration) to minimize artifact formation.

Nucleotide sequencing

Low throughput method. There are number of different nucleotide sequencing methods, but they fall into two broad categories, low- and high-throughput approaches, based on the number of sequences that can be processed. The low-throughput method, often referred to as Sanger sequencing (Sanger et al., 1977), is still used if only a few sequences are being determined. This method is based on the incorporation of dideoxynucleotide analogs, dideoxynucleotide triphosphates (ddNTPs), during DNA synthesis. To use this method, a primer is annealed to single-stranded DNA. Next, standard nucleotide bases (dNTPs), ddNTPs, and DNA polymerase are added to synthesize the complementary DNA strand. Because the ddNTPs lack the 3′-OH group on the ribose moiety, the phosphodiester bond cannot form,

BOX 4.4

Potential issues with enumerating microorganisms based on gene copy numbers

In contrast to culture-based methods, which estimate the number of viable microbial cells in a sample (Chapter 5), DNA-based methods such as quantitative PCR (qPCR) instead determine the number of DNA sequence targets present. The result is usually referred to as the "copy number" of the specific target and expressed as number of copies per gram of soil or some similar measure. The most commonly used microbial targets are ribosomal RNA (rRNA) genes (Box 4.2), but a wide assortment of functional genes (e.g., genes encoding ammonia monooxygenase [*amoA*] and nitrous oxide reductase [*nosZ*]) and even sequences of unknown function have been used as targets. The premise underlying this approach is that there is a direct correlation between the copy number of a target(s) in a sample and the number of organisms that contain the target(s). Although this is generally true, there are several potential issues that can confound interpretation of copy number results:

- **Multiple gene copies per organism**. It is not uncommon for microorganisms to have multiple copies of a specific gene in their genome. For example, different Bacteria and Archaea can have from 1 to over 10 copies of the 16S rRNA gene in each cell. Since rRNA genes are used extensively for microbial community analyses, a database reporting known copy numbers has been developed (*rrn*DB, https://rrndb.umms.med.umich.edu/). There can also be sequence variation between the different copies in an individual cell.
- **Variation in target DNA sequence (i.e., primer-target mismatches) can affect amplification**. The occurrence, number, and location of mismatches between the primer and target sequences can affect the accuracy and relative enumeration of targets from different organisms. Analytical biases resulting from varying degrees of primer-target mismatch are an inherent limitation of many types of PCR-based assays.
- **Detection of live and dead organisms and even extracellular/environmental DNA**. Most DNA-based methods do not differentiate between targets from live and dead organisms. Recent research has also demonstrated that extracellular DNA (a.k.a., relic DNA) can persist in the environment for weeks to years and confound DNA-based analysis. One possible approach to reduce these issues is the use of a modified PCR method that incorporates use of intercalating dyes, such as propidium monoazide. These dyes are excluded from live cells but can penetrate most dead cells, bind with their DNA, and prevent PCR amplification. Another approach to reduce/avoid detection of dead cells is to use RNA (mRNA or rRNA) as the target, although this can be more challenging due to the labile nature of RNA and additional steps required for amplification.
- **Enumeration is based on a standard curve from a reference target**. Due to the availability of synthetic DNA, generating a precise standard curve is less of an

(*Continued*)

BOX 4.4 (*Cont.*)

issue than it was just a few years ago. However, these curves are typically based on a specific sequence from one organism and may not completely match all targets (i.e., does not account for mismatches), especially for environmental samples with mixed populations. Droplet digital PCR is a newer approach that does not require use of a standard curve, but it does not eliminate issues from variable primer-target mismatches among different sequences.

and this failure terminates adding any subsequent nucleotides. Four separate reactions are carried out; each containing all the dNTPs, but only one of the dideoxy analogs. As soon as a specific ddNTP is incorporated during extension, the amplicon will terminate at that base. Because each ddNTP incorporates within a background of the respective dNTP, only a portion of the amplicons in a reaction will terminate after each extension while the rest will continue to subsequent rounds; this results in amplicons of different lengths. During the reactions, fluorescently labeled molecules are incorporated that enable visualization of the different fragments after separation on a polyacrylamide gel or capillary column. With the use of fluorescent labels, detectors attached to computers can be used to automatically obtain nucleotide sequence information. When a limited number of nucleotide sequences are required, this method is still the one that is most commonly used.

High throughput methods. When large numbers of sequences are needed, such as for whole genome sequences or targeted sequences (e.g., small subunit rRNA genes) that aid in identifying communities of organisms, then newer high-throughput sequencing methods are being used. High-throughput sequencing methods have been evolving rapidly over the last couple of decades with new technologies being introduced regularly. High-throughput methods can determine thousands to millions of sequence-reads in a single run. These methods basically fall into two categories in which sequences are either determined by replicating DNA (assembling) or by dissembling DNA strands. The various methods differ mainly by the assembly or disassembly process used and the nucleotide detection method. Fluorescent labels are currently the most commonly used method for detecting and differentiating nucleotides during the replication process. Often PCR is needed prior to nucleotide detection to obtain sufficient fluorescence signal (e.g., Illumina instruments use bridge PCR) but fluorescent molecules with sufficient signal strength for detection during the replication process have been developed (e.g., single molecule sequencing in real time, SMRT from PacBio instruments). Alternatively, differential changes in pH have been detected using ion sensors to distinguish nucleotides during sequence assembly (e.g., Ion Torrent instruments). The disassembly approach uses an enzyme (exonuclease) to release nucleotides from the DNA strands that are being passed through membrane nanopores engineered to have ionic currents. As each nucleotide is released it can be detected by changes in currents as they pass through the nanopores (e.g., Nanopore instruments). All these different approaches and instruments have strengths and weaknesses that need to be considered when choosing sequencing methods.

Additionally, new technology is continually being developed therefore it is important to keep up to date on available technologies if nucleotide sequencing is a major approach that you wish to use. For example, high-throughput sequencing methods have recently been adapted for RNA sequencing (e.g., RNA-Seq).

Applications of genetic methods in soil ecosystems

Previously, the major approach used to study microbes was by cultivation in the laboratory. Although cultivation is still appropriate when one wishes to conduct detailed genetic or physiological studies of a specific microorganism, researchers estimate that only ~1% of the total bacterial species have been cultivated (Yarza et al., 2014). For example, in a forest soil, traditional agar plating methods isolated <100 different bacterial species, while a molecular genetic method showed the presence of over 4000 different bacterial genomes (Torsvik et al., 2002). Now nucleic acids can be extracted directly from soils using a number of different methods or commercial kits specifically designed to maximize cell lysis and to minimize organic acid contamination. Despite the improvement in nucleic acid extraction methods from soil samples, there are still a number of factors that can limit extraction efficiency and impact subsequent molecular methods to characterize soil microbes (Box 4.5).

These modern molecular genetic methods provide insight into microbial composition, richness, and evenness in soils (Chapter 10), and include:

- total DNA isolation and characterization by nucleotide sequencing,
- DNA separation based on G + C content,
- PCR amplification of rRNA or rRNA genes,
- PCR amplification of other functional genes,
- PCR amplification of random regions of the genome, and
- in situ hybridization of fluorescently labeled rRNA oligonucleotide probes.

Early molecular genetics methods to characterize microbial community structure used genetic profiles or fingerprints based on differences in PCR-amplified rRNA sequences in DNA isolated from entire communities; these approaches distinguished fragments or amplicons generated through methods such as terminal restriction fragment length polymorphism (t-RFLP) or denaturing gradient gel electrophoresis (DGGE). Differences between communities were determined by assessing the number of common and different bands found between genetic profiles following analysis with gel electrophoresis (Box 4.6) or capillary columns. Since only a limited number of bands can be distinguished using these genetic fingerprint profiles, communities are only partially characterized.

Instead of fingerprinting techniques, more researchers are now sequencing PCR products (Box 4.7) or using a variety of newer methods to identify and characterize a much larger fraction of soil microbial communities (Chapter 10). This has been made possible by the development of economically feasible, high-throughput "omic" technologies; these methods include, but are not limited to, **genomics, transcriptomics,** and **proteomics** and are based on the molecules being examined—DNA, RNA, and proteins, respectively (Fig. 4.6).

Genomics. The most common of these high-throughput methods is genomics, DNA sequence determination of genome sequences of specific organisms, or of entire communities

BOX 4.5

Soil microbial DNA extraction issues

The basic steps required for extracting microbial DNA from soil include cell disruption/lysis, separation of other cellular and soil components from nucleic acids, and then elution/concentration of DNA. Specific extraction approaches vary in the methods and chemicals used to conduct each of these steps. Numerous publications have shown that different methods for soil DNA extraction can produced different results when used for sequencing or community fingerprinting analysis. Many scientists now choose to use commercial kits, which have somewhat standardized the extraction process and greatly improved the quantity and quality of DNA extracted from most soil samples. When choosing an extraction method, there are several issues that can decrease DNA extraction efficiency and that should be considered:

- **Mechanical lysis to improve cell release of DNA.** There are many methods that use enzymes (e.g., lysozyme) or other chemicals (e.g., detergent) to weaken the cell wall of microbes in order to release the nucleic acids, but many soil microbes have cell walls that make them difficult to lyse using these additives. To maximize cell lysis, it is best to include a mechanical means to disrupt the cells. The most commonly used approach is to use small beads with vigorous shaking perferably in a bead-beating instrument. Speed and time of shaking should be optimized for cell lysis but also to minimize DNA breakage from excessive shaking. The optimized conditions are specific for the instrument and sample source. Various instruments vary in shaking intensity. There are also differences in size and shape of beads available for mechanical lysis likely contributing to some of the differences seen among extraction kits.
- **Community composition can contribute to differential cell lysis.** Some cells, such as Gram-positive bacteria, spores,and dormant cells with tougher cell walls, can be more difficult to lyse. If all cells are not equally lysed from a sample, then the results from analyses will be biased toward more readily lysed groups (e.g., Gram-negative bacteria).
- **Soil solids can reduce the amount of DNA extracted.** Extracted DNA can bind to soil particles. Most problematic are soils high in clay to which the majority of DNA can become adsorbed. Some different techniques have been tested to improve extraction efficiency from high clay soils. One is to separate cells from soil substrate prior to DNA extraction. A second is to treat samples with a compound (e.g., skim milk powder) that can help saturate the reactive surfaces on clay to prevent DNA adsorption.
- **Soil organic matter content can decrease the purity of DNA extracts.** Soils with especially high humic acid concentrations can lead to co-extraction of organic compounds with DNA. This is often evident by the tan to brown color of the final DNA extraction solution. These compounds can inhibit enzymatic reactions in downstream analyses such as PCR. Depending on the chosen method or kit, different chemicals (e.g., charcoal, polyvinylpolypyrrolidone (PVPP), cetyl trimethyl ammonium bromide (CTAB)) have been used to reduce humic contamination.

BOX 4.6

Gel electrophoresis

Gel electrophoresis is a simple and effective technique for viewing and sizing DNA (or RNA) molecules or fragments, including PCR products. The nucleic acid samples are mixed with a dye that includes compounds that will make them sink into wells when loaded into a porous solid matrix, usually an agarose gel. The gel is placed into an ionic buffer (gel is made using the same buffer) and voltage is applied, causing the DNA in the gel to migrate toward the anode because of the negatively charged phosphates along the DNA backbone. The gel is stained with a dye such as ethidium bromide, allowing visualization of the DNA when viewed under ultraviolet (UV) light. Smaller DNA fragments migrate faster through the gel matrix, and larger fragments migrate more slowly. The molecular weight of the DNA (base pairs [bp]) determines the rate of migration through the gel and is estimated from comparison with marker standards of known sizes that are run in parallel on the gel.This approach can also be used to select and/or purify DNA of specific sizes by excising individual bands following electrophoresis and extracting DNA from the agarose. Similarly, gel electrophoresis can also be used to visualize RNA; in that case, the gel should include a denaturant that maintains the RNA in its single stranded form, and the marker should be composed of RNA fragments.

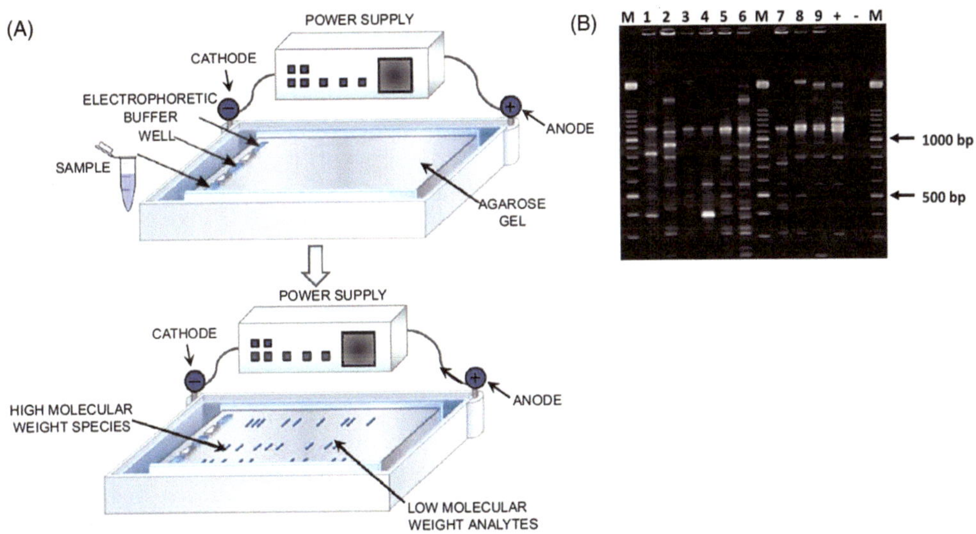

Analysis of DNA using agarose gel electrophoresis. (A) Diagram illustrating use of an agarose gel and electrophoresis system to separate DNA fragments based on size. Note that DNA travels toward the positive electrode with smaller fragments migrating through the agarose more quickly. **(B)** Image of agarose gel after staining and visualization under UV light. Lanes 1–9 are derived from genomic DNA extracts of nine environmental *E. coli* isolates which have been analyzed with a PCR-based fingerprinting method. Lanes labeled "M" include a size marker and lanes " + " and " – " are the positive and negative controls. *Adapted from Hollister et al. (2015); electrophoresis system illustration (***A***) from Drabik et al. (2016); gel photo (***B***) courtesy of E.C. Martin.*

BOX 4.7

Nucleotide sequence analysis of ribosomal RNA genes

Comparing nucleotide sequences of the rRNA genes based on nucleic acids extracted directly from soils show the diversity of microorganisms present and can provide information on the important soil functions they perform (Janssen, 2006; Fierer et al., 2007; Wertz et al., 2006). After amplifying the sequence, usually by PCR using gene-specific primers, the sequence is determined and compared to known sequences using algorithms and databases available on the Internet. The main databank for depositing nucleotide sequences is GenBank (Benson et al., 2013) housed at the National Center for Biotechnology Information (http://www.ncbi.nlm.nih.gov/). A database specifically for rRNA gene sequences is the Ribosomal Database Project (http://rdp.cme.msu.edu/html/) (Cole et al., 2014). This approach is now commonly used to characterize and study the ecology of soil microbial communities. The number of rRNA gene sequences obtained using high-throughput sequencing technology for microbial community analysis has resulted in the development of computational methods that can process millions of reads to identify taxa that they represent as well as perform downstream ecological and statistical analyses (e.g., QIIME2, https://qiime2.org/; Mothur, https://www.mothur.org/). This approach has provided a more complete understanding of soil microbial communities but has also demonstrated that there is much more to learn about soil biota.

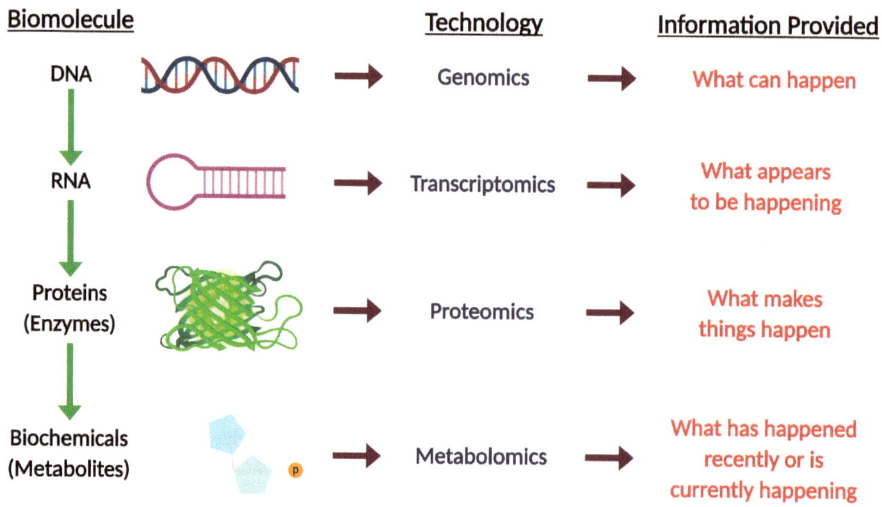

FIGURE 4.6 **Overview of major "omic" approaches for characterizing soil microorganisms.** *Source: Adapted from Zhang et al. (2010). Created with BioRender.com.*

(**metagenomics**). The smaller genome size of prokaryotes and viruses makes it possible to assemble entire genomes of individual taxa within communities. Genomics data can provide genetic information on the potential functions that can be carried out by the organism or, in the case of metagenomics, potential functions of entire communities. However, to identify these potential functions of communities, sequences of genomes from individual organisms, preferably from those in which physiological functions have been tested, are needed for data interpretation. Despite over 220,000 prokaryotic genome sequences available in the National Center for Biotechnology Information (NCBI) database (https://www.ncbi.nlm.nih.gov/), more are needed for complete characterization of soil communities because of the tremendous untapped diversity of soil organisms.

Transcriptomics. Sequencing of all mRNA from a single organism (**transcriptomics**) or from entire communities (**metatranscriptomics**) can reveal genes that are actively being transcribed under specific conditions at a point in time. In this case, RNA is extracted from a sample, but this can be challenging, especially from soil because RNA can be rapidly degraded by enzymes. To stabilize the RNA, it is usually made into double-stranded cDNA using reverse transcriptase to make the complementary strand. After sequencing, typically via high-throughput cDNA methods or RNAseq, bioinformatic analyses aid in determining transcripts with high copy numbers and potential identification of their functional role(s). Here too, a large database of genetic and functional information is needed to aid in data interpretation.

Proteomics. Determination of the proteins produced by an organism (**proteomics**) or from entire communities (**metaproteomics**) identifies the transcripts that are translated into proteins. Basically, it determines which proteins are produced by genes under specific conditions. After protein extraction, proteomes are usually identified using mass spectroscopy methods or by differential separation in gels, but other methods are being developed. Again, bioinformatics and protein databases are essential for data analysis. This is a relatively simplistic explanation of proteomics since all proteins usually cannot be determined in a single analysis and the objective of proteomic studies can differ. This often leads to the need to use multiple analytical instruments and data analysis approaches for charactering soil microorganisms.

The various "omic" technologies produce a tremendous amount of information that has propelled corresponding advancements in the field of **bioinformatics**—defined as the computational approaches to collect, process, analyze, and store this biological data. Together, these technologies are increasingly being used as integrated, multi-method approaches to provide a more complete characterization of soil microbial communities (Chapter 10).

Summary

The basic concepts presented here—nucleic acid biochemistry, replication, transcription, translation, organization of nucleic acids in cells, genetic mutations, and mechanisms of gene transfer—have allowed scientists to develop tools such as the PCR, nucleotide sequencing, and hybridization. These tools have been instrumental in expanding our knowledge of soil microorganisms and soil communities. The chapters ahead will examine microbial diversity and soil community dynamics in greater detail. Certainly, the soil environment is far more diverse than we ever expected.

Study questions

1. Compare and contrast the chemical components and structure of DNA and RNA.
2. Why is the concept of DNA complementation important in the study of nucleic acids?
3. Compare and contrast replication, transcription, and translation.
4. List the three main types of RNA and their role in the synthesis of proteins.
5. Why are genetic mutations important for survival?
6. Describe three mechanisms of gene transfer in the environment.
7. Describe the key components of PCR and the purpose of the three steps in the thermal cycle.
8. Why does the analysis of community DNA probably give a better estimate of microbial diversity than cultivation methods (e.g., dilution plating) alone?
9. What is the purpose of nucleotide sequencing?
10. Discuss the different ways that soil communities can be characterized without cultivation.

References

Amann, R., Fuchs, B.M., Behrens, S., 2001. The identification of microorganisms by fluorescence in situ hybridisation. Curr. Opin. Biotechnol. 12, 231–236.

Benson, D.A., Cavanaugh, M., Clark, K., Karsch-Mizrachi, I., Lipman, D.J., Ostell, J., Sayers, E.W., 2013. GenBank. Nucleic Acids Res. 41, D36–D42.

Brock, T.D., 1990. The Emergence of Bacterial Genetics. Cold Spring Harbor Press, Cold Spring Harbor NY.

Cole, J.R., Wang, Q., Fish, J.A., Chai, B., McGarrell, D.M., Sun, Y., Brown, C.T., Porras-Alfaro, A., Kuske, C.R., Tiedje, J.M., 2014. Ribosomal Database Project: data and tools for high throughput rRNA analysis. Nucleic Acids Res. 42, D633–D642.

Drabik, A., Bodzoń-Kulakowska, A., Silberring, J., 2016. Chapter 7—Gel electrophoresis. In: Ciborowski, P., Silberring, J. (Eds.), Proteomic Profiling and Analytical Chemistry—The Crossroads, second ed. Elsevier, Amsterdam, Netherlands.

Fierer, N., Breitbart, M., Nulton, J., Salamon, P., Lozupone, C., Jones, R., Robeson, M., Edwards, R.A., Felts, B., Rayhawk, S., Knight, R., Rohwer, F., Jackson, R.B., 2007. Metagenomic and small-subunit rRNA analyses reveal the genetic diversity of bacteria, archaea, fungi, and viruses in soil. Appl. Environ. Microbiol. 73, 7059–7066.

Hollister, E.B., Brooks, J.P., Gentry, T.J., 2015. Chapter 13—Nucleic acid based methods of analysis. In: Pepper, I.L., Gerba, C.P., Gentry, T.J. (Eds.), Environmental Microbiology, third ed. Academic Press, NY.

Janssen, P.H., 2006. Identifying the dominant soil bacterial taxa in libraries of 16S rRNA and 16S rRNA genes. Appl. Environ. Microbiol. 72, 1719–1728.

Nakatsu, C.H., Forney, L.J., 1996. Parameters of nucleic acid hybridization experiments. In: Akkermans, A.D.L., van Elsas, J.D., de Bruijn, F.J. (Eds.), Molecular Microbial Ecology Manual. Kluwer Academic Publishers, Netherlands.

Nakatsu, C.H., Marsh, T.L., 2007. Analysis of microbial communities with denaturing gradient gel electrophoresis and terminal restriction fragment length polymorphism. In: Reddy, C.A., Beveridge, T.L., Breznak, J.A., Marzluf, G.A., Schmidt, T.M., Snyder, L.R. (Eds.), Methods for General and Molecular Bacteriology. third ed. ASM Press, Washington D.C..

Sanger, F., Nicklen, S., Coulson, A.R., 1977. DNA sequencing with chain-terminating inhibitors. Proc. Natl. Acad. Sci. USA 74, 5463–5467.

Torsvik, V., Ovreas, L., Thingstad, T.F., 2002. Prokaryotic diversity—magnitude, dynamics, and controlling factors. Science 296, 1064–1066.

Wertz, S., Degrange, V., Prosser, J.I., Poly, F., Commeaux, C., Freitag, T., Guillaumaud, N., Le Roux, X., 2006. Maintenance of soil functioning following erosion of microbial diversity. Environ. Microbiol. 8, 2162–2169.

Woese, C.R., Gutell, R., Gupta, R., Noller, H.R., 1983. Detailed analysis of the higher-order structure of 16S-like ribosomal ribonucleic acids. Microbiol. Rev. 47, 621–669.

Yarza, P., Yilmaz, P., Pruesse, E., Glöckner, F.O., Ludwig, W., Schleifer, K.-H., Whitman, W.B., Euzéby, J., Amann, R., Rosselló-Móra, R., 2014. Uniting the classification of cultured and uncultured Bacteria and Archaea using 16S rRNA gene sequences. Nat. Rev. Microbiol. 12, 635–645.

Zhang, W., Li, F., Nie, L., 2010. Integrating multiple 'omics' analysis for microbial biology: applications and methodologies. Microbiology 156, 287–301.

Further reading

Kowalchuk, G.A., de Bruijn, F.J., Head, I.M., Akkermans, A.D., van, J.D., Elsas, 2004. Microbial Ecology Manual, second ed. Springer Publishers.

Madigan, M.T., Martinko, J.M., Bender, K.S., Buckley, D.H., Stahl, D., 2018. Brock Biology of Microorganisms, fifteenth ed. Pearson Education.

Osborn, A.M., Smith, C.J. (Eds.), 2005. Molecular Microbial Ecology. Taylor & Francis, New York, NY.

Peters, J.E., Henkin, T.M., Champness, W., 2013. Molecular Genetics of Bacteria, third ed. ASM Press, Washington, D.C.

CHAPTER

5

Bacteria and archaea

Mary Ann Bruns

The Pennsylvania State University, University Park, PA, United States

We are all of us walking communities of bacteria. —**Lynn Margulis (1986)**

Bacteria and archaea are prokaryotes, the "unseen majority" of living things on Earth, where a total estimate of 2.6×10^{29} prokaryotic cells inhabit the upper 8 m of soil worldwide (Whitman et al., 1998). Bacteria and archaea are microscopic and single-celled, with high surface-to volume ratios that account for their high sensitivity to changing external conditions. They are the smallest cellular organisms but, just like higher organisms, bacteria and archaea are susceptible to infection by the even smaller acellular viruses. With prokaryotic evolution having proceeded on Earth over billions of years, bacteria and archaea have more highly diversified metabolisms than larger organisms. Many bacteria and archaea are also very long-lived, especially when they enter low-metabolic or dormant states. Bacteria in deep Pacific Ocean sediments have been documented to remain viable for as long as 101 million years. Based on DNA analysis, a single gram of topsoil may contain up to 10,000 species or more. Bacteria and archaea play crucial roles in soil by decomposing organic wastes, oxidizing and reducing redox-active compounds, and interacting with plants, fungi, and other members of the soil biota. Moreover, each higher organism itself harbors its own assemblage of bacteria, archaea, and microbial eukaryotes as a distinctive "microbiome." As the above quotation reminds us, "We are all of us walking communities of bacteria." Becoming aware of the essential functioning of bacteria in and on our own bodies can help us appreciate all the vital processes being carried out by bacteria and archaea dwelling in soils.

Biology of bacteria and archaea

Classification

The two main approaches for identifying archaea and bacteria are phenotypic and phylogenetic classification. Important references on phenotypic classification include *Bergey's Manual of Determinative Bacteriology*, and *Bergey's Manual of Systematic Bacteriology*, 2nd edition. However, both approaches have been combined in a newer, online resource, *Bergey's Manual*

Principles and Applications of Soil Microbiology. http://dx.doi.org/10.1016/B978-0-12-820202-9.00005-8
Copyright © 2021 Elsevier Inc. All rights reserved.

of Systematics of Archaea and Bacteria, which provides phenotypic information within a phylogenetic framework. Recognized phyla are organized alphabetically and provide available information by class, order, family, genus, and type species, if a type species has been designated (Whitman, 2015). A type species or type strain is a descendant of the first-described representative of that species or strain.

Phenotypic classification

Microorganisms can be assigned to groups with similar morphologies and biochemical characteristics (phenotypes) observed in the laboratory. Measurable properties used to categorize microorganisms include morphological, nutritional, physiological, biochemical, antigenic, pathogenic, and ecological characteristics (Table 5.1). Such classification is based on the assumption that microorganisms sharing characteristics are more closely related to each other than to those with contrasting characteristics. This approach is practical because it is based on observable properties employing common laboratory supplies and equipment (e.g., culture media, cell-wall stains, microscopes). Drawbacks to phenotypic classification are that it is restricted to cultured microorganisms and often does not correspond to phylogenetic groupings. Moreover, physiological characteristics within a genus or species may differ in stability and variability.

TABLE 5.1 Phenotypic characteristics commonly used to classify bacteria and archaea.

Characteristic	Description
Morphological characteristics (cell and colonial form and structure)	Size, shape, and grouping of cells; staining reactions; subcellular structures; appendages for motility; colony appearance
Physiological characteristics (cellular function)	Sources of carbon, energy, and reducing equivalents; sources of inorganic nutrients; oxygen requirements; temperature, pH, and osmotic requirements; major metabolic products
Biochemical characteristics (cellular chemistry)	Chemical make-up of the cell wall, cytoplasmic membrane, and other cellular constituents
Antigenic characteristics (serological reactivity)	Distinctive chemical components of cells that react specifically with antibodies produced by an animal
Ecological characteristics (relationships)	Life stages and cycles; normal habitat and distribution of the organism in nature; interactions with the environment and other organisms

Adapted from table by David Alexander, second ed.

Phylogenetic classification

This approach employs DNA sequence similarities to organize microorganisms on the basis of their probable evolutionary relationships. Following years of work on a global catalog of microbial cultures, Carl Woese and his collaborators formally proposed in 1990 a three-domain system of Archaea, Bacteria, and Eukarya based on the nucleotide sequences from RNA molecules found in ribosomes (Woese et al., 1990; Chapters 1 and 4). The underlying hypothesis of their approach was that distantly related microorganisms would show greater differences in ribosomal RNA (rRNA) sequences. The general outline of microbial systematics begins with rRNA-based phylogeny, with one of the most surprising discoveries being that archaea are more closely related to eukaryotes than they are to bacteria. A summary of fundamental differences between the Bacteria and Archaea is given in Table 5.2.

TABLE 5.2 Contrasting characteristics of the domains bacteria and archaea.

Characteristic	Bacteria	Archaea
Membrane lipids	Ester-linked, straight-chain fatty acids	Ether-linked, branched-chain aliphatic compounds
Cell wall	Almost always consists of peptidoglycan containing muramic acid; mycoplasmas lack cell walls	Composition varies; cell walls do not contain muramic acid
Transfer RNA	Thymine present in most tRNAs; initiator tRNA carries *N*-formylmethionine	Initiator tRNA contains methionine
RNA polymerase	Single enzyme Simple subunit structure Sensitive to rifampin	Several enzymes Complex subunit structure Not sensitive to rifampin
DNA polymerase	Primary replication enzyme is DNA polymerase III	Only known DNA polymerase is homologous to eukaryotic polymerase Type 6
Ribosomes	70S size Sensitive to chloramphenicol and kanamycin Resistant to anisomycin	70S size Resistant to chloramphenicol and kanamycin Sensitive to anisomycin[a]
Metabolism	Some species capable of chlorophyll-based photosynthesis No methanogenesis	No chlorophyll-based photosynthesis Some species capable of methanogenesis

[a]*It is interesting to note that eukaryotic cells have antibiotic sensitivities similar to the Archaea.*
Adapted from table by David Alexander, second ed.

In the cell, ribosomes are the sites of protein synthesis, and they depend on the finely tuned molecular structures of their components to ensure cell survival. All ribosomes are composed of multiple subunits containing both protein and RNA molecules. In Archaea and Bacteria, the small and large units of ribosomes are designated 16S and 23S, based on centrifugal sedimentation rates, respectively, while Eukarya have larger subunits (18S and 28S, respectively). The rRNA molecules are encoded by corresponding genes on the organism's chromosome, or **genome**.

Woese chose to compile and compare nucleotide sequences for rRNA molecules, because these crucial genes are more highly conserved (slower-to-change) than other genes (Krieg and Garrity, 2015). The 16S rRNAs in small ribosomal subunits have a convenient length for sequence comparisons (~1540 nucleotides), and prokaryotic phylogeny based on 16S rRNA genes has become widely accepted. At the family and genus levels of phylogenetic classification, 16S rRNA phylogeny correlates well with genome-wide average nucleotide identity (ANI) values based on other conserved genes (Konstantinidis and Tiedje, 2005). However, finer-scale resolution of species and subspecies (e.g., strain, pathovar, genomovar) requires additional sequence information for reliable differentiation (Rodriguez-R et al., 2018)

Phylogenetic classification today is performed routinely by obtaining nucleotide sequences from genomic DNA (from either a pure culture or from an uncultured mixed community in an environmental sample) through PCR amplification of 16S rRNA genes. Sequences are submitted to search programs (e.g., BLAST) in digital repositories like GenBank in NCBI (National Center for Biotechnology Information, National Institutes of Health; https://www.ncbi.nlm.nih.gov/) to identify the most similar sequences. Search results retrieving database sequences with more than 97% sequence similarity may provisionally identify a genus and species related to the unclassified sequence. However, searches may retrieve sequences with much lower nucleotide similarities, so that determination of genus- or family-level relationships is not possible. Such search results indicate that the sequence was likely copied from genomic DNA of a microorganism that has not yet been isolated in laboratory culture. Nevertheless, obtaining close matches to unidentified sequences reported from other research studies is also common. Repeated recovery of highly similar sequences from wide-ranging sources suggests that these sequences are from yet-to-be-cultured "candidate" taxa (Delgado-Baquerizo et al., 2018).

Bacterial phyla. Prior to the advent of 16S rRNA gene-based exploration, 12 bacterial phyla with cultured representatives were recognized. By 2015, cultured isolates had been obtained for 17 additional phyla that were initially identified by 16S rRNA gene sequences. Numerous phyla with no cultured representatives exist, although it has become possible to obtain nearly whole genomes of uncultured bacteria by assembling DNAs into MAGs (metagenome-assembled genomes). Identification of new phyla having no cultured representatives continues, and these are designated as candidate phyla (*Candidatus* phylum) (Hug et al., 2016).

In a 16S rRNA gene survey of 22 surface soils from diverse locations (Fierer et al., 2012), five phyla of bacteria were most abundant in soils (>10% representation): Proteobacteria, Acidobacteria, Actinobacteria, Verrucomicrobia, and Bacteroidetes. Less abundant phyla (0.5%–5%) included the Chloroflexi, Planctomycetes, Firmicutes, Gemmatimonadetes, and Cyanobacteria. Many other phyla were detected at lower frequencies. These phyla are listed in Table 5.3, with means and ranges of percentages and brief descriptions of potential characteristics. Note that the phylum "Actinobacteria" is the phylogenetically appropriate name for "actinomycetes," the more traditional term used in this textbook.

Archaeal phyla. As archaea were a relatively recent discovery and thought to be mostly associated with extreme environments, soil archaea are less well studied than soil bacteria. It is now known that archaea are common and diverse soil inhabitants, although they typically make up only about 1% of total prokaryotes in most soils. In 1990, Woese identified two Archaeal phyla: Euryarcheota, which comprised methanogens and extreme halophiles; and Crenarchaeota, which contained both hyperthermophiles and cold-dwelling representatives. By 2010, four other phyla were recognized: Thaumarchaeota, Aigarchaeota, Korarchaeota, and Nanoarchaeota. As new lineages of Archaea are discovered at all phylogenetic levels, their recognized diversity will continue to expand (Adam et al., 2017). In soils, members of the Thaumarcheota phylum (aerobic ammonia-oxidizers) are consistently detected in oxic samples, even at depths of a meter or more, while members of Euryarcheota (methanogens) are abundant under anoxic conditions. Archaeal members of the microbiomes of plant and animal hosts also are now receiving more attention (Borrel et al., 2020).

TABLE 5.3 Major bacterial phyla detected by their 16S rRNA genes in DNA extracts from surface soils, average percentage representation within communities, and selected group characteristics.

Phylum	Average (range)[a]	Subgroups found in soils and examples of known characteristics (list is not exhaustive)
Proteobacteria Well-represented by cultured isolates, but many remain uncultured	20 (6–31)	Gram-negative cell walls; four classes are abundant in soils (no Epsilonproteobacteria):
	9 (1–15)	Alphaproteobacteria (fix N_2; endophytes; prey on bacteria; oxidize nitrite; oxidize CH_4)
	4 (1–11)	Betaproteobacteria (fix N_2; oxidize ammonia (NH_3); denitrify; prey on bacteria; oxidize CH_4)
	3 (1–10)	Gammaproteobacteria (denitrify; ammonify nitrite; plant pathogens; solubilize phosphate)
	3 (0.1–11)	Deltaproteobacteria (many anaerobes; reduce sulfate; prey on other bacteria)
Acidobacteria Subdivision 1 members are readily cultured	20 (0.1–46)	Gram-negative cell walls; comprise at least eight subgroups, few cultured representatives; Subgroups 1, 4, and 6 most abundant in soils, with Subgroup 3 (Solibacteres) abundant in some soils; all aerobic except for Subgroup 8
Actinobacteria (formerly called "actinomycetes")	14 (2–41)	Gram-positive cell walls; filamentous or pleomorphic morphologies; three subgroups common in soil: Actinobacteridae (most abundant, many cultured, aerobic heterotrophs, produce antibiotics); Acidimicrobidae (few cultured, include acidophiles that oxidize ferrous iron); Rubrobacteridae (few cultured aerobic heterotrophs, possibly other physiologies)
Verrucomicrobia	13 (0.2–40)	Five major classes; Class 2, Spartobacteria most common, aerobic heterotrophs, some are nematode symbionts; many uncultured members with unknown physiologies
Bacteroidetes **CFB** (*Cytophaga, Bacteroides, Flavobacterium*) **group**	11 (2–40)	Gram-negative cell walls; of four subgroups, three common in soil: Sphingobacteria most common (aerobes, anaerobes, or facultative; degrade chitin); Flavobacteriia (aerobic heterotrophs); Cytophagia (aerobic heterotrophs, degrade cellulose); Bacteroidetes (uncommon, but may be symbionts or commensals in soil fauna)
Chloroflexi	2 (0.6–3)	Phylum may be as diverse as Proteobacteria; possibly eight major subgroups; only one isolate from soil (aerobic heterotroph)
Planctomycetes	2 (0.1–4)	Lack cell walls; budding morphology; may possess intracellular, membrane-bound compartments; three major subgroups: Phycisphaerae (rhizosphere) and Planctomycetia (includes soil isolates); Candidatus Brocadiae (anammox bacteria that oxidize NH_3 anaerobically)
Firmicutes	0.5 (0–4)	Gram-positive cell walls; form heat-resistant spores; two major subgroups: Bacilli (readily cultured aerobically; lactic fermenters) and Clostridia (anaerobes)
Gemmatimonadetes	2 (0.1–4)	Gram-negative; aerobic heterotrophs; one major group; few cultured; little known physiologies

(*Continued*)

TABLE 5.3 Major bacterial phyla detected by their 16S rRNA genes in DNA extracts from surface soils, average percentage representation within communities, and selected group characteristics. (*Cont.*)

Phylum	Average (range)[a]	Subgroups found in soils and examples of known characteristics (list is not exhaustive)
Cyanobacteria	3 (0–25)	Photosynthetic; fix N_2; unicellular, colonial, or filamentous morphologies; Gram-negative cell walls; may form biofilms on soil surfaces; symbiotic partners with fungi in lichens
Below are "low-abundance" phyla that are cosmopolitan but detected at frequencies of <0.5% of DNA/RNA reads; also called "rare phyla"[b]		
Nitrospirae		Includes *Nitrospira* spp. that oxidize nitrite and *Leptospirillum* spp. that oxidize ferrous iron for energy
Deinococcus-Thermus		Includes radiation-resistant, thermophilic members; physiologies and functions in soil unknown
Chlorobi		"Green sulfur bacteria" anaerobic photoautotrophs that oxidize sulfide or sulfur; physiologies and functions in soil unknown
Fibrobacteres		Anaerobic, some chitinolytic, includes termite-gut inhabitants
OP11, TM7		Lineages in candidate phyla radiation characterized by small cell size (pass through 0.2-μm filters; small genomes indicative of fermentative physiologies)
Candidate phyla[c]		BRC1, NKB19, OP10, OS-K, SC3, SC4, TM6, WS2, WS3 (no cultured isolates)

[a]*Estimated average percentage representation (range in parentheses) as reported in Fierer et al. (2012).*
[b]*Aanderud et al. (2015).*
[c]*Wu et al. (2009).*
Adapted from Bruns (2017). Used with permission.

Structure and function of prokaryotic cells

Bacteria and Archaea are much smaller than eukaryotic microbes (Fig. 5.1), and they lack prominent membrane-bound organelles so readily observed in eukaryotic cells. Most species range from 0.2 to 2 μm in diameter and from 1 to 10 μm in length. The small size of Bacteria and Archaea endows their cells with high surface-to-volume ratios which enable them to respond and adapt to changing environmental conditions more rapidly than larger, more complex organisms.

Prokaryotic cells consist of cytoplasm enclosed within a cell envelope (Fig. 5.2). The envelope of a prokaryotic cell consists of a cytoplasmic membrane (innermost) and nearly always a cell wall. Prokaryotic cell envelopes may also include additional exterior components such as an outer membrane or a protein S-layer (Albers and Meyer, 2011). The primary function of the cell wall is to protect the cell from osmotic lysis due to excessive turgor pressure because the solute concentration inside a cell is usually higher than the concentration outside it. Nuclear material of prokaryotic cells typically consists of a single DNA molecule that is not separated from the cytoplasm by a nuclear envelope.

Cellular morphology of Bacteria and Archaea. When examined under the microscope, the shapes of many bacterial and archaeal cells are spherical (coccus, singular; cocci, plural) or rod-shaped (bacillus, singular; bacilli, plural). Less common shapes include spirals (spirilla), vibrios (comma-shaped), and lobed forms (Fig. 5.3). Although many archaea exhibit cell shapes similar to those of bacteria, some extremely halophilic archaea are square-shaped

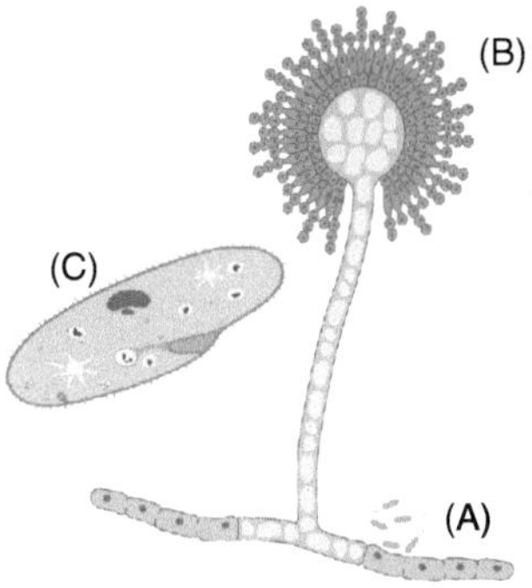

FIGURE 5.1 Relative sizes of common soil microorganisms. (A) bacterial cells with widths of 0.5 μm, or approximately one-tenth of the width of a fungal hyphal filament, (B) asexual spore-bearing, fruiting body of a saprotrophic fungus (*Aspergillus* sp.), and (**C**) protozoan (*Paramecium* sp.) that may range from 50 to 300 μm long. *Source: Created with BioRender.com.*

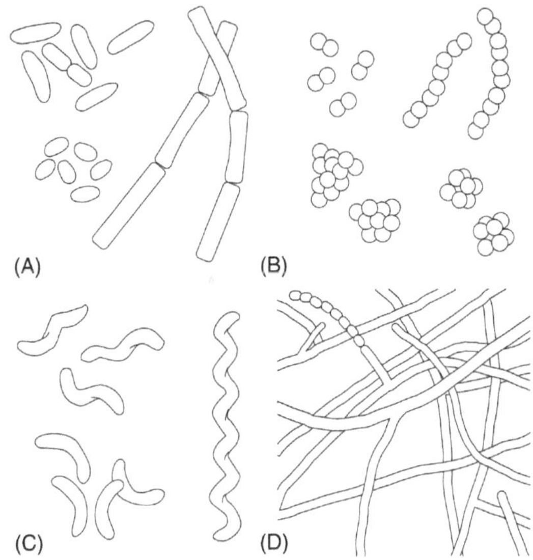

FIGURE 5.3 Common shapes and groupings of soil bacteria. (A) Rods, occurring as single cells, streptobacilli (chains), and coccobacilli (short rods); (B) Cocci, occurring as diplococci (pairs), streptococci (chains), staphylococci (irregular clusters), and sarcinae (cuboidal packets); (C) spirilla (spirals) and vibrios (comma-shaped); and (D) branched or unbranched filaments (hyphae) (actinomycetes, also called actinobacteria). *Source: Original drawing by David Alexander. Used with permission.*

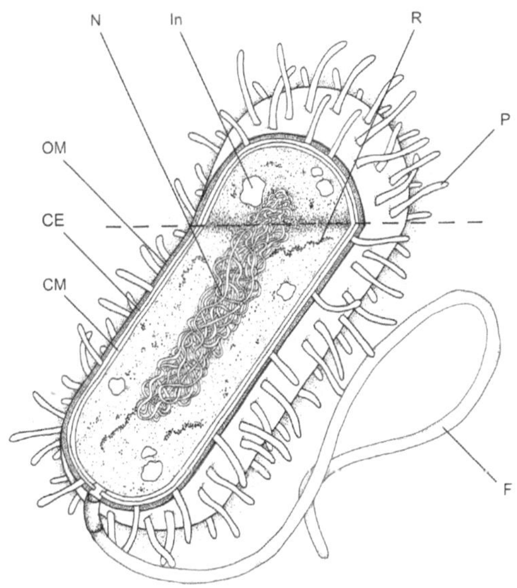

FIGURE 5.2 Generalized structure of a prokaryotic cell. *CM*, cytoplasmic membrane; *CE*, cell envelope; *N*, nucleoid, *In*, Inclusion; *R*, ribosomes (occurring in the form of a polysome); *P*, pilus; *F*, flagellum; *OM*, outer membrane. External structures (cell envelopes and appendages) vary within and between the two prokaryotic domains. Even though external structures of bacteria and archaea appear similar, they are genetically and biochemically distinct. *Source: Original drawing by Kim Luoma. Used with permission.*

and thin like postage stamps (Burns et al., 2007), while one member of the korarchaeal group appears as an ultra-thin filament (0.18 μm) with lengths up to 100 μm (Elkins et al., 2008). Most distinctive within the bacterial domain are actinobacteria that form slender filaments called hyphae (hypha, singular) consisting of individual cells that remain connected to each other. These bacteria were formerly known as actinomycetes (from the Greek: "actino" for ray or beam, and "mycetes" for fungal-like). Actinomycete hyphae may be unbranched or branched and can develop into spore-bearing structures (Fig. 5.3D).

Some bacteria have intermediate shapes, such as short rods (coccobacilli) or comma-shaped (vibrios). Aquatic sediments are often inhabited by corkscrew-shaped bacteria (spirochetes) and chains of rods enclosed within a sheath (sheathed bacteria). Among species that form cocci, individual cells often do not separate after undergoing cell division. Instead, the cells remain attached to one another and exhibit characteristic groupings. Cocci that consistently divide longitudinally form chains of cells called streptococci, while those that divide randomly form irregular cell clusters called staphylococci. A few species of cocci form planar packets of four cells (tetrads) or cuboidal packets of eight or more cells (sarcina, singular; sarcinae, plural).

As mentioned previously, actinomycetes are also called actinobacteria because they have been phylogenetically determined to be bacteria. During their vegetative growth phases, actinomycetes can form dense mats of interwoven hyphae (mycelia) and reproduce by forming asexual spores that are either enclosed in a sac (sporangiospores) or not enclosed (conidiospores). Actinomycete hyphae have diameters similar to those of rod-shaped bacterial cells (1–2 μm), whereas fungal hyphae typically have diameters ranging from 10 to 50 μm.

Bacterial cell envelopes. Although some bacteria, such as the mycoplasmas, do not have cell walls, the cell walls of all other bacteria contain a unique polymer known as peptidoglycan, also called murein. The polysaccharide portion of the polymer consists of long strands of alternating amino sugars, *N*-acetylglucosamine and *N*-acetylmuramic acid. These strands are cross-linked by short peptide chains to form a strong, mesh-like framework that surrounds the cytoplasmic membrane. This arrangement not only gives the cell wall its rigidity but also makes the wall somewhat porous. Water and small, soluble molecules and ions can pass directly through the cell wall to the cytoplasmic membrane. Bacterial cell walls are differentiated by the thickness and composition of their peptidoglycan and associated layers. The two most common types of bacterial cell walls can be differentiated under the microscope using the Gram stain (Box 5.1).

Gram-positive bacteria (purple-staining) have relatively thick peptidoglycan with no outer membrane (Fig. 5.4A). Their peptidoglycan layers contain teichoic acids (negatively charged polysaccharides composed of repeating subunits of glycerol phosphate, ribitol phosphate, or glucosyl phosphate) and lipoteichoic acids that extend outward from the wall surface. These acids may provide a negatively charged cell surface to bind cationic nutrients or function in cell attachment or recognition events. Between the outer peptidoglycan layer and the cytoplasmic membrane lies a relatively thin periplasmic space filled with **periplasm,** a gel-like material containing a variety of proteins and accessory compounds important for transport and energy generation.

Gram-negative bacteria (pink-staining) have an outer membrane in addition to the inner cytoplasmic membrane. Between these two membranes lies a space containing periplasm

BOX 5.1

The Gram stain

Two common types of bacterial cell walls can be distinguished with a staining technique developed in 1884 by the Danish physician Hans Christian Gram. Gram experimented with various combinations of dyes and fixatives in an effort to make bacterial cells more easily visible in infected tissues. In the process, he discovered that different types of bacteria stain different colors when treated in a certain manner. The staining technique that he devised, now called the **Gram stain**, has great practical value for identifying and classifying bacteria.

To prepare a Gram stain, a smear of bacterial cells is first stained with crystal violet, a dye which imparts a deep purple color to the cells. After applying an iodine solution, which forms a complex with the crystal violet in the cell wall (Wilhelm et al., 2015), the smear is carefully rinsed with a solution of acetone and ethanol. This solution removes the crystal violet-iodine complex from the cell wall of Gram-negative bacteria, but not from Gram-positive bacteria because of differences in the molecular structure of the peptidoglycan layer. The smear is then counterstained with a red dye called safranin. The counterstain makes Gram-negative bacteria appear pink, while Gram-positive bacteria appear dark purple.

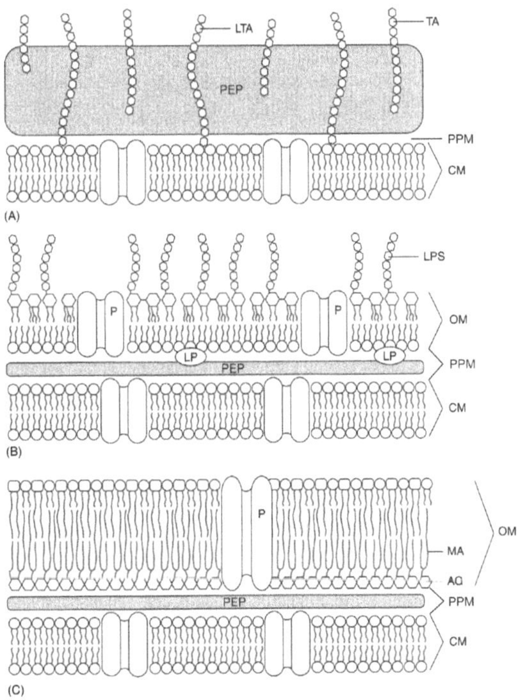

FIGURE 5.4 Structures of (A) Gram-positive, (B) Gram-negative, and (C) acid-fast bacterial cell envelopes. *CM*, cytoplasmic (inner) membrane; *PPM*, periplasm (also called periplasmic space); *PEP*, peptidoglycan layer; *OM*, outer membrane; *TA*, teichoic acid; *LTA*, Lipoteichoic acid; *LPS*, lipopolysaccharide; *p*, porin; *LP*, lipoprotein; *MA*, mycolic acid; *AG*, arabinogalactan. *Source: Adapted from original drawing by David Alexander. Used with permission.*

that surrounds a relatively thin layer of peptidoglycan (Fig. 5.4B). Gram-negative peptidoglycan has less extensive cross-linkage than that of Gram-positive bacteria, which makes the layer more flexible and less resistant to turgor pressure (Koch, 1998). The outer membrane is made of phospholipids and lipopolysaccharides and contains numerous protein channels (porins), allowing water and small molecules to pass into the periplasm. The Gram-negative outer membrane also contains negatively charged polysaccharides that extend outward from the cell surface. As in Gram-positive bacteria, the negatively charged functional groups on cell exteriors interact with cations in the surrounding solution and play important roles in bacterial attachment and host recognition.

Acid-fast bacteria have a highly cross-linked peptidoglycan layer surrounded by an outer membrane composed of glycolipids with long-chain fatty acids (mycolic acids). Bacteria with acid-fast cell walls (Fig. 5.4C) are related to Gram-positive bacteria and include the genera *Mycobacterium* and *Nocardia*. Their outer membranes have high lipid content and also contain porins for nutrient uptake. Because of their unusual cell wall composition, acid-fast bacteria require specialized staining techniques prior to microscopic examination.

Archaeal cell envelopes. Archaeal cell walls contain no peptidoglycan, but cell walls of some archaea are composed of pseudomurein, a polymer similar to peptidoglycan but which does not contain muramic acid. Other archaea possess thick walls of heteropolysaccharides containing more than one type of sugar. Another type of archaeal cell wall is relatively thin and consists of protein or glycoprotein. Some archaeal cell walls are surrounded by proteinaceous S-layers (Albers and Meyer, 2011).

Cytoplasmic membranes of the Bacteria and Archaea. The cytoplasmic membrane is a fluid structure that forms a semipermeable barrier between the interior of the cell and the external environment. In Bacteria, the cytoplasmic membrane is a bilayer of phospholipids made up of straight chain fatty acids linked to glycerols by ester bonds. In Archaea, the lipids consist of branched-chain hydrocarbons linked to glycerols by ether bonds. In some archaea, diether lipids form a bilayer similar to that found in bacteria and eukaryotes, while in others tetraether lipids form a monolayer consisting of long hydrocarbon chains linked to glycerol at both ends. The unique structure and composition of archaeal cytoplasmic membranes may play an important role in their ability to survive in extreme environments.

The cytoplasmic membrane allows passage of water, gases (e.g., O_2, N_2, CO_2, H_2), and a few lipid-soluble molecules by diffusion. Cross-membrane transport of nutrients, wastes, and enzymes is enabled by protein molecules, either embedded within the lipid layer or tightly associated with the inner or outer surface of the membrane. If the concentration of a nutrient in the soil solution is higher than the concentration inside the cell, organisms do not have to expend energy to take it up. Transport proteins known as permeases can conduct nutrients passively by taking advantage of the energetically favorable electrochemical gradient (Fig. 5.5). However, concentrations of dissolved nutrients are rarely high enough for this to occur. Most nutrients must be taken up by proteins that actively transport substances across the membrane by coupling transport with the hydrolysis of ATP (ATPases) or by coupling the movement of one substance down an electrochemical gradient with the movement of another substance against the gradient (co-transporters).

The organic nutrients available to soil microorganisms are often large, complex polymers present in the residues of dead plant or animal tissues. These substances cannot be taken up through small transport proteins in the same manner as simple, water-soluble nutrients. To

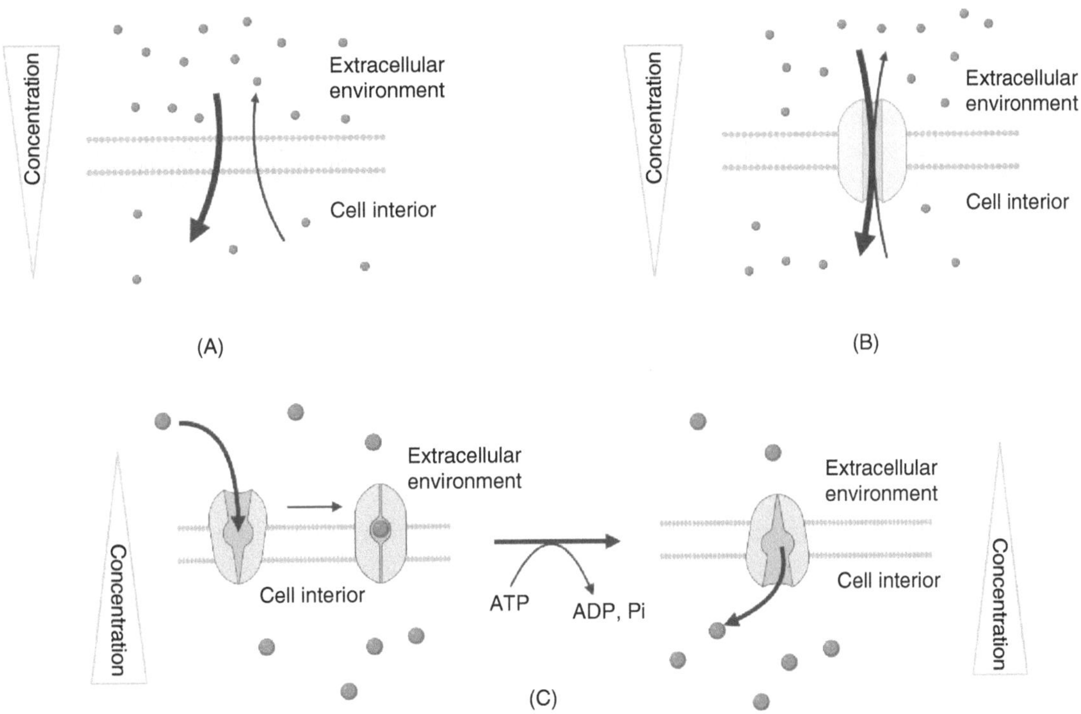

FIGURE 5.5 **Mechanisms by which nutrients pass across bacterial and archaeal cytoplasmic membranes relative to concentration gradients outside and inside cells.** (A) Simple diffusion where nutrients pass through membranes without expenditure of energy; (B) facilitated diffusion where nutrients pass through a protein channel without expenditure of energy; (C) active transport, where nutrients are taken up against a gradient and require energy from ATP generation by proton motive force. *Source: Original figure by Kendall Godwin using BioRender.com.*

utilize these substances, bacteria must excrete hydrolytic (degradative) enzymes (Chapter 13) into the soil solution. Secretory proteins in the cytoplasmic membrane transport extracellular enzymes out of the cell (Green and Mecsas, 2016; Szabo and Pohlschroder, 2012), where they can break down larger molecules to smaller ones that can be taken up by active transporters or permeases. Another means by which bacteria and archaea may release enzymes into the environment is in membrane-bound extracellular vesicles (Gill et al., 2019). In Gram-negative bacteria, these are known as outer membrane vesicles (OMVs), although some produce vesicles from both inner and outer membranes (Kulp and Kuehn, 2010). Extracellular vesicles vary in their molecular cargoes (enzymes, nucleic acids, lipids), and they may play roles in biofilm formation, host interactions, and detoxification. The mechanisms of extracellular vesicle formation are poorly understood but appear to be environmentally controlled (Tashiro et al., 2012). Their significance in organic matter degradation or nutrient acquisition in soils is unknown.

In the cytoplasmic membrane, electron transport proteins and electron carriers like flavins and quinones are responsible for oxidative phosphorylation and generation of the transmembrane proton gradient that generates ATP (Chapter 3). Similar proteins and carriers in the membranes of photosynthetic bacteria catalyze photophosphorylation, the light-activated process by which these bacteria synthesize the ATP and reducing equivalents needed to assimilate CO_2 into organic compounds. The cytoplasmic membrane in bacteria contains sensor proteins, which are an important part of two-component signal transduction systems that enable cells to detect and respond to changes in their environments. These membrane proteins are activated by chemicals, temperature changes, pH, or reduction potential, and they interact with transcriptional regulators in the cytoplasm to elicit cellular responses. While these signaling systems have been studied extensively in bacteria (Stock et al., 2000), less is known about them in archaea, where they have been detected in the genomes of only Haloarchaea and Thaumarcheota (Galperin et al., 2018). In the Archaea, histidine kinase sensor proteins are often located in the cytoplasm rather than in the cell membrane and their response regulators employ protein–protein interactions instead of transcriptional regulation.

Nucleoids. The most prominent structure in the cytoplasm of prokaryotic cells is the nucleoid, which contains the chromosome, a large, double-stranded DNA molecule that encodes the genetic information of the cell. The length of the chromosomal DNA molecule is so large that it is folded into a compact, supercoiled structure by DNA binding proteins (Thanbichler et al., 2005). Nucleoid-associated proteins act to package the chromosome in ways that can accommodate such enzymatic processes as DNA transcription, replication, and repair. In archaea, the types of nucleoid-associated proteins vary among phyla. Euryarchaeota possess histone-like proteins similar to those of eukaryotes, while Crenarcheaota lack histones and employ proteins similar to those of bacteria (Peeters et al., 2015). In both bacterial and archaeal chromosomes, genes are organized in operons which produce functionally related proteins. Although archaeal and eukaryal transcription is carried out by RNA polymerase II, transcriptional regulatory proteins in archaea are more like those of bacteria. Considerable phylogenetic variation thus exists among archaea in nucleoid organization and function.

Plasmids. Many bacterial cells contain one or more plasmids, which are smaller, extrachromosomal DNA molecules containing genes for specialized functions. Such functions include antibiotic resistance, heavy metal tolerance, or ability to degrade specific pollutants. Plasmids are dispersed throughout the cytoplasm and replicate independently. As a result, bacterial cells often contain multiple copies of one or more plasmids, which can be passed from generation to generation (vertical transmission). Some plasmids contain genes that enable them to transfer copies of themselves from cell to cell, sometimes between different species of bacteria (horizontal gene transfer). In the Archaea, plasmids share more genes within phyla than outside phyla. Among Crenarchaeota, plasmids have been studied in Sulfolobaceae, Haloarchaeaceae, and Thermococcaceae, and in various groups of methanogens in Euryarchaeota (Wang et al., 2015). The transfer of plasmids has been shown to occur experimentally in multiple studies using soil microcosms (Neilsen et al., 1994; Fan et al. 2019), but the occurrence of plasmid transfer in natural soils is uncertain.

Ribosomes. The cytoplasm surrounding the nucleoid has a uniform grainy appearance, due to the presence of thousands of ribosomes. Ribosomes are precise assemblages of proteins and non-coding RNAs that translate the genetic code of messenger RNAs (mRNAs) into sequences of amino acids, or proteins (Chapter 4). A **polysome** is a strand of mRNA with many ribosomes attached during active synthesis of proteins. While bacterial and archaeal ribosomes have similar sizes, both are smaller than eukaryal ribosomes and sensitive to different types of chemical agents like antibiotics. Size differences are expressed as Svedberg (S) units, which are based on sedimentation rates during centrifugation. Bacteria and archaea contain 70S ribosomes, while eukaryotes contain 80S ribosomes in their cytoplasm (and 70S ribosomes in mitochondria and chloroplasts that evolved from relict bacteria after their incorporation into eukaryotic cells through endosymbiosis).

Microcompartments or organelles. For many years, it was thought that prokaryotes did not possess membrane-bound organelles typically found in eukaryotes, but it is now recognized that diverse bacteria do possess subcellular organelles of various types. These include bacterial microcompartments consisting of enzyme cores surrounded by protein, the best studied of which is the carboxysome in CO_2-fixing autotrophs (Kerfeld et al., 2018). The microcompartment's membrane protects the enzymes enclosed in the core from the rest of the cytoplasm and permits optimization of enzymatic functions. Other bacterial organelles include thylakoids in cyanobacteria, iron-rich magnetosomes in magnetotactic bacteria, and anammoxosomes in ammonia-oxidizing planctomycetes. Some types of bacterial organelles are enclosed in semipermeable lipid monolayers or bilayers, while others may not have true membranes but rather phase-defined boundaries (Greening and Lithgow, 2020). Diverse organelles have various metabolic and physiological functions and facilitate bacterial adaptation to different environments.

Inclusions. Other structures visible in the cytoplasm of bacterial cells are collectively referred to as **inclusions**, the most common of which are storage granules. When some bacteria access an excess supply of carbon, they can store it as poly-β-hydroxybutyrate or glycogen, which are polymers of fatty acids and glucose, respectively. Other bacteria can store excess phosphorus as polyphosphate granules (also called metachromatic granules due to their red appearance under a brightfield microscope after staining with methylene blue). Some sulfur-oxidizing bacteria produce elemental sulfur granules that can be used as energy sources when external supplies of reduced sulfur become depleted.

Endospores. A few species of soil bacteria, including members of the genera *Bacillus* and *Clostridium*, produce **endospores** in response to nutrient depletion or other environmental stresses. Endospores are the most resistant form of life known. Spore-forming bacteria can withstand high temperature, desiccation (drying), radiation, and exposure to toxic chemicals to a greater extent than any other type of organisms. When nutrients become depleted, vegetative cells transform themselves into metabolically inactive endospores as a survival strategy. When conditions become favorable, endospores germinate to grow and reproduce again as vegetative cells. See Box 5.2 on survival records for endospores and vegetative bacteria. Unlike reproductive spores produced by fungi and actinomycetes, the primary function of bacterial endospores is not to increase the number of individuals in a population, but to enhance survival under adverse conditions. Other bacteria can form resting structures such as spores or cysts, but these are not as resistant to extreme conditions as endospores.

Surface structures. Many bacteria produce a polysaccharide or glycoprotein layer that coats the surface of the cell. The structure and consistency of this surface material vary from a thin slime layer to a thick, gelatinous **capsule**, depending on the species of bacteria and the type and availability of nutrients in the soil. The slime layer or capsule reduces water loss by evaporation from the cytoplasm and protects the bacteria from phagocytosis by protozoa. Thick, sticky capsules may also play a role in the attachment of bacterial cells to surfaces and in the formation of biofilms. They also play an important role, along with fungal and actinomycete hyphae, in binding together clay, silt, and sand particles into soil aggregates that improve soil structure.

BOX 5.2

How long can bacteria live?

How long can bacteria live? A good answer would be, "It depends on environmental conditions." Temperature, humidity, aeration, UV exposure, and radioactivity are all factors influencing the longevity of a bacterial cell and its ability to repair DNA damage. *Endospores*, formed by Gram-positive bacteria in the phylum Firmicutes, likely hold the record for long-term survival in surface soils under ambient conditions. These dormant, dehydrated resting structures, in which all metabolic activity is shut down, require specific physical or chemical triggers to germinate before becoming active again as vegetative bacteria. In surface soils and dust samples, endospores may survive for decades, as was demonstrated by recovery of viable bacilli from dust samples in Charles Darwin's sealed tubes collected during a trip across the Atlantic in 1838 (Gorbushina et al., 2007). It is generally thought, however, that because endospore DNA cannot undergo repair, cumulative radiation impacts over many centuries can make DNA too fragmented and cross-linked to permit germination. Interestingly, survival periods up to 1.1 million years have been documented for non-spore-forming bacteria and vegetative cells of spore-formers as a result of their *very low metabolic activity*. This phenomenon was proposed to have occurred in buried permafrost sediments of northeast Siberia which ranged in burial age (based on depth) from 0.01 to 1.1 million years (Liang et al., 2019). What is the evidence that bacteria in older sediments such as these can remain viable for such long periods of time? In the study by Liang et al. (2019), bacterial cells separated from sediments at both depths contained similar ratios of D- and L- forms of aspartate. This meant that bacteria at both depths had maintained the viability needed to prevent L-aspartate (the amino acid form used by living organisms) from being converted to unusable D-aspartate by natural degradative enzymes called racemases. Further evidence of a low-level maintenance of essential metabolic functions was provided by Stewart Johnson et al. (2007). They examined bacteria from 500,000-year-old deep permafrost samples and found evidence of active DNA repair in *Arthrobacter* spp. (phylum Actinobacteria) exhibiting exceptionally low levels of respiration. Together, these two studies indicated that even non-endospore forming bacteria were capable of long-term survival in low-metabolic states. And in 2020, spore-forming and non-spore-forming bacteria belonging to several major phyla were recovered from 101.5-million-year-old deep marine sediments from the South Pacific Gyre, demonstrating growth on isotope-labeled nutrients within 68 days of incubation (Morono et al., 2020). The dense clay deposits in which these bacteria had been trapped were too nutrient- and energy-poor to enable anything but survival, yet the capability to sustain life for a million years was observed for diverse aerobic species. Thus, regardless of the mechanisms involved (Hoehler and Jørgensen, 2013), research on bacterial longevity has shed light on the profoundly impressive abilities of bacteria to persist in the environment.

Survival in soil often depends on an organism's ability to adhere to soil particles, attach to plant roots, or attach to other soil organisms with which they form symbiotic associations or biofilms. A biofilm is a physically structured community of microbial cells that are attached to a surface, growing near each other, and encased in extracellular secretions. Bacteria often form biofilms on pipelines, pilings, or other objects buried in soil (Fig. 5.6). Attachment may be mediated by two different types of hairlike appendages, pili and fimbriae, which are anchored in the cell envelope and extend outward from the surface of the cell. Some archaea, many Gram-positive bacteria and virtually all Gram-negative bacteria are capable of producing pili (Ng et al., 2008). Some bacteria carry plasmids encoding a special type of pilus, known as a sex pilus, which enables the cell to attach to a recipient cell. In a process known as **conjugation**, a copy of the plasmid passes from the donor to the recipient cell through the pilus (Chapter 4).

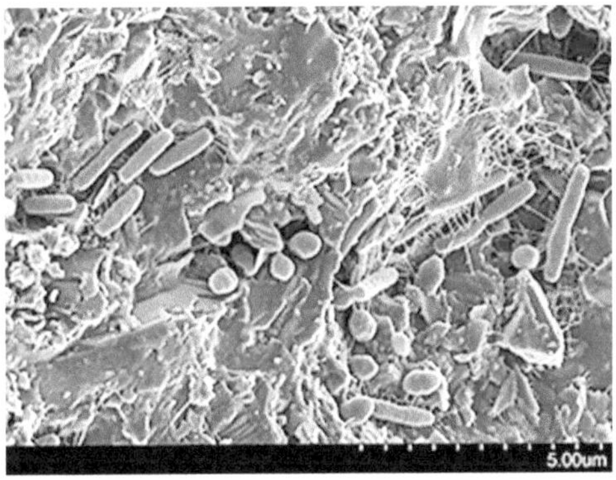

FIGURE 5.6 **Scanning electron micrograph image of a natural community of bacteria on a grain of sand.** Distance between each vertical tick mark on the scale bar is 0.5 µm. *Source: Image provided by the Lewis Lab at Northeastern University, courtesy of Anthony D'Onofrio, William Fowle, Eric Stewart, and Kim Lewis.*

Pili, along with components of the cell envelope, also function in recognition between bacteria and other soil microbes, or between bacteria and plants. Bacteria that colonize plant roots, for instance, bind to specific macromolecules on the surfaces of root cells (Wheatley and Poole, 2017). Pili play a key role in these types of interactions, as do the lipopolysaccharides in the outer membrane of Gram-negative bacteria. These interactions are often very specific because the chemical composition of cell surface structures varies among different species of bacteria. A variation of just one or a few sugars in a polysaccharide or glycoprotein, for instance, can make the difference between recognition and nonrecognition.

The ability to move from one location to another, known as **motility**, is an important characteristic of many soil microbes. It enables microbes to move from sites where conditions are less favorable for growth or survival to sites where conditions are more favorable. Bacteria probably do not travel long distances in soils, but even small movements may enable them to find more suitable microsites. Motile archaea and bacteria "swim" by rotating a long, slender, helically shaped protein filament called a **flagellum** (plural, **flagella**). Some species produce a single flagellum, whereas others produce two or more in various characteristic arrangements (Fig. 5.7). Like pili, flagella are anchored in the cell envelope and usually extend outward from the surface of the cell. A unique group of Gram-negative bacteria, known as spirochetes,

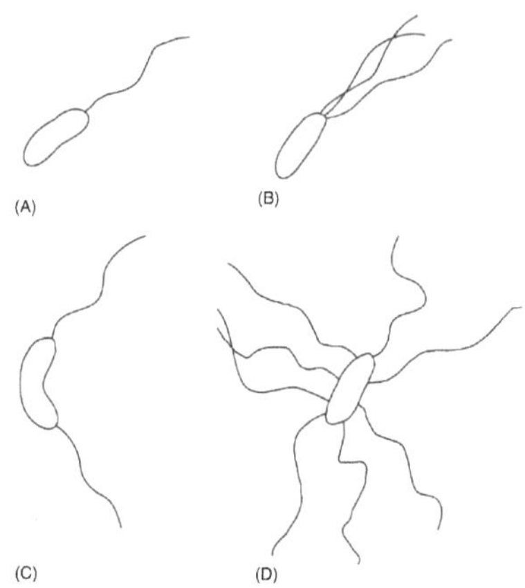

FIGURE 5.7 Common arrangements of bacterial flagella. (A) Monotrichous (single flagellum at one pole of the cell), (B) lophotrichous (tuft of flagella at one or both poles of the cell), (C) amphitrichous (single flagellum or tuft of flagella at both poles of a cell), (D) peritrichous (flagella all around the surfaces of the cell). *Source: Original drawing by David Alexander. Used with permission.*

have flagella that do not extend outward from the cell surface. In these bacteria, the flagella are arranged in a bundle called an axial filament, which wraps around the cell in the space between the cell wall and the outer membrane. Rotation of the axial filament produces a corkscrew type of movement that enables spirochetes to swim through viscous liquids, such as the aquatic sediments that spirochetes typically inhabit.

Water-saturated soils provide more favorable conditions for bacterial movement by flagella than unsaturated soils, where water films on soil particles may be thin or discontinuous, depending on the matric potential of the soil solution. Flagellar motility essentially ceases at matric potentials of -200 kPa but does occur at higher (more positive) matric potentials when water films are at least 1.5 μm thick (Tecon and Or, 2016). Other types of motility enable bacteria to move across soil particle surfaces covered by thin films of moisture. Some species have lateral flagella that enable them to swarm on these surfaces (a phenomenon often observed on agar plates in the laboratory), whereas others exhibit alternate forms of motility described as gliding or twitching. Several models have been proposed to explain these types of movement, but precise mechanisms have yet to be established. It is likely that many different mechanisms operate in bacterial motility (Jarrell and McBride, 2008).

Nutritional strategies

Besides being the most phylogenetically diverse of all soil microorganisms, bacteria and archaea are also the most metabolically diverse, exhibiting the full range of metabolic capabilities introduced in Chapter 3 (Table 5.4). Most are heterotrophic, while some are autotrophic, and some can shift from one type of metabolism to another as environmental conditions change. Aerobes, obligate anaerobes, and facultative anaerobes are found in both domains, which include species that can generate energy and reducing equivalents from oxidations of organic or inorganic compounds.

TABLE 5.4 Major nutritional groups of soil bacteria and archaea.

Nutritional classification	Sources of carbon and energy	Representative groups
Chemoheterotophs	Carbon sources are organic compounds Energy sources are organic compounds	Saprotrophic bacteria, most symbiotic bacteria
Photoautotrophs	Carbon sources are CO_2 gas or (bi)carbonates Energy sources are photons from light	Cyanobacteria, green bacteria, purple bacteria
Chemoautotrophs	Carbon sources are CO_2 gas or (bi)carbonates Energy sources are inorganic compounds (NH_3, NO_2^-, H_2, S^0, Fe^{2+})	Nitrifying bacteria and archaea, hydrogen-oxidizing bacteria, sulfur-oxidizing bacteria and archaea, iron-oxidizing bacteria and archaea

Adapted from table by David Alexander, second ed.

Nutritional requirements

Bacteria and archaea have many of the same nutritional requirements as other organisms. Substances that an organism requires in greater amounts are called **macronutrients**. These substances are structural components of the biological molecules (carbohydrates, proteins, nucleic acids, and lipids) that constitute more than 90% of the dry mass of living cells. Carbon (C) is required in the greatest amount, followed by nitrogen (N), phosphorus (P), and sulfur (S). Water is generally not regarded as a nutrient in the same sense as C, N, P, and S, yet it is essential for all living cells, and it serves as the primary source of hydrogen (H) and (O). Potassium (K), sodium (Na), calcium (Ca), and magnesium (Mg) are also required in substantial quantities. These "secondary" macronutrients function as cofactors for many enzymes and maintain the ionic balance in bacterial cells.

Micronutrients are required in lesser amounts (micromolar concentrations or less). These elements often serve as structural components or activators of specific enzymes in bacterial cells. Iron is the micronutrient that is usually required in the greatest amount. Others include cobalt (Co), zinc (Zn), molybdenum (Mo), copper (Cu), and manganese (Mn). Some bacteria can synthesize all necessary cellular constituents and obtain energy for growth using a single carbon source and a few mineral nutrients. Others have more complex nutritional requirements. Many soil bacteria require specific organic compounds that they are unable to synthesize from simple starting materials. Organic nutrients of this type are called **growth factors** and are usually classified into one of the following categories: amino acids (precursors for protein synthesis), purines and pyrimidines (precursors for nucleic acid synthesis), and vitamins (precursors for the synthesis of certain types of enzymes).

Sources of carbon, energy, and reducing equivalents

Heterotrophic bacteria and archaea obtain carbon, energy, and reducing equivalents (electrons) from soil organic compounds. Most are saprophytes that feed on nonliving plant and animal residues or humus. Organic matter decomposition by these and other microbes is described in Chapter 13. Some bacteria and a few archaea invade the tissues of other living organisms. In some cases, these microbes may harm the host organism (archaea seem to be an exception—none are known to cause disease), whereas in other cases they may benefit or have no effect on the host. The nature of these interactions is discussed more fully later in this chapter and in Chapter 10.

Autotrophic bacteria obtain most or all of their carbon from CO_2 gas or dissolved (bi) carbonate ions. Autotrophs reduce these sources of inorganic carbon for assimilation into carbohydrates or other organic compounds that can be metabolized to support cell growth. Some bacteria are photoautotrophs that use light energy to generate ATP and reducing power needed to fix CO_2, whereas others are chemoautotrophs that generate energy and reducing power through chemical oxidations. While some archaea (*Halobacterium*) use light energy to produce proton gradients for ATP production, no archaea are known to use light energy to fix CO_2 through photosynthesis (Schäfer et al., 1999).

Cyanobacteria are oxygenic photoautotrophs that resemble eukaryotic algae (and plants) in the manner in which they utilize light energy and H_2O to fix CO_2. Their photosynthetic pigments and electron transport proteins are organized into two photosystems (as described in Chapter 3), and their primary photosynthetic pigment is chlorophyll *a*. Like most photosynthetic organisms, they assimilate CO_2 by the Calvin cycle. Chapter 7 provides a more detailed description of cyanobacteria and algae.

Other photosynthetic bacteria differ in several ways from the cyanobacteria (Table 5.5). **Green bacteria** and **purple bacteria** have a single photosystem, produce different photosynthetic pigments, use different primary electron donors, and, in some cases, assimilate CO_2 by different biosynthetic pathways (Blankenship et al., 1995). These bacteria are often found

TABLE 5.5 Major groupings of photosynthetic bacteria[a] and associated metabolic characteristics.

Group	Phylum and representative genera	Electron donors	Oxidized products	CO_2 fixation pathway
Oxygenic, Aerobes				
Cyanobacteria	Cyanobacteria *Nostoc* *Microcoleus*	H_2O	O_2	Calvin-Benson cycle
Anoxygenic, Obligate Anaerobes				
Green sulfur bacteria	Chlorobi *Chlorobium* *Pelodictyon*	H_2S, S^0, H_2	S^0, SO_4^{2-}, H^+	Reverse TCA cycle
Purple sulfur bacteria	Proteobacteria *Chromatium* *Thiospirillum*	H_2S, S^0, H_2	S^0, SO_4^{2-}, H^+	Calvin-Benson cycle
Anoxygenic, Facultative Anaerobes				
Green nonsulfur bacteria	Chloroflexi *Chloroflexus* *Ktedonobacter*	Usually organic compounds,[b] can use H_2, H_2S	H^+, SO_4^{2-} [c]	Hydroxy-propionate pathway
Purple nonsulfur bacteria	Proteobacteria *Rhodospirillum* *Rhodocyclus*	Usually organic compounds,[b] can use H_2, H_2S	H^+, SO_4^{2-} [c]	Calvin-Benson cycle
Anoxygenic, Obligate Anaerobic Photoheterotroph				
Heliobacteria	Firmicutes *Heliobacterium* *Heliobacillus*	Limited carbon sources, not capable of photoautotrophy	Incomplete reductive TCA cycle	

[a]*Among Archaea, only members of the Halobacteria group are known to be phototrophic but they are not autotrophic.*
[b]*Green nonsulfur and purple nonsulfur bacteria grow preferentially as photoheterotrophs under anaerobic conditions, but they can grow in the dark as chemoheterotrophs under aerobic conditions.*
[c]*Products of photoautotrophic growth with H_2 or H_2S as electron donors.*

in riparian soils, mud flats, shallow lakes or ponds, or deep, clear lakes in which light can penetrate to anaerobic depths. Because they do not use water as an electron donor for photophosphorylation, green bacteria and purple bacteria do not evolve O_2. As a result, they are often described as **anoxygenic** photoautotrophs.

Chemoautotrophic (or chemolithoautotrophic) bacteria and archaea oxidize a variety of inorganic compounds to generate ATP and reducing power. Ammonia-oxidizing bacteria and archaea use different pathways for CO_2 fixation, namely the Calvin cycle and hydroxypropionate/hydroxybutyrate cycles, respectively (Könneke et al., 2014). Although not as abundant as heterotrophic species in many soils, chemoautotrophs occupy vital niches in biogeochemical cycles, particularly the nitrogen and sulfur cycles, and they have a profound influence on plant growth in natural and agricultural ecosystems. The ecological roles of these organisms are discussed in more detail later in this chapter and in Chapters 15, 18, and 19.

Oxygen requirements

Oxygen (O_2) concentrations can vary widely from one soil microsite to another. Larger pores filled with air provide relatively high concentrations of O_2, which favors aerobic metabolism. Conversely, smaller pores just a few millimeters away may be filled with water, which can lead to the creation of anaerobic conditions. Oxygen concentrations can also vary widely with time, as soils undergo periods of saturation with water, which may lead to temporary hypoxic or anoxic conditions, followed by the return of oxic conditions as soils dry out. It is therefore not surprising that bacteria and archaea exhibit a full range of adaptations to different O_2 concentrations. Table 5.6 summarizes the oxygen requirements of these microbes and the types of energy metabolism associated with each group. Several of the anaerobes and

TABLE 5.6 Oxygen requirements and energy metabolisms of bacteria and archaea.

Oxygen requirement	Type of energy metabolism	Final electron acceptor	Reduced product of energy metabolism
Aerobes	Aerobic respiration	O_2	H_2O
Facultative anaerobes	Anaerobic respiration[a]	Nitrate (NO_3^-) or Nitrite (NO_2^-)	N_2, N_2O, NO[c]
		Nitrate (NO_3^-)	Nitrite (NO_2^-)[d]
		Nitrate (NO_3^-)	Ammonium (NH_4^+)[e]
		Fumarate	Succinate
	Fermentation	Organic compound	Organic acids, alcohols
Obligate anaerobes	Anaerobic respiration	Sulfate (SO_4^{2-})	Hydrogen sulfide (H_2S)[f]
		CO_2	Acetic acid (CH_3COOH)[g]
		CO_2	Methane (CH_4)[h]
	Fermentation	Organic compound	Organic acids, alcohols
Aerotolerant anaerobes	Fermentation[b]	Organic compound	Organic acids, alcohols

[a] *Facultative anaerobes respire aerobically when O_2 is available.*
[b] *Aerotolerant anaerobes are incapable of either aerobic or anaerobic respiration.*
[c] *Denitrifying bacteria.*
[d] *Nitrate-respiring bacteria which do not reduce nitrite further.*
[e] *Dissimilatory nitrate reduction to ammonium.*
[f] *Sulfate-reducing bacteria.*
[g] *Acetogenic bacteria.*
[h] *Methanogenic archaea.*
Adapted from Alexander (2002). Used with permission.

facultative anaerobes listed in this table play key roles in biogeochemical cycles of carbon, nitrogen, and sulfur, which will be discussed in more detail later in this chapter and in subsequent chapters.

Linking bacteria and archaea to soil

Distribution and abundance of bacteria and archaea

Bacteria and archaea inhabit a variety of habitats and occupy many essential niches in soil ecosystems. Population sizes vary with nutrient availability, temperature, aeration, and other abiotic and biotic factors, but bacteria are generally the most numerous of all cellular soil microbes. Generally, only the acellular viruses are more abundant (Chapter 9). Direct epifluorescence microscopy of soil smears treated with fluorescence-emitting vital stains typically yield counts of 10^8–10^{11} cells g^{-1}, while indirect counts of culturable bacteria often number from 10^7 to 10^{10} **colony-forming units (CFU)** g^{-1} of the same soil. The lower values for indirect counts may reflect improper environmental or nutritional incubation conditions for some community members, as well as lack of detection of some bacteria due to their slow growth on standard laboratory media. Bacterial densities are typically highest in organic matter-rich soil surface layers and in soil immediately surrounding plant roots (a distinctive habitat known as the **rhizosphere**, Chapter 11). Because most soil microbes are heterotrophic, their activity, growth, and survival depend largely on the availability of organic carbon.

Bacteria and archaea exhibit an exceptional variety of adaptations that enable them to survive and grow in habitats that no other organisms can tolerate. We have already seen that these microbes, through their various modes of energy metabolism, can grow in oxic or anoxic environments. Bacteria and archaea also tolerate temperature extremes far beyond those tolerated by other organisms. Most bacteria are **mesophiles**, which grow optimally at temperatures between 20°C and 35°C, or **facultative psychrophiles**, which grow optimally at moderate temperatures but also at temperatures below 5°C. Many bacteria and archaea are **thermophiles** which grow at temperatures in excess of 40°C–50°C, and several, particularly among the archaea, are **hyperthermophiles** that grow at temperatures in excess of 80°C and approaching 100°C. In contrast, most eukaryotes cannot survive temperatures above 45°C. Thermophiles typically produce heat-stable proteins and have high concentrations of saturated fatty acids in their membranes. One important heat-stable enzyme is ***Taq* polymerase** produced by the thermophilic bacterium ***Thermus aquaticus***, for which the enzyme is named. This enzyme is a key ingredient in the **polymerase chain reaction (PCR)** used to copy DNA fragments during molecular analyses of soil DNA extracts (Chapter 4). Thermophiles often play key roles in habitats where intense microbial activity increases temperatures to levels not tolerated by other organisms, such as compost piles (Chapter 24), anaerobic bioreactors, and some mineral deposits used for bioleaching of ores. Hyperthermophiles are found in hot springs and hydrothermal vents, where volcanic activity produces even higher temperatures.

Bacteria in colder climates have adapted to grow optimally at temperatures below 15°C. These psychrophilic microbes produce "cold-active" enzymes and unsaturated fatty acids that maintain the fluidity of their cytoplasmic membranes at low temperatures. Psychrophiles predominate in cold ocean waters and in habitats where soil temperatures remain below 8°C for much of the year (frigid or cryic soils). Research suggests that archaea may constitute as much as one-fifth of oceanic plankton (Bar-On et al., 2018).

Many bacteria and archaea inhabit the highly weathered, acidic soils that predominate in regions with high precipitation, and some can tolerate the extremely low pH of acid springs or other specialized habitats. **Acidophilic** species grow optimally at pH below 5, and extreme acidophiles (such as *Acidithiobacillus thiooxidans*, a sulfur-oxidizing bacterium) tolerate pH as low as 1. Acidophiles maintain a neutral pH in the cytoplasm by actively transporting H^+ out of the cell and often incorporate acid-resistant fatty acids into their cytoplasmic membranes. **Alkaliphilic** species use a similar strategy (ion pumps) to grow at pH as high as 10.5 in arid and semi-arid regions. Microbes that grow well at neutral or slightly alkaline pH are **neutrophilic.** Although actinomycetes are largely characterized as neutrophilic, a few such as *Streptacidiphilus* spp. require acidic conditions for growth. The actinomycetes commonly found in more acidic habitats belong to the genus *Streptomyces* (Poomthongdee et al., 2015). Sensitivity to soil pH can have consequences for the microbial balance of soil ecosystems.

Poor drainage and rapid surface evaporation in arid regions often lead to the formation of salt-affected or **saline** soils (yielding saturated soil extracts having electrical conductivities higher than 4 dS m^{-1}). Many bacteria and archaea adapt to saline or arid environments by accumulating solutes (usually amino acids) in the cytoplasm to compensate for the low water potential of the soil solution or by developing membranes and enzymes that function in solutions of high ionic strength. Microbes that tolerate high salt concentrations are said to be **osmotolerant**. Many of the archaea have adapted so completely to saline environments that they *require* high concentrations of sodium chloride (>1.5 *M* NaCl). These salt-dependent archaea are called **halophiles** or extreme halophiles, depending on the concentration of NaCl that they require. Halophiles are one of the major phylogenetic and ecological groups of archaea. Due to their dependence on high salt concentrations, they are limited to saline or hypersaline environments.

Actinomycetes often predominate in soils subjected to prolonged drying. This is usually attributed to their ability to produce conidia (reproductive spores) that tolerate desiccation, but their filamentous nature could also account for resistance to drying. When more sensitive organisms succumb to low-moisture stress (dry conditions), actinomycetes persist and become the dominant members of the microbial community. Conidia also confer a slight tolerance to elevated soil temperatures. Though most actinomycetes are mesophilic, thermophilic species play an important role in the high-temperature transformations of organic substrates during composting (Chapter 24).

Nutrient transformations mediated by bacteria and archaea

Organic matter decomposition. The primary niche of bacteria and archaea in soils is to function as decomposers. Heterotrophic species play a major role in organic matter decomposition and in bioremediation of soil contaminated with organic wastes. The chemical nature of soil organic matter, mechanisms of decomposition, and practical applications of this process are discussed in Chapters 13, 14, 21, 22, and 24.

Organic substrates decompose most efficiently when a soil, compost heap, or bioreactor is well aerated. Under these conditions, microbial communities respire aerobically and most of the carbon is oxidized to CO_2 or assimilated into microbial biomass. Organic matter decomposes much more slowly in anaerobic microsites or sediments because anaerobic metabolism is less energy efficient than aerobic respiration. Less organic carbon is converted to CO_2 and microbial biomass, so that more organic carbon is partially decomposed to acidic, often foul-smelling waste products.

Anaerobic decomposition can be desirable under certain circumstances. Sewage treatment facilities often use anaerobic digestion to treat the biosolids (sludge) that are generated during various stages of sewage processing. Digestion is required to stabilize the organic matter and reduce the number of pathogens (disease-causing organisms) before the sludge residues (biosolids) can be disposed on agricultural land or further processed for other applications. Anaerobic digestion produces a valuable by-product, methane (natural gas), which can be recovered and used as a source of energy for heat and electricity. Methane is produced by a group of archaea known as **methanogens**, which commonly inhabit anaerobic soil microsites, aquatic sediments, and the rumens of cattle and other ruminant animals. Methanogens are capable of using only a few simple organic substrates, so they typically are members of complex food webs in these habitats. They feed on the H_2, CO_2, acetate, and other products of anaerobic metabolism generated by microbes that decompose more complex organic materials and generate methane by using CO_2 as their final electron acceptor in anaerobic respiration. Some methane produced by methanogens will enter the Earth's atmosphere and contribute to global warming (Chapter 20), if it is not consumed by **methanotroph** bacteria that are also present in soils.

Bacteria rapidly metabolize the easily degradable sugars, starches, fats, and proteins in soil organic matter. Other, more **recalcitrant** substances, such as the lignins, waxes, oils, and resins in plant residues, are decomposed more slowly. Products of organic matter degradation, dead cell material, and polyaromatic compounds formed during decomposition become part of the stable organic fraction of soils known as humus (discussed in Chapters 2 and 13). Humus improves soil structure by binding clay particles together to form stable aggregates and enhances fertility by increasing the water-holding and nutrient-retaining capacity of the soil. As specific groups of microbes gradually breakdown the complex constituents of humus, nutrients are slowly released from organic matter.

Actinomycetes play important roles in decomposing the less easily degradable components of organic matter. These bacteria grow slowly and compete poorly with other soil microbes for easily degradable substrates, but they fill an important niche in soil communities by breaking down an enormous variety of organic compounds (Goodfellow and Williams, 1983). Actinomycetes assist in lignin degradation, and they break down chitin, hemicelluloses, keratin, and other plant, fungal, and animal polymers that many other soil microbes do not decompose. Actinomycetes are known for their production of geosmins, sesquiterpenoid compounds that give soils their characteristic earthy odor (Box 5.3). Geosmins have been shown to attract collembolans (also called springtails), which are some of the most abundant microarthropods in soil (Chapter 8). Collembolans could help disperse actinomycete spores as they move around in soil, thus promoting more extensive colonization and decomposition activity (Box 5.3).

Transformations of nitrogen and sulfur. Bacteria play critical roles in the biological transformations of several key inorganic nutrients in soils, particularly nitrogen and sulfur. Some bacteria and archaea, known as **nitrogen fixers** or **diazotrophs**, use enzymes to convert N_2 from the air into NH_3, which is incorporated into organic molecules as $-NH_2$ groups in

BOX 5.3

Geosmin attracts springtails: an aid in spore dissemination?

After a rainfall, soils have a characteristic smell that can be attributed to a volatile compound called geosmin, produced by actinomycetes, cyanobacteria, and fungi. For some unknown reason, our noses can detect geosmin at very low concentrations (parts per trillion). Apparently, many arthropods, such as insects and arachnids, are also very sensitive to this compound and avoid locations where it is present, perhaps because it might indicate these locations are poorly suited for the survival of their eggs. In contrast, collembolans (also called springtails) are a group of microarthropods that are attracted by geosmin, as demonstrated in a study conducted by Becher et al. (2020). In the laboratory, electrical signals from the antennae of the collembolan *Folsomia candida* were stimulated by the presence of geosmin-producing colonies of *Streptomyces*, a genus that serves as a good model organism for the study of soil actinomycetes. Additionally, the study demonstrated that actinomycete spores are a food source for collembolans and that geosmin is produced by *Streptomyces* during periods of morphological differentiation, thus essentially serving as a signal of active spore production. The production of this chemical attractant may enable spores to "hitch a ride" on the exteriors of collembolan bodies (an example of "phoresy"; the transport of one organism by another), as well as in feces, as a means of dispersal to distant locations in soil.

amino sugars, proteins, and nucleic acids. Dinitrogen fixation is limited to microaerophilic or anaerobic microsites in soils or specialized symbiotic structures (e.g., root nodules of leguminous plants), due to the O_2 sensitivity of the enzyme complex that catalyzes the reaction.

Nitrification refers to the processes by which NH_4^+ is converted to NO_3^-. As dead plant and microbial tissues decompose and become mineralized, NH_4^+ is released into soil solution, becoming available for re-uptake by plant roots and microbes. As a cation, NH_4^+ can be retained by the negatively charged sites of soil colloids if not immediately re-assimilated by roots and microbes. In well-aerated soils, **nitrifying bacteria and archaea** can also utilize NH_4^+ as an energy source, thereby converting the cation to NH_3 inside the cell and oxidizing it to NO_2^-. Rarely detectable in soils, NO_2^- is presumed to be oxidized rapidly to NO_3^-, either abiotically or by two phylogenetically distinct groups of nitrite-oxidizing bacteria (in the Proteobacteria and *Nitrospira* lineages). Plant roots and microbes can also take up NO_3^- to meet cellular N demands, but NO_3^- must first be reduced by assimilatory nitrate reductase enzymes to become incorporated as organic N. Nitrate is soluble in water and, as an anion, is not attracted to negatively charged sites in soils. Nitrate is thus highly susceptible to movement by mass flow and losses by leaching, runoff, and denitrification.

These processes account for significant soil N losses to cause contamination of ground and surface waters.

Denitrification occurs when facultative anaerobes use NO_3^- as a terminal electron acceptor in anaerobic respiration, to produce NO, N_2O, and N_2 gases that diffuse out of the soil. This type of metabolism is often viewed as detrimental to the environment. Substantial portions of fertilizer nitrogen can be lost through denitrification, and the nitrous oxide released into the atmosphere contributes to global warming and ozone depletion (Chapter 20). On the other hand, denitrification is intentionally promoted in municipal wastewater treatment because it provides an inexpensive means to remove nitrate from wastewater and possibly even groundwater.

Sulfate-reducing bacteria have a role in the sulfur cycle similar to that of denitrification in the nitrogen cycle, that is, dissimilatory reduction of SO_4^{2-} and NO_3^-, respectively. These obligate anaerobes make sulfur less available for plant uptake by converting SO_4^{2-} to H_2S in water-saturated soils or anaerobic microsites. The H_2S that they produce usually precipitates as insoluble metal sulfides, rather than diffusing out of the soil in a gaseous form. The ability to respire anaerobically using sulfate as an electron acceptor for respiration is limited to a few genera of anaerobic bacteria, but they are widely distributed in nature.

Sulfur-oxidizing bacteria generate SO_4^{2-} by oxidizing sulfides (H_2S or metal sulfides), elemental sulfur, or thiosulfate ($S_2O_3^{2-}$) in processes analogous to those for nitrification. Most sulfur-oxidizing bacteria are obligate aerobes or microaerophiles (aerobic organisms that grow best at low concentrations of O_2), although at least one species (*Thiobacillus denitrificans*) is a facultative anaerobe that can use nitrate as a terminal electron acceptor. Sulfur-oxidizing bacteria are widely distributed in nature and highly active in well-aerated soils. Mineralization of sulfur-containing organic compounds in soils provides most of the reduced sulfur compounds that these bacteria oxidize.

The roles of bacteria and archaea in biological transformations of nitrogen and sulfur in soils are discussed in greater detail in Chapters 15, 16, and 17 (nitrogen) and Chapter 18 (sulfur). Box 5.4 describes transformations of nitrogen and sulfur in hot springs.

Transformations of other elements. Bacteria are actively involved in transformations of other essential elements in addition to carbon, nitrogen, and sulfur. Bacterial metabolic byproducts include a variety of organic and inorganic acids that dissolve minerals in parent material, thereby releasing soluble materials such as phosphates and metal ions into the soil solution. Some metals and metalloids, including iron, manganese, mercury, arsenic, and selenium, are subject to changes in oxidation state as a result of bacterial activity. Mercury, arsenic, and selenium can also be methylated by bacteria in soils. Methylation increases the toxicity of some metals (organometals are more lipophilic than elemental metals, and therefore tend to accumulate in animal tissues) and increases their mobility in the environment by converting the metals into a more volatile form. The nature of these transformations and the microbes that mediate them are discussed in more detail in Chapter 19.

Interactions with other soil microbes

Bacteria exhibit a broad spectrum of interactions with one another and with other soil microbes. Probably the most common type of interaction among soil microbes is competition,

BOX 5.4

Yellowstone bacteria—taking the heat

Yellowstone National Park is one of the most popular sites in the US National Park Service, attracting more than a million visitors each year. Among the park's most distinctive natural features are its geysers, hot springs, and other geothermal formations. Visitors marvel at the startling beauty of the hot springs' vibrant colors—brilliant hues of red, orange, yellow, purple, and green. The hot springs of Yellowstone owe their vivid colors to thermophilic bacteria and archaea that thrive in the scalding hot, mineral-rich waters (the thermophilic bacterium *Thermus aquaticus*, the source of heat-stable Taq polymerase used in PCR, was isolated from these hot springs). These microbes often form complex, stratified communities called **microbial mats**. The thickness of the mats ranges from a few millimeters to a centimeter or more, and they are characterized by a distinct layering of different microbial populations. The blue-green surface layer consists primarily of cyanobacteria, oxygenic photoautotrophs that serve as primary producers in the mat community. Oxygenic photosynthesis maintains high concentrations of O_2 in the surface layer during the day, but microbial respiration rapidly consumes the oxygen at night. Oxygen depletion provides conditions that allow sulfate-reducing bacteria to reduce sulfate to sulfides and enable dinitrogen-fixing bacteria to capture nitrogen using enzymes that function best when oxygen levels are low. The sulfides and NH_3 generated by these bacteria serve as sources of energy for sulfur-oxidizing and nitrifying bacteria that also reside within the surface layer of the mat. Beneath the cyanobacterial layer, but near enough to the surface to receive sufficient sunlight, is a layer dominated by anaerobic, photosynthetic purple sulfur bacteria. A thin layer of oxidized iron often separates the cyanobacterial layer from the purple layer and may help to maintain a barrier between the aerobic and anaerobic zones of the microbial mat. The purple sulfur bacteria oxidize sulfides in the anaerobic zone, thereby generating elemental sulfur and sulfate that can be used as electron acceptors by anaerobic sulfate-reducing bacteria in the lower layers of the mat. In this way, nearly complete cycles of sulfur and nitrogen occur within the mat community. A thin green layer dominated by green sulfur bacteria often lies beneath the purple zone, though it is not always visible without the aid of a microscope. A thick layer of black sediment containing precipitated iron sulfide typically lies beneath the mat community.

in which bacteria, fungi, and protozoa vie for water, nutrients, shelter, and other vital resources that are limited in supply. On the other hand, decomposition of the complex constituents of soil organic matter requires the cooperative activity of diverse groups of microorganisms possessing different types of degradative enzymes. Successive populations within the microbial community may supply each other's nutritional needs as decomposition progresses. **Syntrophy** occurs when two different species together gain energy from a compound that neither one of them can utilize alone. A good example of syntrophy is that between a methanogen and an alcohol-fermenting, H_2-producing bacterium. Even though fermentation of alcohol is energetically unfavorable, the rapid consumption of H_2 by the methanogen pulls the first metabolic reaction forward, thus enabling alcohol utilization. Bacteria are also subject to **predation** (grazing) by larger protozoa, an interkingdom interaction that limits the size of bacterial populations.

Bacteria may have negative impacts on one another and on other soil microbes by producing antibiotics or other harmful compounds. Actinomycetes are a major source of **antibiotics**, with more than 50 different types of antibiotics being produced by one genus alone (*Streptomyces*), including streptomycin, neomycin, erythromycin, and tetracycline. However, bacteria can develop resistance to an antibiotic themselves (e.g., by genetic mutations) or by taking up plasmids carrying antibiotic resistance genes. The ecological significance of antibiosis in the soil environment remains largely unknown, but the impact on human health has been enormous. The capacity of actinomycetes and other soil microbes to produce antibiotics has spawned a multi-billion-dollar pharmaceutical industry and saved countless human lives in the past half century. Interactions among microbial populations are discussed in greater depth in Chapter 10.

Interactions with plants

Bacteria interact with plants *indirectly* by modifying soil fertility through their roles in organic matter decomposition and transformations of mineral nutrients. Some effects of bacterial activity on nutrient availability were mentioned in the preceding discussion of nutrient transformations. As heterotrophic bacteria decompose soil organic matter, they release NH_4^+, SO_4^{2-}, and HPO_4^{2-}, and other inorganic nutrients through the process of **mineralization**, thereby returning these nutrients to the soil in a form that plants can use. Of course, not all of the nutrients released during decomposition are immediately accessible to plants. Bacteria and archaea use many of the nutrients themselves to support their own growth and survival. **Immobilization** is the assimilation of inorganic nutrients into microbial biomass, which renders these nutrients unavailable to plants. Factors that influence the balance between mineralization and immobilization are discussed in Chapter 15 (nitrogen) and Chapter 18 (sulfur).

Bacteria also interact *directly* with plants, profoundly influencing plant growth and productivity in natural and agricultural ecosystems. Plant roots slough epidermal cells and exude a variety of organic compounds as they extend through the soil in search of water and mineral nutrients. This creates a unique habitat in the vicinity of roots, known as the rhizosphere (Chapter 11), which stimulates the growth of diverse soil microbes. Bacteria grow rapidly when provided with a supply of nutrients and typically colonize plant roots in higher numbers than any other group of soil microbes. Bacteria reach higher densities in rhizosphere soils enriched with root exudates. Many rhizosphere bacteria are harmless saprophytes that feed on root exudates, but some also may protect the plant from infection by disease-causing microbes by competition for nutrients or excreting substances inhibitory to pathogens. Several species of bacteria are currently being investigated as potential biological control agents because of their ability to inhibit pathogen growth or invasion (Chapter 23).

Another way in which many species of bacteria influence plant growth directly is by producing **hormones** or **toxins** that stimulate or impede root function and alter morphology. Several species of N_2-fixing bacteria may form loose **associative symbioses** with plants, feeding on organic substrates exuded from the plant's root system and in turn providing fixed nitrogen to the plant. Other N_2-fixing bacteria, notably those of the genera *Rhizobium*, *Sinorhizobium* (*Ensifer*), *Bradyrhizobium*, and the actinomycete *Frankia*, form intricate mutualistic symbioses with specific plant species. The interactions between plants and associative and symbiotic N_2-fixing bacteria are discussed in Chapters 16 and 17. **Endophytic** bacteria can also grow within plant tissue. Endophytic microorganisms may be introduced through seeds, leaves, or roots and their growth may sometimes have beneficial impacts on the plant, such as synthesizing plant hormones or conferring plant tolerance to stresses (Kandel et al., 2017).

Several species of bacteria are capable of producing disease in plants. Some are well adapted to the soil environment and typically infect plants through their root systems. Others do not compete well with soil saprophytes and rely on some other mode of transmission, such as an insect vector, to move from one host plant to another. Many plant diseases are caused by bacteria that are carried on or within seeds and infect the plant after the seed germinates. Plant pathogens are widely distributed in nature and cause a variety of plant diseases. Some of the more common pathogens and the diseases they cause are listed in Table 5.7.

TABLE 5.7 Representative bacteria that frequently cause disease in plants.

Bacterial pathogen	Host plant	Disease or symptom
Agrobacterium tumefaciens	Various plants	Crown gall
Erwinia amylovora	Apple, pear	Fire blight
Erwinia carotovora	Chrysanthemum	Blight
Erwinia tracheiphila	Cucumber, melon	Vascular wilts
Mycoplasma species	Peach	X-disease, peach yellows
Mycoplasma species	Elm	Phloem necrosis
Pseudomonas syringae pathovars	Diverse trees/crops	Blights, cankers, blisters
Pseudomonas marginalis	Onion	Slippery skin
Pseudomonas syringae pv. *phaseolicola*	Bean	Halo blight
Pseudomonas cichorii	Vegetables/ornamentals	Leaf spot
Streptomyces ipomoeae	Sweet potato	Sweet potato pox
Streptomyces scabies	Potato	Potato scab
Xanthomonas campestris	Crucifers (e.g., cabbage)	Black rot
Xanthomonas citri	Citrus	Canker
Xanthomonas oryzae	Rice	Blight
Xanthomonas pruni	Fruits	Leaf spot

Adapted from table by David Alexander, second ed.

Interactions with invertebrates

Symbiotic interactions between prokaryotic microbes and invertebrate animals are common in nature. Like all living things, invertebrates are colonized by multitudes of microorganisms that adhere to the surfaces of tissues on or within the body of the animal. Invertebrates may also be attracted to the presence specific bacteria through sensing volatile chemicals produced by the microorganisms (Box 5.3). Interestingly, many invertebrates also harbor endosymbionts, which are bacteria or other microbes that live inside the animal's body, often in specialized cells. Relatively little is known about these endosymbionts because few have been cultured successfully outside of their invertebrate host. Microscopic observations of the tissues in which they reside have enabled biologists to study the morphology of these microbes, and research has shown that invertebrates that harbor endosymbionts often depend on the microbes for their survival. When the animals are treated with an antibiotic or other antimicrobial agent that kills the endosymbiont, the animals lose the ability to reproduce and often die prematurely. Antibiotics have no direct effect on invertebrates, so that loss of the endosymbiont is presumably responsible for the death of the host.

Advances in methods to extract DNA and analyze gene sequences from non-culturable bacteria have greatly enhanced microbiologists' ability to understand the interactions between these endosymbionts and their invertebrate hosts. Probably the best characterized of these interactions is the symbiosis of aphids with *Buchnera aphidicola*, a Gram-negative, oval-shaped bacterium that lives in specialized cells called bacteriocytes in the body of the insect (Baumann et al., 1995). Aphids are sucking insects that feed on plant sap, a source of nutrients rich in carbohydrates but deficient in amino acids and vitamins that the insect needs to grow and reproduce. Aphids lack the ability to synthesize 10 "essential" amino acids, but the endosymbiont, *B. aphidicola*, contains the genes needed to synthesize most, if not all, of these essential nutrients. Supplementing an aphid's diet with the required amino acids enhances its ability to survive without *B. aphidicola*, strongly suggesting that one function of this endosymbiont is to provide its host with essential amino acids.

Several other interesting symbioses between bacteria and insects or other invertebrates are known to exist. Termites, for instance, serve as hosts for N_2-fixing bacteria that supplement the nitrogen-deficient diet of the insects. Lice harbor endosymbionts that provide vitamin B and other essential nutrients for their host. Acetogenic bacteria may also play an important role in the nutrition of termites. Acetogens are bacteria that reduce CO_2 to acetate when they use CO_2 as the terminal electron acceptor of anaerobic respiration. Some of these bacteria live in the hindguts of termites, where they convert the CO_2 and H_2 produced by cellulose fermentation into acetate, which can then be absorbed and oxidized by the insect. Many more intriguing interactions between bacteria and invertebrates may soon be discovered as microbiologists develop new methods of studying bacteria in their natural habitats.

Several species of bacteria are capable of producing disease in invertebrates. Some insect pathogens have been used successfully for **biological control** of crop pests. The best known and most widely used are *Bacillus thuringiensis* and *Bacillus sphaericus*. Both species are Gram-positive, endospore-forming bacilli that produce crystalline, insecticidal proteins which have been used to control a variety of insect pests, including silkworm, gypsy moth, tomato hornworm, and many others. Spore products prepared from two other *Bacillus* spp., *B. popilliae* and *B. lentimorbus*, have played a major role in controlling Japanese beetle infestations in the United States, where there are few natural antagonists to control the beetle populations.

Soil-borne pathogens of humans

Soil can serve as a reservoir for microbes that infect humans, although relatively few bacterial pathogens can survive in the soil environment. In this context, a reservoir is a place where disease-causing microbes reside when they are not living in a human host. Soil is a vastly different environment than the human body, so few microbes are capable of surviving in both types of habitats. The soil environment is characterized by a relatively sparse supply of organic nutrients and wide fluctuations in temperature, aeration, and moisture. The human body, on the other hand, offers a warm, moist, nutrient-rich environment, but microbes must be well adapted to contend with a barrage of host defenses designed to eliminate them and protect the body from infection.

The most significant bacterial pathogens that are capable of surviving in the soil environment are listed in Table 5.8, along with the diseases that they cause. All but one of these organisms are endospore-forming bacteria, and the ability to produce endospores contributes to the manner in which they cause disease. *Bacillus anthracis* rarely infects humans under normal circumstances, but it is of great concern because of its potential use as a biological weapon (Box 5.5). *Clostridium tetani* and *Clostridium perfringens* are endospore-forming, anaerobic bacilli that can cause serious infections when soil containing their endospores contaminates deep wounds. *Clostridium botulinum* is another spore-forming anaerobe which causes a deadly form of food-poisoning (botulism). Botulism is usually associated with home-canned, low-acid foods. Spores of *C. botulinum* are widespread in nature and are often present on vegetables and other food crops. If the endospores are not destroyed by heat and pressure during the canning process, they can germinate in the anaerobic environment inside a sealed canning jar. The vegetative bacteria then produce a lethal toxin as they grow in the food.

TABLE 5.8 Soilborne bacteria that are capable of causing disease in humans.

Bacterial pathogen	Description	Disease
Bacillus anthracis	Gram-positive, facultative anaerobe, spore-forming bacillus	Anthrax
Clostridium botulinum	Gram-positive, anaerobic, spore-forming bacillus	Botulism
Clostridium perfringens	Gram-positive, anaerobic, spore-forming bacillus	Gas gangrene
Clostridium tetani	Gram-positive, anaerobic, spore-forming bacillus	Tetanus
Pseudomonas aeruginosa	Gram-negative, aerobic, non-spore-forming bacillus	Infections of the skin, eyes, ears, respiratory tract, and bloodstream

BOX 5.5

Weaponizing anthrax bacteria

In the fall of 2001, an unknown assailant sent letters and packages laced with highly refined *Bacillus anthracis* endospores through the US postal system. Twenty-two people who handled the letters or other materials contaminated with the deadly endospores became infected, five of whom died. Some exposed individuals developed cutaneous anthrax, a less severe disease than the deadly inhalational anthrax. The highly virulent strain of *B. anthracis* used in this bioterrorist act was known as the "Ames strain," because it was thought to have been originally isolated by agricultural scientists in Ames, Iowa. When federal investigators traced the history of the strain in an effort to identify the perpetrator of the attacks, they discovered something surprising—the Ames strain had not originated in Iowa, but in Texas. In 1981, a south Texas rancher had sent samples of infected cow tissue from an animal that he suspected of dying from anthrax to a diagnostic laboratory in College Station, TX. After confirming that the tissues contained *B. anthracis*, the Texas laboratory sent two vials of the anthrax cultures to a US military research laboratory in Fort Detrick, MD, where they were added to a collection of anthrax stains for research purposes. Because the vials were sent in a container bearing a mailing label with the return address of the National Veterinary Services Laboratory in Ames, IA, these cultures were assumed to have been isolated at the Ames laboratory and came to be known as the "Ames strain." It now appears that the strain is naturally and widely distributed in Texas soils, and it is likely to have caused several outbreaks of anthrax in livestock and wild animal populations in Texas. Scaling up its production in a laboratory enables large quantities of spores to be produced and then milled into a fine powder. This process of weaponization makes the agent easy to disseminate and difficult to detect. Because of the danger inherent in its handling, *B. anthracis* is classified as a Biosafety Level-3 (BSL-3) biological agent and, as such, requires that strict protective and containment practices be adhered to at all times by those working with cultures or other viable preparations.

A sobering demonstration of the potential consequences of using *B. anthracis* as a biological weapon can be found in the history of Gruinard Island, a small 196-hectare island off Scotland's west coast. In 1942, during the Second World War, tests were conducted to assess the efficacy of the bacterium in causing disease when aerially dispersed. Sheep were brought to the island as test subjects and exposed to *B. anthracis* as endospores spread by explosive charges, essentially "germ bombs". The sheep died within days following exposure. It was concluded that comparable bombs detonated over German cities would have similar effects on humans and render the affected cities uninhabitable for decades due to the persistence of endospores. After the tests, Gruinard Island was deemed uninhabitable and quarantined against human visitation indefinitely. In response to what were essentially domestic terrorist threats to be carried out if decontamination steps were not taken, the entire island was decontaminated in 1986 by the application of 280 metric tons of formaldehyde and localized excavation and removal of highly contaminated topsoil. Subsequently, sheep were returned to the island and remained healthy. Four years later, in 1990, Gruinard Island was de-quarantined and again opened for human habitation.

The only non-spore-forming bacterium among the soil-borne pathogens listed in Table 5.8 is *Pseudomonas aeruginosa*. This organism causes a variety of infections, typically in patients whose defenses have been weakened by a pre-existing illness or injury. These infections can be very serious and are exceptionally difficult to treat because of the pseudomonad's innate resistance to many antibiotics. It is possible that *P. aeruginosa* owes its remarkable resistance to its natural habitat, soil, where it has coexisted for thousands of years with antibiotic-producing actinomycetes and fungi.

Quantifying bacterial populations

Several methods can be used to estimate the number of bacteria or other microbes in a soil community. Culture-dependent methods involve adding diluted soil suspensions to nutrient media and providing favorable conditions for growth of organisms during an incubation period of suitable length. Culture-independent methods involve estimating cell numbers without growing microbes. The latter methods include direct microscopic counting of fluorescence-stained cells captured on microbiological filters from diluted soil suspensions and quantification of marker genes in microbial community DNA extracts from soils. Each method has certain advantages and limitations. The method chosen depends upon many factors, including the type of microbes studied and the microbes' nutritional and cultural requirements.

Dilution plate count method

This method is often used to enumerate culturable microbial populations in soils. Many soil bacteria have relatively simple nutritional requirements and grow readily on sterile nutrient agar media in Petri plates. The underlying principle of the **plate count method** is that each bacterial cell from a diluted soil sample will produce enough daughter cells on the solid culture medium to form a visible colony during incubation under appropriate conditions. By counting the number of colonies that appear on the medium following incubation, the number of culturable bacteria in the soil sample can be estimated. Bacteria often populate soils in very large numbers, typically ranging from 10^7 to $>10^9$ cells g^{-1} of soil. (Refer to Box 1.2 for additional information on scientific notation). Because it is not possible to isolate and accurately count such large numbers of organisms in a single culture plate, the cells extracted from a soil sample must first be diluted to more manageable numbers before inoculating onto the culture medium.

The bacteria in a soil sample can be rapidly diluted several million-fold with a simple technique known as serial dilution (Fig. 5.8). A liquid suspension of the cells in the soil sample is prepared by mixing 10 g of soil in 95 mL of a sterile diluent (water or a dilute buffer solution). This ratio of soil to diluent produces approximately 100 mL of soil suspension and dilutes the bacterial population by approximately one-tenth, that is, a 1/10 weight/volume dilution. This initial cell suspension is therefore designated a 10^{-1} dilution. The mixture is agitated vigorously to suspend the bacterial cells in the liquid, then allowed to settle briefly so that sand will sediment to the bottom of the dilution bottle.

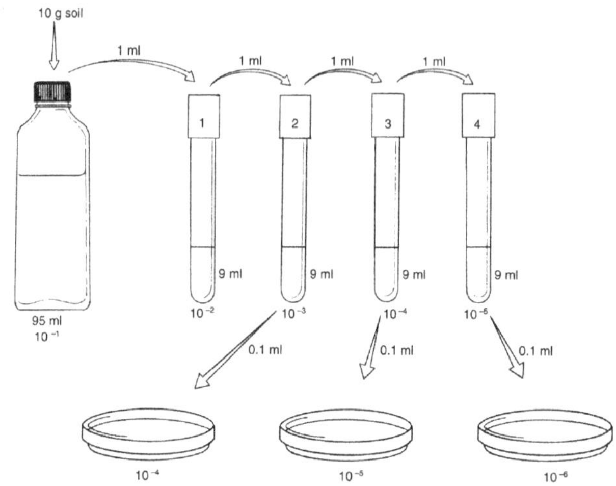

FIGURE 5.8 **The dilution spread plate method for estimating numbers of culturable bacteria.** A transfer of 1 mL soil suspension from the initial bottle to 9 mL sterile buffer in the first tube represents a tenfold dilution. The transfer of 0.1 mL from a dilution tube to a plate also represents a 10-fold dilution because the tube dilution value represents the amount of soil contained in a full milliliter of the suspension. All dilution steps must be taken into account when determining the total dilution factor. *Source: Original drawing by David Alexander. Used with permission.*

Subsequent dilutions are prepared by transferring a defined volume of the bottle cell suspension to test tubes containing defined volumes of sterile diluent. Transfers are made sequentially from tube to tube to obtain increasingly more diluted cell suspensions. In the example illustrated in Fig 5.8, a 10-fold dilution of the original cell suspension is prepared by transferring 1 mL of the cell suspension in the initial dilution bottle to 9 mL of sterile diluent in tube 1. The dilution ratio of this transfer is calculated as follows:

$$\begin{aligned}\text{Dilution ratio} &= \text{Volume transferred} / \text{total volume after the transfer}\\ &= 1\,\text{mL}/(9\,\text{mL}+1\,\text{mL})\\ &= 1/10 \text{ or } 10^{-1}\end{aligned}$$

Given that the suspension in the bottle used to disperse the soil represents a 10^{-1} dilution, tube 1 represents a 10^{-2} dilution of the soil sample. The resulting cell suspension is diluted another 10-fold by transferring 1 mL of the liquid in tube 1 to 9 mL of sterile diluent in tube 2 to obtain a 10^{-3} dilution. Additional 10-fold dilutions are performed to obtain cell concentrations suitable for counting.

At this point, either of two methods may be used to prepare solid culture media to which diluted cell suspensions (inocula) are added. **Pour plates** are prepared by mixing 1 mL of each cell suspension to be enumerated with 15–20 mL of a warm (45°C–50°C) liquid culture medium containing melted agar and pouring the mixtures into sterile Petri plates. After the culture medium solidifies, the plates are incubated, and microbial colonies can be counted as they become visible within the agar medium.

Spread plates are prepared by inoculating 0.1 mL of each cell suspension onto the surface of a sterile culture medium previously poured and solidified in a Petri plate. The suspensions are spread evenly over the surface of the medium with a sterile bent glass rod to disperse the bacterial cells so as to produce well-separated colonies. The colonies are counted after incubating the plates under conditions that allow the microorganisms to grow. Inoculating an agar plate with a 0.1-aliquot in this manner constitutes an additional 10^{-1} dilution of the cells in the suspension. The total dilutions of plates inoculated with 0.1 mL of the cell suspensions in tubes 2, 3, and 4 in Fig. 5.8 would therefore be 10^{-4}, 10^{-5}, and 10^{-6}, respectively.

Soil microbiologists routinely inoculate replicate plates with aliquots from several different cell dilutions and use the average number of colonies on the plates to calculate the number of bacteria in the original soil sample. Plates that yield fewer than 30 or more than 300 colonies are not counted because calculations based on colony counts outside this range are not considered accurate.

The number of cells per gram of soil is calculated by multiplying the average number of colonies by the reciprocal of the total dilution of the plates. For example, if the plates inoculated with 0.1-mL aliquots from tube 3 using the spread plate technique yielded an average of 127 colonies per plate, then the number of cells in the soil sample would be 127×10^4 or 1.27×10^6 cells g^{-1}. Because plate counts are inherently imprecise, some organizations recommend microbial counts be rounded to two significant figures, one before and one after the decimal point, and reported as 1.3×10^6 cells g^{-1}. Note that this estimate is based on grams of moist soil. If counts are to be reported per gram of dry soil, the weight of the moist soil added to the first dilution bottle can be corrected for moisture content. This is done by taking a separate subsample of soil and placing it in an oven for 24–48 h at 105°C to determine oven-dry weight. The calculated weight of dry soil can then be used to adjust the dilution factor in the first dilution step (Fig. 5.8). Correcting microbial counts for moisture content may be relevant if soil and sediment samples are particularly wet at the time of plating.

Usefulness and limitations of the dilution plate method

Soil microbiologists routinely use the dilution plate method to enumerate bacterial populations because the method is simple and reproducible and does not require expensive equipment of supplies. It also enables the microbiologist to count only viable cells (cells that are capable of growing and producing a visible colony) and yields isolated colonies which can be subcultured for further study. Direct microscopic counts often fail to distinguish between viable cells and dead cells, while biochemical detection methods do not lead to the isolation of pure cultures. Estimates of numbers do not necessarily provide information on microbial activity occurring in situ in soil. For example, high numbers of *Bacillus* species may have been derived from dormant endospores in soil.

Compared with other counting methods, however, plate counts often underestimate the actual numbers of viable microorganisms in soils (Amann et al., 1995). Several factors contribute to the relatively low numbers of bacteria recovered with the plant count methods. Only those species that can grow using the nutrients contained in the culture medium will produce a visible colony which can be counted, and no culture medium provides all of the essential nutrients that every species of organism might require for growth. Some organisms grow too slowly to form visible colonies by the end of the incubation period or are overgrown by fast-growing species. In addition, any incubation condition will be unavoidably restrictive. Obligate anaerobes cannot grow on plates that are incubated aerobically, and obligate aerobes cannot grow on plates that are incubated anaerobically. Incubation temperature and the pH of the culture medium may also restrict the growth of certain species and favor the growth of others. For these reasons and others, a plate count detects only a small proportion (typically 1%) of the total microbial community in a soil. Nonetheless it is useful method for determining the relative number of certain types of bacteria in different soils under different environmental conditions. For more information, regarding the dilution plate method, consult Zuberer (1994).

Soil microbiologists often want to count or isolate specific groups of microorganisms within a soil community, rather than studying the community as a whole. They do so with culture media that deliberately exclude or promote the growth of certain types of microbes. **Selective media** contain ingredients which allow certain types of microbes to grow while preventing the growth of others. For instance, a medium can be made selective for acid-tolerant microbes by lowering the pH. **Differential media** contain ingredients which make it possible to distinguish one type of microorganism from another. A pH indicator is often added to a culture medium to differentiate microorganisms that produce acidic metabolic products from those that produce neutral or alkaline products. **Enrichment media** contain ingredients that favor the growth of certain types of microbes over others. A medium which contains cellulose, and no other source of carbon, favors the growth of cellulose-degrading microorganisms over those that cannot decompose cellulose.

Summary

Bacteria and archaea are single-celled, prokaryotic microorganisms that inhabit soils and sediments throughout the world. Bacteria are often the most abundant cellular organisms in soils, outnumbered only by the acellular viruses that infect them. Archaea typically make up about 1% of the prokaryotes living in soil, but some archaea are also present in the microbiomes of higher organisms, and many archaeal species inhabit more extreme environments. Bacteria and archaea are much smaller than eukaryotic microorganisms, and they have a much simpler cellular structure. Their cells consist of cytoplasm that contains a nucleoid, ribosomes, and a variety of catalytic and regulatory proteins, all within a cell envelope that usually includes a cell wall. Many bacteria and archaea also possess a variety of other structures, such as capsules, pili, flagella, plasmids, and inclusions, and recent research has revealed widespread presence of subcellular organelles and extracellular vesicles. These specialized structures may help bacteria and archaea to survive in soils and to adapt to changing environmental conditions.

Despite their small size, bacteria and archaea are extremely versatile metabolically. As a group, they can grow in the presence or absence of oxygen, and many species can shift their metabolism from aerobic respiration to anaerobic respiration or fermentation in response to changing oxygen concentrations. Most are heterotrophs, feeding on nonliving organic matter in soils or forming symbiotic associations with plants, insects, or other soil microorganisms. Many other species are autotrophs, which use either light energy or chemical energy obtained from the oxidation of inorganic substrates to synthesize organic compounds with carbon dioxide as their primary source of carbon.

Through their metabolic activities, bacteria and archaea transform soil minerals and organic matter from one form to another and alter the availability of essential nutrients such as nitrogen, sulfur, and phosphorus for plants and other soil organisms to use. As a result, bacteria and archaea play central roles in organic matter decomposition, nutrient cycling, and soil formation.

At first glance, soil may appear to be a lifeless collection of weathered rocks and minerals, worn down to fine granules by wind and water and time. However, soil is anything but lifeless; it is a habitat teeming with living microorganisms that represent the soil microbiome. A single gram of soil may contain hundreds of millions of microorganisms representing several thousands of different species. It is not an exaggeration to assert that all of the "higher" organisms with which we are more familiar—the plants, mammals, birds, amphibians, reptiles, insects, and others—would cease to exist if not for the activities of the multitudes of invisible organisms in soil.

This chapter begins with a quotation from *Microcosmos* by Lynn Margulis and Dorion Sagan, which asserts that "We are all of us walking communities of bacteria" (Margulis and Sagan, 1986). It may be as difficult to perceive the essential functioning of bacteria in and on our own bodies as it is to be aware of all the vital processes being carried out by bacteria and archaea dwelling in soils. Without microorganisms to decompose dead plants and other organic wastes, higher organisms would be unable to re-capture the nutrients locked inside, and we humans would be unable to cope with the resulting accumulation of debris. Our understanding of these smallest of cellular organisms is being advanced rapidly by DNA-based and other molecular technologies to reveal more of the vast genetic and metabolic diversity which they possess.

Acknowledgment

Portions of this chapter are derived from the work of Dr. David B. Alexander (deceased) who authored a similar chapter for the Second Edition of the textbook. The current author and editors thank his family for permitting the use of his material and graciously supporting production of the Third Edition.

Study questions

1. How are bacteria and archaea similar? How are they different?
2. Identify common shapes and groupings of prokaryotic organisms when viewed under the microscope. Use correct scientific terms to describe cellular morphology.
3. Describe the general structure of a prokaryotic cell and explain the function of its main components.
4. Contrast the structure and chemical composition of Gram-positive and Gram-negative bacterial cell walls.
5. Explain three mechanisms by which cells take up soluble nutrients from the environment. Do all mechanisms require energy?
6. What are plasmids? How might plasmids affect ecological interactions and biogeochemical processes among bacteria in soil?

7. Contrast the types of energy metabolism exhibited by the following groups of bacteria: obligate aerobes, obligate anaerobes, facultative anaerobes, and aerotolerant anaerobes. What types of environment does each group inhabit in soils?
8. Several major phyla of bacteria are consistently found in greatest abundance in surface soils, based on ribosomal RNA gene analyses. List five of these phyla and describe some of their group characteristics.
9. How is the plate count method used to quantify microbial populations in soils? What are advantages and limitations of this method?

References

Aanderud, Z.T., Jones, S.E., Fierer, N., Lennon, J.T., 2015. Resuscitation of the rare biosphere contributes to pulses of ecosystem activity. Front. Microbiol. 6, 1–11.

Adam, P.S., Borrel, G., Brochier-Armanet, C., Gribaldo, S., 2017. The growing tree of Archaea: new perspectives on their diversity, evolution and ecology. ISME J. 11, 2407–2425.

Albers, S.-V., Meyer, B.H., 2011. The archaeal cell envelope. Nat. Rev. Microbiol. 9, 414–416.

Alexander, D., 2002. Soil bacteria. In: Bitton, G. (Ed.), Encyclopedia of Environmental Microbiology. John Wiley & Sons, Inc, New York.

Amann, R.I., Ludwig, W., Schleifer, K.-H., 1995. Phylogenetic identification and in situ detection of individual microbial cells without cultivation. Microbiol. Rev. 59, 143–169.

Bar-On, Y.M., Phillips, R., Milo, R., 2018. The biomass distribution on Earth. Proc. Natl. Acad. Sci. 115, 6506–6511, https://doi.org/10.1073/pnas.1711842115.

Baumann, P., Lai, C.-Y., Baumann, L., Rouhbakhsh, D., Moran, N.A., Clark, M.A., 1995. Mutualistic associations of aphids and prokaryotes: biology of the genus *Buchnera*. Appl. Environ. Microbiol. 61, 1–7.

Becher, P.G., Verschut, V., Bibb, M.J., Bush, M.J., Molnár, B.P., Barane, E., Al-Bassam, M.M., Chandra, G., Song, L., Challis, G.L., Buttner, M.J., Flärdh, K., 2020. Developmentally regulated volatiles geosmin and 2-methylisoborneol attract a soil arthropod to *Streptomyces* bacteria promoting spore dispersal. Nat. Microbiol. 5, 821–829.

Blankenship, R.E., Madigan, M.T., Bauer, C.E., 1995. Anoxygenic Photosynthetic Bacteria. Kluwer Academic, Dordrecht, Netherlands.

Borrel, G., Brugère, J.-F., Gribaldo, S., Schmitz, R.A., Moissl-Eichinger, C., 2020. The host-associated archaeome. Nat. Rev. Microbiol. 18 (11), 622–636, https://doi-org.ezaccess.libraries.psu.edu/10.1038/s41579-020-0407-y.

Bruns, M.A., 2017. Bacteria. Lal, R. (Ed.), Encyclopedia of Soil Science, third ed. Taylor & Francis. doi: 10.1081/E-ESS3-120053846.

Burns, D.G., Janssen, P.H., Itoh, T., Kamekura, M., Li, Z., Jensen, G., Rodriguez-Valera, F., Bolhuis, H., Dyall-Smith, M.L., 2007. *Haloquadratum walsbyi* gen. nov., sp. nov., the square haloarchaeon of Walsby, isolated from saltern crystallizers in Australia and Spain. Intl. J. Syst. Evol. Microbiol. 57, 387.

Delgado-Baquerizo, M., Oliverio, A.M., Brewer, T.E., Benavent-González, A., Eldridge, D.J., Bardgett, R.D., Maestre, F.T., Singh, B.K., Fierer, N., 2018. A global atlas of the dominant bacteria found in soil. Science 359, 320–325.

Elkins, J.G., Podar, M., Graham, D.E., Makarova, K.S., Wolf, Y., Randau, L., Hedlund, B.P., Brochier-Armenet, C., Kunin, V., Anderson, I., Lapidus, A., Goltsman, E., Barry, K., Koonin, E.V., Hugenholtz, P., Kyrpides, N., Wanner, G., Richardson, P., Keller, M., Stetter, K.O., 2008. A korarchaeal genome reveals insights into the evolution of the Archaea. Proc. Natl. Acad. Sci. 105, 8102–8107.

Fan, X.-T., Li, H., Chen, Q.-L., Zhang, Y.-S., Ye, J., Zhu, Y.-G., Su, J.-Q., 2019. Fate of antibiotic resistant *Pseudomonas putida* and broad host range plasmid in natural soil microcosms. Front. Microbiol. 10, 194. doi: 10.3389/fmicb.2019.00194.

Fierer, N., Leff, J.W., Adams, B.J., Nielsen, U.N., Bates, S.T., Lauber, C.L., Owens, S., Gilbert, J.A., Wall, D.H., Caporaso, J.G., 2012. Cross-biome metagenomics analyses of soil microbial communities and their functional attributes. Proc. Natl. Acad. Sci. 109, 21390–21395.

Galperin, M.Y., Makarova, K.S., Koonin, E.V., 2018. Phyletic distribution and lineage-specific domain architectures of archaeal two-component signal transduction systems. J. Bacteriol. 200 (3), e00681–e00617.

Gill, S., Catchpole, R., Forterre, P., 2019. Extracellular membrane vesicles in the three domains of life and beyond. FEMS Microbiol. Rev. 43, 273–303.

Goodfellow, M., Williams, S.T., 1983. Ecology of actinomycetes. Ann. Rev. Microbiol. 37, 189–216.

Gorbushina, A.A., Kort, R., Schulte, A., Lazarus, D., Schnetger, B., Brumsack, H.J., Broughton, W.J., Favet, J., 2007. Life in Darwin's dust: intercontinental transport and survival of microbes in the nineteenth century. Environ. Microbiol. 9, 2911–2922.

Green, E.R., Mecsas, J., 2016. Bacterial secretion systems—an overview. Microbiol. Spect. 4 (1), doi: 10.1128/microbiolspec.VMBF-0012-2015.

Greening, C., Lithgow, T., 2020. Formation and function of bacterial organelles. Nat. Rev. Microbiol 18, 677–689, https://doi-org.ezaccess.libraries.psu.edu/10.1038/s41579-020-0413-0.

Hoehler, T., Jørgensen, B., 2013. Microbial life under extreme energy limitation. Nat. Rev. Microbiol. 11, 83–94.

Hug, L.A., Baker, B.J., Anantharaman, K., Brown, C.T., Probst, A.J., Castelle, C.J., Butterfield, C.N., Hernsdorf, A.W., Amano, Y., Ise, K., Suzuki, Y., Dudek, N., Relman, D.A., Finstad, K.M., Amundson, R., Thomas, B.C., Banfield, J.F., 2016. A new view of the tree of life. Nat. Microbiol. 1, 16048. doi: 10.1038/NMICROBIOL.2016.48.

Jarrell, K.F., McBride, M.J., 2008. The surprisingly diverse ways that prokaryotes move. Nat. Rev. Microbiol. 6, 466–476.

Kandel, S.L., Joubert, P.M., Doty, S.L., 2017. Bacterial endophyte colonization and distribution within plants. Microorganisms 5, 77. doi: 10.3390/microorganisms5040077.

Kerfeld, C.A., Aussignargues, C., Zarzycki, J., Cai, F., Sutter, M., 2018. Bacterial microcompartments. Nat. Rev. Microbiol. 16 (5), 277–290. doi: 10.1038/nrmicro.2018.10.

Koch, A.L., 1998. The biophysics of the Gram-negative periplasmic space. Crit. Rev. Microbiol. 24, 23–59.

Könneke, M., Schubert, D.M., Brown, P.C., Hüglerd, M., Standfest, S., Schwandere, T., von Borzyskowskie, L.S., Erbe, T.J., Stahl, D.A., Berg, I.A., 2014. Ammonia-oxidizing archaea use the most energy efficient aerobic pathway for CO_2 fixation. Proc. Natl. Acad. Sci. 111, 8239–8244.

Konstantinidis, K.T., Tiedje, J.M., 2005. Genomic insights that advance the species definition for prokaryotes. Proc. Natl. Acad. Sci. 102, 2567–2572.

Krieg, N.R., Garrity, G.M., 2015. On Using the Manual. Bergey's Manual of Systematics of Bacteria and Archaea. John Wiley & Sons, https://doi.org/10.1002/9781118960608.bm00019.

Kulp, A., Kuehn, M.J., 2010. Biological functions and biogenesis of secreted bacterial outer membrane vesicles. Annu. Rev. Microbiol. 64, 163–184. doi: 10.1146/annurev.micro.091208.073413.

Liang, R., Lau, M., Vishnivetskaya, T., Lloyd, K.G., Wang, W., Wiggins, J., Miller, J., Pfiffner, S., Rivkina, E.M., Onstott, T.C., 2019. Predominance of anaerobic, spore-forming bacteria in metabolically active microbial communities from ancient Siberian permafrost. Appl. Environ. Microbiol. 85, e00560-19.

Margulis, L., Sagan, D., 1986. Microcosmos: Four Billion Years of Evolution from Our Microbial Ancestors, University of California Press, Berkeley, CA, p. 191.

Morono, Y., Ito, M., Hoshino, T., Terada, T., Hori, T., Ikehara, M., D'Hondt, S., Inagaki, F., 2020. Aerobic microbial life persists in oxic marine sediment as old as 101.5 million years. Nat. Comm. 11, 3626, https://doi.org/10.1038/s41467-020-17330-1.

Neilsen, J.W., Josephson, K.L., Pepper, I.L., Arnold, R.B., DiGiovanni, G.D., Sinclair, N.A., 1994. Frequency of horizontal gene transfer of a large catabolic plasmid (pJP4) in soil. Appl. Environ. Microbiol. 60, 4043–4058.

Ng, S.Y.M., Zolghadr, B., Driessen, A.J.M., Albers, S.-V., Jarrell, K.F., 2008. Cell surface structures of archaea. J. Bacteriol. 190, 6039–6047.

Peeters, E., Driessen, R.P.C., Werner, F., Dame, R.T., 2015. The interplay between nucleoid organization and transcription in archaeal genomes. Nat. Rev. Microbiol. 13, 333–341.

Poomthongdee, N., Duangmal, K., Pathom-aree, W., 2015. Acidophilic actinomycetes from rhizosphere soil: diversity and properties beneficial to plants. J. Antibiotics 68, 106–114.

Rodriguez-R, L.M., Castro, J.C., Kyrpides, N.C., Cole, J.R., Tiedje, J.M., Konstantinidis, K.T., 2018. How much do rRNA gene surveys underestimate extant bacterial diversity? Appl. Environ. Microbiol. 84 (6), e00014–e0018.

Schäfer, G., Engelhard, M., Müller, V., 1999. Bioenergetics of the Archaea. Microbiol. Molec. Biol. Rev. 63, 570–620.

Stewart Johnson, S., Hebsgaard, M.B., Christensen, T.R., Mastepanov, M., Nielsen, R., Munch, K., Brand, T., Gilbert, M.T.P., Zuber, M.T., Bunce, M., Rønn, R., Gilichinsky, D., Froese, D., Willerslev, E., 2007. Ancient bacteria show evidence of DNA repair. Proc. Natl. Acad. Sci. 104, 14401–14405.

Stock, A.M., Robinson, V.L., Goudreau, P.N., 2000. Two-component signal transduction. Annu. Rev. Biochem. 69, 183–215.

Szabo, Z., Pohlschroder, M., 2012. Diversity and subcellular distribution of archaeal secreted proteins. Front. Microbiol. 3, 207.

Tashiro, Y., Uchiyama, H., Nomura, N., 2012. Multifunctional membrane vesicles in *Pseudomonas aeruginosa*. Environ. Microbiol. 14, 1349–1362.

Tecon, R., Or, D., 2016. Bacterial flagellar motility on hydrated rough surfaces controlled by aqueous film thickness and connectedness. Sci. Rep. 6, 19409. doi: 10.1038/srep19409.

Thanbichler, M., Viollier, P.H., Shapiro, L., 2005. The structure and function of the bacterial chromosome. Curr. Opinion Genet. Dev. 15, 153–162.

Wang, H., Peng, N., Shah, S.A., Huang, L., She, Q., 2015. Archaeal extrachromosomal genetic elements. Microbiol. Molec. Biol. Rev. 79, 117–152.

Wheatley, R.M., Poole, P.S., 2017. Mechanisms of bacterial attachment to roots. FEMS Microbiol. Rev. 42 (2018), 448–461.

Whitman, W.B., 2015. Bergey's Manual of Systematics of Archaea and Bacteria. John Wiley & Sons, Inc., in association with Bergey's Manual Trust, doi: 10.1002/9781118960608.bm00022.

Whitman, W.B., Coleman, D.C., Wiebe, W.J., 1998. Prokaryotes: the unseen majority. Proc. Natl. Acad. Sci. 95, 6578–6583.

Wilhelm, M.J., Sheffield, J.B., Sharifian, G.H., Gh, M., Wu, Y., Spahr, C., Gonella, G., Xu, B., Dai, H.-L., 2015. Gram's stain does not cross the bacterial cytoplasmic membrane. ACS Chem. Biol. 10, 1711–1717.

Woese, C.R., Kandler, O., Wheelis, M.L., 1990. Towards a natural system of organisms: proposal for the Domains Archaea, Bacteria, and Eucarya. Proc. Natl. Acad. Sci. 87, 4576–4579.

Wu, D., Hugenholtz, P., Mavromatis, K., Pukall, R., Dalin, E., Ivanova, N.N., Kunin, V., Goodwin, L., Wu, M., Tindall, B.J., Hooper, S.D., Pati, A., Lykidis, A., Spring, S., Anderson, I.J., D'haeseleer, P., Zemla, A., Singer, M., Lapidus, A., Nolan, M., Copeland, A., Han, C., Chen, F., Cheng, J.F., Lucas, S., Kerfeld, C., Lang, E., Gronow, S., Chain, P., Bruce, D., Rubin, E.M., Kyrpides, N.C., Klenk, H.P., Eisen, J.A., 2009. A phylogeny-driven genomic encyclopedia of Bacteria and Archaea. Nature 462, 1056–1060.

Zuberer, D.A., 1994. Recovery and enumeration of viable bacteria. Weaver, R.W., Angle, S., Bottomley, P., Bezdicek, D., Smith, S., Tabatabai, A., Wollum, A. (Eds.), Methods of Soil Analysis, Part 2. Microbiological and Biological Properties, vol. 5, Soil Science Society of America Book Series, Madison, WI, pp. 119–144.

CHAPTER

6

Fungi

Joseph B. Morton

West Virginia University, Morgantown, WV, United States

"I thought a forest was made up entirely of trees, but now I know that the foundation lies below ground, in the fungi."—**Derrick Jensen.**

When the word "fungus" is brought up in conversation, most people think of mushrooms because they have eaten them, marveled at their appearance on forest floors, or have had their imagination fired by legends, myths, and psychedelic properties. The vast majority of fungi are not as visible but no less important. All are involved in a multitude of activities, many of which include decay processes that contribute to Earth as a natural recycling bin. Fungi are unique among microbes in having a filamentous body that promotes long distance foraging for food, the formation of intricate networks and linkages, and indefinite growth with the potential for immortality. For many fungi, batteries of enzymes are released that are able to break down complex molecules to simpler forms, and the filamentous network absorbs these materials for growth and reproduction. While the fungal body may appear simple in structure, it is highly adaptable genetically and physiologically. Not only does a fungus grow locally, but it also produces prolific spores with a wide range of morphologies that can disperse long distances on air currents. With this combination of traits, fungi are able to colonize almost any habitat or organic substrate. As pathogens, they cause diseases in animals, plants, and other microbes. As beneficial symbionts, they help plants and animals to grow better or survive longer. In highly fertile agricultural soils, fungal biomass and activity can equal or exceed that of bacteria; in many natural ecosystems they are even more abundant and can reach ratios of 4:1 (Ananyeva et al., 2006). As cellulose degraders, they are essential players in improving soil quality by mediating organic matter turnover and aggregation. Long ago humans recognized that fungi were important as a food source, for additives such as yeast in bread making and fermented beverages, and for medicinal purposes. They also have a darker side, with some containing poisons and others producing harmful metabolic byproducts known as mycotoxins. At least 70,000 species have been described, but high-throughput DNA sequencing approaches suggest estimates of 5.1 million species or more (Blackwell, 2011). Clearly, fungi are a vital component to ecosystem functionality and vitality, and this chapter explores how they are classified, their biology, and their role in ecosystem processes.

Principles and Applications of Soil Microbiology. http://dx.doi.org/10.1016/B978-0-12-820202-9.00006-X
Copyright © 2021 Elsevier Inc. All rights reserved.

Cell structure

Fungal organisms are **eukaryotic**, in contrast to prokaryotic bacteria discussed in Chapter 5. Membrane-bound organelles present in each cell (Fig. 6.1) are similar to those of insects, **plants**, and animals, but with some important differences. The network of endoplasmic reticulum in fungi is not as extensive, and it has fewer connections with the cell membrane. The membrane system generates many secretory vesicles that substitute for Golgi bodies found in more complex eukaryotes. These vesicles are important in fungi for transport of structural and enzymatic molecules to active metabolic regions near or at the tips of tubular filaments called **hyphae** (singular, **hypha**) that elongate and branch indefinitely to form a **mycelium**.

Fungi do not contain chlorophyll. They have been considered plant-like for a long time, however, because of rigid cell walls, absence of motility, and production of **spores** that are analogous to plant seeds. However, spores lack embryos. Accordingly, Latin binomial names (Chapter 1) of fungi still are based on the Botanical Code of Nomenclature. Expanded morphological and genetic evidence now indicate that fungi are more closely related to animals. They are believed to have evolved about 1 billion years ago as a group of equal rank to plants and animals. Table 6.1 summarizes properties unique to fungi as well as those shared with plants, animals, or both.

Structure and function, with reference to classification

Fungi possess a multitude of traits that can be compared to understand their evolution, which in turn provides the biological basis for organizing them into taxonomic groups. These traits include morphology; chemistry of cell walls; storage carbohydrates and lipids; and sequences of nucleotides and amino acids of DNA and proteins, respectively. These data indicate that most filamentous microbes belong in the Kingdom Fungi (Alexopoulos et al., 1996).

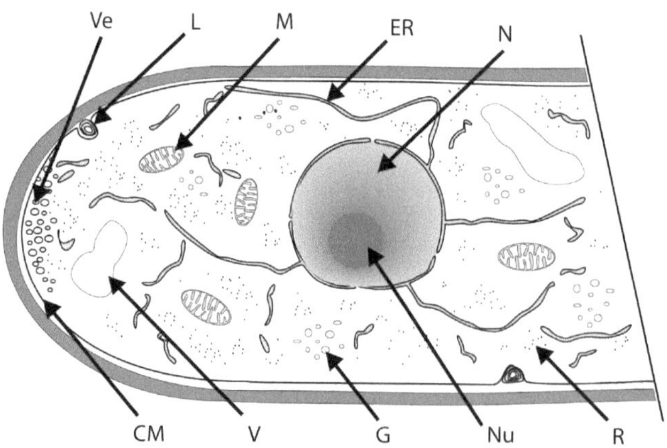

FIGURE 6.1 **General structure and organellar composition of a fungal cell at the apex of a hypha.** *ER*, endoplasmic reticulum; *G*, glycogen granules; *L*, lomasome; *M*, mitochondrion (plural, mitochondria); *N*, nucleus; *Nu*, nucleolus; *CM*, cytoplasmic membrane; *R*, ribosomes; *V*, vacuole; *VE*, vesicles.

TABLE 6.1 General properties of fungal organisms and their similarities to plants and animals.

Fungal property	Comparison with plants and animals
Eukaryotic organization of cells	Similar to animals and plants
Glycogen is a common storage polysaccharide	Similar to animals
Heterotrophic in requiring an external carbon source	Similar to animals, except carbon is obtained by absorption rather than ingestion
Rigid cell walls composed of chitin and non-cellulosic glucans, except in some Oomycota	Similar to some animal groups. Oomycota similar to plants with cellulose and glucans in cell walls
Life cycle includes both sexual and asexual stages	Similar to many plant and a few animal groups
Body (or thallus) made up of filamentous hyphae, except in Chytridiomycota and Oomycota	Similar to primitive plants (algae), except hyphae grow by extension from their tips
The amino acid lysine is produced from α-aminoadipic acid (AAA), except in Oomycota	Similar to the Euglenaceae (Chapter 7), differs from plants, bacteria, and archaea
Asexual and/or sexual spores produced in variable numbers	Similar to some plant groups, except each spore represents a haploid (1N) germ line. Unlike seeds with preformed embryos
Dominant steroid in the cytoplasm is ergosterol	Unique to fungi
Nuclei haploid for most of life cycle, except in Oomycota	Unique to fungi. Oomycota similar to plants and animals in having diploid nuclei
Hyphae often multinucleate and nuclei are migratory	Unique to fungi and Oomycota

Although not true fungi, organisms in the phylum Oomycota traditionally have been treated as fungi because of their filamentous body, formation of similar spore types, and behavior as plant pathogens (Box 6.1).

Advances in genomic analyses have provided vast amounts of data to improve our understanding of evolutionary relationships defining natural classification. At present, this evidence has been used to propose eight phyla and numerous subphyla (Spatafora et al., 2017). Some controversy exists regarding phyla of "lower fungi" because many of them are hard to collect and analyze, so sample sizes are small. The groups most important for this topic are summarized in Box 6.2. "Higher fungi" encompass two well defined and characterized phyla: Ascomycota and Basidiomycota (Box 6.3).

The vegetative body of a fungus is called its **thallus**, and it takes three possible forms:

- **chytrid cells**—solitary globose cells with or without specialized root-like filaments called **rhizoids** (Fig. 6.2A). They are unique to organisms in the phylum Chytridiomycota (Alexopoulos et al., 1996).
- **yeast cells**—spherical to ovoid cells formed by fungi in the phyla Ascomycota and Basidiomycota. They divide by budding or fission (Fig. 6.2B). Many of these fungi are **polymorphic** and can change from a mycelium to a yeast form under conditions where penetration of a substrate is not needed to obtain nutrients, such as in aqueous environments or insect cavities.
- **mycelium**—a filamentous network of hyphae (Fig. 6.2C) that branch and grow only by apical (tip) extension; common to most fungi.

BOX 6.1

Filamentous organisms that aren't true fungi

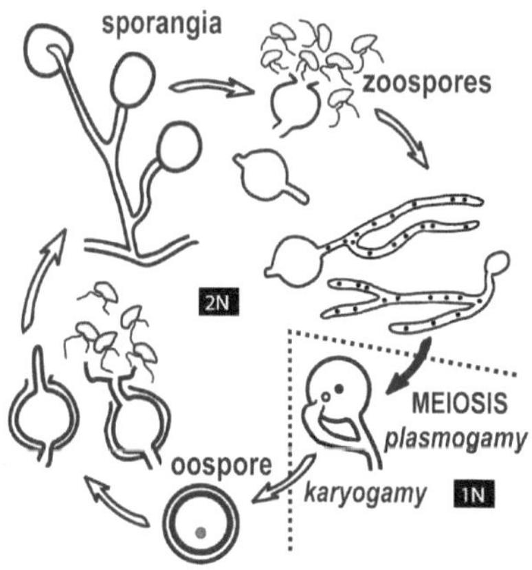

Life cycle of Oomycota.

The phylum **Oomycota** is comprised of filamentous organisms once considered fungi because of similarities in structure and behavior. Genomic differences, a dominantly diploid life cycle (compared to haploid fungi), and a cell wall consisting of cellulose and glycan instead of chitin combine to place them in the Kingdom Protista. These microbes usually are soil- and/or water-borne and have both sexual and asexual stages. Many are saprophytes and degrade organic matter. Some are serious plant pathogens, such as those causing late blight of potatoes (*Phytophthora infestans*), downy mildew of grapes (*Plasmopara viticola*), and damping off of a wide range of hosts (*Pythium* spp.). Their mycelium is coenocytic. Thick-walled **oospores** (teleomorph) are the resting stage, germinating directly under dry warm conditions or producing a vesicle whose contents differentiate into flagellated motile **zoospores** released in wet and cool environments. Asexual **sporangia** develop abundantly for widespread dispersal. Sporangia germinate directly under dry cool conditions, forming motile flagellated zoospores under warm wet conditions.

A mycelium is incredibly resilient because growth is open-ended for as long as a nutrient source is available. The result is that size can be highly variable, ranging from a pinpoint to a large mass covering acres of land. On an agar medium in Petri dishes or on other flat surfaces such as mulch, lawns, and human skin, a mycelium forms circular patches (Fig. 6.3A,B). In a liquid broth or solid medium where no surface boundaries confine growth, a mycelium can form spherical shapes resembling cotton balls (Fig. 6.3C). Soil is more heterogeneous, consisting of zones and microsites that can either benefit or hinder fungal growth. A mycelium provides the ideal mechanism to penetrate soil pores and branch in all directions. Interconnections between hyphae in this network establish a three-dimensional physical continuum capable of spanning nutrient-deficient zones, navigating around physical barriers or gas pockets in the soil matrix, and penetrating herbaceous and woody organic matter. This ability for directed growth sets fungi apart from most bacteria, which cannot do this. Hence, fungi can penetrate materials such as organic residues while most bacteria can only colonize surfaces and are thus more limited in the resources they can access.

The branching of hyphae around and within soil particles of all sizes impacts on soil structure by improving soil aggregation and aeration (Chapter 2). In addition, the mycelium of

BOX 6.2

Lower fungi

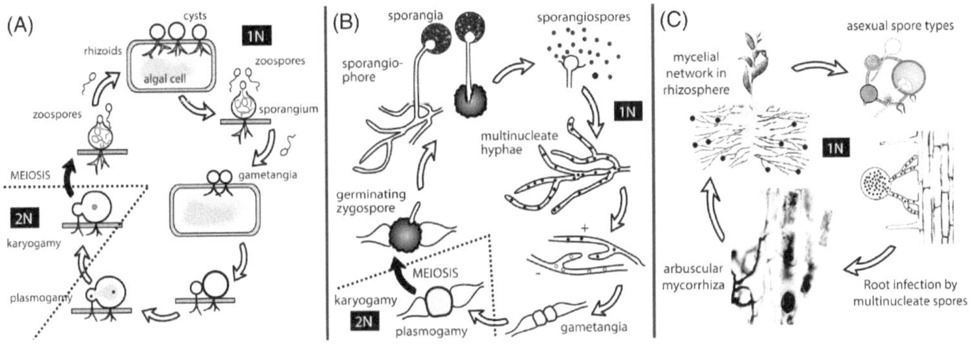

Life cycles of (A) Chytridiomycota, (B) Mucoromycota, and (C) Glomeromycotina.

Of the six phyla now recognized in this grouping, some are poorly understood because so little is known about them. Two are discussed here, in addition to the subphylum Glomeromycotina. All are characterized by a coenocytic mycelium, absent or poorly organized fruiting bodies, and a sexual cycle that is absent or of much shorter duration than the asexual cycle.

(A) Chytridiomycota: Soil- and water-borne; many are saprophytes that decay aquatic vegetation. They can be baited from streams and ponds using pollen grains, hemp seed, and other plant parts. Others are parasites of algae and insects. A few are symptomless plant pathogens. The thallus consists of single cells that vary with how host cells are colonized. Some form rudimentary coenocytic hyphae, **rhizomycelia**, which serve as a channel for movement of nuclei like true hyphae. Some develop totally within the host cell and others remain on the cell surface and are anchored by root-like **rhizoids**. Nuclei and cytoplasm in each sac-like **sporangium** become walled off to form **zoospores**. Each zoospore has a single whiplash flagellum for motility. They swim to new algal cells, penetrate cell walls, and encyst or produce sporangia and zoospores.

(B) Mucoromycota: Terrestrial; many are saprophytes in soil and dung and others are parasites of insects. The mycelium is coenocytic. For the sexual stage to occur, which is not common, two hyphae with genes of opposite mating type (+ and –) establish contact and fuse. A haploid nucleus of each mating type is compartmentalized in gametangia, and they fuse to form a diploid cell that develops into a **zygospore** (teleomorph). The zygospore serves as a resting stage. When it germinates, a **sporangium** develops. Haploid nuclei in each sporangium become partitioned to form **sporangiospores**, which are released into air currents. When conditions are good for growth, numerous sporangia

(*Continued*)

BOX 6.2 (Cont.)

and sporangiospores are produced asexually.

(C) **Glomeromycotina**: For a short time ranked as a phylum, with genomic data now placing these fungi in a subphylum within Mucoromycota. This group is unique in that they co-evolved exclusively with a majority of terrestrial plant lineages over 400 million years. Hyphae are coenocytic, and a variety of asexual spore types are multinucleate. Spores formed singly, in loose clusters, or in compacted aggregates (sporocarps). They are obligate mutualists in roots of >80% of all terrestrial plant species as **arbuscular endomycorrhizal fungi** (Chapter 12). They appear to be strictly asexual, although mating genes have been detected. They exhibit a very wide host range in virtually all habitats globally, but compatibility varies greatly among plant species.

BOX 6.3

Higher fungi

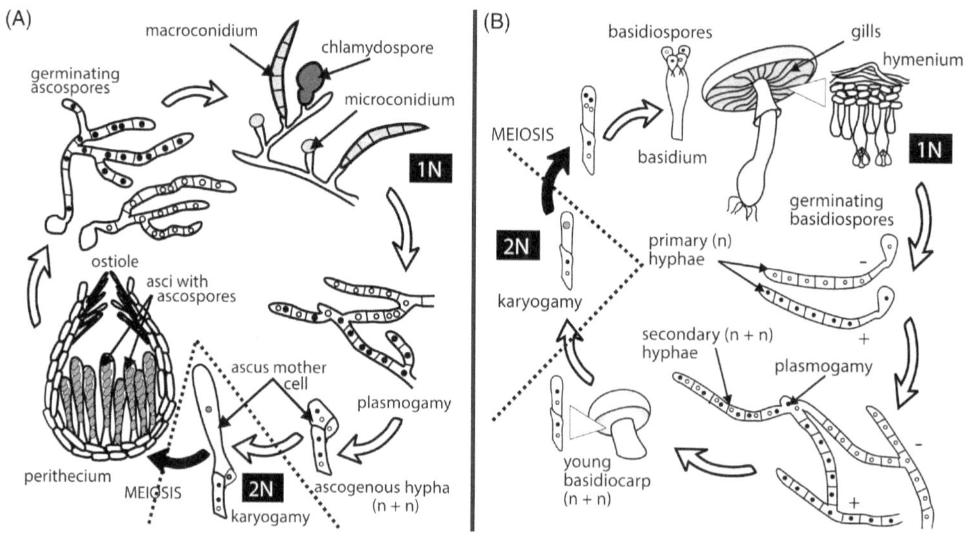

Life cycles of (A) Ascomycota and (B) Basidiomycota.

(Continued)

BOX 6.3 (*Cont.*)

The two phyla in this group are distinguished by a mycelium in which hyphae have cross-walls (septa) and usually are dikaryotic (two nuclei in each compartment) for much of their life cycle.

(A) **Ascomycota**: Terrestrial; many are saprophytes. Others are plant pathogens causing many foliar, root, and canker diseases (e.g., powdery mildew, ergot, chestnut blight, fusarium wilt). Yeasts are important in bread and alcohol production. Morels and truffles are prized delicacies. Most lichens are the result of associations between these fungi and algae or cyanobacteria. Numerous species form only large numbers of asexual spores (**conidia**) in nature, including soil inhabitants such as *Aspergillus*, *Fusarium*, and *Penicillium*. Conidia form in clusters for immediate release or as masses in fruiting bodies of different shapes. When a sexual stage occurs, compatible nuclei fuse to produce a zygote that undergoes meiosis and then one mitotic division to produce eight **ascospores** within a thin "sac" called an **ascus** (plural, **asci**). The asci often are enclosed within fruiting bodies of various shapes.

(B) **Basidiomycota:** Terrestrial; occupying a broad range of habitats. Many are saprophytes capable of degrading cellulose or lignin in soil, leaf litter, and wood. Others are pathogens causing tree root and crown rots and flower and foliar diseases, such as smuts and rusts. Some are beneficial ectomycorrhizal mutualists of trees. Mushroom fruiting bodies can serve as a food source, contain a deadly poison, or have psychedelic/healing properties. The mycelium is septate with complex clamp connections. Sexual reproduction occurs with fusion of compatible nuclei, meiosis, and production of four **basidiospores** on a **basidium** (plural, **basidia**). Basidia often are layered in fruiting bodies such as mushrooms, conks, puffballs, or jelly-like masses. Asexual reproduction is varied, with many forming chlamydospores or mycelial structures (sclerotia, rhizomorphs, mycelial strands) and some forming conidia.

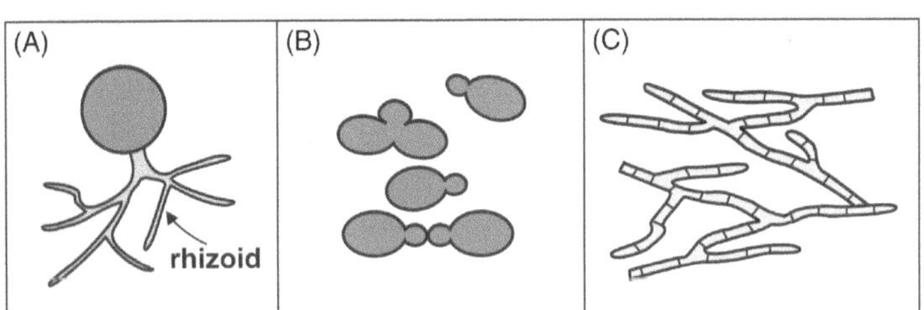

FIGURE 6.2 **Types of cells that make up the vegetative body (thallus) of fungal organisms.** (A) Chytrid cells, (B) yeast cells, and (C) a mycelium containing compartments separated by crosswalls (septa).

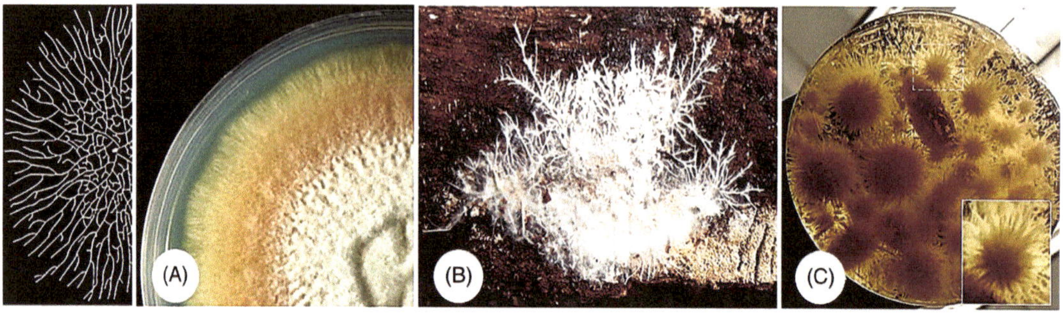

FIGURE 6.3 **Patterns of growth of a mycelium consisting of a branching hyphal network initiated by germination of a spore or a new branch from a hyphal fragment.** (A) Radial spread *Fusarium verticillioides* on the surface of a Sabouraud's agar plate. (B) Radial spread on the surface of decaying wood. (C) Spherical tufts of *Coprinopsis strossmayeri* grown in malt extract liquid medium. *Source: Photo credit: (A) Libero Ajello, Public Domain; (B) Ramjchandran, Wikimedia Commons; (C) Francois Gaascht with permission.*

ubiquitous arbuscular mycorrhizal fungi in the subphylum Glomeromycotina produces a glue-like red-brown glycoprotein called "**glomalin**," which is highly stable and contributes significantly to soil macroaggregate formation. Glomalin-related soil proteins consist of a mixture of macromolecules, so the exact contribution by mycorrhizal fungi is not clear. Nonetheless, these proteins are abundant and cumulative, so they contribute substantially to soil organic matter.

Many soil fungi in the Ascomycota and Basidiomycota can organize hyphae into specialized structures such as mycelial strands, rhizomorphs, and sclerotia. **Mycelial strands** are aggregates of parallel hyphae cemented together by sticky exudates and by fusion of hyphal branches. The cell walls of surface hyphae are thickened and often melanized (containing dark pigments). Mycelial strands of some fungi consist of modified hyphae functioning similar to plant tissues. For example, the wood decay (dry-rot) fungus, *Serpula lacrymans*, produces strands up to 8 mm in diameter consisting of thin-walled hyphae, thick-walled hyphae for structural support, and wide hyphae lacking cytoplasmic contents for mass flow of water much like plant xylem vessels. These strands are usually in close contact with soil to obtain moisture for translocation to the growing edge of a mycelium as it penetrates new wood tissue (Jennings and Watkinson, 1982). Mycelial strands of ectomycorrhizal fungi (in Ascomycetes and Basidiomycetes) not only provide a path for water transport, but also for transfer of nutrients between interconnected plant hosts (Brownlee et al., 1983).

Rhizomorphs are more complex versions of mycelial strands, with a greater degree of multicellular differentiation (Fig. 6.4A, B). They are highly resistant to environmental changes, have greater capability to penetrate soil or organic material, supply oxygen to the leading tips, and transport nutrients and water throughout the fungal organism. Species of basidiomycete fungi in the genus *Armillaria* are aggressive tree root and crown rot pathogens because they form abundant robust rhizomorphs. These rhizomorphs resemble gray to black shoestrings, although they can extend for long distances. They ramify in the soil as much as 8 times faster than a typical mycelium and then persist for years without degradation. The demise of many orchards started in forests cleared of diseased trees can be attributed to persistence of viable rhizomorph networks in soil even after fumigation.

FIGURE 6.4 **Organized structures differentiated by mycelium that enhance survival and other functions.** (A) Rhizomorphs of *Armillaria mellea* on tree roots. (B) Rhizomorph internal structure and function of hyphal tissues. (C) Sclerotia of *Claviceps purpurea* in inflorescence of wild grass, each with a hardened brown to black outer rind. (D) Cross section of a sclerotium showing the cohesive matrix of hyphae and the outer rind with thickened and melanized cell walls. *Source: Photo credit: (A) and (C) from Wikimedia Commons; (A) CC-BY-SA-3.0, (C) Jon Richfield CC-BY-SA-4.0.*

Sclerotia (singular, **sclerotium**) are spherical to oblong aggregates of interwoven hyphae that are thick-walled and melanized on the periphery to form a hard outer rind (Fig. 6.4C,D). Sclerotia often are separated from the parent mycelium and therefore are more readily dispersed than other forms of mycelial aggregates. They usually are 1–3 cm in diameter, but can range from the microscopic to highly visible (25–35 cm across). Sclerotia are better able to resist drought and temperature fluctuations and thus are much longer-lived than mycelia. They also store lipids, carbohydrates, and proteins until soil conditions are favorable enough to germinate and form either a new mycelium or a fruiting body.

Fungi reproduce by forming distinctive spores or fruiting bodies as a result of either sexual or asexual states designated as **teleomorphs** and **anamorphs**, respectively (Box 6.4). Spores are important in life cycles because they not only encapsulate different nuclei and cytoplasmic factors into numerous discrete packets for survival under stress or for dispersal, but they also increase the probability of contact between opposite mating types. Spores and fruiting bodies are important to scientists because they express most of the morphological characters that define taxonomic groups.

In general, the sexual cycle requires four sequential processes:

1. **plasmogamy** is the fusion of cytoplasm of two compatible haploid (1N) sexual organs or **gametangia** (singular, **gametangium**)
2. **karyogamy** is the fusion of two haploid nuclei to establish a diploid (2N) condition
3. **meiosis** is reduction of the diploid nucleus to haploid daughter nuclei, wherein genes can undergo recombination
4. **teleomorph formation** is the encapsulation of each haploid nucleus and its associated cytoplasm in a discrete cell that differentiates into a spore.

Some fungi form morphologically distinctive "male" and "female" gametangia. The gametangia of most fungi are morphologically indistinguishable so that chemical methods are required to recognize each of them. In this case, opposite mating types are designated "+" and "−." In several phyla of lower fungi (i.e., Chytridiomycota) and Oomycota, the teleomorph usually consists of solitary, thick-walled resting spores. In higher fungi (i.e., Ascomycota and

BOX 6.4

Teleomorphs and anamorphs

Fungi are ideal opportunists exploiting almost any habitat in part because of life cycles in which they can alternate between sexual and asexual reproduction. Morphologically distinctive spores produced openly or in fruiting bodies develop from the mycelium in the sexual phase (**teleomorph**) and in the asexual phase (**anamorph**). The teleomorph serves as a primary basis for classification of fungi if and when it is formed (Boxes 6.1–6.3). Many fungi in soil and litter reproduce only asexually under natural conditions, so that Latin binomial names are assigned to the anamorph. Sometimes a sexual phase is discovered in these fungi, at which time it receives a separate name for the teleomorph. This practice has led to dual nomenclature in fungal taxonomy, with separate species names for anamorphs and teleomorphs. However, the name of the whole organism (**holomorph**) is the name given to the teleomorph. Fungi with no known sexual phase were once classified in an artificial phylum, Deuteromycota. Today, they are placed in the phylum of their closest sexual relatives based on molecular, biochemical, physiological, and other morphological or ultrastructural characters. Most are members of Ascomycota or Basidiomycota.

Basidiomycota) spores often are borne on a variety of fruiting bodies like the highly visible and often colorful mushrooms found on the forest floor, in lawns, and in flowerbeds.

The asexual cycle is much simpler than the sexual cycle because it does not require interactions between nuclei or different mating types. The nuclei of one hypha divide repeatedly by mitosis and are partitioned into large numbers of spores called **conidia** (singular, **conidium**). Conidia are formed in a variety of ways, but essentially they arise from hyphal tips. Most soil fungi tend to form conidia freely on a single hypha or hyphae in various degrees of aggregation; a low percentage develop fruiting bodies. The mycelia of some soil fungi are able to transform hyphal compartments into thick-walled, darkly pigmented cells called **chlamydospores** which function as resting or survival spores.

Growth and nutrition

Fungi are **heterotrophs**, in that they must obtain carbon and other nutrients from organic matter in the external environment (see Chapter 3 for further discussion of nutritional types). Nutrients are acquired by absorption through chytrid, yeast, or hyphal cell walls. Fungi that degrade nonliving organic matter are called **saprophytes**, and they are extremely important agents in mineralization and carbon cycling (Chapter 13). In saturated soils or aqueous environments, chytrids and oomycetes are the most common saprophytes, whereas fungi in other phyla are more abundant in drier soils. In general, fungi are obligate aerobes in that they cannot grow without an oxygen supply. They tend to be more abundant in acidic soils, where there is less competition from bacteria.

Saprophytic fungi, together with bacteria, are primary agents in the decomposition of organic substances such as cellulose, hemicellulose, and pectin in plant cell walls. Lignin, the second most abundant cell wall polymer, can be degraded to CO_2 exclusively by "white-rot" fungi that also cause wood decay in trees (Box 6.5). Enzymes that decompose lignin also contribute to the degradation of organic pollutants and pesticides in soil (Aust, 1990). One of the main reasons fungi are better able to degrade many hazardous compounds (Chapter 21) than bacteria is that enzymatic activity occurs outside the mycelium where the chance for mortality to the fungal organisms is greatly reduced.

Saprophytic fungi are capable of producing many other enzymes as well, such as xylanases (to break down xylan, a hemicellulose polymer of plant cell walls) and cutinases (to break down cutin on leaf surfaces). Many of these enzymes are excreted into synthetic media when fungi are cultured in the laboratory. As a result, fungi are excellent candidates for industrial applications such as food processing, waste treatment, and production of alcoholic beverages (Lowe, 1992).

The supreme role of fungi on planet earth is the decomposition of organic matter in a cycle that leads to a restoration of life. Their contribution to decomposition of inorganic compounds is more poorly understood, especially as compared to the activity of bacteria. Much of the definitive results obtained to date are from in vitro experiments of fungi on selective substrates; studies in the field where conditions are more variable and complex have been less conclusive.

The role of fungi in nitrification (Chapter 15) has been hard to document. Reduction of nitrification rates by antifungal compounds such as cycloheximide in acid soils indicates a

BOX 6.5

White rot decay in wood

Fruiting bodies of wood-decay fungi on a log (top) and selective degradation of lignin by white rot fungi (bottom).

Lignin is a major component of cell walls in trees, yet it is one of the most recalcitrant substances to degrade in nature. Of approximately 1700 species of wood-rotting fungi, 94% are able to produce enzymes that selectively decompose lignin. These are called "white rot" fungi because the cellulose that remains behind after selective lignin degradation is white. Example genera in the Basidiomycota are *Ganoderma* and *Heterobasidion*, which with other genera also can cause severe root and collar decay of living trees. The wood of many tree species is low in nitrogen, which stimulates capable fungi to produce white rot. Many of the fungi alter their structure around decaying wood to regulate moisture content that facilitates decay. Lignin degradation (discussed in Chapter 13) has been termed "enzymatic combustion," with two of the most prominent enzymes being lignin peroxidase and manganese-dependent peroxidase.

Fruiting bodies of some of these fungi (e.g., oyster and shiitake mushrooms) also are a food source for animals and insects. For biotechnology, white-rot fungi can degrade persistent materials which otherwise might be hazardous, such as dioxin, creosote, and lindane. The paper industry continues to explore the potential of utilizing efficient lignin-decomposing fungi in chemical pulping to degrade wood chips to cellulose fibers rather than the strong acids typically used (Blanchette, 1991).

fungal contribution, although it is not high. Fungi are able to oxidize a diverse range of compounds, such as ammonium ions, pyruvic oxime, benzohydroxamic acid, cycloserine, and 3-nitroproprionate. It is unclear if these nitrification processes are linked to energy generation or provide maintenance energy to maintain the mycelial network.

Some fungi can produce nitrate reductase under low oxygen conditions in culture, suggesting their possible involvement in denitrification. Most studies, however, show that nitrate is reduced only as far as nitrite or N_2O rather than to dinitrogen gas. These processes, therefore, may have less to do with respiration than with modifying the environment around hyphae, such as transformation of nitrite to less toxic compounds.

Bacteria, especially *Acidithiobacillus* species, have received considerable attention as major players in the oxidation of elemental sulfur, thiosulfate, tetrathionate, and sulfate in soil

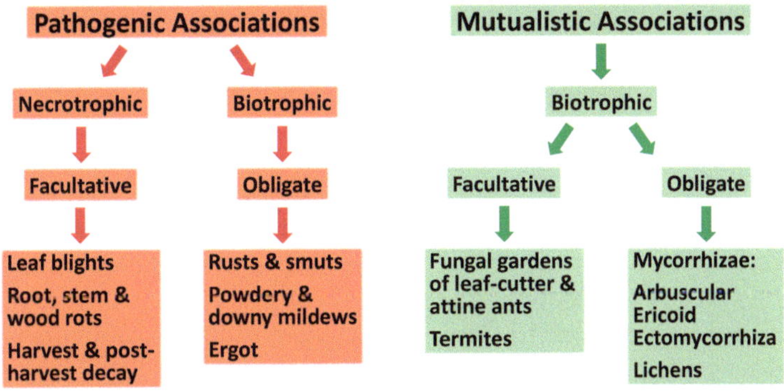

FIGURE 6.5 **Symbiotic associations between fungi and other living organisms to obtain nutrients for growth under natural conditions.** Pathogenic = harmful to one or both associates; mutualistic = beneficial to both associates; necrotrophic = nutrients acquired from decomposition of host cells by enzymes or toxins; biotrophic = no destruction of host cells, with nutrients most often obtained from specialized absorptive organs in association with host; facultative = host-associated and also free-living; obligate = only host-associated. Associations where neither organism in an interaction is affected is commensalistic and therefore not shown.

(Chapter 18). Common fungal saprophytes, such as *Aspergillus*, *Fusarium*, and *Penicillium*, were contaminants in the first isolations of *A. thiooxidans*, indicating they could grow under high sulfur conditions. Another soil saprophyte, *Trichoderma harzianum*, is able to oxidize elemental sulfur to tetrathionate and sulfate even at available carbon as low as 10 to 20 $\mu g\ g^{-1}$ in a synthetic soil (Wainwright and Grayson, 1988). Some yeasts, such as *Rhodotorula*, produce sulfite oxidase and thiosulfate oxidase enzymes, much like *Acidithiobacillus*.

Numerous fungi are not saprophytes but instead form symbiotic associations with other living organisms that are either positive (**mutualistic**), negative (**pathogenic**), or neutral (**commensalistic**) (Fig. 6.5). The relationship is **necrotrophic** when a fungus produces abundant enzymes to degrade and kill tissues of its host. **Biotrophic** relationships are more complex because the fungus establishes close contact with host cell contents via specialized absorptive structures and then induces hormonal changes to channel carbon flow in the host toward those contact sites. Most necrotrophic relationships are **facultative** because the fungi must produce degradative enzymes to obtain nutrients and live freely as a saprophyte for a part of its life cycle. Duration of contact between a fungus and its host is highly variable, often depending on the physiology of each associate and environmental conditions. Most facultative organisms are readily culturable on synthetic nutrient media. **Facultative biotrophic** associations are unique in that the fungal partners are saprophytic. These fungi digest wood as a food source for various ant and termite species. The ants, in turn, forage for plant material which they return to their nests and thus provide a competition-free "home" and source of carbon for the fungus. These fungi are equivalent to "couch potatoes" because they remain confined while their insect partners seek out and transport food resources back to the colony. All other known biotrophic relationships are **obligate** in that fungi need their hosts to grow and reproduce. Some mutualistic fungi, such as ectomycorrhizal and ericoid mycorrhizal fungi (Chapter 12), are obligate symbionts under natural conditions but they still can be cultured slowly on a synthetic medium containing essential nutrients.

Fungi in the Chytridiomycota and Oomycota produce motile zoospores that exhibit chemotaxis (Chapter 5) and swim along a gradient toward increasing concentrations of simple sugars, amino acids, or organic acids exuded by roots of the target host. Ascomycete and basidiomycete fungi do not produce motile spores, but some species have evolved other effective methods to incapacitate their hosts. Some are predatory in that they actively seek out and attack their hosts. For example, ascomycete species of *Arthrobotrys* and *Dactylella* have specialized hyphal structures to "capture" their nematode prey (Chapter 8), such as adhesive knobs, branches, or nets. Other species have constricting and nonconstricting rings as hyphal outgrowths. The mycelia of *Pleurotus* spp. in the Basidiomycota release toxins into the soil to paralyze nematodes long enough to allow hyphal penetration through the tough host cuticle.

The filamentous nature of the mycelium and the three-dimensional branch architecture of component hyphae afford fungi with a reliable and efficient mechanism to obtain patchily distributed nutrients in soils and translocate those nutrients to distant parts of the organisms or wherever they are most needed for continued growth or reproduction. When nutrient depletion extends over a large area, fungi escape by storing available endogenous nutrients in spores and dispersing them to more favorable environments.

Genetics

The main body of a fungus, the mycelium, is structurally simple but complex as a functioning interactive network, providing both genetic and phenotypic plasticity to adapt to changing environments. The nuclei in hyphae of true fungi are haploid during vegetative growth and asexual reproduction. With only one set of chromosomes, phenotypic traits may be expressed immediately and subjected to selection pressures. Structure of the hypha plays a major role in providing some regulation of nuclear movement and how they interact. These behaviors provide some genetic buffering, but also allow different parts of the mycelium to respond to localized environmental differences. A hypha is essentially one long tube with rigid cross-walls known as **septa** (singular, **septum**). These septa are not the same as cell walls because they contain open pores. They provide structural support along the length of a hypha and regulate the flow of cytoplasm, cellular organelles, and nuclei. The hyphae of more ancestral lower fungi (Box 6.2) are **coenocyctic** in that they rarely form septa in actively growing regions and nuclei are free-floating (Fig. 6.6A). When septa are formed, they are located between cytoplasmically active regions of hyphae and senescent or dead regions devoid of cytoplasm. In higher fungi (Box 6.3), septa are an integral part of hyphae. Septal pores can be plugged in hyphae of the Ascomycota by spherical **Woronin bodies** (Fig. 6.6B), thus providing more regulation of flow between hyphal compartments. Basidiomycete fungi have the greatest regulation of flow by means of a structurally complex **dolipore septum**, with flanges at a central pore and a narrow channel capped by perforated membrane domes called **parenthosomes** (Fig. 6.6C). **Clamp connections** between hyphal compartments, if present, further control nuclear division and ensure one pair of haploid nuclei (a **dikaryon**) are present in each compartment.

Hyphae are multinucleate, and associations between nuclei and the genetic differences between them determine expression of phenotypes. For example, a dikaryon consists of two physically separated haploid nuclei. Despite their physical separation, they function together

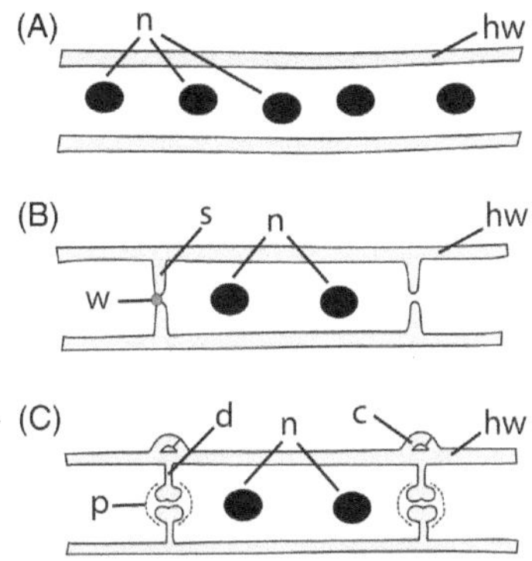

FIGURE 6.6 Structure of a hypha in (A) Oomycota (wall of cellulose) and Mucoromycota (wall of chitin), (B) Ascomycota, and (C) some Basidiomycota. *n*, nucleus; *hw*, hyphal wall; *s*, simple septum; *w*, Woronin body; *c*, clamp connection; *d*, dolipore septum; *p*, parenthosome.

as a diploid, and gene products of both nuclei interact in a common cytoplasm. Genetic flexibility is enhanced by independent movement of each haploid nucleus, especially in branches of hyphae that separate to start new "colonies," in repetitive formation of spores from a hyphal branch, or when there is mixing of cytoplasm of two or more fusing hyphae. The extent to which nuclear migration is controlled varies with the morphology of hyphae of fungi in each phylum, being least in those with coenocytic hyphae and most tightly regulated in Basidiomycetes with dolipore septa and clamp connections.

Genetic variation can occur rapidly with sexual reproduction as a result of gene recombinations. Many soil fungi have only an asexual phase, but genetic flexibility and adaptability still can be considerable with flow of haploid nuclei throughout a mycelium and mutations. Spontaneous mutants can appear at any time (Box 6.6), and they are most noticeable in fungal colonies growing in Petri dish cultures. Common manifestations of a mutant are visible differences in mycelium or spore color, architecture of the hyphae (aerial or adpressed to the agar surface or submerged within agar), spore abundance, or colony size (rate and amount of hyphae growing radially over an agar surface).

Aside from hormonal signals, sexuality in fungi is regulated by the physical origin of gametangia and their mating type. A mycelium on which both gametangia develop is called **homothallic**. It is self-compatible wherein there are no mating type barriers to plasmogamy. When each gametangium develops on a separate mycelium, then the fungus is **heterothallic**. Plasmogamy occurs only when gametangia are sexually compatible, which requires opposite mating types (designated as "+" and "–" when they are visually indistinguishable).

In soil environments, some degree of genetic stability is an asset rather than a liability. With the exception of some species in the Basidiomycota, soil fungi tend to be homothallic. Genetic variation is not lost, since a mycelium still may be heterokaryotic for any number of characters. With karyogamy and meiotic recombination, variation is promoted, but it occurs without too much divergence from the parental phenotype (Raper, 1966). Opposite mating types are more beneficial to fungi producing large numbers of air-dispersed spores which, upon germination, have a greater probability of making contact with each other to begin a sexual cycle (such as the fungus in Mucoromycota illustrated in Box 6.2).

BOX 6.6

Mutations

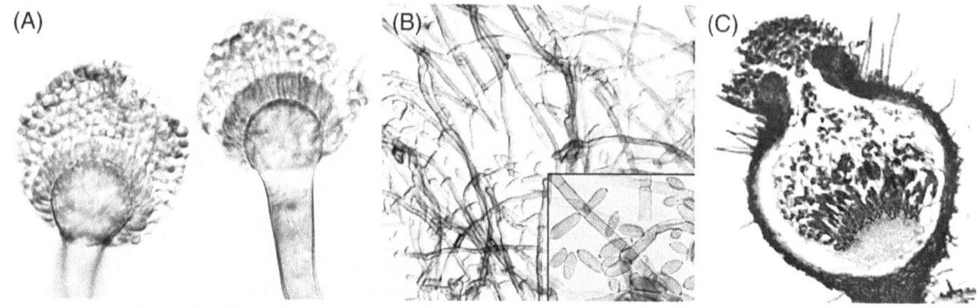

Spores or fruiting bodies of (A) *Aspergillus*, (B) *Cladosporium*, and (C) *Phoma* spp.

Many soil fungi only reproduce by asexual means so that, in the absence of genetic recombination, mutations provide a means to maintain genetic resiliency. Fungi such as *Aspergillus* (A) or *Cladosporium* (B) species produce many thousands of asexual spores from mycelium in 1 kg of soil or from fruiting bodies like that of a *Phoma* species (C). Assuming one nucleus per spore and a rate of one mutation at a gene in each nucleus, as many as 100 million mutants can occur in the top 15 cm of soil in an area of 1 ha. Even if many of the mutations are harmful, they will be detrimental only to those compartments of the mycelium where the nuclei are located. Other compartments of the mycelium lacking these mutants will continue to grow normally. Beneficial or neutral mutations may be inherited immediately if they migrate into mycelium or spores that successfully colonize new substrates and reproduce again. Efficient and widespread dissemination of beneficial mutants via spores increases the chances of their heritability.

Survival

Fungi are like plants in that they cannot move very far or fast during vegetative growth of the mycelium. Therefore, they must rely on large numbers of spores for long-range dispersal. Sporulation often is triggered under conditions where growth becomes limiting (e.g., nutrient deprivation or environmental changes), and thick-walled spores or fruiting bodies often serve as a resting stage. Some fungi inhabiting host tissues do not sporulate, possibly because the niche they occupy affords enough protection. Various mycelial aggregates discussed in the previous section serve as survival structures because they are more resistant than a diffuse mycelial network in resisting parasitism and adverse environmental conditions.

Substrate specialization by saprophytes or host specialization by pathogens or mutualists, are important factors in how long fungi will grow and reproduce in agricultural systems involving crop rotations. A fungus with specialized nutrition is more likely to decline or die out with a nonhost plant (or without a utilizable substrate) in a rotation sequence, whereas a fungus with a wide host range will survive and possibly flourish. Soil fungi often have enough resiliency to convert from one resting structure to another when nutrient levels are too low.

One group of fungi comprising the genus *Fusarium* (Ascomycota) produce up to three types of asexual spores (Box 6.3), with only 20% of species having a sexual cycle (Ma et al., 2013). One spore type can germinate to produce another type of spores as conditions change. If conditions are adverse when a thick-walled chlamydospore germinates, then another new chlamydospore is quickly formed. These patterns may be very important to survival in agricultural management systems that involve fallow periods between crop cycles.

Fungistasis is a general phenomenon in which natural soils suppress germination of fungal spores or other resting structures. It is an important mechanism by which fungi maintain viability when nutrients become limiting. This suppression remains in effect until exogenous nutrients again become available and sometimes even until other environmental conditions are optimal for fungal growth. Fungistasis usually is broken after threshold levels of sugars and nitrogenous compounds are reached in proximity to the dormant spore or resting structure. Plant species-specific stimulants in the rhizosphere also may be involved.

The longevity of fungal organisms has not been measured in many species (Box 6.7), but their genetic/morphological plasticity and open-ended growth habit offers unlimited

BOX 6.7

Fungal longevity

The "individuality" of a single fungal organism is difficult to define because the body consists of a diffuse and indeterminant mycelium and other semi-independent parts, such as sexual and/or asexual spores that disperse and allow "rebirth" of the source colony. To distinguish individuals, researchers measure genetic homogeneity in a mycelium by either testing for vegetative compatibility or comparing amplified molecular markers (Chapter 4). These methods were used on a mushroom-forming basidiomycete fungus, *Armillaria gallica* (reported as *A. bulbosa*), a saprophyte in the root zone of hardwood trees, present in 20 ha of a forest of Michigan's Upper Peninsula. All of the samples collected were found to be part of one organism estimated to be at least 1,500 years old and weighing more than 10,000 kg (Smith et al., 1992). Other fungi also are long-lived. Fairy rings, which result from extensive fungal spread in the root zone of turfgrass, advance in ever-widening circles, much like growth on a Petri plate, for decades, centuries, or even longer geologic periods instead of days or weeks.

Fairy ring in turfgrass. Note the greener grass around the outside of the ring where the fungus is mineralizing nitrogen. *Photo credit: Lee Butler, North Carolina State Univ.*

potential. Fossil evidence, together with measurements of the rate of change in DNA nucleotide sequences, suggests that many fungi have a longevity measured in the millions of years.

Dispersal

Almost any part of the fungal organism can serve as a vehicle of dispersal and renewed growth at a different location, which is why fungi are found in almost any nook or crevice exploiting diverse nutritional niches. Fragmentation of a mycelium does not necessarily kill a fungal organism; instead, it results in a multitude of new propagules that can initiate new growth on a suitable substrate. Researchers take advantage of this potential by fragmenting intact mycelium in water using a blender so as to multiply the number of viable units from one to many. Fragmentation occurs naturally in soils from freezing and thawing cycles between seasons, feeding by microfauna, tillage practices, and other natural or human-driven events. Mycelial fragments or discrete resting structures in soil (such as rhizomorph fragments, sclerotia, chlamydospores, oospores, and zygospores) travel short to moderate distances via water runoff or irrigation. Long-range dispersal of these propagules usually requires aerial mobility in wind currents using vectors such as insects, birds' feet, plant debris, soil particles, seed, and pollen. Airborne sexual spores or asexual conidia usually range in size from 3 to 40 μm, and they are either passively or actively discharged. **Passive discharge** of sporangiospores or conidia produced on individual or aggregated aerial hyphae is implemented by external forces such as splashing raindrops or sporadic wind gusts. **Active discharge** involves a variety of unusual and often dramatic mechanisms. In some ascomycete fungi, spores are forcibly ejected from asci (see Fig. 6.7), much like spitting out a fruit pit. Numerous basidiospores in mushrooms are released ballistically (like projectiles) from each basidium of the gill lining. An average-sized mushroom of *Agaricus campestris* discharges up to 16 billion basidiospores within a 24-h period.

In aquatic environments or moist soils, chytrids and oomycete fungi are dispersed by motile zoospores. These thin-walled spores are fragile and not long-lived, but their motility

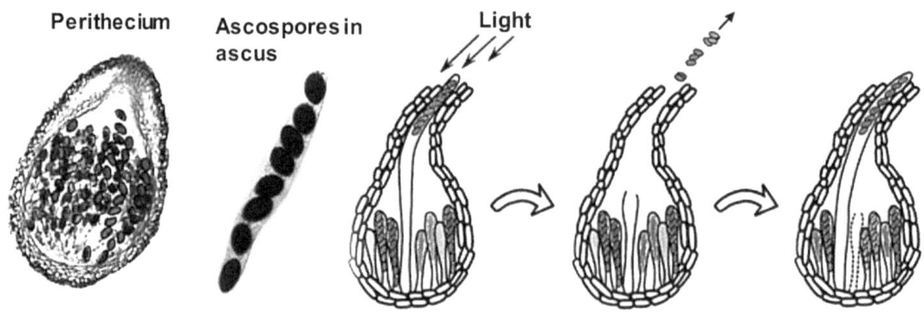

FIGURE 6.7 **Illustration of events in the discharge of ascospores from a perithecium fruiting body of *Sordaria fimicola* (pictured on the left).** The perithecium absorbs water and hydrostatic pressure causes extension of each ascus. When an ascus is ruptured, ascospores within are forcibly discharged into the air. The perithecium neck bends toward light to optimize the probability that discharged ascospores will be picked up by air currents. *Source: Adapted from Ingold (1971).*

allows them to quickly reach and exploit new food sources. Ascomycete and basidiomycete fungi produce spores with elaborate appendages that function to aid in buoyancy and rapid movement in water.

Ecology

Fungal hyphae must be in close contact with a living or dead organic substrate to absorb nutrients, so they rarely grow alone, especially in soils. They compete aggressively among themselves and with other microorganisms for scarce nutrients. Competition and change in available nutrients usually lead to **succession**, or directional changes in composition and abundance of species. Successional changes in fungi are fundamentally different from those in plants in that a stable climax stage rarely is reached. Instead, exploitation of nutrients results in continuous changes in type and abundance of nutrients which then favor one group of species over others.

Competitive interactions among fungi are complex because many species are able to coexist in substrates constantly in a state of flux. For example, Bills and Polishook (1994) carried out extensive culturing procedures on a 5-g sample of leaf litter in a Costa Rican rain forest, from which they recovered between 78 and 134 fungal species in each sample. More than 300 species were found.

Directional taxonomic changes in fungal community structure occur as a plant substratum or a soil undergoes shifts in nutrient availability. In general, fungi in Oomycota, Chytridiomycota, and Mucoromycota tend to precede those in Ascomycota, which in turn precede those in Basidiomycota. Initial colonizers utilize simple soluble sugars, amino acids, and vitamins in plant protoplasm or plant parts such as fruits, seed, and vegetables. These "sugar fungi" produce a rapidly spreading mycelium and abundant asexual structures. The duration of these fungi is short because waste byproducts and other conditions cause growth to cease and resting survival structures to be produced. Examples include *Pythium* species (Oomycota) and common molds like *Rhizopus* species (Mucoromycota). Cellulose degraders intergrade next in this succession, although sugar fungi may grow concurrently by utilizing some of the degradation byproducts. Cellulose degraders constitute the most diverse and competitive group because of the heterogeneity of substrates present. Degradation of straw with a high C/N ratio of 80 (Chapter 13) may cause fungi to parasitize or decompose the mycelium of other fungi to obtain N for growth and enzyme production. Abundant colonizers of straw in no-till crops include many saprophytes or pathogens that reproduce mainly by asexual conidia, such as species of *Chaetomium*, *Fusarium*, and *Trichoderma*, all in the Ascomycota. Degradation of lignin also is stimulated by low nitrogen in wood (Box 6.5). While lignin derivatives constitute up to 60% of the total mass of soil organic matter, the low number of fungal species capable of its degradation reduces competition. Basidiomycete white-rot fungi tend to be slow growing because considerable biomass is devoted to producing rhizomorphs and fruit bodies. These successional trends are generalizations, and so exceptions always can be found.

In semi-arid and arid regions of the world where plant cover is sparse and moisture is limiting, free-living fungi complex with cyanobacteria, lichens, algae, mosses, and liverworts in only a few millimeters of surface soil to form fragile "**biological soil crusts**" (Chapter 7), also called "**cryptobiotic crusts.**" Decades in the making, these thin highly irregular crusts trap seeds, reduce wind and water erosion, and increase water infiltration and nutrient transformations. These crusts are easily disturbed or destroyed by hikers, vehicles, and other human influences, and so there are major efforts to conserve this important natural resource.

Different substrate environments lead to different successions of saprophytic fungi. In compost, for example, successive colonists are increasingly **thermophilic** (heat-tolerant) initially, followed by a decrease in tolerance over time (Chapter 24). In frozen environments, increasingly **psychrophilic** (cold-tolerant) fungi are selected. Succession is likely to be much slower in more specialized niches, such as those inhabited by biotrophic fungi (mycorrhizae and some pathogens) or those involving intricate food partnerships (ants and termites).

Antagonism (Chapter 10) between soil microorganisms also is a common community dynamic. Many soil fauna, especially soil animals such as Collembola, are **mycophagous** (i.e., feed on fungal mycelia). Fungi also produce antibiotics. Antibiotic production by fungal species in ascomycete genera such as *Aspergillus*, *Fusarium*, and *Penicillium* (the latter being the original source of penicillin) provide an ecological advantage in sustaining occupation of a nutrient resource. One of the more clear-cut cases of antibiosis in fungi involves the cereal vascular pathogen, *Cephalosporium gramineum*. It produces a broad-spectrum antibiotic that inhibits other organisms from colonizing infested wheat straw when soil pH is less than 6.5 (Bruehl et al., 1969). This antibiotic appears to play more of a role in survival and abundance than disease development, since its presence or absence has no effect on disease-causing ability.

Summary

Fungi are highly diverse groups of eukaryotic microorganisms that vary greatly in size, shape, and complexity. Unicellular chytrids are well adapted to aquatic environments and moist soils. The branching architecture and indeterminate growth of mycelia of terrestrial fungi optimize foraging ability to explore large soil volumes and exploit heterogeneous and patchily distributed organic substrates. Various mycelial structures optimize this behavior and also provide a hardy resting stage when conditions are adverse. The haploid nuclei common to many fungi are capable of movement in hyphae either singly, in pairs (dikaryons), or even in greater numbers. Cytoplasmic continuity, even with septa, provides a mosaic of genetic and phenotypic possibilities within the same organism. Spores are produced sexually, asexually, or both. Spores not only package nuclei of parental or new genotypes (through recombination or mutation), but they also parcel nutrients, serve as resting structures, and function as vehicles for dispersal. In the soil environment, fungi are important as food sources, pathogens, beneficial symbionts, saprophytes to degrade crop residues, and biotic agents to improve soil structure and aeration. Saprophytes are relatively unspecialized, producing degradative enzymes to reduce complex organic compounds to simple sugars for absorption. Pathogenic or mutualistic associations between fungi and other living host organisms are complex, often producing fungal structures in close contact with host cells to absorb nutrients. Despite the widespread distribution of fungi and their pivotal role in biotic processes, much has yet to be learned about their diversity and the complexity of interactions among each other and with other eukaryotic and prokaryotic organisms.

Study questions

1. What are the unique properties of fungi that distinguish them from prokaryotes, plants, and animals?
2. What are the main differences between fungi considered phylogenetically to be "lower" and those considered to be "higher?"
3. Oomycota are quite different from true fungi, being diploid rather than haploid and their cell walls containing cellulose and glucan instead of chitin. What properties were responsible for this group being considered fungi for so long?
4. What are the advantages of a mycelial body plan for (1) growth in soil with patchy nutrient rich and nutrient poor zones, (2) reproduction, and (3) survival?
5. Why would a fungal organism occupying a soil pore more likely survive loss of moisture or nutrient substrates than a local population of bacteria?
6. What are the main vegetative structures formed to resist environmental fluxes and what other, if any, functions do they serve?
7. What are the advantages of soil fungi in producing numerous asexual spores?
8. Besides genetic recombination, what are important functions of sexual spores in each fungal class?
9. Define heterokaryosis and discuss how this phenomenon benefits genetic plasticity in fungi.
10. In the various specialized symbioses, how do fungi differ in their ability to obtain nutrients?
11. What factors are most important in governing the distribution of fungi in agricultural soils?
12. Define fungistasis and discuss its role in improving survival of fungal resting structures in soil.

References

Alexopoulos, C.J., Mims, C.W., Blackwell, M., 1996. Introductory Mycology, fourth ed. John Wiley & Sons, Inc., NY.

Ananyeva, N.D., Susyan, E.A., Chernova, O.V., Chernov, I.Yu., Makarova, O.L., 2006. The ratio of fungi and bacteria in the biomass of different types of soil determined by selective inhibition. Microbiology 75, 702–707.

Aust, S.D., 1990. Degradation of environmental pollutants by *Phanerochaete chrysosporium*. Microb. Ecol. 20, 197–209.

Bills, G.F., Polishook, J.D., 1994. Abundance and diversity of microfungi in leaf litter of a lowland rain forest in Costa Rica. Mycologia 86, 187–198.

Blanchette, R.A., 1991. Delignification by wood-decay fungi. Annu. Rev. Phytopathol. 29, 381–398.

Blackwell, M., 2011. The fungi: 1, 2, 3…5.1 million species? Am. J. Bot. 98, 426–438.

Brownlee, C., Duddridge, J.A., Malibari, A., Read, D.J., 1983. The structure and function of mycelial systems of ectomycorrhizal roots with special reference to their role in forming inter-plant connections and providing pathways for assimilate and water transport. Plant Soil 71, 433–443.

Bruehl, G.W., Millar, R.L., Cunfer, B., 1969. Significance of antibiotic production by *Cephalosporium gramineum* to its saprophytic survival. Can. J. Plant Sci. 49, 235–246.

Ingold, C.T., 1971. Fungal Spores: Their Liberation and Dispersal. Clarendon Press, Oxford.

Jennings, L., Watkinson, S.C., 1982. Structure and development of mycelial strands in *Serpula lacrymans*. Trans. Br. Mycol. Soc. 78, 465–474.

Lowe, D.A., 1992. Fungal enzymes. In: Arora, D.K., Elander, R.P., Mukerji, K.G. (Eds.), Handbook of Applied Mycology, vol. 4, Marcel Dekker, Inc, NY, pp. 681–706.

Ma, L.-J., Geiser, D.M., Proctor, R.H., Rooney, A.P., O'Donnell, K., Trail, F., Gardiner, D.M., Manners, J.M., Kasan, K., 2013. *Fusarium* pathogenomics. Annu. Rev. Microbiol. 67, 399–416.

Raper, J.R., 1966. Life cycles, basic patterns of sexuality, and sexual mechanisms. In: Ainsworth, G.C., Sussman, A.S. (Eds.), The Fungi: An Advanced Treatise, Academic Press, NY, pp. 473–511.

Smith, M.L., Bruhn, J.N., Anderson, J.B., 1992. The fungus *Armillaria bulbosa* is among the largest and oldest living organisms. Nature 356, 428–431.

Spatafora, J.W., Aime, M.C., Grigoriev, I.V., Martin, F., Stajich, J.E., Blackwell, M., 2017. The fungal tree of life: from molecular systematics to genome-scale phylogenies. Microbiol. Spectrum 5, FUNK-0053-2016.

Wainwright, M., Grayson, S.J., 1988. Fungal growth and stimulation by thiosulphate under oligocarbotrophic conditions. Trans. Br. Mycol. Soc. 91, 149–156.

Further reading

Brundrett, M.C., 2008. Mycorrhizal Associations: The Web Resource. Available from: http://www.mycorrhizas.info.

Carroll, G.C., Wicklow, D.T. (Eds.), 1992. The Fungal Community: Its Organization and Role in the Ecosystem, second ed. Marcel Dekker, Inc, NY.

Hawksworth, D.L., Kirk, P.M., Sutton, B.C., Pegler, D.N., 1995. Ainsworth and Bisby's Dictionary of the Fungi, eighth ed. CAB International, UK.

Henis, Y (Ed.), 1987. Survival and Dormancy of Microorganisms, John Wiley & Sons, NY.

Hudson, H.J., 1986. Fungal Biology. Edward Arnold, Ltd, Baltimore, MD.

Kendrick, B., 1992. The Fifth Kingdom. Focus Texts, Newburyport, MA.

McCoy, P., 2016. Radical Mycology. Chthaeus Press, Portland, OR.

Stamets, P., 2005. Mycelium Running: How Mushrooms Can Help Save the World. Ten Speed Press, Berkeley, CA.

CHAPTER

7

Cyanobacteria and algae

Ferran Garcia-Pichel[a], *Jayne Belnap*[b]

[a]Center for Fundamental and Applied Microbiomics, Biodesign Institute, and School of Life Sciences, Arizona State University, Tempe, AZ, United States; [b]US Geological Survey, Southwest Biological Center, Moab, UT, United States

The role of the infinitely small in nature is infinitely great.—**Louis Pasteur**

Although it might be surprising to some, photosynthetic microorganisms are common soil inhabitants. Most notable are the algae and cyanobacteria. These microbes can occur in or on all surface soils, from arctic deserts to tropical rain forests. Their wide diversity of forms reflects specific adaptations to tolerate the local environmental conditions. In soils, they can be found free-living or in symbiotic associations, for example, with fungi to form lichens. In ecosystems where vascular plant cover is limited (e.g., hot or cold deserts, infertile soils, or areas where vegetation has been temporarily removed), and where sunlight fully illuminates the soil surface, cyanobacterial or algal populations can cover most soil and rock surfaces and become major players in primary production. They have often been described as "pioneer colonizers." Wherever they are found in high numbers, these organisms play many critical ecosystem roles in the carbon, nitrogen, and water cycles, and act as stabilizers against soil erosion.

The term "algae" refers to a functional guild of eukaryotic organisms that, like higher vascular plants, carry out oxygenic photosynthesis and fix carbon using the Calvin cycle (Chapter 3). Despite these similarities, algae are not considered to be plants. Cyanobacteria belong to the same functional guild but are bacteria. Algae do not share an evolutionary common ancestor. Rather, they ultimately derive their capacity to perform oxygenic photosynthesis from ancient endosymbiotic events involving eukaryotic cells with a cyanobacterium, which gave rise to the photosynthetic organelles known as plastids (or chloroplasts). Conversely, cyanobacteria and plastids do share a common ancestor.

Principles and Applications of Soil Microbiology. http://dx.doi.org/10.1016/B978-0-12-820202-9.00007-1
Copyright © 2021 Published by Elsevier Inc.

Classification

In modern taxonomy and systematics, based largely on molecular (DNA) phylogeny, the Cyanobacteria constitute a Phylum within the domain Bacteria (Garcia-Pichel et al., 2020), whereas the algae belong to various Classes or Divisions in the Domain Eukarya (Chapter 1). Of the many algal classes in nature, only a few are common in soils. However, for the purposes of this chapter, we will use a simplified, more traditional classification based on phenotypic characteristic related to cell and **thallus** morphology, photosynthetic pigments, and polymeric storage products (Table 7.1; Fig. 7.1). The term thallus essentially refers to the visible, macroscopic aggregate cells or "body" of an organism.

Cyanobacteria are an evolutionarily well-delimited phylum of prokaryotic organisms that use oxygenic photosynthesis to fix carbon; many can also fix dinitrogen. They are most closely related to non-photosynthetic groups of heterotrophic bacteria (Chapter 5). Cyanobacteria display an unusually high level of morphological diversity, ranging from simple unicells to complex multicellular organisms able to differentiate branching filaments; some display true cellular differentiation. Traditionally, five subsections (or orders) of cyanobacteria have been distinguished based on morphology and modes of cellular division. Molecular phylogenetics, however, indicate that these orders do not always form closely related groups, and that morphological assignments conceal a great deal of genetic diversification, particularly among those that are morphologically simple (Fig. 7.2).

TABLE 7.1 Characteristics used to classify soil cyanobacteria and algae.

Taxon	Example genera	Main photopigments	Storage products
Phylum Cyanobacteria	*Chroococcidiopsis* *Calothrix* *Lyngbya* *Microcoleus* *Nostoc* *Porphyrosiphon*	**Chlorophyll** ***a*** (all species) Chlorophyll *b*, divinyl-*a*, *d*, *f* (some species) **Phycobiliproteins** (phycocyanin, phycoerythrin, phycoerythrocyanin; most species) **Carotenoids** (all species: β-carotene, myxoxanthophyll, echinenone, and others)	Glycogen Poly-β-hydroxybutyrate Cyanophycin Lipids
Eukaryotic algae			
(Division)			
Green algae (Chlorophyta)	*Chlorella* *Chlorococcum* *Klebsormidium* *Trebouxia* *Stichococcus*	**Chlorophyll** ***a*** and ***b*** **Carotenoids** (α- and β-carotene, astaxanthin, lutein, neoxanthin, siphonine, siphonoxanthin, violaxanthin, zeaxanthin)	Starch Lipids
Yellow-green algae (Xanthophyta)	*Botridiopsis* *Geobotrys* *Heterococcus* *Tribonema*	**Chlorophyll** ***a*** and ***c*** **Carotenoids** (β-carotene, lutein, neoxanthin)	Chrysolaminarin Lipids
Diatoms (Bacillariophyta)	*Diploneis* *Fragilaria* *Navicula* *Nitzschia*	**Chlorophyll** ***a*** and ***c***, **Carotenoids** (β-carotene, lutein, fucoxanthin, diadinoxanthin)	Chrysolaminarin Lipids

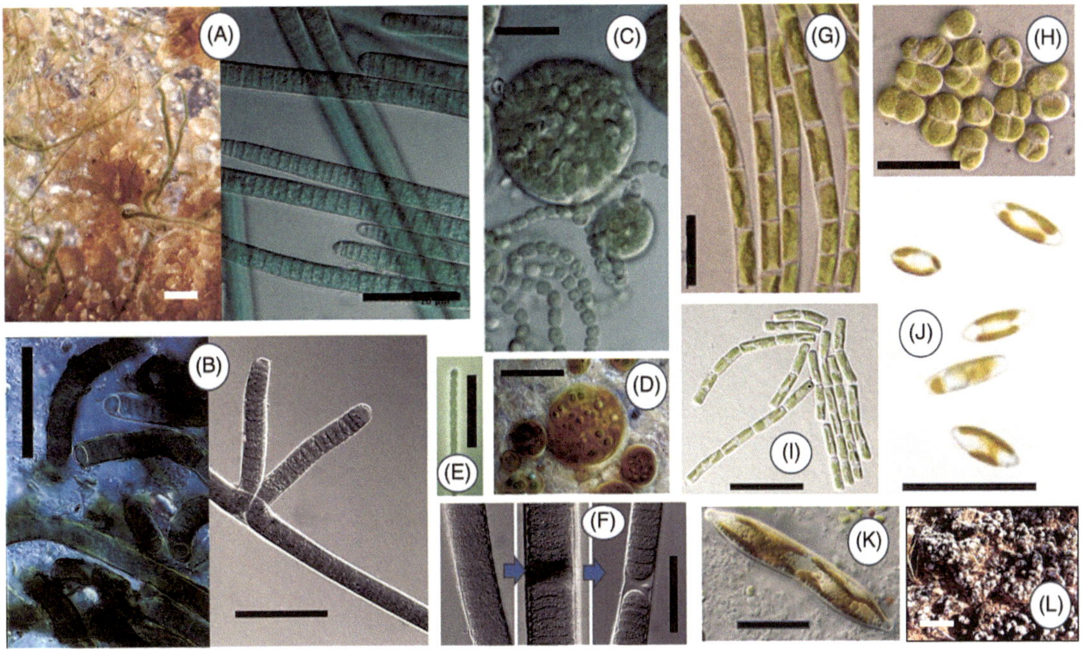

FIGURE 7.1 Some representative soil cyanobacteria (A–F), green algae (G, H), yellow-green algae (I), diatoms (J, K), and a green algal lichen (L). (A) *Microcoleus vaginatus* trichomes in pure culture (right), and bundles of filaments in the soil (left); (B) *Scytonema* sp. field sample (left) and in pure culture (right), showing false branching; (C) spherical microcolonies and single filaments of *Nostoc* sp.; (D) *Gloeocapsa* sp. (large colonies) and *Aphanocapsa* sp. (smaller, faint colonies); (E) *Leptolyngbya* sp.; (F) *Porphyrosiphon notarisii*, showing the process of filament duplication by formation of necridia; (G) *Klebsormidium* sp.; (H) *Interfilum* sp.; (I) *Xanthonema* sp.; (J) *Navicula* sp.; (K) *Hantzschia* sp.; (L) *Psora decipiens*. Scales: A (left) = 100 μm; I = 10 μm; L = 2 cm; all others = 20 μm. *Source: Ana Giraldo-Silva (A, C, D, J), Ulf Karsten (G, H, I, K), Scott Bates (L), Ferran Garcia-Pichel (all others). All with permission.*

Soil algae are eukaryotic organisms that also carry out oxygenic photosynthesis, an ability which they acquired in deep evolutionary time through symbioses with a cyanobacterium. Most soil algae belong to the green algae, diatoms, or yellow-green algae. Euglenoids, dinoflagellates, brown algae, and red algae are also classes of algae, but these groups are largely aquatic and thus will not be further discussed. Furthermore, although diatoms and yellow-green algae can be isolated from soils, only green algae attain large populations unless the soils are flooded for long periods. Green algae, along with cyanobacteria, are also important in soils as the photosynthetic partner (photobionts) in lichen symbioses.

Soil algal morphology and reproduction

Cyanobacteria

Cyanobacteria are one of the oldest life forms on Earth. They are abundant in the fossil record, first appearing as marine stromatolites 2.5–3.5 billion years ago. They dominated

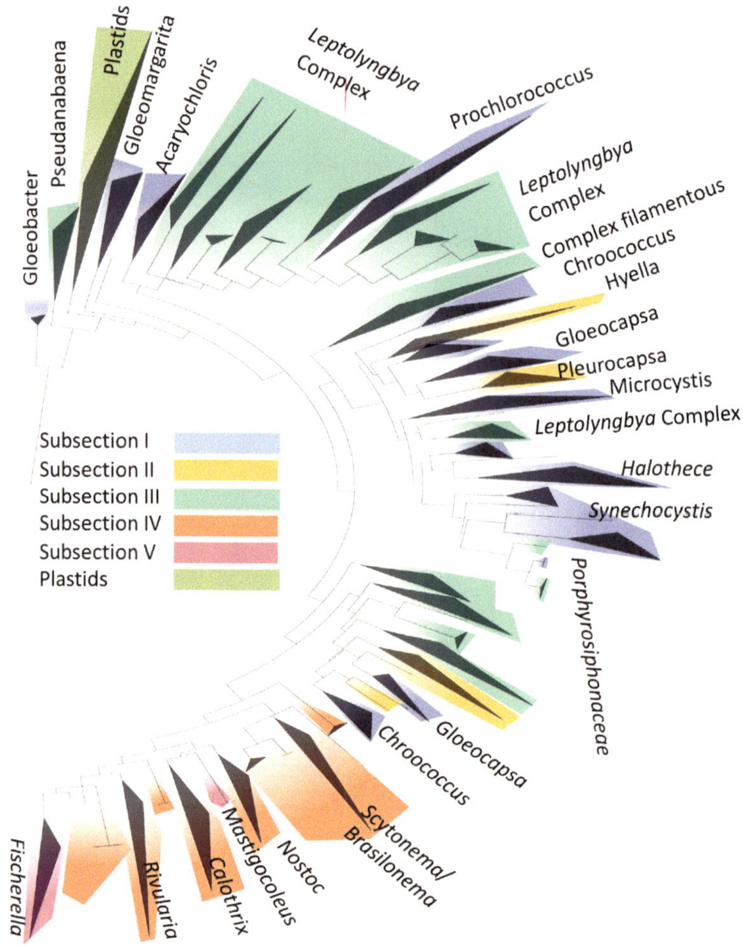

FIGURE 7.2 **Phylogenetic tree of the cyanobacteria and algal plastids (chloroplasts) based on 16S rRNA according to the Cydrasil database** (http://garcia-pichel.lab.asu.edu/labo/cydrasil/). Traditional Subsections are color-coded. *Source: F. Garcia-Pichel's website, with some exemplary genera.*

life on Earth for almost 2 billion years (from 2.5 billion to 570 million years ago) and are considered responsible for oxygenating the atmosphere, thus enabling the evolution of oxygen-breathing life forms (Schopf, 1983). Microfossils of soil cyanobacteria appear in the record as far back as 1.1 billion years ago (Beraldi-Campesi et al., 2014). Through diversification, cyanobacteria have attained a tremendous genetic flexibility that allows them to occupy a wide range of environmental conditions, including extremes in temperature (from hot springs to permanently cold deserts such as those in the Arctic and Antarctic), water availability (aquatic ecosystems to hyper-arid deserts), and salinity (salt-saturated brines to freshwater). As prokaryotes, cyanobacteria have peptidoglycan cell walls and no internal membrane-bound organelles or nuclei (Chapter 5). Instead, the photosynthetic machinery is arranged along multilayered internal membranes that wind throughout the cytoplasm. Cell size is generally between 0.5 and 10 μm in diameter but can be as large as 100 μm.

Most cyanobacteria are obligate phototrophs. They can store carbon either as glycogen, poly-β-hydroxybutyrate, or oil droplets. Some can additionally store organic nitrogen as **cyanophycin**, a non-ribosomally formed polymer of amino acids. Soil-dwelling cyanobacteria often secrete an enveloping mucilaginous sheath or capsule made up of hygroscopic polysaccharides (see Fig. 7.1C,D) that can slow down desiccation and chelate soil nutrient cations. All cyanobacteria contain chlorophylls as their main photosynthetic pigment. Chlorophyll *a* is the most common, but some can also contain chlorophyll *b* or divinyl chlorophyll *a*, and a few have been shown to contain chlorophyll *d* and *f*, helping them extend their photosynthetic range into the infrared and to adapt to low light conditions. Most also contain the light-harvesting **phycobiliproteins,** complex proteins that contain covalently bound tetrapyrrole pigments, phycocyanin being the most common. Phycocyanin imparts a characteristic bluish color to cyanobacteria and is the origin of their name (from the Greek root "cyan") and their older, now obsolete, name "blue-green algae." Other cyanobacteria produce the reddish phycoerythrin that yields various shades red to purple.

Cyanobacteria can be either unicellular, colonial, or filamentous. They do not have flagella, but many move by gliding. Filamentous cyanobacteria travel through moist soils using a gentle corkscrew movement along the long axis, powered mechanically through excretion of polysaccharide sheath material. The term "filaments" refers to the trichomes (rows of cells) plus the extracellular sheaths. In simple filamentous forms, trichomes are made up of independent cells that are held together at their poles (ends) (Fig. 7.1E) but function essentially independently. Complex filamentous forms (Fig. 7.1A,B), however, are truly multicellular organisms and often share a single outer membrane, exchange materials and information readily among cells so as to respond to the environment in a coordinated manner, and undergo cell differentiation and cell specialization for particular tasks. Some cyanobacteria, such as *Nostoc*, present life cycles that can vary from sessile globose colonies, to sessile multicellular filaments (both can be seen in Fig 7.1C), to motile multicellular hormogonia.

Cyanobacterial cells undergo simple cell division (fission). Nonetheless, coordinated changes in the plane of division can give rise to colonies in unicellular forms and to true branches in the filamentous species. Filaments oftentimes split at specific cells called **necridia**, that degenerate sacrificially for this purpose. Continued growth of split filaments within a common sheath can yield false branches (Fig 7.1B). Most cells retain the capacity to divide, with the exception of specialized cells called **heterocysts** (see subsequent section on dinitrogen fixation). Other differentiated cells and structures include **hormogonia**, which are short filaments that differentiate from some sessile cyanobacteria (like *Nostoc*) for the purpose of propagation, and **akinetes** (mostly associated with planktonic forms), which are thick-walled, single-celled resting or survival spores containing high levels of nutritional reserves. The thick walls afford akinetes more resistance to desiccation and freezing compared with other cells, while the storage compounds allow them to survive extended periods in the absence of light for photosynthesis. Terminal hairs are rows of colorless cells specialized in the uptake of phosphorus, and specialized cells called **calcicytes** accumulate calcium intracellularly. Some non-filamentous forms undergo several rounds of division without growth in order to produce very small specialized cells called **baeocytes** (Greek meaning "small cell") that enable dispersal in soil or other environments through tiny soil or rock pores.

Green algae

Green algae are the most diverse of the algal groups, with at least 7000 species. Most are aquatic, but many are found in a variety of habitats that include soils, tree bark, snow, and in symbiotic relationships with a variety of organisms ranging from fungi (forming lichens; Fig. 7.1L) to animals. Like cyanobacteria, green algae are found on all continents and in almost all terrestrial habitats and may be either unicellular, colonial, or filamentous. The earliest evidence of green algae comes from fossils estimated to be a billion years old (Tang et al., 2020). Similar to vascular plant cells, green algae have Golgi bodies, vacuoles, mitochondria, membrane-bound nuclei containing genomic DNA, and plastids. The latter (visible in Fig. 7.1G) contain the photosynthetic machinery, including chlorophylls, electron-transport chains, and the enzymes in the Calvin cycle necessary for carbon assimilation (Chapter 3). Green algae have firm cell walls made of cellulose and other polysaccharides. Solitary green algal cells are generally larger than those of cyanobacteria and can have cell diameters ranging from approximately 1 μm to the single-celled and multi-nucleated seaweed *Caulerpa* sp. which can be up to 3 m long; most cells, however, range from 2 to 7 μm in size. Reproduction can be asexual (through mitotic division or fragmentation) or sexual (involving meiosis and fusion of gametes), the latter producing a **zygospore**.

Diatoms and yellow-green algae

Diatoms first appeared in the fossil record only about 200 million years ago, and many of the fossils resemble living species. Both diatoms and yellow-green algae are predominantly aquatic, but many species are also found in soil. They can be unicellular or filamentous. Both contain chlorophyll *a* and *c* and an abundance of carotenoids that are either fucoxanthin (in diatoms) or the closely related vaucherixanthin (in yellow-green algae). The two groups are closely related to brown algae, having similar chloroplasts, and all three groups store carbon as chrysolaminarin (a glucose polymer) outside of the chloroplast. Yellow-green algae and diatoms additionally store carbon as lipids. The cell walls of yellow-green algae are made of cellulose, whereas diatoms are enclosed in glassy, hard envelopes made of silica having overlapping halves called valves or **frustules** (Fig 7.1J,K; Box 7.1). Diatoms can have either radial or bilateral symmetry; the frustules of the former resemble a Petri dish, and those of the latter a pillbox. Although diatoms are primarily photosynthetic autotrophs, they can also exist heterotrophically.

Yellow-green algae and diatoms have limited mobility in soils. Yellow-green algae reproduce asexually by fragmentation and by either nonmotile or motile spores. Diatoms reproduce primarily asexually by mitotic division, but this brings about a progressive reduction in size that must be overcome by rounds of sexual reproduction resulting in the production of large **auxospores**. The auxospore divides, forming two large vegetative cells, and in this manner the original larger size is regained. Both diatoms and yellow-green algae can form resting cells that maintain low levels of metabolic activity but are otherwise similar to non-resting cells in having limited resistance to harsh environmental conditions. Additionally,

BOX 7.1

Diatomaceous earth

Algae, albeit not soil algae per se, are responsible for the formation of diatomaceous earth. This material consists of diatom frustules (see figure below), both marine and freshwater, that were deposited during the Tertiary and Quaternary Periods of geologic history. Diatomaceous earth (also called diatomite or Kieselguhr) is an almost chemically inert inorganic material containing about 86%–88% silica that has traditionally been used as a mild abrasive material. Diatomite is also commercially available as a filtration aid in sugar refining, the brewing industry, wine making, and antibiotic production. It is also used as insulation and filler for paints and paper products. Historically, diatomite served as the material for lightweight bricks in the construction of the Cathedral of St. Sofia in Constantinople in 532 A.D.

Scanning electron micrographs of diatom frustules. *Source: Cassandra Dudek, with permission.*

diatoms can form a different type of resting cell, called a survival spore that has thickened walls for enhanced protection from the environment and internal stores of energy-rich oils. These spores often have intricate and beautifully ornate exteriors.

Physiology of soil algae

Photosynthesis

Photosynthesis is the pathway by which most chemical energy enters the biosphere. Oxygenic (plant-like, oxygen evolving) photosynthesis is the most important type of photosynthesis, although other types of photosynthesis (anoxygenic) exist among bacteria (Chapter 3). The reducing power and energy derived from light are used to reduce carbon dioxide to sugars for growth. As mentioned previously, other compounds such as lipids can also be produced and stored. Production of the latter has generated interest in using algae and cyanobacteria to produce biofuels (Box 7.2). Most photosynthesis in the ocean is carried out by microbes (algae and cyanobacteria), whereas vascular plants are most important in terrestrial

BOX 7.2

Economic uses of microbial phototrophs

The metabolic characteristics of cyanobacteria and algae have been used to obtain various marketable products through large-scale cultivation of appropriate strains. Harvesting cyanobacteria like *Arthrospira* from natural lake blooms in Africa, or *Nostoc commune* from soils in China, for human consumption has long provided high-protein dietary supplements for traditional communities. *Arthrospira* is currently cultivated in commercial farms and marketed in Western societies under the name "Spirulina." The ability of some algae, including soil strains of the green alga *Chlorella*, to accumulate large amounts of lipids intracellularly has been tapped for the production of diesel-like biofuels. Carotenoids or other pigments produced by various algae and cyanobacteria, for example, the soil cyanobacterium *Leptolyngbya*, have value as food colorants and other uses. There is also interest in using algae for the production of poly-unsaturated fatty acids (similar to vegetable oil) for human consumption. A whole industry sector has developed around these and other specific products, essentially using only carbon dioxide and water as raw materials. Less developed, but perhaps more exciting, is the potential of using algae or cyanobacteria in the production of biohydrogen from water, by re-directing the reducing equivalents produced in the light phase of photosynthesis to the formation of hydrogen gas, or by forcing fermentation of the sugar polymers (such as starch) that accumulate during the dark phase of photosynthesis. In the environmental area, technologies are being developed to scale-up the production of appropriate soil cyanobacteria in order to produce inoculum to promote the growth of biological soil crusts in the restoration of degraded arid land soils.

A microbial nursery to produce soil biocrusts in a greenhouse at Arizona State University (left) and the algal production farm run by Qualitas Health (right). *Source: Photos courtesy of Eneko Ganuza, with permission.*

photosynthesis. However, in desert ecosystems, cyanobacteria and algae can be major contributors to primary productivity through photosynthesis (Figs. 7.3 and 7.4).

In eukaryotic algae and higher plants, photosynthesis occurs in **chloroplasts,** organelles that contain all the photosynthetic pigments, electron-transport systems, and enzymes needed for light reactions and carbon fixation. In fact, chloroplasts are "descended" from ancient cyanobacteria that entered into symbiosis with eukaryotic partners. Cyanobacteria, being prokaryotic, do not have chloroplasts. Instead, photosynthesis occurs on intracellular membranes, and the enzymes used for carbon assimilation are found in the cytoplasm.

Soil cyanobacteria and algae use three types of pigments to capture light: chlorophylls, carotenoids, and phycobiliproteins. As mentioned previously, chlorophyll *a* is the main

FIGURE 7.3 **Populations of the cyanobacterium *Microcoleus vaginatus* greening the plant interspaces of the Chihuahuan Desert of New Mexico after a summer monsoon rain.** *Source: Corey Nelson, with permission*

FIGURE 7.4 **Hypolithic cyanobacteria coating the underside of translucent pebbles in the Namib Desert.** *Chroococcidiopsis,* which usually dominates these environments, is perhaps the most desiccation-resistant cyanobacterium of all. *Source: Heather Throop, with permission*

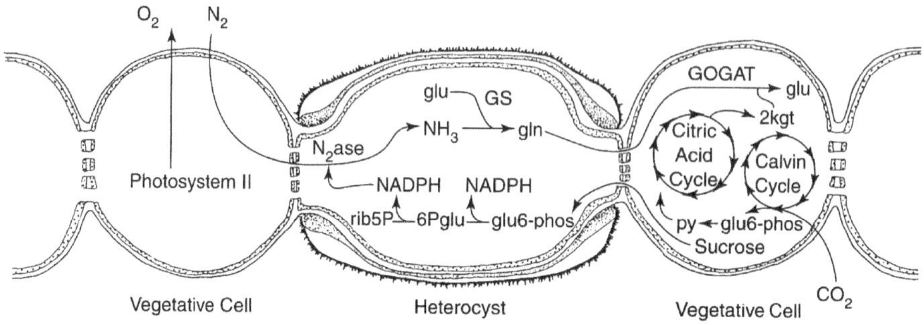

FIGURE 7.5 Diagram of carbon and nitrogen transfer between heterocysts and vegetative cells in filamentous cyanobacteria. Enzymes: *GS*, glutamine synthetase; *GOGAT*, glutamine-oxoglutarate amidotransferase (glutamate synthase); N_2ase, nitrogenase. Metabolites: 6Pglu = 6-phosphogluconate; gln = glutamine; glu = glutamate; glu6-phos = glucose-6-phosphate; NADPH = nicotinamide-adenine dinucleotide phosphate (reduced); py = pyruvate; rib5P = ribose-5-phosphate; 2kgt = 2-ketoglutarate. See Fig. 15.7 for additional details. *Source: Adapted from Haselkorn et al., 1978. Used with permission. Figure redrawn by Kim Luoma. Used with permission.*

reactive pigment in the photosystems of both algae and cyanobacteria, but other types of chlorophylls plus carotenoids (algae and cyanobacteria) and phycobiliproteins (cyanobacteria) can additionally serve to harvest and funnel light energy to the reaction centers in the photosystems. The various chlorophylls (*a*, *b*, c, *d*, and *f*) have different molecular structures that allow them to absorb different wavelengths of light; thus, having more than one type broadens the spectrum of light capture. Nevertheless, chlorophyll *a* is irreplaceable in the photosynthetic process because it is the active pigment in the reaction centers.

As discussed in Chapter 3, the reducing equivalents extracted from water in the light phase of photosynthesis are used to fix carbon from CO_2 in the so-called dark reactions of photosynthesis. While several biochemical pathways for carbon fixation exist in nature, all algae, plants, and cyanobacteria use the Calvin cycle (reductive pentose phosphate pathway). In addition to being used for carbon fixation, light-derived reducing equivalents can be directed to other metabolic needs such as the chemical reduction of nutrients required for growth, an example of which is given in the next section on dinitrogen fixation.

Dinitrogen fixation

Nitrogen is an essential component of all cells. Nitrogen occurs in the atmosphere as gaseous dinitrogen (N_2), a chemically stable form of nitrogen completely unavailable to all eukaryotic organisms (including vascular plants) and most bacteria. To be usable by the vast majority of life forms on Earth, dinitrogen must first be reduced or "fixed" to ammonia (NH_3), a biochemically difficult and metabolically costly process carried out only by certain bacteria (including some cyanobacteria) and archaea (Chapter 16). These organisms are called diazotrophs or dinitrogen fixers. The enzymes responsible for N_2 fixation are destroyed by molecular oxygen (O_2), a situation that requires diazotrophs to provide a low-oxygen environment in which N_2 fixation can occur. Many cyanobacteria accomplish this by producing specialized cells, called **heterocysts** (Fig. 7.5), that are able to restrict oxygen gas from their interiors using one or more of the following strategies:

- thick walls containing glycolipids that slow O_2 diffusion into the cell;
- high levels of cellular respiration to consume any O_2 that does diffuse into the cell;
- elimination of photosystem II from the heterocyst, the biological apparatus that produces oxygen gas during photosynthesis, but keeping photosystem I, which produces energy for the process; and
- importing reduced forms of organic carbon from adjacent cells, thereby avoiding the need for photosynthesis.

The heterocysts export fixed nitrogen in the form of amino acids to the rest of the trichome. Heterocystous genera found in soils include *Tolypothrix, Nostoc, Scytonema,* and *Anabaena.* Some non-heterocystous genera of cyanobacteria can also fix dinitrogen. These forms, like *Lyngbya,* fix nitrogen only at night when they are not producing oxygen. The globally most abundant soil cyanobacterium, *Microcoleus vaginatus,* cannot fix nitrogen but enters into a loose symbiosis with heterotrophic diazotrophic soil bacteria living on its sheath; these bacteria fix nitrogen in exchange for fixed carbon compounds from photosynthesis, which the cyanobacterium "feeds" to them.

Heterocystous cyanobacteria in **lichens** (symbiotic associations between a photosynthetic **photobiont** and a fungal partner; Chapter 17) are important sources of fixed N in some soils. Although not organisms in the strict sense, lichens are given taxonomic names based on the fungal partner identity. A common example of this is the lichen *Collema,* which contains the heterocystous cyanobacterium *Nostoc.* N_2-fixing cyanobacteria can be the dominant source of nitrogen in many tropical, arid, and aquatic ecosystems, as well as rice paddies. Rates of N_2 fixation are dependent on many factors including moisture, temperature, light, pH, carbon stores, and salinity (Belnap and Lange, 2003; Chapter 16). Much of the dinitrogen fixed by cyanobacteria or cyanolichens in soil leaks to the surrounding medium, apparently because of damage to the plasma membrane caused by recurrent wetting and drying cycles. This nitrogen can then be utilized by nearby organisms or transported to the rest of the ecosystem with runoff or percolating water (Johnson et al., 2007).

Ecology of algae and cyanobacteria in soil

Environmental factors affecting soil algae

Many factors affect the distribution and physiological functioning of soil algae and cyanobacteria. These include soil moisture, light, temperature, acidity, nutrient availability, and pollutants.

Moisture. All cyanobacteria and algae, and indeed all organisms generally, require moisture for metabolic activity. Therefore, moisture ultimately controls all activities of these organisms, including photosynthesis, N_2 fixation, respiration, and growth. Both cyanobacteria and eukaryotic algae use hygroscopic, polysaccharide sheaths to retain water, thus prolonging their activity during periods of low moisture availability. Consequently, organisms found in extreme deserts often have thicker sheaths than the same species found in more hospitable climates. Conversely, when these same species are grown in liquid culture in the laboratory, they often have no or minimal sheaths, probably because they have no need to conserve water. Many cyanobacteria and algae can withstand long periods of desiccation, with some species surviving for decades in a dry state. The exact mechanisms of adaptation to an existence under recurrent desiccation are not well known, but they involve the differential expression of large parts of their genetic capacity. Aside from the production of hygroscopic sheaths or capsules, the production of compatible cellular solutes to counteract osmotic stresses brought about by wetting and drying cycles does seem to play a prominent role.

Light. Light is needed for photosynthesis. However, high UV radiation at the soil surface discourages colonization by cyanobacteria and algae, while the visible wavelengths required for photosynthesis drop off quickly within the soil, virtually disappearing within a few millimeters (Garcia-Pichel and Belnap, 2001). Consequently, most cyanobacterial and algal populations concentrate between 1 and 4 mm below the soil surface. The thickness of this photic zone is variable, as it depends on soil characteristics that affect the penetration of light (e.g., mineralogy, water content, soil texture and structure, organic matter content, and plant residues on the soil surface). Exposure to full solar radiation can be very harmful to both algae and cyanobacteria, as photosynthesis, respiration, N_2 fixation, nutrient uptake, and protein synthesis are all negatively affected by UV radiation. Many cyanobacteria produce high levels of sunscreen pigments such as scytonemin to avoid excess UV radiation, which they excrete into their extracellular sheaths or capsules (see the yellow/brown sheaths or capsules in Fig 7.1B,D). Because screening from UV is so important, scytonemin can represent up to several percent of cyanobacterial dry biomass in those strains that produce it. This investment in a passive protective mechanism is particularly effective for soil cyanobacteria that spend long periods of exposure while in an inactive, desiccated state, when alternative protection strategies (for example DNA repair) would be of little use. In *Nostoc commune* colonies, sheath pigments can reduce the amount of UV in the cell by 90%, while not significantly reducing the longer wavelengths needed for photosynthesis. Species with large amounts of UV-protective pigmentation (e.g., *Scytonema, Nostoc, Tolypothrix)* are generally found on the very surface of the soil, above relatively less-pigmented species (e.g., *Microcoleus).* These sunscreen-lacking species glide to the surface only when soils are wet, returning to shaded depths once the soils begin to dry or with prolonged UV exposure. The accumulation of large amounts of scytonemin on the soil surface can impart conspicuous changes to the soil, making the populations of these microbes visible from satellites, and altering the soil albedo sufficiently to result in significant local warming effects (Couradeau et al., 2016). Even with their many adaptations, there is significant mortality of cyanobacteria and algae during times of high UV exposure, such as in summer.

Temperature. Given sufficient moisture and light, temperature is the next most common factor controlling the metabolic activity of cyanobacteria and algae. Although some species can withstand extreme temperatures when active (including cyanobacteria that live in thermal hot springs at temperatures up to 73°C, e.g., *Synechococcus lividus;* Castenholz, 2015) most cyanobacteria and algae prefer more temperate conditions when physiologically active, regardless of the climates where they are found. Therefore, even species from hot deserts or tundra regions generally exhibit maximum carbon and N_2 fixation rates at about 25°C–30°C and die quickly when exposed to high or low temperatures when wet. Even cyanobacteria from Antarctic soils have optimal temperatures at or above 15°C, which are almost never reached in their environment. The optimal temperature range for dinitrogen fixation by soil cyanobacteria and cyanolichens is between 20°C and 30°C. Temperature is a main driver of community composition in soil cyanobacteria, and typical gradients of community dominance in response to temperature have been described for both heterocystous and non-heterocystous forms. For example, *Microcoleus vaginatus* prefers colder temperatures than *M. steentrupii,* and *Scytonema,* typical of hot deserts, is replaced by *Tolypothrix* in colder settings. (Giraldo-Silva et al., 2020). However, when dry, many species can withstand temperatures down to –58°C as experienced during Antarctic winters and over 90°C as experienced during desert summers on the soil surface.

Acidity. Cyanobacteria prefer soils with circumneutral or alkaline pH. Their populations are generally sparse or absent in soils with pH levels of 5–6 and are typically absent at pH levels less than 4. Conversely, although green algae can be found in alkaline soils, they reach their greatest abundance in soils with pH levels of 7 or less. Many are also quite capable of growing in soils with a pH below 4.

Nutrient availability. Cyanobacteria and algae require little in the way of nutrients. As discussed previously, both provide their own fixed carbon and most cyanobacteria fix their own nitrogen. Additional nutrients are obtained from the soil and airborne dust. Especially important among these is phosphorus, which is exceeded only by nitrogen in the amount needed by cells, given that it is an essential part of many cellular constituents including nucleic acids, ATP, NAD(P)H, and membrane phospholipids. Many soil cyanobacteria can accumulate P in the form of intracellular polyphosphate. Secretion of polysaccharides is also important in nutrient acquisition. Polysaccharides are sticky and physically bind nutrients and nutrient-laden soil particles near the cell and, because they are negatively charged, they function similarly to silicate clay minerals in attracting positively charged cations (e.g., potassium and calcium).

Pesticides and pollutants. Oil spills, insecticides, and herbicides can affect individual species of cyanobacteria and green algae (Belnap and Lange, 2003). These effects can be either positive or negative, depending on the compound and species involved. Compounds that uncouple photosystems, like hydrogen sulfide, are the most harmful, with the true algae often showing a greater sensitivity to these compounds than cyanobacteria. Therefore, exposure to these agents can potentially alter the relative abundance of these groups in soils. Cyanobacteria exposed to sulfide can survive by switching from oxygenic to anoxygenic photosynthesis. Exposure to crude oil, oil dispersants, insecticides, herbicides, and phenolic compounds can decrease N_2 fixation by *Nostoc.* Air pollutants also affect cyanobacteria and algae. Nitrogen deposition from the atmosphere and fertilizers increases chlorophyll *a* content and decreases N_2 fixation by soil cyanobacteria. Free-living and lichenized *Nostoc* show stimulation of N_2 fixation at low concentrations of, or short exposure times to, the elements arsenic, nickel, lead, palladium, and zinc. However, longer-term exposures to cadmium, lead, and zinc inhibit N_2 fixation. Alkaline soils often buffer the effects of acidic compounds, minimizing the effect of acid deposition. Areas exposed to intense acid rain undergo community shifts from cyanobacteria to the more tolerant green algae.

Distribution and abundance of soil algae

Algae and cyanobacteria are found almost everywhere light is available, including in or on marine and fresh waters, intertidal zones, the atmosphere, rocks, soils, plants, and animals (e.g., invertebrates and polar bear hair). Because these organisms are metabolically active only when hydrated, their biomass per unit area is greater where the effective precipitation is high. However, their biomass is also limited by the amount of surface area available for colonization. Whereas tropical soils can have very high amounts of algae and cyanobacteria per unit surface area in the absence of vegetation, normally only a small percent of the soil surface can be colonized because of shading from dense plant cover. In contrast, deserts (including the cold arctic deserts) often sustain large populations and algae due to low plant cover. Globally, the soils of semiarid and arid lands alone are estimated to contain 10^9 million metric tons of cyanobacteria. Populations generally range between 10^3 and 10^4 cells g^{-1} of surface soil but can be as much as 10^8 cells g^{-1} soil (Box 7.3). The number of species and individuals can vary widely on a scale of 1 cm or less, due to fluctuations in soil moisture, pH, nutrients, and light.

BOX 7.3

Estimation of soil algal populations

Assessing algal populations or their diversity in the soil using microscopy or cultivation-based methods is difficult and prone to error. Direct microscopic observations do not differentiate between living and dead organisms and are difficult due to the presence of mineral particles and detritus. Culture methods enumerate only those algae able to grow in the particular medium and environmental conditions provided, enriching for those forms by orders of magnitude. Standards for counting soil algae or preparing enrichment media for the isolation of strains are essentially nonexistent. Some forms will not grow well in standard liquid culture, and taxonomic placement is often uncertain. Alternatively, total algal biomass may be estimated by the chlorophyll *a* content of soil, but this too has limitations, since its content can vary with light intensity and species. As a result, estimations of algal populations in soils using these traditional methods may differ greatly among studies.

Modern studies using molecular methods based on DNA have become ubiquitous in this regard and are nowadays the accepted standard. The most common approach is based on the extraction of DNA from soil samples, PCR amplification of marker genes such as the 16S rRNA genes, and their study by sequencing and bioinformatic analyses to yield the relative abundances of specific phylogenetic types (Chapter 4). Absolute determinations of population size can be achieved by using quantitative qPCR (Fernandes et al., 2018). Nevertheless, cultivation and study of individual taxa are still important because much of our knowledge of the physiology and the ecology of soil algae is based on laboratory cultures.

Soil cyanobacterial and algal numbers are seasonally dynamic, with biomass increasing rapidly under favorable conditions (e.g., moderate temperatures and wet soils) and declining slowly during periods of desiccation, high UV incidence, and temperature extremes. For example, both cyanobacteria and algae suffer significant mortality during the fall through spring seasons if they are hydrated and soils freeze. Likewise, exposure to UV radiation and desiccation can result in a 60% decrease in biomass during summer months. Prolonged drought, in general, also reduces numbers, as does the osmotic shock of wetting after prolonged desiccation.

Many of the algae, cyanobacteria, and lichens that occur in desert soils also live on (epilithic) or inside (endolithic) rock. Alternatively, the hypoliths live under surface translucent

BOX 7.4

Two ways to reveal soil cyanobacteria

Many soil cyanobacteria seek refuge under a few mineral grains of soil as a soil dries, so they are usually cryptic and hard to see. But following rainfall, they immediately migrate to the surface of the soil to bask in the sun and photosynthesize. The soil surface then "greens up" (Fig. 7.3), allowing one to easily observe cyanobacterial blooms between plants, as well as their subsequent retreat to below the soil surface as it dries. If one is too impatient to wait for a rain event, by all means create one yourself. This can be accomplished by wetting the soil thoroughly and covering it with thin cloth or piece of screen to simulate the light environment of an overcast sky (these cyanobacteria may not like full sunlight and stay below the soil surface otherwise) and to prevent the soil from drying too quickly. After half an hour or so, re-examine the soil surface for the appearance of a green carpet. Note what happens to this green carpet as the patch of soil dries out.

A simpler experiment is to look for small translucent pebbles sitting on the soil surface. Good candidates are quartz, flint, calcite, or gypsum pebbles. Simply pick them up and turn them over to observe the pebble's underside. You are likely to see an olive-green cover of the hypolithic cyanobacterium *Chroococcidiopsis* (Fig. 7.4). Afterward, remember to put it back in place the way you found it.

pebbles. Being **endolithic** or **hypolithic** (Box 7.4; Fig. 7.4) has many advantages because the habitat is relatively stable, while UV exposure, water loss, herbivory, and accidental damage are reduced. However, nutrients and carbon dioxide may be limiting, and light does not readily penetrate rocks. Therefore, most photosynthetic organisms are within a few millimeters of the rock surface. Some algae and cyanobacteria that inhabit rocks have special adaptations based on calcium uptake and transport to enlarge the cavities they occupy (Guida and Garcia-Pichel, 2016). Endoliths can greatly accelerate weathering rates in carbonate-cemented rocks. Because large expanses of rocks are common in deserts, these organisms are thought to substantially contribute to rates of soil formation in these regions.

Biotic interactions involving soil algae

Cyanobacteria and algae interact with other organisms in many ways, including beneficial symbioses, predator-prey interactions, and pathogenic relationships (Chapter 10). These interactions occur at many scales, ranging from an individual to entire ecosystems. Cyanobacteria and algae can be the dominant primary producers in soils and are also part of the soil food chain. There are many **cyanophages,** viruses that infect and often kill cyanobacteria (Chapter 9). Heterotrophic bacteria, fungi, and protozoa are also major consumers and decomposers of cyanobacteria and algae. The nitrogen provided by the cyanobacteria can stimulate heterotrophic decomposition. Nematodes, mites, tardigrades, isopods, snails, mole crickets, beetles, ants, and termites have all been observed consuming cyanobacteria and algae (Chapter 8), although no quantitative estimations of their effect have been reported. Fecal material from these predatory species can potentially be an important source of nitrogen for the ecosystems in which they occur. Cyanobacteria and algae can also form symbiotic relationships with many organisms, including fungi, plants, and animals. Many algae maintain tight resource exchange partnerships with specialized groups of heterotrophic bacteria in their close proximity, a "**cyanosphere**" or "phycosphere" by analogy to the rhizosphere of plants (Couradeau et al., 2019). **Lichens** are formed when either a cyanobacterium (e.g., *Nostoc)* or a green alga (e.g., *Trebouxia)* is incorporated into the tissue of 1 of the 18,000 fungal species (mostly within the Ascomycetes, Chapter 6) known to establish these symbioses. The cells of the photosynthetic partner are simply enveloped by a layer of fungal tissue and are not found within individual fungal cells. It is not known if this is a "voluntary" association or if the fungi "imprison" their **photobiont**. Once the relationship is formed, the photobiont undergoes significant morphological and physiological changes, including for cyanobacteria an increase in heterocyst number and an increased leakage of carbon and nitrogen to the fungal partner. It is theorized that in return, the fungi provide protection and increased moisture availability to the photobiont and possibly also phosphorus.

Biological soil crusts

Although soil algae and cyanobacteria are often overlooked in many ecosystems, they make many significant contributions to ecosystem functioning. These roles are especially important in arid and semiarid areas where soil formation rates, fertility, and moisture are naturally low, and the sparse vegetation does little to protect the soil surface from erosion. The lack of vascular plant vegetation in deserts provides ample space for colonization by cyanobacteria, green algae, lichens, and mosses. These organisms weave through the top layers of soil, forming a cohesive layer that is collectively referred to as a **biological soil crust** or **biocrust** (Fig. 7.6; Belnap and Lange, 2003). In deserts, biological soil crusts are dominated by

FIGURE 7.6 **Biological crust in southern Utah, with a walking trail cutting through it.** Such remote areas in the Colorado Plateau, historically spared from cattle grazing and other disturbances, display some of the most developed and conspicuous biological soil crusts, owing to their dark pigmentation by the cyanobacterial sunscreen scytonemin and to the highly rugose three-dimensionality imparted by frost-heaving. *Source: Jason Hollinger/Flickr/ Creative Commons License.*

cyanobacterial species, as low rainfall limits the cover of lichens and mosses as well as plants. As rainfall increases, the cover and number of species of lichens and mosses increases, while cyanobacterial and algal cover and species richness decrease. Biological soil crusts contribute substantially to soil fertility in many ways. They can be the dominant source of nitrogen for the plant and soil community, as well as an important source of fixed carbon for the decomposer community. Crusts increase the capture and retention of nutrient-rich dust, increasing the fertility and water-holding capacity of soil. Cyanobacteria secrete powerful metal chelators and other substances that maintain nutrients in bioavailable forms, an important benefit in high-pH desert soils. Crusts reduce soil erosion by binding soil particles together and increasing the formation rate and size of soil aggregates (Fig. 7.7). In areas where soils freeze, the frost-heaving of crusted soils results in a greatly roughened soil surface that further increases capture and retention of water, dust, seeds, and organic matter (Fig. 7.6). Soil crusts increase moisture retention, aeration, and porosity, and their dark color increases soil temperatures. Repeated experiments show that plant productivity and the nutritional content of plant tissue are higher when plants grow in biologically crusted soils compared to adjacent soils without crusts.

Soil crusts are highly susceptible to physical damage, particularly to compressional forces due to foot traffic, off-road vehicles, and livestock grazing. Crushing of the soil surface decreases cover and the biomass of soil crust organisms and also flattens the roughened soil surface. This in turn decreases dust capture, water infiltration, and soil surface temperature. Because broken crusts have less resistance to raindrop impact and the smaller soil aggregates are easily moved, erosion susceptibility is increased many-fold after disturbance. Dinitrogen fixation is greatly reduced or halted when crusts are disturbed, and carbon fixation is reduced as well. Burial of these photosynthetic organisms often means death; thus, blowing or washing sediment can often kill crusts adjacent to those that are directly disturbed. Anthropogenic pressure has resulted in a loss of biocrusts in many arid lands, and scientists are currently developing strategies to restore natural crusts in these degraded areas, particularly given that natural re-colonization can be very slow in the absence of an inoculum source (Box 7.2). That said, colonization of disturbed areas by cyanobacteria and algae can

FIGURE 7.7 **Scanning electron micrograph of biological soil crusts (bar = 100 μm).** The strands are sheaths of *Microcoleus vaginatus* winding between and connecting sand grains. *Source: Original photo by J. Belnap.*

be rapid, as long as propagules are available nearby and precipitation events are sufficient to allow growth. However, recovery of full, pre-disturbance biomass and soil structure is often fairly slow in environments where effective soil moisture is low. Natural recovery of cyanobacterial and algal populations after severe disturbance has been estimated to range from 10 years in cool deserts to over 100 years in hot deserts. Lichen recovery, however, can take centuries

Summary

The dominant groups of photosynthetic microorganisms in soils are the prokaryotic cyanobacteria and the eukaryotic green algae, diatoms, and yellow-green algae. Cyanobacteria are bacteria and therefore have no membrane-bound organelles; many can fix atmospheric dinitrogen. Algae are eukaryotic and have cellular organelles. Both have photosynthetic systems similar to those of vascular plants. Cyanobacteria divide by fission, whereas algae undergo both sexual and asexual reproduction. Both are able to withstand extreme environments, some forming spores or resting cells in response to especially harsh conditions, while others simply modulate the resistance of vegetative cells to the environment. Most cyanobacteria and algae are found just below the soil surface to a depth of 4 mm. This distribution strikes a balance between harmful UV radiation, which is too great at the soil surface, and the need for photosynthetically active radiation, which is too low at greater depths. Cyanobacteria and algae require soil moisture for metabolic activity; once wet, rates of activity are determined primarily by soil temperatures. These organisms provide carbon and nitrogen to soils, chelate nutrients, increase mineral weathering rates, influence local hydrologic cycles, and help retain nutrient-rich dust. They also form the base of the soil food chain, providing a food source for other soil biota (e.g., heterotrophic bacteria, nematodes, protozoa, and microarthropods). In desert ecosystems, cyanobacteria, algae, lichens, and other biota form biological soil crusts. These soil crusts are essential in stabilizing soils and increasing soil structure and fertility. Crust organisms have low resistance to crushing, and thus are disrupted by many common types of land use.

Study questions

1. What are the advantages of spore formation among the algae and cyanobacteria? Are there any disadvantages?
2. List some characteristics of cyanobacteria and algae that allow them to be successful pioneer organisms.
3. Some cyanobacteria fix nitrogen from N_2 in specialized vegetative cells called heterocysts. In which ways are heterocysts well-suited to housing the enzymes involved in N_2 fixation?
4. What factors affect the depth to which algae will grow in the soil?
5. What are the major ecological roles of biological soil crusts? How does disturbance affect these ecological roles?

6. Why do algae often have more than one type of photosynthetic pigment?
7. Land plants appeared in the fossil record some 0.4 billion years ago. However, terrestrial ecosystems existed much earlier. Can you imagine what these earlier systems might have looked like?

References

Belnap, J., Lange, O.L., 2003. Biological soil crusts: Structure, function, and management. Ecological Studies Series. Springer-Verlag, New York.

Beraldi-Campesi, H., Farmer, J.D., Garcia-Pichel, F., 2014. Modern terrestrial sedimentary biostructures and their fossil analogs in Mesoproterozoic subaerial deposits. Palaios 29, 45–54.

Castenholz, R.W., 2015. Portrait of a geothermal spring, Hunter's Hot Springs, Oregon. Life 5, 332–347.

Couradeau, E., Karaoz, U., Lim, H.C., Da Rocha, U.N., Northen, T., Brodie, E., Garcia-Pichel, F., 2016. Bacteria increase arid-land soil surface temperature through the production of sunscreens. Nat. Commun. 7, 1–7.

Couradeau, E., Giraldo-Silva, A., De Martini, F., Garcia-Pichel, F., 2019. Spatial segregation of the biological soil crust microbiome around its foundational cyanobacterium, *Microcoleus vaginatus*, and the formation of a nitrogen-fixing cyanosphere. Microbiome 7, 55.

Fernandes, V.M., Machado de Lima, N.M., Roush, D., Rudgers, J., Collins, S.L., Garcia-Pichel, F., 2018. Exposure to predicted precipitation patterns decreases population size and alters community structure of cyanobacteria in biological soil crusts from the Chihuahuan Desert. Environ. Microbiol. 20, 259–269.

Garcia-Pichel, F., Belnap, J., 2001. Small-scale environments and distribution of biological soil crusts. In: Belnap, J., Lange, O.L. (Eds.), Biological Soil Crusts: Structure, Function, and Management. Springer, Berlin, Heidelberg, pp. 193–201.

Garcia-Pichel, F., Zehr, J.P., Bhattacharya, D., Pakrasi, H.B., 2020. What's in a name? The case of cyanobacteria. J. Phycol. 56, 1–5.

Giraldo-Silva, A., Fernandes, V., Bethany, J., Garcia-Pichel, F., 2020. Niche partitioning with temperature among heterocystous cyanobacteria (*Scytonema* spp., *Nostoc* spp., and *Tolypothrix* spp.) from biological soil crusts. Microorganisms 8 (3), 396.

Guida, B.S., Garcia-Pichel, F., 2016. Extreme cellular adaptations and cell differentiation required by a cyanobacterium for carbonate excavation. Proc. Natl. Acad. Sci. 113, 5712–5717.

Haselkorn, R., Mazur, B., Orr, J., Rice, D., Wood, N., Rippka, R., 1978. Heterocyst differentiation and nitrogen fixation in cyanobacteria (blue-green algae). In: Newton, W.E., Orme-Johnson, W.H. (Eds.), Nitrogen Fixation, vol. 2, University Park Press, Baltimore, MD, pp. 259–278.

Johnson, S.L., Neuer, S., Garcia-Pichel, F., 2007. Export of nitrogenous compounds due to incomplete cycling within biological soil crusts of arid lands. Environ. Microbiol. 9, 680–689.

Schopf, J.W., 1983. Earth's Earliest Biosphere: Its Origin and Evolution. Princeton University Press, NJ.

Tang, Q., Pang, K., Yuan, X., Xiao, S., 2020. A one-billion-year-old multicellular chlorophyte. Nat. Ecol. Evol. 4, 1–7.

Further reading

Garcia-Pichel, F., 2009. Cyanobacteria. In: Schaechter M. (Ed.), Encyclopedia of Microbiology, third ed., Elsevier Inc. https://doi.org/10.1016/B978-012373944-5.00250-9.

Madigan, M.T., Bender, K.S., Buckley, D.H., Sattley, W.M., Stahl, D.A., 2017. Brock Biology of Microorganisms, fifteenth ed. Pearson, New York.

Nash, III, T.H., 1996. Biology of Lichens. Cambridge University Press, England.

Overmann, J., Garcia-Pichel, F., 2006. The phototrophic way of life. In: Dworkin, M., Falkow, S., Rosenberg, E., Schleifer, K.-H., Stackebrandt, E. (Eds.), The Prokaryotes (vol. 2): Handbook on the Biology of Bacteria: Ecophysiology and Biochemistry, pp. 32–85.

Van den Hoek, C., Mann, D., Jahns, H.M., 1995. Algae: An Introduction to Phycology. Cambridge University Press, England.

Vymazal, J., 1995. Algae and Element Cycling in Wetlands. Lewis Publishers, Boca Raton, FL.

Whitton, B.A., Potts, M., 2000. The Ecology of Cyanobacteria. Kluwer Academic Publishers, Dordrecht.

CHAPTER

8

The soil fauna

Josef H. Görres[a], *Jose A. Amador*[b]

[a]Plant and Soil Science Department, University of Vermont, Burlington, VT, United States; [b]Department of Natural Resources Science, University of Rhode Island, Kingston, RI, United States

Earthworms, though in appearance a small and despicable link in the chain of nature, yet, if lost, would make a lamentable chasm…worms seem to be the great promoters of vegetation, which would proceed but lamely without them—***Gilbert White of Selborne.***

Members of the soil fauna (animals) are ascribed all manner of attributes, from plant pests to predators to soil enhancers. In this chapter, we describe both **invertebrate** animals that live in the soil and their habitat and additionally examine their activities in the context of the soil food-web (Fig. 8.1). We focus on the role of soil animals in controlling microbial pathogens, mineralizing nutrients, changing microbial community composition, and enhancing primary production. Like aboveground fauna, the physical structure of the ecosystem places constraints on the activities of the soil fauna, especially in relation to the microflora. Because of these constraints, the activities and interactions of the soil fauna are also discussed in the context of the soil physical environment. Climate change—through its effects on soil moisture and temperature dynamics—also impacts faunal activities and interactions and is addressed throughout the chapter.

The organisms that use the soil as habitat at some point in their life are numerous and span all three domains of life on Earth. As with archaea, bacteria, fungi, cyanobacteria, and viruses, only a fraction of the soil fauna has been described. For example, only about 4000 earthworm species have been described of the estimated 8000 species on Earth. For our purposes, the soil fauna includes those animals that spend a significant portion of their life cycle within the soil profile or at the soil–litter interface. In this chapter, we will focus on the **earthworms, microarthropods** (mites and collembolans), **nematodes**, and **protozoa**. Phylogenetically, organisms in the first three groups belong in the kingdom Animalia, whereas protozoa belong in the kingdom Protozoa.

Despite the obvious morphological differences among them, soil animals share a number of similarities:

- They are **eukaryotes** (Chapter 1).
- With few exceptions, they are **aerobes** and use oxygen as a terminal electron acceptor. The exceptions are some species of nematodes and protozoa that are **facultative anaerobes** (i.e., they can tolerate anoxic conditions but use oxygen preferentially).

Principles and Applications of Soil Microbiology. http://dx.doi.org/10.1016/B978-0-12-820202-9.00008-3
Copyright © 2021 Elsevier Inc. All rights reserved.

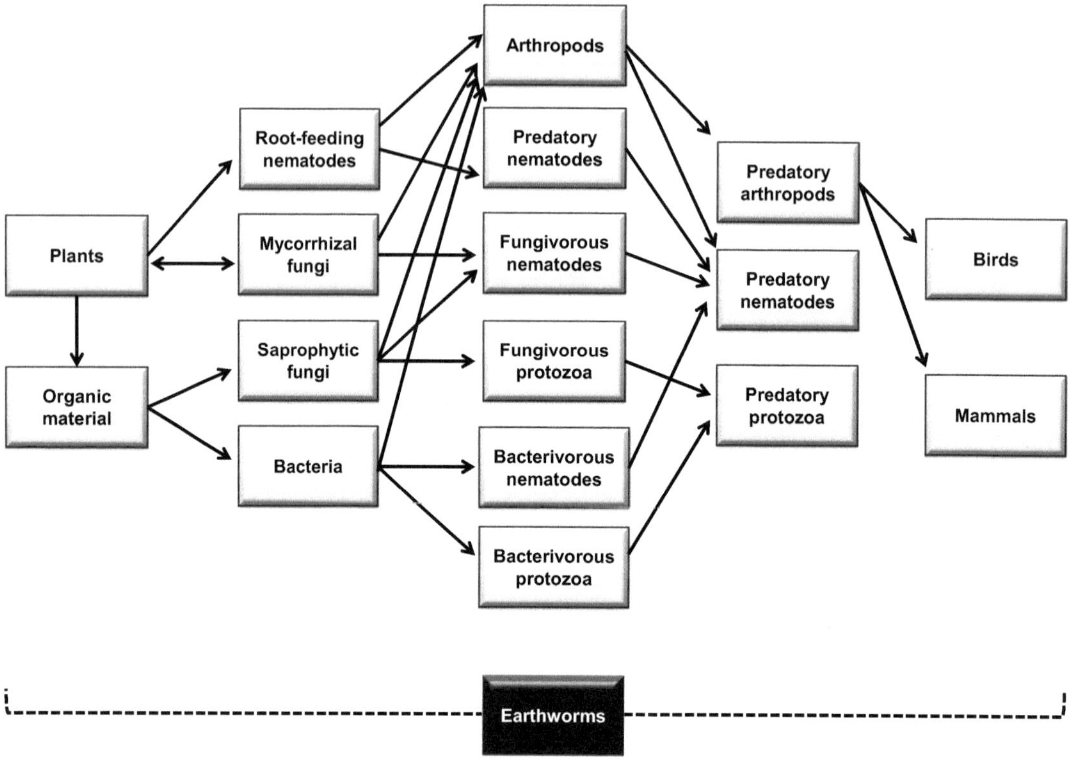

FIGURE 8.1 **The soil foodweb showing multiple interactions among plants, soil fauna, and soil microorganisms.** Earthworms interact with various groups of organisms within the foodweb.

- They are **chemoheterotrophs** because they meet their carbon and energy needs by metabolizing organic compounds (Chapter 3).
- They are **motile** and travel distances proportional to their size by propelling themselves on their own power. Generally, earthworms travel meters, collembolans (springtails) centimeters, and nematodes and protozoa a few millimeters.

Habitat

Soil animals differ in the ways that they inhabit the soil, partly dictated by differences in their size. The length and width of an organism constrains its ability to take advantage of habitat and food resources. The soil fauna can be divided into three size groups, based on their body width: (1) **macrofauna** (>2 mm), represented by earthworms; (2) **mesofauna** (0.1–2 mm), represented by mites, collembolans, and nematodes; and (3) **microfauna** (<0.1 mm), represented by protozoa (Fig. 8.2). The soil faunal community is also divided into organisms that are essentially **aquatic** (living in water-filled pores; e.g., protozoa and nematodes), organisms that are **aerial** (living in the air-filled spaces; e.g., microarthropods), and

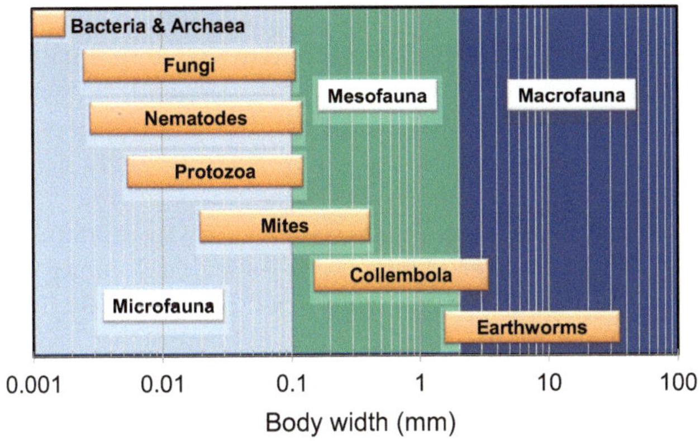

FIGURE 8.2 **Body width ranges for soil organisms.**

those that engineer their own space (e.g., earthworms; Heal and Dighton, 1985). As noted in Chapter 2, soils comprise three states of matter: solid, liquid, and gaseous. From an ecological perspective, this is a good description of the partitioning of habitat available to soil organisms (see Fig. 1.1) that are not engineers. Soil pore spaces vary in size, water content, and air, resulting in variations in habitat:

- Members of the microfauna are generally sufficiently small to fit into **mesopores**, which are 30–100 μm in diameter, and **micropores**, which have diameters less than 30 μm. These pores are found within soil **aggregates** (or **peds**) and retain water at matric potentials less than field capacity (< −0.03 MPa). Thus, the basic unit of habitat for aquatic organisms like nematodes and protozoa is the water-filled pores of aggregates (Hattori, 1994). Relatively small aggregates (5 mm in diameter) can be home to hundreds of protozoa and tens of nematodes.
- Members of the mesofauna fit within pores with a diameter larger than 100 μm (**macropores**), especially the spaces between aggregates. Interaggregate pores drain quickly, making good habitat for these generally aerial organisms.
- Members of the macrofauna do not fit within the existing pore structure of soil, so they either create their own habitat by burrowing into the soil, use burrows created by other organisms, or live on the soil surface.

The activities of burrowing fauna, such as earthworms, alter the soil habitat for many other types of organisms by changing its physical structure and the distribution of resources. Burrows increase aeration and water movement within the soil profile. Greater gas exchange between the soil and the atmosphere increases the availability of oxygen for aerobic soil microorganisms and plant roots. Earthworm casts and soil lining earthworm burrows (the so-called "**drilosphere**") have a different porosity from bulk soil. The smaller pores created in the burrow walls may provide more habitat for nematodes (Görres and Amador, 2010).

A given pore space may be filled with water or air, depending on the current soil water potential. When soil dries, the habitat of the aerial soil community expands, and shrinks when soil becomes wet. The reverse is true for the aquatic soil fauna. These changes in water-filled pore space take place in response to rainfall, irrigation, snowmelt, and evapotranspiration. By altering the magnitude, intensity, and frequency of precipitation events and/or evapotranspiration, climate change can impact the habitat of both aerial and aquatic soil organisms. Thus, the availability of habitat for different kinds of animals depends on water dynamics as well as soil structure and texture, which control the distribution of pore sizes in the soil. This in turn affects decomposition/C mineralization rates, which are partially determined by soil matric potential (Davidson et al., 1998).

Soil animals have a number of adaptations to survive environmental changes in the soil. For example, some mites—which are aerial organisms—survive flooding of their habitat by clustering together. The air bubbles trapped between their bodies satisfy their demand for oxygen when soil is saturated. Other aerial organisms, like collembolans, avoid saturated conditions by migrating. Some aquatic species, such as nematodes and protozoa, form **cysts**—life stages with thick membranes or walls—or enter a state called **anhydrobiosis** when soil dries. The latter is a strategy to survive desiccation as well as freezing temperatures that is characterized by an almost complete loss of moisture from the organism.

In addition to controlling habitat availability and suitability for soil organisms, the pore and aggregate structure of soil, in conjunction with soil moisture, controls the ecological interactions among soil fauna within the soil foodweb (Fig. 8.1). For example, as soil dries bacterivorous nematodes that prey on bacteria can become trapped in pores with their bacterial prey, increasing predation as well as excretion of mineral N by the predators (Savin et al., 2001). In contrast, increased soil wetting can increase movement of organisms between soil aggregates, allowing them to move more freely among aggregates to quest for food resources, as is the case with some protozoans and nematodes (Görres et al., 1999).

Functional groups of soil fauna

Earthworms

Earthworms (Animalia: Annelida: oligochaetes) exhibit external and internal segmentation of the body. They share this characteristic with other annelids including polychaetes, or bristle worms, found mostly in aquatic sediments, and leeches, as well as enchytraeids, also known as potworms—small, unpigmented soil worms.

Earthworms are either **hermaphroditic** or **parthenogenic**; although most species reproduce by cross-fertilization, parthenogenic species reproduce without sex, producing clones. Ova are contained in cocoons that are deposited at or near the soil surface. Adult earthworms range from a few centimeters to almost 2 m in length for several giant earthworm species and from 2 to 20 mm in diameter. Most earthworms mature within 2 years of emerging from the cocoon, although there are exceptions. For example, earthworms used for composting have much shorter maturation periods, because high fecundity is desirable in vermicomposting beds to replace worms removed with compost. Typical cocoon production varies among species but ranges from 0.1 to 2 cocoons per adult per day for *Lumbricus rubellus*, a temperate species, and *Perionyx excavatus*, a tropical worm and popular vermicomposting species, respectively. Cocoons may contain between 1 and 4 ova (Dominguez, 2004).

Earthworms are found in most ecosystems throughout the world except for deserts, areas with permafrost, areas without soil or vegetation, and mountain ranges. Based on current

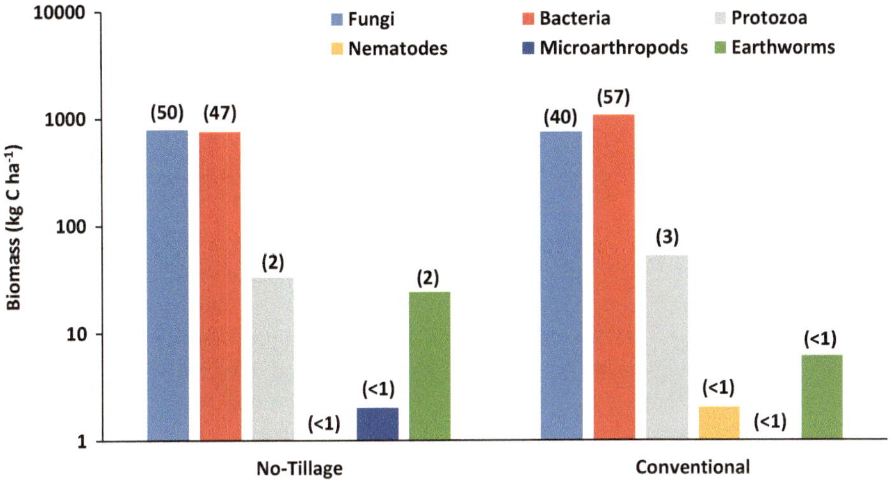

FIGURE 8.3 **Relative biomass of microbial and faunal groups in conventional (CT) and no-tillage (NT) soils in summer/fall at Horseshoe Bend Experimental Area, Georgia, United States.** Values in parentheses are percentage of total biomass. *Source: Adapted from Beare (1997).*

climate change predictions, their range is likely to expand into areas now covered in permafrost and contract in areas undergoing desertification. The number and species distribution vary considerably with the type of ecosystem, with numbers ranging from less than 10 m^{-2} in agricultural fields to more than 700 m^{-2} in forest soils. They are generally more abundant in undisturbed than disturbed environments. In agricultural ecosystems, their numbers depend on management practices employed. For example, earthworms constitute a greater proportion of the biomass in soil under no-tillage than under conventional tillage (Fig. 8.3).

To enumerate or identify earthworms, they must be driven out of the soil by drenching the soil with an irritant, such as a mustard or formaldehyde solution, or by saturating the soil with water. Electrical fields, vibration, hand sorting, and wet sieving are also used. As is true for other soil animals, estimates of the size and composition of earthworm communities are dependent on the method used (Box 8.1). The various methods introduce different biases into enumerations. For example, using mustard as a vermifuge was more effective for extracting anecic earthworms than applying an electrical field, while no difference was found for epigeic earthworms (Eisenhauer et al., 2008). When ecological experiments are conducted, the effect of a chosen method on non-target species needs to be considered (Table 8.1). For instance, chemical extraction methods may affect other species groups such as arbuscular mycorrhizae and other microorganisms. Destructive methods like hand sorting would preclude studying other species or processes in the same location where earthworms were enumerated.

Earthworms are classified into three ecological groups based on food source, pigmentation, adult size, burrow type, mobility, longevity, generation time, drought survival, predation, and other factors (Bouché 1971, 1977). These groups are:

- **Epigeic (upon the earth).** Worms in this group are small to medium in size (2–7 cm) and are heavily pigmented. They live at the soil–litter interface, do not form burrows, and

BOX 8.1

Extraction and enumeration—What are we counting?

The **extraction efficiency** of methods used to count soil animals is invariably less than 100%. For example, extraction of nematodes from soil is at best 50% efficient. Furthermore, the efficiency may vary within a group of organisms. For example, one problem with extracting earthworms from soil with chemical irritants is that estimates of community composition may be skewed toward those groups that can get to the soil surface relatively quickly (anecic and epigeic), leading to undercounting of endogeic species. Different methods to count the same organism may yield varying results because they rely on different attributes of the organism. The Baermann funnel (Fig. 8.7) method yields numbers of live nematodes because it requires that the nematodes swim out of the soil to be counted. By contrast, the sugar flotation method does not discriminate between dead and live nematodes. Finally, many soil animals have resting stages (e.g., nematodes can go into anhydrobiosis and protozoa form cysts) that allow them to survive unfavorable environmental conditions, and thus they may go undetected.

TABLE 8.1 Efficacy of methods for the recovery of earthworms and their associated long-term side effects on non-target organisms in a temperate meadow soil.

Methods	Earthworm extraction efficacy:			Non-target effects on:					
	Total	Adult	Juv.	Collembola	Gamasida	Oribatida	G−	G+	AM
							--------	PFLA	--------
Chemical methods									
AITC [mmol]									
0.5	+	++	+	–	+	–	–	–	–
1.0	+	++	+	–	–	–	–	–	–
Formalin [%]									
0.2	+	++	+	–	–	–	+	–	+
0.4	+	++	+	–	–	–	+	+	+
Physical methods									
Octet	++	+	++	–	–	–	–	–	–
Handsorting	++	+++	++	/	/	/	/	/	/
Kempson	+++	+++	+++	/	/	/	/	/	/

AITC, allyl isothiocyanate; AM, arbuscular mycorrhizal fungi; G−, Gram-negative bacteria; G+, Gram-positive bacteria; PLFA, phospholipid fatty acids. Extraction efficacy: (+) = low; (++) = medium; (+++) = high. Nontarget effects: (–) = no negative long-term effects; (+) = negative long-term effects; (/) = destructive method, no possibility to observe non-target effects. Effects on bacteria and fungi were inferred from changes in the presence of their corresponding PLFAs.

Modified from Čoja et al. (2008).

FIGURE 8.4 (A) Anecic earthworms make vertical burrows. Burrow opening is indicated by *white arrow*. The earthworms feed on litter from the surface, which they drag into the burrow. (B) Nightcrawler midden clearly raised above soil and assembled from leaf litter *(yellow circle)*. (C) Same midden with leaf debris removed showing casting materials *(blue circles)* and a burrow lined with plant material (yellow circle). *Source: Photo credit: (A) E. Nicosia, (B and C) Maryam Nouri Aiin.*

ingest little or no soil. They eat decomposing litter or manure found on the soil surface. Birds, mammals, and arthropods prey heavily on epigeic worms. Red or blood worms (*Lumbricus rubellus*) are representative of this ecological grouping.

- **Anecic (out of the earth).** The common nightcrawler, *Lumbricus terrestris*, is an anecic earthworm. It makes permanent or semi-permanent vertical burrows that have an opening to the surface (Fig. 8.4). Anecic earthworms have some pigmentation and their adults are the largest of the three groups (typically 10–30 cm) if one excludes the various species of giant earthworms. They feed on decomposing litter on the soil surface, which they pull into their burrows, where further decomposition takes place. They deposit their **casts** (excreta) within their burrows and on the soil surface. A **midden** consisting of the casts along with organic debris is often found as a raised pile near the opening to burrows. Although burrows provide some protection, anecic earthworms are still subject to predation by birds at the soil surface and by moles.
- **Endogeic (in the earth).** These subsoil species are represented by the genus *Diplocardia*, the most abundant indigenous species in the southeastern United States. Other common species in this group belong to the genus *Aporrectodea*, which are exotic in North America. Endogeic earthworms make extensive horizontal or semi-horizontal, continuous burrows within a few centimeters of the soil surface. Endogeic worms generally lack pigmentation and are intermediate in size between anecic and epigeic earthworms. They feed on the organic matter fraction of mineral soils and therefore prefer those with high organic matter content. Because they reside within the soil profile, the predatory pressure is low.

Earthworm classification by ecological group can broadly be done by color. Pigmented worms tend to spend time at the surface whereas the unpigmented worms tend to be endogeic worms. This level of classification is at a very broad scale, and earthworm species that belong to one genus can be part of all three groups. *Lumbricus*, the genus to which the common night crawler belongs, is one such genus. From an ecological point of view, the community composition according to both species and ecological group is important. However, identifying species is often hampered by lack of easily identifiable taxonomic characteristics

BOX 8.2

Using DNA to identify soil fauna

Classifying and counting soil organisms in a soil sample requires that they be identified taxonomically and that we account for all individuals in the sample. Traditional morphological identification of fauna is a rare skill among ecologists, and the methods of extraction can be particularly ineffective. Conversely, more recently developed methods employing nucleic acid analyses avoid the need for morphological identification. This is based on the principle that a species has unique DNA sequences of taxonomic value. Once we establish which sequences are uniquely associated with a morphologically identified specimen, the presence of these sequences, or **barcodes**, in the DNA of an unknown specimen allows us to identify it by comparison to barcodes deposited in a database. Similarly, the presence of a barcode in a DNA extract from a soil sample indicates the presence of particular species of fauna.

Different regions of the genome may be used to identify fauna, including the 18S ribosomal RNA region, which is unique to eukaryotes, as well as genes coding for mitochondrial cytochrome oxidase and microsatellite DNA found interspersed throughout the genome. These regions exhibit sufficient genetic variation to allow us to distinguish among species. The process involves either extracting DNA from a soil sample (for meso and microfauna) or from organisms isolated from a soil sample. The extracted DNA is mixed with primers targeting a particular region of the genome and the enzyme DNA polymerase. This process, known as the **polymerase chain reaction** (PCR), produces thousands of copies of a sequence, which can subsequently be used for quantification of the number of copies of the sequence (quantitative PCR). Alternatively, the PCR product can be sequenced using high-throughput technology (Chapter 4) and compared to a database of sequences for previously identified fauna.

This approach has been used successfully to identify novel species of earthworms (Boyer et al., 2011) and to assess soil faunal diversity (Horton et al., 2017). It has led, for example, to the discovery that the earthworm *Dendrodrilus rubidus*, which was regarded as an exotic species of European origin, may have originated in America. It is now called *Bimastos rubidus* (Csuzdi et al., 2017), reassigned to a genus found in America based on genetic analysis.

and by the fact that there appear to be species that are sufficiently different to be distinguished genetically but not morphologically. For example, DNA sequencing (Box 8.2) has determined that the morphological species *L. terrestris* (night crawler) includes a cryptic look-a-like called *L. herculeus* (James et al., 2010).

Because earthworms cannot regulate their water content or body temperature, their activities are dependent on soil moisture and temperature. When adverse conditions occur, they migrate to areas of the soil profile compatible with their needs. For example, during dry, hot conditions, anecic earthworms migrate farther into the soil profile, seeking greater soil moisture and milder temperatures. They come back to the surface to feed at night when the soil has cooled. If hot and dry conditions persist for long periods of time, they will go into a dormant state known as **aestivation**. When it is cold, they burrow below the frost line and remain dormant until the soil warms up sufficiently for them to resume their activities. The well-known phenomenon of earthworms emerging to the surface during a rainstorm may have several explanations. The most credible of these are: avoidance of predators which may cause vibrations in the soil similar to those set up by raindrop impact; ability to migrate further on the surface; and greater ability of hermaphrodites to more easily find a mate at the surface. The explanation that they avoid drowning in saturated soils is not convincing since rainwater contains sufficient dissolved oxygen.

At a much larger geographic scale, a group of earthworms known as **cosmopolitan worms** (Box 8.3) may be expanding their range at the northern boundary because of global climate change. For example, several earthworms in the megascolecid family originating in Japan appear to have extended their range from southern to northern New England over the past 30 years. This may have been due to the extension of the frost-free period during the last few decades, which has lengthened the growing season by 20 days. If this trend continues, Canada will be susceptible to a megascolecid invasion.

The effects of earthworms on the soil ecosystem depend to a large extent on the ecological group to which they belong, because this determines their level of interaction with soil and the organisms that live within the ecosystem. Thus, epigeic worms have little direct effect on what goes on in the soil because they live and feed at the soil–litter interface. Their main effects may be **comminution** (grinding) of litter or competing for plant detritus with other litter-feeding organisms, including anecic worms. By contrast, anecic worms can have a number of direct effects on the soil ecosystem, from creating habitat for microorganisms to causing shifts in the composition of microbial communities. They also repackage organic matter into small micro-aggregates where it is protected from decomposition (Knowles et al., 2016), depriving microorganisms of access to substrates. This mechanism of carbon sequestration may have broader implications by helping to balance out increased greenhouse gas emissions (Lubbers et al. 2013; Zhang et al., 2013) associated with the activities of other types of earthworms.

Microarthropods

Arthropods have segmented bodies, jointed legs, and a cuticle that acts as an exoskeleton. Insects, mites, and spiders are arthropods. The two most abundant groups of soil microarthropods are the **collembolans** (Animalia; Arthropoda, Insecta) and **mites** (Animalia, Chelicerata: Arachnida). They are found on or near the surface of the soil and in plant litter. Populations range between 10^3 m^{-2} in agricultural soils and 10^6 m^{-2} in forest soils.

Collembolans, also known as springtails, are wingless insects a few millimeters in length and 0.2–2 mm in width (Fig. 8.5). They get their common name from the ability of springtails that live above-ground to catapult considerable distances using their furcula, a tail-like, forked organ. However, this organ is reduced or not present in soil-dwelling collembolans, an adaptation to subterranean life. They are aerial organisms but are susceptible to dry as well as saturated soil conditions. They are sufficiently small to fit within interaggregate spaces in soil where water drains freely. They are thought to be **omnivorous**, feeding on bacteria, fungi, nematodes, algae, other collembolans, and decaying plant material. Ecologically, springtails that consume plant detritus are regarded as shredders that fragment the litter and increase

BOX 8.3

Exotic soil fauna

Conservation biology places importance on the dispersal of species to places beyond their original range. These introduced or exotic species can become a threat to native ecosystems when they become invasive. Plants and aboveground vertebrates are familiar examples of invasive organisms, which are often regarded as weeds or pests. Soil fauna can also be invasive—earthworms, ants, flatworms, and springtails are examples of soil organisms that have spread beyond their original range.

Earthworms can be invaders of soil ecosystems, particularly in areas where native species were extirpated during the last glaciation. In North America, earthworms from Europe (Lumbricidae) and Asia (Megascolecidae) have invaded the formerly glaciated areas of the northern tiers of the United States and southern Canada but can also be found south of the extent of the last glaciation. Their presence initially changes forest soil structure through the consumption of the thick folistic O-horizon that developed in the absence of earthworms. This sets off a cascade of events that may begin with the reduction of forest plant biodiversity, thereby depriving other decomposer species of resources and habitat and additionally affecting populations of potential predators, for example, salamanders, and of some ground-nesting bird species whose nest are deprived of vegetative cover and become vulnerable to nest predators. Earthworms are dispersed by commercial plant material, composts, and fishing bait. Earthworm invasions are culturally cryptic because they are regarded as beneficial organisms in agriculture and gardening. Thus, they go undetected as invaders of forests.

Flatworms (or hammerhead worms) from Asia and New Zealand have been introduced to the United States and most of Europe. In parts of the United Kingdom, *Artiopsothia triangulata* (the New Zealand flatworm)—which was probably imported with ornamental plants—has decimated earthworm populations. The two most abundant flatworm species invading American ecosystems, *Bipalium adventitium* and *Bipalium kewense*, are widespread in New York and are found in several other states. Tetrodotoxin, the same neurotoxin found in pufferfish, is present in these flatworm species and is likely the substance that paralyses their earthworm prey.

Unlike earthworms and flatworms, smaller exotic organisms such as springtails (Collembola) are likely to go unnoticed. For example, the table here shows that collembolans on the remote Macquarie Island, located between New Zealand and Antarctica, originated in Europe and North and South America. They likely entered the island with supplies (mainly plants) to the local research station around which these organisms are commonly found.

Exotic springtails (collembola) on remote sub-Antarctic Macquarie Island.

Family	Genus	Species	Orginal locality
Entomobryodae	*Lepidocyrtus*	*violaceus*	France
Hypogastruridae	*Hypogastrura*	*viatica*	Scandinavia
Isotomidae	*Desoria*	*tigrina*	Switzerland
Isotomidae	*Isotomurus*	*palustris*	Europe
Katiannidae	*Sminthurinus*	*quadrimaculatus*	North America
Katiannidae	*Sminthurinus*	*mime*	South America
Onychiuridae	*Protaphorura*	*finata*	Switzerland
Sminthurididae	*Sminthurides*	sp. cf *malmgreni*	Northern Hemisphere
Tullbergiidae	*Mesaphorura*	*macrochaeta*	Canada

Modified from Convey et al. (1999).

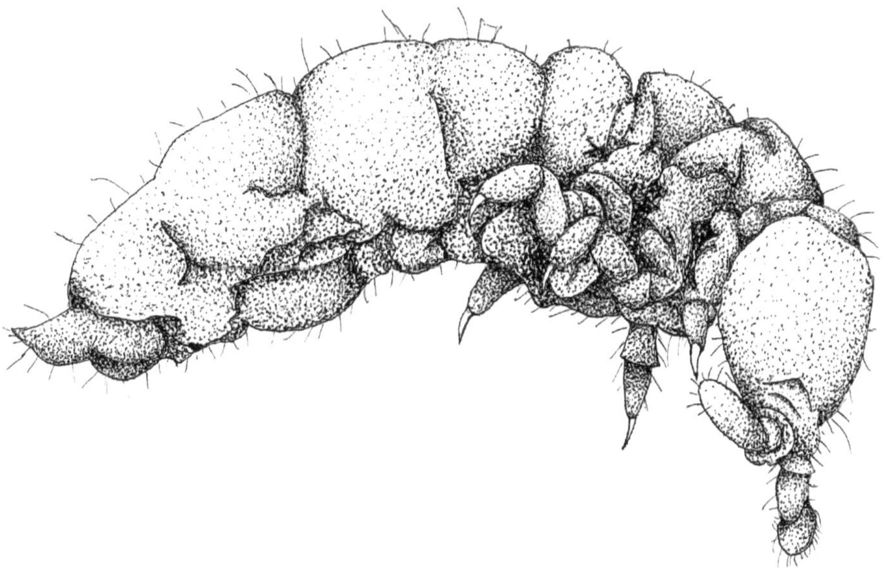

FIGURE 8.5 **A soil collembolan (*Onychiurus* sp.).** Collembolans that live within the soil tend to be more compact than those that live aboveground. *Source: Drawing by Laura Amador, based on Eisenbeis and Wichard (1987).*

the susceptibility of the litter to microbial colonization by increasing its surface-to-volume ratio.

Mites are the other important group of microarthropods in the soil (Fig. 8.6). They have oval bodies with four pairs of legs. Oribatid mites feed on fungi, decomposing plant detritus, or both. They tend to be the most abundant group of mites in many soils. The mesostigmatid mites are primarily predators of small arthropods and nematodes. Prostigmatid mites include some fungal feeders, but most species appear to be predators of small arthropods

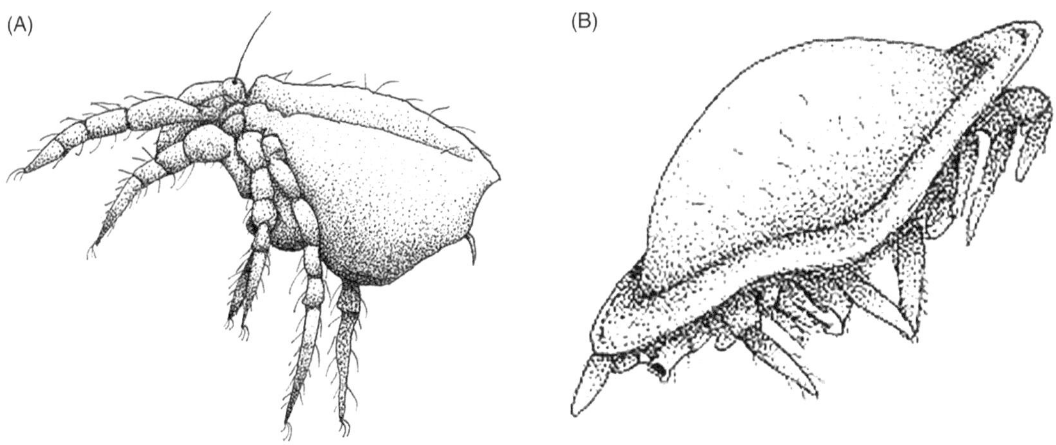

FIGURE 8.6 Examples of (A) oribatid (*Nothrus* sp.) and (B) mesostigmatid (*Uropoda cassidea*) soil mites. *Source: Drawings by Laura Amador, based on Eisenbeis and Wichard (1987).*

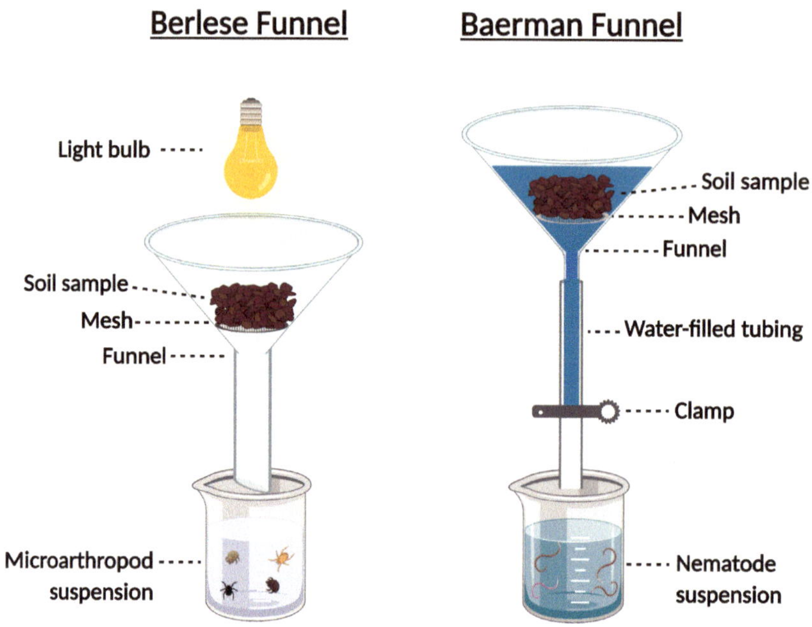

FIGURE 8.7 **Schematic diagrams of a Berlese funnel, used to determine the number of microarthropods in soil, and a Baerman funnel, used to extract nematodes from soil.** *Source: Created with BioRender.com.*

and nematodes. They can be distinguished from oribatid and mesostigmatid mites by the presence of a constriction between the anterior and posterior part of the body that is absent in the other two.

Determining the number of microarthropods in soil involves driving or floating them out of a soil sample (Moldenke, 1994). The *Berlese funnel* method involves placing the soil sample on a wire mesh within a funnel and applying an environmental stimulant (light, heat, or a chemical irritant) that forces the organisms out of the soil (Fig. 8.7). The mites fall through the mesh and are collected in a bottle placed below the funnel. Microarthropods can also be extracted by shaking a soil sample in a dense salt solution. The organisms of interest float to the surface while the soil particles fall to the bottom of the container.

Nematodes

Nematodes (Animalia, Nematoda: Adenophorea and Animalia: Nematoda: Secernetea), also known as roundworms, are cylindrical, unsegmented worms with tapered ends. They are small enough (100–1000 μm in length and 5–100 μm in diameter) to fit within the existing pore structure of soil, especially within soil aggregates. Like other soil microfauna, nematodes are aquatic organisms. Their habitat is the water films and water-filled pores with diameters similar to their body diameter.

The mouth and esophagus of nematodes show a wide variety of shapes that are generally related to their feeding habits. There appear to be two general strategies of feeding. One group has a **stylet** or a tooth to puncture living cells and suck out the cytoplasm, while the

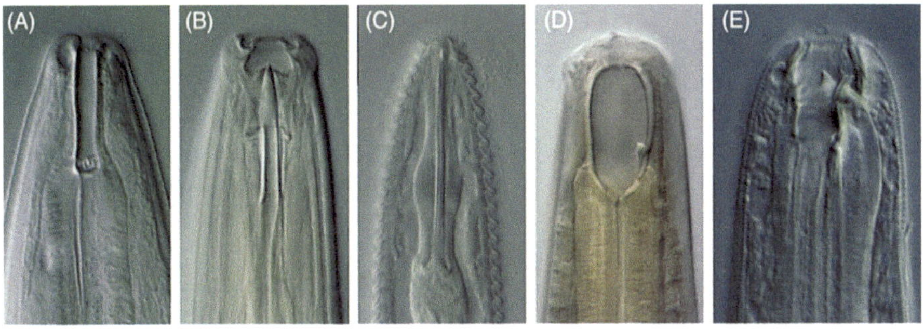

FIGURE 8.8 **Nematodes in different trophic groups are identified by the shape of their mouth and esophagus.** (A) *Rhabditoides inermis*, a bacteriovore, uses its stoma to feed on bacteria; the plant-feeding nematodes *Paractinolaimus* sp. (B) and *Mesocriconema* sp. (C) use a hollow stylet to puncture and feed on the contents of plant roots, as do nematodes that feed on fungi; predatory nematodes such as *Coomansus zschokkei* (D) and *Monochoides* sp. (E) use their large teeth to puncture their prey. *Source: Adapted from Kiontke and Fitch (2013). Photo credit: (A, E) K. Kiontke, (B-D) De Ley Lab. Used with permission.*

second ingests food whole. Nematodes can, as a first approximation, be classified into five major trophic groups (Yeates et al., 1993), depending on the morphology of the esophagus and mouthparts (Fig. 8.8):

- Bacteriovores are relatively small and have a **stoma**, an orifice that allows them to feed on bacteria.
- Predators use a very large stoma and tooth-like structure to prey on other nematodes.
- Fungivores use a slender stylet or a small tooth to puncture fungal hyphae and suck out the contents of the cell.
- Plant feeders puncture plant roots with a large stylet. Like fungivores, they feed on cytoplasm, sucking out the contents of the root cell.
- Omnivores are less specialized in their feeding habits, eating plant roots, protozoa, algae, fungi, and bacteria. Omnivores are sometimes combined with predators into a single category because many predaceous nematodes also appear to feed on other types of organisms.

A distinction is often made between free-living and **parasitic** nematodes. Free-living nematodes include bacteriovores, fungivores, predators, and omnivores, whereas parasitic nematodes are the plant feeders. Parasitic nematodes may live outside their hosts (ectoparasitic) or within them (endoparasitic) and can be free-living in their infectious, larval stage.

Nematodes are found in every ecosystem on earth. Their numbers range from less than 10^6 m^{-2} in desert soils to 10^7 m^{-2} in grasslands. Nematodes from all trophic groups are most abundant in surface soil horizons, with their numbers diminishing exponentially with depth. Within surface soils, free-living nematodes tend to be more numerous near patches of decomposing organic matter (e.g., plant litter, decaying insect remains), whereas plant feeders are most abundant in **rhizosphere** soil (Chapter 11). The rhizosphere and drilosphere are also regions of the soil containing high numbers of microbes. These in turn are resources for **microbivorous** nematodes and protozoa; therefore, populations of bacteriovores and fungivores are also more abundant in those habitats than in bulk soils.

Nematodes can be extracted from soil with a *Baermann funnel* (Fig. 8.7). The apparatus consists of a funnel fitted with a mesh to which a length of plastic tubing is attached. The tubing is clamped at the bottom and filled with water, the soil sample is placed on the mesh, and the funnel is filled with water to saturate the soil. Nematodes migrate out of the soil into the water-filled plastic tubing. Because they cannot swim effectively outside the narrow passages of soil voids, they collect at the clamped end of the tube. The clamp is then opened to collect the nematodes for enumeration and identification. Nematodes can also be extracted by wet sieving soil through successively finer mesh. The sieved soil is suspended in a concentrated sucrose solution and centrifuged. Because the specific gravity of nematodes is less than the concentrated sucrose solution, the nematodes float to the top while soil particles and other debris fall to the bottom of the centrifuge tube. This is known as the sugar flotation method.

Protozoa

Soil protozoa (Protozoa: Ciliophoran, Discomitochondria, Rhizopoda) are unicellular organisms ranging in size from 6 to 100 μm in diameter. Like nematodes, they are aquatic. They are sufficiently small to reside within the existing water-filled pores of soil aggregates. Many protozoa are **bacteriovorous**, but their diet may also include other protozoa, fungi, and organic detritus. Some ingest their food by engulfing it, whereas others brush food toward their mouth or **cytosome**. Protozoa are classified based on their morphology and means of locomotion:

- **Flagellates** (Fig. 8.9A). These protozoa rely on single or multiple **flagella**, long, whip-like appendages, for their motility. Flagellates feed primarily on bacteria and tend to be the most numerous of the soil protozoa.
- **Amoebae**. Amoebae extend and contract their protoplasm (the extensions are called **pseudopodia**) to move within soil pores. Two types of amoeba are recognized: *testate* and *naked*. Testate amoebae are encased in a rigid siliceous or chitin-based shell, or **test** (Fig. 8.9B), whereas the naked amoebae lack such a structure (Fig. 8.9C). Because of

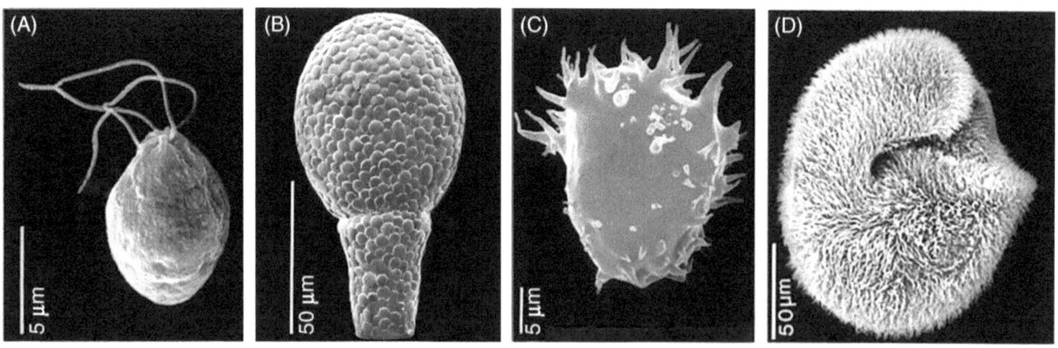

FIGURE 8.9 **Examples of soil protozoa.** (A) *Polytomella* sp., a common heterotrophic soil flagellate. (B) *Nebela vas*, a testate amoeba. The shell is made of circular platelets originating from its prey, another testate amoeba, *Trinema lineare*. (C) *Mayorella* sp., a naked amoeba with cone-shaped pseudopodia mainly in the anterior body half. (D) *Bresslauides discoideus*, a large ciliate. *Source: From Foissner (1999).*

their ability to change shape, naked amoebae can explore smaller soil pores than testate amoebae, ciliates, or flagellates. Amoebae are common in the rhizosphere and at root surfaces, where they graze on the bacterial population.

- **Ciliates** (Fig. 8.9D). Ciliated protozoa move by beating the numerous **cilia** that cover the cell surface. Cilia are much shorter than flagella. Ciliates are well known for their ability to graze on bacteria, but about half of the known soil ciliate species prey on protozoa, including other ciliates.

Protozoa can make up a significant fraction of the faunal biomass in soil in agricultural ecosystems (Fig. 8.3). Their numbers range from 10^9 to 10^{10} m^{-2}, and they are more abundant in cultivated than in uncultivated soils. The most common way to enumerate protozoa is the **most probable number** (**MPN**) method, which involves serial dilution to the point where protozoa are no longer present (Box 8.4).

BOX 8.4

The most probable number (MPN) method of enumeration

Like the dilution plate count method (Fig. 5.8), the MPN method is a way to count organisms. It is the method of choice for enumerating protozoa in soil samples. To do an MPN assay, a series of increasingly dilute soil suspensions are prepared, to the point at which protozoa are no longer present (i.e., dilution to extinction). The dilutions are incubated with a growth substrate (e.g., bacteria) that allows the protozoa to reproduce. Samples are then examined microscopically for the presence of particular groups of protozoa. Replicate dilutions are inoculated into tubes or microplate wells in order to ensure accuracy. Positive replicates at particular dilutions are used to calculate the number of protozoa in the original soil sample using an MPN table or software program.

For example, assume that a number of 10-fold dilutions (10^{-2}, 10^{-3}, 10^{-4}, 10^{-5}, 10^{-6}, and 10^{-7}) are made of a sample, and each dilution is inoculated into four replicate wells of a 24-well microplate. After a suitable incubation period, a sample from each well is examined, and the presence of protozoa is recorded ("+" = protozoa present; "–" = no protozoa). In this example, the number of positive replicates is 4-4-3-2-0-0 for dilutions of 10^{-2}–10^{-7}. Using these values to read the corresponding table indicates that the MPN is 21,000 individual protozoa g^{-1} in the original soil sample (Woomer, 1994).

The MPN method is also useful for enumerating many other microorganisms, such as cyanobacteria and algae that are challenging to grow on solid media.

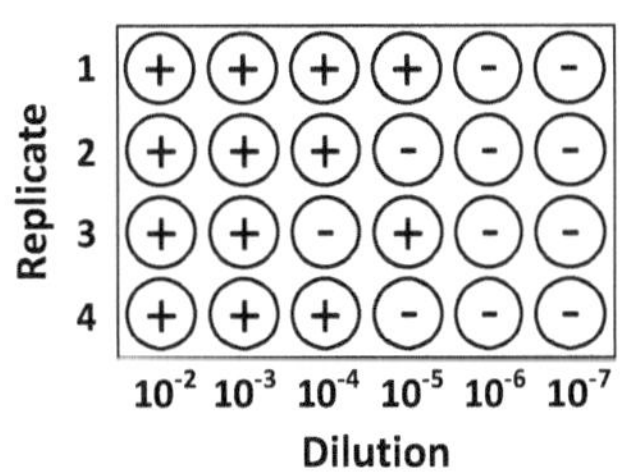

Fauna in the context of the soil food web

The activities of soil fauna take place in the context of the **soil food web** (Fig. 8.1). Because soil animals constitute only a small fraction of the biomass of heterotrophic soil organisms (Fig. 8.3), their importance in soil ecosystems results more from their ecological interactions with other microbes rather than their total biomass. These interactions mediate the transfer of carbon and nutrients, which originate in plant material, among **trophic groups** in the soil. Transfer of matter from one trophic level to the next becomes more inefficient as one moves up the trophic levels of the food web. One of the important consequences of such interactions is the release of nutrients through mineralization, especially phosphate and ammonium, from organic materials that form the detritus and the biomass of soil organisms. Plants require a source of inorganic (mineralized) nutrients, thus mineralization by the soil fauna contributes to plant productivity. This mineralization is particularly important for maintaining soil ecosystems, which are driven by the inputs of energy and nutrients entering the soil from plant residues.

The connectivity between trophic levels of the below-ground food web depends in part on the soil water dynamics. For aquatic fauna like nematodes and protozoa, accessibility of prey may become limited when soils dry. At the pore scale, plants can easily remove moisture from the relatively large pores of the nematode and protozoan habitat, thus reducing the spaces that can be explored for food by these aquatic animals. This will affect mineralization rates and plant nutrient availability.

Differences in soil habitat translate into differences in ability of fauna to interact with soil microorganisms within the soil foodweb. Members of the microfauna, many of which prey on bacteria, inhabit the same intraaggregate spaces that bacteria do but can be excluded from some of the bacterial habitat when soils dry. Fungal hyphae grow inside and outside aggregates, where **fungivorous** micro- and mesofauna can eat them. The macrofauna, on the other hand, interact with soil microorganisms at different levels, ranging from fortuitous ingestion to "engineering" their habitat to favor soil microorganisms that are part of their diet.

Direct effects of the soil fauna on microbial communities

Grazing. The soil fauna and microflora form part of food webs in which they are both consumers and are themselves consumed (see Chapter 10). Soil food webs are driven by the chemical energy contained in detritus that is harnessed by soil microorganisms. Energy and nutrients are passed through the trophic levels of a food web. The consumption of microorganisms by the fauna is called grazing. Grazing accelerates the detrital microbial pathway of nutrient mineralization. There are three main aspects of grazing that affect functioning of the soil ecosystem:

- Reduced microbial biomass due to predation
- Changes in community composition
- Mineralization and release of nutrients from microbial biomass that is consumed

Grazers directly affect microbial biomass by removing individual microorganisms from the population. Yet predicting the effect of grazing on microbial populations is not as straightforward as it may appear. Intuitively, one expects the population size to decrease with grazing, but grazing can also accelerate microbial growth rates (Bengtsson et al., 1993; Traunspurger et al., 1997). In fact, microbial population size may increase, decrease, or remain the same in response to grazing. What determines the outcome depends on factors such as the intensity of grazing, the initial size of microbial populations, and nutrient and energy limitations.

The accelerated growth rate that results when bacteria and fungi are grazed by collembolans, protozoa, and nematodes is called **compensatory growth**. Compensatory growth is a well-known effect in aboveground ecosystems. A familiar example is mowing a lawn periodically. Mowing removes some biomass and adds detritus, making a mown ecosystem more productive than one that is not. Similarly, in the grasslands of the Serengeti, grazing by zebras and antelopes stimulates plant growth.

Nutrient cycling. The soil fauna contributes considerably to nutrient cycling. Indeed, soil fauna may contribute as much as 30% of the total soil mineralizable nitrogen (Chapter 15). The contributions can be even higher in environments where microorganisms predominantly immobilize the nitrogen, particularly where detrital C/N ratios are high, and are partly a result of compensatory growth. The fauna remobilizes nutrients as metabolic waste products, which benefits both plant nutrition and microbial growth. Grazing results in net nutrient mineralization because feeding and digestive processes are never 100% efficient and C/N ratios of microorganisms and grazers differ.

In general, both the elemental compositions of grazers and microorganisms, and the physiological efficiencies of grazers, affect carbon and nutrient mineralization. Consider bacteria with a C/N ratio of 5 and bacteriovorous nematodes with a C/N ratio of 6. For every 6 units of bacterial biomass C consumed, nematodes receive 1.2 units of N. Nematodes only need 1 unit of N per 6 units of carbon. Thus, there is an excess of 0.2 units of N that may be excreted by the nematode as ammonia or readily mineralizable N compounds.

The discrepancy between C/N ratios of microbivorous animals and microorganisms is only half the mineralization story. Consider protozoa and bacteria, both with C/N ratios of 5. Protozoa, like all heterotrophs, meet their energy and carbon requirement by metabolizing organic carbon compounds and assimilate only a fraction of the carbon they consume into their biomass; the rest is lost as carbon dioxide. Assuming a C assimilation efficiency of 50%, protozoa would have to metabolize 10 units of C from the bacterial pool, corresponding to 2 units of N, to assimilate 5 units of C. This would leave 1 unit of N that cannot be used by the protozoa and may be mineralized. This example applies equally to microbivorous nematodes.

The effects of grazing on nutrient transformations also need to be placed in the context of soil physical constraints. There are two main hypotheses on the effect of changing soil moisture on grazing pressure by protozoa and nematodes. On the one hand, the *exclosure hypothesis* (Hassink et al., 1993) states that as soils dry and water films of soil particle become thinner, fewer of the most abundant types of aquatic grazers (nematodes and protozoa) can access their microbial prey and nutrient turnover decreases as a result. On the other hand, the *enclosure hypothesis* (Görres et al., 1999) states that as soils dry, aquatic grazers are concentrated in pockets of remaining soil water where they can consume microorganisms more readily, because they do not have to search for them as extensively. Because the coverage of soil particle surfaces with bacteria may be as low as 8% (Postma and van Veen, 1990) and searching for bacteria may be an important energy expenditure for the microbivorous fauna, "enclosure" would result in more rapid nutrient turnover, albeit in a smaller space. Both hypotheses are probably valid depending on the prevailing soil moisture content.

Microbial community composition. The feeding habits of soil fauna can affect not only the size but also the species composition of the microbial community. Microarthropods, free-living nematodes, and protozoa can be selective in their feeding, grazing preferentially on particular microorganisms. Selection may be based on size or other, often unidentified, characteristics of the microorganism. For example, some fungivorous nematodes and collembolans feed preferentially on plant pathogenic fungi. Furthermore, grazing can alter the outcome of competition for food resources between groups of microorganisms. For example, grazing of fungi by collembolans gives bacteria a competitive advantage. Not only do fewer fungi compete for food resources, but grazing additionally releases nutrients from fungal biomass for use by the resident bacterial community.

Indirect effects of the soil fauna on microbial communities

Primary production. Grazing of soil microorganisms by soil fauna can have both negative and positive effects on primary production. On the one hand, plant parasitic nematodes are best known for their negative effects on crop production, which can potentially reduce the amount of plant detritus deposited on the soil surface. Because the soil food web is detrital, reduced litter inputs can decrease the numbers and activities of soil microorganisms. On the other hand, the mineralization of bacterial biomass by the soil fauna, as well as moderate levels of faunal feeding on plant roots, increases plant biomass production by increasing the size of the pool of above- and below-ground plant litter that fuels the soil food web. For example, small or moderate levels of root consumption by nematodes may enhance the growth rate of roots, an effect analogous to compensatory growth exhibited in grazed microbial populations and savannah grass. Injury to roots caused by root-feeding nematodes can result in additional carbon exudation from roots as well as root death, providing additional resources to fungi and bacteria.

As mentioned earlier, burrowing by earthworms alters the physical structure of soil, which in turn affects the availability of habitat and the movement of air and water in soil. Earthworms also impinge on the soil microbial community by translocating resources. Endogeic earthworms translocate and enrich the mineral soil they ingest by mixing it with intestinal secretions, enhancing microbial activity and nutrient turnover in the casts they produce. Similarly, anecic earthworms translocate large amounts of mineral soil from within the soil profile to the surface. They also drag plant litter into their burrows and line their burrows with mucus as well as nutrient-enriched castings. This enrichment generally increases microbial respiration and nitrogen mineralization, as well as numbers of autotrophic nitrifiers (Chapter 15). Conditions within the burrow can also promote the proliferation of fungi. Populations of microbivorous fauna, especially nematodes and protozoa, tend to be higher in the soil surrounding earthworm burrows than in soil without such burrows. Thus, the combined activities of earthworms, soil microorganisms, and the microbivorous fauna enhance decomposition of plant detritus and accelerate the release of nutrients in plant-available form.

Transport. In combination with their interaction with soil microorganisms, soil fauna of all sizes, including earthworms, microarthropods, nematodes, and protozoa can transport microorganisms through the soil profile. Transport can have positive effects for soil microbial populations, such as providing opportunities for the colonization of new habitat. Transport begins by soil animals ingesting cells and reproductive or survival structures, such as fungal and bacterial spores. These structures are not always digested by soil animals and may instead be activated by passage through the gut. As soil animals move through the soil and excrete their wastes, they release bacteria and various types of spores away from where they were originally ingested. A second mechanism involves transport on the surface of soil animals, where spores and bacterial cells can attach themselves. As soil animals travel through the soil, microorganisms and their spores may become detached from their surfaces and deposited.

The ability of soil fauna to transport microorganisms has been used to inoculate large volumes of soil with bacteria for specific purposes. For example, earthworms can enhance the distribution of fungi and bacteria that are pathogens of insects, as well as bacteria that have been genetically engineered to enhance biodegradation of organic contaminants. However, not all microorganisms transported by the soil fauna have a positive effect on the soil. Soil animals can just as easily transport plant pathogenic fungi and their spores, and this transport may increase the likelihood of encountering a susceptible host plant.

Summary

The soil fauna include earthworms, collembolans, mites, nematodes, and protozoa. These are eukaryotic, heterotrophic, motile organisms that require oxygen for their metabolism. Their physical range, habitat, and food resources are constrained by their respective sizes and the availability of pores of appropriate size within the soil. Nematodes and protozoa are microbivorous aquatic organisms, living within water-filled pores in aggregates and water films on particle surfaces, where they can access microbial and faunal prey. Collembola and mites are aerial organisms that live in interaggregate spaces or at the soil surface, feeding on fungi, plant detritus, microfauna, and other microarthropods. Earthworms are organisms that can live at the soil–litter interface or in burrows within the soil, feeding on soil or plant detritus. The interactions of the soil fauna with microorganisms take place in the context of the soil food web. Through grazing on bacteria and fungi, microbivorous faunal species directly affect the size and composition of microbial communities, populations of plant pathogenic microorganisms, and the mineralization of nutrients. They act to accelerate decomposition and nutrient cycling in soil. As a result of their feeding, burrowing, and movement, the soil fauna also engineer the habitat for the soil microflora, transport beneficial as well as pathogenic microorganisms, and affect the production of detrital resources from plants. The contribution of the soil fauna to the flow of nutrients and energy through soil ecosystems is disproportionately larger than their contribution to the total biomass of soil.

Study questions

1. How would trophic interactions between soil fauna and microorganisms change as the soil water potential changes from saturated (0 MPa) to field capacity (−0.03 MPa) to permanent wilting point (−1.5 MPa)?
2. A nematicide is applied annually to an agricultural field. What may be the long-term consequences of this practice to (1) free-living nematodes, (2) microbial biomass, (3) N availability, and (4) the farmer's profits?

3. Farmer Ebe's fields are next to an old industrial site contaminated with 2,4,5-trichlorophenoxyacetic acid (2,4,5-T). The soil at the site is being bioremediated using anecic earthworms to distribute a genetically modified microorganism (GMO) within the soil profile. The GMO is a *Pseudomonas* species containing a plasmid that codes for the enzymes involved in the biodegradation of 2,4,5-T. Farmer Ebe is concerned that the activities going on at the neighboring bioremediation site are affecting his farm. He's noticed that the 2,4-dichlorphenoxyacetic acid (2,4-D) he has used for the past 20 years to control broadleaf weeds isn't working anymore. His raspberry plants have started to develop galls. He blames the bioremediation project for his problems and is considering suing the large conglomerate that owns the property. Develop scenarios that support Farmer Ebe's interpretation of the facts, as well as alternative scenarios.
4. As part of a demonstration project, a farmer has managed nutrient cycling in a tomato field by adding microbivorous nematodes annually (For examples of the successful use of nematodes in managing soil fertility, see Ferris, 1998). This year he has received a contaminated nematode culture that includes 5% predatory nematodes. Should he be concerned about this year's nutrient management success?
5. Septic system drainage fields remove pathogens from household sewage. One obvious mechanism involves mechanical removal by filtration, especially in conventional systems that drain in the C horizon. However, in novel systems, drainage is closer to the soil surface because treatment relies on the "biologically active zone" of the soil profile. What sorts of interactions might take place between microbial pathogens in the sewage and the soil fauna?
6. Climate change activists demand that all organic soils (northern forests, thawing permafrost in the Tundras) be inoculated with earthworms so the organic matter can be physically protected in earthworm castings. Argue for and against this idea.

References

Beare, M.H., 1997. Fungal and bacterial pathways of organic matter decomposition and nitrogen mineralization in arable soils. Brussaard, L., Ferrera-Cerrato, R. (Eds.), Soil Ecology in Sustainable Agricultural Systems. Lewis Publishers, Boca Raton, FL, pp. 37–70.

Bengtsson, G., Helund, K., Rundgren, S., 1993. Patchiness and compensatory growth in a fungus-Collembolan system. Oecologia 93, 296–302.

Bouché, M.B., 1971. Relationship between the spatial and functional structures of ecosystems, as illustrated by the pedobiological role of earthworms. Pesson, P. (Ed.), The Life of the Soil: Novel Aspects, Experimental Studies-Gauthier-Villars, Paris, pp. 187–209.

Bouché, M.B., 1977. Earthworm strategies. Ecol. Bull. 25, 122–132.

Boyer, S., Blakemore, R.J., Wratten, S.D., 2011. An integrative taxonomic approach to the identification of three new New Zealand endemic earthworm species (Acanthodrilidae, Octochaetidae: Oligochaeta). Zootaxa 2994, 21–32.

Čoja, T., Zehetner, K., Bruckner, A., Watzinger, A., Meyer, E., 2008. Efficacy and side effects of five sampling methods for soil earthworms (Annelida, Lumbricidae). Ecotoxicol. Environ. Safety 71 (2), 552–565.

Convey, P., Greenslade, P., Arnold, R.J., Block, W., 1999. Collembola of sub-Antarctic South Georgia. Polar Biol. 22 (1), 1–6.

Csuzdi, C., Chang, C.H., Pavlícek, T., Szederjesi, T., Esopi, D., Szlávecz, K., 2017. Molecular phylogeny and systematics of native North American lumbricid earthworms (Clitellata: Megadrili). PLoS ONE 12 (8), e0181504.

Davidson, E.A., Belk, E., Boone, R.D., 1998. Soil water content and temperature as independent or confounded factors controlling soil respiration in a temperate mixed hardwood forest. Global Change Biol. 4, 217–227.

Dominguez, J., 2004. State-of-the-art and new perspectives on vermicomposting research. Edwards, C. (Ed.), Earthworm Ecology. CRC Press, Boca Raton, FL, pp. 401–424.

Eisenbeis, G., Wichard, W., 1987. Atlas on the Biology of Soil Arthropods. Springer–Verlag, New York.

Eisenhauer, N., Straube, D., Scheu, S., 2008. Efficiency of two widespread non-destructive extraction methods under dry soil conditions for different ecological earthworm groups. Eur. J. Soil Biol. 44 (1), 141–145.

Ferris, H., 1998. The role of nematodes in soil fertility [online]. Available from: http://nemaplex.ucdavis.edu/Ecology/fertil.htm (accessed 23.01.2020.).

Foissner, W., 1999. Soil protozoa as bioindicators: pros and cons, methods, diversity, representative samples. Agric. Ecosyst. Environ. 74, 95–112.

Görres, J.H., Amador, J.A., 2010. Partitioning of habitable pore space in earthworm burrows. J. Nematol. 42, 68–72.

Görres, J.H., Savin, M.C., Neher, D.A., Weicht, T.R., Amador, J.A., 1999. Grazing in a porous environment: 1. The effect of soil pore structure on C and N mineralization. Plant Soil 212, 75–83.

Hassink, J., Bouwman, L.A., Zwart, K.B., Brussaard, L., 1993. Relationships between habitable pore space, soil biota and mineralization rates in grassland soils. Soil Biol. Biochem. 25, 47–55.

Hattori, T., 1994. Soil microenvironment. Darbyshire, J.F. (Ed.), Soil Protozoa. CAB International, Wallingford, England, pp. 43–64.

Heal, O.W., Dighton, J., 1985. Resource quality and trophic structure in the soil system. Fitter, A.H. (Ed.), Ecological Interactions in Soil. Blackwell Scientific Publication, Oxford, pp. 339–354.

Horton, D.J., Kershner, M.W., Blackwood, C.B., 2017. Suitability of PCR primers for characterizing invertebrate communities from soil and leaf litter targeting metazoan 18S ribosomal or cytochrome oxidase I (COI) genes. Eur. J. Soil Biol. 80, 43–48.

James, S.W., Porco, D., Decaëns, T., Richard, B., Rougerie, R., Erséus, C., 2010. DNA barcoding reveals cryptic diversity in *Lumbricus terrestris* L., 1758 (Clitellata): resurrection of *L. herculeus* (Savigny, 1826). PloS ONE 5 (12), e15629.

Kiontke, K., Fitch, D.H., 2013. Nematodes. Curr. Biol. 23 (19), R862–R864.

Knowles, M.E., Ross, D.S., Görres, J.H., 2016. Effect of the endogeic earthworm *Aporrectodea tuberculata* on aggregation and carbon redistribution in uninvaded forest soil columns. Soil Biol. Biochem. 100, 192–200.

Lubbers, I.M., Van Groenigen, K.J., Fonte, S.J., Six, J., Brussaard, L., Van Groenigen, J.W., 2013. Greenhouse-gas emissions from soils increased by earthworms. Nat. Clim. Change 3 (3), 187–194.

Moldenke, A.R., 1994. Arthropods. Weaver, R.W., Angle, S., Bottomley, P., Bezdicek, D., Smith, S., Tabatabai, A., Wollum, A. (Eds.), Methods of Soil Analysis, Part 2. Microbiological and Biochemical Properties. Soil Science Society of America Book Series, No. 5, Madison, WI, pp. 512–542.

Postma, J., van Veen, J.A., 1990. Habitable pore space and survival of *Rhizobium leguminosarum* biovar *trifolii* introduced into soil. Microbial Ecol. 19, 149–161.

Savin, M.C., Görres, J.H., Neher, D.A., Amador, J.A., 2001. Uncoupling of carbon and nitrogen mineralization: role of microbivorous nematodes. Soil Biol. Biochem. 33 (11), 1463–1472.

Traunspurger, W., Bergtold, M., Goedkoop, W., 1997. The effect of nematodes on bacterial activity and abundance in freshwater sediment. Oecologia 112, 118–122.

Woomer, P.L., 1994. Most probable number counts. Weaver, R.W., Angle, S., Bottomley, P., Bezdicek, D., Smith, S., Tabatabai, A., Wollum, A. (Eds.), Methods of Soil Analysis, Part 2. Microbiological and Biochemical Properties. Soil Science Society of America Book Series, No. 5, Madison, WI, pp. 59–79.

Yeates, G.W., Bongers, T., de Goede, R.G.M., Freckman, D.W., Georgieva, S.S., 1993. Feeding habits in soil nematode families and genera—An outline for soil ecologists. J. Nematol. 25, 315–331.

Zhang, W., Hendrix, P.F., Dame, L.E., Burke, R.A., Wu, J., Neher, D.A., Li, J., Shao, Y., Fu., S., 2013. Earthworms facilitate carbon sequestration through unequal amplification of carbon stabilization compared with mineralization. Nat. Comm. 4, 2576.

Further reading

Berry, E.C., 1994. Earthworms and other fauna in the soil. Hatfield, J.L., Stewart, B.A. (Eds.), Soil Biology: Effects on Soil Quality. CRC Press, Boca Raton, FL, pp. 61–90.

Coleman, D.C., Callaham, M., Crossley, Jr., D.A., 2017. Fundamentals of Soil Ecology, third ed. Academic Press, New York.

Darbyshire, J.F., 1994. Soil Protozoa. CAB International, Wallingford, England.
Edwards, C.A., Bohlen, P.J., 1996. Biology and Ecology of Earthworms, third ed. Chapman & Hall, London.
Edwards, C.A., Arancon, N.Q., Sherman, R.L., 2010. Vermiculture technology: earthworms, organic wastes, and environmental management. CRC Press, Boca Raton, FL.
Eisenhauer, N., Partsch, S., Parkinson, D., Scheu, S., 2007. Invasion of a deciduous forest by earthworms: changes in soil chemistry, microflora, microarthropods and vegetation. Soil Biol. Biochem. 39 (5), 1099–1110.
Ferris, H., 2010. Form and function: metabolic footprints of nematodes in the soil food web. Eur. J. Soil Biol. 46 (2), 97–104.
Geisen, S., Koller, R., Hünninghaus, M., Dumack, K., Urich, T., Bonkowski, M., 2016. The soil food web revisited: diverse and widespread mycophagous soil protists. Soil Biol. Biochem. 94, 10–18.
Ingham, R.E., Trofymow, J.A., Ingham, E.R., Coleman, D.C., 1985. Interactions of bacteria, fungi, and their nematode grazers: Effects on nutrient cycling and plant growth. Ecol. Monogr. 55, 119–140.
Lavelle, P., Spain, A.V., 2001. Soil Ecology. Kluwer Academic Publisher, Boston.
Margulis, L., Schwartz, K.V., 1998. Five Kingdoms: An Illustrated Guide to the Phyla of Life on Earth, third ed. W.H. Freeman and Co., New York.
Ruiz, N., Lavelle, P., Jiménez, J., 2008. Soil Macrofauna Field Manual. FAO of the United Nations, Rome, Italy.
Tugel, A.J., Lewandowski, A.M., Happe-vonArb, D. (Eds.), 2000. Soil Biology Primer. Soil and Water Conservation Society, Ankeny, IA.

CHAPTER

9

Viruses

Kurt E. Williamson

Biology Department, The College of William & Mary, Williamsburg, VA, United States

"[Viruses], not lions, tigers or bears, sit masterfully above us on the food chain of life, occupying a role as alpha predators who prey on everything and are preyed upon by nothing."—***Claus Wilke and Sara Sawyer***

Viruses are molecular genetic parasites of cellular systems. Viruses require cellular systems for either copying nucleic acids, transcribing nucleic acids, translating nucleic acids into protein, or all three of these processes in order to replicate. Because viral reproduction is different from that of cells, this process is typically called **replication**. The replication cycle for most viruses involves an extracellular stage, but some viruses copy and insert themselves within the host genome or replicate without producing extracellular particles. This is an important detail when it comes to defining what exactly viruses are. A narrow definition that focuses on physical or structural properties of virus particles cannot encompass the diversity of viruses, some of which only exist as genes encoded in nucleic acids.

As we will see in this chapter, virus particles are incredibly abundant on earth, outnumbering cells by a factor of at least 10. Furthermore, viruses are important in the evolution of life on earth, playing important roles in selection and in the transfer of alleles between organisms. Finally, viral genomes constitute the greatest reservoir of largely uncharacterized genetic diversity on Earth.

Virus particle structure

For most viruses, replication involves an extracellular stage during which the viral genome is released from the cell in a protective package. An intact virus particle, consisting of a nucleic acid genome and protein coat, is called a **virion**. In this form, viruses are metabolically inert (Box 9.1). Unlike all other organisms on earth that encode their genetic information in double-stranded DNA, the specific nature of the viral genome that resides within this package can vary considerably among viruses. Viral genomes can be composed of either DNA or RNA and can be double-stranded, single-stranded positive sense, or single-stranded negative sense (the concept of "positive sense" or "negative sense" is discussed in more detail later), and it can be

Principles and Applications of Soil Microbiology. http://dx.doi.org/10.1016/B978-0-12-820202-9.00009-5
Copyright © 2021 Elsevier Inc. All rights reserved.

BOX 9.1

Are viruses alive?

Scientists are not in complete agreement about how to define the state known as life, or what is alive. But most agree on five key elements:

1. Life is made out of cells.
2. Living organisms generate and use energy.
3. Living organisms process information—both genetic information and information about their environments (stimulus and response).
4. Living organisms reproduce.
5. Living organisms are products of evolution, and populations of organisms continuously evolve.

Viruses generally fail to meet the first two requirements on this list—viruses are not cellular, and they lack the enzymes and systems for independently synthesizing energy (ATP). However, viruses process genetic information and can even respond to environmental stimuli as evidenced by **induction** (Fig. 9.10). Viruses are highly effective at replicating themselves, given a cellular host to parasitize. Lastly, viral populations evolve. Like seeds or spores, viruses behave as inert particles until the right conditions present themselves. Inside a compatible host cell, the viral program that is encoded in its genome is carried out by host cell machinery, enabling viruses to live a kind of "borrowed life" (van Regenmoretel and Mahy, 2004).

packaged as a linear or circular molecule. The structural arrangement of the package containing the genome can vary but typically involves a rigid protein structure known as a **capsid**. Most virus capsids are constructed using either **helical symmetry** or **icosahedral symmetry**.

Capsids with helical symmetry are assembled by stacking wedge-shaped protein subunits around a central axis, similar to a spiral staircase (Fig. 9.1A) and result in particle morphologies resembling rods or filaments (Fig. 9.1C). Capsids with icosahedral symmetry are assembled by fitting triangular protein complexes together edge-to-edge to form a closed shell (Fig. 9.1B,D). A perfect icosahedron consists of 20 identical triangular faces. But the number, size, and grouping of proteins that make up these triangular faces varies, depending on the virus, leading to different sized capsid shells that can accommodate smaller or larger genome sizes. In general, viruses with smaller genomes tend to use capsids that are true icosahedrons, while viruses with larger genomes tend to use additional protein subunits in their construction, forming capsids that more closely approximate a sphere and therefore have more volume to contain the larger genome. While most virus particles are smaller than cells, viruses span a wide size range from smaller than 20 nm in diameter to dimensions rivaling those of cells (Table 9.1).

The particles produced by some viruses include a lipid envelope derived from the host, while other viruses produce "naked" capsids. Some viruses that infect bacteria (bacteriophages, or simply "phages") produce particles with distinct tail structures that are built using radial symmetry, in which ring-shaped protein complexes are stacked along a central (radial) tube. Not all bacteriophage particles include tails, but currently, the only virus particles known

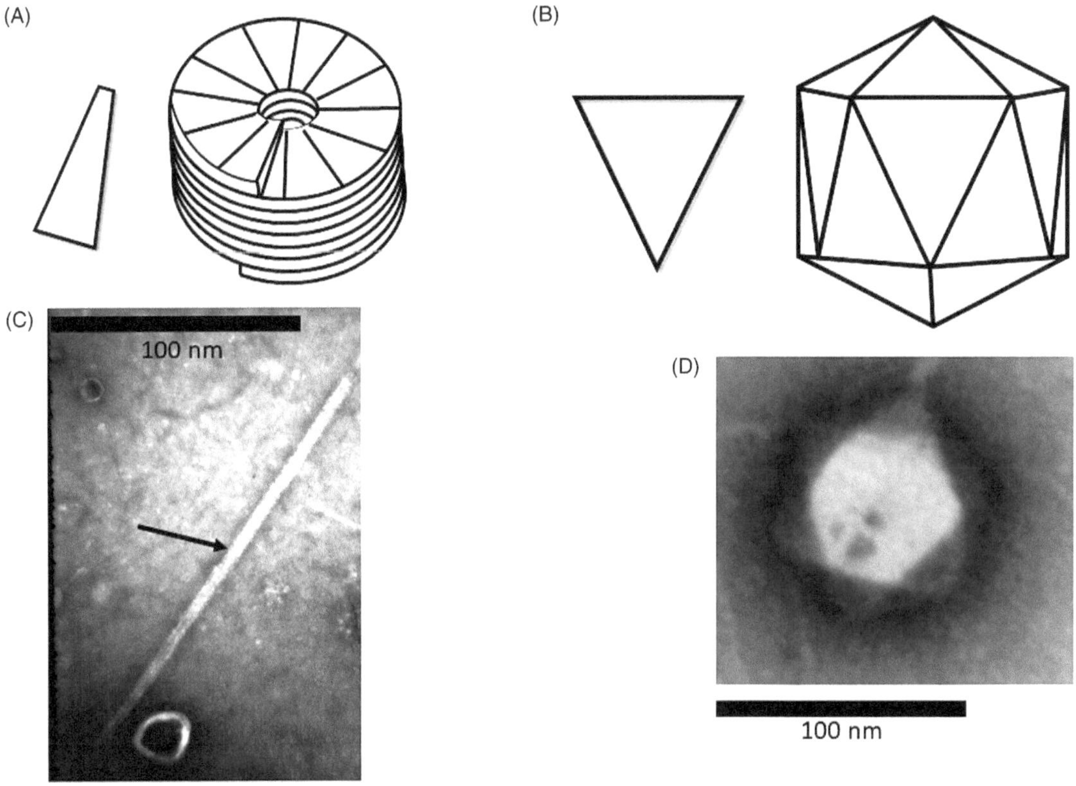

FIGURE 9.1 **Examples of virus particles with icosahedral and helical symmetry.** (A) Diagrammatic representation of helical symmetry, showing how wedge-shaped subunits are stacked; (B) diagrammatic representation of icoshahedral symmetry, showing how triangular subunits are assembled; (C) example electron micrograph of a virus particle with helical symmetry *(arrow)*, scale bar represents 100 nm; (D) example electron micrograph of a virus particle with icosahedral symmetry, scale bar represents 100 nm. *Source: Diagrams and electron micrographs by Kurt Williamson.*

to include tails are bacteriophages. For tailed bacteriophages, the tail structure is important in attachment and infection of host cells. Each virus particle that has been observed thus far has been composed of some type of nucleocapsid, some with tails, and some with envelopes. Possibilities for typical virus particle morphologies are shown in Fig. 9.2.

Virus genome structure

All cellular organisms, from unicellular bacteria to multicellular blue whales, encode their genomic information in double-stranded (ds) DNA. Some virus genomes are linear dsDNA, but as a group, viruses demonstrate much more diversity in their genome arrangements. Viral genomes may be composed of DNA or RNA, but not both. The DNA comprising virus genomes may be ds or single-stranded (ss) DNA.

TABLE 9.1 Sizes of representative viruses.

Virus	Particle size (nm)	Genome size (bp[a])
Circovirus *Example*—porcine circovirus 1	17	1,700
Parvovirus *Example*—canine parvovirus 2	20	5,000
Microvirus *Example*—phage ΦX174	32	5,400
Tobamovirus *Example*—tobacco mosaic virus	40 × 300	6,400
Podovirus *Example*—phage T7	58	40,000
Myovirus *Example*—phage T4	90 × 200	169,000
Adenovirus *Example*—human adenovirus 7	90–100	26,000–48,000
Herpesvirus *Example*—herpes simplex 1	125	153,000
Poxvirus *Example*—smallpox	325 × 250	186,000
Mimivirus (a giant virus)	400–600	1,181,000
Pandoravirus (a giant virus)	500 × 1000	2,800,000

[a]*bp, base pairs of DNA or RNA. Note that for viruses with single-stranded genomes, size is in bases rather than base pairs.*

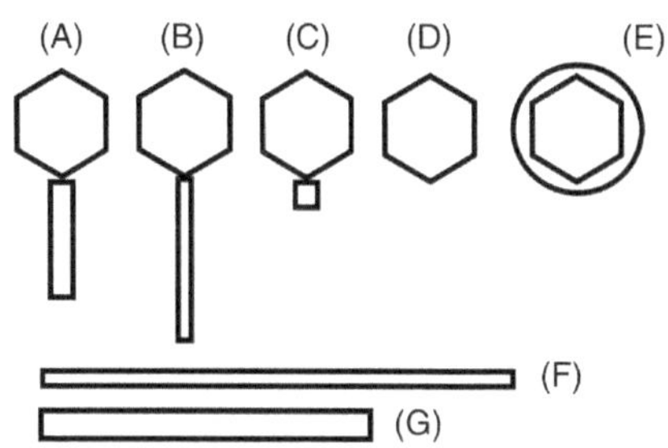

FIGURE 9.2 **Typical virus particle morphologies.** (A) Myoviridae (phage with contractile tail); (B) Siphoviridae (phage with flexible non-contractile tail); (C) Podoviridae (phage short tail); (D) naked icosahedron with no tail; (E) enveloped icosahedon; (F) filamentous (flexible, radial symmetry); (G) rod-shaped (rigid, helical symmetry). *Source: Diagrams by Kurt Williamson.*

RNA virus genomes may be single-stranded, positive-sense, which is to say it is the same sense or orientation as cellular messenger RNA. For many such viruses, like the picornaviruses, this arrangement enables the viral genome to be immediately translated by cellular ribosomes. The retroviruses also have single-stranded, positive-sense genomes, but rely on a viral reverse transcriptase enzyme that converts this single-stranded RNA genome into

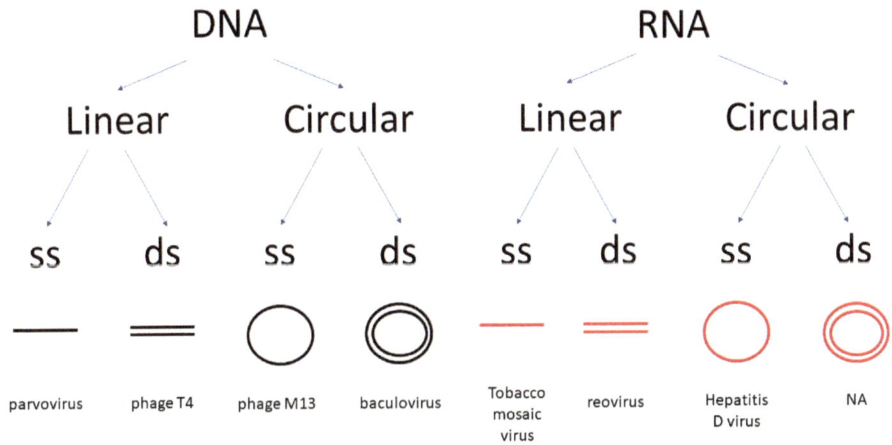

FIGURE 9.3 **Viral genomes may be either DNA or RNA.** The genome may be double stranded or single stranded and may also be linear or circular. Further, viral genomes may be monopartite (a single piece of nucleic acid) or segmented (multiple pieces)—this aspect of genome structure is not represented in this figure. Examples of specific viruses with a particular genome arrangement are listed below each specific arrangement; for example, parvoviruses have linear ssDNA genomes. *Source: Diagram by Kurt Williamson.*

dsDNA that is later inserted into the host cell's dsDNA genome. RNA virus genomes may be single-stranded, negative-sense, meaning that the viral genome is not in the correct orientation for immediate translation by host ribosomes. Single-stranded, negative sense viral genomes require an intermediate transcription step carried out by a specialized viral RNA-dependent RNA polymerase to convert the genome to positive or message sense before viral proteins can be made. Certain viruses, like the coronaviruses, have ambisense RNA genomes that include regions of both positive and negative sense RNA in the same genome. Finally, RNA virus genomes may be dsRNA. Most dsRNA viruses transcribe one of the strands into a single, positive-sense strand that can be translated in order to produce viral proteins.

In addition to nucleic acid type and strandedness, viral genomes also vary in topology—they can be circular or linear and may also be monopartite (consisting of one piece of nucleic acid) or segmented (consisting of several pieces). Example viruses for each genome type are given in Fig. 9.3.

Viral replication

Viral infection and replication can be broken down into five basic steps: attachment, entry, expression of early genes, expression of late genes, and release of progeny. When an infectious virus particle comes into contact with a susceptible host cell, a chain reaction of events is initiated. In attachment, surface groups on the virus particle interact with and bind to specific surface groups or **receptors** on the host cell. This is one of the means by which the specificity of viral infection is controlled—hosts lacking the correct receptor cannot be infected by a particular virus. Attachment of the virion to the host cell receptor typically induces a conformational (physical, three-dimensional) change in either the virus capsid (or tail), the host cell, or

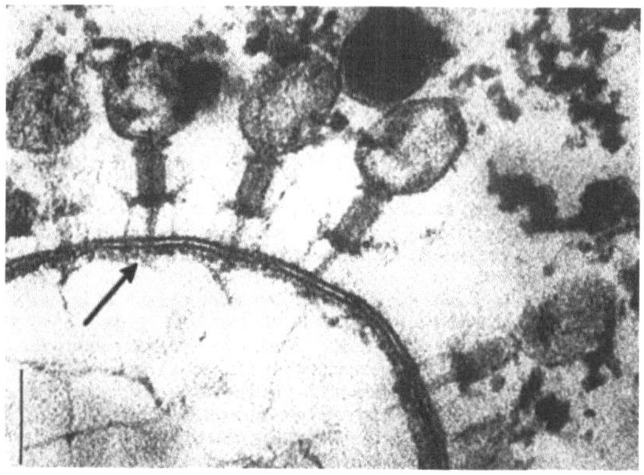

FIGURE 9.4 **Bacteriophage T4 infecting its host cell, *Escherichia coli*.** Phage particles are bound to the bacterial outer membrane by protein groups found on their tails. The tail structures of phage T4 are constructed using a tube-in-tube configuration. After attachment, the outer tube of the tail structure contracts, forcing the inner tail tube through the bacterial cell wall *(arrow)*, allowing entry of the phage genome into the cytoplasm. *Source: Micrograph from Simon and Anderson (1967). Used with permission.*

both, resulting in viral entry. The process of entry differs dramatically depending on whether the virus infects prokaryotes or eukaryotes. Prokaryotic cell entry typically involves either physically penetrating (Fig. 9.4) or enzymatically degrading the cell wall and then transporting the viral genome to the host cell's cytoplasm. For tailed bacteriophages, attachment of the particle and entry of the phage genome is mediated by the tail structure. The virus capsid remains outside the cell. Eukaryotic cell entry typically involves the movement of the entire virus capsid into the cell; after entry, the capsid breaks down and the viral genome enters the nucleus (for most DNA viruses) or remains in the cytoplasm (for most RNA viruses). Once the viral genome is inside of the cell, viral genes are expressed. Like the cell itself, the virus must regulate the expression of its genes so that viral proteins are produced in the correct order and the needed quantities. Typically, early gene expression is focused on the production of viral proteins required for genome replication, while late gene expression is focused on the production of viral proteins needed for genome packaging and release of progeny virions. Mechanisms for release of progeny virus particles vary by virus. These typically involve degradation of the peptidoglycan cell wall for viruses that infect prokaryotes or activation of the apoptotic pathway (programmed cell death and disintegration) for viruses that infect eukaryotes. For both eukaryotic and prokaryotic hosts, some viruses may be released through shedding of virus particles through the cell membrane with little immediate damage to the host cell.

Measuring viral abundance in soils

Viruses can be detected and enumerated within soil samples using a number of techniques, each with a particular set of strengths and weaknesses.

Plaque assays and infection assays

Plaque assays and infection assays require the infection of a host that can be maintained in a laboratory setting. Most infection assays involving viruses from soil samples have focused on microbial hosts, particularly bacteria. However, laboratory cultures of animal or plant cells may also be used. Infection assays are performed by mixing susceptible host cells with a viral **inoculum** (sample that is known or suspected to contain infectious virions) and a molten growth medium containing a low percentage of agar. This viral inoculum is typically an aqueous extract from a soil sample. The host-virus-agar mixture is then poured into a Petri dish containing a solidified bottom agar layer also containing nutrients for host growth. Once the molten agar layer hardens, the virus particles and host cells are "frozen" in place, and viral infections may only spread by diffusion to host cells immediately adjacent to an initial infection. Once a contiguous layer of cell growth—a "lawn"—develops, the macroscopic evidence of localized viral infections can be observed as clear areas called plaques. Plaques, the areas where the growth of host cells has been clearly impeded or completely lysed by viral infection, are the macroscopic evidence of viral infection and lysis (Fig. 9.5). By counting the number of plaques on the plate and accounting for any dilution, it is possible to calculate the concentration of infectious virus particles in the original inoculum or sample (see discussion of the dilution plate count method in Chapter 5). This concentration is typically reported as plaque-forming units per milliliter (PFU mL^{-1}).

The key strength of infection assays is their ability to detect and enumerate *infectious* virus particles; the molecular and microscopic techniques described later do not generally distinguish between infectious and non-infectious (e.g., defective) viruses. Enumeration of infectious viruses is often vital when assaying for plant, animal, or human pathogens in soil samples. Infectivity assays are also easily adapted for isolation of novel virus strains and propagation of viruses in laboratory settings. However, infection-based assays are limited to those virus strains specifically capable of infecting the host organisms under consideration, and these hosts must be first isolated and maintained under laboratory conditions. In

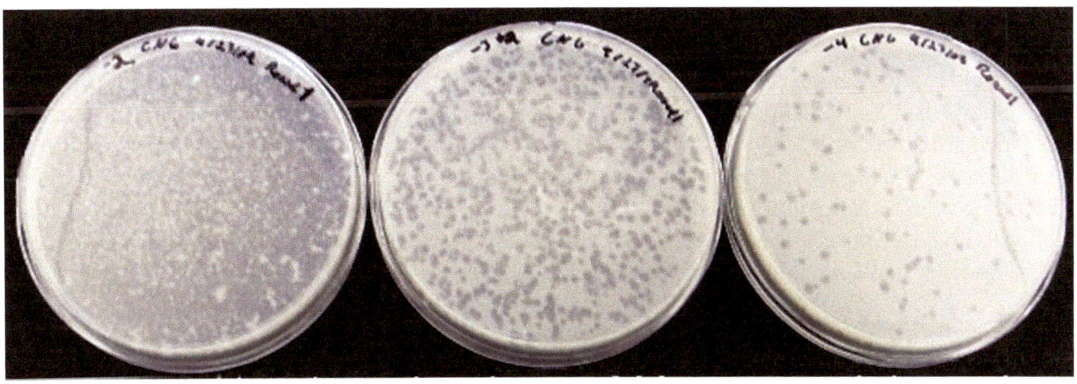

FIGURE 9.5 **Plaques (zones of clearing) in a lawn of *Mycobacterium smegmatis* cells.** The phage inoculum has been serially diluted 10-fold, from left to right: 10^{-2}–10^{-4}. By counting plaques and accounting for the dilution factor, the number of plaque forming units (PFU) in a sample can be determined. *Source: Photograph by Kurt Williamson.*

this regard, plaque and infection assays provide a limited and narrow view of viruses in soil environments and provide no information about total viral abundance or community-level diversity of viruses in soils.

Molecular detection assays

Specific virus types can be detected and often enumerated using molecular assays that focus on the capsid proteins or nucleic acid genomes of the virus of interest. Proteins are usually detected using antibodies that recognize specific proteins on the virus surface. In some assays, such as enzyme-linked immunosorbent assays (ELISAs), antibodies may be bound to an enzyme that generates a colored product. In this way, the degree of color generated by the assay is directly proportional to the amount of virus protein present in the sample (i.e., the number of virus particles). Detection of virus particles using antibody-based assays is specific and relatively sensitive, with a limit of detection of approximately 10^4 viruses mL^{-1}.

Viral nucleic acids are usually detected using the polymerase chain reaction (PCR; Chapter 4, Fig. 4.5). PCR uses short, single stranded pieces of DNA called **primers** that are complementary to a sequence of interest in the viral genome. If the target sequence of interest is present in a particular sample, then the primers will enable the copying of this sequence by heat-stable DNA polymerase (e.g., Taq polymerase). For viruses with RNA genomes, a reverse transcriptase (Chapter 4) can be used to create a complementary DNA copy that can then be amplified using PCR, a method referred to as reverse-transcriptase PCR (RT-PCR). The presence of the target sequence, and therefore the virus of interest, in a given PCR is confirmed using gel electrophoresis. The PCR products are loaded onto an agarose gel and separated by an electrical current. If an amplification product was generated, it will be visible as a band of the predicted size in the agarose gel, thereby confirming the presence of nucleic acids from the virus of interest. The lack of a band of the expected size is interpreted as an absence of the virus of interest. PCR-based assays can also be combined with fluorescent dyes or fluorescently labeled probes in order to estimate the number of viral genome templates in the original sample, a modification known as quantitative PCR (qPCR; Chapter 4). PCR-based methods are highly sensitive and can detect as few as one virus particle in a sample.

Molecular detection assays have many strengths. They are rapid (providing results within hours), relatively cheap, high throughput (many samples can be tested in a short amount of time), highly sensitive, and highly specific for the viruses that are targeted. This same specificity is a key weakness, however. For molecular detection assays to work, the virus of interest must be isolated in culture and either specific antibodies raised against its surface proteins, or PCR primers must be developed based on known gene or genome sequences. In other words, we can only develop targeted molecular assays against viruses that are well characterized. Therefore, these methods are effective for a very narrow and specific range of viruses—typically, those of agronomic or medical interest.

Microscopic techniques

Total viral abundance in soils can be determined by either transmission electron microscopy (TEM) or epifluorescence microscopy (EFM). Both techniques first require the dispersion of soil samples in an aqueous extraction buffer and the removal of cells and larger soil particles, typically by passing the extracts through a filter that retains particles and cells ≥0.22 μm in size. (The smallest bacteria are approximately 0.3 μm, although clay particles can be much smaller. Some larger viruses are >0.22 μm in size and will be excluded by this filtration step; see Box 9.2 and the section on viruses of amoebae and protozoa later in this chapter.) For EFM, the filtered soil extracts are typically then passed through a filter that retains particles ≥ 0.02 μm in size, thereby capturing viruses on the surface of the filter. The virus particles are subsequently stained with a dye that binds to nucleic acids and fluoresces

BOX 9.2

Giant viruses

The first virus of any kind was described in 1898. Its discovery was the culmination of years of work by multiple researchers who were trying to identify the causative agent of a disease that caused tobacco plants to develop yellow, mottled leaves. The causative agent could pass through a device known as a Chamberlain filter candle, a porcelain double-cylinder with pores too small to allow the passage of bacteria and other cells (see below). Therefore, whatever was causing the disease in the tobacco plants must be something altogether different from a bacterium or other cellular pathogen. The agent, dubbed tobacco mosaic virus, was classified as a "filterable agent" (subsequently as a virus), and this filtration step became a key method for distinguishing viral from bacterial pathogens. Around the mid-20th century, the porcelain filter was replaced by a 0.22 μm membrane filter (typically composed of polymers like cellulose esters or nylon). But filtration has remained a key method in separating viruses from cells in every microbiology and virology laboratory, up to present day.

In 2003, however, another discovery was made—some viruses are too big to pass through the 0.22 μm filters that have been a mainstay of virus isolation for over 100 years. This so-called "giant virus" was initially mistaken for a bacterium and was finally recognized as a virus 10 years after its first description. The discovery of this first giant virus has led to the search for—and discovery of—many others, all of which have large, dsDNA genomes. Examples of giant viruses are described in the following table. These giant viruses pose challenges to viral ecology. Due to their large size, they cannot be separated from cells using filtration. Most researchers rely on a combination of antibiotics and repetitive, painstaking dilutions to eliminate bacterial contaminants. Up to 85% of the genes in giant virus genomes cannot be identified because they do not match to sequences currently cataloged in any database. Many giant virus genomes contain genes that are found only in that particular virus, and nowhere else.

The discovery of giant viruses has led to major changes in viral classification and important questions remain regarding the evolutionary origins of giant viruses, and, indeed, the evolutionary relationship between giant viruses and cellular life (Brandes and Linial, 2019). Giant viruses have been isolated from soils, but their abundance and overall ecological impacts in soils have yet to be determined.

(*Continued*)

BOX 9.2 (Cont.)

Virus	Year isolated	Particle size (nm)	Genome size (Mbp)[a]
Acanthamoeba polyphaga mimivirus	2003	400–600	1.18
Cafeteria roenbergensis virus	2003	300	0.73
Megavirus chiliensis	2011	680	1.26
Pandoravirus salinus	2013	1,000	2.80
Pithovirus sibericum	2014	700	1.25

[a]Mbp, Mega (million) base pairs.

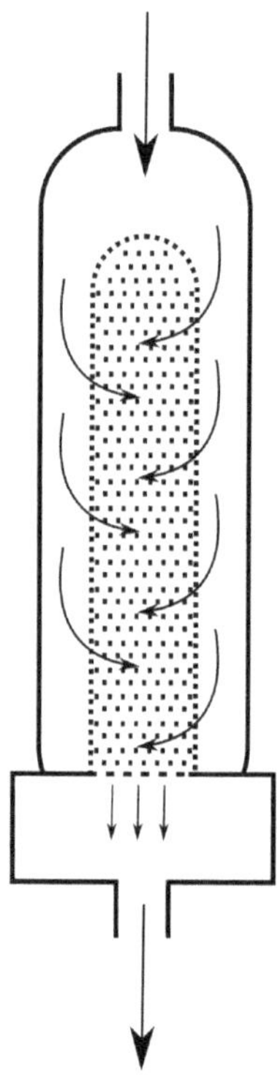

Diagram of a Chamberlain filter candle. The device was made of unglazed porcelain, flow direction of fluid sample is indicated by *arrows*. Fluid was forced into the inlet *(arrow at top)* and through the many small pores *(represented by dots)* in the center plug. Bacteria and other cells were too large to pass through these pores, but virus particles smaller than approximately 0.2 μm (200 nm) would pass through to the outlet *(arrow at bottom)* where the filtrate could be collected in a sterile vessel. The presence of viruses in the filtrate was then confirmed by inoculating live organisms or performing plaque assays.

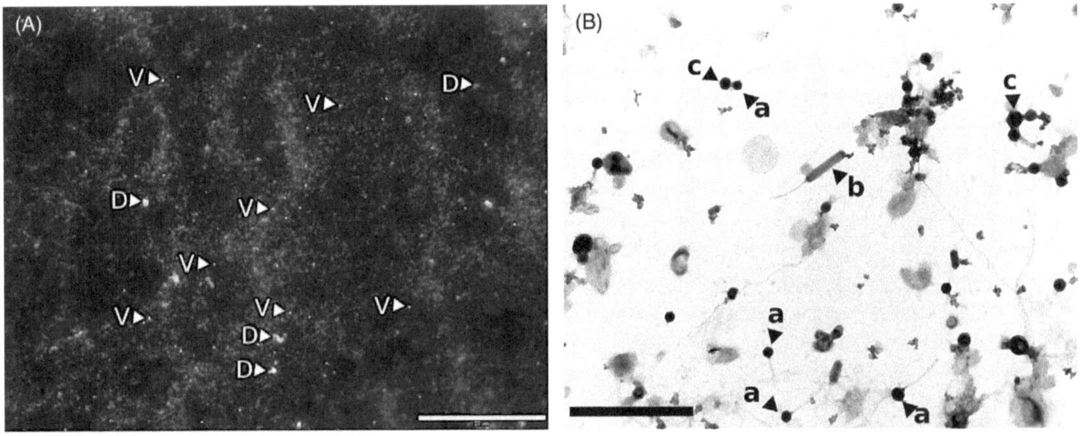

FIGURE 9.6 (A) Example epifluorescence micrograph of viruses extracted from soil. Soil extracts were filtered through 0.22 μm to remove bacteria prior to immobilization on 0.02 μm filters. Virus particles (V) appear as small, *circular points*; additional fluorescent signals originate from cell and other debris (D), as well as filter background. Scale bars indicate 20 μm. (B) Transmission electron micrograph of virus particles extracted from soil: (a) icosahedral capsids with long, flexible tails (Siphoviridae); (b) particle with elongated capsid; (c) icosahedral capsids lacking obvious tails. Particles were purified using density gradient centrifugation and stained with uranyl acetate prior to visualization. Scale bar indicates 500 nm. *Source: Micrographs courtesy of Kurt Williamson.*

in the presence of ultraviolet radiation. Using an epifluorescent microscope equipped with a UV source and digital camera, the fluorescence from the virus particles can be viewed and photographed as microscopic points of light, even though the particles themselves are too small to be viewed directly. The fluorescing virus particles are then counted, typically with assistance from image processing software (Fig. 9.6A). For TEM, additional purification and concentration, for example, by density gradient centrifugation, is usually needed prior to loading samples onto electron microscopy grids, otherwise virus particles may be obscured by soil particles and other co-extracted debris. For electron microscopy, grids are stained with an electron-dense heavy metal salt, such as uranyl acetate, and the stained virus particles are visualized in a transmission electron microscope (Fig. 9.6B).

A key strength of TEM is that permits visualization of particle morphology in addition to enumeration. A disadvantage is that electron microscopes are costly and expensive to maintain. Additionally, it takes longer to prepare samples for electron microscopy, and the capture of virus particles on electron microscopy grids is not as efficient as on 0.02 μm filters, the latter leading to lower abundance estimates as compared to EFM. Dilution errors may also be introduced during purification and concentration. Epifluorescence microscopy has higher particle-capture efficiency and is generally cheaper and achieves higher throughput, but it does not allow for visualization of virus-particle structure. Furthermore, EFM is sensitive to false positives through the detection of fluorescent signals of non-viral origin.

Estimates of viral abundance in soils

Viral abundance in soils ranges from 2.2×10^3 gdw^{-1} (per gram dry weight) in desert sands collected from Saudi Arabia to 5.8×10^9 gdw^{-1} in forest soil collected from eastern Virginia

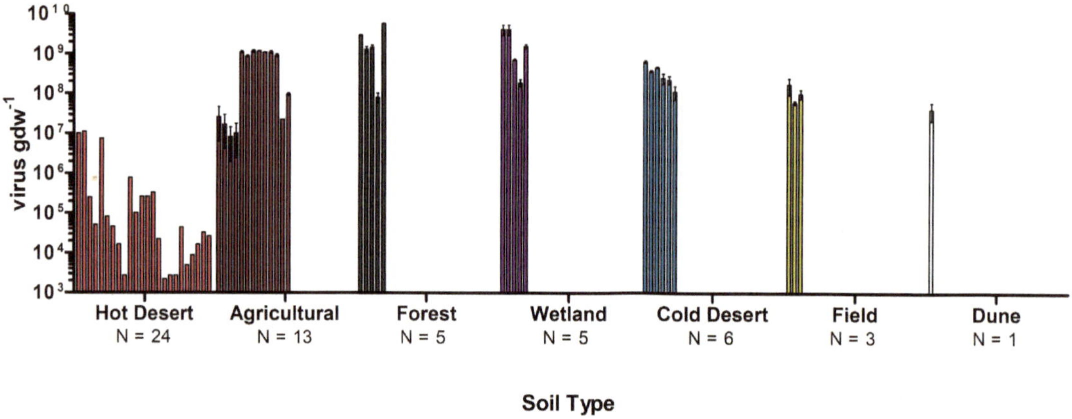

FIGURE 9.7 **Viral abundance grouped by soil type.** Hot Desert soils include samples from the Mojave, Sonoran, Chihuahuan (United States), Saharan, Namib (Africa), Arabian (Middle East), and Registan (Afghanistan) deserts; agricultural soils include samples from England, Scotland, United States, China, and Vietnam; forest soils include samples from Delaware and Virginia, United States; wetland soils include samples from Delaware and Virginia, United States; Cold Desert soils include samples from the McMurdo Dry Valleys, Antarctica; Field (fallow) soils include samples from Virginia, United States; Dune soils include samples from the Outer Banks, North Carolina, United States. *Source: Data for figure from Williamson et al. (2017).*

(Williamson et al., 2017) (Fig. 9.7). For a sense of how many virus particles this translates to at a global scale, see Box 9.3. Some of the differences in viral abundance reported across studies stem from the different methods used for extraction and enumeration of viruses. For example, some studies used TEM while the majority used EFM. Even accounting for these differences in enumeration methods, however, viral abundance seems to be strongly correlated with soil biome or land use. In general, viral abundance is lowest in hot deserts, intermediate in agricultural soils, fallow fields, and cold deserts, and highest in forested and wetland soils. Soil viral abundance is typically correlated with bacterial abundance, strongly suggesting that many if not most viruses in soils are bacteriophages. Viral abundance is negatively correlated with soil pH; virus capsids become positively charged at lower pH, and this influences viral attachment to negatively charged soil surfaces and therefore viral persistence and extraction efficiency. Temperature is also known to influence viral persistence and abundance in soils, with lower temperatures likely preserving virions for longer periods of time, and higher temperatures likely promoting thermal degradation of viral particles.

Measuring viral diversity in soils

Estimates of viral diversity in soils can provide important insights to the activity and impacts of viruses in soil microbial ecosystems. Until the mid-2000s, most surveys of soil viral diversity focused on the genetic diversity within agronomically or medically important viruses, but few studies attempted to quantify community-level viral diversity in soils. In subsequent years, multiple methods have been employed to estimate how many different kinds of viruses can be found in a given soil sample.

BOX 9.3

Getting a sense of global viral abundance in soils.

It has been estimated that the global ocean, which covers about 70% of the Earth's surface, contains roughly 10^{30} virus particles (Suttle, 2007). But what about soils? Using approaches similar to those taken to obtain estimates for the global ocean (Calculations and approaches derived from: http://www.phagehunter.org/2008/09/how-far-do-those-phages-stretch.html) (which involve several assumptions), we can obtain a rough estimate for soils:

Land masses cover approximately 5×10^{18} cm^2 (5×10^8 km^2) of Earth's surface.

In most studies, soil has been sampled to a depth of 10 cm, so we can estimate viral abundance in the upper 5×10^{19} cm^{-3} of soil.

The average density of loamy soils (accounting for pore spaces) is 1.33 g cm^{-3}. Multiplying total soil volume by this average density yields 6.8×10^{19} g of soil, total.

The average viral abundance across all soils can be estimated by averaging the highest and lowest values reported in the literature. This yields an average of 3×10^9 virus particles g^{-1}.

Finally, if we multiply the average viral abundance by the total mass of soil on land, we get 2×10^{29} virus particles, total.

How many viruses is that?

We can get a sense of how big this number is by considering how long a chain of viruses, placed end-to-end, would stretch.

The average diameter of a phage particle is between about 20 and about 200 nm (Table 9.1). The average of these two values is 110 nm but let us use 100 nm for the sake of simplicity.

2×10^{29} virus particles $\times$ 100 nm per particle = 2×10^{31} nm or 2×10^{19} km.

But how long is this? At this scale, many scientists measure things in light years.

There are 9.46×10^{12} km in a light year.

Dividing 2×10^{19} km by 9.46×10^{12} km [light year]$^{-1}$ yields a combined length of **2.1 $\times$ 10^6 light years.**

For comparison:

Distance from Earth to the moon: 0.00000004 light years.

Distance from Earth to the sun: 0.000016 light years.

Distance from Earth to Pluto: about 0.0005 light years.

Distance from the sun to the center of the Milky Way: about 2.6×10^4 light years.

Diameter of the Milky Way: 1×10^5 light years.

Distance from Earth to the Andromeda Galaxy: 2.5×10^6 light years.

Distance from Earth to the M81 Local Group: 1.1×10^7 light years.

Additionally, note that modern rockets can travel about 8 km s^{-1}. At this velocity, it would take us about 37,200 years to travel 1 light-year.

Morphological diversity (TEM)

Initial characterizations of viral diversity in soils used TEM to examine and compare viral particle morphologies across different soils. In spite of the limitations of this approach, TEM-based morphological comparisons suggest that different soils contain distinctly different viral assemblages. In particular, the viral community of some soils appear to be dominated by tailed phages (Williamson et al., 2005), while other soils are dominated by small, spherical particles that lack tails (Swanson et al., 2009). A main limitation of this approach is that the specific identity of most virions cannot be determined. Nevertheless, the large differences in morphological diversity within virus communities across soils suggest important differences in host-community structure and function in different soils. While electron microscopy was the first tool to be applied in estimating viral diversity in soils, its use has been declining in favor of the higher-resolution and higher-throughput techniques described later.

Marker gene analysis

Viruses lack a universally conserved phylogenetic marker gene analogous to the small subunit ribosomal gene, or 16S rRNA gene, that has enabled a comprehensive understanding of bacterial diversity in soils. This lack of a universally conserved marker gene has meant slow progress toward characterizing viral diversity in soils. However, genes that are conserved within specific viral families have enabled researchers to characterize phylogenetic diversity within these families. A key success story for this approach comes from the use of the g23 gene, which encodes the major capsid protein for T4-like bacteriophages (Filée et al., 2005). In this approach, PCR primers are used to amplify the g23 gene from DNA that has been extracted from soil samples. The amplified DNA is then sequenced, and the sequences are compared through nucleotide alignment and the building of phylogenetic trees. This approach has been successfully used in multiple soil types and has revealed wide within-family diversity across soils, as well as across related viruses from other environmental sources (Wang et al., 2011; Liu et al., 2012). Despite this success, any given marker gene (including g23) may not be found in all soils, and this limits the applicability of the marker gene approach for understanding viral diversity across soils.

Community-level profiling

The main community-level profiling (CLP) technique that has been applied to soil viral communities thus far is randomly amplified polymorphic DNA-PCR, or RAPD-PCR. This technique uses a single primer (typically 10 nucleotides long) to amplify DNA fragments from a mixture of purified viral DNA templates. Separation of these amplified fragments on a high-resolution gel generates a pattern of bands, wherein each band is presumed to have originated from a unique viral taxon (Fig. 9.8). Thus, the banding pattern indicates the underlying structure of the viral community from which the template DNA was obtained. Early applications of RAPD-PCR to soil viral assemblages were promising but difficult to reproduce. More recently, the method has been modified by using fluorescently labeled primers to generate fluorescent amplification products that are then treated with a restriction endonuclease. The fluorescently labeled restriction fragments are separated using capillary electrophoresis. This resulted in a series of peaks, rather than a banding pattern, that likewise reflect the viral community composition. The main strengths of CLP techniques are their high throughputs, which enables large numbers of samples to be processed simultaneously, their low cost per sample, and their ability to discern viral community-level changes over time and space. However, the major weakness of CLP techniques is that they only enable the characterization of

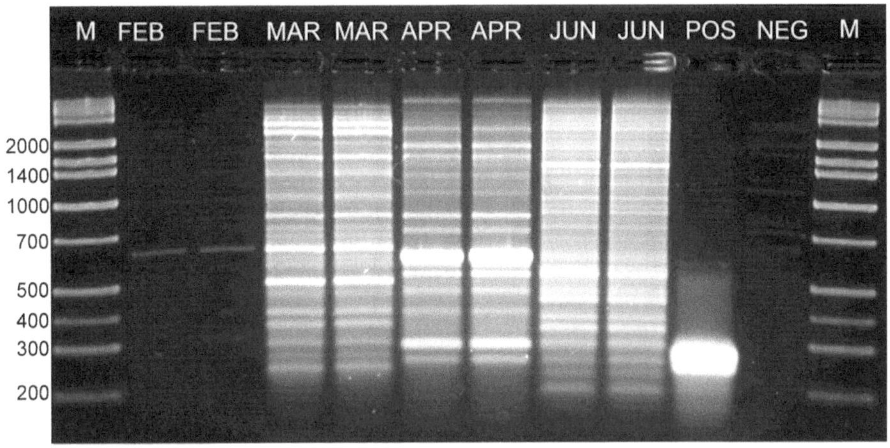

FIGURE 9.8 **Example of RAPD-PCR gel.** This gel demonstrates how temporal change in viral community composition can be documented; the duplicate samples show reproducibility. Each band is assumed to have originated from a particular viral taxon, and the pattern of bands denotes a particular viral community composition. It is possible to correlate these changes with environmental drivers such as temperature or host nutrient availability. M denotes molecular weight markers, with band sizes in bp of DNA; POS denotes positive control; NEG denotes negative control. *Source: Photograph by Kurt Williamson.*

relative changes over time, space, or in response to a stimulus or disturbance, and cannot identify the specific taxa that are responding to these changes.

Metagenomic analysis (viromics)

Metagenomics typically involves the sequencing of the collective genomic DNA from the organisms found within a specific environmental sample. Viral metagenomics, frequently shortened to "viromics," involves steps to extract and enrich for virus particles and to reduce or eliminate cellular (e.g., bacterial) sources of nucleic acids prior to sequencing. Most approaches involve dispersion of soil and elution of virus particles in an aqueous buffer, removal of soil solids through sedimentation, and removal of cells through filtration. Additional steps such as density gradient centrifugation and nuclease treatment are frequently used to further reduce non-viral sources of nucleic acids. The latter degrades free DNA (e.g., from lysed bacteria), whereas viral DNA is protected within capsids. Subsequently, the nuclease is inactivated, and viral capsids are removed to release viral DNA. This ensures that the majority of data obtained from sequencing efforts reflects the viruses found in the sample.

The resulting DNA sequences, known as "reads," may be assembled into larger contiguous sequences, known as "contigs." Reads or contigs are then analyzed using a number of bioinformatic tools to determine taxonomic identities (i.e., which reads came from which virus), to determine the possible functions of proteins encoded by the DNA sequences, to determine the richness (how many different viral taxa are represented in the sample), and to determine evenness (how the number of reads or contigs is distributed across the viral taxa represented in the sample).

In order to determine the taxonomic or functional identity of a particular read, the read is compared to a set of known sequences stored in a reference database, typically the National Center for Biotechnology Information (NCBI) non-redundant public database (https://www.ncbi.nlm.nih.gov/). However, within the soil viromes that have been published, 54.5%–97.3% of all reads cannot be matched with a specific taxonomic source, a specific protein function, or both, rendering these reads unusable in most downstream analyses. This problem highlights the importance of isolating and characterizing individual viruses from soil samples, as sequence data alone is of limited value if it cannot be matched to a known virus. Based on the identity of reads that can be matched to database entries, some soils are dominated by dsDNA phages while others are dominated by small, ssDNA viruses. But it is not yet clear to what degree these differences in the representation of virus genome types across soils is due to actual differences in viral community composition as opposed to differences in the methods used to obtain the sequence data.

The estimated number of distinct viral genotypes (richness; Chapter 10) in soils varies between 1,000 and 1,000,000 and appears to vary according to soil type. For comparison, viral richness estimates vary from 532 to 129,000 for marine samples and from 400 to 40,000 for freshwaters. Note that both the lower and upper estimates of viral richness in soils exceed the corresponding lower and upper bounds for aquatic systems. Given the high numbers of viral metagenomic sequences that cannot be identified and the high viral richness estimates, the soil virome may represent the largest reservoir of genetic diversity on the planet.

The chief strength of viromics is that sequence data can be obtained for virtually all of the viruses in a given soil sample without relying on laboratory cultivation or prior knowledge about what kinds of viruses might be in the sample. In addition, the relatively low cost of sequencing, combined with improved methods for processing multiple samples in parallel, means that a large amount of sequence data can be generated quickly and cheaply for a variety of samples. Paradoxically, this strength of rapid data generation is also a weakness of viromics. As mentioned earlier, determining the identity and predicted function of a particular sequence depends on matching it to a known reference sequence (i.e., one whose taxonomy and function are known), and the rapid rate at which novel viral reads have been generated and deposited into databases has far outmatched the rate for known reference sequences. Furthermore, the development of appropriate bioinformatic tools for analyzing sequence data, and for comparing reads within and between very large sequence data sets, has also been outpaced by the rate at which viral sequences can be generated, leading to a situation known as the bioinformatics bottleneck.

Viruses by host

Animals

Historically (until the turn of the 21st century) most of our understanding regarding the ecology of viruses in soils had been from a public health perspective, dealing with the fate, transport, and detection of pathogenic viruses introduced to soils from outside sources. Exogenous bacteria and viruses, mostly from surface wastewater application and septic tank discharges, are often purposely introduced to soils with the belief that any pathogens will be rapidly degraded in the soil environment. In theory, wastewater is purified upon passage through soil before returning to underground aquifers. However, this is not always the reality, leading to the historic focus on plant and animal pathogens in soil. Among the animal viruses, most studies focused on enteric viruses that are commonly shed in feces and so likely to be found in wastewater.

The objective of these early investigations of viruses in soil was to determine the fraction of land-applied viruses that are degraded or inactivated by passage through the soil profile versus the fraction transported deeper into groundwater. While the fate and transport of human and animal pathogens remains relevant today, the 21st century has witnessed a blossoming of interest in the natural viral ecology of soils—that is, in viruses that are autochthonous (indigenous) to soils—and how viruses impact hosts involved in various soil processes. The earlier fate and transport studies are valuable to this newer focus on soil viral ecology, as they provide critical information regarding the behavior of viruses in soils under a variety of conditions. We currently understand that factors influencing viral adsorption include:

- Characteristics of the soil solution, including ionic strength and composition, pH, and presence of dissolved organic matter.
- Characteristics of the virus, including isoelectric point (charge characteristics) and hydrophobicity.
- Soil features, including water content, clay and organic matter content, cation exchange capacity, and presence of organic or metal oxide coatings.

We also know that viruses adsorbed to soil surfaces can remain infectious and are even protected from degradation. This discovery has had strong implications for the interactions between naturally occurring soil viruses and the soil microbial community. If adsorption of virus particles to soil solids does not inactivate virions, then soils can give rise to active communities of naturally occurring viruses, which may then play important roles in soil ecosystems.

Plants

While plant viruses are likely common to most soils that support herbaceous or woody plants, few plant viruses are directly transmitted from soil to plant host. Because plant cells include tough cell walls, viruses are not typically transmitted directly from infected individuals and generally require arthropod or nematode vectors (Chapter 8). These vectors have piercing mouthparts capable of penetrating the plant cell wall and delivering a viral payload to the cytoplasm of the cell during feeding. Researchers that study plant viruses have capitalized on this aspect of plant virus transmission, adapting a "vector enabled genomics" approach, in which arthropod vectors are collected and their gut contents analyzed for plant virus nucleic acids to learn more about plant virus diversity and ecology (Ng et al., 2011).

Insects

Multiple families of viruses infect insects, including viruses with RNA genomes and those with DNA genomes. Of particular relevance to soils, however, are the baculoviruses. Part of the baculovirus replication cycle within their insect hosts includes the production of viral

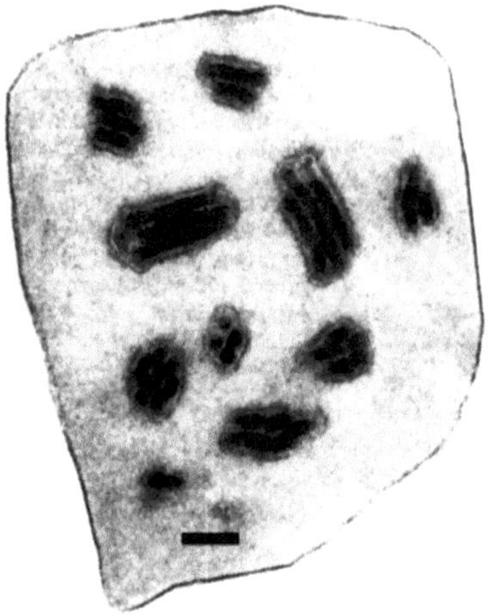

FIGURE 9.9 **Example of occlusion bodies generated by baculoviruses.** Scale bar represents 20 nm. *Source: Micrograph by Drion Boucias. Used with permission.*

nucleic acids embedded in protein crystals known as **occlusion bodies**. While baculoviruses use rod-shaped virus particles to spread from cell to cell in an infected host, occlusion bodies (Fig. 9.9) are released into the environment after the death of the host, which usually involves liquefaction of the insect. These occlusion bodies are resistant to environmental damage and can persist in soils for anywhere from 5 to 40 years. When occlusion bodies are ingested by susceptible insect hosts, either directly from soil or from soil-contaminated plant matter, the infection cycle begins again. Baculoviruses, as a group, infect a broad range of insect hosts, encompassing at least 3 taxonomic orders and over 500 different species. Understanding of baculovirus ecology has been instrumental in their development as biological control agents for managing insect pests (Fuxa, 2004).

Bacteria

Bacteriophage life cycles

Viruses of bacteria, or bacteriophages (or simply, phages) are probably the best-characterized group of viruses in soil ecosystems. In order to fully appreciate the ecological impacts of viruses in soil microbial communities, however, it is important to understand that most phages replicate through one of two major life cycles: lytic or lysogenic (Fig. 9.10).

Lytic phage infections are characterized by rapid production of intracellular virus particles followed by host cell lysis and release of progeny virions. Lytic infections tend to quickly eliminate local populations of susceptible hosts, and extracellular virus particles must survive long enough to propagate infection if the virus strain is to remain successful. By contrast, **temperate phages** may switch between the lytic replication pathway described earlier and **lysogenic** replication. Lysogenic replication involves the silencing of lytic gene expression by a phage-encoded repressor protein and frequently also the integration of the phage genome into the bacterial chromosome mediated by a phage-encoded enzyme (integrase). The specific replication pathway of temperate phages, lytic or lysogenic, is determined primarily by host

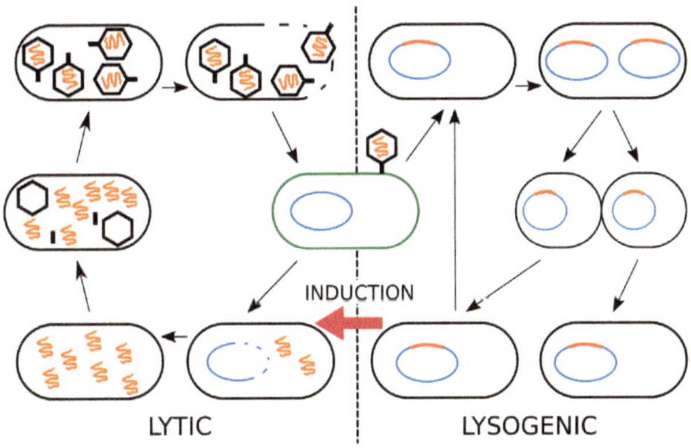

FIGURE 9.10 **Bacteriophage replication.** Start in the center of the figure, with the green cell infected by a phage particle. The bacterial chromosome is represented by *blue ellipses* and the phage genome is represented by the *orange squiggles* or *lines*. The lytic cycle proceeds downward and to the left; the lysogenic cycle proceeds upward and to the right. In lysogeny, the virus inserts into the host chromosome as a "prophage". Lysogenized cells replicate a copy of the prophage to each daughter cell. Lysogenized cells can also be induced *(red arrow)* to begin the lytic cycle. *Source: Diagram by Kurt Williamson.*

metabolic status. The state in which the host cell has been colonized by a repressed phage genome, termed **lysogeny**, can be maintained indefinitely. During cell division, the phage genome (**prophage**) is regularly assorted to daughter cells either as an integrated part of the host chromosome or as an extrachromosomal element. In this fashion, the phage is maintained within the host population. While lysogeny is highly stable, this relationship is not necessarily terminal for the infecting phage. Environmental signals, particularly damage to host DNA, can initiate a cascade of cell responses that inactivate the phage repressor protein. Following this event, in a process known as **induction**, the virus shifts into lytic replication and progeny virus particles are released into the environment through cell lysis (Fig. 9.10).

The biological, chemical, and physical properties of soil can change over time and space, and specific soil properties may favor specific viral replication strategies. In particular, lysogenic replication appears to hold a distinct advantage for viruses in soils, as lysogeny supports the long-term maintenance of phages within host populations when host distribution is patchy, when local host densities are low, or when host cells are starved for nutrients. It is estimated that approximately 30%–60% of the bacterial population in any given soil contain inducible prophages. A predominance of lysogeny may result in significant impacts of soil phages on the genetic diversity and evolution of hosts. This is because lysogenic relationships increase the probability of both genetic exchange between phage and host and horizontal gene transfer between hosts. However, soils also contain appreciable numbers of obligately lytic viruses, whose impacts on the ecology of soils should not be underestimated.

Bacterial mortality and community succession

Phages in soils appear to play important roles in mediating bacterial mortality and controlling bacterial community composition and diversity. Field experiments focusing on the sugar beet rhizosphere have shown that phage and host dynamics exhibit recurring annual cycles, suggesting that viruses are important in microbial community succession in soils. Microcosm experiments using naturally occurring soil microbes have shown that both bacterial and viral communities are highly dynamic and responsive to perturbations. These experimental results suggest that viral-mediated mortality is a significant factor controlling bacterial abundance and, by association, biomass turnover and rates of nutrient cycling in soil ecosystems. However, soils are notorious for their extremely high spatial heterogeneity and temporal variability, and it is not yet clear whether these results apply at global or even landscape scales.

Phage impacts on host fitness and evolution

Temperate phages are known to affect host fitness through two different processes, known as **superinfection immunity** and **lysogenic conversion**. In superinfection immunity, the repressor protein that is encoded by the prophage can also silence the gene expression of any new infections (superinfections) if the superinfecting phage genome has a similar binding sequence for the pre-existing repressor protein. In this regard, a bacterium that is infected by a prophage may be granted protection from infection and lysis by other closely related phages. In lysogenic conversion, the prophage may contain genes that increase the bacterial host's fitness under specific conditions, providing a selective advantage over non-lysogenized hosts. Some of these genes may have originated from previously infected hosts as described later. The high frequency of lysogeny among soil bacteria implies that temperate soil viruses influence host fitness and evolution through these processes.

Beyond the impacts of superinfection immunity and lysogenic conversion, which are limited solely to temperate phages, viruses in soils may have significant impacts on host population genetics by acting as vectors for the horizontal transfer of genes between different bacterial strains. Generally speaking, bacteria may acquire new alleles via three mechanisms: (1) transformation, the uptake of naked DNA from the environment by competent cells; (2) conjugation, the transfer of genetic material from one cell to another through a pilus; and (3) transduction, phage-mediated transfer of DNA (Chapter 4). Transduction begins as an error during the phage replication process in which host genomic DNA is mistakenly packaged into virus capsids. Two types of transduction are possible, generalized or specialized, depending on the replication mode of the phage mediating the transfer. Generalized transducing phages, which may be virulent or temperate, obtain essentially random fragments of the host genome through packaging errors. By contrast, specialized transduction is mediated only by temperate phages that integrate into the host chromosome. Only those bacterial genes closely flanking the prophage insertion site are transferred due to improper excision of the phage genome from the host chromosome. In either case, these defective phages, now termed **transducing particles**, act as shuttles capable of transferring DNA fragments between bacterial hosts. While transduction is a rare event leading to a low frequency of gene transfer, the short generation times and fast growth rates of many bacteria means that even rare recombination events may quickly alter the genotypic diversity of a bacterial community.

In laboratory-based experiments performed in liquid culture, bacterial hosts and phages have been used to estimate the frequency of transduction in aquatic environments. In these liquid culture experiments, the presence of particulate matter increased transduction frequencies by providing surface area on which phages and host cells could aggregate and interact. This suggests that soils, especially those which have extremely high specific surface areas (80–800 $m^2\ g^{-1}$; Chapter 2), can provide many potential sites for phage and host adsorption and interaction. However, only a few experiments using laboratory-gown hosts and phages have demonstrated the potential for transduction in soils.

Auxiliary metabolic genes

Beyond transduction, or phage-mediated transfer of genes from one bacterium to another, phages may acquire novel genes from the genomes of their hosts through recombination. In some cases, these acquired genes may be expressed during viral infection and increase the fitness of the phage by altering the metabolism of the infected host. Such genes are called **auxiliary metabolic genes** (AMGs). The best understood and most heavily documented example of this phenomenon is found in cyanophages, phages that infect photosynthetic cyanobacteria in aquatic environments (Mann et al., 2003). The majority of sequenced cyanophage genomes include a gene called *psbA*, which encodes the cyanobacterium major light harvesting protein designated D1. When cyanophages infect their cyanobacterial hosts, cellular metabolism shuts down as most cellular functions are diverted to the production of progeny phage, and this includes cellular expression of the D1 protein. D1 is labile, and once existing copies of the protein have broken down, photosynthesis in the infected cell ceases. This places a fundamental limit on the raw materials available in the cell for the production of viral progeny. However, phages that have acquired their own copy of the cyanobacterial *psbA* gene, and can thus express the D1 protein during infection, are able to reinstate cellular photosynthesis. This increases the raw materials available to the phage, and thus the number of progeny that can be made per infected cell, demonstrating the fitness benefit that results from this AMG.

Since the dramatic discovery of *psbA* in cyanophages in the early 2000s, dozens of additional AMGs have been identified that affect host carbon, nitrogen, and phosphorus metabolism, impacting cellular processes such as the Calvin cycle and nucleoside biosynthesis (Thompson et al., 2011; Warwick-Dugdale et al., 2019). However, the vast majority of described AMGs come from marine phages. Much less is known regarding the prevalence or diversity of AMGs in soils. The earliest documented example of AMGs in soil phages came from molecular detection assays. Phage particles were obtained from the induction of bacteria that had been extracted from soils treated with the broad-leaf herbicide atrazine. Using specific primers, PCR screening detected the presence of the atrazine catabolic gene *trzN* within the induced phage community, suggesting its status as an AMG (Ghosh et al., 2008). The acquisition of this gene by the phage may increase phage fitness by enabling the host to use atrazine as a nitrogen source during phage infection. Similar to the example of *psbA* in cyanophages, this arrangement could enable phages that have acquired the *trzN* gene to generate more progeny than competitors lacking the *trzN* gene. More recently, viromics approaches analyzing the viral diversity in thawing permafrost soils in Sweden has identified 14 putative glycoside hydrolase genes within the sampled viral genomes (Emerson et al., 2018). AMGs and their potential impact on biogeochemical cycles in soils remain an active area of research.

Archaea

The archaea are phylogenetically distinct from the bacteria. In terms of cell structure, archaea use ether phospholipids, as opposed to ester phospholipids found in bacteria, and archaeal cell walls lack peptidoglycan. Two of the four archaeal phyla (Thaumarchaeota and Korarchaeota) appear to be widely distributed in many environments, including fresh and marine waters and soils, where they are believed to carry out the aerobic oxidation of ammonia, a key step of the nitrogen cycle. Only a few Thaumarchaeota have been isolated and characterized, and no Korarchaeota have been isolated in laboratory culture. As a result, archaeal viruses have so far been isolated only on crenarchaeal or euryarchaeal hosts, all originating from hot spring, hypersaline, or hydrothermal environments. There are currently no cultured isolates of archaeal viruses from soils. Based on the presence of archaeal hosts, it is highly likely that soils harbor archaeal viruses. Currently, however, nothing is known regarding the abundance, genetic diversity, or ecological roles of archaeal viruses in soil environments.

Fungi

Fungal viruses, also known as mycoviruses, are likely found in most soils that support the growth of their hosts. Based on the presence of cytoplasmic virus-like particles and dsRNA within fungal cells, mycovirus infections appear to be common in all major fungal groups. The majority of the known mycoviruses have dsRNA genomes, but about a third of all known representatives have positive-sense ssRNA genomes. The mechanism by which mycoviruses infect and enter cells from the extracellular environment is unknown, and mycoviruses do not appear to cause lysis of infected host cells. This lack of virulence among most mycoviruses precludes plaque assay methods of detection. While most mycovirus infections are cryptic, producing no apparent changes associated with host fitness, infection by certain mycoviruses can cause dramatic and noticeable changes in host phenotype as described later (see also Williamson, 2018).

Killer phenotype

The killer phenotype, or so-called "killer viruses," encode a secreted toxin as well as an antidote molecule that confers immunity to the infected host. The mechanism of action of the secreted toxin varies by virus but appears to fall into two main classes. Some killer viruses encode ionophoric toxins that disrupt normal cell membrane function by forming cation-selective channels. The resulting ionic imbalance in the cytoplasm causes osmotic distress leading to cell death. Other killer viruses encode proteins that interact with transcription factors that block the cell cycle, resulting in eventual cell death. The killer phenotype conferred by these viruses upon their hosts enables the infected host to kill off competitor cells lacking the toxin antidote, resulting in clear fitness gains since the death of competitors ensures more resources for the host, and its parasitic mycovirus, for survival and reproduction. Because only infected hosts contain the toxin antidote molecules, killer viruses can control mating success between infected and non-infected fungal partners. Mating will be more successful between two infected partners, ensuring that surviving progeny carry the virus, and potentially driving evolution of the host species. This toxin-antidote system may represent an important survival strategy for the mycoviruses, potentially leading to their evolutionary success in the highly variable soil environment containing a patchy distribution of hosts.

Hypovirulence

Hypovirulence refers to a reduction in the severity of disease caused by the fungal host, typically in fungal infection of plants, and results from mycovirus infection of the fungus. Hypovirulence has been observed in several plant-parasitic fungi but is best characterized in the hypoviruses of chestnut blight fungus, *Cryphonectria parasitica*. *C. parasitica* typically enters its host plant through physical wounds or breaks in the bark, infecting the actively growing cells. The fungus then secretes toxic compounds that kill the exposed host plant cells. Eventually the fungal growth or canker will girdle the entire trunk, killing everything above it. Infection of *C. parasitica* fungus by hypoviruses can greatly reduce the virulence of the fungal infection, in many cases allowing the host plant to survive and retain living tissues (e.g., trunk, branches, and leaves) above the fungal infection site. Because of their role in reducing the pathogenicity of fungi toward their host plants, there has been significant interest in the use of hypoviruses as biological control agents of chestnut blight since the 1950s. It is not clear the extent to which viruses of other soil microbes may alter host function. But given the impacts of temperate phages and hypoviruses on their respective hosts, it is very likely that we have yet to discover other important ways in which viruses impact their hosts, leading to differential success in the heterogeneous soil environment.

Thermotolerance

In the only known example of its kind among plants, a mycovirus has been shown to alter the fitness of a host plant that has been colonized by the mycovirus' endophytic fungal host. (An endophyte is an endosymbiont that resides within a plant host during at least part of its life cycle without causing discernable disease.) The mycovirus, *Curvularia* thermal tolerance virus (CThTV), alters the phenotype of its fungal host *Curvularia protuberate* such that, when the roots of the panic grass *Dichanthelium lanuginosum* are colonized by the fungal endophyte, the tripartite interaction allows both fungus and host plant to grow at high soil temperatures (Márquez et al., 2007). This results in significant niche expansion, as evidenced by the host plant growing near geothermal features in Yellowstone National Park, United States. Laboratory experiments showed that when grass root zones were heated to 65°C, non-symbiotic plants were not able to survive, while symbiotic plants survived. Additional experiments involving virus-infected fungi and virus-free fungal isolates revealed that the mycovirus was a crucial element in the symbiosis, as only those plants inoculated with fungal isolates carrying the mycovirus were able to survive heat stress. In the absence of heat stress, symbiotic plants exhibited enhanced growth rate and drought tolerance compared with non-symbiotic plants. The molecular mechanisms underlying thermotolerance and other phenotypic changes in this symbiosis have not yet been established. Nevertheless, the existence of this tripartite relationship raises important questions regarding how common such complex interactions might be among the mycoviruses as well as among viruses of other soil microbes.

Cyanobacteria, algae, and diatoms

Algae, diatoms, and cyanobacteria are common inhabitants of most soils with sufficient moisture-holding capacities to support their growth, and they can impart many benefits to soils and associated ecosystem services (Chapter 7). Yet, despite their relatively broad global distribution, almost nothing is known regarding the viruses that infect soil algae and cyanobacteria. In aquatic systems, viruses are well-studied and are known to have significant impacts on host abundance, host population dynamics, host genetic diversity, nutrient cycles, and even cloud formation and atmospheric conditions (Arnold et al., 2013). Several studies have examined the dynamics and potential impacts of cyanophages in rice paddy soils. Paddy systems, however, are typically flooded most of the year, and therefore it is difficult to determine whether the cyanobacteria and associated phages found in these systems should be considered aquatic or terrestrial. For the time being, the ecology of viruses that infect algae, cyanobacteria, and diatoms in soil ecosystems remains unknown.

Protozoa

The discovery of Mimivirus, an unusually large virus (400–800 nm) infecting the protozoan host, *Acanthamoeba polyphaga*, started a revolution in virology, spawning concerted efforts to determine the prevalence of these so-called giant viruses in nature. In the past decade, isolation efforts have produced several additional giant viruses, with at least 20 isolates originating from soils (Pagnier et al., 2013). The mimiviruses, as a group, hold much interest due to their large genomes and array of novel and unusual genes. Indeed, the genetic diversity of this group is still being characterized and compared. Currently, however, the abundance of amoebal viruses in soils, as well as their specific ecological impacts on host population structure and function, remain unknown.

Summary

Viruses are molecular genetic parasites of cellular systems. Most viruses found in soils are composed of a protein capsid that protects a genome comprised of various arrangements of either DNA or RNA. In order to replicate, viruses must attach to, enter, and express their genes within susceptible host cells. In many cases, viral replication leads to host cell death and lysis, although in some cases, viral replication can result in transfer of viral genetic information to host daughter cells.

The persistence of infectious virus particles in soil depends on temperature, soil water content, and microbial activity in soil. Although most of the studies that determined these factors were concerned with fate and transport of human and animal viruses, it is highly likely that most viruses are similarly affected by changes in biological, chemical, and physical conditions.

Much remains unknown regarding the roles and impacts of autochthonous viruses on host population dynamics, host-mediated process rates, and host fitness and evolution in soil ecosystems. Bacteriophages are currently the best understood viruses in soils, but even here, many questions remain. For some virus groups, such as those infecting soil archaea, fungi, and algae, almost nothing is known. Important advances in molecular techniques, particularly viromics, have enabled us to learn more about viruses in soil environments, and no doubt, continued advances will allow further elucidation of viral diversity and viral impacts on host activities. We continue to move slowly forward toward an improved understanding of the role of viruses in soil ecology.

Study questions

1. Are viruses alive? Which aspects of viruses support their inclusion with other living organisms? Which aspects of viruses result in their exclusion from being considered living organisms?

2. Which step of virus replication determines the species of host(s) it can infect?
3. You are working in a state public health laboratory monitoring water quality at a popular lake that is used for swimming, boating, and recreational fishing. You are using a PCR method to keep track of human adenovirus, a common indicator of fecal contamination. (A) What are the advantages and disadvantages of using this detection method? (B) One particular set of water samples shows an alarmingly high amount of human adenovirus using your PCR-based method. Do those same samples contain any infectious human adenovirus? How could you tell?
4. A small start-up company is seeking to develop an antibody-based assay (ELISA) for a novel human coronavirus. What information is needed so that such a test can be developed?
5. Why is lysogeny predicted to be a more common replication strategy for bacteriophages in soil than lytic replication?
6. What is hypovirulence (observed in fungal viruses)?
7. What are the advantages and disadvantages of using plaque assays to determine viral abundance in soil samples?
8. Explain why auxiliary metabolic genes are of such great interest to researchers.
9. What is the purpose of occlusion bodies observed in baculoviruses?
10. Predict two possible roles or impacts of bacterial viruses in soil ecology.

References

Arnold, H.E., Kerrison, P., Steinke, M., 2013. Interacting effects of ocean acidification and warming on growth and DMS-production in the haptophyte coccolithophore *Emiliania huxleyi*. Glob. Change Biol. 19, 1007–1016, https://doi.org/10.1111/gcb.12105.

Brandes, N., Linial, M., 2019. Giant viruses—big surprises. Viruses 11 (5), 404. doi: 10.3390/v11050404.

Emerson, J.B., Roux, S., Brum, J.R., Bolduc, B., Woodcroft, B.J., Jang, H.B., Singleton, C.M., Solden, L.M., Naas, A.E., Boyd, J.A., Hodgkins, S.B., Wilson, R.M., Trubl, G., Li, C., Frolking, S., Pope, P.B., Wrighton, K.C., Crill, P.M., Chanton, J.P., Saleska, S.R., Tyson, G.W., Rich, V.I., Sullivan, M.B., 2018. Host-linked soil viral ecology along a permafrost thaw gradient. Nat. Microbiol. 3, 870–880, https://doi.org/10.1038/s41564-018-0190-y.

Filée, J., Tétart, F., Suttle, C.A., Krisch, H.M., 2005. Marine T4-type bacteriophages, a ubiquitous component of the dark matter of the biosphere. Proc. Natl. Acad. Sci. USA 102, 12471–12476, https://doi.org/10.1073/pnas.0503404102.

Fuxa, J.R., 2004. Ecology of insect nucleopolyhedroviruses. Agric. Ecosyst. Environ. 103, 27–43, https://doi.org/10.1016/j.agee.2003.10.013.

Ghosh, D., Roy, K., Williamson, K.E., White, D.C., Wommack, K.E., Sublette, K.L., Radosevich, M., 2008. Prevalence of Lysogeny among Soil Bacteria and Presence of 16S rRNA and trzN Genes in Viral-Community DNA. Appl. Environ. Microbiol. 74, 495–502, https://doi.org/10.1128/AEM.01435-07.

Liu, Junjie, Wang, G., Wang, Q., Liu, Judong, Jin, J., Liu, X., 2012. Phylogenetic diversity and assemblage of major capsid genes (*g23*) of T4-type bacteriophages in paddy field soils during rice growth season in Northeast China. Soil Sci. Plant Nutr. 58, 435–444, https://doi.org/10.1080/00380768.2012.703610.

Mann, N.H., Cook, A., Millard, A., Bailey, S., Clokie, M., 2003. Marine ecosystems: bacterial photosynthesis genes in a virus. Nature 424, 741, https://doi.org/10.1038/424741a.

Márquez, L.M., Redman, R.S., Rodriguez, R.J., Roossinck, M.J., 2007. A virus in a fungus in a plant: three-way symbiosis required for thermal tolerance. Science 315, 513–515, https://doi.org/10.1126/science.1136237.

Ng, T.F.F., Duffy, S., Polston, J.E., Bixby, E., Vallad, G.E., Breitbart, M., 2011. Exploring the diversity of plant DNA viruses and their satellites using vector-enabled metagenomics on whiteflies. PLoS ONE 6, e19050, https://doi.org/10.1371/journal.pone.0019050.

Pagnier, I., Reteno, D.-G.I., Saadi, H., Boughalmi, M., Gaia, M., Slimani, M., Ngounga, T., Bekliz, M., Colson, P., Raoult, D., Scola, B.L., 2013. A decade of improvements in mimiviridae and marseilleviridae isolation from amoeba. Intervirology 56, 354–363, https://doi.org/10.1159/000354556.

Simon, L.O., Anderson, T.F., 1967. The infection of *Escherichia coli* by T2 and T4 bacteriophages as seen in the electron microscope. Virology 32, 279–297.
Suttle, C.A., 2007. Marine viruses—major players in the global ecosystem. Nat. Rev. Microbiol. 5, 801–812.
Swanson, M.M., Fraser, G., Daniell, T.J., Torrance, L., Gregory, P.J., Taliansky, M., 2009. Viruses in soils: morphological diversity and abundance in the rhizosphere. Ann. Appl. Biol. 155, 51–60, https://doi.org/10.1111/j.1744-7348.2009.00319.x.
Thompson, L.R., Zeng, Q., Kelly, L., Huang, K.H., Singer, A.U., Stubbe, J., Chisholm, S.W., 2011. Phage auxiliary metabolic genes and the redirection of cyanobacterial host carbon metabolism. Proc. Natl. Acad. Sci. USA 108, E757–E764, https://doi.org/10.1073/pnas.1102164108.
van Regenmoretel, M.H.V., Mahy, B.W.J., 2004. Emerging issues in virus taxonomy. Emerg. Infect. Dis. 1, 8–13.
Wang, G., Yu, Z., Liu, J., Jin, J., Liu, X., Kimura, M., 2011. Molecular analysis of the major capsid genes (g23) of T4-type bacteriophages in an upland black soil in Northeast China. Biol. Fertil. Soils 47, 273–282, https://doi.org/10.1007/s00374-010-0533-1.
Warwick-Dugdale, J., Buchholz, H.H., Allen, M.J., Temperton, B., 2019. Host-hijacking and planktonic piracy: how phages command the microbial high seas. Virol. J. 16, 15, https://doi.org/10.1186/s12985-019-1120-1.
Williamson, K.E., 2018. Viruses of microorganisms in soil ecosystems. In: Hyman, P., Abedon, S.T. (Eds.), Viruses of Microorganisms. Caister Academic Press, Norfolk, UK, pp. 77–94.
Williamson, K.E., Fuhrmann, J.J., Wommack, K.E., Radosevich, M., 2017. Viruses in soil ecosystems: an unknown quantity within an unexplored territory. Annu. Rev. Virol. 4, 201–219, https://doi.org/10.1146/annurev-virology-101416-041639.
Williamson, K.E., Radosevich, M., Wommack, K.E., 2005. Abundance and diversity of viruses in six Delaware soils. Appl. Environ. Microbiol. 71, 3119–3125, https://doi.org/10.1128/AEM.71.6.3119-3125.2005.

Further reading

Enquist, L.W., 2009. Virology in the 21st century. J. Virol. 83, 5296–5308.
Williamson, K.E., 2011. Soil phage ecology: abundance, distribution, and interactions with bacterial hosts. In: Witzany, G. (Ed.), Biocommunication in Soil Microorganisms. Springer Berlin Heidelberg, Berlin, Heidelberg, pp. 113–136.
Ghabrial, S.A., Castón, J.R., Jiang, D., Nibert, M.L., Suzuki, N., 2015. 50-plus years of fungal viruses. Virology 479, 356–368.

CHAPTER

10

Microbial ecology

Stephanie A. Yarwood

Department of Environmental Science and Technology, University of Maryland, College Park, MD, United States

A clod of earth seemingly simple and lifeless ... its particles most elaborate in composition... inhabited by prodigious numbers of living organisms inconceivably small... yet somehow linked up with our lives. ...

—***Sir E. John Russell***

Soil microbial ecology is the study of the properties and behavior of microorganisms in their soil environment. We seek to understand how soil physical and chemical properties and organismal interactions influence the microbially mediated processes that drive soil ecosystems. Because of the interplay between the physicochemical and biological properties of soil, many different approaches can be used to study soil microbial ecology. For example, we may wish to identify the soil properties that determine the rate of a specific microbially mediated process in different soil types. Alternatively, we might be more interested in determining if different soil management practices influence the overall composition of the soil microbial community. These changes in community composition might then be correlated with soil health or resistance and resilience in the face of environmental perturbations. At the other end of the spectrum, we might wish to identify how one member of a bacterial soil population can prevent a specific plant pathogen from infecting a crop species. The purpose of this chapter is to introduce the subject of soil microbial ecology and to point out the interactions that must occur between microorganisms and their environment if soil ecosystems are to function effectively.

Soil organisms and their interactions

Many kinds of organisms spend all, or at least a part, of their lives in the soil. These include plants, animals, fungi, bacteria, and viruses. A typical gram of soil contains billions of individual microorganisms representing thousands of different species. Genetically similar individuals (e.g., same species) are often referred to collectively as a population (Fig. 10.1). Different populations which perform similar functions and compete for resources are grouped into guilds. All microbial guilds, and thus all populations, inhabiting a given environment comprise the microbial community, or **microbiome**, for that soil. The microbiome together with

Principles and Applications of Soil Microbiology. http://dx.doi.org/10.1016/B978-0-12-820202-9.00010-1
Copyright © 2021 Elsevier Inc. All rights reserved.

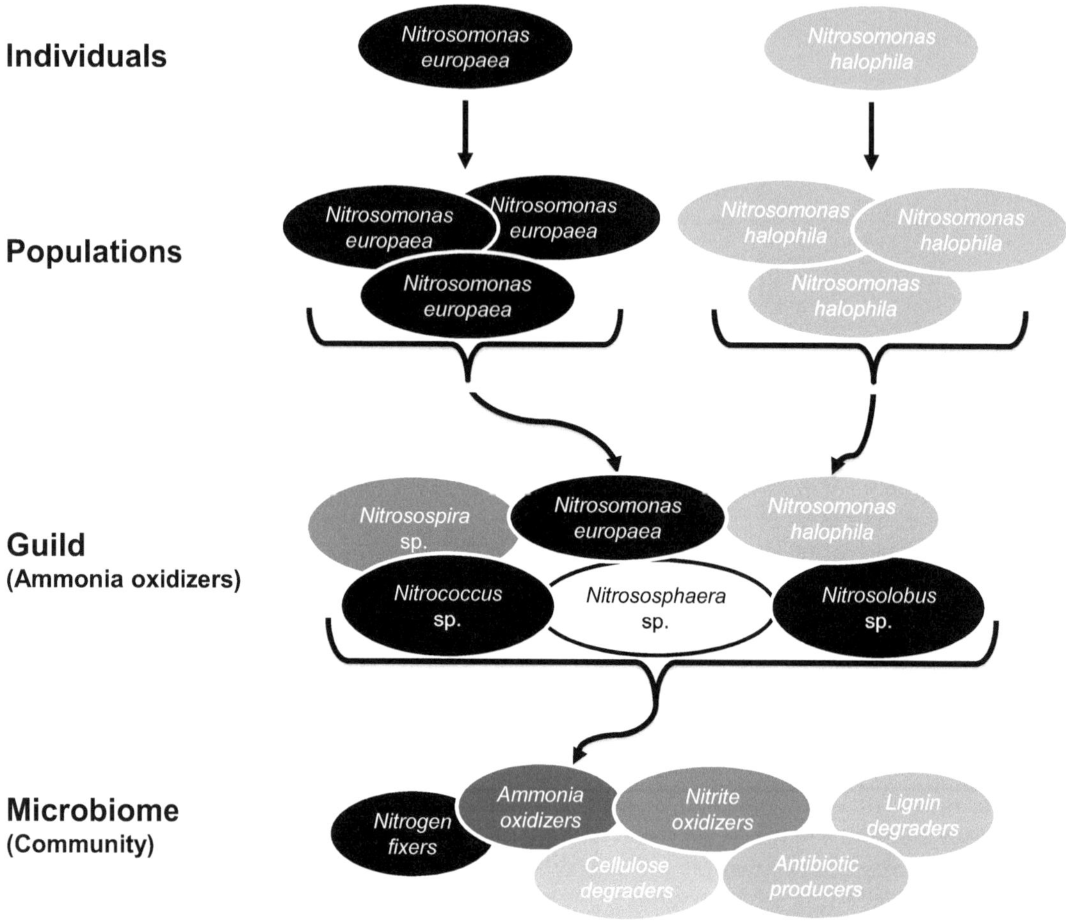

FIGURE 10.1 **Ecological organization within a microbial community.** In this example, populations of *Nitrosomonas* spp. along with other bacteria comprise the ammonia oxidizer guild. The various guilds (e.g., ammonia oxidizers, nitrogen fixers) collectively comprise the microbial community, or microbiome, for the ecosystem. Note that guilds are not mutually exclusive in that a given organism may be capable of performing more than one ecological function, for example, both cellulose degradation and nitrogen fixation.

other organisms (e.g., plants) and abiotic components (e.g., soil solids) interacting within an environment forms an ecosystem.

Most soil microbiomes contain large populations with nutritional versatility and fast metabolism as the hallmark features of many (but not all) microorganisms. These nutritional and metabolic characteristics permit microorganisms to play a major role in decomposing organic materials and returning essential plant nutrients to the soil.

Nevertheless, it is difficult for microorganisms to achieve their metabolic potential in the soil matrix. Accessing nutrients is a major problem for soil microorganisms because soil is a discontinuous matrix of vast surface area (Chapter 2). Despite seemingly high densities of microorganisms, it has been estimated that less than 0.4% of soil pore space is occupied by microorganisms. Moreover, because a large percentage of the pore diameters in soil aggregates are less than 2 μm, particulate substrates in these voids are inaccessible to all but the smallest cells. The bioavailability of hydrophobic substrates is limited by adsorption to soil surfaces, whereas diffusion of soluble substrates is often limited by the lack of continuous water films or water-filled pores. To further complicate matters, most bacteria possess a unicellular growth habit that restricts their ability to efficiently colonize massive, solid, low surface area-to-volume food sources (e.g., plant residues) and prevents them from moving effectively between spatially separated food sources. Because of their small size, bacteria and many of the smaller fungi occupy micropores that limit their interaction with organisms only a few millimeters away and physically exclude the larger organisms that prey upon them.

If decomposition of organic materials is to proceed effectively and if nutrients are to move along the soil **food chain** from microorganisms into the higher **trophic levels** (Fig. 10.2), the limits on substrate availability must be circumvented. The highly branched nature of plant

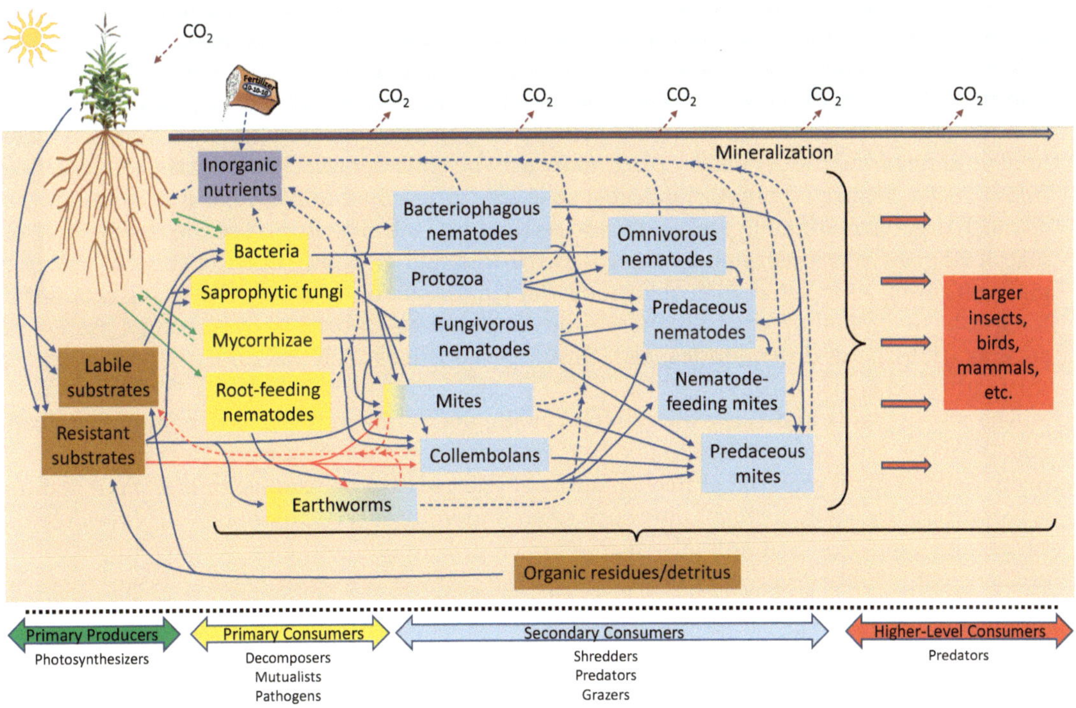

FIGURE 10.2 **A simple soil food web showing the movement of carbon and nutrients from plant roots and aboveground litter through soil trophic levels.** For *blue* and *green lines*, line color indicates movement of compounds/nutrients between organisms in soil *(blue)* or directly between plants and microbes *(green)*; line style indicates movement of organic *(solid)* or inorganic *(dashed)* forms of compounds/nutrients. *Red lines* illustrate activity of shredders with *solid lines* indicating movement of resistant substrates through shredders and *dashed lines* indicating return of digested material to the labile substrate pool. *Arrows* at bottom of figure indicate producer and consumer levels and representative organisms at various stages of the soil food web. *Source: Adapted from (Hunt et al., 1987).*

root systems and the mycelia of fungi are examples of spreading growth habits that permit access to spatially separated food sources. Many fungi and soil animals possess the mobility or mass, or both, to physically disrupt both the structure of soil and its massive nutrient sources (e.g., plant residues). For example, like plant roots, fungal hyphae display directed growth, allowing them to penetrate solid organic substrates that are largely unavailable to most bacteria. These activities circumvent some of the substrate bioavailability problems by permitting fungi and some bacteria (e.g., Actinobacteria) to gain access to and more efficiently colonize and decompose their nutrient sources. In addition, they enable predators to gain access to their microbial prey.

Soil food web: producers and consumers

Plant life is the primary energy source that drives most terrestrial soil ecosystems (there are few exceptions such as deserts where algae and cyanobacteria are substantial contributors to primary productivity; Chapter 7). Because soils contain low concentrations of biologically available mineral nutrients, terrestrial plants need large root systems to access a sufficient volume of the soil to obtain an adequate supply of essential minerals and water. However, plant roots also release various organic compounds into the adjacent soil volume. As a consequence, plant roots provide a supply of relatively labile organic compounds (photosynthate) to microorganisms in their immediate vicinity (i.e., the **rhizosphere** effect, Chapter 11). The total volume of soil influenced by plant roots may be considerable. An estimated 30%–90% of fine roots (< 1 mm in diameter) in forest soils are turned over annually and provide at least as much organic material to soil organisms as does leaf fall. The impact of the rhizosphere on microbial growth is illustrated by the 10- to 100-fold higher population densities of microorganisms closely associated with plant roots than in the surrounding soil. As a consequence of the greater microbial productivity near roots, populations of bacterial and fungal predators, such as protozoa and nematodes, also thrive in the rhizosphere, as do their predators, the microarthropods.

The role of plants in driving the soil food chain, that is, the movement of energy and nutrients from one type of organism to another, also occurs at the soil surface. Leaf fall and plant death deposit plant material at the soil surface. The greatest percentage of this plant residue is located in physically resilient, low surface area-to-volume structures such as leaves, stems, twigs, and branches. Soil animals, such as microarthropods and earthworms, improve the efficiency of microbial decomposition of the large plant debris by increasing its surface area-to-volume ratio through their shredding activities and by distributing the debris throughout the soil volume. In addition, the burrowing activity of moles and other earthworm predators causes further disturbance of the soil environment. These activities result in the redistribution of mineral nutrients and organic matter throughout the soil volume and create more uniform microbial activity and population growth. The interactions of the various members of the soil community and their associated roles in processing energy and essential elements through different chemical and physical forms are often conceptualized schematically in a **food web** (Fig. 10.2). As energy is transferred between trophic levels, some is lost as heat via cellular metabolism.

Growth characteristics of soil microoganisms

Although essential for soil ecosystem function, microbial growth in soil is transitory and depends on the availability of nutrients and adequate physicochemical conditions. Different scenarios of nutrient availability occur in soil and, as a result, different growth strategies are required. For example, under dry surface soil conditions, biodegradable nutrients (fecal material, animal and microbial corpses, plant residues, seeds, and fruits) might accumulate because there is insufficient water to support microbial activity. Upon the return of water, a "flush" of rapid microbial growth occurs. This scenario of transient, luxuriant nutrient availability can be contrasted with the rhizosphere where a more consistent supply of nutrients is available or with the portion of soil organic matter that is less labile and which can be degraded only by the subset of microorganisms equipped with the enzymes necessary to attack these complex substrates (Burns et al., 2013; Chapter 13). Each environment provides microorganisms with an opportunity to express their individual growth characteristics and to compete for nutrients with whatever unique attributes they can bring to bear (such as substrate versatility, fast growth potential, high affinity nutrient transport systems, and production of antimicrobial compounds).

Mathematical concepts of microbial growth

Under favorable environmental conditions of initially nonlimiting nutrients, microbial growth proceeds through various phases (Fig. 10.3). This section provides a basic overview of the quantification of archaeal and bacterial growth, both of which reproduce via binary fission (division to produce two daughter cells of equal size). Briefly, after an adaptive or lag phase (Fig. 10.3A), the organism enters a period of unlimited (exponential or log phase) growth (Fig. 10.3B). Eventually, nutrients become limiting or toxic metabolites accumulate, and growth slows down at the end of the exponential phase. Finally, net growth ceases signaling the start of the stationary phase (Fig. 10.3C) which is subsequently followed by cell death (Fig. 10.3D). It should be noted that there can be new cell growth during both the stationary

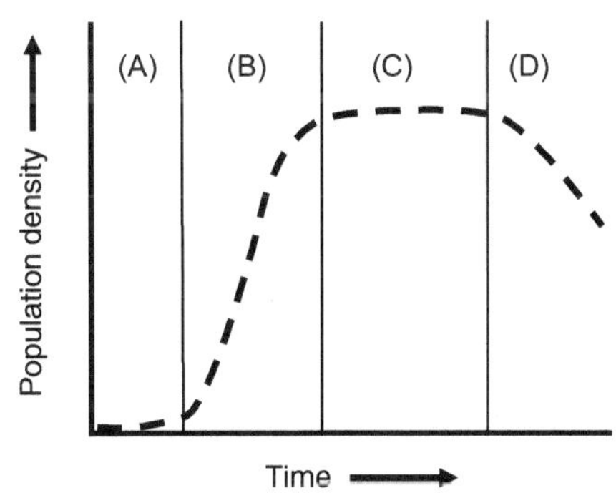

FIGURE 10.3 (A) Adaptive or lag phase, (B) logarithmic or exponential phase, (C) stationary phase (growth and death in equilibrium), and (D) death phase. *Source: Adapted from Bottomley (2005).*

and death phases, but it is matched or exceeded by death; death = growth during stationary phase, and death > growth during death phase.

During the exponential growth phase, bacterial cell numbers double per constant interval of time. This time interval is referred to as the **doubling time,** or **generation time** (often abbreviated to t_d). The number of cells per unit volume of growth medium (N_t) after a given period of time (t) depends on the number of cells per unit volume in the initial population of cells (N_0) and the number of doublings (n) that have occurred during the time interval under consideration (t). Mathematically this can be expressed as:

$$N_t = N_0 \times 2^n \tag{10.1}$$

Because $n = t/t_d$, Eq. (10.1) can be rearranged as:

$$N_t = N_0 \times 2^{t/t_d} \tag{10.2}$$

It is more convenient to consider the linear form of Eq. (10.2):

$$\ln N_t = \ln N_0 + \frac{\ln 2 \times t}{t_d} \tag{10.3}$$

Because it is also more convenient to work with logarithms of base 10 than with natural logarithms, the latter are divided throughout by 2.303 to give:

$$\log_{10} N_t = \log_{10} N_0 + \frac{\log_{10} 2 \times t}{t_d} \tag{10.4}$$

During the unlimited growth phase, a plot of $\log_{10}N_t$ against t will produce a straight line with an intercept on the *y*-axis at $\log_{10}N_0$ and a slope of $\log_{10}2/t_d$.

This type of bacterial and archaeal cell growth is referred to logarithmic or exponential growth, with the slope representing the growth rate constant (often abbreviated to "μ"). The magnitude of the growth rate constant is inversely proportional to the doubling time. Some bacteria can achieve amazingly short generation times (<1 h) when nutrients are non-limiting. As a result, they can produce astronomical populations in short periods of time (Box 10.1).

It is important to note that these growth characteristics differ for viruses (Chapter 9), which create numerous viral particles (i.e., some bacteriophages produce bursts of 100–300 viral particles per replication cycle), and fungi, which often reproduce by sporulation. Under unlimited conditions they too can achieve large populations. Nevertheless, despite these impressively high growth potentials, most microbial growth in soil probably occurs under substrate limited conditions. In fact, ~80% of the soil microbial community may be dormant at any given time (Lennon and Jones, 2011).

Effect of substrate concentration on growth rate

Microbial growth is the outcome of hundreds of chemical reactions that occur in the cell and are catalyzed by enzymes (Chapter 3). The rate at which an enzyme carries out its reaction is controlled in part by the concentrations of the reactants. The relationship between the rate of an enzyme reaction (v) and the concentration of its reactant or substrate (S) is often expressed by the *Michaelis-Menten equation*:

BOX 10.1

Calculating microbial growth

Example 1. Using Eq. (10.4), you can calculate that if a bacterium has a doubling time of 1 h, it would produce 16 million progeny in 24 h.

Rearrange Eq. (10.4) to

$$\log_{10} N_t - \log_{10} N_0 = \frac{\log_{10} 2 \times t}{t_d}$$

Substitute numerical values into the equation,

$$\log_{10} N_t - \log_{10} N_0 = \frac{0.30 \times 24}{1}$$

Rearrange the equation,

$$\log_{10} N_t \times 1 = 7.2 + \log_{10}(1)$$

Therefore, $\log_{10} N_t = 7.2 + 0$ and antilog 7.2 equals 1.6×10^7 cells.

Example 2. Often, we know the values of N_t and N_0 and we want to calculate the doubling time. For example, if N_t and N_0 are 10^8 and 10^6 cells mL^{-1} and the length of the unlimited growth interval is 24 h, we can determine that the doubling time is 3.6 h:

$$\log_{10}(10^8) - \log_{10}(10^6) = \frac{\log_{10} 2 \times 24}{t_d}$$

Therefore, $8 - 6 = \frac{0.3 \times 24}{t_d}$ and $2 \times t_d = 7.2$ and $t_d = 3.6$ h.

$$v = \frac{V_{max} \times S}{K_m + S} \tag{10.5}$$

Here, v is the reaction velocity when substrate is limited. V_{max} is the maximum velocity when substrate concentration is optimal, and K_m, the Michaelis-Menten constant, represents the concentration of substrate needed to drive the enzyme reaction at one-half of its maximum velocity.

An analogous quantitative relationship exists between the growth rate of a microorganism and substrate concentration. This relationship can be expressed mathematically in an equation referred to as the *Monod equation:*

$$\mu = \frac{\mu_{max} \times S}{K_s + S} \tag{10.6}$$

where K_s represents the substrate concentration at which the growth rate of the microorganism is one-half of its maximum rate under specified environmental conditions. The growth response to substrate concentration is illustrated graphically as a hyperbolic curve (Fig. 10.4). When $S \gg Ks$, then $\mu = \mu_{max}$ for all practical purposes. When $S = K_s$ then μ is $0.5\ \mu_{max}$. The lower the value of Ks, the greater is the affinity of the microorganism for the substrate and

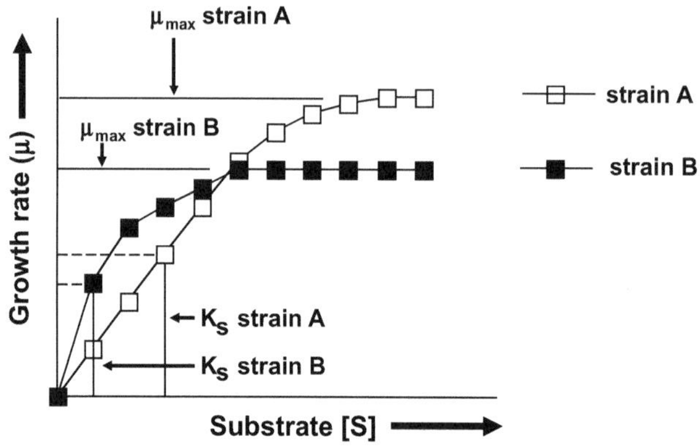

FIGURE 10.4 **Growth as a function of substrate concentration for two bacteria, A and B, with different maximum growth rate potentials (μ_{max}) and with different affinities (K_s) of their substrate uptake.** *Dashed lines* indicate growth rates corresponding to respective K_s values. *Source: From Bottomley (2005).*

the less is the influence of substrate concentration on growth rate. Values of K_s have been determined for a variety of substrates, and generally they are low (Table 10.1). When all substrate is consumed, the final microbial biomass is proportional to the product of the initial concentration of the limiting substrate and the **growth yield coefficient** (Y), which is defined as the quantity of biomass carbon formed per unit of substrate carbon consumed (sometimes also called carbon use efficiency, or CUE; Chapter 13). Heterotrophic microorganisms typically exhibit growth yield coefficients of 0.30–0.55 g (30%–55%) of biomass-carbon g^{-1} of substrate carbon consumed, but some values as high as 0.85 have been observed for more reduced substrates under aerobic conditions (Geyer et al., 2016). This value depends on the energy content of the substrate, oxygen availability, and the nitrogen source available for growth.

In contrast to pure-culture laboratory conditions, microbial growth in soil occurs under mixed population conditions. When substrates are nonlimiting, the growth rates of two organisms are influenced by their relative maximum growth rates. In contrast, when competitive growth occurs at low substrate concentrations, K_s plays a more significant role in the relative success of the two competing microorganisms. Under these conditions, it is theoretically

TABLE 10.1 Affinity constants (Ks) of the uptake transport systems of bacteria for various substrates.

Substrate	K_s (μM)
Glutamate	1–10
Lactate	20–60
Succinate	5–300
Maltose	1–50
Glucose	5–10

From Bottomley (2005).

possible that an organism with the smaller K_s will outgrow another organism with a greater μ_{max} (Fig. 10.4). An example of this is the potential for ammonia-oxidizing archaea to out-compete ammonia-oxidizing bacteria when soil ammonium concentrations are low (Chapter 15).

Classification of soil microorganisms based on growth characteristics

The idea that soil microorganisms can be classified on the basis of their growth characteristics, nutritional versatility, and affinity for substrates has surfaced in various forms and at various times throughout the last century. For example, Winogradsky (1924) referred to soil microorganisms that grow rapidly when high-energy-containing substrates are readily available as **zymogenous.** In contrast, he referred to organisms that colonized the more recalcitrant material remaining after a primary attack had subsided as the **autochthonous** microflora.

Despite the logical intuitiveness of Winogradsky's concept, little experimental evidence has surfaced to indicate that soil bacteria can be concretely separated on these grounds. Another attempt to group soil bacteria based upon growth characteristics arose from the work of researchers in Japan (Ohta and Hattori, 1983). They isolated soil bacteria, referred to as **oligotrophs**, that grow much better at low concentrations of substrates than at high concentrations. In contrast, many other soil bacteria grow better at relatively high concentrations of nutrients, such as those provided in most standard laboratory growth media; these bacteria are often referred to as **copiotrophs.**

Andrews (1984) discussed the possibility of linking microbial classification based on growth properties with the r-K theory of colonization and succession. Briefly, the theory suggests that a species needs to adopt a different strategy to colonize an environment in which it is initially present at low density **(r-strategy)** than to persist in an environment in which it is already present near to its carrying capacity **(K-strategy).** The r-strategy species place a large percentage of their available energy into reproduction and attempt to occupy the niche as quickly as possible. Species specializing in this behavior are considered to be poorly adapted in dealing with environmental stress and periods of low nutrient availability. As a consequence, populations of r-strategists are thought to undergo large fluctuations over time. Many of the soil bacteria that have been cultured and subsequently described, particularly those exhibiting zymogenous or copiotrophic traits, probably belong to the r-strategist category. In the case of K-strategists, growth rate and reproduction are finely tuned to the limited resources available, and these species are adept at dealing with periods of starvation and stress. Traditionally, actinobacteria and fungi have been considered K-strategists because they possess slower growth rates, produce long-lived spores and other survival structures, and metabolize complex nutrients such as lignin.

The wide adoption of molecular tools to characterize soil microbial communities has shifted some views related to r-K theory and led to qualifications of the generalizations presented earlier. For example, DNA sequencing has revealed large populations of uncultured soil bacteria, many suspected to have slow growth rates and better adaptation to oligotrophic conditions. These include members of the Acidobacteria and Verrucomicrobia. Similarly, fungi have been shown to vary considerably in their growth rates and nutrient requirements. Ernebjerg and Kishony (2012), observing the rate of colony growth, concluded that rather than a clear dichotomy of oligotrophs and copiotrophs, soil bacteria exhibit a wide range of growth strategies. They further observed that colony formation was somewhat dependent on the proximity of cells to each other on culture media, underscoring that pure culture characterization gives us only limited information when dealing with the mixed communities of soil.

Interactions between organisms and soil

Despite tremendous advances in our understanding of soil microbiology, we still lack knowledge about the conditions under which most organisms exist within soil. This is partially because the scale at which we study soil properties is often mismatched with the scale of conditions impacting soil microorganisms. For example, we might determine soil pH by homogenizing 5 g of a soil sample in 10 mL of water and then measuring pH of the slurry, but these 5 g contain numerous pores (microsites) that can vary in pH. Water in a pore containing ammonia-oxidizers may have a lower pH because these organisms exude protons. In contrast, water in pores containing sulfate reducers could have a higher pH because their metabolism consumes protons. This discrepancy between analytical and microbially relevant scales limits our ability to elucidate linkages between organisms and processes in soil. Adding to the spatial variability, soil microorganisms are unevenly distributed in soil, often occurring as microcolonies attached to soil solids within biofilms rather than as single cells floating in water films (Fig. 10.5). A **biofilm** is a collection of organisms that excrete, or release upon cell death and lysis, extracellular polymeric substances (e.g., polysaccharides, DNA) that aid adherence to surfaces. (Fig. 10.6). As biofilms develop, cells are initially able to detach and move to other locations; however, as the biofilm matures, cells become permanently embedded. This is one of the ways that microaggregates form (Chapter 2). Biofilms are important to survival, as they buffer against desiccation and help to trap nutrients. Some are composed of a single species and others are multispecies.

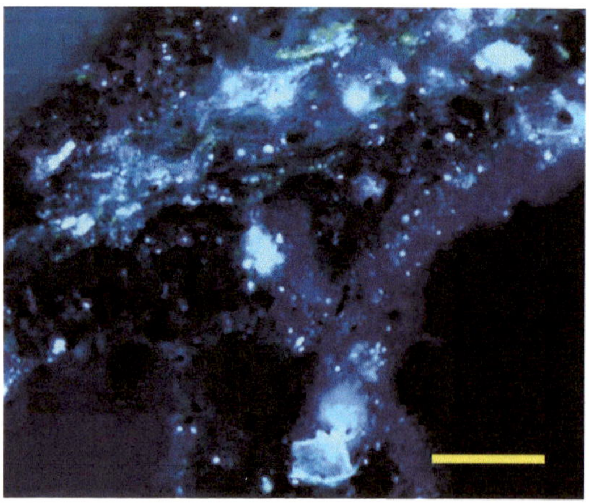

FIGURE 10.5 **Example of patchy bacterial colonization *(light blue)* of soil *(dark background)* as shown by fluorescence microscopy.** Scale bar represents 20 μm. *Source: From Nunan et al. (2003). Spatial distribution of bacterial communities and their relationships with the micro-architecture of soil. FEMS Microbiol. Ecol. 44(2), 203–215. By permission of Federation of European Microbiological Societies.*

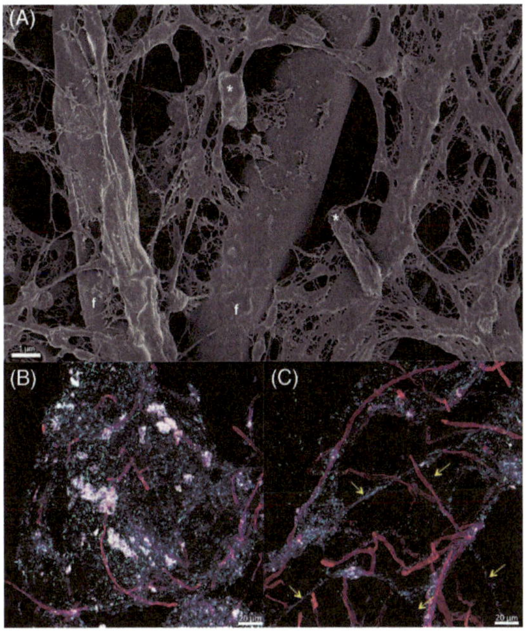

FIGURE 10.6 **Characterization of the matrix components of *Pseudomonas fluorescens* BBc6 biofilms on the surface of *Laccaria bicolor* S238N (a mushroom).** (A) Scanning electron microscopy image of 24 h-old biofilm showing bacterial cells (*) and fungal hyphae (f) encased in a matrix made of aggregates and filaments. (B) Confocal microscopy image showing the presence of protein aggregates *(white)* and eDNA (extracellular DNA; *blue*) in the matrix of 16 h old *P. fluorescens* BBc6 *(green)* biofilm on *L. bicolor* S238N hyphae *(red)*. Proteins and eDNA were stained with SYPRO Ruby and DAPI, respectively. Magnification 40×. (C) Confocal microscopy image showing the presence of filaments stained by DAPI *(yellow arrows)* connecting hyphae to bacterial cells. Magnification 40×. *Source: From (Guennoc et al., 2018). By permission of Federation of European Microbiological Societies.*

Interactions between soil organisms

Given the constraints that soil places on its biota in terms of nutritional limitations and environmental stresses, it is not surprising that many soil organisms interact with each other to circumvent some of these difficulties. Scientists have attempted to define these interactions on the basis of whether one or both organisms gain benefit from the association and if there is any specificity in the association (Table 10.2). However, these interactions cannot be rigidly defined. For example, the specificity of the organisms in a mutualistic association may vary from being highly specific (e.g., legume-*Rhizobium* symbiosis) to promiscuous (e.g., some mycorrhizal symbioses).

Positive interactions

Commensalism. This term refers to an association in which one population, or organism, benefits from the interaction and the second remains unaffected. The organism(s) that benefit can obtain nutrients, protection, or transport. In many cases, this involves one population metabolizing a metabolic by-product of another population. A common example of this is nitrification where ammonia-oxidizing bacteria and archaea oxidize ammonia to nitrite and

TABLE 10.2 Types of interactions among soil microorganisms.

Name of interaction	Effect of interaction on: Population 1	Population 2
Commensalism	+	0
Protocooperation	+	+ or 0
Mutualism	+	+
Predation (Parasitism)	−	+
Competition (Antagonism)	−	−
Amensalism	−	+ or 0

"+," positive effect; "−," negative effect; "0," no effect.
Adapted from Bottomley (2005).

other bacteria then oxidize the nitrite to nitrate (Chapter 15). A classic example of commensalism that can occur in hydrocarbon-polluted soils is the incidental conversion of cyclohexane to cyclohexanol by one microbial population, while trying to metabolize propane, followed by a second population metabolizing the cyclohexanol for energy. The first population gains no energy from the conversion of cyclohexane to cyclohexanol and cannot metabolize the produced cyclohexanol. This overall process is known as **cometabolism** and occurs due to the non-specific activity of some enzymes (Chapter 21). Given the vast diversity of soil microorganisms and the enzymes they produce, it is likely this type of **cross-feeding** occurs far more frequently within the soil microbiome than we currently recognize.

Another, indirect, example is the respiratory activity of aerobic soil microorganisms that lower the oxygen concentration allowing an anaerobic process to occur in close proximity. Commensalism is central to the redox dependent interactions that occur in wetlands (Fig. 10.7). At high redox values (Eh $>$ 400 mV), oxygen drives aerobic respiration. As the depth from the surface or the distance from the wetland plant root (which also leak oxygen) increases, oxygen is depleted, and alternative electron acceptors are used. Under highly reduced conditions, methanogenesis, a strictly anaerobic process, is made possible by the depletion of oxygen and other electron acceptors (Chapter 13).

Protocooperation. Both populations (or microorganisms) benefit from the association. The interaction is not essential for the two microorganisms to exist in that environment, and membership in the association is not usually so specific that one microorganism cannot be replaced by another.

Wetland soils are also a location for protocooperation. Many methane-producing archaea and sulfate-reducing bacteria obtain their energy from acetate, hydrogen gas, or both. Under the same conditions, fermentative bacteria metabolize sugars, fatty acids, and amino acids to generate energy and produce acetate and hydrogen gas as by-products. The methane-producing archaea and sulfate-reducing bacteria benefit from the acetate and hydrogen gas produced by the fermenters, and the fermentative bacteria benefit because the concentration of hydrogen gas and acetate is kept low, which permits them to couple thermodynamically marginal biochemical reactions to cell growth (Chapter 20). This is a classic example of **syntrophy** or cross-feeding where one species lives off of the products of another species.

Evidence has accumulated that commensalistic or synergistic associations of microorganisms are required to completely mineralize some soil-applied chemicals, such as pesticides

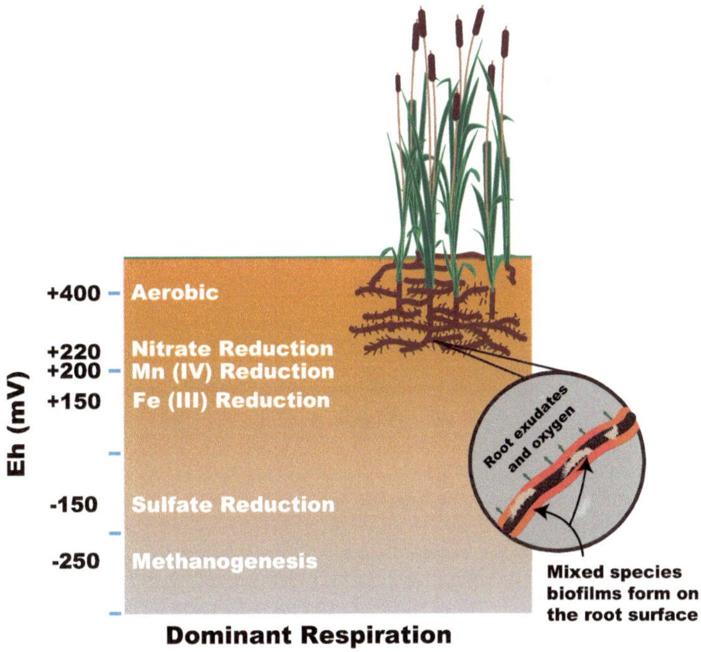

FIGURE 10.7 **Commensalism is exemplified in wetland soils where aerobic microorganisms consume oxygen allowing for a series of alternative electron acceptors that become more favorable as depth or distance from the plant root increases.** The inset to the right illustrates the oxic zone *(orange)* around the root of a wetland plant due to the release of oxygen into an otherwise anoxic *(gray)* soil environment. The release of root exudates stimulates microbial use and depletion of oxygen and promotes facultative and obligate anaerobic processes near the root. For example, some bacteria can use oxidized iron (Fe^{3+}) as an electron acceptor during degradation of root exudates under anaerobic conditions. This produces reduced iron (Fe^{2+}) which can then be used as an energy source by iron-oxidizing bacteria in the oxic environment around the root.

and hazardous waste. This type of association is described as a microbial **consortium**. For example, many locations have soils contaminated with diesel fuel. A number of bacteria have been recognized for their ability to degrade diesel, but they often work better in consortia when cross-feeding facilitates increased degradation. Although a consortium of bacteria expressing a broad suite of enzymes may degrade diesel, the hydrophobic nature of hydrocarbons makes them difficult to access. Researchers have been able to improve bioremediation by adding bacteria that do not cross-feed on the diesel, but rather produce a biosurfactant. They operate synergistically with the degrading consortium and benefit from small carbon compounds that are produced at the end of degradation (Fig. 10.8).

Mutualism. Also referred to as **symbiosis,** this type of association usually occurs at the level of individuals organisms (e.g., a legume plant being colonized by a nitrogen-fixing bacterium). Both organisms benefit from the association. The relationships can be highly specialized between the two organisms and allow them to thrive in habitats that neither could occupy alone. The metabolic activities of the partners are often modified in the association, and occasionally the partners cannot grow in the absence of the association.

Many examples of mutualism exist between soil microorganisms and plants and animals. Although plants are the primary source of reduced carbon for the soil microbial community,

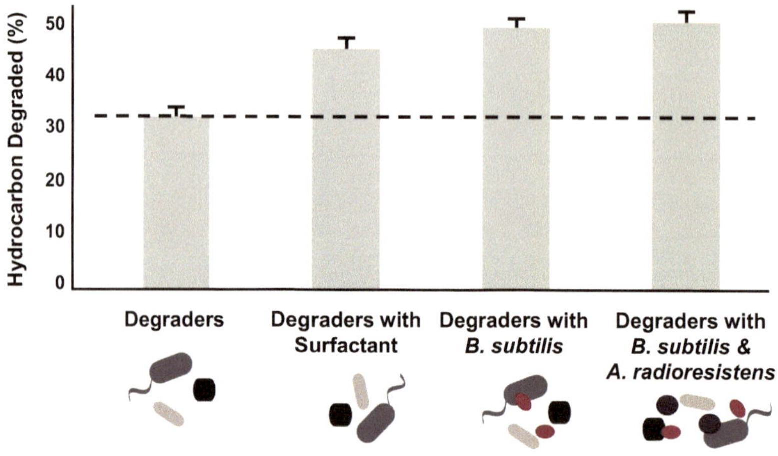

FIGURE 10.8 **Hydrocarbons can be more efficiently degraded when a microbial consortium (in this case composed of three diesel degraders) is aided by the addition of the surfactant-producing bacteria *Bacillus subtilis* and *Acinetobacter radioresistens*.** This is an example of a synergistic interaction where the hydrocarbon-degrading bacteria benefit from the surfactant increasing hydrocarbon bioavailability, and the surfactant-producing bacteria benefit by catabolizing metabolites produced by the hydrocarbon-degrading bacteria. *Source: From Hays et al. (2015); data from Mnif et al. (2015).*

plant growth is often limited by the availability of nitrogen and phosphorus. Many plants have solved this problem by forming associations with various soil microorganisms. Certain fungi invade plant roots and form an elaborate network of hyphae that penetrate either between or into the cortical cells and also into the surrounding soil matrix. These associations are collectively referred to as mycorrhizae (Chapter 12). The fungus benefits by obtaining a readily available supply of reduced carbon from the plant. Various mechanisms have been proposed to explain how the fungus benefits the plant, including:

- The mycorrhizal roots enhance the uptake of mineral nutrients (especially phosphorus) and water from soil because the external mycelia explore a greater volume of soil than do nonmycorrhizal roots.
- The fungus sequesters mineral nutrients at low concentrations more efficiently than does the nonmycorrhizal root.
- The fungal hyphae produce various hydrolytic enzymes that release nitrogen and phosphorus from organic compounds that are otherwise unavailable to the plant.

Leguminous plants form mutualistic symbioses with bacteria, primarily of the genera *Bradyrhizobium* and *Rhizobium* (Chapter 17). A somewhat random array of plant species forms symbiotic associations with the actinobacterial genus *Frankia* and are called actinorhizal. These bacteria invade the plant via root hairs or by passage between epidermal and cortical cells of the root. As a result of chemical signals passed between the bacterium and the plant, the latter is stimulated to construct a specialized multicellular structure called a **root nodule.** Inside the nodule, the bacteria "fix" atmospheric dinitrogen gas into ammonia (Chapter 16), which the plant then assimilates into amino acids. The plant supplies the bacteria with reduced carbon to use as the energy source for metabolic processes including N_2 fixation. As a result of these symbioses, leguminous and actinorhizal plants can grow in environments that are extremely deficient in mineral nitrogen.

Mutualistic associations with microorganisms play an essential role in the well-being of soil animals that feed primarily on plant debris. Some soil animals have protozoa and bacteria in their gut that decompose the cellulose component of plant cells into metabolites that the animal can assimilate. Furthermore, many gut bacteria fix dinitrogen into ammonia, which is then incorporated into protein via amino acids. As a result, the nitrogen-deficient diet of the soil animal is supplemented. In another form of mutualism, various ants, beetles, and termites cultivate specific kinds of fungi (Box 10.2). The insects provide the fungi with nutrients in the form of plant debris and feces. By feeding on the fungal biomass, the insects circumvent the problem of digesting cellulose.

Negative interactions

Predation. This interaction occurs when one population has a negative impact on the size of another population by feeding on it and reducing its numbers.

BOX 10.2

Leafcutter ants are fungal farmers

Humans are not the only animals that create complex societies. Hundreds of species of leafcutter ants organize their communities around farming fungi, with different members of their society fulfilling specific jobs. Some ants are charged with collecting leaf material that is carried back to the colony where it is transferred to workers that feed the leaves to fungi. Although adult ants eat leaf sap, their larvae rely on the fungi that grow in specialized chambers. The fungal farmers grow specific strains of fungi and make sure to eliminate waste products and any potential contaminants. The ants harbor bacteria that can be transferred to the surface of the fungi and produce antibiotics to further ensure the purity of the farmed hyphae. The ants, fungi, and bacteria compose a fascinating mutualistic relationship.

Tiny nurse ants tending to white ant larvae are dwarfed by the queen ant in the upper right. All the ants feed upon protein-rich food produced by a white-gray fungus that they cultivate underground. *Photo by Karolyn Darrow, Smithsonian Institution.*

There are many predator-prey interactions within soil, occurring across multiple trophic levels (Fig. 10.2). For example, soil animals, such as nematodes and protozoa, feed on bacterial and fungal populations and reduce their numbers. As mentioned earlier in this chapter, this activity is an essential part of soil ecosystem function. Conversely, several species of fungi use specialized hyphal structures or paralyzing toxins to capture nematode prey (Chapter 6). Furthermore, various viruses parasitize all other organisms. Although such activity is detrimental to the size of the prey population, the activity of the survivors can be stimulated because the predators release excess (waste) nutrients during their feeding activities.

Competition. In this interaction, both populations compete for the same limited resources and have a negative effect on each other.

For example, heterotrophic microorganisms and ammonia-oxidizing bacteria compete for ammonium. The former needs ammonium to make amino acids and proteins, and the latter needs ammonium primarily as an energy source. If ammonium exists in limiting quantities, then neither population will grow to the extent that it would in the absence of competition. In another example of competition, different strains of a symbiosis-forming species of *Rhizobium* compete with each other to establish a symbiosis with a leguminous plant. In many cases, the number of root nodules formed by each of the competing strains will be less than if only one strain is present in the soil. Furthermore, if the more competitive strain is less effective at fixing dinitrogen, then the plant does not grow as vigorously as it would if all nodules were occupied by the more effective strain.

Amensalism. In this type of interaction one population gains a competitive advantage over another by producing a growth-inhibiting substance.

Many fungi and bacteria have the potential to produce antimicrobial substances (e.g., antibiotics) that may serve to inhibit other nearby microorganisms (Box 10.3). When amensalism occurs in the rhizosphere, the growth of soilborne plant pathogens can be inhibited and plant disease lowered. This phenomenon, known as **biological control,** is described in Chapter 23. Amensalism can also be caused by more general effects of metabolism. For example, sulfur-oxidizing microorganisms can cause soil acidification, which may adversely affect acid-sensitive *Streptomyces* species in their immediate vicinity (Chapter 18).

Microbial communication

Many microorganisms lack mobility but can still interact and communicate through the production of chemical signals. Often this involves diffusible chemicals such as antibiotics, lipochitooligosaccharides involved in mycorrhizal colonization (Chapter 12), "nod factors" involved in the rhizobia-legume symbiosis (Chapter 17), and N-acyl homoserine lactones (AHLs) used in **quorum sensing** by many Gram-negative bacteria (Chapter 25).

One important microbial activity directed by signal molecules and quorum-sensing is the formation of biofilms. When microbial populations are low, these quorum-sensing compounds are not detectable or are at low concentrations. For example, bulk soils may have lower concentrations of AHLs compared to rhizosphere soils. Accumulation of AHLs and other signaling molecules is dependent on population density. As populations increase, cells produce and detect increasing concentrations of signaling compounds leading to coordinated responses. If in the vicinity of other bacteria producing signal molecules, a bacterial cell may detect these signaling molecules and change its own cellular metabolism to produce even more signal molecules thus further increasing their concentration. In response to these signals, the same cell may increase production of extracellular polysaccharides and other molecules

BOX 10.3

Antibiotic nursery

There are many reasons that scientists may want to culture a soil microorganism. The most fundamental reason is that it allows microorganisms to be better described in terms of growth rate and nutrient requirements. It can also lead to the discovery of new compounds, such as antibiotics. The isolation (i) Chip was developed as a way to culture previously uncultured soil microorganisms. Many soil bacteria may be slow growing or rely on metabolites or secondary compounds that we cannot easily mimic in laboratory culture media. The iChip is designed to physically isolate cells within a permeable membrane; this allows them to be cultured under somewhat in situ conditions where chemical signals can be exchanged between various populations. This technique has helped to expand the number of soil microbes that are currently in culture and led to the discovery of novel antibiotics (Berdy et al., 2017).

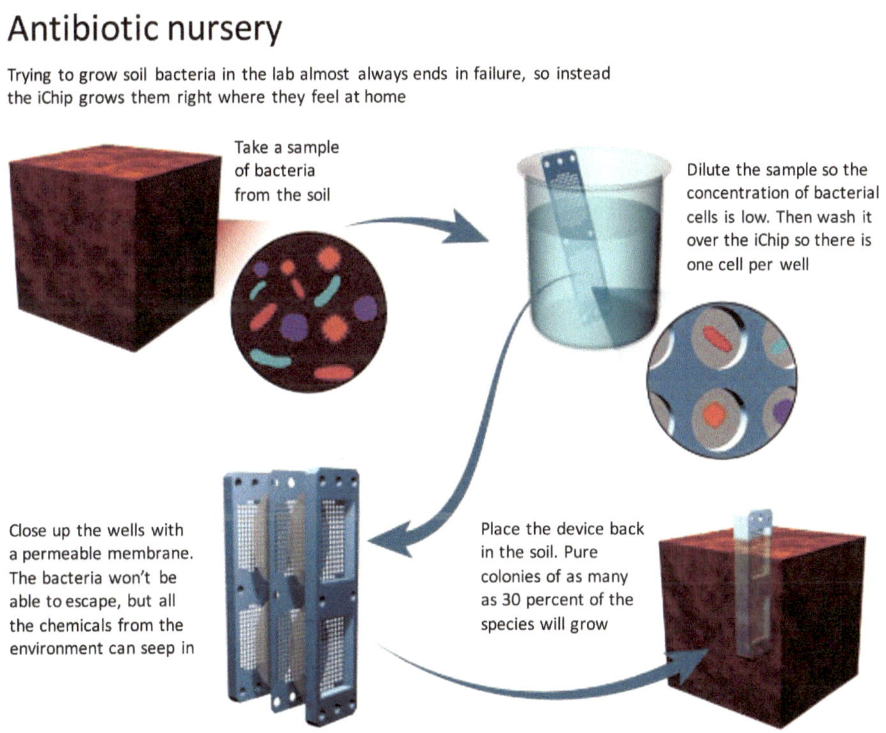

Use of iChip for discovery of novel antibiotics produced by soil microorganisms (Graber, 2016).

needed to form the biofilm (Fig. 10.9). Quorum sensing via AHLs and other compounds has also been linked to expression of extracellular enzymes and virulence genes, contributing to the potency of some pathogens.

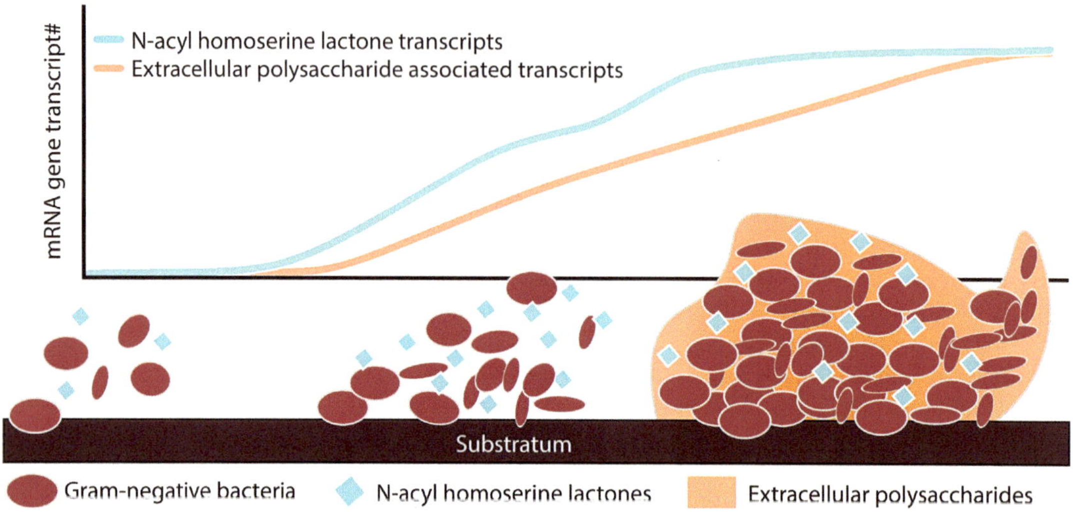

FIGURE 10.9 **The establishment of a biofilm is accompanied by an increase over time in the expression of genes associated with signaling molecules such as N-acyl homoserine lactones involved in quorum sensing and genes governing extracellular polysaccharide production.**

A wide variety of microbially produced volatile organic compounds (mVOCs) can also serve as signaling molecules. These volatile compounds are produced as secondary metabolites and may enable longer distance communication, as compared to soluble chemicals (Schulz-Bohm et al., 2017). Recent research has demonstrated an assortment of positive or negative impacts on organisms including bacteria, fungi, protozoa, and plants (Fig. 10.10). An example mVOC is geosmin which is responsible for the earthy odor from soil after a rainfall.

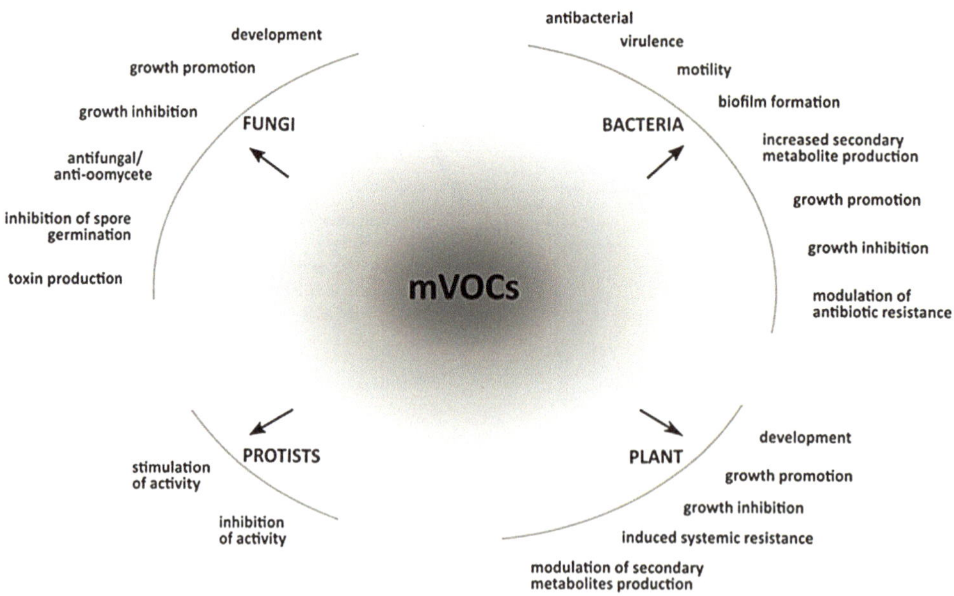

FIGURE 10.10 **Responses in bacteria, fungi, protists, and plants caused by microbial volatile organic compounds (mVOCs).** *Source: From Schulz-Bohm et al. (2017).*

Geosmin is produced by certain actinobacteria, cyanobacteria, and fungi. As discussed in Box 5.3, geosmin appears to either attract or repel different groups of arthropods. The effects of mVOCs on plants can be dramatic leading to changes in root development, plant physiology, and plant growth. Although most of the current research on mVOCs focuses on model plants like *Arabidopsis thaliana*, this is an emerging research area (Schulz-Bohm et al., 2017).

Microbial succession

Succession is a fundamental ecological theory to describe how ecosystems develop. In a new environment (e.g., lava flow) or following a disturbance in an established ecosystem (e.g., forest fire), colonization/recolonization by plants, animals, and microbes occurs in a predictable sequence of primary followed by secondary communities. For example, the regeneration of a forest following fire typically begins with the growth of small shrubs. Succession arises because as organisms (from microbes to trees) grow, they alter the environment around them. Eventually large trees shade out smaller trees and shrubs on the forest floor. Similarly, microbes can alter their environment as they grow, an example being the addition of fixed C by pioneering algae and cyanobacteria (Chapter 7). In both instances, these events cause changes that pave the way for development of different populations.

We sometimes use the term *primary microflora* to describe the early populations. A primary microflora community may colonize fresh crop residues or leaf litter in the soil. As these microbes grow, they alter their environment and the substrate and, thereby, change plant-derived products into microbial biomass and metabolic waste products. Soon a *secondary microflora* develops and feeds on the dead cells of the primary flora and its products. Thus, we see a wave-like progression of change in microbial populations as the substrates are metabolized and the abiotic environment (e.g., pH, oxygen in the soil atmosphere, redox potential) changes.

The succession of microbes on a piece of plant litter would be considered short term succession, but studies have also observed differences in microbial communities along succession gradients corresponding to different-aged forests (Cline and Zak, 2015) and in soils exposed by receding glaciers (Nicol et al., 2005). Taxonomic differences in community composition do not necessarily translate to differences in the function of a microbial community, however. Many of the billions of microbial cells in a gram of soil have similar metabolic capability, a phenomenon termed **functional redundancy** (Fig. 10.11). For example, numerous bacteria and fungi can breakdown cellulose, and more still can use the small C molecules in root exudates.

Functional redundancy contributes to the maintenance of ecosystem functioning. For example, a drought may select for one species of denitrifying bacteria over another, but denitrification may still occur when it eventually rains, as long as at least one of the species remains viable. In macroecology, we often hear about **keystone species**, a species that when removed from the ecosystem leads to dramatic changes in ecosystem health and function. Although microbial communities as a whole are fundamental to ecosystem functioning, there are likely few to no keystone species within the soil microbiome due to its amazing functional diversity.

Resistance and **resilience** (Fig. 10.12) of the microbial community also ensures that key functions are maintained even when a disturbance occurs. For example, a forest fire may damage above ground vegetation but may leave soil bacterial communities largely intact because of their resistance. The fungal community, on the other hand, may be affected by the same fire but may return to a similar community shortly after in terms of composition (resilience) or function (functional redundancy).

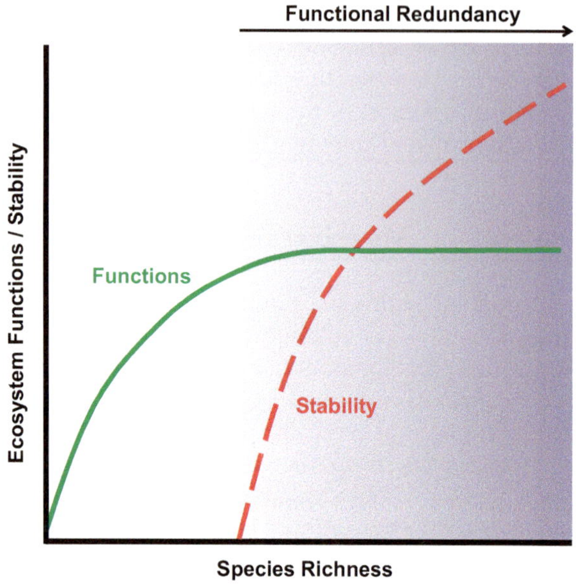

FIGURE 10.11 **Relationship between functional redundancy and ecosystem stability.** As microbial species are added to an ecosystem, this increases the functional capability of the microbiome to a point. Beyond this, additional species do not add new functions, but they do serve to increase the functional redundancy and thus stability of the ecosystem. *Source: Adapted from Pepper et al. (2015), (Konopka, 2009).*

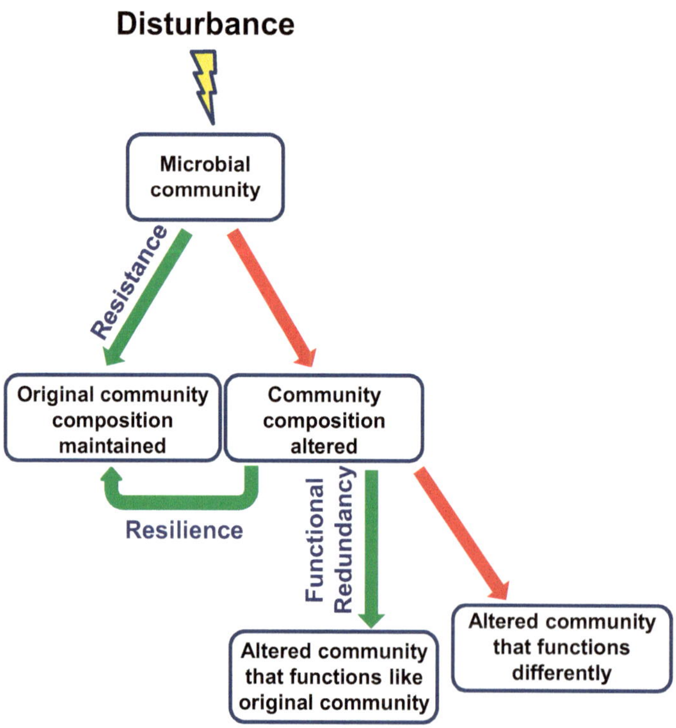

FIGURE 10.12 **Illustration of potential microbial community responses to disturbance and the resulting impacts on ecosystem processes.** Note that multiple mechanisms (e.g., resistance, resilience, and functional redundancy) can help maintain microbial function(s) following a perturbation. *Source: Adapted from Pepper et al. (2015); (Allison and Martiny, 2008).*

Measurement of microbial diversity and interactions

For many years, soil ecosystem modelers assumed that the quantity of soil microbial biomass was the major driver of decomposition and nutrient cycles, and that substrate availability, temperature, and moisture were the major controlling variables. After discovering the vast diversity and dynamic nature of the soil microbiome, microbiologists began evaluating methods capable of detecting links between changes in community composition and ecosystem functions (Box 10.4). Due to the rapid expansion in microbiological methods, especially nucleic acid-based assays, over the past few decades, it is not possible to cover all available methods in a single chapter. Instead brief descriptions of a select number of these methods follow as examples of approaches for interrogating the soil microbiome.

Enzymes. The intracellular and extracellular enzymes produced by microorganisms are ultimately responsible for metabolism (Chapter 3). Assays have been developed to measure the potential enzyme activity for processes including denitrification, nitrification, cellulose degradation, and numerous other C, N, and other nutrient transformations (Burns et al., 2013). In the case of denitrification, a soil sample is added to a slurry containing a buffered solution with supplemental C and NO_3^-. The headspace is purged of oxygen and acetylene is added along with N_2. The acetylene blocks the last step of denitrification resulting in a buildup of N_2O (Chapter 15). The accumulated N_2O is then used to calculate the denitrification rate. The

BOX 10.4

Soil health

Soils are incredibly diverse in terms of their physical and chemical properties. Soil scientists have long attempted to define soil types in terms that reflect their ability to function in natural and managed settings. The term "soil quality" was coined in this context, but more recently the term "soil health" has become common (Chapters 14 and 25). Soil quality is the ability of a soil to function within its surroundings, support plant and animal productivity, and maintain or enhance water and air quality. Soil health extends this idea to acknowledge that just as humans function best when they are healthy, so too do soils. Soil microbial communities play an enormous role in soil health; they need to be morphologically and phenotypically diverse enough to successfully occupy the numerous micro-niches found in soil, and to process and partition the wide range of substrate inputs that occur at different times of the year. Several of the critical indicators of soil health include level of organic matter, aggregate stability, water infiltration, crusting, erosion, and acidification. These parameters are strongly influenced directly and/or indirectly by microbial activity. Changes in these soil properties invariably modify the niches occupied by soil microorganisms, resulting in changes in the content, composition, and activities of the microbial community. As part of the effort to define soil health, soil microbiologists have developed procedures to gain insight into community composition and its dynamics at both the taxonomic and functional levels. One important goal is to identify those practices that improve and sustain soil health in agriculture.

activity of enzymes involved in the breakdown of C molecules typically makes use of colorimetric or fluorometric assays, such as the phosphatase assay described in Chapter 19. In those cases, when the targeted substrate is broken down, a color or light signal is produced. The accumulation of those signals again can be translated into the rate of reaction and potential for conversion of the substrate.

Cellular Fatty Acids. Lipids are essential components of living organisms. Several classes of lipids exist in microorganisms, including phospholipids, glycolipids, lipopolysaccharides, and lipoproteins. Many different fatty acids exist in microbial lipids, and they are used in the identification of microorganisms. In general, soil microbiologists use two relatively straightforward procedures to examine the fatty acid profiles of soils. In one approach, fatty acids are liberated from soil lipids by heating a soil sample at 100°C in a solution of NaOH and methanol. The fatty acids are converted into their methyl ester derivatives (FAME) by heating under hot acidic conditions, whereupon the FAMEs are extracted into an organic solvent such as hexane and analyzed by gas chromatography. Alternately, lipids can be extracted from soil into a mixture of water, chloroform, and methanol, followed by separation of the different classes of lipids (neutral, glycolipids, and phospholipids). Usually, the phospholipid fraction (PLFA) receives the most attention and is analyzed after hydrolysis and methylation under alkaline conditions. The PLFAs are extracted with an organic solvent and analyzed by gas chromatography. Typically, 20–50 fatty acids can be detected and differentiated (more than 100 fatty acids can be identified when more sophisticated extraction and analytical procedures are used). Since some PLFA markers (and other fatty acids) are considered diagnostic for specific microbial groups, these data can help distinguish between microbial communities in different soils, or in a specific soil under different management conditions. This analysis can also be done on a quantitative basis and be used as a measure of microbial biomass. Because both bacteria and fungi contain ester-linked phospholipids, both groups can be simultaneously examined and the relative abundance of select fungal and bacterial PLFAs used to determine the fungal:bacterial ratio. Two drawbacks of common whole-soil lipid-based analyses are (1) they can be used to identify broad groupings such as Gram-positive bacteria and mycorrhiza but have limited taxonomic resolution and cannot distinguish between species and (2) Archaea have ether linkages and are not detected using the same extraction protocol.

DNA Sequencing. New molecular technologies and computing resources have changed many fields within biology, but perhaps none more fundamentally than microbial ecology. The limitations of culturing and microscopy to capture the diversity of the soil microbial community have largely been overcome by increased capacity to sequence millions of stretches of DNA simultaneously. For example, numerous studies have targeted regions of the 16S rRNA genes using PCR (Chapter 4) and then applied bioinformatics to match those sequences against publicly available databases for sequence identification and calculation of diversity metrics.

This is the primary way that researchers currently survey soil microbial community composition across soil types, land management, and disturbance. For example, Williams et al. (2013) sequenced 16S rRNA bacterial genes and used multivariate statistics to visualize differences in community composition in soils that had developed from sand deposits along the coast of Lake Michigan, United States (Fig. 10.13). Further investigation of this data set revealed that bacteria belonging to Acidobacteria increased but Cyanobacteria decreased in abundance with increasing age (Fig. 10.14).

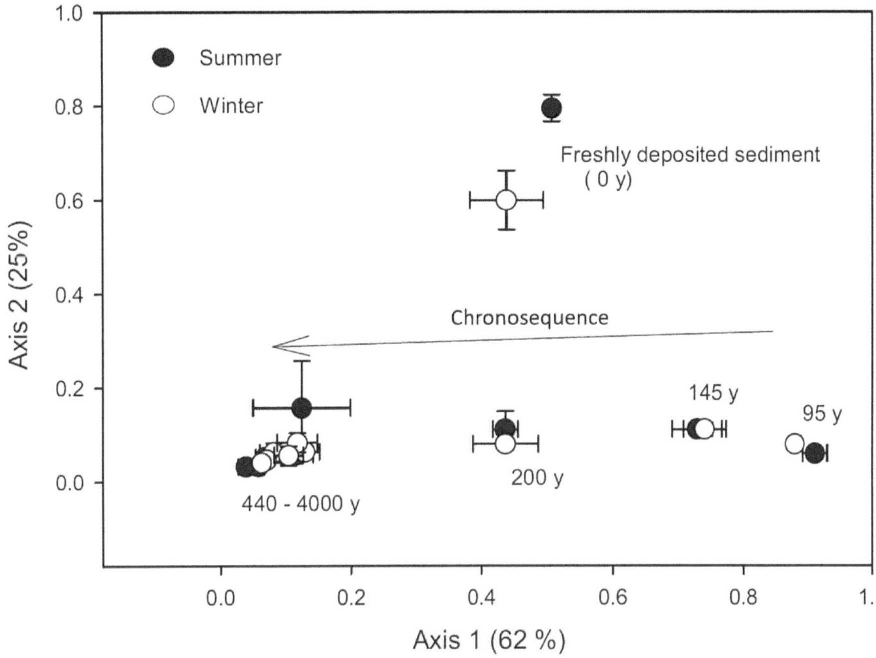

FIGURE 10.13 **Multivariate ordinations are a way to distill information from thousands of species so as to visualize differences between microbial community compositions.** In this example, bacterial communities were characterized using 16S rRNA gene sequencing, and each point on the ordination represents the bacterial community at different ages of soil development (chronosequence) in sand deposits along the coast of Lake Michigan, United States. *Source: From Williams et al. (2013).*

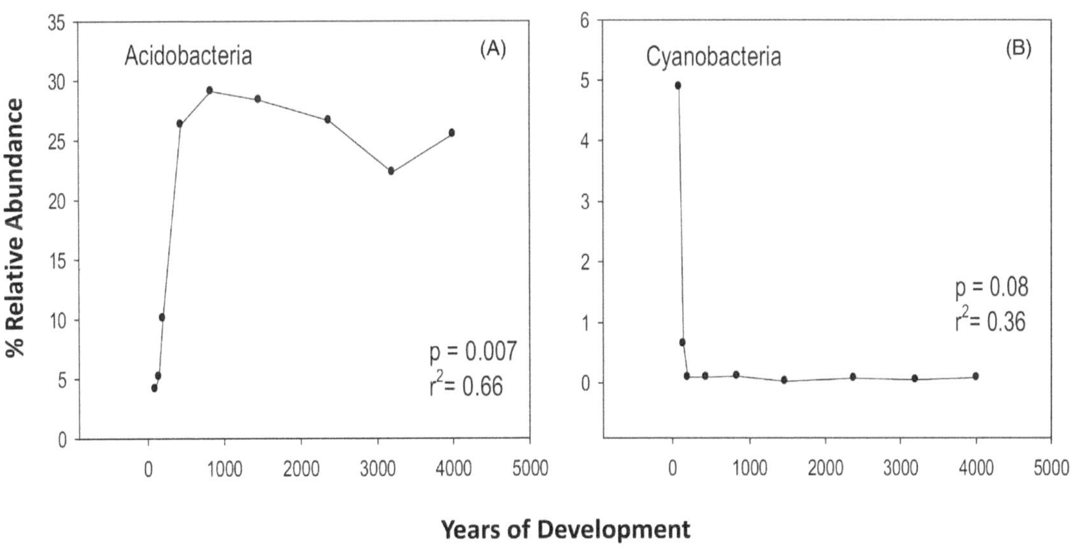

FIGURE 10.14 Relative abundance of 16S rRNA gene sequences belonging to Acidobacteria (A) and Cyanobacteria (B) at different ages of soil development in sand deposits along the coast of Lake Michigan, United States. *Source: From Williams et al. (2013).*

In addition to amplicon sequencing, where PCR is used to create numerous copies of a target gene, researchers may decide to apply shotgun metagenomics. In this case, soil DNA is directly sequenced without the preliminary PCR step. The resulting sequences typically represent 300–900 bp stretches of DNA that can code for any part of the microbial genome. This requires a larger sequencing capacity to sufficiently sample the community and also requires significantly more computing power. The advantage of this approach is the elimination of PCR bias and the potential to sequence yet unknown genes for which PCR primers have not been developed.

RNA Analysis. Although DNA sequencing has been a method of choice for characterizing microbial communities, it has limitations. One of the most significant being that we cannot easily distinguish between DNA from active versus dormant microorganisms. Consider if we went to a meadow to characterize the plant community but could not tell the difference between plants that were growing when we visited and seeds that were in the soil yet to sprout. Understandably, community composition and the amount of DNA in the soil have been found to rarely correlate to function at a specific time point. A more direct measure of activity is the mRNA transcribed when cells are manufacturing the enzymes needed to carry out a particular function. A growing number of studies have targeted mRNA that codes for specific activities such as denitrification (Chapter 15). Compared to DNA-based analyses, far fewer studies have targeted mRNA because it is less stable in the soil, and therefore difficult to extract. However, mRNA-based analyses can provide a more accurate measure of specific microbial activities than DNA-based analyses because mRNA is only transcribed in large quantities when a process is being induced. For example, when soils initially become anaerobic, we may observe increased transcripts associated with denitrification, but those transcripts might be present in lower numbers when the system has been anaerobic for an extended period because the enzymes have already been translated. For example, Theodorakopoulos et al. (2017) examined mRNA transcript abundance in an agricultural field following a rain event. Their goal was to understand what enzymatic processes may contribute to increased N_2O (Fig. 10.15). Because N_2O can be generated during nitrification and denitrification, this approach allowed the researchers to identify which pathway was likely contributing the most N_2O and in what stage of wetting and drying.

Stable Isotope Probing. Stable isotopes are useful tools for tracking the fate of molecules through the microbial community and determining which organisms are responsible for specific processes. Biogeochemists use isotopes to determine the rate of processes such as nitrogen mineralization (Chapter 15). More recently these tracers have been added to soils as a way to label the microbial community, in a method known as **stable isotope probing** (SIP). For example, ^{13}C-labeled cellulose can be added to soil, and the ^{13}C allowed to be incorporated into the microbial biomass during metabolism. This approach is based on the principle that organisms which catabolize the cellulose will incorporate a portion of the ^{13}C into their cells. The label can then be detected in cellular lipids and nucleic acids to determine which organisms are catalyzing the process of interest. In the case of DNA- and RNA-SIP, the labeled and unlabeled nucleic acids are separated using ultracentrifugation (^{13}C being heavier than ^{12}C) and the different fractions then analyzed by sequencing or another approach.

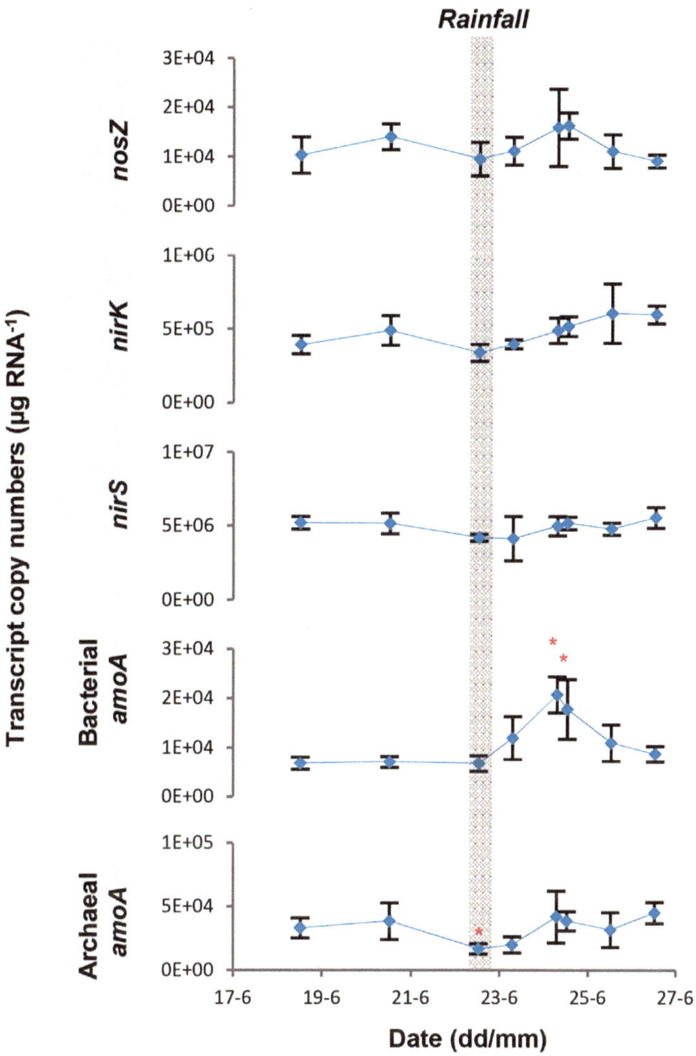

FIGURE 10.15 Quantification of gene expression by quantitative PCR for bacterial and archaeal genes involved in nitrification (*amoA*) and denitrification (*nirS, nirK, nosZ*) before and after a rainfall event *(gray shading)*.Differences are indicated by *(*)* when the value was significantly different from those obtained at the two first sampling dates (prior to the rain event). *Source: From Theodorakopoulos et al. (2017).*

Emerging Technologies. Numerous microbial characterization methods have been developed over the last 2 decades that take advantage of the molecules produced by soil microorganisms. Methods development for **metaproteomics** is on-going but would allow us to describe all of the proteins (including enzymes) in the soil in a manner similar to DNA-based metagenomics. Another approach is **metabolomics** where microbial metabolites including amino acids, sugars, and alcohols are extracted and described. Each approach has advantages and disadvantages when it comes to methodological constraints and interpretability.

Due to the complexity of the microbial community, a multi-pronged approach likely yields the most accurate information. For example, Mbuthia et al. (2015) compared microbial community composition, activity, and soil quality in response to differences in farming systems that included tillage versus no-till practices and the presence of cover crops. They concluded that long-term differences in farming practices altered the microbial community and its function, with most analytical methods indicating benefits (e.g., increased microbial activity and biomass) from use of no-till practices and/or cover crops. However, treatment benefits varied depending on the parameter being measured. For example, a vetch cover crop produced higher β-glucosaminidase and basal microbial respiration rates but lower mycorrhizal biomass levels compared to wheat and no cover crop.

Another study examined the effects of drought and rainfall on the soil microbial communities in native tall grass prairies (Roy Chowdhury et al., 2019). Soil samples were collected and subjected to either wetting or drying. mRNA sequencing was used to determine how bacteria responded to these potential stressors, by identifying the decrease or increase of gene expression (Fig. 10.16A). Additionally, metabolomics was used to examine the change in metabolites following moisture shifts (Fig. 10.16B). Along with DNA sequencing, these data were compiled to create reaction-metabolite integrative bipartite networks, in other words, metabolic maps of how soils react to wetting and drying (Fig. 10.16C). This basic information about how the soil microbiome changes during wetting and drying can help us understand the short- and long-term effects of climate change on soil functioning and health.

(A)

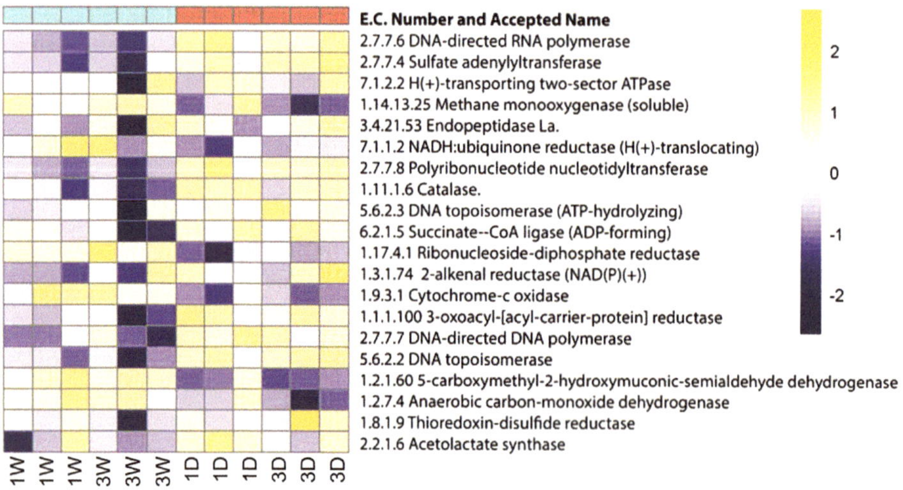

FIGURE 10.16 **Multi-omics characterization of soil microbiome response to wetting and drying.** (A) Metatranscriptome data with heat map showing the top 20 most abundant transcripts (enzyme name listed) observed under dry compared to wet conditions. The *x*-axis indicates soil sample (1 or 3) and treatment (W, wet; D, dry). Moisture conditions are also indicated by the header row in *blue* (wet) or *red* (dry). The color gradient for each cell is scaled to a $\log_2$ fold change of −2 to 2.

(Continued)

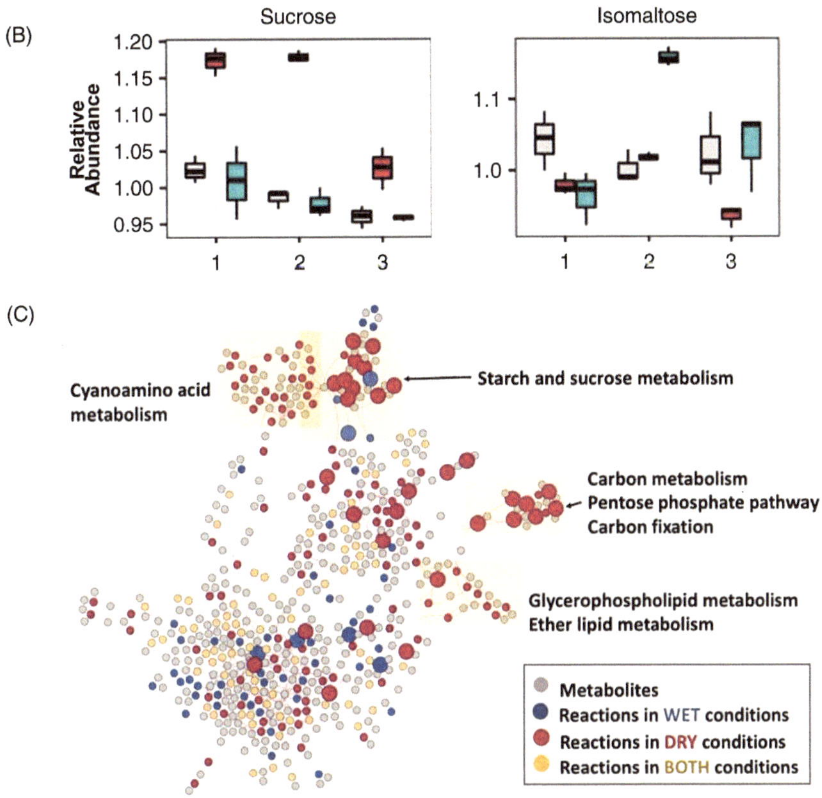

FIGURE 10.16 (*Cont.*) (B) Bar charts showing the metabolome (relative abundance of selected metabolites) from three native prairie soils (1, 2, and 3) (*gray*, control; *blue*, wet; *red*, dry). (C) Metabolite networks for soil 1. Colored symbols indicate reactions that are more prevalent under specific incubation conditions: *blue*, wet; *red*, dry; *yellow*, both. *Yellow shading* highlights specific pathways that are more prevalent under dry conditions. *Larger nodes* represent the commonly predicted reactions in both soils (1 and 3), while *small nodes* represent the uniquely predicted reactions. *Source: From Roy Chowdhury et al. (2019).*

Summary

This chapter places soil microorganisms in the context of their soil environment and conceptually describes their roles in soil ecosystems. Soil organisms play crucial and often complementary roles in nutrient cycling. However, the complexity of soil structure and fluctuations in environmental conditions and nutrient availability combine to make it difficult for soil microorganisms to achieve their metabolic potential. Organismal interactions circumvent some of these problems and thereby help to maintain soil health and ecosystem productivity. Some interactions between soil microorganisms benefit one or both of the participating populations, whereas other interactions have a negative impact on one of the participants. Clearly, soil contains an enormous diversity of microbial populations, yet we remain woefully ignorant of the forces that control microbiome composition. New technologies have expanded our capacity to study soil microorganisms, and increased computing resources allow us to integrate large data sets to uncover important interactions and metabolic pathways.

Acknowledgments

Portions of this chapter, with permission, are derived from the work of Dr. Peter J. Bottomley who authored a similar chapter for the Second Edition of the textbook. The current author and editors thank him for the use of his material and for graciously supporting production of the Third Edition.

Study questions

1. List the complementary roles of the bacteria, fungi, and soil animals that contribute to efficient nutrient cycling in soil.
2. If 54 bacterial cells were to divide every 45 min for a period of 20 h, what would be the final number of cells in the population?
3. List the benefits of organismal interactions in ecosystem function, thinking both about positive and negative interactions.
4. Why might it be difficult to culture a soil microorganism, especially into pure culture?
5. What are the theoretical differences between r- and K- strategists?
6. Describe an interaction between two microorganisms and an interaction between one macroorganism and one microorganism. Explain how the organisms immediately benefit from the interaction, and how the soil ecosystem ultimately benefits from it.
7. Describe the advantages of soil possessing a highly diverse microbial community. Consider the roles of resistance, resilience, and functional redundancy.
8. Explain the differences in information that can be gathered from soil enzymes, cellular lipids, and nucleic acid sequencing.

References

Allison, S.D., Martiny, J.B.H., 2008. Resistance, resilience, and redundancy in microbial communities. Proc. Natl. Acad. Sci. U.S.A. 105, 11512–11519.

Andrews, J.H., 1984. Relevance of r- and K-theory to the ecology of plant pathogens. In: Klug, M.J., Reddy, C.A. (Eds.), Current Perspectives in Microbial Ecology. American Society for Microbiology, Washington, DC, pp. 1–7.

Berdy, B., Spoering, A.L., Ling, L.L., Epstein, S.S., 2017. In situ cultivation of previously uncultivable microorganisms using the ichip. Nat. Protoc, 12, 2232–2242.

Bottomley, P.J., 2005. Microbial ecology. In: Sylvia, D.M., Hartel, P.G., Fuhrmann, J.J., Zuberer, D.A. (Eds.), Principles and Applications of Soil Microbiology, second ed. Pearson Prentice Hall, Upper Saddle River, NJ.

Burns, R.G., DeForest, J.L., Marxsen, J., Sinsabaugh, R.L., Stromberger, M.E., Wallenstein, M.D., Weintraub, M.N., Zoppini, A., 2013. Soil enzymes in a changing environment: current knowledge and future directions. Soil Biol. Biochem. 58, 216–234.

Cline, L.C., Zak, D.R., 2015. Soil microbial communities are shaped by plant-driven changes in resource availability during secondary succession. Ecology 96, 3374–3385.

Ernebjerg, M., Kishony, R., 2012. Distinct growth strategies of soil bacteria as revealed by large-scale colony tracking. Appl. Environ. Microbiol. 78, 1345–1352.

Geyer, K.M., Kyker-Snowman, E., Grandy, A.S., Frey, S.D., 2016. Microbial carbon use efficiency: accounting for population, community, and ecosystem-scale controls over the fate of metabolized organic matter. Biogeochemistry 127, 173–188.

Graber, C., 2016. Make 'em happy. New Scientist 229, 36–39.

Guennoc, C.M., Rose, C., Labbé, J., Deveau, A., 2018. Bacterial biofilm formation on the hyphae of ectomycorrhizal fungi: a widespread ability under controls? FEMS Microbiol. Ecol. 94 (7), fiy093.

Hays, S.G., Patrick, W.G., Ziesack, M., Oxman, N., Silver, P.A., 2015. Better together: engineering and application of microbial symbioses. Curr. Opin. Biotechnol. 36, 40–49.

Hunt, H.W., Coleman, D.C., Ingham, E.R., Ingham, R.E., Elliot, E.T., Moore, J.C., Rose, S.L., Reid, C.P.P., Morley, C.R., 1987. The detrital food web in a shortgrass prairie. Biol. Fertil. Soils 3, 57–68.

Konopka, A., 2009. What is microbial community ecology? ISME J. 3, 1223–1230.

Lennon, J.T., Jones, S.E., 2011. Microbial seed banks: the ecological and evolutionary implications of dormancy. Nat. Rev. Microbiol. 9, 119–130.

Mbuthia, L.W., Acosta-Martínez, V., DeBruyn, J., Schaeffer, S., Tyler, D., Odoi, E., Mpheshea, M., Walker, F., Eash, N., 2015. Long term tillage, cover crop, and fertilization effects on microbial community structure, activity: implications for soil quality. Soil Biol. Biochem. 89, 24–34.

Mnif, I., Mnif, S., Sahnoun, R., Maktouf, S., Ayedi, Y., Ellouze-Chaabouni, S., Ghribi, D., 2015. Biodegradation of diesel oil by a novel microbial consortium: comparison between co-inoculation with biosurfactant-producing strain and exogenously added biosurfactants. Environ. Sci. Pollut. Res. 22, 14852–14861.

Nicol, G.W., Tscherko, D., Embley, T.M., Prosser, J.I., 2005. Primary succession of soil *Crenarchaeota* across a receding glacier foreland. Environ. Microbiol. 7, 337–347.

Nunan, N., Wu, K., Young, I.M., Crawford, J.W., Ritz, K., 2003. Spatial distribution of bacterial communities and their relationships with the micro-architecture of soil. FEMS Microbiol. Ecol. 44, 203–215.

Ohta, H., Hattori, T., 1983. *Agromonas oligotrophica* gen. nov., sp. nov., a nitrogen-fixing oligotrophic bacterium. Antonio van Leeuwenhoek 49, 429–446.

Pepper, I.L., Gerba, C., Gentry, P., T, J. (Eds.), 2015. Environmental Microbiology, third ed. Academic Press, New York.

Roy Chowdhury, T., Lee, J.-Y., Bottos, E.M., Brislawn, C.J., White, R.A., Bramer, L.M., Brown, J., Zucker, J.D., Kim, Y.-M., Jumpponen, A., Rice, C.W., Fansler, S.J., Metz, T.O., McCue, L.A., Callister, S.J., Song, H.-S., Jansson, J.K., 2019. Metaphenomic responses of a native prairie soil microbiome to moisture perturbations. mSystems 4, e00061-19.

Schulz-Bohm, K., Martín-Sánchez, L., Garbeva, P., 2017. Microbial volatiles: small molecules with an important role in intra- and inter-kingdom interactions. Front. Microbiol. 8, 2484.

Theodorakopoulos, N., Lognoul, M., Degrune, F., Broux, F., Regaert, D., Muys, C., Heinesch, B., Bodson, B., Aubinet, M., Vandenbol, M., 2017. Increased expression of bacterial *amoA* during an N_2O emission peak in an agricultural field. Agric. Ecosyst. Environ. 236, 212–220.

Williams, M.A., Jangid, K., Shanmugam, S.G., Whitman, W.B., 2013. Bacterial communities in soil mimic patterns of vegetative succession and ecosystem climax but are resilient to change between seasons. Soil Biol. Biochem. 57, 749–757.

Winogradsky, S., 1924. Sur la microflore autochthone de la terre arable. C. R. Acad. Sci. (Paris) 178, 1236–1239.

CHAPTER

11

Rhizosphere

Lindsey Slaughter

Department of Plant and Soil Science, Texas Tech University, Lubbock, TX, United States

"To own a bit of ground, to scratch with a hoe, to plant seed, and watch the renewal of life-this is the most commonest delight."—***Charles Dudley Warner***

To a soil microorganism, the rhizosphere is like a lush oasis in the desert. In comparison to the near-starvation conditions of the bulk soil, the rhizosphere is a place where nutrients are plentiful and microorganisms flourish. The plant root is an adaptive organ that both affects and is affected by the physical, chemical, and biological properties of the surrounding soil. The **rhizosphere** is the zone of soil immediately influenced by the root with altered microbial diversity, increased activity and number of organisms, and complex interactions between soil microorganisms and the root. The significance of the rhizosphere arises from the release of organic material from the root and the subsequent effect of increased microbial activity on nutrient cycling and plant growth. The unique assemblage of microorganisms in the rhizosphere, known as the rhizosphere **microbiome**, where microbial community composition, abundance, and functional attributes are distinct from the bulk soil microbiome of the surrounding environment, can influence plant growth in beneficial, neutral, variable, or harmful ways (Table 11.1).

The term **rhizosphere** was first used by Hiltner in 1904 to describe the zone of soil under the influence of roots (Hiltner, 1904). The rhizosphere can extend more than 5 mm from the root and, more importantly, is an area of increased microbial activity (Fig. 11.1). The zone of influence and area of increased microbial activity are the key components that define the rhizosphere. The spatial boundaries of the rhizosphere, although commonly extending only a few millimeters from the root surface as noted earlier, are not fixed due to the massive range and combinations of soil types and plant root architectures that moderate the volume of soil that is physically, chemically, and biologically altered by the presence of the plant root. As the direct interface between plants, microbes, and soils, understanding the complex interactions that take place in the rhizosphere is critical to sustainably manage plant and soil ecosystems in a changing world (Philippot et al., 2013).

Principles and Applications of Soil Microbiology. http://dx.doi.org/10.1016/B978-0-12-820202-9.00011-3
Copyright © 2021 Elsevier Inc. All rights reserved.

TABLE 11.1 Effects of the rhizosphere microbiome on plant growth.

Impact	Effect
Beneficial	Growth promotion
	Symbiosis
	N_2 fixation/nutrient acquisition
	Biocontrol
	Water uptake
	Soil stabilization
Neutral or Variable	Nutrient flux
	Free enzyme release
	Attachment
	Allelopathy
Harmful	Competition
	Phytotoxicity
	Disease

Adapted from Lynch (1990).

The root environment

Spermosphere

The relationship between plants and soils begins with the seed prior to root development, including exudation of compounds and interactions with soil microorganisms in what will become the rhizosphere (Lemanceau et al., 2017). Microorganisms can be found both in and on seeds prior to contact with the soil through internal or floral pathways or through external contact with microbes prior to entering the soil environment. As soon as the seed enters the soil and begins the germination process, microbial activity increases around the seed. Much like the rhizosphere, the area of heightened microbial activity around germinating seeds is known as the **spermosphere** (Verona, 1958). Seeds exude organic compounds such as carbohydrates and amino acids during imbibition, or adsorption of water to rehydrate the seed, at the onset of germination and as the seed coat ruptures to form the primary root. Soil microbes immediately begin to colonize and compete for these resources in the spermosphere. Other antimicrobial and signaling compounds are also exuded to protect the seed against pathogenic microbes. This initial web of interactions in the spermosphere forms the foundation for the developing root and rhizosphere environment.

Root systems

Plant root systems are complex organ systems comprised of different root types that have evolved to be highly plastic and environmentally responsive. The first organ to emerge from the seed is the primary (or tap) root, and most plants then soon develop basal (or seminal)

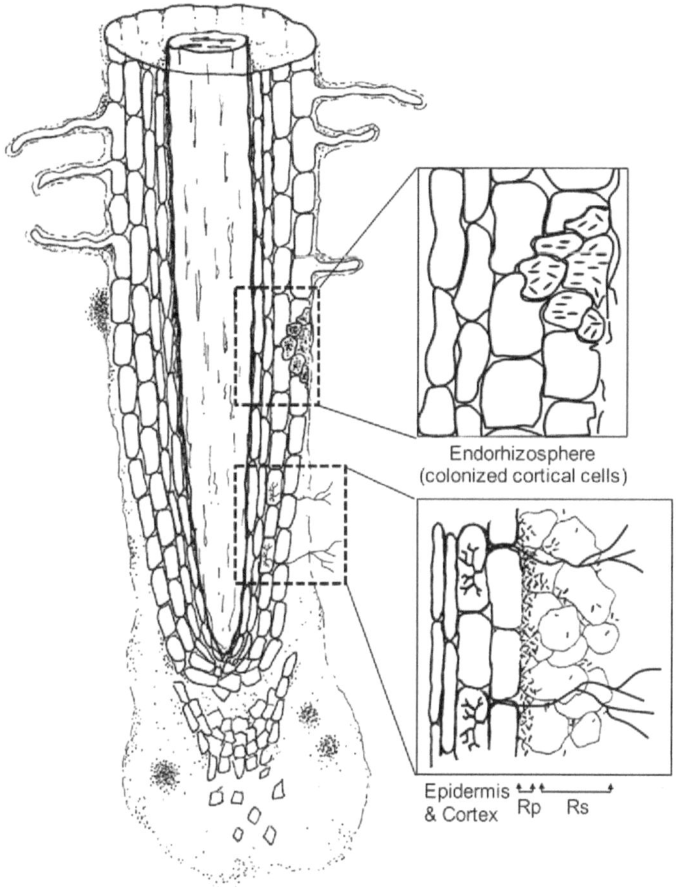

FIGURE 11.1 **A root and corresponding rhizoplane and rhizosphere.** Root hairs, the root cap, and other structures are shown along with microorganisms colonizing the root environment. The top inset shows the endorhizosphere consisting of cortical cells colonized by bacteria. The bottom inset illustrates the rhizoplane (Rp) immediately adjacent to the epidermis and cortex (colonized by mycorrhizal fungi) followed by the rhizosphere (Rs) extending a short distance (i.e., a few millimeters) into the surrounding soil. *Source: Original drawing by Kim Luoma.*

roots that become part of the embryonic root system. In addition, many plants can also develop adventitious or shoot-borne roots that originate from the stem (Box 11.1). The plant may deploy root types differently (deep roots versus shallow scavenging roots). Eventually all of these root types may form branches (or lateral roots). As described by York and Lobet (2017), the resulting root system is a complex tree-like structure composed of a large range of root types, ages, and developmental stages, which leads to a heterogeneity in functional properties within each root system. The root tips at the apex of each root type are responsible for directional growth of the root and are the primary location for water and nutrient absorption and rhizodeposition. The success of the plant depends in large part on the ability of the root tips to sense environmental signals and direct root tip growth in response. The massive

BOX 11.1

Roots from shoots

Not all roots are hidden beneath the soil. Adventitious roots originate from the non-root (e.g., stem) tissue of a plant and may be produced during normal growth or in response to stress or tissue damage. This is why we are able to vegetatively propagate certain plants, such as roses, from stem or leaf cuttings. Adventitious roots are a natural feature of many plants that take a variety of forms, from the short aerial roots along the stems of climbing plants that allow them to attach to surfaces, to the "stilt" roots extending from mangrove tree trunks into the soil, to the "prop" roots of corn (maize) that help anchor the plant in the soil and provide extra stability. In addition to structural support, shoot-borne adventitious roots can play a role in plant stress responses and nutrition. The adventitious roots of an indigenous maize landrace (Sierra Mixe) have been found to secrete a carbohydrate-rich mucilage that harbors beneficial N_2-fixing bacteria (Van Deynze et al., 2018). This can provide a substantial portion of the nitrogen needs of the maize plant.

The aerial roots of Sierra Mixe maize secrete large quantities of carbohydrate-rich mucilage between 3 and 6 months after planting. *Source: From Van Deynze et al. (2018).*

number of root tips in the soil accounts for a considerable portion of root surface area and is the most significant location for root interactions with rhizosphere microbial communities. Because of the importance of root tips for root development and function, they are also the site where microbes may have the most significant effect on plant root systems and ultimately the development and physiology of the entire plant.

The growing root tip is a site of rapid cell growth, elongation, differentiation (which includes the production of root hairs essential for water and nutrient uptake), and finally suberization with maturation. Because the root meristem is essential for the continued growth of each root tip, it must be carefully protected. The root cap at the apex of the root tip protects the meristematic region of the tip from the considerable compaction forces exerted by the growing root. The root cap and border cells released from the root cap also protect the notoriously susceptible newly synthesized root tissues from attack by soil-borne organisms that are attracted to the root tip in response to root exudates.

Root growth through soil

The elongating tip of a root is sterile but, as the root grows into the soil, contact is made with colonies of microorganisms distributed throughout the soil (Box 11.2). Roots normally grow into air-filled pores larger than 10 μm in diameter. Properties of the soil affect root diameter, root hairs, and the branching pattern of lateral roots. The distribution of roots is influenced by species and cultivar, soil texture, physical and chemical properties, and temporal changes.

Root hairs

Root hairs are extensions of the epidermal cells of the plant root. The functions of root hairs are still debated among researchers, but root hairs are known to be important in ion uptake from soil, especially with relatively immobile ions like $H_2PO_4^-$ and Zn^{2+}. More root hairs are found in coarsely aggregated soils (aggregate size of 3–5 mm) than in finely aggregated soils (less than 1 mm). Root hairs may penetrate adjacent soil aggregates and take up nutrients within them (Fig. 11.2).

Root cap

The primary function of the root cap is to protect the meristem as the root moves through the soil. The cap produces mucilage and may release substances that enhance plant cell growth. This nutrient-rich area provides a lush region for microbial growth. Root cap cells may elongate and persist in the soil surrounding the root for many centimeters behind the root tip. The mucilaginous sheath (sometimes called mucigel layer) functions as a nutrient absorptive area

BOX 11.2

Overview of scale

Roots are much larger than bacteria and fungi. For example, a soil bacterium may be 0.5 μm wide and 5 μm long; a fungal hypha may be 10 μm wide and more than 10,000 μm long. It is difficult to estimate the length of individual fungal hyphae in soil because most sampling techniques break these filamentous structures. In contrast, plant roots are about 1,000 μm wide and extend 1,000,000 μm (1 m) down the soil profile. Compared to an average soil bacterium and fungus, roots may be approximately 10–1,000 times wider and 100–100,000 times longer. In contrast, an average root hair is 10 μm wide and 500 μm long and more similar in size to fungal hyphae.

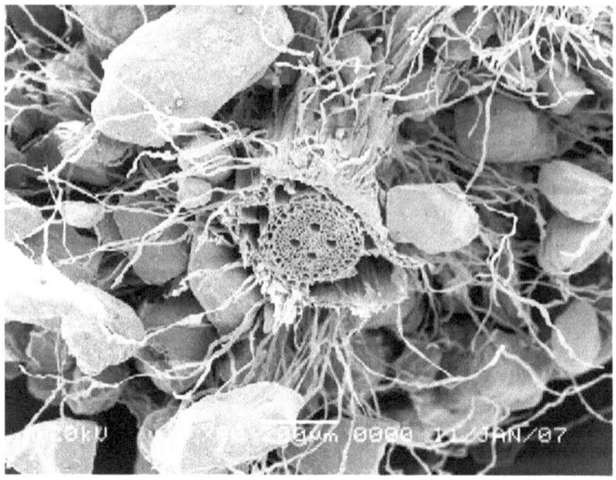

FIGURE 11.2 **Scanning electron micrograph of *Pascopyrum smithii* root cross section showing root hairs and attached soil particles.** *Source: From Bergmann et al. (2009). Reprinted with permission. Copyright 2009, Brigham Young University, Provo, UT 84602.*

and protects the root tip from chemicals in the soil that could injure the root. This covering reduces desiccation and abrasion damage to the root tip or apical region of the root.

Border cells

Cells sloughed from the root cap, known as border cells, are constantly being released into the rhizosphere. Although lysed cells sloughed from the root cap and other areas along the root are major contributors of carbon into the rhizosphere, border cells are typically intact and metabolically active when released. Researchers are beginning to explore the potential functions and importance of border cells in root signaling and pathogen defense, such as their use of extracellular DNA and antimicrobial proteins to lure and trap invading pathogens of the root (Hawes et al., 2016).

Root zone of influence

The root system is a dynamic and adaptive plant organ that physically and chemically influences the surrounding soil, both creating a habitat for, and responding to, increased microbial activity in cyclical relationship. The rhizosphere is a spatially and temporally heterogeneous zone created by active roots. Physical effects of root growth increase microscale compaction and impact aggregation as soil is pushed aside by the root or enmeshed by root hairs. Nutrient absorption, respiration, and exudation by plant roots create zones of nutrient depletion or enrichment as well and changes in pH and aeration. The vast range of possibilities in physical root zone characteristics such as root architecture and surrounding soil type, or chemical characteristics such as the amount and types of compounds absorbed or exuded from roots, make spatially defining the zone of influence around the root difficult. Nutrients with different absorption rates can create different zones of influence (Fig. 11.3). Nutrients such as potassium can be depleted within several millimeters of the root, while the zone of influence for nitrate can extend for several centimeters, thus complicating the quantification of rhizosphere size and extent as defined by the zone of root influence (Hinsinger et al., 2005). Gases

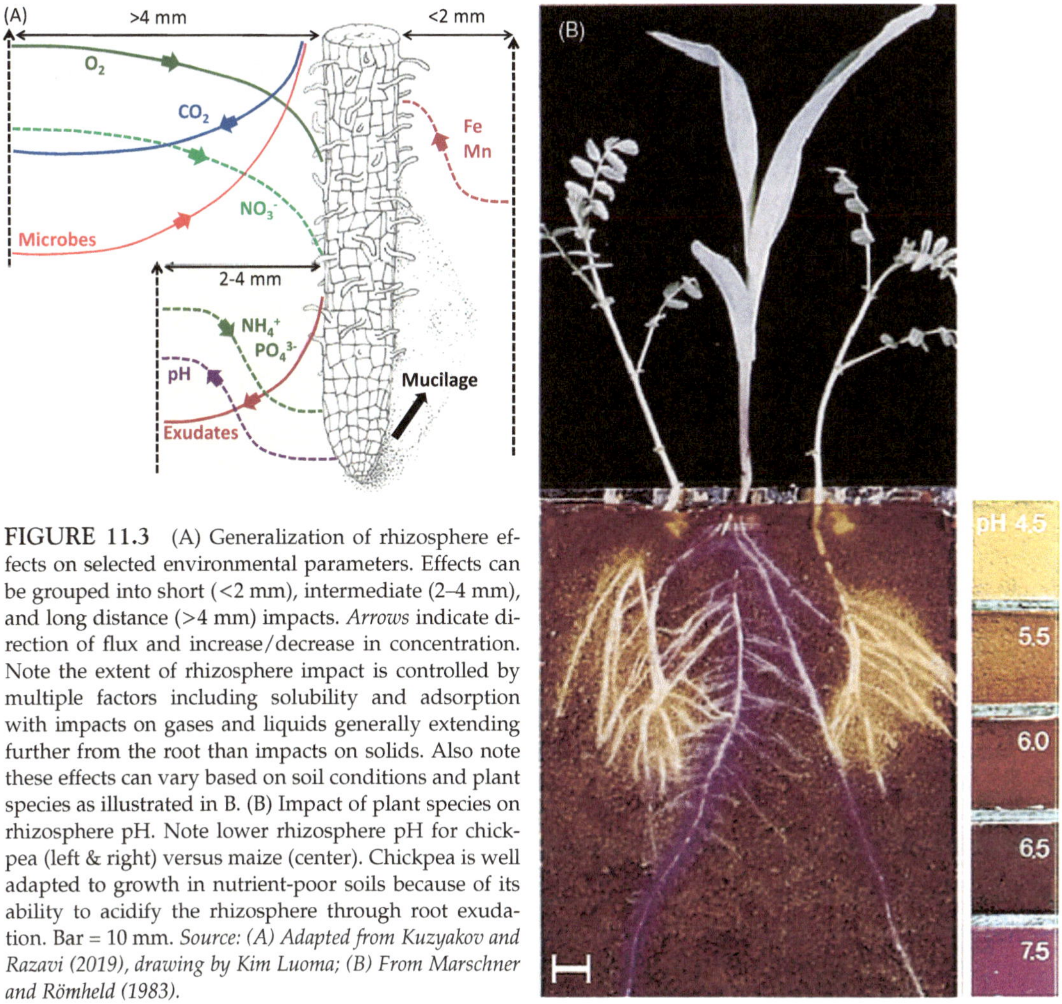

FIGURE 11.3 (A) Generalization of rhizosphere effects on selected environmental parameters. Effects can be grouped into short (<2 mm), intermediate (2–4 mm), and long distance (>4 mm) impacts. *Arrows* indicate direction of flux and increase/decrease in concentration. Note the extent of rhizosphere impact is controlled by multiple factors including solubility and adsorption with impacts on gases and liquids generally extending further from the root than impacts on solids. Also note these effects can vary based on soil conditions and plant species as illustrated in B. (B) Impact of plant species on rhizosphere pH. Note lower rhizosphere pH for chickpea (left & right) versus maize (center). Chickpea is well adapted to growth in nutrient-poor soils because of its ability to acidify the rhizosphere through root exudation. Bar = 10 mm. *Source: (A) Adapted from Kuzyakov and Razavi (2019), drawing by Kim Luoma; (B) From Marschner and Römheld (1983).*

produced during root and microbial metabolism also diffuse several centimeters beyond the conceptualized zone of influence (de la Porte et al., 2020). However, the area of increased microbial activity around the root relative to bulk soil is another definitive characteristic of the rhizosphere that helps to delineate these rhizosphere boundaries. The complexity and heterogeneity of the zone of influence around growing roots that creates the rhizosphere is challenging to define and study, but understanding and managing this dynamic environment holds massive implications for ecosystem functioning and productivity.

Rhizosphere boundaries

The boundaries of the various sections of the rhizosphere are also difficult to demarcate. A simple solution is to divide this area into the rhizosphere, rhizoplane, and endorhizosphere. As previously defined, the rhizosphere is the zone (volume) of soil influenced by roots and their exudation of substances that affect microbial activity. The zone of rhizosphere soil extending beyond the root surface into the bulk soil can be alternatively designated as the **ectorhizosphere**, although in many studies the overall term rhizosphere is used to represent this zone. The **rhizoplane** is the immediate surface of a plant root and any strongly adhering soil particles. The area just inside the root that includes portions of the root cortex and endodermis is commonly referred to as the **endorhizosphere**, which may at times be simplified to "endosphere." The microbial populations that colonize the interior of the root and form intimate associations with the root are considered **endophytes** that inhabit the endorhizosphere.

The rhizosphere or ectorhizosphere, rhizoplane, and endorhizosphere are easier to conceptualize than to study (Box 11.3). For example, if the rhizosphere can extend from 1 to 4 mm or more from the root surface and the rhizoplane with adhering soil can be up to 2 mm in width, then the rhizoplane and the rhizosphere are difficult to differentiate. Furthermore, inhabitants in the rhizosphere may be difficult to distinguish from those of the rhizoplane because of mucilage secreted by the root. Methodologies for distinguishing endorhizosphere, rhizoplane, rhizosphere, and bulk soil microbial communities are continually being refined to develop standardized sampling protocols, especially when sequencing-based analyses are to be used (McPherson et al., 2018). Studies that distinguish these boundaries typically find increased microbial diversity and altered community composition when moving from the endorhizosphere to the rhizosphere, and sometimes subtler differences between rhizosphere and bulk soil (Figs. 11.4 and 11.5). The soil-root interface is the area with the greatest number and variety of plant-microbial associations. As standardized methodologies develop and become more accessible to determine functional microbiome characteristics in addition to composition, researchers find that even small changes in microbial communities between the different rhizosphere boundaries can have large implications for ecosystem processes such as nutrient cycling.

BOX 11.3

Practical rhizosphere analysis

How does one determine which organisms inhabit the various areas of the rhizosphere when these regions are indistinct? Rhizosphere analyses are often limited by the procedures to identify the microorganisms of these zones, as well as by variety in plant root architecture and soil structure that result in widely different rhizosphere characteristics between plant types and environments. For example, rhizosphere soil may not adhere well to plants with thick taproot systems in coarse-textured or poorly-aggregated soils, which poses a challenge for excavation and sampling. For measurement purposes, rhizosphere microorganisms are often considered to be the microorganisms removed when the roots and adhering soil are shaken gently in water. Rhizoplane microorganisms are those removed when the washed roots are transferred to fresh diluent or sterile phosphate-buffered saline solution (PBS), and shaken vigorously. Finally, endophytes are the microorganisms that are recovered after the root is surface sterilized and macerated.

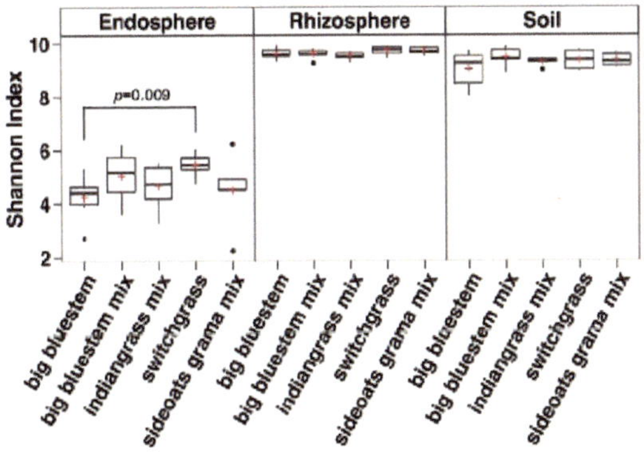

FIGURE 11.4 **Bacterial diversity (Shannon index) within the endorhizosphere (endosphere), rhizosphere, and bulk soil of five field-grown grass species determined via 16S rRNA gene amplification and sequencing.** A higher Shannon index in rhizosphere and bulk soil samples can be attributed to greater overall richness (number of species) and/or evenness (relative abundance of species) of these communities compared to the endosphere. Note that various grass species impacted bacterial diversity in the endosphere but not the rhizosphere or bulk soil. *Source: From McPherson et al. (2018).*

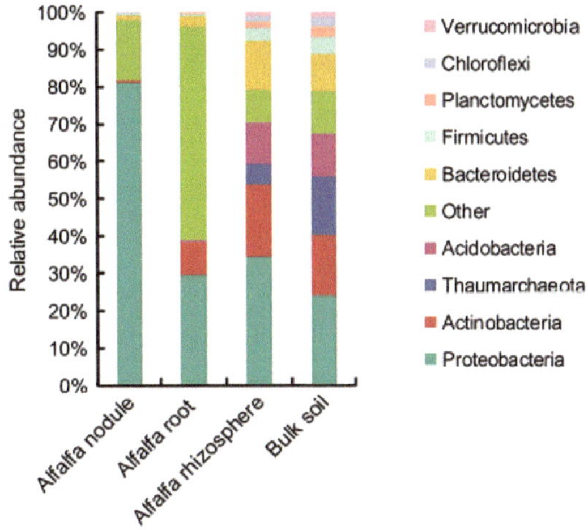

FIGURE 11.5 **Relative abundance (%) of archaeal and bacterial phyla detected within the endorhizosphere (nodule and root), rhizosphere, and bulk soil of alfalfa determined via 16S rRNA gene amplification and sequencing.** Note the similarity between the rhizosphere and bulk soil communities in contrast to the nodule and root. *Source: From Xioa et al. (2017).*

The rhizosphere environment

The physical and chemical properties of soil were discussed in Chapter 2. This chapter focuses on how differences or management-induced changes in some of these soil properties in the vicinity of a plant root affect microbial activity. The nutritional aspect of the rhizosphere will be covered in the next section.

Rhizosphere moisture will affect microbial colonization directly or indirectly as it affects plant growth. Microbial cells at the surface of the root may experience severe changes in water potential (Fig. 11.6). On the one hand, soils above or near field capacity (approximately –30 kPa or a less negative potential) have little change in soil water potential even though plant transpiration can move a large amount of the soil water. On the other hand, in soils having –200 kPa or a more negative potential, the soil water potential near the root can stress microbial cells in the rhizosphere. At low soil water potential, motility and diffusion of nutrients can be reduced, dramatically influencing microbial growth. This does not mean that high soil water potentials are necessarily beneficial, because a large percentage of pore space is water-filled and oxygen may be limiting due to restricted oxygen diffusion combined with plant and microbial respiration.

Soil texture can affect microbial transport and root colonization by microorganisms. In general, bacteria can move greater distances in coarse-textured soils like sands than in fine-textured soils like clays, due to lower permeability and increased potential for adhesion to clay particles in the latter. Texture also influences the extent and physical characteristics of the rhizosphere. For example, increased porosity and connectivity between pores in soils with higher clay contents can increase the radial extent of the rhizosphere compared to sandy soils.

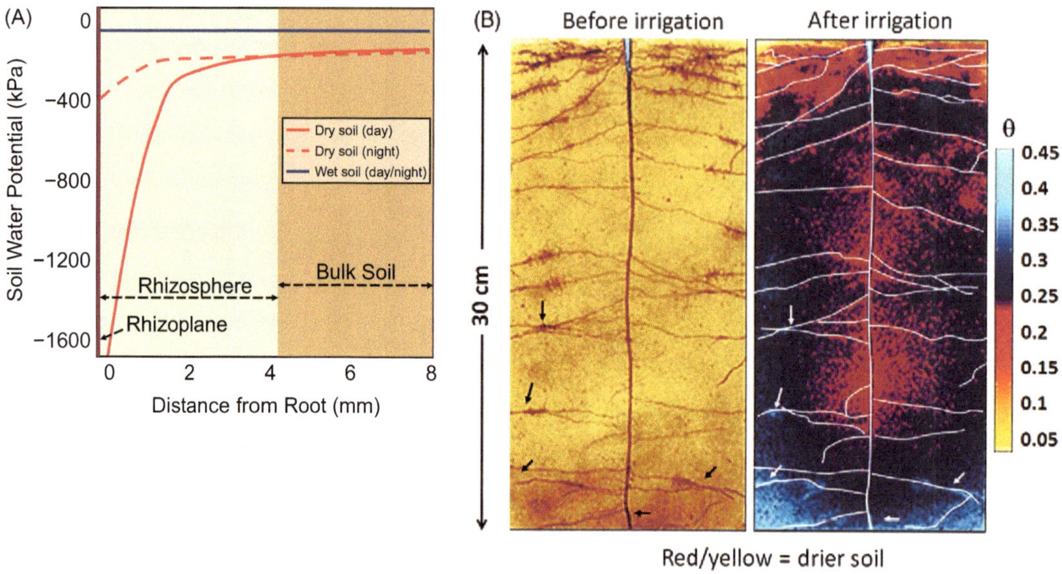

FIGURE 11.6 (A) Soil water potential in the rhizoplane, rhizosphere, and bulk soil illustrating the impacts of day versus night and dry versus wet conditions at varying distances from the plant root. (B) Spatial impacts of rhizosphere during soil drying. Volumetric soil water content (θ) before and 30 min after irrigation of *Lupinus albus*. Following irrigation, the rhizosphere of the upper roots, tap root, and proximal parts of the lower roots remained markedly drier than the rhizosphere of the root tips in the lower parts of the sample. This is shown by high water content visible around the root tips *(white arrows). Source: (A) Adapted from Papendick and Campbell (1975); (B) Adapted from Carminati (2013).*

Temperature can play a role in colonization rates of the rhizoplane. Often microorganisms colonize best at temperatures below their optimum growth temperature. This phenomenon may result from reduced competition from the indigenous microflora or slowed root growth at the lower temperatures, thus allowing greater root colonization.

Soil pH can alter the survival and competitive abilities of microorganisms. Root-induced production of H^+, HCO_3^-, or organic compounds and their subsequent release into the rhizosphere affects ion uptake and thus pH (Box 11.4). In addition, both microbial production of

BOX 11.4

Effect of nitrogen source on rhizosphere pH

Nitrogen is an important, and typically limiting, resource that is taken up in large quantities by the plant compared to other nutrients. The source of nitrogen influences ion uptake by the root and thus affects rhizosphere pH. When nitrate (NO_3^-) is supplied to the plant, the rhizosphere pH increases because more HCO_3^- is released than H^+. When ammonium (NH_4^+) is supplied, the reverse is true. Ammonium-based fertilizers such as urea, anhydrous ammonia, and aqua ammonia can reduce rhizosphere pH. In acid soils, rhizosphere acidification may reduce bacterial numbers.

organic acids and oxidation reactions during mineralization of organic matter can decrease the pH. Rhizosphere pH is typically lower than the bulk soil. The magnitude and direction of pH change depends mainly on nitrogen nutrition, where ions released from plant roots during uptake of cationic nutrients (e.g., NH_4^+) can acidify the rhizosphere while uptake of anionic nutrients (e.g., NO_3^-) can instead increase pH. However, symbiotic relationships and microbial activity can also influence pH.

The plant community influences the rhizosphere microbial community by the exudates it releases from the root and the composition of its residues. Plant species, and even cultivars of the same species, release different amounts and types of carbon to the rhizosphere. Cultivars differing in disease susceptibility via plant defense signaling and exudation may also differ in their populations of rhizosphere microbes. For example, susceptible lines may support more abundant and diverse rhizosphere microbial communities than non-susceptible lines (Doornbos et al., 2012). Genetically modified plants often have distinct exudate patterns and residue composition compared to the wild type plant. Although these differences may result in changes in the microbial community, they may not be any greater than those seen among various cultivars.

Residue quality and management will influence the composition of the microbial community. Decaying residue and root systems serve as a source of nutrients for the surrounding microorganisms, and the quality of these crop residues may vary with species and cultivar. Similarly, crop species and varieties differ in their decomposition rates and in the microbial populations they support. Crop rotation may allow for greater carbon inputs and diversity of plant material added to soils. Crop rotation favors beneficial microbes, increases microbial diversity, interrupts the cycle of pathogens, and reduces weed and insect populations. Studies have long shown the positive effects of crop rotation on crop growth, attributing this effect to changes in microbial community composition.

Genetically modified organisms

As more genetically modified (GM) plants and microorganisms are introduced into agricultural systems, their impact on native soil populations and plant productivity is being questioned. Genetic engineering of plants and microbes to enhance crop health, nutrition, and productivity, as well as for bioremediation of contaminated soils, is a continually developing area of research as sequencing and gene editing technologies improve and become more widespread and affordable. Although genetically engineered microorganisms have shown promising applications for bioremediation, the largest focus in GM use has been directed at plants. Plants influence rhizosphere microbiota through residue composition, resource demands, root system dynamics, exudate composition and patterns, and plant defense signaling. As such, genetic modifications that alter one or more of these characteristics have the potential to impact rhizosphere biota. Differences in rhizosphere biota have been detected in GM crops, for example, those that produce Bt toxins aimed at reducing insect damage to roots. However, many studies have shown inconsistent or transient effects. How these differences affect the soil microbial communities and nutrient cycling, plant productivity, soil quality, and organic matter accumulation on a long-term basis is unclear (Mandal et al., 2020). In addition, application of genetically modified plants and microbes are often accompanied by changes in pesticide, herbicide, and tillage and rotation management, which may obfuscate effects of genetically modified organisms on soil microbial communities and ecosystem functioning. The possibility exists that DNA from engineered plants or microorganisms could be transferred to indigenous bacterial populations through a variety of gene transfer mechanisms. However, transfer of recombinant genes between genetically modified bacteria and indigenous microbial communities has not been consistently demonstrated despite some apparent effects of these introduced organisms on microbial community structure (Kong et al., 2016). There is evidence that antibiotic resistance genes can be transferred to rhizosphere bacteria from transgenic plants, but this may be preventable during plant sequence selection and development. Overall, observed changes to rhizosphere microbial communities and functioning caused by GM plants and microbes have been inconsistent, and perturbations due to their introduction are often no larger than those resulting from other management practices. More studies on how genetically modified organisms affect the soil environment and indigenous microbial community, as modified by management strategies, are needed to better understand the risks and ecosystem outcomes of genetically modified plants and soil-applied microbes intended to enhance N_2-fixation, promote plant growth, and assist in bioremediation and soil conservation.

Plant-derived compounds

Plant roots excrete various organic substances into the rhizosphere, and these substances provide a rich source of nutrients for the microbial community. The term **rhizodeposit** is used to encompass all C-containing compounds that are leaked or exuded into the rhizosphere from the root. The organic fractions of rhizodeposits from the root are of various forms and composition and are further categorized as exudates, secretions, mucilages, mucigel, and lysates from sloughed root cap and border cells (Table 11.2). Because most soils are low in readily available carbon for microbial growth, C-containing plant rhizodeposits, which are often

TABLE 11.2 Carbon compounds originating from plant roots.

Exudates	Compounds slowly oozing directly from epidermal cell walls into the soil or oozing indirectly into the intercellular spaces of the plant root and then into the soil. The release of these low-molecular-weight compounds is not the result of metabolic processes.
Secretions	Compounds released that are the results of metabolic processes. There compounds are both low- and high-molecular weight.
Lysates	Compounds released from sloughed cells and when epidermal cells are damaged or autolyzed.
Mucilages	Compounds (1) secreted by the Golgi apparatus in the root cap, (2) secreted by epidermal cells that still have only primary walls, (3) released from hydrolyzed polysaccharides of primary cell wall (between epidermal cells and sloughed root cap cells), and (4) released from bacterial degradation of multilamellate primary cell walls of dead epidermal cells. These gelatinous products, together with natural mucilages, bacterial cells and their metabolic products, colloidal minerals, and organic matter, form the mucigel—the gelatinous material on the surface of roots grown in soil. The mucigel maintains contact between the root and the soil as the root shrinks during daytime water stress, permitting continuous uptake of nutrients and water.

From Rovira et al. (1979).

estimated to be as much as 20% of photosynthetically-produced C, are of major importance to heterotrophic microbial growth. Some plants can deposit over 40% of photosynthetically-fixed carbon into the soil each day through their roots, depending on plant physiology and environmental conditions such as wet-dry cycles (Odell et al., 2008). Most of this is deposited via sloughed root cap or border cells and other extracellular substances.

Exudate composition

Exudate composition varies with the soil environment and the growth stage of the seed and root. It is difficult to determine which compounds are true exudates and which compounds are released from lysed root cells, although methodological developments in metabolomics and transcriptomics increasingly allow researchers to identify common metabolites in root exudates, as well as begin to investigate the transport mechanisms and plant-microbial signaling involved in import and export of metabolites (Sasse et al., 2018). Compounds in exudates include sugars, sugar alcohols, amino acids, organic acids, nucleotides, peptides, fatty acids, inorganic nutrients, secondary metabolites, and other compounds including hormones, sterols, flavonoids, vitamins, tannins, alkaloids, phosphatides, and unidentified substances (growth factors, fluorescent substances, nematode cyst or egg-hatching factors, and fungal growth stimulants and inhibitors) (Table 11.3). The composition and dynamics of root exudates are important in shaping the rhizosphere microbiome, both as a source of nutrients and raw materials for microbial metabolism and as chemical signals between plant roots and microbial symbionts. Alterations in the quantity and composition of metabolites, including sugars and amino acids, exuded from roots in response to plant stress such as drought also play a role in shaping the root and rhizosphere microbiome. Advances in techniques to study root exudates and their interactions with root and rhizosphere microbes offer exciting avenues of research for understanding and managing plant-soil interactions to improve plant health and productivity in stressed environments.

Sugars provide readily available sources of carbon for microbial growth. Amino acids, whose absolute and relative percentages vary with plant species, are a readily available source of nitrogen for microbial growth. These compounds are the most-studied root exudates and are thought to be associated with susceptibility or resistance of plants to root-infecting fungi. Organic acids can chelate metals, affecting pH and the absorption and translocation of nutrients. The range in quantity and variety of vitamins in exudates can account to some extent for differences in bacterial populations. Although much research has focused on water-soluble and nonvolatile exudates, water-insoluble exudates and volatile organic compounds (VOCs) are also produced. A diverse range of VOC's emitted from roots are involved in pathogen interactions and defense in addition to their use as a source of carbon for microbes.

Exudation rate

The actual rate of exudation from roots is difficult to calculate. Exudation rates and release patterns likely vary from different root portions and cell types, with the greatest amount of exudate arising from those cells involved in cell elongation and lateral root formation. The concentration of carbon in the rhizosphere declines exponentially as the distance from the root increases. Increased exudation can result from extremes in temperature, water stress,

TABLE 11.3 Compounds detected in plant root exudates.

Class	Examples
Sugars	Arabinose, deoxyribose, fructose, galactose, galactinol, glucose, mannose, maltose, oligosaccharides, ribose, raffinose, rhamnose, sucrose, xylose
Sugar alcohols	Arabitol, erythritol, glycerol, inositol, myo-inositol, ribitol, threitol, xylitol
Sugar phosphate	Glucose-6-phosphate, glucose-1-phosphate
Amino acids	α-Alanine, β-alanine, γ-aminobutyric acid, α-aminoadipic acid, arginine, asparagine, aspartic acid, alanine, 4-aminobutryric acid, citrulline, cystathionine, cysteine, cystine, deoxymugineic acid, 3-epihydroxymugineic acid, glutamic acid, glutamine, glycine, isoleucine, homoserine, histidine, lysine, leucine, methionine, ornithine, proline, phenylalanine, proline, serine, threonine, tryptophan, tyrosine, valine
Organic acids	Acetic, aconitic, ascorbic, aldonic, aspartic, butyric, *trans*-benzoic, caffeic, p-coumaric, citric, erythronic, ferulic, formic, fumaric, gluconic, glutaric, glyceric, glycolic, glyoxylic, *p*-hydroxybenzoic, α-ketoglutaric, lactic, malic, malonic, oxalacetic, oxalic, piscidic, propionic, pyruvic, succinic, syringic, tartaric, tetrionic, valieric, vanillic
Nucleotides, purines	Adenine, guanine, uridine/cytidine, thymine
Peptides	Various hormones/signaling peptides, antimicrobial peptides (defensins, thionins)
Fatty acids	Linoleic, linolenic, oleic, palmitic, stearic
Inorganics	Nitrate, phosphate, sulfate, potassium
Enzymes	Amylase, invertase, peroxidase, phenolase, acid/alkaline phosphatase, polygalacturonase, protease
Growth factors, vitamins	*p*-Amino benzoic acid, biotin, choline, inositol, *N*-methyl nicotinic acid, niacin, pathothenic, pantothenate, pyridoxine, riboflavin, strigolactones, thiamine
Sterols	Campesterol, cholesterol, sitosterol, stigmasterol
Flavonoids	Chalcone, coumarine, flavones, flavonols, flavanones, flavonones, isoflavones
Other compounds	Al-induced polypeptides, alcohols, alkyl sulfides, auxins, esculetin, esculin, scopoletin, scopolin, 4-methylumbelliferone, camalexin, dihydroquinone, ethanol, glucosides, glucosinolates, glycinebetaine, hydrocyanic acid, inorganic ions and gaseous molecules (e.g., CO_2, H_2, H^+, OH^-, HCO_3), isothiocyanates, unidentified ninhydrin positive compounds, unidentifiable soluble proteins, reducing compounds, scopoletin, sorgoleone, strigolactones

Adapted from Sasse et al. (2018) and Vives-Peris et al. (2020).

phosphorus deficiency, increased light intensity, and increased populations of microorganisms. Exudation rate is also influenced by herbicides, pathogens, foliar treatments, symbiotic associations, and root and seed injury (Box 11.5). Decreased exudation can result from nitrogen deficiency, decreased light intensity, and decreased microbial populations.

Plant exudates are substrates for microbial growth, and the level of microbial activity in turn affects root and plant growth. Greater populations of microorganisms in the rhizosphere competing for nutrients may create a nutrient deficiency for the plant. In contrast, microorganisms may make nutrients from poorly-soluble sources, such as phosphates and trace metals, available to plants. In addition, symbiotic associations of roots with rhizobia

BOX 11.5

Effect of seed scarification

Over a 24-h period, unblemished, moistened soybean seeds with intact seed coats exude an estimated 5.3 μg glucose equivalents (C) h^{-1} per seed. If the seeds are scarified (which breaks or injures the seed coat), then the seeds exude up to 217 μg C h^{-1} per seed (Schlub and Schmitthenner, 1978). Thus, damaging the seed coat can vastly change the exudation pattern and the microbial community around that seed.

(Chapter 17) or mycorrhizal fungi (Chapter 12) increase root exudates. Mycorrhizal roots have enhanced nutrient-absorbing capacity, which subsequently increases plant growth. Plant pathogenic microorganisms (both foliage-infecting and root-infecting) can alter the quality and amount of exudate and result in greater populations of microorganisms in the rhizosphere (Chapter 23).

Mucilage and mucigel

The mucilage exuded by roots forms a sheath of slime on the external surface of the root. The principal site of mucilage release is the root tip or apical zone of a root, particularly the meristematic zone immediately behind the root cap. By definition, mucigel contains plant mucilages and plant products as well as bacterial cells and their products. The mucigel creates an intimate contact between roots or root hairs and soil particles and is principally found on the root and root hairs compared to the root tip. Organic materials excreted from living root cells or released from cortical cells, as well as senescent root hairs or border cells sloughed off from the root, are contained in this sheath. The combined production and mixing of mucilage from plant roots and extracellular polymeric substances (EPS) from bacteria is a critical mechanism involved in holding water in the rhizosphere and maintaining connections between pore spaces as soils dry (Benard et al., 2019).

Organisms inhabiting the rhizosphere

A vast number of species of microorganisms are present in the rhizosphere, and their numbers are typically higher than in bulk soil (Figs. 11.3, 11.7, and 11.8; Tables 11.4 and 11.5). The abundance and activity of bacteria and fungi in the rhizosphere tends to be greatest directly next to the root and decreases farther away from the root, but the opposite is true for microbial diversity (Fig. 11.8). Organisms in the rhizosphere closest to the root, and thus closest to exudates and signaling compounds, tend to be exhibit more characteristics of r-type strategists with rapid turnover rates rather than K-type strategists with slow turnover rates (Fig. 11.8; Chapter 10). Comparing the number of microorganisms in the rhizosphere (R) and in the bulk soil (S), resulting in an R:S ratio, can be used to indicate how strongly the rhizosphere

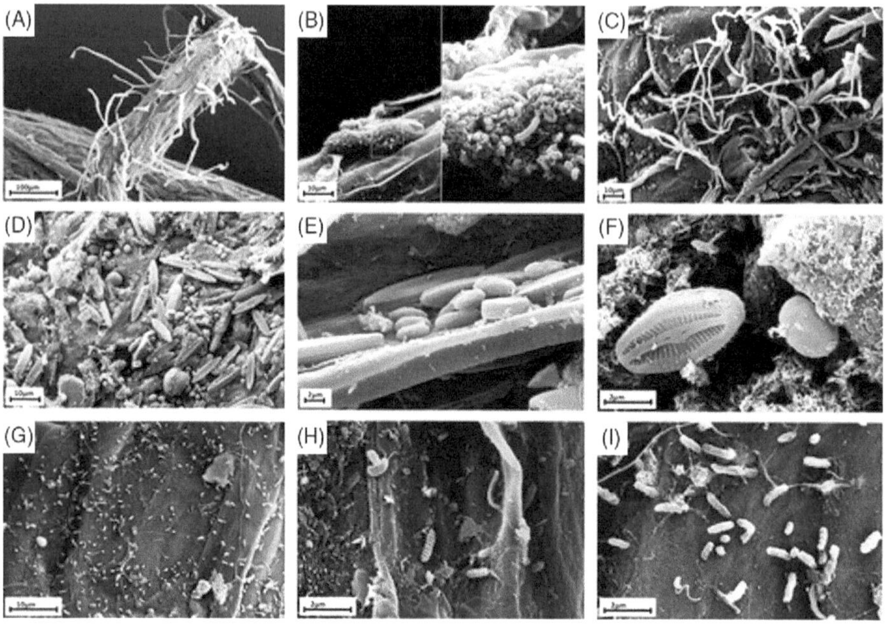

FIGURE 11.7 **Scanning electron microscopy pictures of *Arabidopsis thaliana* root surfaces showing the complex microbial networks formed on roots.** (A) Overview of primary root with numerous root hairs; (B) Biofilm-forming bacteria; (C) Fungal or oomycete hyphae surrounding the root surface; (D) Primary root densely covered by spores and protists; (E,F) Protists; (G) Bacteria and bacterial filaments; and (H,I) Different bacterial individuals showing a variety of shapes and morphological features. *Source: From Hassani et al. (2018).*

affects general microbial abundance (e.g., bacteria and fungi) or even particular groups (e.g., denitrifiers) (Table 11.4). Both the microbiome composition and the abundance of specific microbes found in the rhizosphere depend on a variety of plant and environmental factors, including plant identity, growth stage, bulk soil microbiome, and soil physical and chemical characteristics (Fig. 11.9). For example, bacterial composition and diversity in the rhizosphere microbiome can fluctuate throughout the growing season based on plant resource needs and the availability and types of plant root exudates and signaling compounds.

Bacteria

Bacteria, including actinomycetes, are the most numerous inhabitants of the rhizosphere, typically numbering 10^6–10^9 organisms g^{-1} of rhizosphere soil, although they account for only a smaller portion of the total biomass because of their small size. The dominant bacterial phyla found in the rhizosphere include Proteobacteria, Actinobacteria, Acidobacteria, Bacteroidetes, and Firmicutes, as well as Chloroflexi, Gemmatimonadetes, and Planctomycetes. In general, nonsporulating rods are abundant in the rhizosphere. Pseudomonads and other Gram-negative Proteobacteria are especially competitive in the rhizosphere and typically occupy a large portion of the total bacterial population on the root (Figs. 11.5 and 11.10A). Actinobacteria can account for approximately 10%–30% of the total microflora in the rhizosphere, depending on

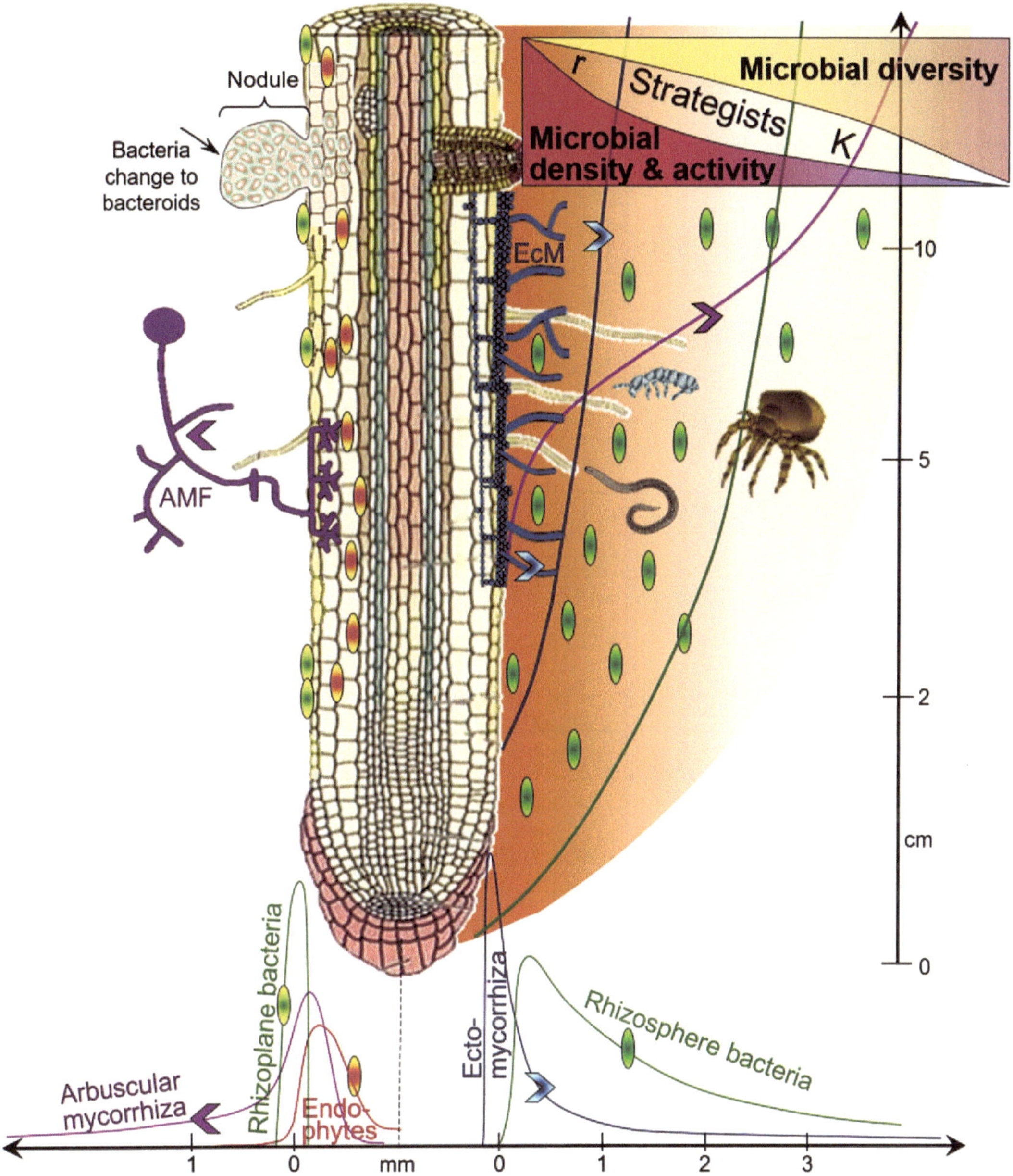

FIGURE 11.8 **Distribution of life in the rhizosphere.** The abundance of various microbial groups across (*x*-axis at the bottom, in millimeter from the root surface) and along (*y*-axis at the right, in centimeters, not proportional) the root is presented by continuous color curves. Microbial groups include: Arbuscular mycorrhiza (*violet*, AMF) and Ectomycorrhiza (*blue*, EcM); Endophytic *(red)*, Rhizoplane *(green)*, and Rhizosphere bacteria *(green)*. The gradients of microbial density, activity, and diversity, as well as the dominance of r and K strategists are presented at the top right. The overall microbial density is presented with orange shading on the right. Note that the size of illustrated microorganisms is not proportional to their real size. *Source: From Kuzyakov and Razavi (2019).*

TABLE 11.4 Colony-forming units (CFU) of microorganisms in the rhizosphere (R) of wheat (*Triticum aestivum* L.) and bulk soil (S) and their resultant R:S ratios.

Microorganisms	Rhizosphere	Bulk soil	R:S ratio
	----------CFU g^{-1} soil----------		
Bacteria	1.2×10^9	5.0×10^7	24.0
Fungi	1.2×10^6	1.0×10^5	12.0
Protozoa	2.4×10^3	1.0×10^3	2.4
Ammonifiers	5.0×10^8	4.0×10^6	125.0
Denitrifiers	1.3×10^8	1.0×10^5	1260.0

From Rouatt et al. (1960).

TABLE 11.5 Phospholipid fatty acid (PLFA) biomass in the rhizosphere of tall fescue (*Lolium arundinaceum*) and bulk soil.

	Rhizosphere	Bulk soil
	---------- nmol PLFA g^{-1} soil ----------	
Bacteria	255.8 ± 14.6	203.7 ± 13.3
Fungi	28.1 ± 3.1	16.4 ± 3.2
16:1ω5c	17.5 ± 0.2	12.4 ± 0.1
Bacteria: Fungi	9.2 ± 0.7	12.7 ± 1.4

Total bacteria and fungi represent the sum of PLFA biomarkers (± standard deviation) associated with each group. Abundance of the 16:1ω5c biomarker shown in addition to total fungi is often associated with arbuscular mycorrhizal fungi (AMF).
Adapted from Rojas et al. (2016).

the season or when the nutrients are added. These modifying factors cause difficulty in characterizing populations in the rhizosphere.

Fungi

Although plate counts of rhizosphere fungi are generally less than bacteria, these counts are based on spores and mycelial propagules and may be misrepresentative. Direct microscopy shows considerable fungal growth along root surfaces (Fig. 11.7C); there may be as much as 12–14 mm of hyphae mm^{-1} of root length. Population abundances may best be reported as ranges, most frequently 10%–20% of the total microflora or 10^5–10^6 g^{-1} rhizosphere soil. The rhizosphere initially favors colonization by fungi that grow on simple sugars (Chapter 6). The predominant fungi represented in the rhizosphere are typically members of the Ascomycota, Basidiomycota, and Zygomycota phyla (Fig. 11.10B). Several rhizosphere-inhabiting fungi are pathogenic or saprophytic, whereas others form symbiotic (mycorrhizal) associations with roots. Mycorrhizal fungi may be a major component of the microbial biomass in the rhizosphere. Because many of these fungi do not grow on artificial media in the laboratory, they have been historically overlooked, yet advances in imaging, mass-spectrometry, and

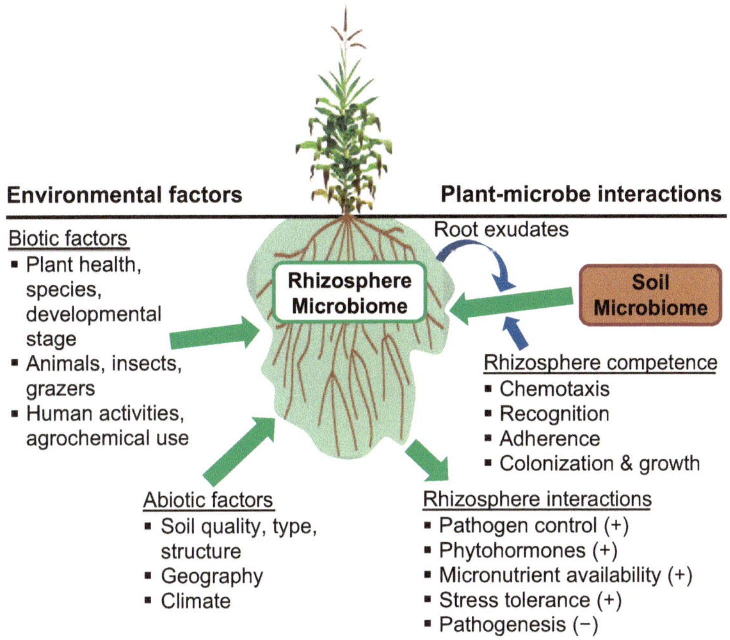

FIGURE 11.9 **Factors that influence selection, composition, abundance, and function of the rhizosphere microbiome.** *Source: Adapted from Berg and Smalla (2009).*

sequencing-based methodologies have revealed much about the composition and function of the fungal rhizosphere microbiome.

Faunal populations

As with the microfloral populations, micro-, meso-, and macrofauna are affected by the rhizosphere. Protozoa (i.e., amoebae, flagellates, and ciliates), nematodes, mites (Acarina), and springtails (Collembola) exhibit the greatest rhizosphere effect (Chapter 8). Unsurprisingly, these organisms are not uniformly distributed in soil, but are found where food is most plentiful. In the case of protozoa, they can feed selectively and therefore influence the composition of the bacterial populations in the rhizosphere.

Root exudates influence the abundance and behavior of nematodes directly or indirectly in a variety of crops, and their populations are much greater in the rhizosphere than in root-free soils. Root exudates can induce cyst- or egg-hatching of nematodes. Some plants stimulate hatching; others do not. Nematode larvae are usually found in the highest concentration at the region of greatest cell elongation behind the root tip. This enhanced population is not only a response to an increased food supply, but also in response to the release of substances that attract nematodes. Conversely, nematodes may also be repelled by some plant exudates.

Because many microarthropods are **mycophagous** (fungal-eating), they are more abundant near root surfaces. Mites and springtails predominate in the rhizosphere and feed on

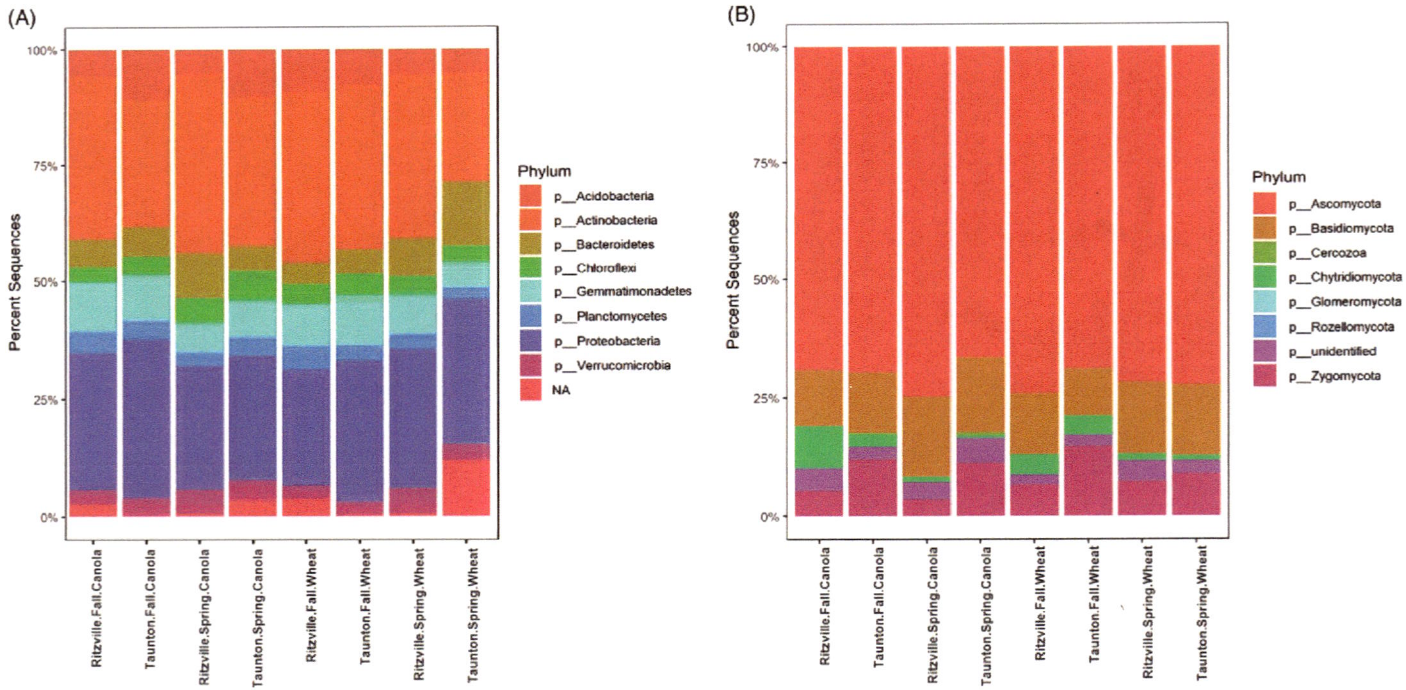

FIGURE 11.10 Relative abundance (%) of rhizosphere bacterial (A) and fungal (B) phyla in different seasons (Fall, Spring) and soils (Ritzville, Taunton) of semi-arid field-grown canola and wheat. *Source: From Schlatter et al. (2019).*

woody tissue, leaf tissue, fungi, bacteria, or algae. They follow plant roots for their available food supply. Like protozoa, mites and springtails can feed selectively on certain plants and microbes. The high populations of bacteria and fungi in the rhizosphere encourage the buildup of many of these faunal species that graze on microorganisms. The faunal component is often overlooked in rhizosphere studies, and yet it is a dynamic, integral factor in rhizosphere functioning.

Movement of microflora in the rhizosphere

Dispersal

Microorganisms can move through the soil along the root as they colonize the rhizosphere. Microorganisms are thought to move along the root by three different mechanisms:

- Motility of the organism (active transport)
- Water movement on the surface of the root (passive transport)
- Movement of microorganisms on the root apex as the root cells elongate (passive transport)

Although bacteria inoculated on the seed will colonize roots within a few centimeters of the seed in the absence of percolating water, their further dispersal along the roots is increased by water movement through the soil.

Root colonization

A number of biotic and abiotic factors affect the ability of an organism to colonize the rhizosphere, or its **rhizosphere competence**. A microorganism with strong rhizosphere competence is a good candidate for use as a microbial inoculant. The biotic and abiotic factors involved in determining rhizosphere competence include the microorganisms themselves, the plant, and the environment (Table 11.6). Although most scientists report colonization on a per root length or root dry mass basis, electron micrographs show that microorganisms occupy only 7%–15% of the root surface. For example, rhizoplane bacteria are typically scattered across the root in small colonies associated with only a small portion of the root surface (Figs. 11.11 and 11.12). Nor are these colonies uniformly scattered along the root but rather are usually associated with holes, tears, crevices, and junctions of epidermal cells. The greatest colonization (approximately 10^7 CFU cm^{-2}) is found behind the root tip and on older root surfaces, with the least ($<10^3$ CFU cm^{-2}) at the root tip. Bacteria may attach to the surface of the root cell wall, whereas fungal hyphae attach to the surface and also penetrate the wall. Recognition signals may play an important role in root colonization. For example, bacteria that agglutinate root exudates can better follow the growth of the root in the soil and colonize the rhizosphere than those with no agglutination activity. In such a case where microbes have high agglutination activity, transport (active or passive) is not a problem. There are, however, many cases where neither bacteria nor fungi applied as an inoculant to the seed are transported along the root surface. Thus, transport of microorganisms in the rhizosphere can be highly variable and depends on the microorganism, the plant, and the environment.

TABLE 11.6 Factors influencing rhizosphere competence.

Microorganisms	
	Chemotaxis toward root exudates
	Nutritional versatility
	Early growth rate
	Cellulase production
	Antibiotic tolerance or production
	Siderophore production
	Unique physiological attributes
	Tolerance to fungicides or other chemicals
Plants	
	Species
	Age
	Altered genetics
	Foliar treatments
	Abundance of unoccupied binding sites or niches
Soil	
	Type
	Texture
	Moisture
	Atmosphere
	Temperature
	Fertility

From Kennedy (2005).

Microbe-plant interactions in the rhizosphere

The growth of a root through the soil increases the abundance of many microorganisms. Of interest and possible use are the microbial interactions that benefit plant growth and regulate terrestrial nutrient cycles such as C and N cycling (Table 11.7). The most intensely studied of these interactions are N_2-fixation, mycorrhizal associations, plant growth promotion, decomposition, and nutrient cycling.

Microorganisms play a large part in the decomposition and subsequent mineralization of organic matter in the soil (Chapter 13). Their activity yields various compounds, including nitrates, phosphates, sulfates, carbon dioxide, and water. Microbial activities are greater in the rhizosphere than in non-rhizosphere or root-free soil. Much research focuses on how soil organisms, roots, and ecological interactions in the rhizosphere influence **ecosystem functions** related to water and nutrient cycling, especially when subjected to stressors such as drought, salinity, and nutrient deficiency (Williams and de Vries, 2020). This is increasingly important due to rising global water and nutrient scarcity, widespread soil degradation,

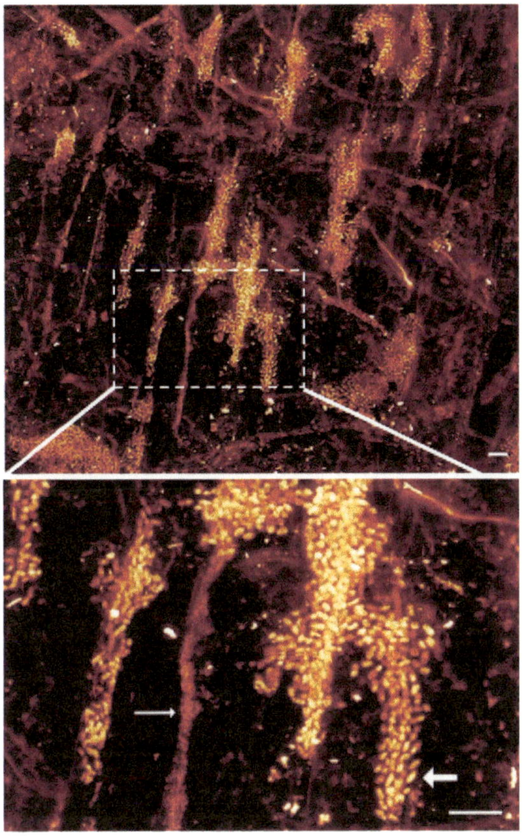

FIGURE 11.11 **Colonization of sugar beet root by native soil bacteria (stained with fluorescent probe).** Note microcolonies of small- *(thin arrow)* and large-sized *(thick arrow)* bacteria that occur in string-forming microcolonies along the epidermal root cells. Scale bar = 10 μm. *Source: From Lübeck et al. (2000). By permission of Federation of European Microbiological Societies.*

greater frequency of extreme weather events and disturbances, and the need for remediating contaminated soils due to human activities. To face the challenges of a changing climate and growing world population, researchers continue to produce a vast body of knowledge regarding rhizosphere microbes and their role in ecosystem functioning and plant stress responses.

Mycorrhizal fungi form symbiotic associations with plant roots and are discussed in more detail in Chapter 12. They are typically regarded as mutualistic symbionts with plants, although the direction (beneficial vs. parasitic) and strength of their relationship with host plants depends on environmental conditions and availability of resources such as light and nutrients. Mycorrhizae are involved in nutrient cycling processes, especially in stressed environments (e.g., phosphorus- and water-deficient soils). Mycorrhizal associations enhance nutrient solubilization and nutrient uptake in the rhizosphere and expand the volume of soil the root can explore for water and nutrients. Potential benefits to plants include increased rate of nutrient absorption, selective ion uptake, and protection from environmental extremes. This relationship may also impart some protection to plant roots from pathogens. Mycorrhizal

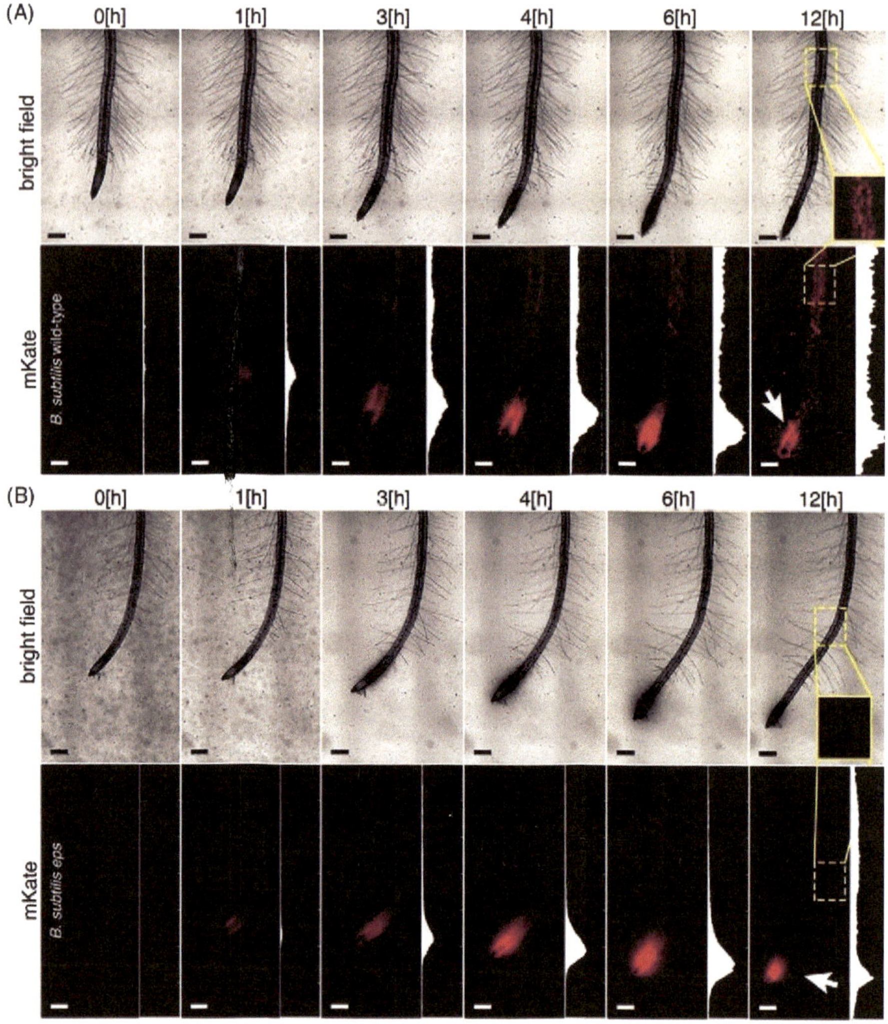

FIGURE 11.12 **Colonization of growing *Arabidopsis thaliana* roots by *Bacillus subtilis*.** (A) Selected images from time-lapse confocal microscopy of a root incubated with red fluorescent-labeled (mKate), wild-type (WT) *B. subtilis* for 12 h. (B) A *B. subtilis* biofilm defective mutant (eps) shows similar attraction as the WT strain but with little or no subsequent biofilm formation. The *white area plots* (on the right side of the mKate fluorescent images) show normalized bacterial quantification. Higher magnification view of the area marked by a *dashed box* in the mKate view at 12 h shows biofilm formation at a mature part of the root. Note greatest bacterial abundance behind actively growing root tip. Although this growth was similar for both WT and mutant bacteria, the mutant was less capable of colonizing mature root surfaces. *White arrows* in A and B indicate lower bacterial accumulation, likely because of fewer motile cells at 12 h (compared with the same area at 6 h). Scale bars = 200 μm. *Source: From Massalha et al. (2017).*

TABLE 11.7 Summary of beneficial activities of microorganisms in the rhizosphere.

Decompose plant residues and organic materials	
	Produce and consume organic matter
	Mineralize organic nitrogen, sulfur, and phosphorus
Increase plant nutrient availability of phosphorus, manganese, iron, zinc, and copper	
	Increase root area (symbiotic mycorrhiza)
	Produce siderophores
	Cause oxidation-reduction reactions
	Solubilize phosphorus
Biologically fix dinitrogen (symbiotic and asymbiotic bacteria)	
	Promote plant growth
	Produce plant growth hormones
	Protect against root pathogens
	Enhance nutrient use efficiency
Promote biological control	
Increase biodegradation of synthetic pesticides or contaminants	
Enhance plant drought tolerance	
Improve soil aggregation	

From Kennedy (2005).

associations are enhanced by crop rotation and management practices favoring minimum disturbance.

Plant-growth-promoting rhizobacteria (PGPR) are specific bacteria that colonize the rhizosphere and can enhance seed germination and plant growth (Chapter 23). Many different mechanisms are responsible for plant growth promotion. Many PGPR have been used in biocontrol to protect against plant pathogens, biofertilization to increase P solubilization and fix atmospheric dinitrogen asymbiotically, and phytostimulation, which directly improves plant growth through the production of hormones. Use of PGPR has the potential to reduce application of inorganic fertilizers and pesticides, and thus reduce pollution. For example, *Azotobacter* and *Azospirillum*, as well as other bacterial genera (e.g., *Bacillus*), produce plant growth-stimulating hormones such as gibberellic and indoleacetic acid. Plant growth promotion may also be an indirect consequence of reduced pathogen colonization of the seed and root due to antagonistic activities of PGPR that increase their competitive ability in the rhizosphere. Some PGPR have shown promise as biocontrol agents due to antibiotic production to deter pathogens or by triggering "Induced Systemic Resistance" (ISR) in plants to stimulate plant defenses against pathogens (Beneduzi et al., 2012). Iron-chelating compounds called **siderophores** produced by some microbes can make iron more available to a PGPR and less available to a plant pathogen. Because of the role iron plays in cellular metabolism (e.g., as a co-factor in some enzymes), this siderophore production may affect microbial growth.

PGPR may be exophytic or endophytic. The microbial populations that colonize the interior of the root and form intimate associations with the root are considered endophytes.

Endophytic bacteria may increase plant growth, confer disease resistance, and aid the plant in withstanding stresses such as drought. Populations of endophytic bacteria can be introduced and manipulated in agricultural systems through the use of plant tissue culture, crop rotations, tillage systems, and chemical inputs that encourage bacterial growth and additionally through cultivar selection, seed treatments, and genetic modification. A key to the success of endophytes may be the ability of the organism to compete with other bacteria on the external or internal portions of plant roots. Bacteria are thought to increase mycorrhizal symbiotic efficiency. Inoculation with a helper bacterium (e.g., *Pseudomonas fluorescens*) may aid mycorrhizal colonization of tree seedlings and increase plant growth. Introduction of PGPR often modifies the overall composition of the microbial community, especially in the presence of plants, due to enhanced competitive ability of PGPR (Mawarda et al., 2020).

Dinitrogen-fixing associations can add otherwise unavailable nitrogen to the plant (Chapter 17). Symbiotic rhizobia and related bacteria form nodules on the roots of plants and transform atmospheric N_2 to plant-available ammonia. The plant provides nutrients and a competition-free home for the bacteria. Associative N_2-fixing bacteria (Chapter 16) colonizing nonlegumes can be classified into three groups: (1) rhizosphere organisms, such as *Azotobacter paspali*; (2) facultative endophytes that colonize the rhizosphere or root interior of forage grasses and cereals, such as *Azospirillum* spp.; and (3) obligate endophytes that occur within plant tissues, such as *Gluconacetobacter diazotrophicus*, *Azoarcus* spp., *Herbaspirillum* spp., and *Burkholderia* spp. In low nitrogen soils, such as agricultural systems in the tropics, these organisms may contribute significant amounts of nitrogen.

Plant growth-inhibiting microorganisms (Chapter 23) may colonize the rhizosphere and affect plant development in several ways. The traditional plant pathogen attacks plant tissue and causes disease symptoms. Some plant pathogens may parasitize root cells without causing distinct disease symptoms. In other cases, organisms do not parasitize the plant but instead colonize the rhizosphere or rhizoplane and produce compounds that may harm root development and inhibit plant growth. Their detrimental effects on plant growth are due mainly to the production of metabolites that disturb plant physiological processes and include phytotoxins, plant growth regulators, volatile substances, and antibiotics. In some cases, these plant-microbial interactions are host-specific enough to have potential for biocontrol suppression of weeds.

The principal interface for many plant-microbe interactions is the rhizosphere. The dynamics of these interactions will determine the health of the plant. Microorganisms inhabiting the rhizosphere, such as pathogens or microorganisms that produce phytotoxins, may be detrimental to plant health. The challenge is to enhance beneficial relationships and minimize harmful interactions.

Inoculants

Recognizing the impact that microorganisms have on plant growth has spurred decades of research and development of ways to manipulate rhizosphere community composition to enhance plant growth (Boxes 11.6, 11.7). Researchers are continually refining methods to identify and isolate new microbes for potential use in inoculants, improving strategies for deployment and inoculation success, and evaluating their effects on plant growth, soil microbial

BOX 11.6

Inoculation

Farmers have known for centuries about the benefits of adding a small amount of soil from a legume field to the seed or soil of another leguminous crop. *Rhizobium* inoculants have been successfully used for more than 100 years, and attempts to improve inoculation technology for PGPR is ongoing. For inoculation to be successful, the organism must be consistently delivered, and it must be sufficiently competitive to survive and persist in the soil, rhizosphere, and rhizoplane. Care must also be taken to avoid laboratory acculturation, where repeated laboratory culturing reduces the field survivability of a microorganism, much in the same way a pampered house pet would not have what it takes to survive in the wilds of Africa. Because inoculants contain live organisms, appropriate storage and handling is also critical to ensure survival and efficacy. Inoculant carriers such as peat, alginate, clay, plant bioproducts, or synthetic polymers can increase the likelihood of successful establishment. Co-inoculation or mixtures of organisms may have a synergistic effect by providing protection, increasing nutrients, or stimulating microbial growth.

BOX 11.7

Engineering the rhizosphere microbiome

Many attempts are being made to develop tools and technologies that optimize beneficial microbiome composition and function in the rhizosphere, effectively "engineering" the rhizosphere microbiome. This includes novel identification, isolation, and delivery technologies for inoculants, as well as using gene editing tools to enhance desired traits in introduced organisms (Qiu et al., 2019). Another promising strategy is to use **host-mediated microbiome engineering (HMME)** to indirectly select for beneficial rhizosphere microbiomes using desired plant phenotypic traits as a guide.

In a laboratory-based study, rhizosphere inoculum was created from bermudagrass (*Cynodon* spp.) samples collected from a semi-arid environment subjected to frequent and extreme drought. This served as the starting inoculum that underwent several rounds of HMME in wheat (*Triticum aestivum* subsp. *aestivum* cultivar TAM111), where sterilized seeds were grown in pots containing the inoculum and subjected to extreme drought stress. Rhizosphere inoculum was collected from the best-performing plants and used in the next round of selection with new seeds. This process was repeated until no further improvements to drought resistance were observed. Applying a 10^{-1}or 10^{-2} dilution of the final HMME inoculum conferred this drought resistance to new seedlings, lending support to the idea that rhizosphere microbiomes can be effectively engineered to help plants adapt to stressful conditions such as drought (Jochum et al., 2019).

(*Continued*)

BOX 11.7 (Cont.)

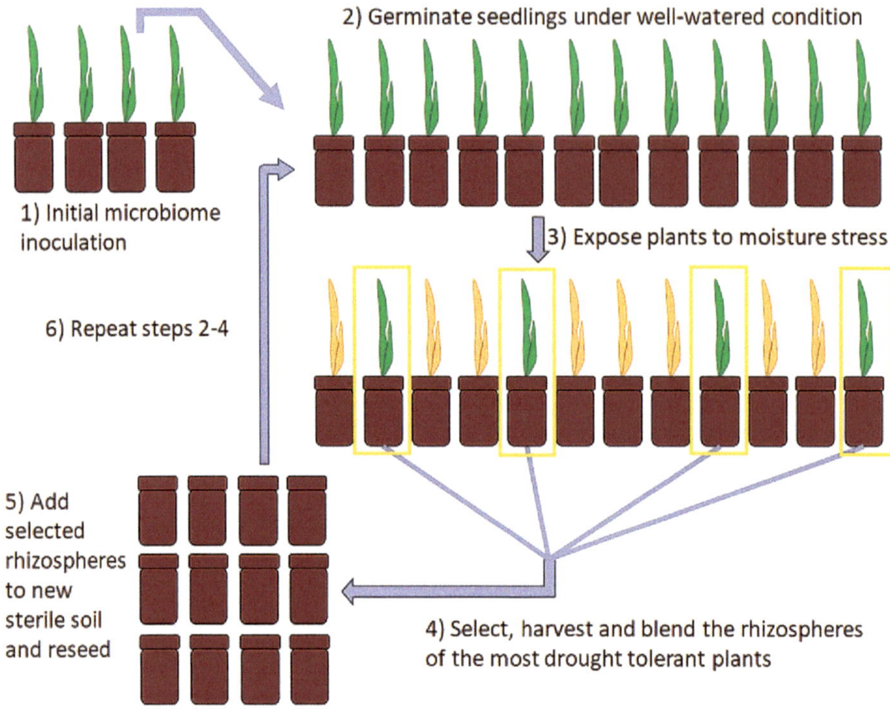

Concept of host mediated microbiome engineering (HMME) for drought tolerance. *Source: From Jochum et al. (2019).*

communities, and the environment (Martínez-Hidalgo et al., 2018). Species of N_2-fixing bacteria, a wide range of PGPR, mycorrhizal fungi, and bacterial and fungal biological control agents are often evaluated and developed as commercial microbial inoculants (Table 11.8). Most inoculants are developed to enhance agricultural productivity through plant growth promotion or pest control and also to improve soil properties such as aggregate stability or water and nutrient dynamics. Introduced microbes may also be used to enhance restoration of degraded land or to remediate contaminants or toxins in soil (Chapter 22). Inoculation will only be successful, however, if the microorganisms flourish when they are introduced into the soil or on the seed. When introduced with seeds, the subsequent exposure of the microorganism to exuded nutrients will enhance microbial growth and establishment of the organism. Yet, introduced microorganisms face several challenges to successful inoculation, including compatibility with plant hosts, competition with existing microbial populations, and suitability for survival in the soil habitat into which they are introduced. Further, successful inoculation and plant growth promotion does not necessarily mean that inoculated organisms will persist in soils and lead to beneficial changes in soil properties or ecosystem

TABLE 11.8 Microorganisms with use or potential as inoculants for the rhizosphere.

Benefit	Example organisms
Nutrient availability, uptake	*Rhizobium* spp., *Bradyrhizobium* spp., *Glomus intraradices, Glomus mosseae, Glomus* spp., *Rhizophagus* spp., *Azotobacter* spp., *Azospirillum brasilense, Bacillus* spp., *Burkholderia* spp., *Pseudomonas syringae* pv. *Tabaci, Trichoderma virens, Penicillium bilaii*
Drought tolerance	*Glomus* spp., *Pseudomonas putida, Pantoea aglomerans, Trichoderma harzianum*
Pathogen control	*Achromobacter xylosoxidans, Agrobacterium tumefaciens, Alcaligenes* spp., *Bacillus* spp., *Bacillus subtilis, Bacillus amyloliquefaciens, Pseudomonas* spp., *Pseudomonas fluorescens, Pseudomonas chlororaphis, Streptomyces* spp., *Trichoderma harzianum*
Pest control	*Bacillus thuringiensis, Bacillus firmus, Pseudomonas fluorescens, Beauveria bassiana, Klebsiella pneumoniae*
Weed control	*Pseudomonas putida, Stenotrophomonas maltophilia, Enterobacter taylorae, Fusarium oxysporum* f. sp. *strigae*

functioning. Some introduced microbes may be excellent promoters of plant growth under controlled conditions (e.g., in a greenhouse) but produce no effects in field soils, while others may have short- or long-term effects on soil fertility and microbial diversity depending on the environmental context and manner in which they are applied. The overall effects of inoculated organisms for agricultural use on soil ecosystems, and the mechanisms and contexts that regulate these ecosystem outcomes, remain unclear.

Understanding and modeling rhizosphere function

Studying and modeling the spatial dynamics, metabolism, and growth of microbes in the rhizosphere and their interactions with plant roots is a much-needed area of research given the impacts of rhizosphere microbes on soil structural development, root growth, and water and nutrient dynamics. Creating conceptual and mathematical models of the rhizosphere based on research findings allows us to better understand and predict the complex interactions in the rhizosphere. However, this is extremely challenging given the size of many plants and the spatial and temporal heterogeneity of the rhizosphere. Furthermore, many of the characterization techniques are destructive (preventing repeated measurements) and suffer from limited resolution (i.e., do not measure at microbially relevant scales). For rhizosphere projects in the field, researchers typically excavate plant roots and separate the bulk soil from rhizosphere and endorhizosphere components using shaking and washing methods. This captures many of the plant roots, but deeper roots (>25 cm) are often not obtained due to the difficulty in collection, and very fine roots can be lost in processing. The large variation in soil and environmental conditions at the field-scale can also mask any treatment-specific impacts on the rhizosphere. For more controlled conditions, researchers often use hydroponic or soil-based microcosms in greenhouses or growth chambers to collect and characterize rhizosphere soil or root exudates using a variety of destructive or non-destructive means. Microcosms can also be constructed using membranes that separate root mats or rhizosphere components from the surrounding soil. A common method in rhizosphere ecology that can be adapted for both lab and field settings is the use of rhizoboxes or rhizotrons, which are constructed using flat boxes or half-pipes that are held at about a 45 degrees angle and often re-packed with soil or artificial media, which allow monitoring and measurement of root growth characteristics (Schmidt et al., 2018). Rhizoboxes can be customized during construction to facilitate use of additional imaging and sampling techniques.

Recent methodological developments that use combined chemical, isotopic, or microbial imaging techniques provide new ways to study rhizosphere processes and microbial functioning in controlled settings (Oburger and Schmidt, 2016). For example, using microfluidic approaches combined with imaging techniques allow researchers to make real-time spatial and temporal observations of root-microbe interactions in lab-grown seedlings (Massalha et al., 2017). Advances in digital imaging to characterize excavated roots, or in-situ scanning of roots in soil using X-ray computed tomography (CT) or magnetic resonance imaging (MRI), has helped researchers investigate and understand how root growth and root system architecture respond to environmental conditions and soil biota in field soils, although it is not yet possible to image root-microbe interactions at the microscopic scale using these techniques. Using data collected from lab- and field-based studies to construct mathematical models at both the macroscopic and microscopic scales enables researchers to model specific properties or processes such as root architecture and growth dynamics or chemical and water movement in the rhizosphere. Greater use of image-based modeling has further strengthened our ability to examine and predict rhizosphere processes and interactions (Roose et al., 2016). Modeling may assist investigations on the overlapping boundaries of the various regions of the rhizosphere, which may lead to further understanding in delineating these controversial, complex regions. Studies of rhizosphere ecology can use these models as a tool for testing subsequent hypotheses about the zone of soil around the root.

Summary

The rhizosphere is the volume of soil influenced by the root, which is an adaptive organ that both affects and is affected by the physical, chemical, and biological characteristics of the surrounding soil. Organic materials are released from the root or seed that alter microbial diversity and increase numbers of organisms, microbial activity, and interactions among microorganisms, the root, and the soil. The microbial community in the rhizosphere can influence plant growth in beneficial, neutral, or detrimental ways, ultimately influencing the health, vigor, and productivity of the plant. The complex interactions between microbes and roots in the rhizosphere regulate processes such as nutrient cycling and plant responses to stresses such as drought. Given the small size, relative to the surrounding bulk soil, and diffuse nature of plant rhizospheres, microbial ecology in the rhizosphere holds disproportionately large implications for soil properties and ecosystem functioning.

Acknowledgments

Portions of this chapter, with permission, are derived from the work of Dr. Ann C. Kennedy who authored a similar chapter for the Second Edition of the textbook. The current author and editors thank her for permitting the use of this material and for graciously supporting production of the Third Edition. Dr. Elizabeth Pierson provided critical direction during chapter revision and contributed material regarding root structure and environmental influence.

Study questions

1. Why is the rhizosphere a zone of increased microbial activity?
2. Why is the rhizosphere an important area of study?
3. What role do exudates play in the rhizosphere?
4. Distinguish among rhizosphere, rhizoplane, and root.
5. How does microbial activity compare among the rhizosphere, rhizoplane, and soil?
6. By what mechanisms can bacterial populations be beneficial, detrimental, or neutral to plant growth?
7. What factors affect root colonization?

References

Benard, P., Zarebanadkouki, M., Brax, M., Kaltenbach, R., Jerjen, I., Marone, F., Couradeau, E., Felde, V.J.M.N.L., Kaestner, A., Carminati, A., 2019. Microhydrological niches in soils: how mucilage and EPS alter the biophysical properties of the rhizosphere and other biological hotspots. Vadose Zone J. 18, 180211.

Beneduzi, A., Ambrosini, A., Passaglia, L.M.P., 2012. Plant growth-promoting rhizobacteria (PGPR): their potential as antagonists and biocontrol agents. Genet. Mol. Biol. 35, 1044–1051.

Berg, G., Smalla, K., 2009. Plant species and soil type cooperatively shape the structure and function of microbial communities in the rhizosphere. FEMS Microbiol. Ecol. 68, 1–13.

Bergmann, D., Zehfus, M., Zierer, L., Smith, B., Gable, M., 2009. Grass rhizosheaths: associated bacterial communities and potential for nitrogen fixation. West. N. Am. Nat. 69, 105–114.

Carminati, A., 2013. Rhizosphere wettability decreases with root age: a problem or a strategy to increase water uptake of young roots? Front. Plant Sci. 4, 298.

de la Porte, A., Schmidt, R., Yergeau, É., Constant, P., 2020. A gaseous milieu: extending the boundaries of the rhizosphere. Trends Microbiol. 28, 536–542.

Doornbos, R.F., van Loon, L.C., Bakker, P.A.H.M., 2012. Impact of root exudates and plant defense signaling on bacterial communities in the rhizosphere. A review. Agron. Sustain. Dev. 32, 227–243.

Hassani, M.A., Durán, P., Hacquard, S., 2018. Microbial interactions within the plant holobiont. Microbiome 6, 58.

Hawes, M., Allen, C., Turgeon, B.G., Curlango-Rivera, G., Tran, T.M., Huskey, D.A., Xiong, Z., 2016. Root border cells and their role in plant defense. Annu. Rev. Phytopathol. 54, 143–161.

Hiltner, L., 1904. Ober neuere Erfahrungen und Probleme auf dem Gebiet der Bodenbakteriologie und unter besonderer Beriicksichtigung der Griindiingung und Brache. Arb. Dtsch. Landwirtsch. Ges. Berlin 98, 59–78.

Hinsinger, P., Gobran, G.R., Gregory, P.J., Wenzel, W.W., 2005. Rhizosphere geometry and heterogeneity arising from root-mediated physical and chemical processes. New Phytol. 168, 293–303.

Jochum, M.D., McWilliams, K.L., Pierson, E.A., Jo, Y.K., 2019. Host-mediated microbiome engineering (HMME) of drought tolerance in the wheat rhizosphere. PLoS One 14, e0225933.

Kennedy, A.C., 2005. Rhizosphere. In: Sylvia, D.M., Hartel, P.G., Fuhrmann, J.J., Zuberer, D.A. (Eds.), Principles and Applications of Soil Microbiology, second ed. Pearson Prentice Hall, Upper Saddle River, NJ.

Kong, H.G., Kim, N.H., Lee, S.Y., Lee, S.-W., 2016. Impact of a recombinant biocontrol bacterium, *Pseudomonas fluorescens* pc78, on microbial community in tomato rhizosphere. Plant Pathol. J. 32, 136–144.

Kuzyakov, Y., Razavi, B.S., 2019. Rhizosphere size and shape: Temporal dynamics and spatial stationarity. Soil Biol. Biochem. 135, 343–360.

Lemanceau, P., Barret, M., Mazurier, S., Mondy, S., Pivato, B., Fort, T., Vacher, C., 2017. Chapter Five—Plant communication with associated microbiota in the spermosphere, rhizosphere and phyllosphere. In: Becard, G. (Ed.), Advances in Botanical Research, vol. 82, Academic Press, New York, pp. 101–133.

Lübeck, P.S., Hansen, M., Sørensen, J., 2000. Simultaneous detection of the establishment of seed-inoculated *Pseudomonas fluorescens* strain DR54 and native soil bacteria on sugar beet root surfaces using fluorescence antibody and in situ hybridization techniques. FEMS Microbiol. Ecol. 33, 11–19.

Lynch, J.M., 1990. The rhizosphere. John Wiley and Sons, Chichester, UK.

Mandal, A., Sarkar, B., Owens, G., Thakur, J.K., Manna, M.C., Niazi, N.K., Jayaraman, S., Patra, A.K., 2020. Impact of genetically modified crops on rhizosphere microorganisms and processes: A review focusing on Bt cotton. Appl. Soil Ecol. 148, 103492.

Marschner, H., Römheld, V., 1983. In vivo measurement of root-induced pH changes at the soil-root interface: effect of plant species and nitrogen source. Z. Pflanzenphysiol. 111, 241–251.

Martínez-Hidalgo, P., Maymon, M., Pule-Meulenberg, F., Hirsch, A.M., 2018. Engineering root microbiomes for healthier crops and soils using beneficial, environmentally safe bacteria. Can. J. Microbiol. 65, 91–104.

Massalha, H., Korenblum, E., Malitsky, S., Shapiro, O.H., Aharoni, A., 2017. Live imaging of root–bacteria interactions in a microfluidics setup. Proc. Natl. Acad. Sci. 114, 4549–4554.

Mawarda, P.C., Le Roux, X., Dirk van Elsas, J., Salles, J.F., 2020. Deliberate introduction of invisible invaders: A critical appraisal of the impact of microbial inoculants on soil microbial communities. Soil Biol. Biochem. 148, 107874.

McPherson, M.R., Wang, P., Marsh, E.L., Mitchell, R.B., Schachtman, D.P., 2018. Isolation and analysis of microbial communities in soil, rhizosphere, and roots in perennial grass experiments. J. Vis. Exp. 137, 57932.

Oburger, E., Schmidt, H., 2016. New methods to unravel rhizosphere processes. Trends Plant Sci. 21, 243–255.

Odell, R.E., Dumlao, M.R., Samar, D., Silk, W.K., 2008. Stage-dependent border cell and carbon flow from roots to rhizosphere. Am. J. Bot. 95, 441–446.

Papendick, R.I., Campbell, G.S., 1975. Water potential in the rhizosphere and plant and methods of measurement and experimental control. In: Bruehl, G.W. (Ed.), Biology and Control of Soil-borne Pathogens. The American Phytopathological Society, St. Paul, MN, pp. 39–49.

Philippot, L., Raaijmakers, J.M., Lemanceau, P., van der Putten, W.H., 2013. Going back to the roots: the microbial ecology of the rhizosphere. Nat. Rev. Microbiol. 11, 789–799.

Qiu, Z., Egidi, E., Liu, H., Kaur, S., Singh, B.K., 2019. New frontiers in agriculture productivity: Optimised microbial inoculants and in situ microbiome engineering. Biotechnol. Adv. 37, 107371.

Rojas, X., Guo, J., Leff, J.W., McNear, D.H., Fierer, N., McCulley, R.L., 2016. Infection with a shoot-specific fungal endophyte (*Epichloë*) alters tall fescue soil microbial communities. Microb. Ecol. 72, 197–206.

Roose, T., Keyes, S.D., Daly, K.R., Carminati, A., Otten, W., Vetterlein, D., Peth, S., 2016. Challenges in imaging and predictive modeling of rhizosphere processes. Plant Soil 407, 9–38.

Rouatt, J.W., Katznelson, H., Payne, T.M.B., 1960. Statistical evaluation of the rhizosphere effect. Soil Sci. Soc. Am. Proc. 24, 271–273.

Rovira, A.D., Foster, R.C., Martin, J.K., 1979. Note on terminology: Origin, nature and nomenclature of the organic materials in the rhizosphere. In: Harley, J.L., Russell, R.S. (Eds.), The Soil-Root Interface. Academic Press, London, pp. 1–4.

Sasse, J., Martinoia, E., Northen, T., 2018. Feed your friends: do plant exudates shape the root microbiome? Trends Plant Sci. 23, 25–41.

Schlatter, D.C., Hansen, J.C., Schillinger, W.F., Sullivan, T.S., Paulitz, T.C., 2019. Common and unique rhizosphere microbial communities of wheat and canola in a semiarid Mediterranean environment. Appl. Soil Ecol. 144, 170–181.

Schmidt, J.E., Lowry, C., Gaudin, A.C.M., 2018. An optimized rhizobox protocol to visualize root growth and responsiveness to localized nutrients. J. Vis. Exp. 140, e58674.

Schlub, R.L., Schmitthenner, A.F., 1978. Effects of soybean seed coat cracks on seed exudation and seedling quality in soil infested with *Pythium ultimum*. Phytopathology 68, 1186–1191.

Van Deynze, A., Zamora, P., Delaux, P.-M., Heitmann, C., Jayaraman, D., Rajasekar, S., Graham, D., Maeda, J., Gibson, D., Schwartz, K.D., Berry, A.M., Bhatnagar, S., Jospin, G., Darling, A., Jeannotte, R., Lopez, J., Weimer, B.C., Eisen, J.A., Shapiro, H.-Y., Ané, J.-M., Bennett, A.B., 2018. Nitrogen fixation in a landrace of maize is supported by a mucilage-associated diazotrophic microbiota. PLoS Biol. 16, e2006352.

Verona, O., 1958. The spermosphere. Ann. Inst. Pasteur. 95, 795–798.

Vives-Peris, V., de Ollas, C., Gómez-Cadenas, A., Pérez-Clemente, R.M., 2020. Root exudates: from plant to rhizosphere and beyond. Plant Cell Rep. 39, 3–17.

Williams, A., de Vries, F.T., 2020. Plant root exudation under drought: implications for ecosystem functioning. New Phytol. 225, 1899–1905.

Xioa, X., Chen, W., Zong, L., Yang, J., Jiao, S., Lin, Y., Wang, E., Wei, G., 2017. Two cultivated legume plants reveal the enrichment process of the microbiome in the rhizocompartments. Mol. Ecol. 26, 1641–1651.

York, L., Lobet, G., 2017. Phenomics of root system architecture: Measuring and analyzing root phenes. The Plant Cell 29 (9), https://academic.oup.com/plcell/article/29/9/tpc.117.tt0917/6100632.

Further reading

Cardon, Z.G., Whitbeck, J.L. (Eds.), 2007. The Rhizosphere: An Ecological Perspective. Elsevier Academic Press, London.

Curl, E.A., Truelove, B., 1986. The Rhizosphere. Springer-Verlag, Berlin.

Keister, D.L., Cregan, P.B., 1991. The Rhizosphere and Plant Growth. Kluwer Academic Publishers, Dordrecht, The Netherlands.

Parniske, M., 2008. Arbuscular mycorrhiza: the mother of plant root endosymbioses. Nat. Rev. Microbiol. 6, 763–775.

CHAPTER

12

Mycorrhizal symbioses

Rhae A. Drijber, Morgan R. McPherson

Department of Agronomy and Horticulture, University of Nebraska-Lincoln, Lincoln, NE, United States

"The study of plants without their mycorrhizas is the study of artefacts."

As the American plant pathologist, Steven Wilhelm, said:

"...Plants, strictly speaking, do not have roots; they have mycorrhizas."

La Banque Européene des Glomales (BEG) Committee, 25th May 1993

Mycorrhiza (*pl* -zae or -zas) refers to a variety of symbiotic associations between plants and fungi that colonize the cortical tissue of roots during periods of active plant growth. The association is ancient (~450 million year (Myr) ago; Strullu-Derrien et al., 2018) and paralleled the evolution of land plants in three distinct waves, the most recent and ongoing wave being linked to substantial climate change in the past 50 Myr coupled with plant diversification in biological "hotspots" (Brundrett and Tedersoo, 2018). Regardless of the type of mycorrhizal association, the symbiosis is characterized by the movement of (1) plant-produced carbon to the fungus and (2) fungal-acquired nutrients to the plant. The term **mycorrhiza** (literally meaning fungus-root) was first applied to fungus-tree associations described in 1885 by the German forest pathologist A.D. Frank. Since then we have learned that more than 80% of all land plant species form symbiotic associations with fungi (see Figure 1 in Brundrett and Tedersoo, 2018, https://doi.org/10.1111/nph.14976). Indeed, the mycorrhizal condition is the rule among plants, not the exception.

The benefits afforded plants from mycorrhizal symbioses can be characterized either agronomically (i.e., increased growth and yield) or ecologically (i.e., improved fitness or reproductive ability). In either case, the benefits accrue primarily because mycorrhizae form a critical linkage between plant roots and soil. Mycorrhizae usually have a proliferation of fungal biomass both in the root and in soil. The soilborne **extraradical hyphae** take up nutrients from the soil solution and transport them to the root thereby increasing the effective absorptive surface area of the plant root system and volume of soil explored. Another function attributed to mycorrhizal fungi is the ability to access pools of soil phosphorus (Chapter 19) not readily available to the plant (Miyasaka and Habte, 2001). Many mycorrhizae produce organic acids (e.g., oxalic and citric acids) that can liberate phosphate from phosphate-containing minerals

Principles and Applications of Soil Microbiology. http://dx.doi.org/10.1016/B978-0-12-820202-9.00012-5
Copyright © 2021 Elsevier Inc. All rights reserved.

or produce enzymes to hydrolyze P from soil organic matter. In nutrient-poor or moisture-deficient soils, nutrients taken up by the extraradical hyphae lead to improved plant growth and reproduction. As a result, mycorrhizal plants are often more competitive and better able to tolerate environmental stresses than are nonmycorrhizal plants.

Global perspective

Mycorrhizal associations vary widely in structure and function. There are four main types of mycorrhiza: arbuscular mycorrhizae (AM), ectomycorrhizae (EcM), ericoid mycorrhizae, and orchid mycorrhizae. These differ in host preference, extent of mutualism (obligate to weak saprophytes), morphology, habit preference, ecology and dominant function(s) (Table 12.1). As summarized by Brundrett and Tedersoo (2018) and others, over 80% of vascular plants

TABLE 12.1 Plant hosts, fungi, and important characteristics of major types of mycorrhizal associations.

	Endomycorrhizae (AM)	Ectomycorrhizae (EcM)	Ericoid	Orchid
Host plant	~78% vascular plants Angiosperms Gymnosperms Bryophytes Pteridophytes	Mostly woody plants Angiosperms Gymnosperms Some Liverworts	Ericaceae Epacridaceae Empetraceae Some Bryophytes Diaspensiaceae	All Orchidaceae
Fungal taxa	Glomeromycotina ~300–1600 species; Endogonomycetes	Basidiomycota and Ascomycota; ~20,000 species	Ascomycota ~150 species; some Basidiomycota	Basidiomycota ~25,000 species
Fungal structure(s)	Intracellular arbuscules; vesicles; axillary cells; aseptate multinucleate extraradical hyphae	Intercellular Hartig nets; sheathing mantle; rhizomorphs	Intracellular hyphal coils; some form undifferentiated mantle; septate extraradical hyphae	Intracellular hyphal pelotons or coils; septate extraradical hyphae
Fungal function(s)	Acquisition of nutrients for host plant; soil aggregation; source of soil C; plant stress resistance	Acquisition of nutrients for host plant; mineralization of soil organic matter; soil aggregation; source of soil C; plant protection	Mineralization of soil organic matter and subsequent transfer between plants; acquisition of nutrients for host plant	Supplies C and vitamins to plant embryo/ seedling; acquisition of nutrients for host plant
Nutrients acquired	Mostly P, some N; Poor C biodegradation capacity	N and P; moderate C biodegradation capacity	N and P; strong C biodegradation capacity	N and P, some C; strong C biodegradation capacity
Habitat	Tropical and temperate forests; grasslands; savannas; croplands; deserts	Boreal, temperate and tropical forests; tundra; N-poor soils; agroforestry	Boreal and temperate forests; tundra; highly acidic soils	Mainly tropical and subtropical forests; some temperate forests; includes epiphytic habitats

Information for table compiled from multiple sources (Sylvia, 2005; Brundrett and Tedersoo, 2018; Johnson and Jansa, 2017; Tedersoo et al., 2020; van der Heijden et al., 2015).

are mycorrhizal: 71% are AM, 2.0% are EcM, 1.4% are ericoid, and 10% are orchid. Only 8% are completely nonmycorrhizal (NM) and are confined to a few plant families such as mustards and cabbages (family *Brassicaceae*), succulent stonecrops (family *Crassulaceae*), parasitic broomrapes (family *Orobanchaceae*), and flowering shrubs and trees with short proteoid roots (family *Proteaceae*). Approximately 7% of plants have variable NM–AM associations and function as nutritional specialists (carnivores, e.g., Venus fly traps) and parasites (e.g., Mistletoe) or habitat specialists (e.g., hydrophytes and epiphytes) (see Figure 1 in Brundrett and Tedersoo, 2018, https://doi.org/10.1111/nph.14976).

The global distribution of mycorrhizal plants varies strongly by type across biomes (Fig. 12.1; Soudzilovskaia et al., 2019). In Read's model of the biogeography of mycorrhizas in the Northern Hemisphere (Read and Perez, 2003), nutrient stoichiometry plays an important role along with climate, vegetation, and soil processes. Note the wide global dominance of AM plants (Fig. 12.1A) while EcM plants, and to a lesser extent ericaceous plants, dominate mainly northern latitudes (Fig. 12.1 B,C).

Regions dominated by grasslands, deciduous trees, and shrubs have high percentages of AM plants (Fig. 12.1A) and are characterized by arid or semi-arid climates. In these ecosystems nutrient use is high and phosphorus is frequently a limiting nutrient for growth. Grasses (as well as a wide variety of other plants, including major food crops) are colonized by fungi that form arbuscules (highly branched structures—the term literally means little trees (Fig. 12.2A,B) *within* (i.e., intracellular) root cortical cells, giving rise to the **arbuscular**

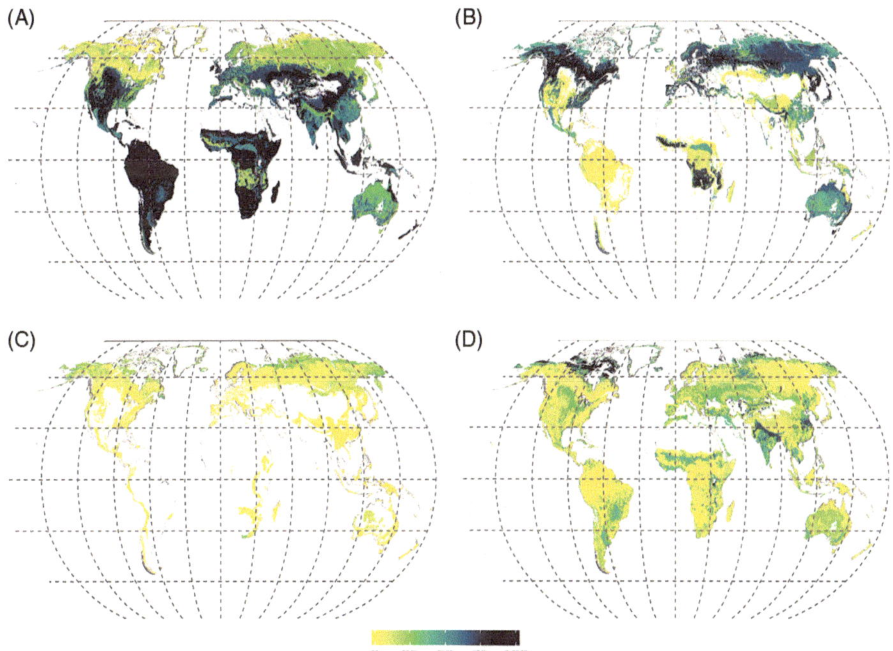

FIGURE 12.1 **Percentage of aboveground plant biomass of mycorrhizal vegetation.** (A) Arbuscular mycorrhizal plants, (B) ectomycorrhizal plants, (C) ericoid mycorrhizal plants, and (D) non-mycorrhizal plants. The map resolution is 10 arcmin. *Source: Figure from Soudzilovskaia et al. (2019). This article is licensed under a Creative Commons Attribution 4.0 International License http://creativecommons.org/licenses/by/4.0/.*

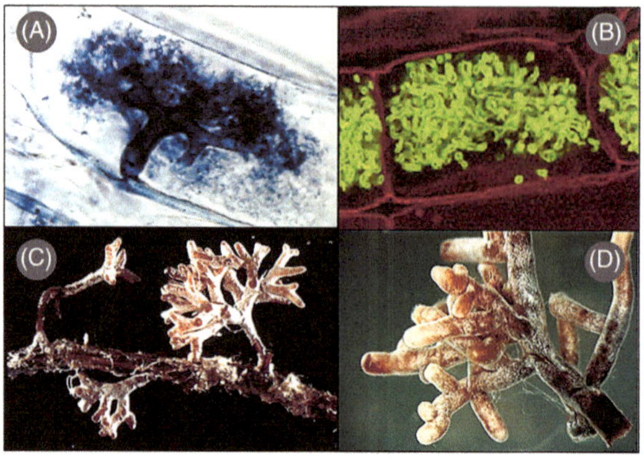

FIGURE 12.2 (A) Stained root section showing the "tree-like" form of an arbuscule; (B) Fluorescence photomicrograph of an arbuscule in a root of *Medicago truncatula* colonized by *Glomus versiforme*. (C, D) Examples of ectomycorrhizal root tips, 1–2 mm in diameter. *Source: Figure (A) provided by D. Zuberer; Figure (B) courtesy of Dr. Maria Harrison, Boyce Thompson Institute, used with permission; Figure (C) by Randy Molina, Public Domain, http://www.fs.fed.us/pnw/mycology/photo/content7.html, and Figure (D) by Ellen Larsson/CC BY (https://creativecommons.org/licenses/by/2.5).*

type of mycorrhiza (sometimes called endomycorrhizae). These fungi may produce extensive extramatrical hyphae to significantly increase uptake of phosphorus and other limiting nutrients by the plants they colonize. In addition to providing carbon to the soil through hyphal necromass, AM fungi also produce glomalin-related soil proteins (GRSPs), an important stabilizer of soil aggregates.

As we move into northern latitudes, coniferous trees replace grasslands and EcM fungi become dominant (Fig. 12.1B). These trees are colonized by a wide range of mostly basidiomycetous fungi (Chapter 6) that grow *between* (i.e., intercellular) root cortical cells forming the **ectomycorrhizal** type of mycorrhiza (Fig. 12.2C,D). Ectomycorrhizal fungi may produce large quantities of hyphae on the root and in soil, providing a major source of belowground soil carbon (Soudzilovskaia et al., 2019). Similar to AM hyphae, these hyphae function in the absorption and translocation of inorganic nutrients and water but also release nutrients from litter layers by production of enzymes involved in mineralization of organic matter.

Ericaceous plants, found in high-organic soils of subarctic and subalpine heathlands (Fig. 12.1 C), are colonized by a group of ascomycetous fungi (Chapter 6) giving rise to the **ericoid** type of mycorrhiza. This mycorrhizal type is characterized by extensive growth *within* (i.e., intracellular) cortical cells, but less hyphal extension into the soil compared to other mycorrhizal types. The fungus produces extracellular enzymes that mineralize organic matter, allowing the plant to access nutrients tied up in the organic materials surrounding the root. The Orchid family is one of the largest in the plant kingdom with more than 20,000 species identified globally, primarily in tropical and subtropical forests where they often form part of the epiphytic community. All orchids, whether they contain chlorophyll or not, have a growth stage where they are non-photosynthetic (notably the seedling stage) and therefore are dependent on external sources of sugars and nutrients. Orchids are colonized exclusively by Basidiomycetes, often from the genus *Rhizoctonia*. Although globally ubiquitous, nonmycorrhizal plants (Fig. 12.1 D) are more common and diverse in arid and degraded habitats as well as arctic and alpine regions (Brundrett and Tedersoo, 2018). The diversity of these mycorrhizal associations affords plants a range of strategies for efficient functioning in an array of plant-soil systems.

Mycorrhizal types

Ectomycorrhizae (EcM)

The diagnostic feature of EcM is the presence of hyphae between the root cortical cells producing a net-like structure termed the **Hartig net,** after Robert Hartig who is considered the "founder of forest biology" (Fig. 12.3A). Many EcM also have a sheath or **mantle** of fungal tissue that may completely cover the absorbing root (usually the fine feeder roots). The mantle can vary widely in thickness, color, and texture depending on the particular plant-fungus combination. The mantle increases the surface area of absorbing roots and often affects

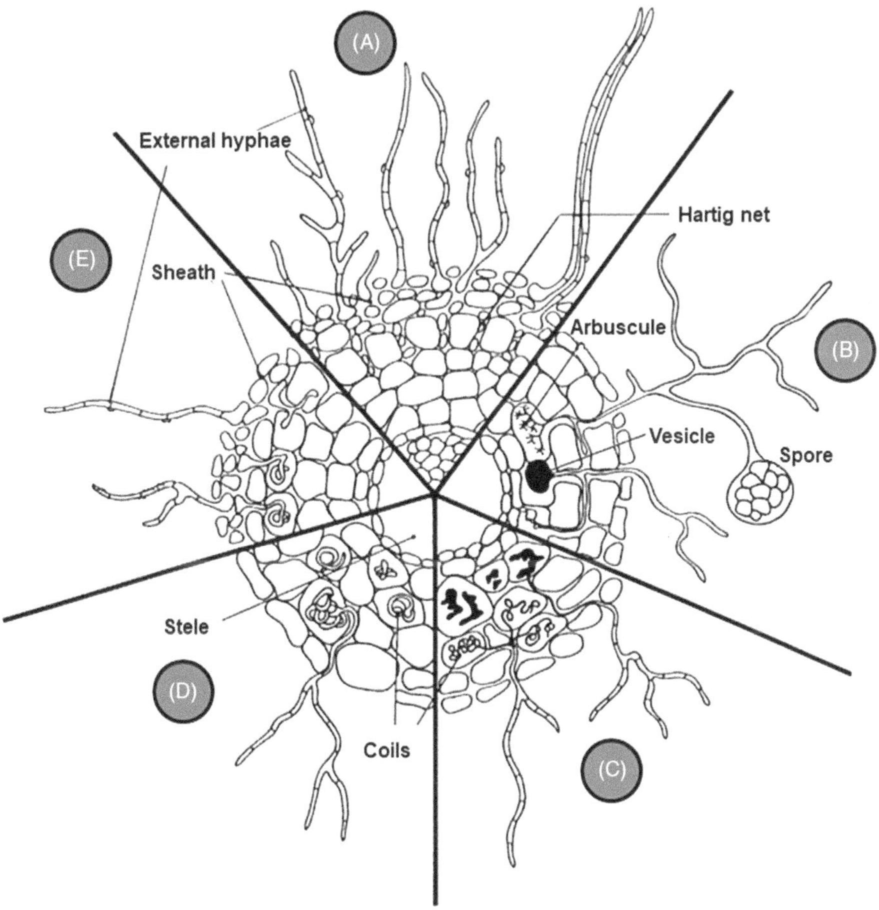

FIGURE 12.3 **Diagrammatic cross-section of a fine root illustrating key features of several types of mycorrhizae.** (A) Ectomycorrhiza, (B) Arbuscular mycorrhiza, (C) Orchid mycorrhiza, (D) Ericoid mycorrhiza, (E) Arbutoid Ectendomycorrhiza. *Source: Figure from Selosse and Le Tacon (1998). License #:4872730167028.*

fine-root morphology, resulting in root bifurcation, clubbing, and clustering (Fig. 12.2C,D). Contiguous with the mantle are hyphal strands that extend into the soil. Often the hyphal strands will aggregate to form **rhizomorphs** (see Fig. 6.4A) that may be visible to the unaided eye. The internal portion of rhizomorphs can differentiate into tube-like structures specialized for long-distance transport of nutrients and water. Ectomycorrhizal associations are found predominantly in superficial layers of the forest floor high in organic matter where their enzymatic capabilities release organically-bound nutrients, primarily nitrogen, for uptake and transport to the plant host.

Ectomycorrhizae are found on woody plants ranging from shrubs to trees and are readily recognized by their well-known fruiting bodies (truffles, chanterelles, and the very poisonous amanitas to name a few). Many of the host plants belong to the families Pinaceae, Fagaceae, Betulaceae, and Myrtaceae. Currently, ~20,000 fungal species, belonging primarily to the Basidiomycota and fewer to the Ascomycota, are known to form ectomycorrhiza (van der Heijden et al., 2015). Some fungi have a narrow host range (e.g., *Boletus betulicola* on *Betula* spp.) while others have a very broad host range (e.g., *Pisolithus tinctorius* forms ectomycorrhiza with more than 46 tree species belonging to at least 8 genera).

Arbuscular mycorrhizae (AM)

The diagnostic feature of AM fungi is the development of highly branched **arbuscules** within root cortical cells (Fig. 12.2A,B). The fungus initially grows between cortical cells, but soon penetrates the host cell wall and grows within the cell. The general term for all mycorrhizal types where the fungus grows within cortical cells is **endomycorrhiza.** In this process neither the fungal cell membrane nor the host cell membrane is breached. As the fungus grows, the host cell membrane invaginates and envelops the fungus, creating a new compartment where material of high molecular complexity is deposited. This **apoplastic** or **periarbuscular** space prevents direct contact between the plant and fungal cytoplasm and allows for efficient transfer of nutrients between the symbionts. The arbuscules are relatively short lived (<15 days) and may be difficult to see in field-collected and stained root samples. A second key feature of AM fungi is the production of **coenocytic** (multinucleate) **aseptate hyphae** *external* to the root, the so-called **extraradical** mycelium (pl. mycelia). The form and extent of this extraradical mycelium in the soil can also vary across genera: *Glomus* hyphal networks show profuse branching close to the root while *Scutellospora* networks are more diffuse with fewer branches (Parniske, 2008).

Other structures produced by some, but not all, AM fungi include vesicles, auxiliary cells, and asexual spores (Fig. 12.3B). **Vesicles** are thin-walled, lipid-filled structures that usually form in root intercellular spaces. Their primary function is thought to be storage; however, vesicles can also serve as reproductive propagules for the fungus. **Auxiliary cells** are formed in the soil and can be coiled or knobby. The function of these structures is not known. Reproductive **spores** can be formed either in the root or more commonly in the soil (i.e., hypogeous). Spores produced by fungi forming AM associations are asexual, formed by the differentiation of vegetative hyphae. In some fungi (e.g., *Glomus intraradices;* renamed *Rhizophagus intraradices*) spores may also arise from vesicles but more often spores are borne from hyphal swellings in the soil.

Originally, all symbiotic associations formed by fungi in the subphylum Glomeromycotina were called **vesicular-arbuscular mycorrhiza** (VAM); however, since a major family (Gigasporaceae) lacks the ability to form vesicles in roots; AM is now the preferred acronym. Until recently, morphology has been the mainstay of AM taxonomy. Key features used to distinguish among groups are still relevant today, such as presence of vesicles in the root and formation in soil of **chlamydospores** (thick-walled, asexual spores, borne from subtending hyphae), auxiliary cells, and **azygospores** (spores resembling zygospores but developing asexually from subtending hyphae resulting in distinct bulbous attachments). The spores of AM fungi are very distinctive. They range in diameter from 10 μm for *Glomus tenue* to >1000 μm for some *Scutellospora* spp. The spores can vary in color from hyaline (clear) to black and in surface texture from smooth to highly ornamented. *Glomus* forms spores on the ends of hyphae, *Acaulospora* forms spores *laterally* from the neck of a swollen hyphal terminus while *Entrophospora* forms spores *within* the neck of the hyphal terminus. Internal features of spores can also be diagnostic: spores of *Scutellospora* contain inner membranous walls and a **germination shield** (wall structure from which germ tube can arise) that is absent from *Gigaspora* a close relative.

Over the years, the taxonomy of AM fungi has transcended from a morphology-based system to one of phylogenetics with the advent of molecular tools (Redecker et al., 2013). Currently, there are over 200 species of AM fungi within the subphylum Glomeromycotina, divided among 4 orders, 11 families, and 25 genera (Table 12.2; Modified from Souza, 2015). The AM type of symbiosis is very common as the fungi involved have the ability to colonize a vast taxonomic range of both herbaceous and woody plants, indicating a general lack of host **specificity** among this type. It is, however, important to distinguish between specificity (innate ability to colonize), **infectiveness** (amount of colonization), and **effectiveness** (plant response to colonization). Most plants are colonized simultaneously by multiple species of AM fungi, with varying degrees of root colonization over time and space. Control occurs at the plant cellular level in response to the spatial structure of resources in the environment (e.g., distribution of nutrients in the soil) with varying outcomes on nutrient uptake and plant growth. Plant mutants with defective arbuscular-specific phosphate transporters (PT_4) showed early arbuscular death suggesting that arbuscule lifespan was contingent on supplying phosphate and likely other nutrients to the plant (Parniske, 2008). This may have important consequences for the development of plant communities (van der Heijden et al., 2015).

Orchidaceous mycorrhizae

Mycorrhizal fungi have a unique role in the life cycle of plants in the Orchidaceae. Orchids typically have very small seeds with little nutrient reserve.. The plant becomes colonized shortly after germination and during a prolonged seedling stage, the mycorrhizal fungus supplies carbon and vitamins to the developing plant. For **achlorophyllous** species that cannot photosynthesize, the plant depends on the fungal partner to supply carbon throughout its life. Infection of the seed occurs after the seed takes up water and swells thereby rupturing the seed coat. Germination of the embryo produces a protocorm with structures similar to root hairs (e.g. rhizoids) that become rapidly colonized by fungal hyphae (Yeung, 2017). The fungus grows into the plant cell, invaginating the cell membrane and forming hyphal coils

TABLE 12.2 Taxonomic relationships among the subphylum Glomeromycotina within the Mucoromycota as adapted from Souza (2015).

Order	Family	Genera	Main features
Diversisporales	Diversisporaceae	**Tricispora*, **Otospora*, *Diversispora*, **Corymbiglomus*, *Redeckera*	Produce glomoid, acaulosporoid, and entrophosporoid propagules. Germination not accompanied by formation of a germination shield.
	Acaulosporaceae	*Acaulospora*	Spores have bi-layered outer wall. Spores formed on or within a cylindrical or funnel-shaped hypha terminating in a sporiferous saccule: spores become sessile after detachment from the saccule neck.
	Sacculosporaceae	**Sacculospora*	Spores have outer and inner walls. Spores formed within the hyphal neck of closely adherent, terminal, or intercalary sporiferous saccules.
	Pacisporaceae	*Pacispora*	Glomoid spores with a combination of amyloid inner wall components. Germination by means of a germination shield, coupled with the formation of vesicular–arbuscular mycorrhizas.
	Gigasporaceae	*Scutellospora*, *Gigaspora*, **Intraornatospora*, **Paradentistucata*, *Dentiscutata*, *Cetraspora*, *Racocetra*	Produces single spores in soil (rarely in roots). Spores formed on bulbous sporogenous cells arising from subtending hyphae (sporophore) that differentiates from mycelial hyphae in soil. Spore contents separated by plug and rarely by septum.
Glomerales	Claroideoglomeraceae	*Claroideoglomus*	Single spores or small clusters (rarely in sporocarps) form in soils and rarely in roots. The subtending hyphae is hyaline to white, rarely subhyaline, and often conspicuously bill-shaped. Spores contain mono- to multi-layered outer walls.
	Glomeraceae	*Glomus*, *Funneliformis*, *Septoglomus*, *Rhizophagus*, *Sclerocystis*	Spores have one mono-to-multiple layered outer walls. Spores produced terminally on intercalary hyphae in soil and sometimes in roots. Can be formed singly, in clusters, or in sporocarps (with or without peridium).
Archaeosporales	Ambisporaceae, Archaeosporaceae, Geosiphonaceae	*Ambispora*, *Archaeospora*, *Geosiphon*	Oldest order within Phylum Glomeromycota. Dimorphic propagules (acaulosporoid and glomoid morphotypes). Spores lack pigmentation.
Paraglomerales	Paraglomeraceae	*Paraglomus*	Mycorrhizal morphology very similar to Archaeosporaceae

Members of each family (names ending in aceae) and genera (names in italics) differ in mode of spore formation, organization, and germination processes. Species are distinguished by differences in subcellular properties of the spore wall or internal flexible walls usually associated with germination (not shown). *Tentative due to insufficient evidence.

Table based on Souza (2015) and references cited within.

(pelotons) within the cell (Fig. 12.3C) to greatly increase the interfacial exchange surface between fungus and host. These coils are active for only a few days after which they lose turgor and degenerate, and the developing orchid absorbs the nutrient contents. The fungi participating in the symbiosis are basidiomycetes (~25,000 species; Johnson and Jansa, 2017) similar to those involved in decaying wood (e.g., *Coriolus*, *Fomes*, *Marasmius*) and pathogenesis (e.g., *Armillaria* and *Rhizoctonia*). In mature orchids, orchidaceous mycorrhizae have similar roles to other mycorrhiza in nutrient uptake and translocation.

Ericoid and other ericaceous mycorrhizae

The term ericaceous mycorrhizae is applied to mycorrhizal associations found on plants in the order Ericales. The hyphae in the root are able to penetrate cortical cells (endomycorrhizal habit); however, no arbuscules are formed (Fig. 12.3D). Three major forms of ericaceous mycorrhizae have been described: ericoid (most common), arbutoid, and monotropoid. In the **ericoid** form, cells of the inner cortex become packed with fungal hyphae. A loose welt of hyphae grows over the root surface, but a true mantle is not formed. The ericoid mycorrhizae are found on plants such as *Calluna* (heather), *Rhododendron* (azaleas and rhododendrons), and *Vaccinium* (blueberries) that have very fine root systems and typically grow in acid, peaty soils. The fungi involved are ascomycetes of the genus *Hymenoscyphus*. In the **arbutoid** form (Fig. 12.3E) characteristics of both EcM and endomycorrhizae are found. Intracellular penetration can occur, a mantle forms and a Hartig net is present. These associations are found on *Arbutus* (Pacific madrone), *Arctostaphylos* (Bearberry, Hairy Manzanita), and several species of the family Pyrolaceae (evergreen herbs, or wintergreens). The fungi involved in the association are basidiomycetes and may be identical to fungi that colonize EcM tree hosts in the same region. In the **monotropoid** form, the fungi colonize achlorophyllous plants in the family Monotropaceae (*Monotropa* species; Indian Pipes) producing a Hartig net and mantle. The same fungi also form EcM associations with trees and thereby form a link through which carbon and other nutrients can flow from the autotrophic host plant to the heterotrophic parasitic plant.

Mixed infections

Several fungi can colonize the roots of a single plant but the type of mycorrhiza formed is usually uniform for a host. There are, however, examples where a host can support more than one type of mycorrhizal association. *Alnus* (alders), *Salix* (willows), *Populus* (poplars), *Juniperus* (junipers), and *Eucalyptus* can have both AM and EcM associations on the same plant. Some ericaceous plants have also been shown to have occasional EcM and AM colonization.

An intermediate mycorrhizal type can be found on coniferous and deciduous hosts in nurseries and burned forest sites. The **ectendomycorrhiza** type forms a typical EcM structure except the mantle is thin or lacking and hyphae in the Hartig net may penetrate root cortical cells (Fig. 12.3E). The ectendomycorrhiza is replaced by EcM as the seedling matures. The fungi involved in the association were initially designated "E-strain" but were later shown to be ascomycetes and placed in the genus *Wilcoxina*.

Mycorrhizal biomass and community profiling

Assessing mycorrhizal biomass in plant roots and soils

Knowing the amount of biomass produced by an organism or group of organisms is necessary to understand their function and impact on soils, plant communities, and/or ecosystems. Historically, mycorrhizal biomass in both roots and soils has been quantified microscopically using various staining techniques or through direct visualization (e.g., minirhizotron) (Wallander et al., 2013). More recently, the discovery of a fatty acid biomarker (C16:1*cis*11 or C16:1ω5) abundant in the neutral lipids of AM fungi has made quantification of this group in soils much more reliable than hyphal measurements (Olsson, 1999). Additionally, it provides a useful measure of plant photosynthate carbon released into the soil via the extraradical mycelium and indicates the extent of the mycorrhizosphere, the latter being key to plant nutrient uptake. Unfortunately, no such biomarker exists for EcM or other mycorrhizal types. Therefore, quantification of these groups typically relies on amplicon sequencing (see later; Chapter 4) and reporting of relative abundances or enumeration of these sequences by quantitative PCR (Wallander et al., 2013).

Molecular approaches to assessing mycorrhizal diversity and function

Mycorrhizal diversity and function can be characterized using amplicon sequencing, whole genome sequencing, and meta-omics approaches (Chapter 4), a significant step up from mycorrhizal spore-based taxonomic diversity. Amplicon sequencing relies on the use of taxon-specific primers for either the 18S ribosomal RNA subunit (AM fungi) or internal transcribed spacers (ITS; for EcM and other mycorrhizal types). These techniques assess diversity by profiling the taxonomic composition of entire communities (Lekberg et al., 2018). While amplicon sequencing is often used for community analysis, whole-genome sequencing is applied to individual organisms. Sequencing is performed on isolates in order to characterize genomic potential, that is, enzymatic products and processes. Using this technique, putative function(s) can be assigned to specific mycorrhizal groups. Similarly, genetic potential of entire communities can be obtained using meta-omics techniques. This includes transcriptomics, proteomics, and metabolomics which provide a snapshot of the potentially expressed and active molecules (transcripts, proteins, and metabolites) that could contribute to mycorrhizal and/or plant host function (Kamel et al., 2017). Molecular approaches give insight into two main questions: who is there (diversity), and what might they be doing (functional potential). These techniques are necessary to unravel complex symbiotic associations and their ecological functions.

Features of the symbiosis

Signaling and forming the AM association

Signaling between plant hosts and their associated symbiotic soil microbes, including EcM, AM, and N_2-fixing rhizobia, is a remarkable co-evolutionary relationship (Box 12.1). These intimate interactions include cellular communications and extensive host-cell remodeling to accommodate entry of the symbiont and subsequent lifestyle (Martin et al., 2017).

BOX 12.1

Plant signaling similar for both AMF and rhizobium

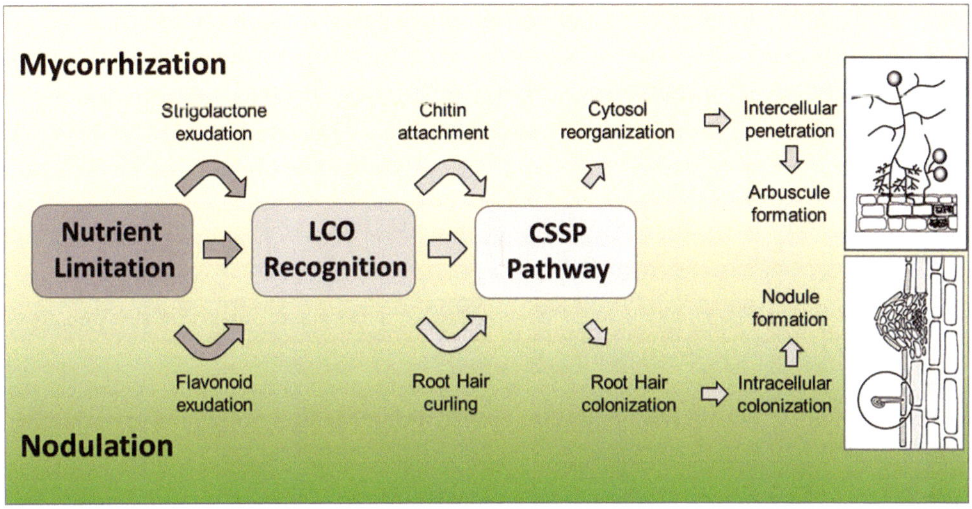

A perplexing question is how does the host plant distinguish between friend and foe? Many microorganisms find their homes within plant roots and co-occurrence of multiple symbionts (or endophytes) is common. In model legume species, a unique signal transduction pathway is activated in plant host cells on recognition of diffusible lipochito-oligosaccharide (LCO) signals produced by legumes (Nod factors) and AM fungi (Myc factors) that allows for microbial entry into the root and subsequent steps to establish the symbiosis (Oldroyd, 2013). This pathway is now referred to as the "common symbiotic signaling pathway" or CSSP (Martin et al., 2017). By keeping the invading microbe within the apoplastic space (EcM symbioses) or within intracellular infection compartments (AM and rhizobium symbioses), and thus separate from the host's cytoplasm, activation of host antimicrobial defenses is minimized. Figure redrawn from Sasse et al. (2018).

Establishing the AM symbiosis (Fig. 12.4) begins with presymbiosis involving spore germination, conservative hyphal growth, host recognition, and attachment. Spore germination, although not dependent on the plant, occurs more readily in the presence of host root exudates (triggered by nutrient stress in the host) containing specialized plant signals known as strigolactones (Fig. 12.5). Host recognition of the symbiont occurs through detection of AM secreted oligosaccharide signals, known as mycorrhizal lipochito-oligosaccharides (Myc-LCOs) that are chemically similar to other microbial signals that cause root hair branching and calcium spiking (Gutjahr and Parniske, 2013) in the *Rhizobium*-legume symbiosis (Box 12.1; Chapter 17). Upon root contact, the fungus forms a thick, flattened hyphal tip, called a hyphopodium or appressorium. The host cell then initiates formation of the prepenetration apparatus (PPA) to allow hyphal entry into the root and establishment of the symbiosis. Altogether, unraveling these symbiotic relationships will provide insight into how plants differentiate between symbiotic and pathogenic intentions, respond to abiotic and biotic stressors, and guide our predictions on how best to manage plant-microbe relationships

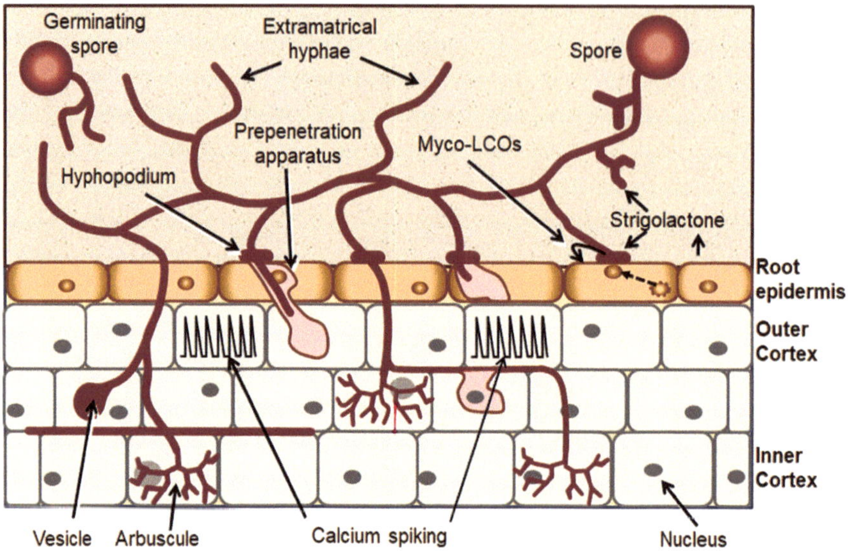

FIGURE 12.4 **Life cycle of AM symbiosis.** Under optimal abiotic conditions, fungal spores germinate and in the presence of plant root exudates, navigate toward the host root. Strigolactones exuded by the host root induce fungal hyphal branching and production of lipochitooligosaccharides (Myc-LCOs) signaling molecules to activate the common symbiotic signaling pathway (CSSP; see Box 12.1) and subsequent development of the arbuscule. *Source: Figure redrawn from Gutjahr and Parniske (2013) and Parniske (2008).*

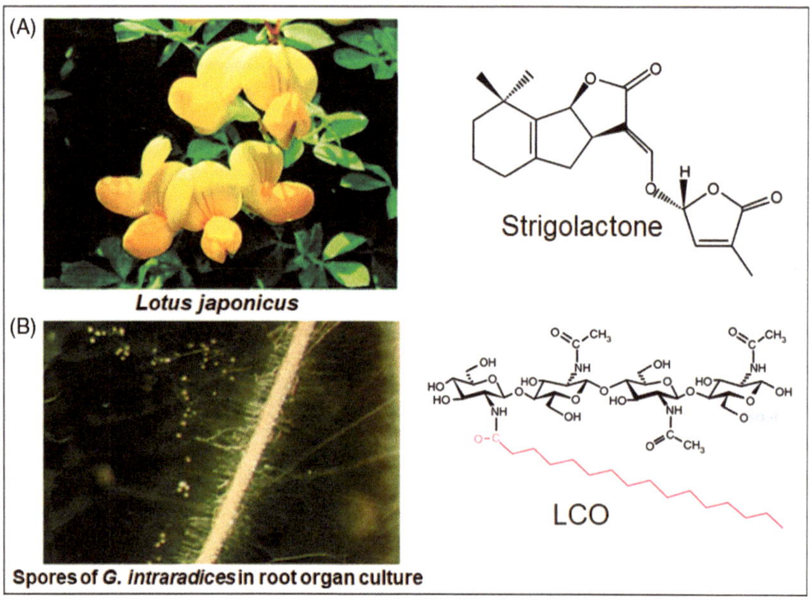

FIGURE 12.5 **Two main groups of bioactive molecules produced by the plant and the fungal partner have been identified so far as being crucial in initiating the molecular dialog in arbuscular mycorrhiza.** (A) The strigolactone group, carotenoid-derived metabolites produced by many plants including the AM host plant *Lotus japonicus*; (B) lipochitooligosaccharides (LCOs) produced by *G. intraradices* grown in axenic conditions in root organ cultures. *Source: Figure from Bonfante and Requena (2011). With permission.*

to maintain agronomic productivity and ecosystem sustainability under a changing climate (Martin et al., 2017).

Carbon fluxes in mycorrhizal plants

Mycorrhizal fungi range from **obligate symbionts** (e.g., AM) that only obtain carbon from the plant host to **facultative symbionts** (e.g., EcM) that can also mineralize organic carbon. In nature, however, mycorrhizae obtain all or most of their carbon from the autotrophic host plant. Photosynthetic carbon in the form of sucrose is transported to the roots where it is cleaved into glucose and fructose (Garcia et al., 2016). Colonized roots form a strong sink for sugars and the flux of sugars is controlled by membrane transport systems: sucrose transporters (SUTs), monosaccharide transporters (MSTs), and the SWEET (sugars will be eventually exported transporter) family. Glucose is now thought to be the main source of carbon taken up by the fungus at the plant-fungus interface (Garcia et al., 2016). In AM colonized plants, host carbon in the form of fatty acids is also transferred across the plant-fungal interface within arbusculated plant cells (Wipf et al., 2019). Lipid bodies (e.g., vesicles) are a major storage form of C in AM fungi and are a primary source of lipids transported to the external mycelium to facilitate growth, nutrient uptake, and later spore formation. As much as 20% of the total carbon assimilated by plants may be transferred to the fungal partner. This transfer of carbon to the fungus has sometimes been considered a drain on the host. However, the host plant may increase photosynthetic activity following mycorrhizal colonization, thereby compensating for carbon "lost" to the soil. Occasionally plant growth suppression has been attributed to mycorrhizal colonization, but usually this occurs only under low-light (photosynthate limiting) or high-phosphorus conditions.

Uptake of soil nutrients and transfer to host plant

When a nutrient is deficient in the soil solution, the critical root parameter controlling its uptake is surface area. Hyphae of mycorrhizal fungi have the potential to greatly increase the effective absorbing surface area of the root. For example, Rousseau et al. (1994) found that while extramatrical hyphae accounted for <20% of the total nutrient absorbing surface *mass*, they contributed nearly 80% of the absorbing surface *area* of pine seedlings. It is also important to consider the distribution and function of the extramatrical hyphae.

For the mycorrhiza to be effective in nutrient uptake, the hyphae must be distributed beyond the **nutrient depletion zone** that develops around the root (Fig. 12.6). A nutrient depletion zone develops when nutrients are removed from the soil solution more rapidly than they can be replaced by diffusion. For a poorly-mobile ion such as phosphate, a sharp and narrow depletion zone develops close to the root. Hyphae can readily bridge this depletion zone and grow into soil regions having a greater ability to provide phosphorus. Uptake of micronutrients such as zinc and copper are also improved by mycorrhizae since these elements are also diffusion limited in many soils. For more mobile nutrients such as nitrate the depletion zone is wider (Kuzyakov and Razavi, 2019), but it has been shown that AM hyphae can take up and translocate nitrogen from distances quite remote to the root (i.e., 5–10 cm or more). Another factor contributing to the effective absorption of nutrients by mycorrhizae is their narrow diameter relative to roots. The steepness of the diffusion gradient for a nutrient is

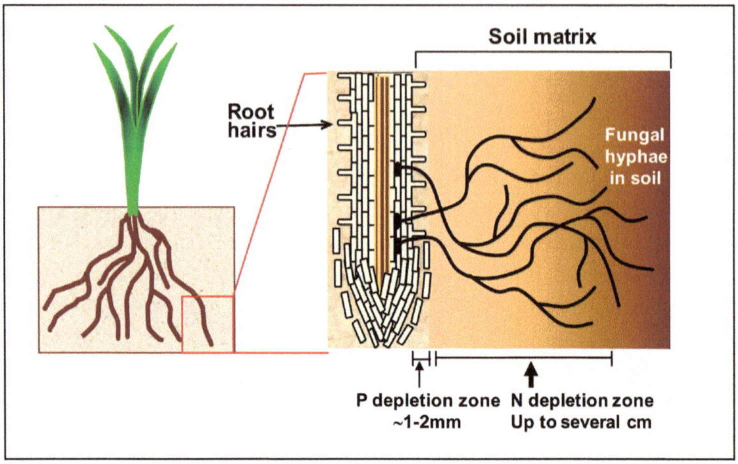

FIGURE 12.6 **Diagrammatic representation of the development of a depletion zone adjacent to a plant root.** A narrow depletion zone (e.g., 1–2 mm) usually develops for phosphate whereas a wide depletion zone (10 mm) develops for nitrate. Hyphae of mycorrhizal fungi can generally grow beyond the phosphorus depletion zone. Evidence also indicates that hyphae of the extraradical mycelium (ERM) can take up and translocate nitrogen from distances quite remote to the root (i.e., 5–10 cm or more). *Source: Original drawing by D.A. Zuberer. Used with permission.*

inversely related to the radius of absorbing unit; therefore, the soil solution will be less depleted at the surface of a narrow absorbing unit such as a hypha. Furthermore, narrow hyphae can grow into small soil pores not accessible to roots or even root hairs. It should be noted that depletion zones can also form around individual hyphae as well.

In the soil, mycorrhizal extramatrical hyphae take up nutrients via membrane transporters located in hyphal tips. These membrane transporters are specific to each nutrient and form. Nitrate (NO_3^-) is the most important source of N in soil solution and is more mobile than ammonium (NH_4^+). Nitrate is taken up by a highly regulated, energy-dependent family of nitrate/peptide transporters (NRT and PTR families). However, microbes including mycorrhizae generally prefer direct assimilation of NH_4^+ as a less costly alternative to NO_3^-, as the latter has to be reduced to NH_4^+ for amino acid biosynthesis (Wipf et al., 2019). Phosphorus in soil exists largely in organic matter, adsorbed to clay minerals, or as inorganic salts (e.g., calcium phosphates) of low solubility. Inositol phosphates (e.g., phytic acid, Chapter 19) constitutes the largest pool of soil organic P and originates largely from seeds and grains (Wang et al., 2017). Phosphate is released from this phosphorylated cyclic sugar by phytase-mediated hydrolysis of organic phosphate (C-O-P) ester bonds. Soil microbes can help solubilize phosphate salts by release of oxalic or citric acids. The low solubility of Pi in soil solution, coupled with direct uptake of Pi by plants via membrane specific transporters, sets up a depletion zone as described earlier. Mycorrhiza extend beyond this depletion zone to acquire Pi using highly regulated, energy-driven phosphate transporters of the Pht family (Garcia et al., 2016). Nutrients are then transported via the extraradical mycelium to the colonized root for subsequent transfer across the fungal-host interface using nutrient-specific fungal and host membrane transporters.

Interactions with other organisms

Mycorrhizal fungi interact with a wide assortment of organisms in the **rhizosphere** (Chapter 11). The effect on the mycorrhizal association or a particular component of the rhizosphere can be positive, neutral, or negative. An important consequence of the carbon flow to the fungal partner is the development of a unique rhizosphere microbial community termed the **mycorrhizosphere** (Hodge, 2000) and is distinct from the communities associated with nonmycorrhizal roots. This altered community may have important impacts on nutrient uptake. For example, in contrast to many elements, the uptake of manganese is often reduced in mycorrhizal plants which may be beneficial in soils prone to manganese toxicity. Manganese is predominantly taken up in the reduced form (Mn^{2+}); however, the abundance of Mn-reducing bacteria has been found to be higher in the rhizosphere of nonmycorrhizal plants compared to mycorrhizal plants, resulting in lower acquisition of Mn by mycorrhizal plants (Posta et al., 1994). Furthermore, specific bacteria have been shown to associate with EcM and stimulate mycorrhizal formation in conifer nurseries. These "mycorrhization helper bacteria" may even preclude the need for soil fumigation (Garbaye, 1994).

The interaction between rhizobia (Chapter 17) and AM fungi has received considerable attention because of the relatively high phosphorus demand of N_2 fixation. The two symbioses typically act synergistically, resulting in greater nitrogen and phosphorus content in combination than when each is inoculated onto the legume alone. Legumes are typically coarse-rooted and therefore not efficient in extracting phosphorus from the soil. The AM fungi associated with legumes are an essential link for adequate phosphorus nutrition, leading to enhanced nitrogenase activity (Chapter 16) for N_2 fixation, which in turn promotes root and mycorrhizal growth.

Mycorrhizal fungi colonize feeder roots and thereby interact with root pathogens (Chapter 23) that parasitize these same tissues. Therefore, it is not surprising that mycorrhizae have been shown to reduce the incidence and severity of root diseases. Mycorrhizae may stimulate root colonization by selected biocontrol agents, but our understanding of these interactions is meager. Much more research has been conducted on the direct effects of mycorrhizal colonization on root pathogens. The mechanisms proposed to explain this protective effect include:

- development of a mechanical barrier—especially the mantle of EcM—to infection by pathogens
- production of antibiotic compounds by EcM that suppress the pathogen
- competition for nutrients in the rhizosphere, including production of siderophores (Chapter 19)
- induction of generalized host defense mechanisms

Mycorrhizae in ecosystems

By this point it should be apparent that mycorrhizae have important multi-functional roles in natural ecosystems (Fig. 12.7). Besides improving uptake of poorly mobile nutrients such as P, mycorrhizae contribute to soil quality by channeling plant photosynthate C into the soil as living fungal biomass, labile exudates, and ultimately to SOM largely as fungal necromass (Chapter 13). Extramatrical hyphae of mycorrhizae also function to bind soil particles together into aggregates thereby physically protecting SOM from losses through decomposition or

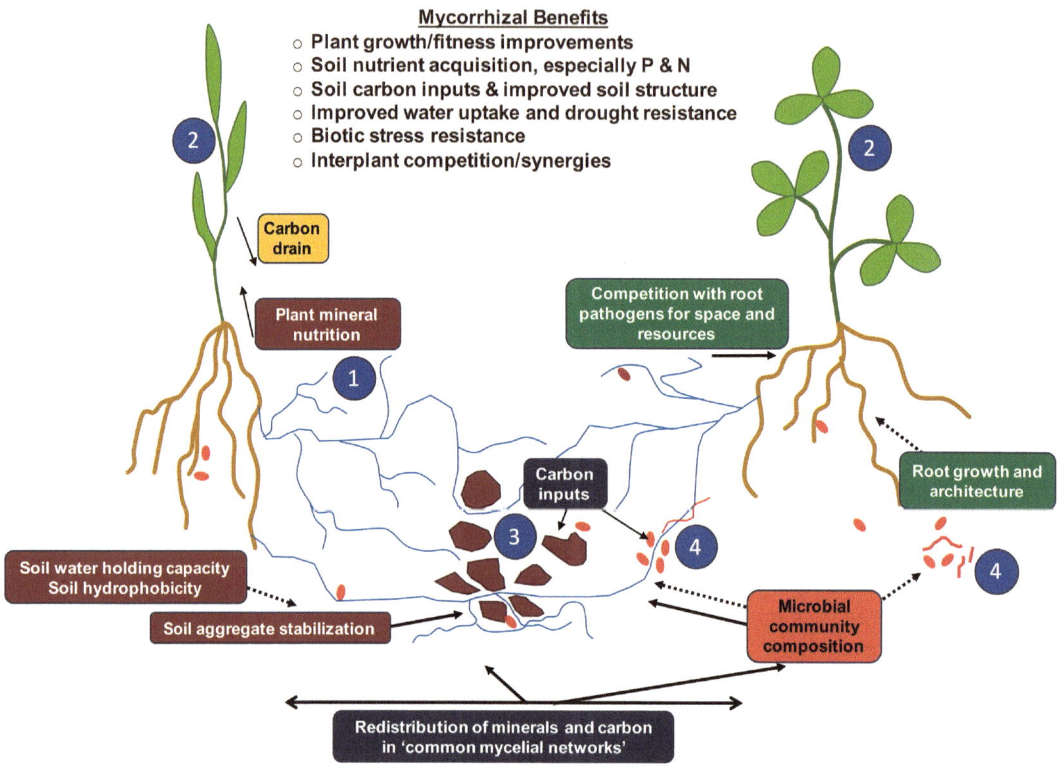

FIGURE 12.7 **Schematic representation of the different functions played by arbuscular mycorrhizal (AM) fungi in the physiology and community ecology of their host plants.** (1) Extraradical AM hyphae linking plant communities within the soil and providing inputs to SOM; (2) Grass and legume host plants; (3) Mycorrhizal hyphae bind soil particles to form soil aggregates and interconnect roots across plant species to redistribute nutrients within "common mycelial networks"; (4) Mycorrhizae and roots affect soil microbial communities. *Solid lines* represent direct and the *dotted lines* indirect effects of the AM fungi on the plants, soil, and soil microbes. *Source: Modified from Jansa et al. (2013). CC BY 3.0 https://creativecommons.org/licenses/by/3.0/.*

soil erosion. Typically, there are between 1 and 20 m of AM hyphae g^{-1} of soil. Furthermore, high-molecular-weight glycoproteins (GRSPs) are produced in abundance by AM fungi. This material accumulates in soil and is positively correlated with aggregate stability (Wright and Upadhyaya, 1998). Thus, the impact of mycorrhizal fungi on SOM storage will depend on their relative roles in mycorrhizal-mediated SOM formation, stabilization, and destabilization processes (Frey, 2019).

Ericoid and EcM have a special role in the mineralization of nitrogen (Martin et al., 2001). Most plant litter entering the soil has a high C/N ratio and is rich in lignin and tannins. Only a few mycorrhizal fungi can mobilize nutrients from these primary sources. However, a wide range of ericoid and EcM fungi can obtain nitrogen (and other nutrients) from secondary sources of organic matter such as recalcitrant SOM and dead microbial biomass because they produce hydrolytic and oxidative enzymes capable of depolymerizing organic nitrogenous polymers such as proteins and chitin. AM fungi are more limited in their capacity to mineralize SOM and thus rely on N mineralized by other microorganisms or external sources (e.g., fertilizers) to supply the plant host. This dichotomy underpins the "mycorrhizal-associated nutrient economy" or MANE framework of temperate forests (Box 12.2) with applicability to other

BOX 12.2

Mycorrhizal-associated nutrient economy (MANE)

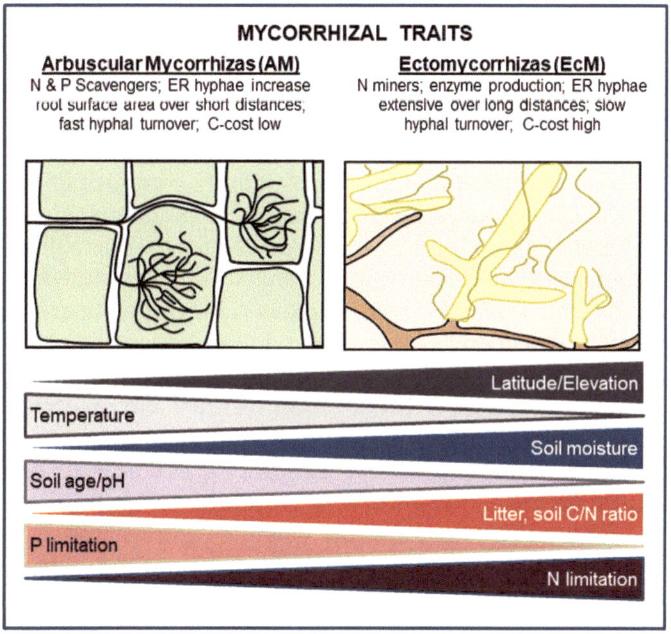

Mycorrhizal traits and ecosystem characteristics are intimately linked (Brzostek et al., 2017) with the goal of providing nutrients (N, P and others) to the host plant at the lowest relative C cost. The mycorrhizal-associated nutrient economy (MANE) is a conceptual framework that differentiates carbon and nutrient flow through temperate forest environments depending on the dominant mycorrhizal symbiont; endo- or ectomycorrhizae (Phillips et al., 2013). Forest communities dominated by ectomycorrhizae (EcM) have an organic 'nutrient economy,' while endo- or arbuscular mycorrhizae (AM) dominated forests have an inorganic 'nutrient economy'. The 'organic' nutrient economy is characterized by slower decomposition of lower quality plant inputs, greater conservation of soil inorganic nutrients, and possibly greater soil C storage. A larger pool size of dissolved organic soil C, increased availability of organic P and lower inorganic N concentrations help retain nutrients in the system. A temporal study conducted in 2009 found that as the percentage of EcM trees increased, net nitrification rates decreased thereby reducing nitrate leaching (Phillips et al., 2013). Total fungal biomass, including mycorrhizal hyphal biomass, is also higher in EcM-dominated forests, and may contribute to greater C storage (Frey, 2019). In contrast, the 'inorganic' nutrient economy of AM-dominated temperate forests is characterized by inputs of high-quality plant material (litter, root exudates) and fungal carbon (chitin) resulting in rapid mineralization of organic compounds (proteins, chitin, amino acids, etc.) into inorganic forms (ammonium and nitrate; phosphate) by a diverse microbial community (Phillips et al., 2013; Frey, 2019). Because EcM and AM use different nutrient acquisition strategies their impact on SOM storage may also differ. AM fungi are scavengers of inorganic nutrients due to lack of extracellular SOM-degrading enzymes and thus are more dependent on other members of the soil microbial community to mineralize nutrients from plant residues and SOM. EcM are miners of SOM, either directly through enzymatic breakdown or indirectly via oxidative mechanisms, leaving C-rich, nutrient poor substrates behind (Frey, 2019). The MANE concept provides an ecological tool to address factors of SOM stabilization versus destabilization within temperate forests, and likely other ecosystems. By coupling this model to spatial and temporal features of the global carbon cycle, such as nitrogen deposition and elevated atmospheric CO_2, we may be better equipped to predict environmental changes driven by climate change (Phillips et al., 2013). Figure modified from Brzostek et al. (2017). Used with permission.

mycorrhizal systems. We now realize that carbon flow to the soil is critical for the maintenance of a healthy plant-soil system and should be considered an important benefit of mycorrhizal colonization.

There is also mounting evidence that mycorrhizal fungi are important determinants of plant community structure and plant succession (Parniske, 2008; Tedersoo et al., 2020; van der Heijden et al., 2015). Plant communities compete for limiting soil resources within (intra-) and among (inter-) species as influenced by the supply of those resources via the mycorrhizal community. Furthermore, coexisting members of the mycorrhizal community also compete for host C and soil resources. Coexistence of mycorrhizal genotypes can occur when genotypes are limited by different resources (Johnson, 2010). In early successional plant communities, such as in reclamation of mine spoil material, plant community composition appears more important for structuring AM communities than soil properties (Kruger et al., 2017). As ecosystems evolve, AM communities respond to plant-induced changes in soil habitat (e.g., pH and nutrient content/allocation) and altered microclimate. This was clearly shown for AM communities in roots of eastern red cedar as stand density increased in temperate grasslands thereby altering soil pH and resource allocation between soil and litter layers (Liang et al., 2017).

Uncovering the functions of mycorrhizal fungi in managed agroecosystems has gained prominence in recent years as concerns over food security and environmental stewardship became mainstream. A rising global population, associated land degradation and scarcity, coupled with water quality and quantity issues have necessitated realignment of our cropping systems to reduce inputs and capitalize on crop diversification, especially in the face of climate variability. More reliance on managing existing resources, such as mycorrhizae, will be essential as we move forward to address these challenges. For AM dominated systems, it is becoming clear that the balance of nutrients is particularly important to the outcome of the symbiosis. This also applies to agroecosystems where *"the nutritional benefits provided by mycorrhiza in field crops are assumed to be a mosaic of the functions of diverse AMF species"* (Kobae, 2018). This has been conceptualized as the "trade balance model" where the success of the symbiosis depends on the interaction of host C supply with fungal C demand as regulated by the availability of soil N and P (Box 12.3; Johnson, 2010). How the trade balance model functions as we diversify our cropping systems with novel rotations, cover crops, intercropping, etc., will impact productivity and land/water stewardship. Nonetheless, when one takes a more holistic view of the plant-soil continuum, the "cost" of the symbiotic association may turn out to be an important benefit to soil health and overall ecosystem services.

Conventional agronomic practices common to large-scale industrial agriculture may adversely affect the diversity and abundance of AM fungi in agroecosystems (Abbott and Johnson, 2017; Rillig et al., 2019). Excessive application of phosphate fertilizers, including manures high in phosphorus, may decrease mycorrhizal biomass in soils as well as root colonization. Annual tillage also decreases mycorrhizal biomass in soils due to disruption of mycelial networks that then have to be rebuilt. Over time, mycorrhizal taxa more resistant to these practices become dominant and may lead to decreased mycorrhizal colonization of roots and loss of function. Crop rotations that include non-host plants such as mustards or long-term fallow may also reduce AM fungal populations in soils. Fungicides, especially fumigants, substituted aromatic hydrocarbons and benzimidazoles may adversely affect the activity of AM fungi in soil. Strategies to enhance indigenous populations of mycorrhizal fungi in agroecosystems include reduced tillage, reduced fertilizer and pesticide use, as well as careful management of crop rotations, crop sequences, and cover crops (Abbott and Johnson, 2017).

BOX 12.3

The trade balance model

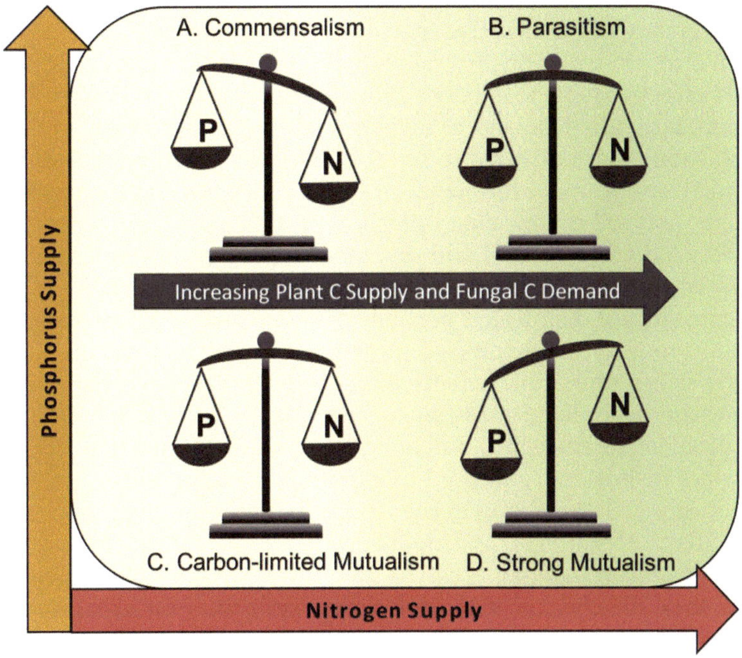

The trade balance model is based on fungal acquisition of soil P in exchange for C from the host plant in AM symbioses. Strong mutualisms (scenario D) are favored when the host plant is productive despite a limited supply of soil P. This occurs when N is non-limiting to plant growth and photosynthesis is sufficient to support the high C demand by the fungal symbiont in exchange for soil P. This scenario may be found in highly weathered soils with high P fixation capacity or in highly fertile soils where soil P supply cannot keep up with plant demand making the C-for-P trade favorable (Tian et al., 2013). Commensalism (scenario A) is predicted when P is not limited, so plants have nothing to gain from C-for-P trade, but C demand by AM fungi is kept in check because fungal growth is N-limited. Under conditions of both N and P limitation, a limited mutualism is predicted (scenario C) despite a favorable C-for-P trade balance because N limitation restricts plant photosynthetic acquisition of C. This may occur in some natural grasslands when conditions are suboptimal for mineralization of N from SOM. When neither P nor N is limited plants gain no benefit from the C-for-P trade but a high supply of photosynthetic C coupled with high fungal demand for C may lead to parasitism (scenario B). To evaluate the trade balance model, one would calculate a mycorrhizal growth response [MGR = ln (AM/NM) where AM and NM are total plant dry weights of mycorrhizal and nonmycorrhizal plants, respectively] in conjunction with measured nutrient concentrations in plant tissue. Although this approach works well in controlled experiments where NM controls are easily established, it becomes problematic in the field. Figure modified from (Johnson, 2010).

Biotechnological applications

In situations where native mycorrhizal inoculum is low or ineffective, consideration should be given to introducing appropriate fungi into the plant-production system. With the current state of technology, inoculation is most feasible for transplanted crops and in areas where soil disturbance has greatly reduced native inoculum potential.

The first step in any inoculation program is to obtain **infective** (able to penetrate and spread in the root) and **effective** (able to enhance growth or stress tolerance of the host) isolate(s) for a given situation. Individual isolates of mycorrhizal fungi vary widely in these properties so screening trials should be conducted to select isolates that will perform successfully. Ideally, screening should be done under the actual cropping conditions since indigenous mycorrhizal fungi, pathogens, and soil chemical and physical properties will influence success.

Isolation and production of EcM and AM inocula present very different problems. Many EcM fungi can be cultured on artificial media; therefore, isolates of EcM fungi can be obtained by placing surface-disinfested portions of sporocarps (e.g., mushrooms, puffballs) or mycorrhizal short roots on an agar-based growth medium. The resulting fungal biomass can be used directly as inoculum but, for ease of use, inoculum often consists of fungal material mixed with a carrier or bulking material such as peat or biochar. Obtaining isolates of AM fungi is more problematic since they will not grow apart from their host. Spores can be sieved out of soil, surface disinfested, and used to initiate "pot cultures" on susceptible host plants in sterile soil or artificial plant-growth media. Pot cultures are then used to "scale up" production of inocula. Alternatively, hydroponic or aeroponic culture systems can be used—a benefit of these systems is that plants can be grown without a supporting substrate, allowing colonized roots to be sheared into inoculum of high propagule number. Methods for working with AM inocula have been summarized elsewhere (Sylvia, 1994).

The goal of inoculation is to introduce *viable* propagules of selected mycorrhizal fungi into the rhizosphere of the target plant. The most common method is to place inoculum below the seed or seedling prior to planting. Alternative strategies include seed coatings and somatic embryo encasement with inoculum. A key concern for commercially produced inocula is viability. Further development of technologies for the production, storage, and delivery of viable and effective inocula should allow for more widespread use of mycorrhizal fungi in the future.

Summary

Mycorrhizal associations are symbiotic relationships that form between plant roots and fungi. Bi-directional movement of nutrients characterizes these symbioses where photosynthetic carbon flows to the fungus and inorganic nutrients move to the plant. Mycorrhizae occur on most plant species where they provide a critical linkage between the plant root and soil. In infertile soils, nutrients taken up by the mycorrhizal fungi can lead to improved plant growth and reproduction. As a result, mycorrhizal plants are often more competitive and better able to tolerate environmental stresses than are nonmycorrhizal plants. Mycorrhizal associations vary widely in form and function. Ectomycorrhizal fungi are mostly basidiomycetes that grow between root cortical cells of many tree species, forming a Hartig net. Arbuscular mycorrhizal fungi belong to the subphylum Glomeromycotina and form highly branched structures (arbuscules) within root cortical cells of many herbaceous and woody plant species. Plant responses to colonization by mycorrhizal fungi can range from dramatic growth promotion to growth depression. Factors affecting this response include the mycorrhizal dependency of the host crop, the balance of nutrients in the soil, and the inoculum potential of the mycorrhizal fungi. Mycorrhizae add significant quantities of carbon to the soil, improve soil aggregation, and contribute to SOM. Management practices such as tillage, crop rotation with non-hosts, and fallowing may adversely affect populations of mycorrhizal fungi in the field. Where native inoculum potential is low or ineffective, inoculation strategies may be considered. With the current state of technology, inoculation is most feasible for transplanted crops and in areas where soil disturbance has greatly reduced the native inoculum potential.

Acknowledgment

Portions of this chapter are derived from the work of Dr. David Sylvia who authored a similar chapter for the Second Edition of the textbook. The current authors and editors thank him for permitting the use of his material and graciously supporting production of the Third Edition.

Study questions

1. The term symbiosis literally means living together and usually implies a mutually beneficial association. Describe how each partner in a mycorrhizal symbiosis gains benefit from the association?
2. Describe the differences in structure and function between hyphae of mycorrhizal fungi that grow in the root (intramatrical) and the hyphae that grow in the soil (extramatrical).
3. When grown in phosphorus-limited soil, citrus is a mycorrhizal-dependent plant. However, when citrus is grown under high-phosphorus conditions (such as found in a nursery), inoculation with mycorrhizal fungi may lead to growth depression. Provide an explanation for this phenomenon.
4. What are some of the consequences of host carbon that is transferred to the soil via mycorrhizal structures?
5. A tripartite association often forms between legumes, N_2-fixing rhizobial bacteria, and arbuscular mycorrhizal fungi. Why does the legume especially benefit from colonization by these two symbiotic microorganisms?

References

Abbott, L.K., Johnson, N.C., 2017. Introduction: Perspectives on mycorrhizas and soil fertility. Johnson, N.C., Gehring, C., Jansa, J. (Eds.), Mycorrhizal Mediation of Soil. Elsevier, pp. 93–105.

Bonfante, P., Requena, N., 2011. Dating in the dark: how roots respond to fungal signals to establish arbuscular mycorrhizal symbiosis. Curr. Opin. Plant Biol. 14, 451–457.

Brundrett, M.C., Tedersoo, L., 2018. Evolutionary history of mycorrhizal symbioses and global host plant diversity. New Phytol. 220, 1108–1115.

Brzostek, E.R., Rebel, K.T., Smith, K.R., Phillips, R.P., 2017. Integrating mycorrhizas into global scale models: A journey toward relevance in the Earth's climate system. Johnson, N.C., Gehring, C., Jansa, J. (Eds.), Mycorrhizal Mediation of Soil. Elsevier, pp. 479–499.

Frey, S.D., 2019. Mycorrhizal fungi as mediators of soil organic matter dynamics. Ann. Rev. Ecol. Evol. Syst. 50, 237–259.

Garbaye, J., 1994. Helper bacteria: a new dimension to the mycorrhizal symbiosis. New Phytol. 128, 197–210.

Garcia, K., Doidy, J., Zimmermann, S.D., Wipf, D., Courty, P.E., 2016. Take a trip through the plant and fungal transportome of mycorrhiza. Trends Plant Sci. 21, 937–950.

Gutjahr, C., Parniske, M., 2013. Cell and developmental biology of arbuscular mycorrhiza symbiosis. Ann. Rev. Cell Dev. Biol. 29, 593–617.

Hodge, A., 2000. Microbial ecology of the arbuscular mycorrhiza. FEMS Microbiol. Ecol. 32, 91–96.

Jansa, J., Bukovská, P., Gryndler, M., 2013. Mycorrhizal hyphae as ecological niche for highly specialized hypersymbionts – or just soil free-riders? Front. Plant Sci. 4, 134. doi: 10.3389/fpls.2013.00134, Published 2013 May 16.

Johnson, N.C., 2010. Resource stoichiometry elucidates the structure and function of arbuscular mycorrhizas across scales. New Phytol. 185, 631–647.

Johnson, N.C., Jansa, J., 2017. Mycorrhizas. Johnson, N.C., Gehring, C., Jansa, J. (Eds.), Mycorrhizal Mediation of Soil. Elsevier, pp. 1–6.

Kamel, L., Keller-Pearson, M., Roux, C., Ane, J.M., 2017. Biology and evolution of arbuscular mycorrhizal symbiosis in the light of genomics. New Phytol. 213, 531–536.

Kobae, Y., 2018. High-resolution functional analysis of arbuscular mycorrhizal symbiosis in field crops. J. Integr. Field Sci. 15, 18–21.

Kruger, C., Kohout, P., Janouskova, M., Puschel, D., Frouz, J., Rydlova, J., 2017. Plant communities rather than soil properties structure arbuscular mycorrhizal fungal communities along primary succession on a mine spoil. Front. Microbiol. 8, 719.

Kuzyakov, Y., Razavi, B.S., 2019. Rhizosphere size and shape: temporal dynamics and spatial stationarity. Soil Biol. Biochem. 135, 343–360.

Lekberg, Y., Vasar, M., Bullington, L.S., Sepp, S.-K., Antunes, P.M., Bunn, R., Larkin, B.G., Öpik, M., 2018. More bang for the buck? Can arbuscular mycorrhizal fungal communities be characterized adequately alongside other fungi using general fungal primers? New Phytol. 220, 971–976.

Liang, Z.B., Lee, D.J., Dweikat, I.M., Wedin, D.A., Yuen, G.Y., Drijber, R.A., 2017. Molecular diversity of arbuscular mycorrhizae in roots of *Juniperus virginiana* invasive to grasslands. Soil Sci. Soc. Amer. J. 81, 526–536.

Martin, F., Cliquet, J.B., Stewart, G., 2001. Nitrogen acquisition and assimilation in mycorrhizal symbioses. Lea, P., Morot-Gaudry, J.M. (Eds.), Plant Nitrogen Springer-Verlag, Heidelberg, pp. 147–166.

Martin, F.M., Uroz, S., Barker, D.G., 2017. Ancestral alliances: plant mutualistic symbioses with fungi and bacteria. Science 356 (6340)doi: 10.1126/science.aad4501, eaad4501.

Miyasaka, S.C., Habte, M., 2001. Plant mechanisms and mycorrhizal symbioses to increase phosphorus uptake efficiency. Commun. Soil Sci. Plant Anal. 32, 1101–1147.

Oldroyd, G., 2013. Speak, friend, and enter: signalling systems that promote beneficial symbiotic associations in plants. Nat. Rev. Microbiol. 11, 252–263.

Olsson, P.A., 1999. Signature fatty acids provide tools for determination of the distribution and interactions of mycorrhizal fungi in soil. FEMS Microbiol. Ecol. 29, 303-301.

Parniske, M., 2008. Arbuscular mycorrhiza: The mother of plant root endosymbioses. Nat. Rev. Microbiol. 6, 763–775.

Phillips, R.P., Brzostek, E., Midgley, M.G., 2013. The mycorrhizal-associated nutrient economy: a new framework for predicting carbon–nutrient couplings in temperate forests. New Phytol. 199 (1), 41–51.

Posta, K., Marschner, H., Römheld, V., 1994. Manganese reduction in the rhizosphere of mycorrhizal and nonmycorrhizal maize. Mycorrhiza 5, 119–124.

Read, D.J., Perez-Moreno, J., 2003. Mycorrhizas and nutrient cycling in ecosystems–a journey towards relevance? New Phytol. 157 (3), 475–492.

Redecker, D., Schussler, A., Stockinger, H., Sturmer, S.L., Morton, J.B., Walker, C., 2013. An evidence-based consensus for the classification of arbuscular mycorrhizal fungi (Glomeromycota). Mycorrhiza 23, 515–531.

Rillig, M.C., Aguilar-Trigueros, C.A., Camenzind, T., Cavagnaro, T.R., Degrune, F., Hohmann, P., Lammel, D.R., Mansour, I., Roy, J., van der Heijden, M.G.A., Yang, G., 2019. Why farmers should manage the arbuscular mycorrhizal symbiosis. New Phytol. 222, 1171–1175.

Rousseau, J.V.D., Sylvia, D.M., Fox, A.J., 1994. Contribution of ectomycorrhiza to the potential nutrient-absorbing surface of pine. New Phytol. 128, 639–644.

Sasse, J., Martinoia, E., Northen, T., 2017. Feed your friends: Do plant exudates shape the root microbiome? Trends Plant Sci. 23, 25–41, http://dx.doi.org/10.1016/j.tplants.2017.09.003.

Selosse, M-A., Le Tacon, F., 1998. The land flora: a phototroph-fungus partnership? Trends Ecol. Evol. 13, 15–20. doi: 10.1016/s0169-5347(97)01230-5.
Soudzilovskaia, N.A., van Bodegom, P.M., Terrer, C., Zelfde, M.V., McCallum, I., McCormack, M.L., Fisher, J.B., Brundrett, M.C., de Sa, N.C., Tedersoo, L., 2019. Global mycorrhizal plant distribution linked to terrestrial carbon stocks. Nat. Commun. 10, 5077, https://doi.org/10.1038/s41467-019-13019-2.
Souza, T., 2015. Glomeromycota classification. Handbook of Arbuscular Mycorrhizal Fungi. Springer International Publishing, Switzerland, pp. 87–128.
Strullu-Derrien, C.M.A., Selosse, P., Kenrick, F.M., Martin, F.M., 2018. The origin and evolution of mycorrhizal symbioses: from palaeomycology to phylogenomics. New Phytol. 220, 1012–1030.
Sylvia, D.M., 1994. Vesicular-arbuscular mycorrhizal (VAM) fungi. In: Weaver, R.W. et al., (Ed.), Methods of Soil Analysis, Part 2. Microbiological and Biochemical Properties. Soil Science Society of America, Madison, WI, pp. 351–378.
Sylvia, D.M., 2005. Principles and Applications of Soil Microbiology, second ed. Pearson Prentice Hall, Upper Saddle River, N.J.
Tedersoo, L., Bahram, M., Zobel, M., 2020. How mycorrhizal associations drive plant population and community biology. Science 367, 1–9.
Tian, H., Drijber, R.A., Zhang, J.L., Li, X.L., 2013. Impact of long-term nitrogen fertilization and rotation with soybean on the diversity and phosphorus metabolism of indigenous arbuscular mycorrhizal fungi within the roots of maize (*Zea mays* L.). Agri. Ecosyst. Environ. 164, 53–61.
van der Heijden, M.G.A., Martin, F.M., Selosse, M-A., Sanders, I.R., 2015. Mycorrhizal ecology and evolution: the past, the present, and the future. New Phytol. 205, 1406–1423.
Wallander, H.A., Ekblad, D.L., Godbold, D., Johnson, A., Bahr, P., Baldrian, R.G., Björk, B., Kieliszewska-Rokicka, R., Kjøller, H., Kraigher, C., Plassard, M., Rudawska, M., 2013. Evaluation of methods to estimate production, biomass and turnover of ectomycorrhizal mycelium in forests soils—A review. Soil Biol. Biochem. 57, 1034–1047.
Wang, X.-X., Hoffland, E., Gu Feng, G., Kuyper, T.K., 2017. Phosphate uptake from phytate due to hyphae-mediated phytase activity by arbuscular mycorrhizal maize. Front. Plant Sci. 8, 684.
Wipf, D., Krajinski, F., van Tuinen, D., Recorbet, G., Courty, P.E., 2019. Trading on the arbuscular mycorrhiza market: from arbuscules to common mycorrhizal networks. New Phytol. 223, 1127–1142.
Wright, S.F., Upadhyaya, A., 1998. A survey of soils for aggregate stability and glomalin, a glycoprotein produced by hyphae of arbuscular mycorrhizal fungi. Plant Soil 198, 97–107.
Yeung, E.C., 2017. A perspective on orchid seed and protocorm development. Bot. Stud. 58, 33.

Further reading

Bender, S.F., Wagg, C., van der Heijden, M.G., 2016. An underground revolution: biodiversity and soil ecological engineering for agricultural sustainability. Trends Ecol. Evol. 31, 440–452.
Bonfante, P., Genre, A., 2015. Arbuscular mycorrhizal dialogues: do you speak 'plantish' or 'fungish'? Trends Plant Sci. 20, 150–154.
Bucking, H., Mensah, J.A., Fellbaum, C.R., 2016. Common mycorrhizal networks and their effect on the bargaining power of the fungal partner in the arbuscular mycorrhizal symbiosis. Commun. Integr. Biol. 9, e1107684.
Johnson, N.C., Gehring, C., Jansa, J. (Eds.), 2017. Mycorrhizal Mediation of Soil. Elsevier.
Koltai, H., Kapulnik, Y. (Eds.), 2010. Arbuscular Mycorrhizas: Physiology and Function. Springer Science & Business Media.
Peay, K.G., 2016. The mutualistic niche: mycorrhizal symbiosis and community dynamics. Annu. Rev. Ecol. Evol. Syst 47, 143–164.
Rillig, M.C., Sosa-Hernandez, M.A., Roy, J., Aguilar-Trigueros, C.A., Valyi, K., Lehmann, A., 2016. Towards an integrated mycorrhizal technology: harnessing mycorrhiza for sustainable intensification in agriculture. Front. Plant Sci. 7, 1625.
van der Heijden, M.G., Bardgett, R.D., van Straalen, N.M., 2008. The unseen majority: soil microbes as drivers of plant diversity and productivity in terrestrial ecosystems. Ecol. Lett. 11, 296–310.

CHAPTER

13

Carbon transformations and soil organic matter formation

Jeffry J. Fuhrmann[a], *David A. Zuberer*[b]

[a]Department of Plant and Soil Sciences, University of Delaware, Newark, DE, United States; [b]Department of Soil and Crop Sciences, Texas A&M University, College Station, TX, United States

Without death and decay, how could life go on?—**John Burroughs**

Soil microorganisms play a vital role in the cycling of carbon (C) on Earth. The plants responsible for nearly all terrestrial CO_2 fixation are soil inhabitants and, as such, they deposit a wide variety of C-containing materials to soils including shoot and root structural tissues, storage compounds, and various root exudates. Additionally, soils contain diverse microbial and animal communities, members of which die and contribute their own unique C-containing constituents. Humans add organic materials to soils both purposefully (e.g., manures, composts, pesticides) and accidently (e.g., oil spills, industrial solvents). All of these organic materials are subject to decomposition by soil microbes; in fact, humans and higher lifeforms in general depend absolutely on the degradative abilities of soil microbes. Knowing how microbes perform this essential task is critical for understanding how soils function to produce food and fiber, cycle C and other nutrients, and impact the quality of soil, water, and air resources. As such, aspects of C in soils are presented in several chapters in this textbook. In particular, this chapter will discuss the fundamental aspects of C in soil, focusing primarily on the types of plant, animal, and microbial residues added to soil, the specific microbiological processes responsible for their decay, and the nature of the C-containing compounds that tend to persist in soil and contribute to soil organic matter (SOM) reserves. Chapter 14 (Applied Aspects of Soil Carbon) will build on these topics by presenting examples of how land management practices affect organic C inputs, decomposition, and storage in soil and the resulting consequences for overall soil health. Still other chapters will discuss C in additional contexts including nutrient cycling (Chapters 15–19), pollution (Chapters 21 and 22), climate change (Chapter 20), and composting (Chapter 24).

Principles and Applications of Soil Microbiology. http://dx.doi.org/10.1016/B978-0-12-820202-9.00013-7
Copyright © 2021 Elsevier Inc. All rights reserved.

Terrestrial carbon cycle

Soils are a major pool of stored C and contain an amount of C at least equal to that in the terrestrial living biomass and atmosphere combined (Table 13.1). Some estimates of the soil C pool are even greater (Weil and Brady, 2019; Scharlemann et al., 2014). In natural systems, the relative flux of C between net photosynthesis (C fixation minus plant respiration) and decomposition processes is roughly balanced (Table 13.1; Fig. 13.1). As such, the increasing amounts

TABLE 13.1 Soil carbon pools in relation to atmospheric carbon pools and annual fluxes.

		Carbon (Pg)[a]
Pool		
	Soil organic matter	1500–2400
	Living biomass	450–650
	CO_2-C in atmosphere	860
Annual flux		
	Net primary production (net photosynthesis—plant respiration)	63
	Soil respiration (release of CO_2-C to atmosphere)	60
	Mineralization of eroded soil C[b] (release of CO_2-C to atmosphere)	1.1

[a]*Pg, Petagrams (10^{15} g); Pool values from Friedlingstein et al. (2019). Flux values from Lal (2018), who estimated that an additional 10.7 Pg C year^{-1} is released to the atmosphere due to anthropogenic activities.*
[b]*Mineralization of soil C dispersed to various environments due to erosional processes.*

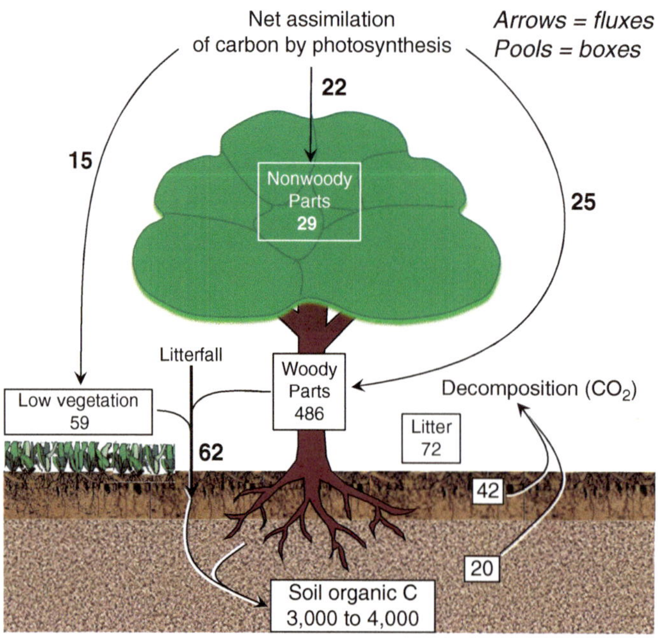

FIGURE 13.1 **Terrestrial carbon cycle showing fluxes and reservoirs of the various components.** Photosynthesis results in net assimilation of atmospheric CO_2-C, which is approximately balanced by the decomposition of plant residue and litter, and soil organic matter. The sizes of the compartments are indicated in units of Petagrams (Pg = 10^{15}g). *Source: Adapted from Post et al. (1990).*

of atmospheric CO_2 occurring over the last 100 years or so is due primarily to the burning of fossil fuels (Chapter 20), although land clearing and agricultural practices, especially tillage, have also contributed significantly (Chapter 14).

Significance of microbes to carbon cycling

A basic activity of soil microbes, like that of other life forms, is survival through reproduction. Soil microbes use residue components as substrates for energy and also as carbon sources in the synthesis of new cells. Energy is furnished to the microbial cells through the oxidation of the organic compounds (Chapter 3). The major end product of respiration is carbon dioxide, which is released back into the atmosphere. This release of CO_2 results from the mineralization of the organic C present in the original material, **mineralization** referring to the conversion of an organic form of an element to an inorganic form. A related, more general, term is **decomposition**, which is defined as the chemical breakdown of a compound into simpler compounds, often accomplished by microbial metabolism.

Microorganisms are generally short-lived and may themselves be decomposed by successive populations that find the resulting dead cells to be more suitable growth substrates than the initial residues. The overall process of decomposition generally involves a broad spectrum of complementary microbes that act in concert on a substrate. They differ in the kinds of enzymes they produce to disassemble the wide array of organic compounds present. Through their combined efforts, the carbon present in various organic compounds is progressively transformed to its most oxidized form, that of carbon dioxide.

One of the more persistent transformation products of residue decay is SOM, traditionally referred to as humus. SOM has long been recognized as a valuable constituent because it imparts many desirable biological, chemical, and physical properties to soil. The nature of organic residues and the subsequent microbial decomposition of their components results in SOM, a collection of substances that enhances soil quality (Chapter 14).

Organic materials added to soil

The first step in our discussion of carbon transformations in soil is to focus on the nature of organic materials added to soil, beginning with the kinds, amounts, and composition of plant residues from cultivated crops or native vegetation.

Plant residues

In agricultural ecosystems, the residues that decompose in soil are generally derived from crop plants and consist primarily of leaf, stem, and root tissues that remain after harvest. The residues may comprise an annual production of several megagrams per hectare (Mg ha^{-1}) that, with time, become fully incorporated into the soil. Because of their location in the soil, roots are very susceptible to decay. Even during growth, roots supply exudates and sloughed-off cell materials that serve as carbon substrates for the microbial population in the same manner as post-harvest root residues (Chapter 11).

In natural ecosystems, the residues that undergo decomposition are derived primarily from native prairie and forest vegetation. In the native prairie, the aboveground plant debris accumulates year after year on the soil surface. In addition, considerable belowground residue is deposited each year from root systems that die back periodically during periods of cold- or drought-induced dormancy. In the forest system, leaf litter and other plant debris fall to the soil surface and accumulate as layers in which decay occurs.

Residue quantities

Total carbon inputs (above- and belowground) to terrestrial systems vary widely depending on corresponding plant biome, ranging from approximately 900 kg C ha^{-1} $year^{-1}$ for tundra and desert ecosystems to 20,300 kg C ha^{-1} $year^{-1}$ for tropical forests (Paul, 2016). In comparison, representative values for croplands and temperate forests globally are 4,800 and 8,500 kg C ha^{-1} $year^{-1}$, respectively. Carbon inputs from the above-ground residues of different agricultural crops vary considerably (Table 13.2). Moreover, these values can be expected to vary for different regions as modified by soil, climate, and agricultural management.

TABLE 13.2 Estimated average carbon content of residues for selected crops for the year 2000 in the United States.

Crop	Carbon content of crop residues (kg C ha^{-1} $year^{-1}$)
Barley	1,930
Corn	3,725
Oat	1,405
Sorghum	2,245
Soybean	1,500
Sunflower	1,040
Wheat	1,710

Carbon contents calculated from residue values obtained from Johnson et al. (2006) assuming an average C content in residues of 50%. (Cellulose and lignin contain approximately 44% and 65% C, respectively.) Original residue values were estimated based on average grain yields.

Residue yields for some Missouri crops, fairly representative of central US agriculture, are given in Table 13.3. The values illustrate the quantitative relationships involved in the decomposition process and serve as reference points for the study of specific cases in other regions and land uses. For wheat and soybean, the total annual residues are about 10,000 kg ha^{-2} with the root contribution being somewhat less than half that value (note that g $m^{-2} \times 10 =$ kg ha^{-1}). This amount of net annual production is similar to that for native prairie in this region. Corn production is considerably greater, with a mean annual value of about 27,000 kg ha^{-1}. The more productive forests of the region would show an annual litter drop similar to this value. These residues, in combination with root exudates, supply the total annual substrate input for the soil organisms to a depth of approximately 20 cm. The corresponding microbial biomass is around 150 g m^{-2} (1,500 kg ha^{-1}), and the SOM content ranges from 2% to 2.5% on a soil weight basis. A portion of the substrate carbon is used by the microbial community for reproduction, but a larger share (approximately 70%) is liberated as carbon dioxide derived

TABLE 13.3 Annual residue production (kg ha^{-1}) for three agronomic crops in Missouri, United States.

Residue type	Corn	Soybean	Wheat
Stems and leaves	12,620	4,000	5,900
Cobs or pod hulls	1,580	1,360	–
Roots	13,040	4,780	5,210
Total	27,240	10,140	11,110

Values calculated from Buyanovsky and Wagner (1986). Residue estimates based on direct measurements.

from energy-yielding metabolic processes. Soil organisms work to decompose the residue substrates at a rate that generally maintains an equilibrium level of SOM year after year for a given long-term crop production system.

Residue composition

The crop residues added to and decaying in soil have various compositions and are comprised of complex polymers such as cellulose and lignin as well as soluble cellular constituents. As described below, these compounds differ in their susceptibility to degradation. Representative compositions of aboveground residues of corn, soybean, and wheat are shown in Table 13.4.

TABLE 13.4 Representative percentage compositions of above–ground corn, soybean, and wheat residues.

Component	Corn	Soybean	Wheat
		%	
Soluble components	29.3	55.7	28.8
Hemicellulose	26.8	9.0	18.4
Cellulose	28.4	22.2	36.1
Lignin	5.6	22.9	14.1
Ash[a]	9.3	6.4	8.4
Nitrogen	1.0	2.2	0.9

[a]*Material remaining following ignition of residues.*
Values calculated from Broder and Wagner (1988).

Microbial residues

Remnants of dead soil microbes are also substrates for living microorganisms. Bacterial cell walls are composed primarily of peptidoglycan, a polymer of N-acetylglucosamine and N-acetylmuramic acid, along with other polysaccharide materials, that serve as substrates. Fungal cell walls composed of chitin can also be degraded. Like those of higher plants, algal cell walls often contain cellulose (Chapter 7), as do those of the Oomycota (Chapter 6). These too differ in their susceptibilities to degradation, with fungal cell walls being more difficult to degrade than those of bacteria (Kästner and Miltner, 2018). As discussed later, recent evidence indicates that a large portion of the persistent organic C in soils is derived from microbial cell walls.

Other residues and substrates

Land application of biosolids (sewage sludge) and animal manure is a common practice, aspects of which are covered in Chapter 14. Additionally, synthetic pesticides applied to agricultural fields and residential areas routinely end up in soils, as do various other synthetic organic compounds such as industrial solvents from accidental spills. The degradation of these compounds is described in Chapter 21.

Decomposition of naturally occurring organic materials

Factors influencing degradation

Incorporation of residues

Decomposition of residues is enhanced by mixing with soil. In agricultural settings, decay of aboveground residues begins and progresses in a manner characteristic of the particular agricultural management system. Conventional tillage operations, such as plowing and disking, incorporate straw and stalks within the soil where contact with moisture and microbes hasten decay. Minimum and no-till operations result in residues on the soil surface, and the decay process progresses in a manner similar to that for natural ecosystems (Chapter 14). At the interface between residue litter and moist soil, the microbial community becomes very active. The decay progresses primarily from beneath this litter layer, although in moist environments the entire residue layer may be actively under attack by soil microbes. During the decay process, some of the litter may be transported beneath the soil surface by the soil fauna, where active decay accelerates as it becomes mixed into the moist upper mineral horizon of the soil. Additionally, the activities of soil fauna fragment residues into smaller particles, thereby increasing the surface area available for microbial attack. Variations in moisture, temperature, nutrient supply, and pH modify the rate of decay. Limitations imposed by these influences may cause greater fluctuation in the rate of decay for surface residues than for residues residing in the bulk soil.

Composition effects

Composition influences decay because residues do not decompose as whole units. Rather, the various groups of organic compounds comprising plant tissues are selectively attacked by a range of soil microbes, each of which produces a particular suite of degradative enzymes active on the insoluble polymers and soluble, low molecular weight, compounds. Fungi are thought to be particularly important in the degradation of polymeric substances, whereas bacteria are more active on smaller, more readily degradable, compounds contained in the original residue or produced by the degradation of polymers (Kästner and Miltner, 2018).

The general processes involved in decomposition of residues may occur under aerobic or anaerobic conditions (Fig. 13.2). For most soils, the aerobic pathway is of greatest importance. Residue decay progresses through the interrelated activities of many different microorganisms. Simple substrates that are readily assimilated may have a community of microorganisms feeding in competition with one another as the material comes under attack. More complex and resistant substrates, such as carbohydrate polymers, may initially attract one or more microbial groups that break down the polymer into simpler components that can be used by a wide range of microorganisms. In nearly all cases, the final step is the assimilation of decay

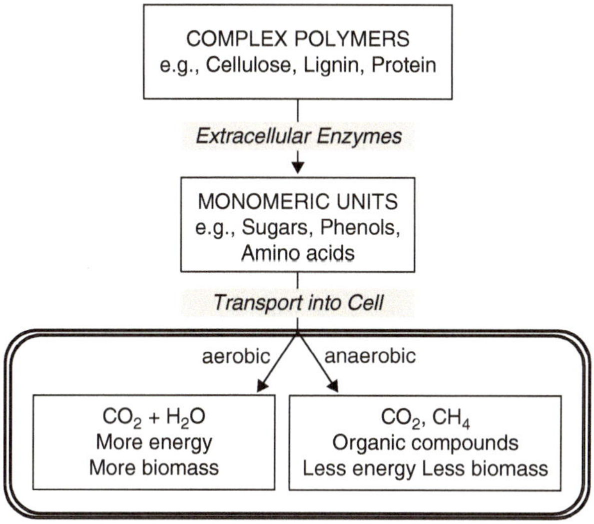

FIGURE 13.2 **Major steps in the decomposition of polymers by soil microorganisms.**

products by diverse microbes that oxidize the compounds to obtain energy and carbon for the production of new cell tissue. A generalized summary of the microbial decomposition process under aerobic conditions is presented in Fig. 13.3. This figure shows a progressive decrease in substrate accompanied by an increase in evolved carbon dioxide along with a temporary increase in microbial biomass.

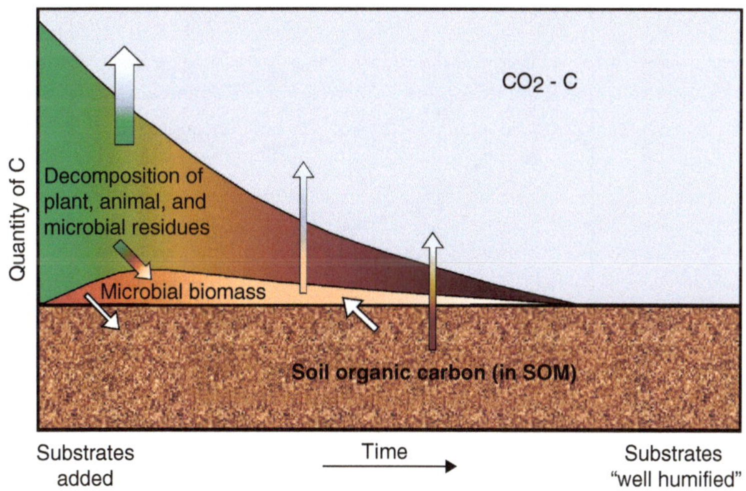

FIGURE 13.3 **General reactions in microbial decomposition of plant, animal, and microbial residues.** When residue is added to the soil and undergoes decomposition, the amount of original residue decreases with time. The microbial biomass increases as does the carbon evolved as carbon dioxide. Note the initial increase in microbial biomass and then the steady decline. As the soil returns to a steady-state condition, the microbial populations and soil organic matter return to approximately the same levels prior to adding residues. Width of the *arrow* indicates the relative amount of carbon exchanged between carbon pools. *Long upward arrows* indicate relative magnitudes of carbon dioxide fluxes from the soil. *Source: Adapted from Weil and Brady (2019).*

Although agricultural residues consist primarily of various complex polymers, they also contain smaller amounts of simple, often soluble, organic constituents that can be utilized by a vast array of soil microbes. In favorable environments, these simple constituents may be completely converted to new cell structures and carbon dioxide within a few days. For example, studies with a common simple sugar like glucose show that this substrate is completely metabolized by soil microorganisms in 1 or 2 days. Free amino acids also are readily used by soil microbes.

The selective action of microbes in attacking the various components of the residue was demonstrated by Broder and Wagner (1988), who studied the decay of wheat straw buried in nylon bags in a field environment. Their experimental system allowed periodic removal of one or more samples to examine the progression of decay. The results shown in Fig. 13.4 demonstrate the rapid rate at which at least a part of any crop residue can decompose. Of the various polymeric carbohydrates (polysaccharides) examined, hemicelluloses were degraded with relative ease by numerous soil microorganisms and, as a result, this component was essentially exhausted within 160 days. Cellulose, the most abundant polysaccharide, decomposed more slowly but still more rapidly than the residue as a whole, and approximately 20% of this component remained at the end of the study. Lignin decayed at a markedly slower rate than the other wheat straw components. This persistence reflects its polymerized aromatic character and possibly also its physical stabilization on mineral surfaces and within soil aggregates due to the greater time required for its degradation.

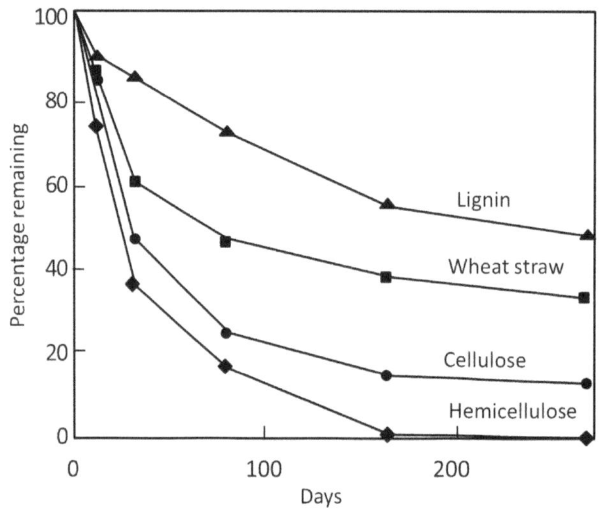

FIGURE 13.4 **Decomposition of wheat straw and its major constituents in a silt loam with the residue buried in nylon bags to facilitate periodic removal and analysis.** The initial composition of wheat straw was 50% cellulose, 25% hemicellulose, and 20% lignin (Wolf and Wagner, 2005). *Source: Adapted from Broder (1985).*

Although this study did not examine degradation of root residues, other studies have shown that wheat roots decompose more rapidly than straw. This most likely occurs because the roots are already in intimate contact with an active soil population, and their decomposition may begin quickly once plant senescence begins. An active heterotrophic population surrounding the roots would have already developed and been nurtured by simple organic compounds exuded by the roots and other cellular debris that is sloughed off as the plant root grows. It should be reiterated that residues derived from aboveground crop tissues, such as straw and stubble, may begin to decompose at the soil surface if soil moisture is adequate, but their decay will accelerate following soil incorporation by tillage or animal activities.

The degradation of microbial cells follows similar patterns. Soluble cellular constituents degrade quickly but, as in the case of lignin in plant residues, some microbial cell components can be very resistant to decay. An example is melanized (darkly pigmented) fungal cell walls whose resistance may lead to their eventual appearance in SOM. By contrast, hyaline (nonpigmented) fungal cell wall tissues, such as those from the fungus *Aspergillus niger*, consist predominantly of carbohydrate polymers, which are easily decomposed. Representative decomposition curves for a number of different microbial cells are presented in Fig. 13.5.

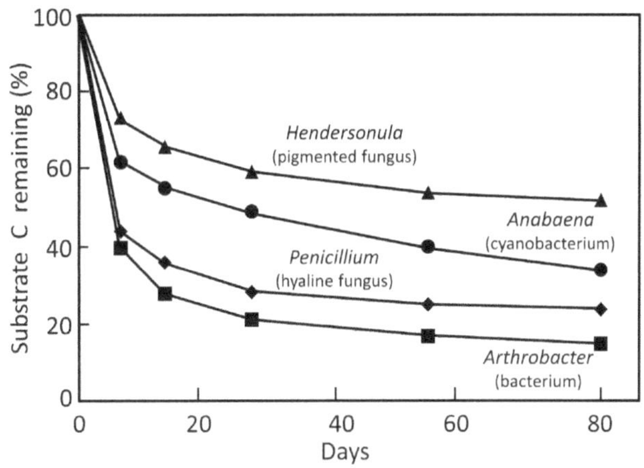

FIGURE 13.5 **Decomposition of representative microbial cells added to soil and incubated under aerobic conditions.** The percentage substrate carbon remaining in the soil does not consider formation of microbial biomass or humus (Wolf and Wagner, 2005). *Source: Adapted from Burns and Martin (1986).*

Seasonal influences

The annual cycle of climate greatly influences the progression of residue decay. For example, microbial activity may be restricted by cold temperatures in winter or by dry conditions in summer. In the case of wheat in the central United States, about 50% of the original residue mass carries forward through winters, and the warming that occurs in spring marks the beginning of a second phase of decay (Fig. 13.6). Corn and soybean crops, which mature later in the fall, may have a greater proportion of the total residue carried into the next spring, at which time a flush of decomposition activity occurs. This carryover for soybean is largely stem residues given that leaf materials drop to the ground in the summer as the plant matures, and the leaves degrade almost completely by the end of the fall season.

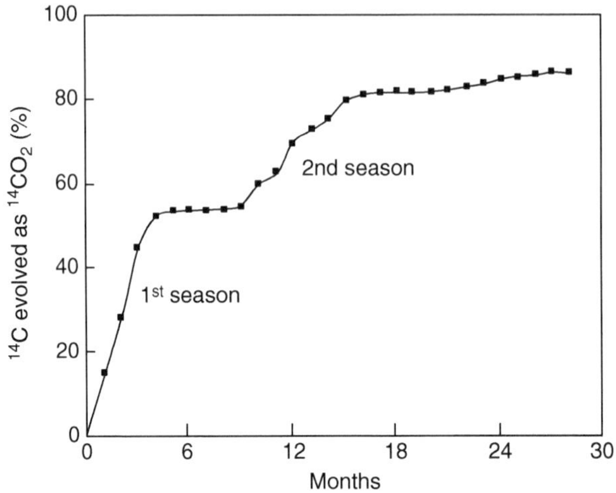

FIGURE 13.6 **Cumulative curve of $^{14}CO_2$ evolution from labeled wheat residues following a June harvest with residues disked into soil** (Wolf and Wagner, 2005). *Source: Adapted from Buyanovsky and Wagner (1987).*

The overall process of decomposition in soil, therefore, should be looked upon as an annual cycle. New residues are produced each year, a large percentage of which is completely mineralized to carbon dioxide during that year. However, residues generally persist into the second year so that most materials have a residence time of 2 years or more (Fig. 13.6). These materials, along with substantial amounts of dead microbial debris (necromass), can become physically protected by various mechanisms (Kästner and Miltner, 2018; see section on SOM formation), and persist in the soil for tens or hundreds of years. Nevertheless, even these materials are slowly mineralized at rates of 2%–5% year^{-1}, especially in response to soil disturbance which makes them positionally available for enzymatic attack.

Temperature and soil moisture strongly modify residue decay within the framework of the seasonal pattern just described. Wheat residues decay rapidly in late summer and fall when soil temperatures in the central United States approach 24°C, with soil moisture at a satisfactory level at least for some part of that period. The rate of microbial response to increasing temperature is greatest over the range of 3°C–19°C, but more modest responses occur if temperature increases from 20°C to 25°C.

The soil water content for maximum decomposition is about 50%–60% of the water holding capacity of the soil or −20 to −50 kPa moisture potential. This level helps to ensure good aeration for oxidative processes and provides sufficient moisture for all groups of heterotrophic microbes that are most active in the decomposition process.

Degradation of specific compounds

Soluble components

Plant residue contains an appreciable amount of water-soluble organic compounds (Table 13.4), such as free amino acids, organic acids, and sugars that are readily available for decomposition by the vast majority of soil microbes. These materials are rapidly taken up by the microbes for catabolic and anabolic activities. These water-soluble compounds are generally utilized by bacteria and "sugar fungi" (Mucoromycota such as *Mucor* and *Rhizopus*; Chapter 6), all of which exhibit rapid growth rates. These organisms are called zymogenous (Chapter 10).

Numerous water-soluble compounds also pass from the plant root into the soil as exudates. As with the water-soluble compounds in plant residue, root exudates are rapidly used by the microbes in the rhizosphere.

Proteins

Proteins are generally a small but significant component of plant residue added to soil. However, proteins can be a sizable portion of dead microbial cells and plant material used as a green manure. Proteins are polymers of amino acids linked by peptide bonds which are readily broken by hydrolytic enzymes. However, recent studies have shown that proteins and shorter peptides tend to persist in soil as a result of their adsorption to the surfaces of soil particles as well as through other mechanisms (Kästner and Miltner, 2018). Nevertheless, a wide variety of microbes produce proteolytic enzymes (proteases, peptidases) that hydrolyze proteins into individual amino acids. These products are then transported into the cell for further catabolism or for synthesis of new proteins required by the microbe (Fig. 13.7). Proteins are rich in nitrogen (~16% N) and therefore release nitrogen to soil when mineralized.

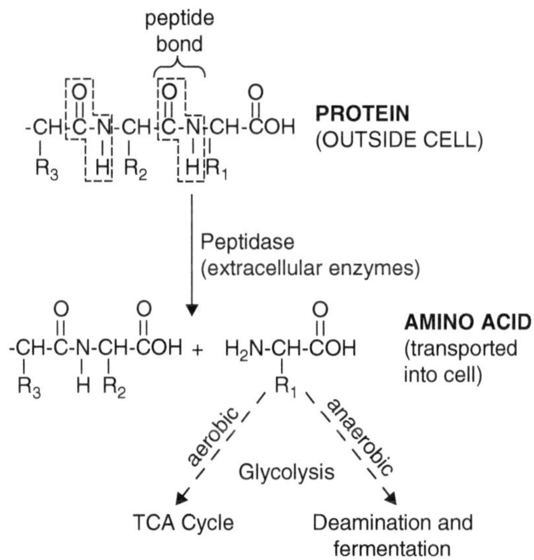

FIGURE 13.7 **Degradation of protein by hydrolysis of the peptide bond.**

Cellulose

Most carbon in plant residues is in the form of complex carbohydrates, such as the structural polysaccharides. Structural polysaccharides are so named because they are generally the constituents responsible for conferring structural rigidity to cell walls, thereby allowing plants to grow as erect structures. The most common structural polysaccharide in residues is cellulose, which is a linear chain of glucose units joined by β-1,4 linkages (Fig. 13.8). Cellulose is therefore a glucan, meaning a polymer of glucose, and each cellulose molecule may contain up to 10,000 glucose units. The cellulose content of plant material generally increases as the plant matures; it may be as low as 15% on a dry-mass basis for young plants but greater than 50% in wood, straw, and leaves. As stated previously, cellulose is also an important component of some algal cell walls and those of the Oomycota.

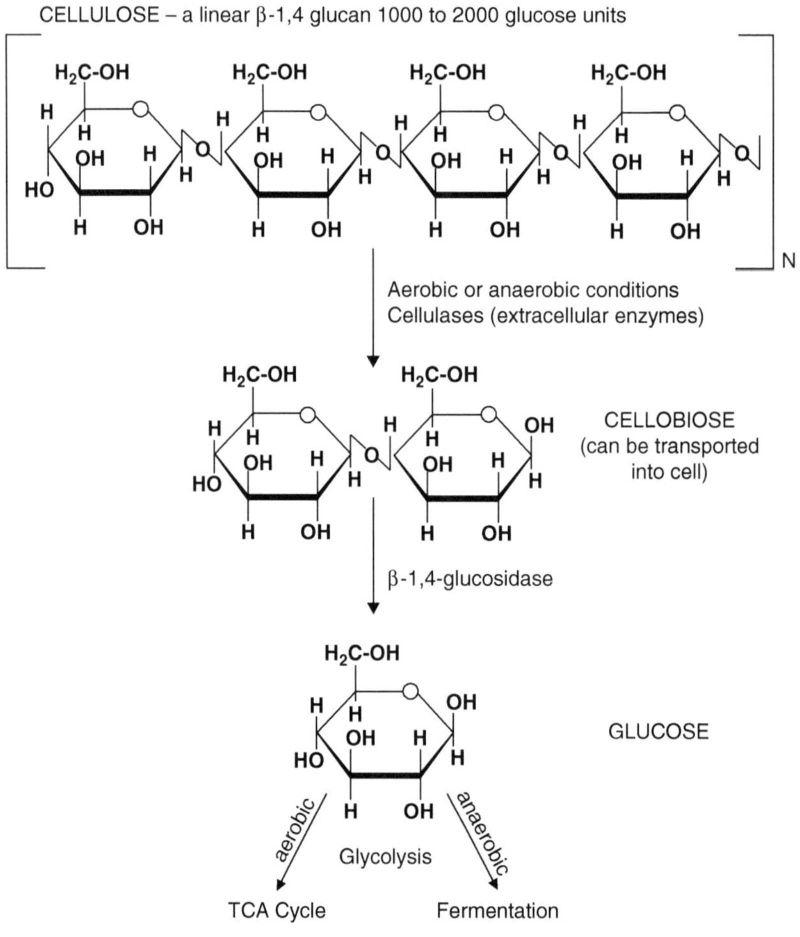

FIGURE 13.8 **Decomposition of cellulose under aerobic and anaerobic soil conditions.**

Because cellulose is a large polymer that is insoluble and too large to enter the microbial cell, it must first be cleaved by extracellular enzymes into smaller units that can be transported into the microbial cell for subsequent catabolism (Fig. 13.2 and Box 13.1). Cellulose decomposition is initiated by a diverse group of enzyme complexes known as cellulases. The cellulase enzyme complex generally contains three types of enzymes: a β-1,4-endoglucanase, a β-1,4-exoglucanase, and a β-1,4-glucosidase; the latter is also called cellobiase. The first

BOX 13.1

Extracellular enzymes determine the rate of decomposition of polymeric compounds

Polymeric substances are typically too large to be taken up intact by microorganisms. They must first be degraded by extracellular enzymes to produce smaller units that can be transported into a microbial cell to meet its nutrient and energy requirements. This extracellular degradation is generally the rate-limiting step in the decomposition of polymers. Extracellular enzymes are classified as either endoenzymes or exoenzymes. Endoenzymes operate along the internal portions of the polymer chain, whereas exoenzymes act at the end of the chain, cleaving off monomers and dimers. The fragments produced by endoenzymes serve to increase the number of "ends" available for the action of exoenzymes and thus facilitate degradation. Extracellular enzymes are typically synthesized in the cell and excreted into the soil but may alternatively be located on the external surfaces of cells (Traving et al., 2015). One might ask how microbes can sense the presence of insoluble polymers in order to produce the proper degradative enzymes (unnecessary enzyme production constitutes a drain on a cell's energy resources and must therefore be avoided). Because a portion of the extracellular enzymes produced during a prior decomposition event can persist in soil, it is thought that these residual enzymes may produce degradation products that serve as signals to a cell that a particular polymer is nearby, thereby inducing the cell to produce more of the corresponding enzyme(s).

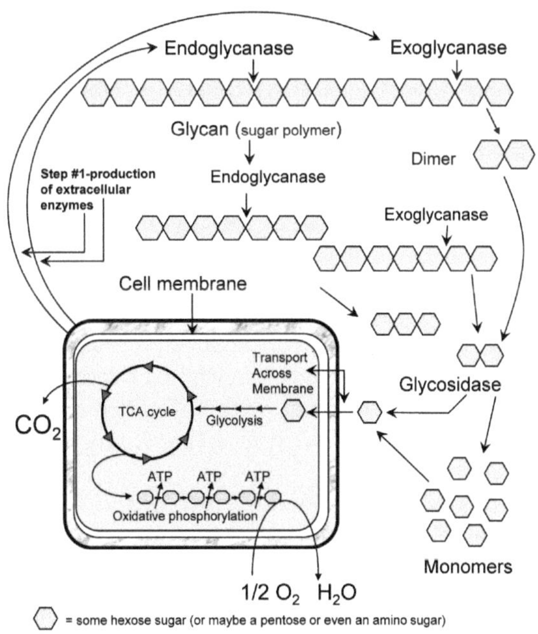

Depiction of the extracellular degradation of a polysaccharide and its subsequent intracellular processing by catabolic metabolism. Extracellular enzymes degrade the polymer into small sugar molecules (monomers) able to cross the cellular membrane. Inside the cell, the sugar is metabolized via glycolysis, the tricarboxylic acid (TCA) cycle, and oxidative phosphorylation to produce CO_2, H_2O, and metabolic energy (ATP) (Chapter 3).

step in cellulose mineralization involves the loss of crystalline-like structure, followed by depolymerization. The resulting linear chains of two or three glucose units are called cellobiose and cellotriose, respectively. Before the carbon can be metabolized by the cell for energy production, cellobiose and cellotriose must be hydrolyzed to single glucose units by β-1,4-glucosidase. The individual glucose molecules can then be metabolized for energy and biomass production.

Soil fungi, such as *Trichoderma*, *Aspergillus*, *Penicillium*, and *Fusarium*, and bacteria, such as *Streptomyces*, *Pseudomonas*, and *Bacillus*, are important in the initial extracellular depolymerization of cellulose. Once cellulose is hydrolyzed into smaller units that can be transported into a cell, the entire glucose-metabolizing microbial community of the soil can participate in the decomposition processes. Particularly noteworthy are the "brown rot" fungi, because they decompose the cellulose and hemicellulose in wood, leaving lignin and phenolic materials behind. Rotting wood under attack by these organisms takes on a brown color. In contrast, the "white rot" fungi can decompose all of the wood including the lignin, but a greater quantity of cellulose accumulates, leading to a white residue. Recent interest in the white rot fungi, notably *Phanerochaete chrysosporium*, has been stimulated by the observation that they may be important in bioremediation of soils contaminated with toxic organic chemicals (Chapters 21 and 22).

Hemicelluloses

Hemicelluloses are the second most common carbohydrate constituents in plant residues and consist of polymers containing hexoses (6-C sugars), pentoses (5-C sugars), and uronic acids. They constitute a diverse group of structural polysaccharides that comprise up to 30% of the dry mass of plant residue. Most frequently, hemicelluloses in plant residues contain from 50 to 200 sugar units. Hemicelluloses are glycans, a more general term than glucan, indicating they are polymers of any of several monosaccharides (i.e., not only glucose). The polymers are a collection of homoglycans and heteroglycans whose components are joined primarily by β-1,4 and β-1,3 linkages and may be branched or unbranched. The presence of uronic acid indicates that the $-CH_2OH$ group of the 6-C sugar has been oxidized to $-COOH$.

Pectic substances, an important component of plant cell walls, are chains of galacturonic acid, which is often esterified with a methyl group (i.e., methylgalacturonic acid). Pectins differ from other hemicelluloses in that the units are joined primarily by α-1,4 linkages. They occur especially in the middle lamella, where they help to "cement" the plant cell walls together. An example of pectin decomposition is shown in Fig. 13.9. The decomposition process involves several enzymes often referred to collectively as pectinases. Many of the same fungi and bacteria that degrade cellulose are active in the decomposition of pectic substances. Plant root invasion by plant pathogens and symbionts, such as the N_2-fixing *Bradyrhizobium* and *Rhizobium*, and mycorrhizal fungi, such as *Glomus*, has been related to production of pectinases.

Decomposition of hemicellulose is generally rapid and exceeds the rate for cellulose (Fig. 13.4). However, high levels of available simple sugars reduce hemicellulose decomposition, as the microbes will preferentially use more readily available substrates before attacking more complex materials. As with cellulose, the initial hydrolytic disintegration of the polymer occurs outside the cell; when the monomeric sugar or uronic acid residues are produced, they can be transported into the cell for subsequent metabolism.

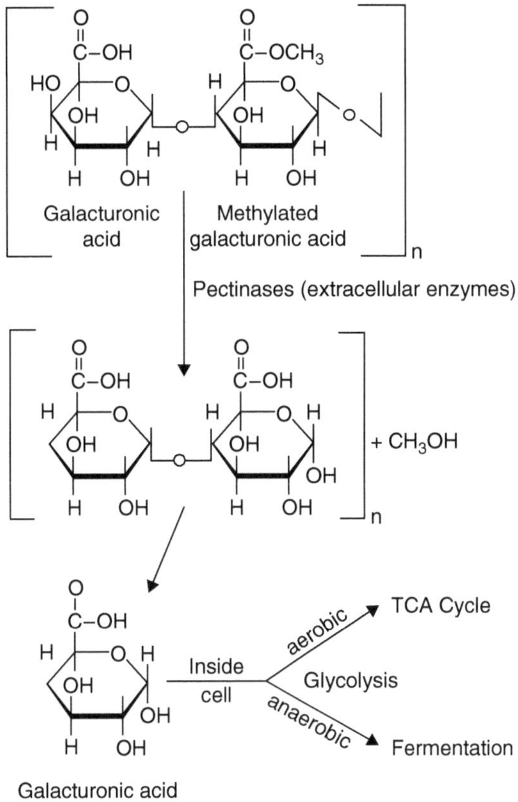

FIGURE 13.9 **Decomposition of the hemicellulose pectin in soil.**

Starch

Starch is a common plant storage polysaccharide that serves as an energy storage compound in cereal grains (seeds), stems, roots, rhizomes, and tubers. Starch can consist of both linear and branched polymers of glucose. When glucose is linked in the α-1,4 position, the linear polymer is known as amylose. The α-1,4 linkage facilitates a more rapid breakdown rate than the β-1,4 linkage found in cellulose. As a branched polymer, the glucose is linked in the α-1,4 and α-1,6 positions to produce a material known as amylopectin (Fig. 13.10).

Extracellular enzymes known as amylases (α-1,4-exo- and endoglucanase) are produced by numerous bacteria and fungi and hydrolyze amylose and amylopectin to the disaccharide (two-glucose units) maltose. Subsequent hydrolysis of maltose by an α-1,4-glucosidase (maltase) yields monomeric glucose units that can enter the glycolytic pathway. Inulin, a linear β-2,1 fructosan (fructose polymer), is another common storage polysaccharide. It is degraded by inulinase, an extracellular fructosanase.

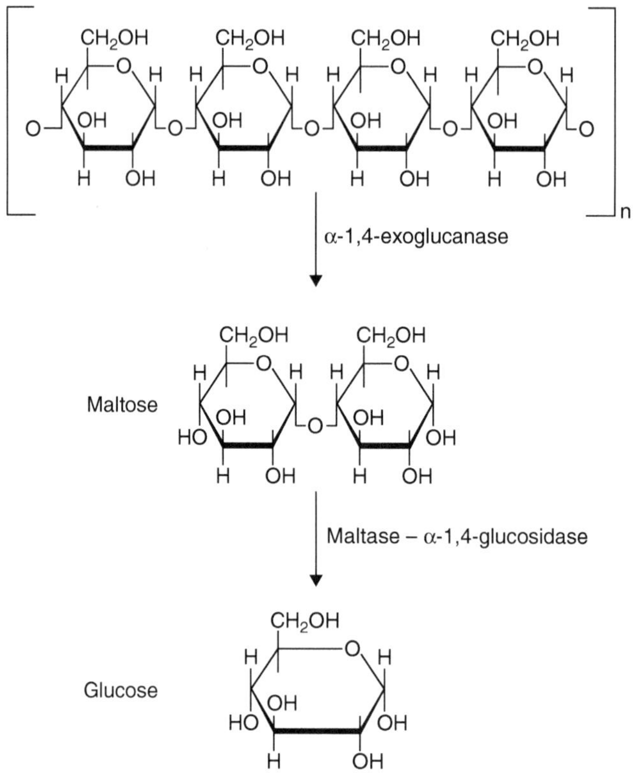

FIGURE 13.10 **Structure and decomposition of the linear polymer of starch, amylose.**

Lignin

Lignin is generally the third most common component of plant residue. It is especially prominent in woody tissues and is thus particularly important in forest ecosystems. Lignin is an integral component of the cell walls of plants, and its content increases as the plant ages. The lignin content of young plants is often less than 5% by mass whereas mature plants may contain 15%, and the wood in trees may have levels near 35%. Because the chemical composition of lignin varies among plant species, it is not possible to give a specific structure that is representative of all lignins. The basic building block of lignin is the phenylpropene unit that consists of a hydroxylated 6-C aromatic benzene ring (phenol) and a 3-C linear side chain. Three common phenyl alcohols used by plants in lignin synthesis are shown in Fig. 13.11. Lignin is typically composed of 500–600 phenylpropene units with randomly condensed units having substantial cross-linkage (Fig. 13.12). The linkage can involve ether linkages (C-O-C) or C-C bonds and may occur between benzene rings, side chains, or rings to side chains. Fig. 13.13 illustrates an example of possible degradation pathways for lignin.

Because of the size, complexity, and aromatic nature of lignin, its decomposition rate is generally slow compared to that of cellulose and hemicellulose (Fig. 13.4), although some studies have reported relatively rapid rates for some lignins (Olk et al., 2019). The decomposition process generally begins with oxidation and removal of the exposed side chains. The

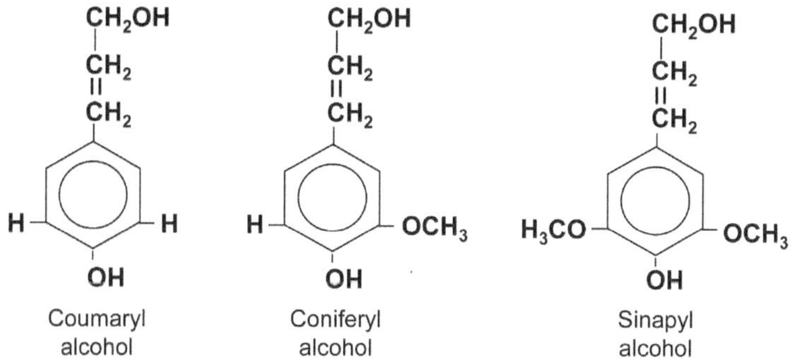

FIGURE 13.11 Examples of phenylpropene units that are the basic building blocks of lignin.

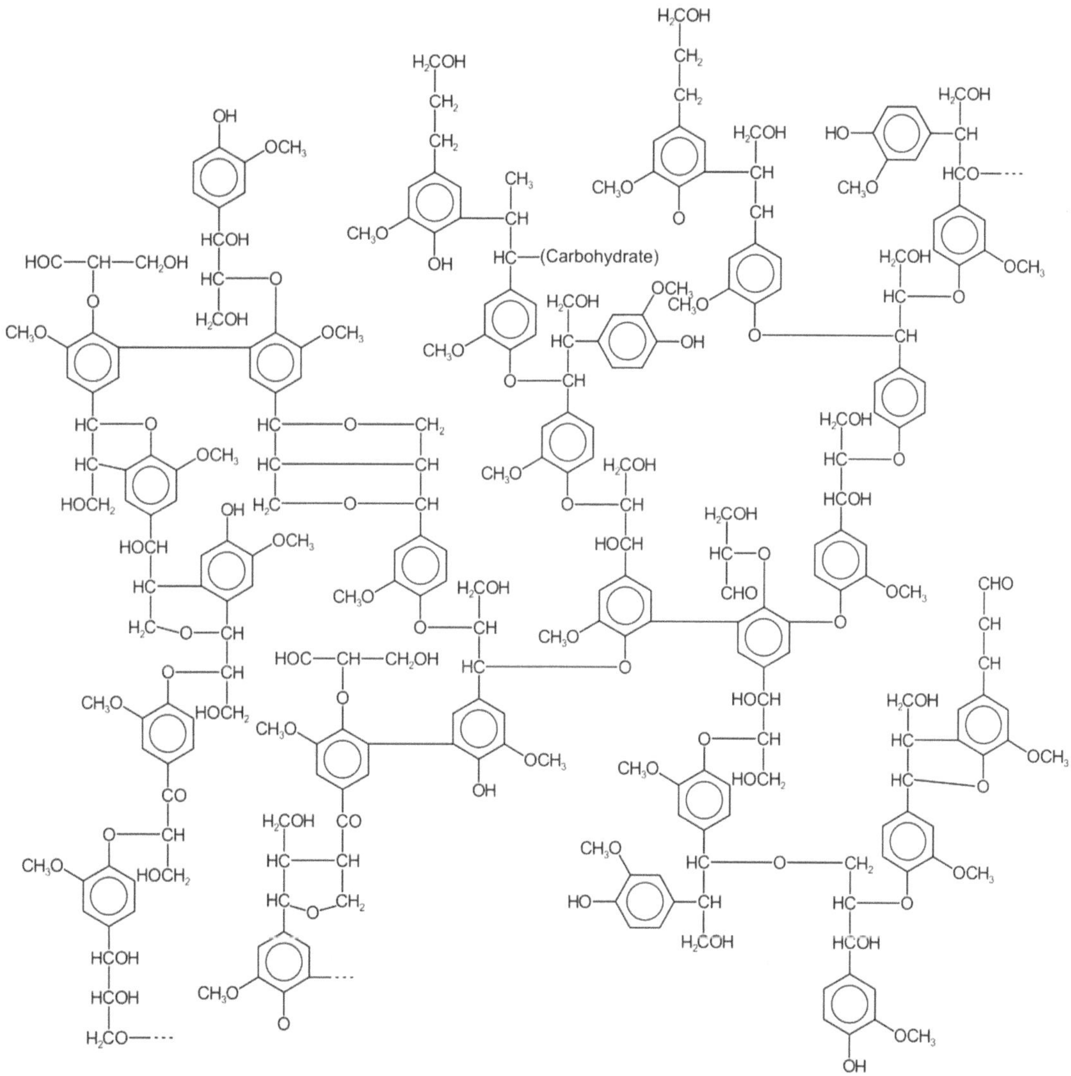

FIGURE 13.12 Generalized structure of lignin. *Source: Karol Głąbpl.wiki: Karol007commons: Karol007e-mail: kamikaze007 (at) tlen.pl / CC BY-SA (http://creativecommons.org/licenses/by-sa/3.0/)(https://commons.wikimedia.org/wiki/File:Lignin_structure.svg)*

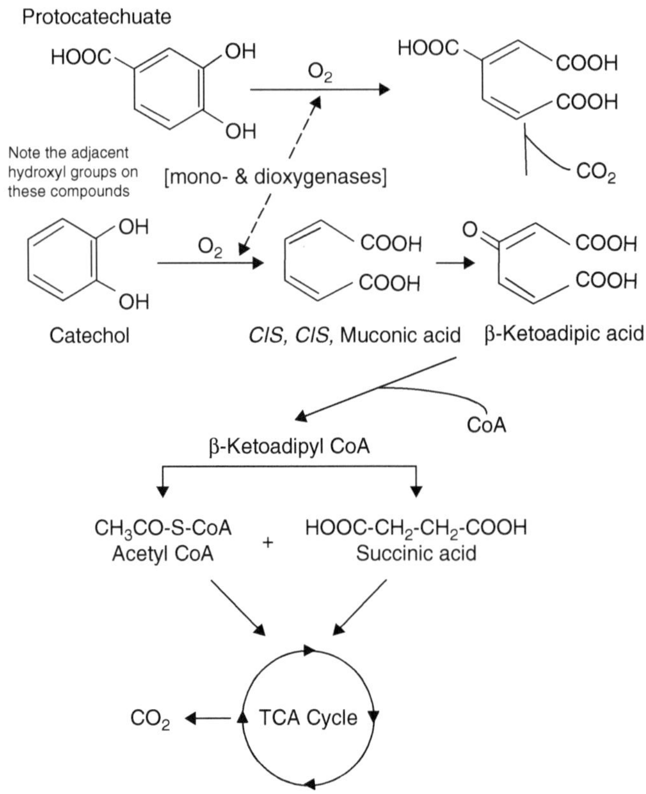

FIGURE 13.13 **Generalized pathway for the metabolism of the aromatic decomposition products of lignin via oxygenases and entry into the Krebs (TCA) cycle.**

methoxyl (CH_3O-) group is oxidized to carbon dioxide and a hydroxyl (-OH) group is left on the ring. The second step in lignin degradation is depolymerization, which results in the liberation of individual phenolic units with side chains. Due to the structural variability of lignin, microorganisms use relatively non-specific enzymes, such as lignin peroxidase, manganese peroxidase, and laccase, to catalyze oxidation and depolymerization. Once these units are released, they are available to much of the soil microbial community and may be transported inside microbial cells for subsequent metabolism. The side chains are then removed and, if hydroxyl groups exist on adjacent carbons on the phenolic ring, the ring can be cleaved. Ring cleavage via ortho or meta (position on the benzene ring) fission occurs under aerobic conditions (Chapter 21). Degradation of lignin can occur under anaerobic conditions, but the process is limited and ligninaceous materials tend to accumulate in these environments (e.g., peat bogs).

Fungi, especially basidiomycetes, are generally recognized as the major microbial group responsible for lignin degradation. However, actinomycetes, such as *Streptomyces*, and certain other bacteria are also involved. Research suggests that the efficiency of carbon utilization for microbial biomass production is low with most of the carbon being either evolved as carbon dioxide or incorporated into SOM.

Chitin

Chitin is a constituent of the cell walls of common soil fungi (e.g., *Aspergillus* and *Penicillium*), with from 3% to 25% of fungal biomass consisting of chitin on a dry-weight basis. Chitin is also added to soil as part of the exoskeletons of arthropods and crustaceans (Chapter 8). As a long-chain, linear biopolymer of N-acetylglucosamine units joined by β-1,4 linkages, chitin has a chemical structure similar to cellulose.

N-acetylglucosamine units are also present in the polysaccharide portion of the peptidoglycan in bacterial cell walls (Chapter 5). If the acetyl group is absent from N-acetylglucosamine, the resulting polymer of glucosamine is known as chitosan and is also found in fungal cell walls.

As a large, water-insoluble polymer, the first step in chitin degradation is hydrolysis into N-acetylglucosamine units by the extracellular enzymes chitinases (Fig. 13.14). The soluble N-acetylglucosamine is subsequently transported into the microbial cell where it is hydrolyzed to yield acetyl-CoA and glucosamine which are easily oxidized under aerobic

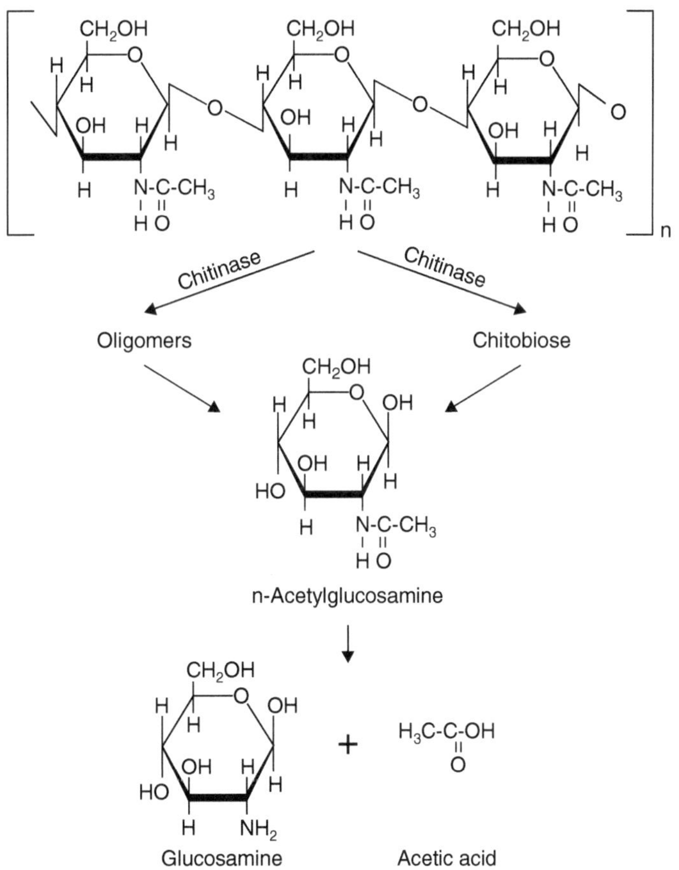

FIGURE 13.14 **Decomposition of chitin under aerobic soil conditions.**

conditions. In its naturally occurring state, chitin is often physically protected from degradation by formation of a protein–chitin complex. Chitin is important in the formation of SOM and, with a C/N ratio of 7 (~6.3% N), it provides a source of slowly mineralizable nitrogen.

The degradation rates of chitin and chitosan in soil appear to be similar to rates reported for cellulose. In aerobic soils, the dominant chitin degraders appear to be actinomycetes (Actinobacteria) of the genera *Streptomyces* and *Nocardia*. Fungi, such as *Trichoderma* and *Verticillium*, and bacteria, such as *Bacillus* and *Pseudomonas*, have also been shown to degrade chitin but appear to be less important than actinomycetes. It is common to use finely-ground chitin as the carbon source in selective media for the isolation of actinomycetes. As the particulates of chitin are degraded in the media, a "zone of clearing" forms around the colonies of degradative actinomycetes or other microorganisms.

Peptidoglycan (murein)

Bacterial cell walls contain varying amounts of peptidoglycan (also called murein), a polymer consisting of an alternating arrangement of the amino sugars N-acetylglucosamine and N-acetylmuramic acid (Fig. 15.4). It is degraded by lysozyme, an enzyme that hydrolyzes the β-1,4 glycosidic bonds linking the amino sugars. Peptidoglycan is relatively rich in nitrogen (5.5% N; C/N~8), and its mineralization therefore releases available nitrogen to plants and nearby microbes. The cell walls of archaea do not contain peptidoglycan, but some possess a similar polymer called pseudomurein (Chapter 5). Degraded fragments of bacterial cell walls, including peptidoglycan, are thought to be a major form of persistent organic C in soils.

Decomposition activity of microorganisms

Magnitude of the microbial community

The dynamic community of heterotrophic microbes engaged in residue decomposition is regulated primarily by aeration, moisture, temperature, pH, and nutrient status in the soil. Conditions favoring rapid decomposition of plant residues and multiplication of microorganisms include:

- plant residue with a low lignin content and small particle size
- adequate available nitrogen, either in the residue itself (i.e., low C/N) or as soil NH_4^+ or NO_3^-
- near neutral soil pH to allow diverse microbial populations to be active
- adequate soil moisture and aeration
- warm soil temperature with an optimum of 30°C–45°C

Population increases following addition of fresh residues may result in a doubling of the microbial biomass. Subsequent declines in populations result from the onset of unfavorable environmental conditions, such as cold temperatures or dry conditions, or from depletion of the more readily available substrates. Predation and parasitism by other microbes can also contribute to declines. Typical estimates of soil microbial biomass are around 150 g m^{-2} (or 1,500 kg ha^{-1}) and are influenced by the method used in the determination. Common methods involve fumigating the soil with chloroform and measuring the amount of carbon (or nitrogen) released by chemical analysis or by determination of CO_2 evolved during a subsequent incubation (Box 13.2).

BOX 13.2

Determining soil microbial biomass by fumigation-incubation method

Many approaches have been proposed to measure microbial biomass in soils. One method, called the fumigation-incubation method (Jenkinson and Powlson, 1976), has often been used as a standard by which other approaches are compared. It employs a 24-h laboratory incubation of a soil sample in the presence of chloroform vapor (fumigation) to lyse the resident microbial community. The vapor is removed, and a subsequent 10-day incubation is performed during which the CO_2 produced by the decomposition of the killed biomass is measured. (A sufficient number of microbes generally survives the incubation to grow and decompose the dead biomass; in some cases, the fumigated sample is reinoculated with a small amount of fresh soil.) The carbon liberated as carbon dioxide during the 10-day incubation, minus that produced by similarly incubated but nonfumigated (control) soil sample, is quantitatively related to the initial microbial biomass. The flush of carbon dioxide generally accounts for 40%–45% of the microbial biomass carbon, although this value can vary depending on the nature of soil sample and the incubation conditions.

For example, the amount of CO_2-C evolved during 10-day incubation was determined to be 35.8 mg CO_2-C 100 g^{-1} dry soil. A nonfumigated sample of the same soil evolved 23.5 mg CO_2-C 100 g^{-1} dry soil during the period.

The amount of microbial biomass C is calculated as:

Microbial biomass C = (A − B)/k

where A = CO_2- C evolved from the fumigated soil,
B = CO_2- C evolved from the nonfumigated soil,
and k = fraction of biomass C mineralized to CO_2 = 0.41

$$\text{Therefore, microbial biomass C} = \left(35.8 - 23.5 \text{ mg } CO_2 - C \text{ } 100 \text{ g}^{-1} \text{ dry soil}\right)/0.41$$
$$= 30.0 \text{ mg C} 100 \text{ g}^{-1} \text{ dry soil}$$

This value corresponds to approximately 670 kg biomass C ha^{-1}.

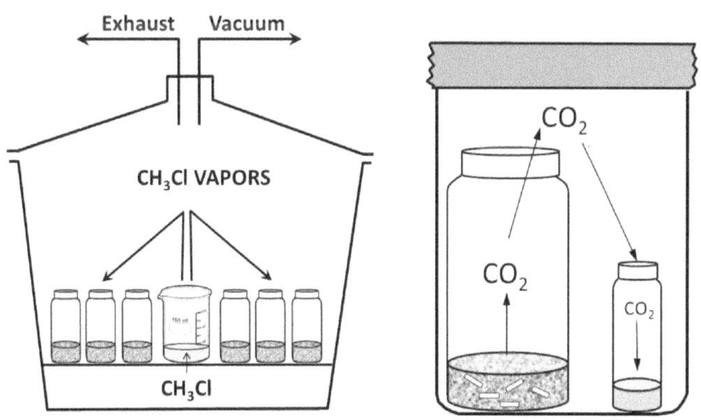

Left: Vacuum desiccator containing soil samples to be fumigated and a beaker of chloroform. Drawing a vacuum saturates the atmosphere in the desiccator with chloroform vapor. Right: Sealed chamber containing a fumigated soil sample (left) and a vial of NaOH solution (right). Carbon dioxide evolved from the soil during the 10-day incubation is trapped by the NaOH and subsequently measured.

Monitoring the decomposition process

As alluded to previously, the decomposition of organic material added to soil involves successive generations of numerous microbial groups that work in concert and continually respond to changing nutritional and environmental conditions. It is therefore a complex process and one that is difficult to monitor with precision. Also as discussed earlier, the decomposition and eventual mineralization of organic C added to soil can take many months or even years (Fig. 13.6) and first involves decomposition of the more easily degraded compounds contained in the original material (Fig. 13.4). Initially, decomposition produces primarily CO_2 and new microbial biomass (Fig. 13.15). As decomposition continues, however, more resistant compounds are degraded along with some of the early-produced microbial biomass. With time and after successive generations of microbial populations (microbial succession), the remaining plant-derived material and accumulated dead microbial debris (necromass) become physically protected from active decomposition processes (see section of soil organic matter [SOM]).

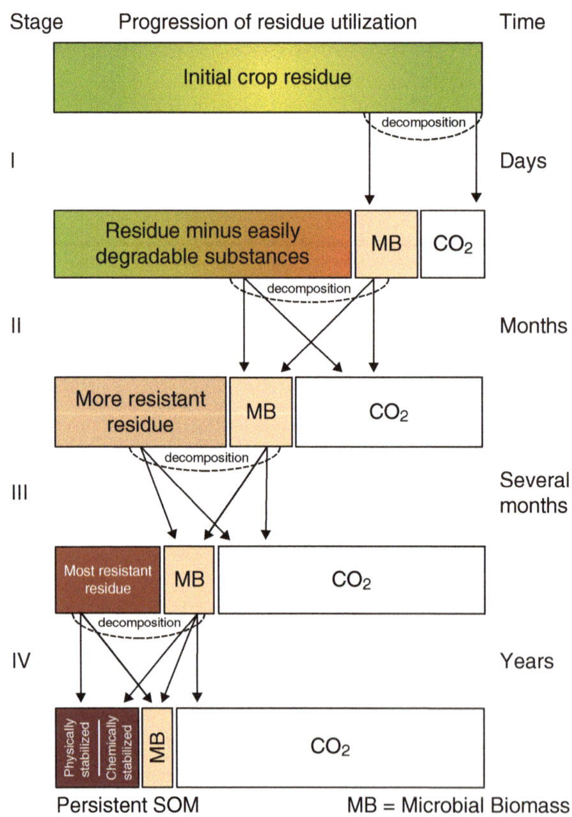

FIGURE 13.15 **Stages of residue decomposition in soil.** Changing box sizes indicate the relative quantities of carbon in the various fractions as decomposition progresses toward completion.

A number of factors make it difficult to unambiguously determine the fate of organic C added to soil or even the extent to which the original material has been degraded. A common approach is to measure the cumulative evolution of CO_2 over time, CO_2 representing the complete mineralization of the original organic C. However, CO_2 measurements underestimate true rates of residue degradation because a significant portion of the organic C present in the original residue is not mineralized but is instead converted to microbial biomass (Fig. 13.16). Conversely, some of the CO_2 evolved may arise from the mineralization of preexisting microbial biomass or other, preexisting, plant-derived compounds unrelated to the material under consideration. Additionally, the amount of biomass C produced per amount of C consumed, referred to as carbon use efficiency or CUE, decreases over time due to the increasingly recalcitrant nature of the compounds being degraded. Conversion efficiencies also vary depending on the microbial groups involved, for example, bacteria typically exhibit efficiencies lower than those for fungi. In general, soil microbes exhibit CUEs between 0.30 and 0.55 (Soares and Rousk, 2019). An additional consideration is the differing requirements among microbial groups for nitrogen and other nutrients for the synthesis of new microbial biomass. For a given amount of residue carbon incorporated into microbial biomass, the nitrogen requirement of bacteria is approximately twice that for fungi. For materials that are low in nitrogen, the decay process does not yield sufficient nitrogen from the residue itself to meet the nitrogen requirement of the developing microbial biomass. The inorganic nitrogen pool of the soil is then drawn upon to meet the nitrogen needs of microorganisms. This removal of available nitrogen from the soil and its assimilation into an organic form in microbial tissues, which renders it less readily available to plants and other microbes, is called **immobilization** (Chapter 15). Thus, limited N availability can indirectly alter overall

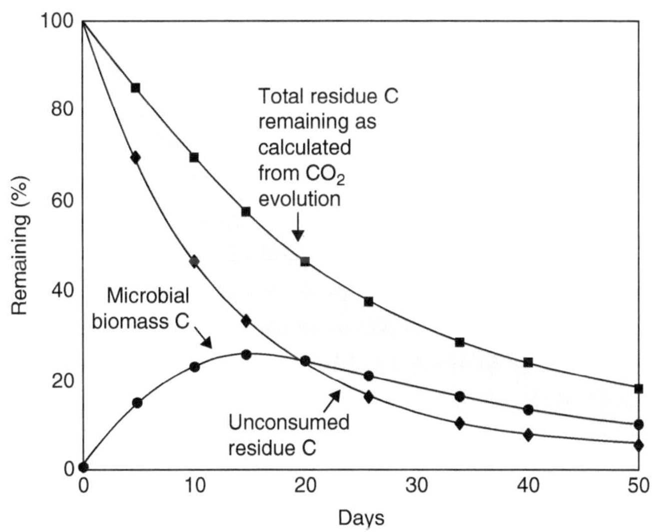

FIGURE 13.16 **Total residue carbon remaining in soil and microbial biomass carbon produced during residue decomposition.** The values were calculated using a microbial efficiency of 60%. Note that the *shaded area* equals the C in microbial biomass. *Source: Adapted from Paul and van Veen (1978).*

CUEs by, for example, favoring fungi versus bacteria as the primary degraders. In general, low availability of non-C nutrients such as N lead to lower CUEs, that is, increased evolution of CO_2 versus production of microbial biomass (Soares and Rousk, 2019). However, it is important to note that both bacterial and fungal cells are rich in nitrogen relative to most other residues being decomposed and are therefore inherently more susceptible to degradation upon cell death. Alternative approaches to simply measuring CO_2 evolution for monitoring organic C decomposition typically involve quantifying changes in the amount of C residing in different pools (e.g., CO_2; microbial biomass, Box 13.2; extractable non-living organic C) often in combination with stable (^{13}C) or radioactive (^{14}C) isotopic labeling of the original organic material. Regardless of the method used, modeling the behavior of C in soils is a formidable challenge.

Given the persistent nature of the organic compounds produced toward the end of the decomposition process, one might logically ask why SOM levels do not continually increase in response to annual organic C inputs. One factor contributing to the degradation of SOM is the **priming effect**. The priming effect refers to the well documented phenomenon of enhanced degradation of SOM in response to the addition of labile (fresh) organic residues to soil. Originally, it was thought to arise from the resultant increased microbial activity and associated production of enzymes capable of degrading otherwise persistent SOM. More recent studies show that other factors are involved, the exact nature of which are the subject of on-going research (Chen et al., 2019; Lal, 2018).

Soil organic matter (SOM)

Plant residues serve as the major substrates for the formation of SOM. Various definitions of SOM exist, but many soil scientists would define **SOM** as the nonliving organic fraction of the soil exclusive of undecayed animal and plant residues. Other scientists use broader definitions that include living biomass, undecayed substances, and even surface residues (e.g., O horizons; Chapter 2) (Weil and Brady, 2019). To a certain extent, these differing definitions reflect that organic matter in soils consists of a myriad of compounds representing a continuum of decomposition states, with new compounds continually being formed as a result of biological and abiotic processes. The reader is referred to Tan (2014) for a detailed discussion of terms used to describe SOM and related compounds. Regardless of how it is defined, all soil scientists would agree that SOM is critical to the proper functioning of natural soils.

Traditionally, SOM has been divided into nonhumic and humic substances. **Nonhumic substances** include those products of decomposition that retain the general chemical nature of the original residues (e.g., fragmented polymers) as well as microbial biomass and identifiable microbial byproducts (e.g., extracellular polysaccharides). The remaining **humic substances**, historically called **humus**, are characterized as being high in molecular weight, brown to black in color, and highly resistant to decomposition. Using a standardized extraction procedure, humic substances are divided into fractions of decreasing solubility and increasing molecular weight—**fulvic acid**, **humic acid**, and **humin**, respectively (Box 13.3). These extracted fractions have been well characterized and differ greatly from the various biopolymers, such as lignin and cellulose, produced by plants and soil microbes. According

BOX 13.3

Classical and novel schemes for characterization of SOM

At the time of this writing, there is considerable research investigating how best to characterize various components of soil organic matter, in particular those persistent compounds that typically accumulate in soil. Described here are the classical alkaline extraction method used for many decades in SOM research and an example of the newer methods aimed at avoiding sole reliance on harsh alkaline extraction. The newer method presented is intended to both reduce the possibility of producing artifactual compounds differing from those actually present in soil and simultaneously provide more detailed information on the composition of SOM.

Classical extraction and fractionation (Stevenson, 1994)

Following pretreatment of soil with 0.1 *M* HCl to remove cations and iron and aluminum hydroxides to increase extraction efficiency, humic substances are extracted with 0.5 *M* NaOH under an N_2 atmosphere. The extracted substances are adjusted to pH 1–2 with HCl, and the flocculated and soluble materials are separated by centrifugation. This procedure yields the following fractions:

- *Humic acid*—dark brown to black organic material that is alkali extractable and which precipitates upon acidification; molecular weight mostly > 10^5 g mol^{-1}; high in polyphenolic components; low in polysaccharides
- *Fulvic acid*—yellow material that is alkali extractable and which remains in solution upon acidification; molecular weight mostly < 10^3 g mol^{-1}; low in polyphenolic components; high in polysaccharides
- *Humin*—fraction of soil organic matter that is not alkali extractable and remains associated with soil particles, particularly clay surfaces; relatively poorly characterized; higher molecular weight than humic and fulvic acids; may represent 50%–70% of total SOM

"Humeomics" extraction and fractionation (Drosos et al., 2017)

Following pretreatment of soil with 0.1 *M* HCl and washing to neutrality, the following extractions are performed sequentially under N_2 atmosphere to produce five fractions:

- *Unbound fraction*—washed soil is extracted with solution of dichloromethane and methanol; consists primarily of alkanes (acyclic saturated hydrocarbons) and fatty acids
- *Weakly bound ester fraction*—residual SOM is extracted with solution of boron trifluoride and methanol; consists primarily of fatty acids and sugars
- *Strongly bound ester fraction*—residual SOM is extracted with a solution of KOH and methanol; consists primarily of fatty acids and phenolic acids (aromatic hydrocarbons)
- *Strongly bound ether fraction*—residual SOM is extracted with highly acidic hydroiodic acid (HI); consists primarily of amides and nitrogen-containing heterocyclic compounds
- *Residual extractable OM*—residual SOM is extracted with a solution of NaOH and sodium pyrophosphate; relatively poorly characterized but may consist largely of various linear and cyclic organic acids

(Continued)

BOX 13.3 (*Cont.*)

The non-extractable SOM that remains in the soil following the above extractions may represent approximately 25%–30% of total SOM; there are indications that this material is highly hydrophobic in nature and tightly bound to soil minerals, which may contribute to its non-extractability. Nevertheless, results to date suggest that the "Humeomics" approach may extract a greater proportion of total SOM than is possible with the traditional alkaline extraction.

Given the current efforts invested in SOM research that use novel approaches, including sophisticated instrumentation for physical and chemical analyses of extracted SOM components as well as analyses of whole soils (Box 13.4), one can confidently conclude that the final chapter on SOM characterization is yet to be written. However, the future looks bright for increased understanding of the nature and dynamics of soil C, information which will ideally lead to improved insights for the management and preservation of this precious soil resource.

to this characterization of SOM, the humic fractions represent the products of chemical condensation and synthesis reactions and persist in the soil primarily due to their inherent resistance to microbial decomposition processes and intimate association with mineral surfaces.

Contrary to the historical view, evidence over the past few decades has pointed to a very different perspective of the nature, formation, and persistence of SOM (Kästner and Miltner, 2018). Although some aspects of the traditional view perhaps hold true and have value (Olk et al., 2019), it is increasingly thought that the bulk of SOM in soil consists not of humic material as defined previously but rather the partially oxidized and physically protected remnants of dead microbial cells and lesser amounts of plant-derived polymeric substances. An important distinction relative to the traditional view is that these remnants are not inherently recalcitrant and can be degraded readily should they become unprotected, for example, by disruption of soil microaggregates or dissolution of protective iron, aluminum, or other metallic oxides. This view posits that soil microbes are central to SOM formation by virtue of their extensive processing of plant residues (to produce microbial biomass) and by eventually contributing their own cellular polymers (especially cell wall constituents) upon their death (Fig. 13.17). One emerging view is that the persistence of SOM involves the accumulation of hydrophobic components, largely in association with mineral surfaces (Piccolo et al., 2019). Additionally, rather than being high-molecular-weight materials, these sequestered compounds are relatively small molecules non-covalently arranged in larger units that, due to their hydrophobicity, have little association with water, thereby making them less accessible for microbial degradation. Thus, the validity of the process of "humification" itself has been brought into question, with some scientists urging discontinued use of "humus" and related terms (Kleber and Lehmann, 2019).

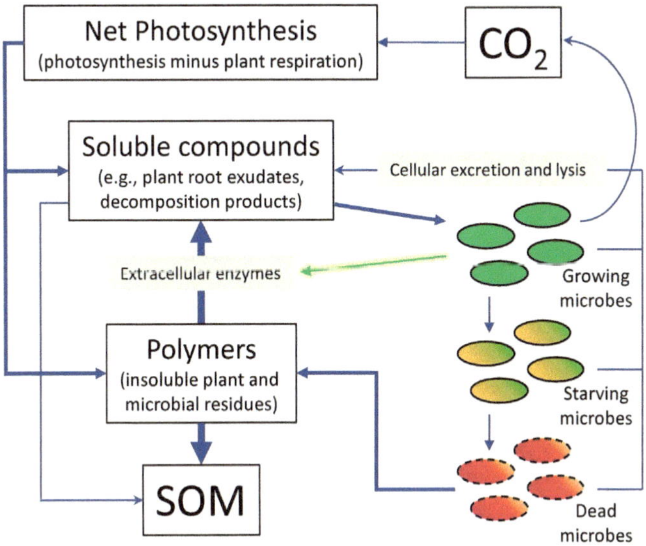

FIGURE 13.17 **Generalized scheme for the formation of soil organic matter (SOM) from plant and microbial residues, emphasizing the central role of soil microorganisms.** Active soil microbes grow on soluble compounds derived largely from the degradation of polymeric substances by extracellular enzymes. Fragments of polymers of both plant and microbial origin are thought to be major components of persistent SOM. *Source: Based on Kästner and Miltner (2018).*

As one might anticipate, these divergent views on the nature of SOM have fueled a vigorous and, importantly, healthy debate among scientists researching C cycling in soils (Box 13.4).

BOX 13.4

The humus debate (aka "How Science Progresses")

The nature of SOM has been the subject of study essentially since the origin of soil science as a discipline. Thus, one might be tempted to think the topic to be settled within the scientific community, but one would be mistaken. Recent developments, fueled largely by advances in the analytical techniques available for SOM characterization, provide an excellent insight to the progression of science in the light of new experimental evidence.

For more than 200 years, the traditional view of SOM, based on the study of alkaline extractions (Stevenson, 1994; Box 13.3), prevailed and provided a basis for formulating models of organic matter persistence and associated properties in soil. These models have proven to be useful over the years and into the present day (Olk et al., 2019). However, over the past few decades, concerns have increasingly been expressed by some scientists that the traditional models were insufficient to account for some observed behaviors of organic C in soils. Alternative methods of SOM extraction have been developed. For example, Drosos et al., 2017 employed a se-

(Continued)

BOX 13.4 (*Cont.*)

quential extraction procedure, which they called "Humeomics" (Box 13.3), which led to the proposal that SOM exists largely as suprastructures of smaller molecules bound together by various non-covalent and covalent linkages. Other researchers have explored the potential value of employing modern instrumentation techniques (e.g., infrared and nuclear magnetic resonance (NMR) spectroscopy, atomic force microscopy, X-ray spectromicroscopy) when applied directly to soil, thereby circumventing the use of extractants entirely (Kästner and Miltner, 2018). These newer approaches, combined with concerns that alkaline extractions may produce chemical artifacts, has led some to challenge aspects of the traditional models of SOM (Kleber and Lehmann, 2019).

Thus, there is currently a stimulating and healthy debate among scientists that is sure to produce new revelations and improved models of carbon behavior in soils. The timing of these developments is fortuitous, given increasing concerns about climate change and the role of land management in altering CO_2 flux from terrestrial landscapes. Additionally, the role of SOM in promoting soil sustainability is a major concern with regard to food security and environmental quality. Clearly, the true nature of many of the organic compounds in soils remains largely a mystery with much left to be discovered.

Beneficial properties imparted to soil

SOM imparts many desirable biological, chemical, and physical properties to soil (Box 13.5; see also Chapter 14). Although it represents only around 5% of the total soil volume, it exerts a disproportionally large influence on soil properties.

The major chemical effect of SOM in most mineral soils is that it contributes 20%–80% of the cation exchange capacity (CEC). The source of the charge is pH dependent, that is, as the pH increases, the CEC also increases. Physically, SOM improves soil **aggregation**, which is also encouraged by extracellular polysaccharides (EPS) and other compounds (e.g., glomalin-related soil proteins, or GRSPs; Chapter 12) produced by soil microbes. A soil in which the smaller soil particles are bound into water-stable aggregates is highly desirable because it results in:

- higher infiltration rates that reduce water runoff and associated erosion
- less detachment of individual particles that can be eroded by water and wind
- more rapid gas exchange with the atmosphere and thus better soil aeration
- easier cultivation and penetration by plant roots because the soil offers less resistance

In general, organic residues that have the highest levels of readily decomposable carbon exert the greatest and most rapid increase in soil aggregation. However, the effect is generally of short duration. Organic residues that are more resistant to decomposition require more time to exert maximum effect on aggregation, but the influence persists for a longer period.

Not to be confused with soil humic substances, there are commercially available products called "humates" that are advertised to provide many of the benefits of associated with SOM. As discussed more fully in Chapter 24, these products differ compositionally from soil humic

BOX 13.5

Beneficial properties imparted to soil by organic matter

This list illustrates the importance of adopting soil management systems (e.g., no-tillage, crop rotation, manure addition, high biomass cover crops) to maintain and augment the level of organic matter in soil.

Biological properties

- Provides a slowly available carbon and energy source to support a large, diverse, and metabolically active microbial community
- Source of certain compounds that may exert plant growth-promoting effects

Chemical properties

- Increases cation exchange capacity of soil (often 20%–80% of the total CEC is due to organic matter)
- Buffers pH change
- Provides a slow-release supply of organically bound nutrients such as nitrogen, phosphorus, and sulfur (soil organic matter decomposes at 2%–5% $year^{-1}$)
- Enhances chelation and thus bioavailability of trace elements to plants
- Accelerates mineral weathering and aids in solubilization of plant nutrients (such as phosphorus) from otherwise insoluble minerals
- Has a high adsorptive capacity for organic compounds and thus reduces the bioavailability of toxic xenobiotics

Physical properties

- Contributes to improving soil structure and aggregation
- Decreases bulk density and thus increases percentage pore space, particularly macroporosity, thereby enhancing water infiltration and reducing runoff and erosion
- Increases total water-holding capacity but may also increase the strength with which water is held
- Increases heat absorption in early spring due to dark color, but soil often holds more water, which requires more energy to heat

substances and, when applied to most soils at the recommended rates, would not appear to contain sufficient amounts of the components required to produce the beneficial effects often advertised.

Characterization of SOM

Elemental composition

The carbon content of SOM is generally given as 58% (1% organic-C = 1.7% SOM), whereas nitrogen values range from 3% to 6%. The carbon/nitrogen/phosphorus/sulfur ratio is approximately 100/10/1/1. Thus, being nutrient rich relative to its C content, any portion of SOM that becomes accessible to and is decomposed by the microbial community will result in the net mineralization of N, P, S, and other nutrients that it may contain.

Functional groups

The CEC of SOM ranges from 150 to 300 cmol kg^{-1} and is derived from such functional groups as carboxyls (-COOH) and phenolic hydroxyls (-OH) which characteristically bear a negative charge in soil. As noted in Chapter 2, this range is considerably higher than the exchange capacity of many clay minerals. Note, however, that SOM is less dense than minerals, thereby somewhat inflating its CEC when compared to clay minerals. Nevertheless, SOM represents an extremely significant source of CEC. Sites for anion exchange reactions also exist and are due in part to protonated amino groups ($-NH_3^+$). A noted property of organic functional groups is the influence of pH on their ionization. As the pH increases, more negatively charged sites ($-COO^-$ and $-O^-$) are available for cation exchange and fewer sites exist for anion exchange.

Complexes

Many organic compounds found in the soil form complexes with various metal ions in soil and water. When a metal ion such as Zn^{2+} combines with an electron donor (certain organic compounds), the substance formed is known as a complex or chelate. Examples of Zn^{2+} chelates are shown in Fig. 13.18. The bonds between the electron-accepting metal (Zn^{2+}) and the electron-donating group may be ionic or covalent. Organo-metallic complexes are important in micronutrient and heavy metal availability to plants. It is generally thought that most metal complexes are more bioavailable to plants and other organisms and more mobile in soil

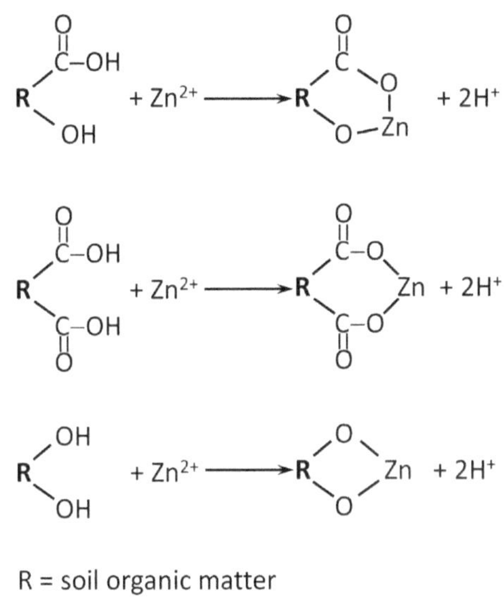

FIGURE 13.18 **Possible Zn^{2+} chelates formed with soil organic matter functional groups** (Wolf and Wagner, 2005).

than the corresponding ionized form, primarily because they are less susceptible to precipitation and adsorption reactions. Micronutrient availability is facilitated by chelate formation, and it has been demonstrated that up to 75% of the copper in soil solution is in a chelated form. However, chelate formation by heavy metals such as nickel and cadmium would be undesirable if it resulted in increased plant uptake. Certain organic compounds in SOM form chelates with iron and aluminum. The chelates facilitate the downward movement of iron and aluminum through the soil profile in a process known as cheluviation. The chelates are subsequently deposited in the lower soil horizons, and the resulting redistribution of iron and aluminum is important in soil profile development.

Adsorption

Adsorption of pesticides by SOM plays a major role in determining pesticide activity, effectiveness, and leachability (Stevenson, 1994; Werner et al., 2013). The mechanism of adsorption varies depending upon the chemical form of the pesticide. A common observation is that the application rate of certain herbicides must be increased as the SOM level increases in order to maintain adequate weed control. Higher SOM levels adsorb larger amounts of the herbicide and result in a lower concentration of the herbicide available for uptake by weeds. Similarly, adsorption of toxic organic contaminants can influence transformation and transport of xenobiotics in soil.

Decomposition of SOM

Aerobic conditions

SOM decomposes at a rate of only 2%–5% year^{-1} depending upon climatic conditions. Increasing soil temperatures will result in increasing decomposition up to about 45°C. Tropical conditions favor rapid SOM decomposition, and nitrogen deficiencies commonly result following SOM depletion. Adequate soil aeration favors oxidative decomposition and results in a faster decomposition rate than under anaerobic conditions. The presence of adequate available nitrogen will increase the rate of decomposition. Cultivation or mechanical mixing of the soil exposes new surfaces of organic matter (aggregate disruption, etc.) to microbial oxidation and thus increases the rate of SOM decomposition. The practice of no-till or minimum-till crop production increases SOM levels compared to conventional tillage management due to less soil cultivation and inherently lower levels of residue incorporation (Chapter 14).

Anaerobic conditions

When plant residue is added to an anaerobic environment (e.g., wetland, swamp, peat bog, or marsh), decomposition is greatly reduced and organic residues typically accumulate, often in layers. These layers represent various successions of vegetation types and various degrees of decomposition of the plant residue. Eventually an organic soil, or Histosol, is formed. In the past, Histosols were called peats or bogs to indicate little plant-residue decomposition, or mucks to indicate advanced plant-residue decomposition.

Histosols have often been drained for growing vegetables and other crops. A unique management problem that arises when Histosols are drained is the resulting rapid decomposition of the SOM, which results in a drop in land elevation. This decline in soil volume, known as **subsidence**, occurs at rates up to 10 cm year^{-1} (Fig. 13.19).

FIGURE 13.19 **A demonstration of soil subsidence in the Everglades Agricultural Area, Florida, United States.** The marker post was driven into the ground flush with the surface of the soil when water flow patterns were changed due to conversion of wetlands to agricultural usage primarily for sugarcane production. Eighty-five years later the soil surface has subsided nearly 6.5 feet. The loss of soil volume is due to the oxidation of soil organic matter in the highly organic soils (Histosols) due to enhanced aeration. *Source: Photo courtesy of Dr. Alan Wright, Univ. of Florida.*

Summary

Decomposition of plant residues, microbial biomass, and SOM is a major function of soil microorganisms. In agricultural ecosystems, above- and belowground crop residues are the primary carbon substrates for the microbial community. Plant residues consist of readily decomposable fractions such as soluble components, starch, and protein, less available components such as cellulose and hemicellulose, and the slowly available fraction typified by lignin. Microbial remains contribute the polymers associated with their cell walls (primarily peptidoglycan and chitin), protein, and soluble compounds. Decomposition rates differ for each of the various components, and the rates are influenced by soil abiotic factors of temperature, moisture, and aeration. The levels of nitrogen in the soil and in the residue are also critical.

During the decomposition of the plant residues, a portion of the carbon is evolved as carbon dioxide and an additional fraction is incorporated into microbial biomass. The efficiency with which the microbes incorporate the residue carbon into biomass may be as high as 65% for soluble components but may be only 10% for lignin. The newly formed biomass also serves as substrate for other microorganisms upon lysis or predation.

As the plant residues and microbial biomass undergo biochemical alteration by soil microorganisms, some of the compounds are incorporated into the SOM pool. The formation of SOM and the decomposition of residue are intimately associated. SOM accounts for a small percentage of the soil volume, but it exerts a large influence on the biological, chemical, and physical properties of soil. This decomposition of organic materials in the soil also releases carbon dioxide into the atmosphere and is a major factor in the carbon cycle.

Acknowledgment

Portions of this chapter, with permission, are derived from the work of Drs. Duane C. Wolf and George H. Wagner who authored a similar chapter for the Second Edition of the textbook. The current authors thank them for the use of their material and for graciously supporting production of the Third Edition.

Study questions

1. What are the relative quantities of carbon in the Earth's atmosphere and in soil organic matter compared to annual photosynthetic fixation by terrestrial plant life?
2. What are the major organic constituents of crop residues and how do they differ in ease of decomposition by the soil microbial community?
3. In the decomposition of wheat straw in soil over time, is the loss of carbon from the soil as carbon dioxide more or less than the assimilation of carbon in the residues for production of new cells by the microorganisms? Why?
4. One gram of ^{14}C glucose is added to 100 g of soil, and approximately 80% of the glucose carbon is evolved as $^{14}CO_2$ during 4 weeks of soil incubation. After 4 weeks, in what general forms would you expect to find the ^{14}C remaining in the soil?
5. What are the difficulties encountered when measuring decomposition rates of organic matter added to soil? Describe the effects of environmental conditions on decomposition rates.
6. Describe the different approaches used to characterize SOM. What are the major organic fractions obtained by these approaches? To what extent do you think scientists have successfully described the compounds comprising SOM?
7. List some of the major biological, chemical, and physical benefits to soil attributable to SOM.

References

Broder, M.W., 1985. Changes in the chemical composition and microbial population on corn, wheat, and soybean residue decomposing in Sanborn Field. Ph.D. diss. University of Missouri, Columbia (Diss. Abstr. 86-07891).

Broder, M.W., Wagner, G.H., 1988. Microbial colonization and decomposition of corn, wheat, and soybean residue. Soil Sci. Soc. Am. 52, 112–117.

Burns, R.G., Martin, J.P., 1986. Biodegradation of organic residues in soil. In: Mitchell, M.J., Nakas, J.P. (Eds.), Microfloral and Faunal Interactions in Natural and Agroecosystems. Martinus Nijhoff/Dr. W. Junk Publications, Boston, pp. 137–202.

Buyanovsky, G.A., Wagner, G.H., 1986. Post-harvest residue input to cropland. Plant Soil 93, 57–65.
Buyanovsky, G.A., Wagner, G.H., 1987. Carbon transfer in a winter wheat (*Triticum aestivum*) ecosystem. Biol. Fert. Soils 5, 76–82.
Chen, L., Liu, L., Qin, S., Yang, G., Fang, K., Zhu, B., Kuzyakov, Y., Chen, P., Xu, Y., Yang, Y., 2019. Regulation of priming effect by soil organic matter stability over a broad geographic scale. Nat. Comm. 10, 5112. doi: 10.1038/s41467-019-13119-z.
Drosos, M., Nebbioso, A., Mazzei, P., Vinci, G., Spaccini, R., Piccolo, A., 2017. A molecular zoom into soil Humeome by a direct sequential chemical fractionation of soil. Sci. Total Environ. 586, 807–816, http://dx.doi.org/10.1016/j.scitotenv.2017.02.059.
Friedlingstein, P., Jones, M.W., O'Sullivan, M., Andrew, R.M., Hauck, J., Peters, G.P., Peters, W., Pongratz, J., Sitch, S., Le Quéré, C., Bakker, D.C.E., Canadell, J.G., Ciais, P., Jackson, R.B., Anthoni, P., Barbero, L., Bastos, A., Bastrikov, V., Becker, M., Bopp, L., Buitenhuis, E., Chandra, N., Chevallier, F., Chini, L.P., Currie, K.I., Feely, R.A., Gehlen, M., Gilfillan, D., Gkritzalis, T., Goll, D.S., Gruber, N., Gutekunst, S., Harris, I., Haverd, V., Houghton, R.A., Hurtt, G., Ilyina, T., Jain, A.K., Joetzjer, E., Kaplan, J.O., Kato, E., Klein Goldewijk, K., Korsbakken, J.I., Landschützer, P., Lauvset, S.K., Lefèvre, N., Lenton, A., Lienert, S., Lombardozzi, D., Marland, G., McGuire, P.C., Melton, J.R., Metzl, N., Munro, D.R., Nabel, J.E.M.S., Nakaoka, S.I., Neill, C., Omar, A.M., Ono, T., Peregon, A., Pierrot, D., Poulter, B., Rehder, G., Resplandy, L., Robertson, E., Rödenbeck, C., Séférian, R., Schwinger, J., Smith, N., Tans, P.P., Tian, H., Tilbrook, B., Tubiello, F.N., van der Werf, G.R., Wiltshire, A.J., Zaehle, S., 2019. Global carbon budget 2019. Earth Syst. Sci. Data 11, 1783–1838.
Jenkinson, D.S., Powlson, D.S., 1976. The effects of biocidal treatments on metabolism in soil—V. A method for measuring soil biomass. Soil Biol. Biochem. 8, 209–213.
Johnson, J.M-F., Allmaras, R.R., Reicosky, D.C., 2006. Estimating source carbon from crop residues, roots and rhizodeposits using the nation grain-yield database. Agron. J. 98, 622–636.
Kästner, M., Miltner, A., 2018. SOM and microbes—what is left from microbial life. In: Garcia, C., Nannipieri, P., Hernandez, T. (Eds.), The Future of Soil Carbon: Its Conservation and Formation. Elsevier Science & Technology, Saint Louis, pp. 125–163.
Kleber, M., Lehmann, J., 2019. Humic substances extracted by alkali are invalid proxies for the dynamics and functions of organic matter in terrestrial and aquatic ecosystems. J. Environ. Qual. 48, 207–216. doi: 10.2134/jeq2019.01.0036.
Lal, R., 2018. Digging deeper: A holistic perspective of factors affecting soil organic carbon sequestration in agroecosystems. Global Change Biol. 24, 3285–3301. doi: 10.1111/gcb.14054.
Olk, D.C., Bloom, P.R., Perdue, E.M., McKnight, D.M., Chen, Y., Farenhorst, A., Senesi, N., Chin, Y.-P., Schmitt-Kopplin, P., Hertkorn, N., Harir, M., 2019. Environmental and agricultural relevance of humic fractions extracted by alkali from soils and natural waters. J. Environ. Qual. 48, 217–232. doi: 10.2134/jeq2019.02.0041.
Paul, E.A., 2016. The nature and dynamics of soil organic matter: Plant inputs, microbial transformations, and organic matter stabilization. Soil Biol. Biochem. 98, 109–126, http://dx.doi.org/10.1016/j.soilbio.2016.04.001.
Paul, E.A., van Veen, J.A., 1978. The use of tracers to determine the dynamic nature of organic matter. In: Transactions of the Eleventh International Congress Soil Science, vol. 3, June 19-27, 1978, Edmonton, Canada. pp. 61–102.
Piccolo, A., Spaccini, R., Cozzolino, V., 2019. The soil humeome: Chemical structure, functions and technological perspectives,. Vaz, Jr., S. (Ed.), Sustainable Agrochemistry. Springer Nature Switzerland, pp. 183–222, https://doi.org/10.1007/978-3-030-17891-8_7.
Post, W.M., Peng, T.H., Emanuel, W.R., King, A.W., Dale, V.H., DeAngelis, D.L., 1990. The global carbon cycle. Am. Scientist 78, 310–326.
Scharlemann, J.P.W., Tanner, E.V.J., Hiederer, R., Kapos, V., 2014. Global soil carbon: understanding and managing the largest terrestrial carbon pool. Carbon Manag. 5, 81–91. doi: 10.4155/cmt.13.77.
Soares, M., Rousk, J., 2019. Microbial growth and carbon use efficiency in soil: Links to fungal-bacterial dominance, SOC-quality and stoichiometry. Soil Biol. Biochem. 131, 195–205. doi: 10.1016/j.soilbio.2019.01.010.
Stevenson, F.J., 1994. Humus chemistry: Genesis, composition, reactions, second ed. John Wiley and Sons, New York.
Tan, K.H., 2014. Humic Matter in Soil and the Environment: Principles and Controversies, second ed. CRC Press.
Traving, S.J., Thygesen, U.H., Riemann, L., Stedmon, C.A., 2015. A model of extracellular enzymes in free-living microbes: which strategy pays off? Appl. Environ. Microbiol 81, 7385–7393. doi: 10.1128/AEM.02070-15.
Weil, R.R., Brady, N.C., 2019. Elements of the Nature and Properties of Soils, fourth ed. Pearson, NY.
Werner, D., Garratt, J.A., Pigott, G., 2013. Sorption of 2,4-D and other phenoxy herbicides to soil, organic matter, and minerals. J. Soils Sediment. 13, 129–139. doi: 10.1007/s11368-012-0589-7.

Wolf, D.C., Wagner, G., 2005. Carbon transformations and soil organic matter formation. In: Sylvia, D.M., Fuhrmann, J.J., Hartel, P.G., Zuberer, D.A. (Eds.), Principles and Applications of Soil Microbiology, second ed. Pearson, New Jersey, pp. 285–332.

Further reading

Garcia, C., Nannipieri, P., Hernandez, T. (Eds.), 2018. The Future of Soil Carbon: Its Conservation and Formation. Elsevier Science & Technology, Saint Louis.

Jenkinson, D.S., 1988. Soil organic matter and its dynamics. In: Wild, A. (Ed.), Russell's Soil Conditions and Plant Growth, eleventh ed. John Wiley and Sons, New York, pp. 564–607.

Kononova, M.M., 1966. Soil Organic Matter, second ed. Pergamon Press, New York.

Tate, R.L., 1987. Soil Organic Matter: Biological and Ecological Effects. John Wiley and Sons, New York.

Waksman, S.A., 1936. Humus: Origin, Chemical Composition, and Importance in Nature. Williams and Wilkins Co, Baltimore, Md.

CHAPTER

14

Applied aspects of soil carbon

Alan Joseph Franzluebbers

USDA—Agricultural Research Service, Raleigh, NC, United States

"As microbiologists, may we add a new element to the monuments of the century, monuments more significant than metals, rocks, and plastics mingled in the impressive dams, high-energy accelerators, and space rockets. May we dream dreams of perfection and make our contribution that of ideas. Hopefully, these ideas will spill over from our Science and help to imbue our fellow man with more desire for accomplishment than wealth, more eagerness for involvement than ease, and greater regard for intelligence and wisdom than things." **– Robert L. Starkey, 1963**

One of the most fundamental activities of soil microorganisms is to decompose organic materials that are so bountifully produced from the diversity of plants thriving on the essential natural resources of sun, air, water, and soil. Carbon (C) forms the "backbone" biochemical structure of all living things. As described in the preceding chapter, soil microorganisms transform C from these plant- and animal-derived organic molecules, as well as from dead microbes themselves, into various byproducts of decomposition and a myriad of complex substances collectively called soil organic matter (SOM; Chapter 13). This chapter expands on those fundamental processes and describes how soil can be managed to enhance soil C reserves.

Importance of soil carbon management

Soil organic matter provides numerous benefits to plant growth, environmental quality, and to the health of soil itself, including resident microorganisms. While it is beyond the scope of this chapter to discuss the entire breadth of topics related to soil C, those presented are representative of ways in which human activities impact this important soil resource, both positively and negatively.

An overriding negative effect that humans have had on SOM over the past several thousand years is through the conversion of natural landscapes such as forests and grasslands to food and feed production. This has resulted in immense losses of SOM in many regions of the globe. The introduction of soil tillage, in particular, has accelerated decomposition processes

Principles and Applications of Soil Microbiology. http://dx.doi.org/10.1016/B978-0-12-820202-9.00014-9
Copyright © 2021 Elsevier Inc. All rights reserved.

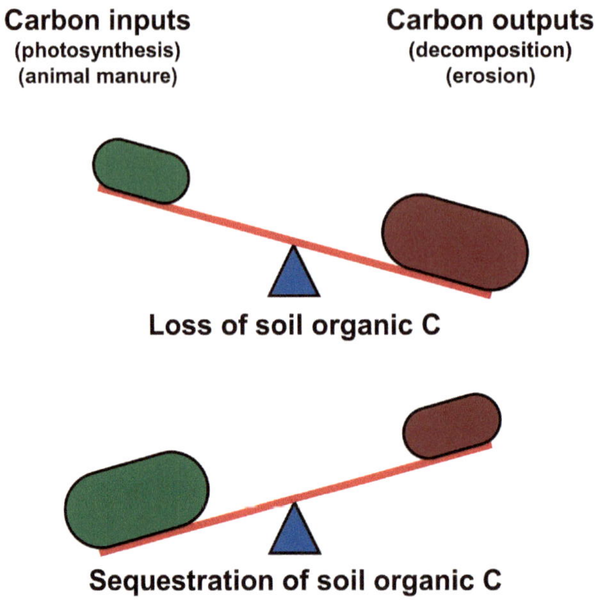

FIGURE 14.1 **Schematic diagram of the core principle underlying retention and loss of soil organic C.** Soil organic C storage is a balance between inputs and outputs.

and, in the absence of adequate C inputs from plants or other sources, has led to a net loss of SOM (Fig. 14.1). Globally, soil organic C loss from conversion of native vegetation to crop cultivation has been estimated as 44–537 Pg (Petagram is 10^{12} kg) from the total of 1550 Pg estimated still stored as soil organic C (Lal, 2004).

It is difficult to exaggerate the negative consequences of SOM losses, as they can include loss of soil fertility, decreased soil aggregation leading to increased soil erosion, reduced soil biodiversity, and global warming. Fortunately, recognition of the seriousness of these effects has spurred many soil scientists, farmers, conservationists, and politicians to identify, employ, and appeal for measures to restore SOM to more sustainable levels. The following sections describe impacts of tillage and organic amendments on soil organic C and soil microbial activity.

Tillage effects

Tillage systems

Soil tillage is an ancient practice that was necessary to eradicate weeds and loosen soil for planting seeds. In the past, aggressive tillage methods were the norm and largely involved inversion or "flipping" of topsoil layer(s), for example, by a moldboard plow (described later), such that the original soil surface was buried. Modern agriculture, particularly on upland landscapes, increasingly uses either no-tillage or conservation-tillage systems at planting to minimize negative consequences on soil erosion, greenhouse gas emissions, and water pollution. However, some forms of inversion tillage are still performed for a number of worthy reasons, including for:

- controlling weeds, insects, and diseases, when cultural and chemical approaches are not feasible or not effective
- improving the soil's physical condition by loosening compacted layers, enhancing soil warming in spring, and/or ameliorating subsoil conditions with various amendments
- occasionally incorporating lime, fertilizer, herbicides, plant residues, or green manure crops for specific goals of soil mixing, nutrient redistribution, sequestering chemical residues, or eliminating pathogens
- conserving soil and water, primarily in arid environments by creating a dust mulch to limit evaporation
- preparing a fine-grained seedbed for sensitive seeding operations

Since the 1970s, conservation tillage and particularly no tillage have been increasingly utilized, as the need for inversion tillage has been reevaluated in light of its many negative consequences on erosion, water quality, and SOM oxidation. In particular, the susceptibility of inverted soil to wind and water erosion has highlighted the threats of inversion tillage to environmental quality and sustainable food production. The term "conservation tillage" includes a variety of systems, all of which are designed to minimize residue incorporation and soil disturbance with the intent of abating soil erosion. According to the USDA definition of the term, >30% residue cover must be on the soil surface immediately after planting for it to meet the conservation tillage standard.

Tillage practices range from the very simple to the very complex, and a complete discussion of these is beyond the scope of this chapter. However, three types that illustrate the differing effects of tillage on SOM dynamics are the moldboard plow, shallow tillage (e.g., chisel plow), and no tillage.

- The moldboard plow (Fig. 14.2A,B,D), developed in the early 1800s, was perhaps the most widely used primary tillage implement during the early part of the 20th century. The moldboard plow inverts soil to a depth of 15–30 cm, resulting in nearly complete burial of aboveground crop residues.
- Shallow tillage uses any of a wide variety of implements to kill weeds and scarify the soil surface. For example, the working depth of the chisel plow (Fig. 14.2C) is similar to the moldboard plow, but the degree of soil inversion with the chisel plow is much less.
- No tillage involves cutting a slit in the soil for planting seeds but otherwise leaves the soil undisturbed. As such, application of chemical herbicides is often viewed as necessary to control weeds. However, organic growers are increasingly interested in using modified versions of no-tillage management that use cover crops, mowers, flamers, and other cultural methods to smother or kill weeds.

Tillage effects on depth distribution of soil organic matter

Inversion tillage mixes residues dispersed at the soil surface into deeper soil layers. The type of tillage tool greatly influences the eventual location of aboveground residues within the soil profile. Moldboard plowing to a depth of 25 cm buries ~70% of the aboveground

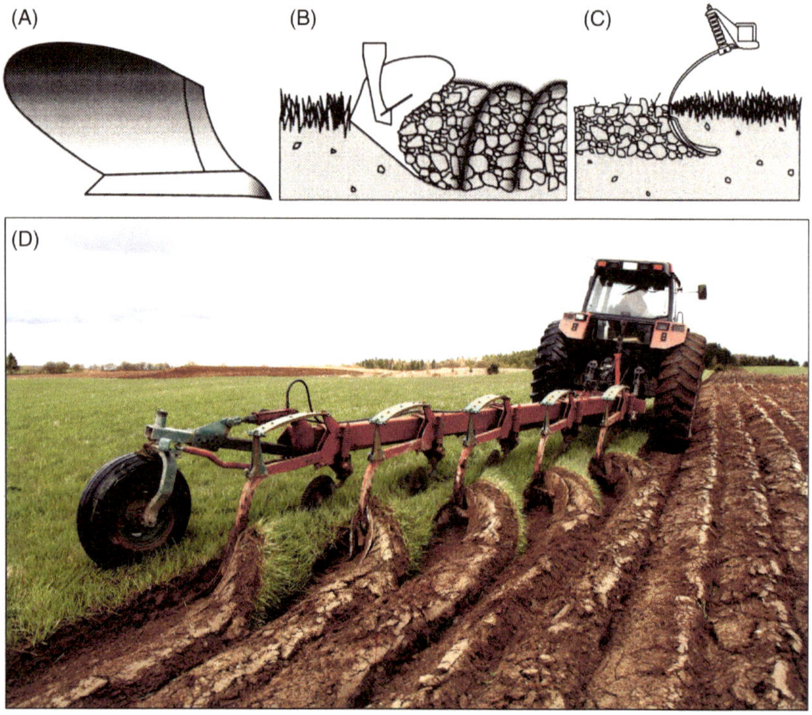

FIGURE 14.2 A moldboard plow viewed from the side (A) and rear (B). (D) A moldboard plow in action. Note the complete inversion of the soil to plow depth as opposed to the chisel plow (C) which results in less soil disturbance. *Source: Photo credit: Lurin 123RF.com.*

oat residue at a depth of 12–24 cm, whereas chisel plowing to a depth of 15 cm leaves nearly 60% of the residue at a depth of 0–6 cm (Fig. 14.3). No tillage leaves nearly all of the non-root residues at or above the soil surface. Since plant residues contribute to subsequent SOM formation, placement of plant residues with different tillage practices is highly important for understanding depth distribution of SOM.

With repeated inversion tillage, SOM becomes more uniformly distributed within the plowed layer. The fate of organic matter that is mixed into soil versus the organic matter left on the soil surface depends on the prevailing climatic conditions. In general, the environment within soil is more buffered against extremes in moisture and temperature than that at the soil surface. More favorable moisture conditions *in* the soil than *on* the soil surface is probably the primary factor that leads to greater decomposition of organic matter in tilled soil.

Surface-placed plant residues typical of conservation tillage systems experience frequent drying and rewetting events, depending upon frequency of precipitation. Although decomposition of surface-placed residues is slower than that of buried residues, concentration of N in these decomposing plant residues generally increases with time relative to buried residues. This seems counter to the common observation that higher N concentration of plant tissues leads to faster decomposition. This apparent contradiction suggests that frequent drying and rewetting of surface-placed residues increases the resistance of certain N compounds to

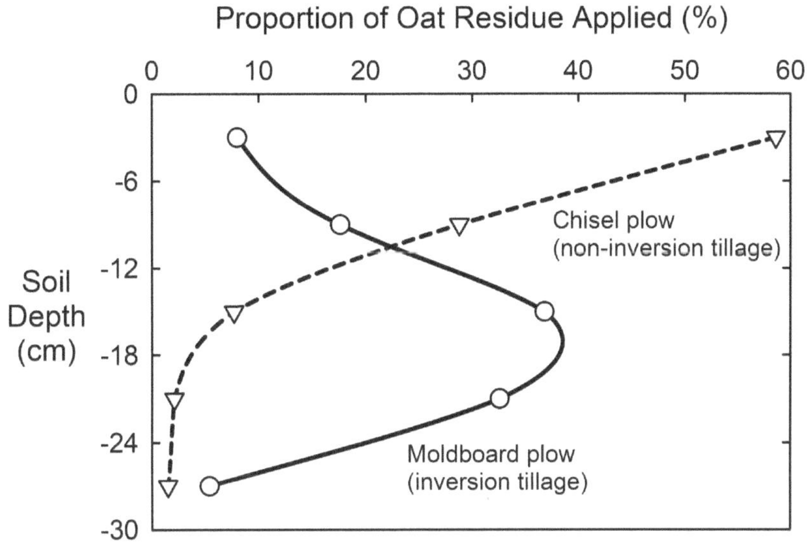

FIGURE 14.3 Crop residue distribution as affected by tillage implement. *Source: Data from Allmaras et al. (1996). Used with permission.*

microbial decomposition, potentially leading to greater total N accumulation in the surface of no-tillage soils. Alternatively, N concentration in surface residues could increase as a result of fungi colonizing residues and subsequently taking up and accumulating N from adjacent soil. Research continues to explore how C and N transformations vary during decomposition of various crop residues under different micro- and macro-climatic conditions.

Soil organic matter impacts the bulk density of soil and, therefore, the amount of organic C within a given volume of soil (Box 14.1). Since conservation tillage systems leave residues near the soil surface, most investigations report a substantial increase in soil organic C content in surface layers, as compared with inversion tillage systems. However, in order to make valid comparisons of total C storage in soil across tillage systems, it is necessary to determine net changes in soil organic C to at least the depth of the deepest tillage tool used.

By concentrating various C and N fractions near the soil surface, conservation tillage systems have beneficial consequences for soil functioning beyond that of simply sequestering more C. The soil surface receives fertilizer and pesticide applications, is subjected to the intense impact of rainfall events throughout the year, and additionally partitions the fluxes of gases and water into and out of soil. Therefore, protection provided by surface residues, combined with more open structure afforded by high levels of SOM, becomes essential for erosion control, water infiltration, and conservation of nutrients, all of which are vitally important soil functions. The degree to which these functions are improved by this stratification depends upon the inherent level of SOM, intensity of soil disturbance, type of agricultural production system, and duration of management. Soils with low inherent levels of organic matter (such as in warm, semi-arid, and humid regions of the subtropical and tropical regions of the world) may be the most functionally improved with conservation tillage, despite modest or no change in total standing stock of soil organic C within the rooting zone. Alternatively,

BOX 14.1

Soil organic C stock calculation with different tillage systems

A study determined organic C distribution with soil depth after 4 years of tillage (working depth of 15 cm) or no tillage on a sandy loam in Georgia. At a depth of 0–2.5 cm, organic C under no tillage was 81% greater on a volume (content) basis and 95% greater on a weight (concentration) basis, when compared to the disk tillage treatment (see figure below). Conversely, at a depth of 2.5–7.5 cm, corresponding values were 2% and 14% lower with no tillage than with disk tillage. As a result, summation of soil C on an area (content) basis to a depth of 15 cm indicated no difference between tillage systems. However, including surface residue C along with soil organic C to a depth of 15 cm did result in significantly greater storage of C under no tillage compared with disk tillage.

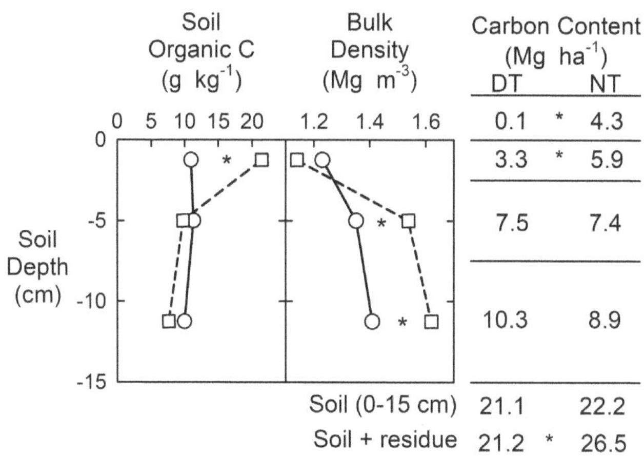

Soil organic C depth distribution as affected by tillage system (*DT*, disk tillage [circle]; *NT*, no tillage [square]) and calculation of soil organic C stock within the sampled profile. *, Means are significantly different at $P < 0.05$. *Source: Data from Franzluebbers et al. (1999).*

those soils with inherently high SOM may obtain relatively little additional soil functional benefit with adoption of conservation tillage because the functional properties of soil might already be at a high level.

Stratification ratio of soil organic C and N fractions has been suggested as a simple and rapid indication of soil health improvement with conservation management. This approach simplifies numerical comparisons among sites. It also recognizes that surface soil properties are most responsive to management and that high stratification ratios would be uncommon under degraded conditions (Fig. 14.4). Soil microbial biomass and potential C mineralization are directly linked to the concentration of SOM at the soil surface in conservation agricultural systems.

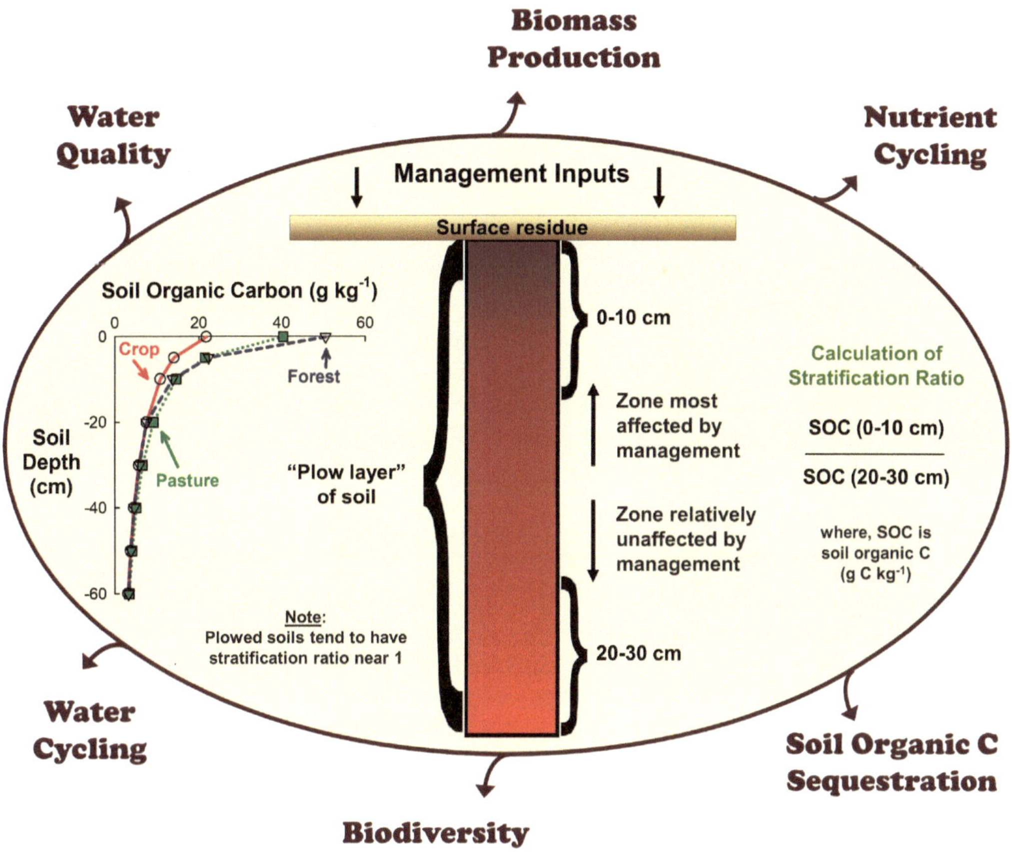

FIGURE 14.4 **Concept of stratification ratio of soil organic C and N fractions and its impact on environmental outcomes.** The central portion of the figure depicts the soil depth affected by tillage and the depth intervals to be sampled for calculating organic C stratification ratio (formula shown in inset on the right). Inset on the left shows example C distributions with soil depth for different land management systems (Perkins, 1987): undisturbed forest, long-term pasture, and long-term conventional tillage. *Source: Based on Franzluebbers (2002).*

Tillage effects on total organic carbon stock

Conservation tillage systems often have greater soil organic C levels in the upper 30 cm of soil than conventional, inversion tillage systems. This is mainly due to placement and preservation of surface soil organic C at a relatively dry position near the soil surface and relative lack of disturbance to soil that would otherwise lead to SOM oxidation. Some debate has occurred whether redistribution of organic matter with inversion tillage actually enhances soil organic C and N storage. Increased storage could occur in environments that have high clay concentrations at the bottom of the plow layer, provided those clays are not already fully saturated with adsorbed C and are thus able to sequester additional C. There are also likely to be soil type- and climate-specific responses to tillage management with regard to SOM dynamics and other important ecosystem properties. Long-term, multidisciplinary studies are necessary to address these site-specific issues because complex relationships are likely to occur among management, environment, and genetic resources.

The magnitude of differences in soil organic C storage between no tillage and conventional tillage systems can be as great as 20 Mg C ha^{-1}, but more typical differences are around 3 Mg C ha^{-1} after 10 years of no tillage. However, there are a number of field studies that show no appreciable difference in soil organic C and N stocks in the upper 30 cm between conservation and conventional tillage systems. Although C and N contents in surface residues are not always accounted in agricultural system comparisons, this "trash" component at the soil surface can be significant for C accounting (Box 14.1) and is certainly important for soil health improvement. Only recently has there been much effort to quantify the surface residue component of soil management systems. As an illustration of the value of surface residues in corn cropping systems in the Midwest United States, the amounts of corn crop residue needed to control wind erosion, water erosion, and maintain SOM content were estimated to be 1.2, 3.2, and 8.3 Mg ha^{-1}, respectively (Wilhelm et al., 2007).

Organic amendments

Organic amendments were the essential source of fertility for food production systems prior to the industrial revolution and the development of inorganic sources of nutrients. Organic amendments depend on soil microorganisms to decompose molecules and release inorganic constituents for uptake by plants. Despite wide availability of inorganic fertilizers, organic amendments are still widely utilized in agricultural systems in the form of crop residues, cover crops, animal manures, and municipal biosolids. **Organic agriculture** still relies almost solely on organic amendments for adding fertility to soil. Decomposition of organic amendments is a natural process in forests and grasslands through root, leaf, and litter inputs, so utilization of organic amendments in agriculture should also be considered normal and be appreciated despite the abundance of inorganic fertilizers available today.

Animal manures

Animal manures are a large source of organic materials feeding soil biology worldwide. Animal manures are typically high in organic C. Following digestion of available substrates in feedstuffs by livestock, a significant proportion of remaining C in manure is low in susceptibility to further microbial degradation. However, transformation of the original feedstuff by gastro-intestinal and rumen microflora leads to considerable transport of microbial biomass and byproducts that may activate soil microbial communities. Composition of animal manures is often variable due to different feeding rations on different farms, as well as varying environmental conditions among seasons (Table 14.1). Nutrient concentrations are nevertheless considerable, making manures valuable fertilizers for agricultural lands. Most of the manure produced on livestock farms is spread back on adjacent farmland or sold to neighbors. The development of large concentrated animal feeding operations has led to unique challenges to process and effectively utilize such large quantities of animal manure.

TABLE 14.1 Characteristics of animal manures.

Source	Moisture	Total N	NH_4^+–N	Phosphorus	Potassium
	%	g kg^{-1}			
Beef solids	62–81	3–7	1–3	1–3	3–12
Dairy slurry	75–97	5–15	1–7	1–4	4–14
Swine slurry	94–99	2–6	1–3	1–3	1–2
Layer hen	58–72	21–26	10–15	7–10	5–10

Manitoba Agriculture Food and Rural Development (Manitoba Agriculture, 2015).

Composting and biogas digestion facilities have been utilized to reduce volume by driving off C through decomposition, concentrate nutrients, and capture value from these large volumes of manure.

Animal manures contain nutrients in addition to the primary elements of C, N, and P (Table 14.2). A complete elemental analysis of poultry litter shows its broad nutritional value and justification for widespread utilization. An imbalance of nutrients in some animal manure suggests that different combinations are needed to avoid unnecessary application to target plants and fields. Animal manures are an ideal C substrate to feed the soil microbial community and maintain SOM. Long-term research in Rothamsted, England and the Morrow Plots in Illinois has shown the possibility to maintain SOM with repeated animal manure application.

TABLE 14.2 Nutrient composition of poultry broiler litter.

Component	Concentration (g kg^{-1})	Component	Concentration (mg kg^{-1})
Moisture	180–400	Aluminum (Al)	300–700
Carbon (C)	260–340	Iron (Fe)	300–800
Total nitrogen (N)	25–33	Manganese (Mn)	300–500
Phosphorus (P)	9–13	Zinc (Zn)	200–350
Potassium (K)	12–25	Copper (Cu)	100–300
Calcium (Ca)	12–18	Nitrate (NO_3^-N)	100–200
Ammonium (NH_4^+-N)	4.1–6.5	Arsenic (As)	0–20
Sulfur (S)	3.5–5.1	Boron (B)	0–20
Magnesium (Mg)	2.4–4.4	Cadmium (Cd)	Trace
Sodium (Na)	3.4–4.4	Chromium (Cr)	Trace
		Molybdenum (Mo)	Trace
		Nickel (Ni)	Trace

Franzluebbers et al. 2007.

Biosolids

Municipal wastewater-treatment facilities produce a de-watered residual termed **biosolids** or sewage sludge. The composition and quality of biosolids can vary depending on the size of the local community and its associated industries producing wastewater for treatment. A variety of stabilization processes can be used at the treatment plant to help control odors, pathogens, levels of biodegradable toxins such as hydrocarbons, and human disease vectors such as rodents and flies. These processes can also affect the concentration of heavy metals derived from industrial sources in the waste stream. Composting of biosolids (Chapter 24) is an effective strategy to stabilize organic C and nutrients and simultaneously reduce pathogens, resulting in a potentially valuable soil conditioner and source of slow-release nutrients. For example, some municipal solid waste facilities have composted mixtures of biosolids and yard wastes (e.g., trimmings from plants). US-EPA (1999) estimated that 6.3 Pg of biosolids were generated in the United States in 1998, of which 60% was land applied, composted, or used as landfill cover. The remaining 40% was disposed of in landfills or incinerated and represents a large unused supply of this potentially valuable resource. Beneficial use of biosolids is mostly hindered by public opposition. Important factors dictating public acceptance include the possible pollutants in the biosolids, risk of disease, and odors. When properly treated and managed in accordance with regulations and standards, biosolids are considered safe for the environment and human health.

Biosolids are regulated by US-EPA, and their utilization is restricted by a variety of real and perceived threats to humans and the environment. Class A biosolids contain virtually no detectible pathogens, whereas Class B biosolids are treated but still contain detectible levels of pathogens. There are buffer requirements (e.g., proximity to surface waters) and restrictions regarding public access and crop harvesting for virtually all forms of Class B biosolids that are land applied. Additionally, the US-EPA has established regulations regarding the amounts of trace metals applied to land as contaminants in biosolids (Table 14.3); these include maximum concentrations in the biosolids on a weight basis and maximum amounts

TABLE 14.3 Pollutant limits for land application of biosolids.

Element	Ceiling concentration (mg kg^{-1})	Annual loading (kg ha^{-1} yr^{-1})
Arsenic (As)	75	2.0
Cadmium (Cd)	85	1.9
Chromium (Cr)	3000	150
Copper (Cu)	4300	75
Lead (Pb)	840	15
Mercury (Hg)	57	0.85
Molybdenum (Mo)	75	–
Nickel (Ni)	420	21
Selenium (Se)	100	5.0
Zinc (Zn)	7500	140

US-Environmental Protection Agency US-EPA, 1994.

that can be applied annually and cumulatively on a land area basis. Reduction or elimination of pathogens and vectors are required prior to land application. Common treatments are heat (>52°C for 12 h) and/or high pH (raising pH to 12 for 72 h) and various other approved means to meet classification standards. Recently, unregulated pollutants have been identified in biosolids, including pharmaceuticals (e.g., ciprofloxacin, diphenhydramine, and triclocarban); steroids and hormones (e.g., campesterol, cholestanol, and coprostanol); and flame retardants (e.g., perfluoroalkyl substances [PFASs]; Chapter 22). An inspector general's report mandated that the US-EPA conduct risk assessments on 352 pollutants found in biosolids across the United States (OIG, 2018). Additional research is needed for making these assessments. It is also relevant to consider whether individuals and societies in general should re-evaluate the necessity of producing and using products potentially harmful to their health and environment, including those that enter municipal waste streams.

Industrial and commercial by-products

Industrial and commercial by-products are diverse in origin and composition and include such materials as wastes from the manufacture of paper and gelatin, restaurant waste, and ash from the burning of coal or biofuels. While some of these can stimulate soil biological activity, others may contain substances that are inhibitory. Potential beneficial effects are varied. Organic byproducts can provide C and other nutrients that promote plant growth, stimulate soil microbial communities, enhance soil health, or aid in the restoration of degraded soils. Nutrients and particulates contained in inorganic materials, such as ash from the burning of coal or biofuel, can also have beneficial effects when applied to soils at appropriate rates. Nevertheless, detrimental effects can arise from excessive application rates or the presence of inherently harmful constituents such as heavy metals and various organic contaminants. Thus, a key issue in utilization of industrial byproducts is obtaining a complete analysis and knowing whether harmful elements and/or constituents are present that could threaten soil life and surrounding ecological habitats, not to mention the eventual poisoning of wildlife and humans through ecological transfer via air, water, and biological mechanisms.

Biochar

Burning of biomass for heating of living quarters and for fueling various smelting and metallurgy operations has been a part of civilization for thousands of years. Alternatively, biomass and other organic materials can be pyrolyzed under conditions of low O_2 to produce charcoal. **Biochar** is a special type of charcoal produced for use as a soil amendment. Interest in biochar is due to its resistance to decomposition (thereby increasing soil C sequestration) and several other potential benefits to soil. These characteristics are especially relevant in the tropics, which typically have high rates of organic matter decomposition. Biochar is thought to persist in soil for more than 1000 years due to its high aromatic C content. It is also highly adsorptive of other substances. "Terra preta" or black earth was reported by Prof. Wim Sombroek during his explorations in the Amazon basin in the 1950s; plants growing on terra preta were more productive, leading Sombroek and others to investigate these unusual soils. Indigenous people in the region burned forest vegetation in piles, and a portion of the material would pyrolyze under the heaps of biomass that limited O_2 entry. In addition to producing biochar, modern pyrolysis techniques capture otherwise harmful gases historically released from open pits. Biochar produced from modern pyrolysis operations can be useful as a soil amendment, partly because of its high cationic and anionic charge, as well as its high specific surface area. The potential benefits of biochar applied to soil include:

- capturing toxins and excess N, thereby protecting our waterways and groundwater quality
- creating alternative sources of energy (syngas and bio-oil) for generating heat and power, thereby reducing CO_2 emissions
- sequestering C from photosynthesis (CO_2) for thousands of years, thereby slowing climate change
- reducing the need for fossil fuels used to manufacture commercial fertilizers
- converting biomass and biowaste into value-added products while creating jobs

Indeed, many scientists are researching multiple aspects of biochar production, including its chemical composition when produced from different feedstocks, that is, raw biomass. Biochar additions to soil can increase nutrient and water retention in soils that have been degraded by conventional agricultural techniques and, thereby, could help improve crop yields. Biochar production may also promote recycling of organic waste and remediation of soil contaminants. Composition of biochar and the specific characteristics it imparts on soil will be determined to a large extent by the source of the feedstock, pyrolysis temperature, oxygen conditions, and duration of pyrolysis. Table 14.4 shows some of the variation in biochar composition that can be expected.

TABLE 14.4 Composition of biochars as affected by source of feedstock and pyrolysis temperature.

Source of feedstock	Pyrolysis temperature	Mass recovered after pyrolysis	C	H	O	N	P	Aromatic C[a]
	°C		------	------	%	------	------	------
Peanut hull[b]	–	–	51	6	38	1.7	–	–
	400	40	75	5	10	2.7	0.3	57
	500	35	82	3	3	2.7	0.3	82
Pecan shell	–	–	52	6	41	0.3	–	–
	350	50	65	5	28	0.3	<0.1	42
	700	30	91	2	2	0.5	<0.1	58
Poultry litter	–	–	36	5	24	4.1	–	–
	350	57	46	4	9	4.9	2.9	57
	700	36	44	<1	<1	2.8	4.3	–
Switchgrass	–	–	48	6	43	0.5	–	–
	250	78	55	6	36	0.4	0.1	29
	500	29	84	2	4	1.1	0.2	82

[a]*For Aromatic C, the values are percentages of the total C percentages (column 4).*
[b]*First row for each source represents values for raw feedstock.*
Novak et al. (2009).

More often than not, biochar leads to improvement in a variety of soil physical and chemical properties and, therefore, to an improvement in conditions for soil microbial communities. Stimulation of soil microbial activity with biochar application can occur, but its resistant aromatic composition suggests that any direct value as a C or energy source is limited to the small fraction of leachable (and more degradable) constituents that are present; thus, indirect stimulatory effects via changes in soil properties are more likely. Various changes in microbial communities have been reported in response to biochar amendment, and these changes can differ depending on the nature of the biochar applied. There are studies that show negative impacts of biochar on plant growth, and these have mostly been attributed to either very high rates of biochar addition that cause imbalances in nutrient conditions, presence of polycyclic aromatic hydrocarbons from unweathered materials, or various other toxic components. With a wide variety of feedstocks available to produce biochar, there is potential to design biochars with different nutrient contents, abilities to absorb various molecules, and susceptibilities to degradation. There is also interest in co-composting biochar with traditional organic materials to produce a compost product having the combined benefits of traditional compost and biochar and possibly other synergistic benefits. As is the case with many soil amendments, research is needed to assess the long-term effect of biochar and associated management on soil microbial communities and processes.

Cover crops

Cover crops can be any non-harvested crop used primarily to protect soil from erosion during off-season periods, provide actively growing roots to extract and stabilize nutrients that might be otherwise vulnerable to leaching or volatile loss, and increase levels of SOM to promote soil physical properties and C sequestration. Cover crops have other values to farmers, as some crops can also be harvested for forage or seed or to diversify the cropping system to suppress diseases (Chapter 23), obtain other crop rotation benefits, improve off-season access to fields, or extract water during wet periods.

As a source of additional C delivered to soil during non-cash-crop growing periods (e.g., in fall and winter in many temperate regions), cover crops are particularly effective in supplying soil microorganisms with readily available C sources from both root exudates during growth and C-rich crop residues upon termination. Several studies have found greater soil organic C sequestration with implementation of cover crops (Poeplau and Don, 2015). Often combined with no-tillage, management of cropland with cover cropping can enhance soil organic C sequestration due to addition of organic materials growing directly on land rather than imported from another location.

The type of cover crop used can have important effects on soil microbial communities and net C storage in soil. Typical cover crops in the southeastern United States, which is a favorable environment for all sorts of cover crops due to the generally mild winter conditions and abundant rainfall, have been single-species rye, wheat, or crimson clover. With the blossoming of the soil health movement across the United States, multi-species cover crops have also been used effectively by many growers. Many of these mixtures have three major botanical components of annual grass, broadleaf (typically *Brassica* species), and leguminous plants. The value of each depends on soil, cropping systems, and growing conditions. Diversity of species can provide balance to fill spatial gaps of establishment of any one particular species, fix N by leguminous plants (Chapter 17), provide dense rooting and above-ground structural support by cereal grasses, and provide a suite of biochemically active plant inputs to feed soil microbial communities.

Managing crop residue removal for biofuel production

Technology and infrastructure to produce ethanol and other forms of energy from a variety of plant biomass sources (i.e., **biofuel**) have been developing in the United States since the turn of the millennium. Some of the plants and associated materials considered for feedstocks are forest slash, fast-growing woody plantations of willow or poplar, perennial grasses like switchgrass and *Miscanthus*, crop residues from corn and wheat, and corn grain. Crop residues have historically been considered as an effective armor against soil erosion. Farmers are specifically required to keep >30% surface residue cover to meet conservation compliance requirements in US government assistance programs, although the removal of crop residues as fodder or silage for feeding livestock can be practiced if conservation compliance can be met in other ways. Similarly, removal of some portion of crop residues for biofuel production has been proposed for high-yielding regions where it might be possible to still meet conservation compliance guidelines. However, in either case, meeting minimum requirements for conservation compliance (i.e., to reduce erosion) may not be sufficient to avoid loss of soil organic C over time. Soil organic C is important for retaining and recycling nutrients, improving soil structure, maintaining sufficient water infiltration and storage, and sustaining soil microbial life. An analysis of several studies conducted across multiple locations throughout the US Corn Belt found that greater C input is needed to maintain soil organic C than is needed to provide protection from water and wind erosion (Fig. 14.5). Therefore, sustainable harvest of corn residue should only be expected on those fields that use the best conservation tillage approaches and have highest yield potential.

Soil biological changes with crop residue removal are mostly negative because crop residues are a significant source of organic substrates that fuel soil faunal and microbial activities. Across a number of studies, removal of crop residue has led to a reduction in earthworm density. Long-term crop residue removal can dramatically reduce soil microbial biomass and potential microbial activity. In those systems that include crop residue removal for biofuel, stimulation of soil biological activity could be achieved with organic amendment application and/or use of cover crops.

An alternative biofuel approach that can stimulate soil biological activity and avoid SOM loss is use of perennial crops, such as switchgrass. Compared with annual cropping systems, utilizing perennial crops often leads to greater soil organic C, even when the aboveground plant biomass is removed. Enhanced retention of C results from both greater amount of stubble that remains following harvest and constant presence of actively growing root systems, the latter continually providing soil and associated biological communities with soluble and particulate organic C.

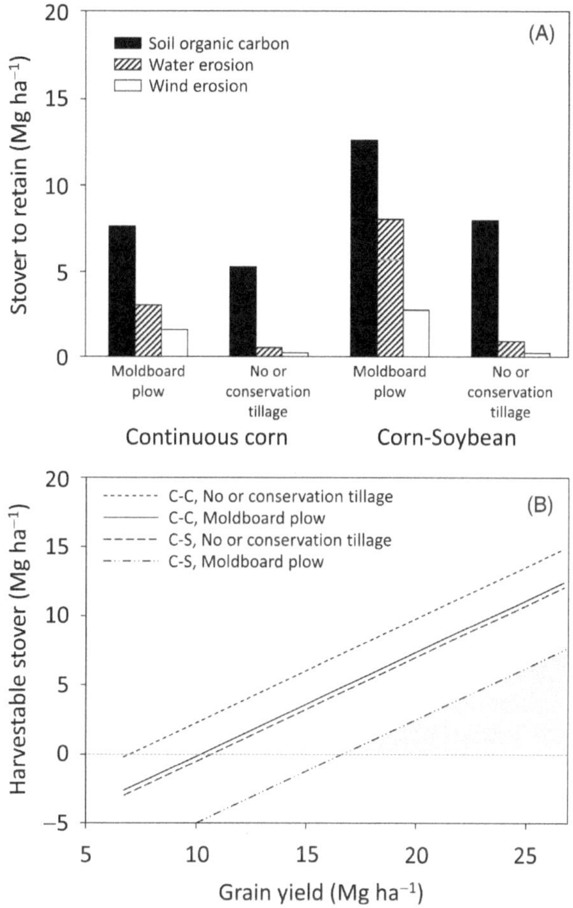

FIGURE 14.5 (A) Estimated corn residue requirements to maintain soil organic C and protect against water and wind erosion under different tillage and cropping systems. (B) Estimated amount of sustainably harvestable residue under different management systems as a function of grain yield; as an example, the shaded area represents sustainably harvestable stover for a corn-soybean rotation with moldboard plow tillage. C-C, continuous corn; C-S, corn-soybean rotation. *Source: Figure redrawn from Wilhelm et al. (2007). Used with permission.*

Carbon sequestration

Soil C sequestration is defined as an increase in soil organic C stock over time. It can be achieved with adoption of a variety of conservation agricultural systems or any other change in management that leads to greater balance of soil organic C (i.e., C inputs exceeding C outputs; Fig. 14.1). All conservation agricultural systems aim to achieve high crop or other production goals (yield and profitability), while simultaneously enhancing soil and environmental resources. These approaches generally rely on four guiding principles:

- minimizing soil disturbance
- maximizing soil cover

- maintaining presence of living roots as long as possible
- promoting biodiversity and associated carbon inputs

It is important to note that conservation systems are not universal recipes, because each environmental and socio-economic setting offers unique opportunities and challenges for optimizing the mix of principles. Conservation agricultural systems often include minimum- or no-tillage management of crops but may also rely on a diversity of cropping strategies to maximize photosynthesis and provide a diversity of substrates to ecologically control pests. Additionally, they may include pasture management approaches to integrate crop and livestock production. These strategies provide practical methods for achieving the goals of other holistic soil management systems. One of these is "**regenerative agriculture**" which is aimed at sustainably enhancing overall soil health. Promoting biodiversity, optimizing soil biological activity, balancing landscape interactions, and fostering robust ecosystem services are core goals of regenerative agriculture. Just as sustainable agriculture is a process for the continual improvement of soil services, so too is regenerative agriculture a vision for a better future with many milestones along the way. Other goals of holistic approaches include managing soil C to develop climate change resilience, protect human health, and provide opportunities for sustainable agro-ecosystem business development.

Soil C sequestration is a process that helps farmers on their own land, but it also contributes collectively to the global issue of climate change. Reducing concentrations of CO_2 and other greenhouse gases in the Earth's atmosphere is one of the most pressing modern-day environmental issues (IPCC, 2007) due to the enormous and largely unknown, but potentially cataclysmic, effects of climate change on ocean acidification and many other important ecosystem functions. There is no doubt that historical land clearing and other management practices have caused significant erosion and oxidation of SOM by soil microorganisms, greatly reducing C stocks in soils and contributing to this rise in atmospheric CO_2 concentration. Estimates of soil organic C loss vary. However, if we assume 20% loss of soil organic C stock (i.e., 400 Pg from an original level of 2000 Pg to depth of 2 m), there is an enormous potential to recapture most or all of this soil organic C through technological innovations and restoration activities. This would have to be achieved primarily by manipulating the activity of soil microorganisms and their accessibility to organic substrates. Assuming that aggressive global restoration efforts occur throughout the remainder of the 21st century, nearly all of the recent rate of CO_2 increase in the atmosphere (i.e., 4.1 Pg C yr^{-1}) could be mitigated. How this mitigation potential can be realistically achieved is shown by calculating that the needed total C sequestration of 400 Pg C/5 billon ha of agricultural land/75 year would require an annual soil C sequestration rate of 1.1 Mg C ha^{-1} (Mg, or Megagram, is 10^3 kg). This is certainly a tremendous goal but also plausible, based on empirical estimates obtained from more complex cropping systems that use perennials in combination with annuals. Of course, restoration efforts may take longer than 75 years, and the continued release of additional fossil fuel-derived CO_2 might trigger other greenhouse gas emissions to the atmosphere (e.g., boreal emissions of CH_4). Nevertheless, the potential for soil organic C sequestration could have a major impact on the atmosphere. Meeting these bold targets will require enactment and implementation of socio-economic policies, but the sooner restorative actions are initiated, the fewer the risks to the global ecology.

Storing C in soil as organic matter is not just a viable strategy to sequester C from the atmosphere; it is also essential in improving the health and vitality of soil. Maximizing C input to the terrestrial biosphere from the atmosphere is possible in agricultural production systems through a variety of management options, including plant selection, tillage, fertilization, and integrated management. Minimizing SOM decomposition and C loss from soil to the atmosphere can be achieved by reducing soil disturbance, more fully utilizing available soil water, and maintaining surface residue cover. Some key agricultural activities to sequester soil C and counter land degradation include tree plantings, conservation-tillage cropping, effective animal manure application and utilization, greater reliance on biological N fixation, novel green-manure cropping systems, improved grassland management, cropland-pasture rotation, and optimal fertilization strategies. A key aspect of plant-based strategies is the benefit derived from maximizing rhizosphere effects. The rhizosphere has much greater microbial biomass and activity than surrounding bulk soil due to the release of readily available C substrates and sloughing of plant cells and other organic materials (Chapter 11).

Fields managed to achieve greater soil organic C content over time will often have greater water infiltration, less water runoff, and potentially lower loss of nutrients in runoff, depending on how nutrients are applied. Soil organic matter provides C substrates needed for soil macro- and micro-fauna to thrive and for soil microbial activity to rework soil particles into **water-stable aggregates**. An abundance of water-stable aggregates at the soil surface provides resistance to the impacts of external energy forces, such as rainfall, that can otherwise lead to disaggregation and surface sealing, thereby restricting water infiltration. Water-stable aggregates are a key feature promoting a greater number of continuous, open macropores at the soil surface, allowing water to infiltrate and percolate to lower depths where deep roots of plants can extract a larger volume of available water. The extent of water-stable aggregation within a particular soil type is often directly related to the soil microbial biomass present as a result of the integration of soil organic C embedded within these aggregates (Fig. 14.6).

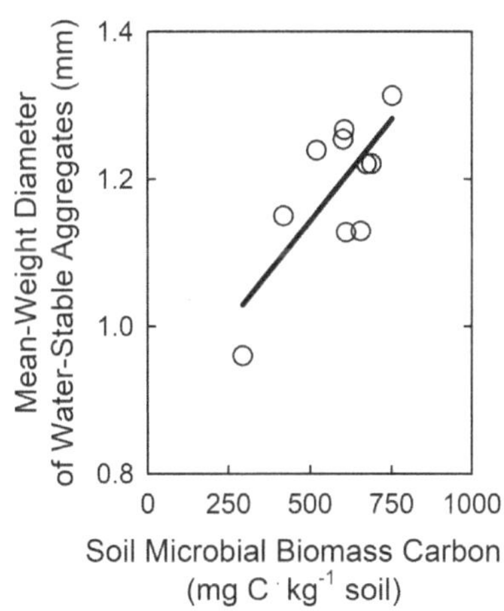

FIGURE 14.6 **Association of water-stable aggregation [expressed as mean-weight diameter (MWD), that is, the average diameter of aggregates based on mass distribution among several size classes] with soil microbial biomass C concentration from a silt loam (Mollic Cryoboralf) at depth of 0–10 cm in Alberta, Canada.** *Source: Data from Arshad et al. (2004).*

Soil health

Soil and the science of soil are vitally important for solving many of the global issues society is facing now and in the future, including providing for food security, mitigating and adapting to climate changes, providing clean water and its availability, managing nutrient recycling, and promoting effective nutrient utilization. Soil organic C powers many of the ecosystem services derived from well-functioning soil, including these key processes:

- Water and nutrient cycling
- Climate regulation
- Food, feed, fiber, and fuel production

Soil health (Chapter 25) is a concept borne from its predecessor—**soil quality**. In practice, little difference exists between the two, yet "soil health" has become a more popular title for the movement to counter practices considered harmful to soil resources. Conceptually, soil quality is concerned primarily with those soil properties and services most directly important to humans, whereas soil health is more holistic and focused on the biological integrity of the soil community itself as a living entity. Important indicators of soil health are those that relate to critical soil services in the field such as organic matter decomposition, cycling water and nutrients, controlling gas emissions, and harboring biodiversity.

Healthy soils are needed so that agriculture can continue to produce abundant food, feed, and fiber in the face of climate change uncertainties and ever-increasing input costs, and soil C is a key factor in meeting these goals. An oft-repeated question by agricultural advisors is "how should we measure the health of soil?" Chemical, physical, and biological dimensions of soil health are important to consider, because soil is a holistic body that has multi-purpose processes and functions. Indeed, a key biological indicator is the "breath" of soil itself. Like us, soil is living and therefore breathes—the more it breathes, the more it needs to be fed—the more it is fed, the more work it can do. Soil tests of biological activity, as determined from the flush of CO_2 following rewetting of dried soil, provide strong indications of:

- how well soil provides crops with N and other nutrients from mineralization
- the degree to which soil becomes aggregated and able to filter and cycle water effectively in the environment
- the magnitude of readily utilizable substrates available to support a diversity of soil microorganisms
- the potential of soil to sequester C from the atmosphere, as well as store plant nutrients in slowly accessible forms of SOM

To effectively and efficiently characterize soil functions, soil health indicators should (1) be easy to measure, (2) detect changes in soil function, (3) integrate soil physical, chemical, and biological properties and processes, (4) be accessible to many users and applicable to field conditions, (5) be sensitive to variations in management and climate, (6) encompass ecosystem processes and relate to process-oriented modeling, and (7) where possible, be components of existing soil data bases. The following minimum dataset could be considered for soil health evaluation, focusing on important associations with C in soil (see also Chapter 25).

- **Total organic C**—reflecting the functional capability of soil to supply nutrients to plants, serve as an organic nutrient reserve, mitigate greenhouse gas accumulation, provide organic resources for stabilizing the soil surface against erosion, purify water, buffer against nutrient extremes, and promote a biologically diverse and healthy microbial population—assessed by combusting soil to release all C for analysis

- **Water-stable aggregation**—reflecting the functional capability of soil to provide physical stability and resistance against wind and water erosion—assessed by immersing soil in water and determining extent of aggregate breakdown
- **Soil-test biological activity—**reflecting the functional capability of soil to cycle nutrients, decompose organic amendments, and catalyze and stabilize ecosystem processes through the interactions among a diversity of organisms – assessed by measuring the flush of CO_2 following rewetting of dried soil
- **Microbial substrate utilization**—reflecting the functional capability of soil to provide biodiversity and habitat, cycle nutrients, decompose organic amendments, and provide biological resilience to soil disturbance—assessed by measuring the ability to degrade a diversity of C sources
- **Inorganic N, extractable P and K, and soil pH**—reflecting the functional capability of soil to supply readily available nutrients, as well as to foster microbial function under optimum pH—assessed by routine soil testing for nutrients

Conservation and regenerative management systems are capable of restoring soil functions for the benefit of society. Soil microbiologists, ecologists, and agronomists must collaborate and work with local, regional, and national policy makers and community leaders to enhance soil functioning for the benefit of both current and future inhabitants of our global environment, not just those that temporarily own the land. Although producing plants and food have been at the forefront of civilization historically, it is not the only soil function of importance (Fig. 14.7). The following quotation perhaps says it best:

> "We abuse land because we regard it as a commodity belonging to us. When we see land as a community to which we belong, we may begin to use it with love and respect." –*Aldo Leopold,* **A Sand County Almanac,** *1949*

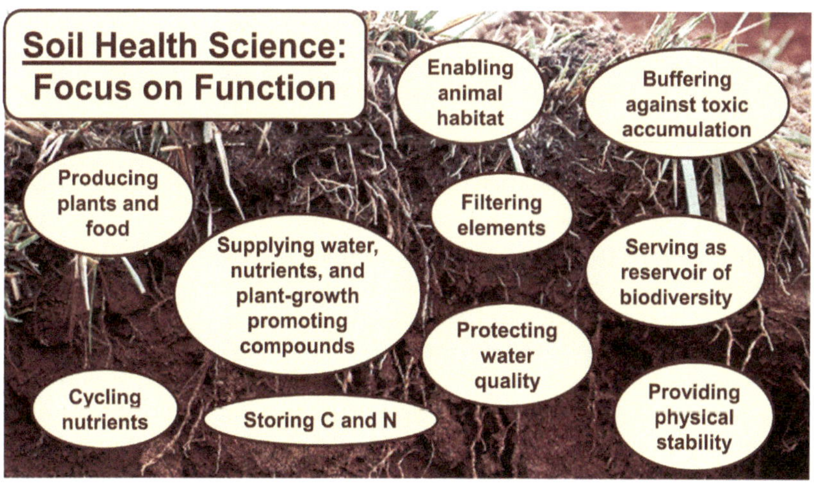

FIGURE 14.7 **Some of the functional attributes of soil that should be addressed by a robust soil health indicator.**

Summary

Soil organic matter provides numerous benefits to plant growth, environmental quality, and to the health of soil itself, including resident microorganisms. Agricultural production has for the most part contributed greatly to the loss of SOM, most notably due to the historical removal of native vegetation and disturbance of the soil with tillage. Food production is vital for our existence, but conservation agricultural approaches are now recognized as necessary to balance the need for food and preservation of environmental quality through regeneration of SOM.

Stratification of SOM with depth, that is, the concentration of SOM at or near the soil surface, under conservation tillage systems has positive consequences on important soil functions, including control of erosion, providing concentrated habitat for soil organisms, preserving nutrients in organic form, and cycling water more effectively. This depth stratification of SOM provides a good indication of soil health improvement with conservation management.

Effectiveness of organic amendments for agricultural production depends on the activity of soil microorganisms to decompose substrates and release inorganic constituents to the root zone for uptake by crops. Animal manures are a major source of organic materials capable of re-invigorating soil biology. Biosolids provide an abundant source of organic C and nutrients, but usage is regulated to ensure environmental protection due to potential concentration of heavy metals and pharmaceuticals during wastewater treatment. Industrial byproducts can also be beneficial, depending on their composition. Cover crops are highly effective to supply soil microorganisms with readily available C sources from root exudates during growth and nutrient-rich crop residues upon termination. Biochar addition to soil can provide a long-term source of C to retain nutrients and water, particularly in highly degraded soils and, therefore, could help improve crop yields. Soil biological changes with crop residue removal are mostly negative because crop residues are a significant source of organic substrates that fuel soil faunal and microbial activities.

Soil organic C sequestration can be achieved with adoption of a variety of conservation agricultural systems. Minimizing C loss from decomposition by soil microorganisms is possible by reducing soil disturbance, more fully utilizing available soil water through crop rotation and cover cropping, and maintaining surface residue cover. Soil with greater organic C content will often have greater water infiltration, less water runoff, and potentially lower loss of nutrients in runoff, depending on how nutrients are applied.

Soil and the science of soil are vitally important to solving many of the global issues facing society now and in the future, including providing for food security, mitigating and adapting to climate changes, providing clean water and its availability, nutrient recycling, and more effective nutrient utilization. Healthy soil with a diverse and active microbial community contributes to these essential soil functions. A variety of chemical, physical, and biological properties and processes can be used as indicators of soil health, including the simple measure of soil biological activity. Our health is directly linked to the health of the soil. Management of soil and SOM should be for now and for the future. Humanity depends on the orderly functioning of soil—driven by the soil organisms that inhabit it.

Study questions

1. If you had a farm, what are three things you would do to improve its soil organic carbon content?
2. How does tillage impact soil organic matter and soil microbial communities?
3. Why is stratification of soil organic matter important for soil microbial communities?
4. What are common sources of organic amendments and how do they differ in their contributions to soil microbial characteristics?
5. What are two important characteristics needed to calculate rate of soil carbon sequestration?
6. What do you think is a good indicator of soil health? Why?

References

Allmaras, R.R., Copeland, S.M., Copeland, P.J., Oussible, M., 1996. Spatial relations between oat residue and ceramic spheres when incorporated sequentially by tillage. Soil Sci. Soc. Am. J. 60, 1209–1216.

Arshad, M.A., Franzluebbers, A.J., Azooz, R.H., 2004. Surface-soil structural properties under grass and cereal production on a Mollic Cryoboralf in Canada. Soil Tillage Res. 77, 15–23.

Franzluebbers, A.J., 2002. Soil organic matter stratification ratio as an indicator of soil quality. Soil Tillage Res. 66, 95–106.

Franzluebbers, A.J., Langdale, G.W., Schomberg, H.H., 1999. Soil carbon, nitrogen, and aggregation in response to type and frequency of tillage. Soil Sci. Soc. Am. J. 63, 349–355.

Franzluebbers, A.J., Schomberg, H.H., Endale, D.M., 2007. Surface-soil responses to paraplowing of long-term no-tillage cropland in the Southern Piedmont USA. Soil Tillage Res. 96, 303–315.

Intergovernmental Panel on Climate Change (IPCC), 2007. Climate change 2007: The physical science basis. Summary for policymakers. Working Group I to Fourth Assessment Report of the IPCC. IPCC Secretariat, Geneva, Switzerland, p. 21.

Lal, R., 2004. Soil carbon sequestration impacts on global climate change and food security. Science 304, 1623–1627.

Manitoba Agriculture, Food and Rural Development, 2015. Properties of manure, pp. 37. Available from: https://www.gov.mb.ca/agriculture/environment/nutrient-management/pubs/properties-of-manure.pdf.

Novak, J.M., Lima, I., Xing, B., Gaskin, J.W., Steiner, C., Das, K.C., Ahmedna, M., Rehrah, D., Watts, D.W., Busscher, W.J., Schomberg, H.H., 2009. Characterization of designer biochar produced at different temperatures and their effects on a loamy sand. Ann. Environ. Sci. 3, 195–206.

Office of Inspector General (OIG), US-EPA, 2018. Cleaning up and revitalizing land: EPA unable to assess the impact of hundreds of unregulated pollutants in land-applied biosolids on human health and the environment. Report No. 19-P-0002. Washington, DC. Available from: https://www.epa.gov/sites/production/files/2018-11/documents/_epaoig_20181115-19-p-0002.pdf.

Perkins, H.F., 1987. Characterization data for selected Georgia soils. Spec. Publ. Georgia Agric. Expt. Sta., Univ. Georgia. 43, 545.

Poeplau, C., Don, A., 2015. Carbon sequestration in agricultural soils via cultivation of cover crops – A meta-analysis. Agric. Ecosyst. Environ. 200, 33–41.

US-Environmental Protection Agency (US-EPA), 1994. A Plain English Guide to the EPA Part 503 Biosolids Rule. EPA/832/R-93/003, Office of Wastewater Management, US-EPA, Washington, DC, p. 177. Available from: https://19january2017snapshot.epa.gov/sites/production/files/2015-05/documents/a_plain_english_guide_to_the_epa_part_503_biosolids_rule.pdf.

US-EPA, 1999. Biosolids Generation, Use, and Disposal in the United States. EPA530-R-99-009, Office of Solid Waste, Municipal and Industrial Solid Waste Division, US-EPA, Washington, DC, p. 74. Available from: https://www.epa.gov/sites/production/files/2018-12/documents/biosolids-generation-use-disposal-us.pdf.

Wilhelm, W.W., Johnson, J.M.F., Karlen, D.L., Lightle, D.T., 2007. Corn stover to sustain soil organic carbon further constrains biomass supply. Agron. J. 99, 1665–1667.

Further reading

Coleman, D.C., Crossley, D.A., 1996. Fundamentals of Soil Ecology. Academic Press, San Diego, CA, p. 205.
Doran, J.W., Coleman, D.C., Bezdicek, D.F., Stewart, B.A. (Eds.), 1994. Defining Soil Quality for a Sustainable Environment. Soil Science Society of America Special Publication No. 35, SSSA-ASA, Madison, WI, p. 244.
Fuhrer, J., Gregory, P.J. (Eds.), 2014. Climate Change Impact and Adaptation in Agricultural Systems. CABI Climate Change Series, vol. 5, CABI, Wallingford, UK, p. 285.
Magdoff, F., Weil, R.R. (Eds.), 2004. Soil Organic Matter in Sustainable Agriculture. Advances in Agroecology Series. CRC Press, Boca Raton, FL, p. 398.

CHAPTER

15

Transformations of nitrogen

David D. Myrold

Department of Crop and Soil Science, Oregon State University, Corvallis, OR, United States

Nature works only in cycles, there are no straight lines.
The forward movement is provided by time. Everything within it must revolve.—**Anonymous**

Perhaps more time and effort have been invested in studying the nitrogen (N) cycle than any other topic in soil microbiology. Nitrogen is an essential nutrient for all life on earth. Thus its fixation into useable forms by bacteria and subsequent transformations and recycling through organic and inorganic forms are of great practical interest. Indeed, nitrogen is the nutrient most often limiting plant growth in terrestrial ecosystems, and the use of synthetic nitrogen fertilizers boosts agricultural productivity worldwide. The nitrogen cycle affects the environment as well (Fowler et al., 2013; Battye et al., 2017). Current concerns include: greenhouse warming (Chapter 20), ozone depletion, high concentrations of nitrate (NO_3^-) in ground and surface waters, and the contribution of gaseous nitrogen oxides, such as nitric oxide (NO) and nitrous oxide (N_2O), to large-scale environmental problems of nitrogen deposition, and changes in plant biodiversity. The large diversity of nitrogen-containing compounds, which exist in numerous oxidation states, and the wide array of microbial transformations make the nitrogen cycle an extremely interesting intellectual challenge.

The intent of this chapter is to present the generally accepted workings of the nitrogen cycle and its associated organisms. Such generalizations inherently leave out some of the details and deviations from the accepted norm. But for those who desire more of the nuances of the nitrogen cycle, a few exceptions to the rules will be presented in some of the Boxes.

Nitrogen cycle

The terrestrial nitrogen cycle can be viewed at different levels of detail. We will first consider the global scale in terms of inputs and outputs of nitrogen, with an emphasis on how human activities have altered the nitrogen balance (Fig. 15.1).

Prior to the industrial revolution, natural inputs of nitrogen to the terrestrial biosphere were about 125 Tg N year^{-1}, consisting of biological nitrogen fixation (80%), atmospheric deposition (8%), and rock weathering (12%), a more recently recognized source (Houlton

Principles and Applications of Soil Microbiology. http://dx.doi.org/10.1016/B978-0-12-820202-9.00015-0
Copyright © 2021 Elsevier Inc. All rights reserved.

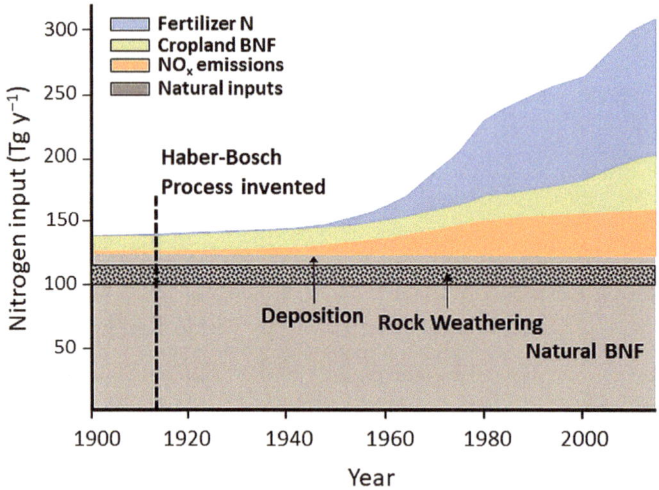

FIGURE 15.1 **Global inputs of nitrogen to the terrestrial ecosystem.** Natural inputs are dominated by biological nitrogen fixation (BNF). Anthropogenic inputs have approximately tripled the total amount of nitrogen entering the terrestrial ecosystem during the past century. *Source: Anthropogenic nitrogen input data provided by William Battye (Battye et al., 2017); natural nitrogen input data from Houlton et al. (2018).*

et al., 2018). These inputs changed little until the mid-20th century when use of inexpensive synthetic nitrogen fertilizer became widespread as part of the Green Revolution. This change was made possible by the invention of the Haber-Bosch process in 1913, which produced ammonia (NH_3) from dinitrogen (N_2) and hydrogen gases. Although synthetic nitrogen fertilizer has allowed food production to keep pace with the growing population, it has nearly doubled the amount of nitrogen entering the terrestrial biosphere annually, by an additional 110 Tg N year^{-1}. When added to the nitrogen from biological nitrogen fixation in agroecosystems (43 Tg N year^{-1}) and deposition of atmospheric nitrogen oxides (NO_x) from the combustion of fossil fuels (37 Tg N year^{-1}), more than two times the amount of nitrogen enters terrestrial ecosystems each year than it did in 1950. The unintended consequences of this excess nitrogen include: contamination of ground and surface waters through losses of nitrate, greater production of nitrous oxide—a potent greenhouse gas, and changes in plant biodiversity. Outputs from the terrestrial biome currently include losses to the atmosphere of dinitrogen (100 Tg N year^{-1}) and nitrous oxide (13 Tg N year^{-1}) from biological processes, and transport to the ocean (80 Tg N year^{-1}), both of which have increased from preindustrial levels, when they would have roughly balanced natural inputs. Other short-lived gases, such as ammonia and nitric oxide, are released into the atmosphere but they are redeposited within a few weeks and thus don't contribute to net outputs on an annual basis. Anthropogenic nitrogen inputs have resulted in some accumulation of N in terrestrial ecosystems through enhanced biomass production and soil organic matter accumulation (9 Tg N year^{-1}).

As we move down to the plot-scale, we become more interested in the pools of nitrogen in the soil and the transformations among those pools. This concept is known as the **nitrogen cycle.** A greatly simplified version is shown in Fig. 15.2.

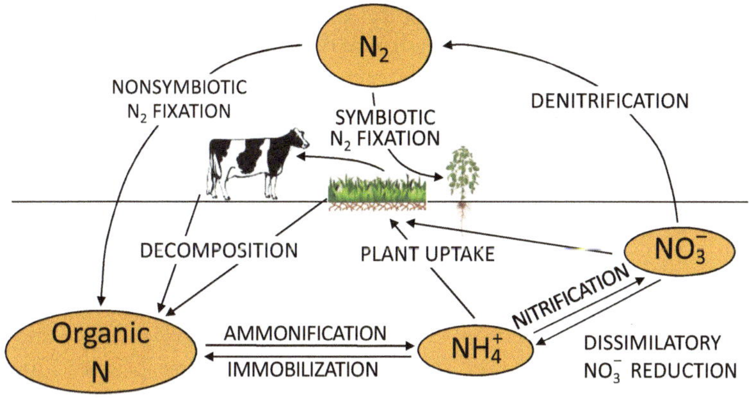

FIGURE 15.2 **Overview of the N cycle, showing major pools *(shaded ovals)* and transformations *(arrows)* of N.**

TABLE 15.1 Pool sizes of terrestrial nitrogen based on mineral soil to a depth of 1 m.

Pool	**Typical size (range) ----- g N m^{-2} -----**	**Remarks**
N_2 (Dinitrogen)	1,150 (230–27,500)	Minimum based on 0.25 m^3 air-filled pore space in the soil; maximum based on soil air plus a 30–m tall cylinder of air above the soil surface, for example, a tall forest stand.
Total organic N	725 (100–3,000)	From Post et al. (1985); typical value is median of reported soil N contents. Histosols are not included and would likely contain 3,000–8,000 g N m^{-2}.
Plant N	25 (1–240)	Minimum based on desert regions, maximum based on agricultural crops (Olson and Kurtz, 1982) and forest systems (Waring and Schlesinger, 1985; Anderson and Spencer, 1991).
Particulate organic N (PON)	15 (2–100)	Assumes 1 m^3 soil at a bulk density of 1.25 Mg m^{-3} and typical PON concentrations for soils.
Dissolved organic N (DON)	5 (1–50)	Assumes 1 m^3 soil at a bulk density of 1.25 Mg m^{-3} and typical DON concentrations for soil solutions.
NH_4^+ (Ammonium)	1 (0.1–10)	Assumes 1 m^3 soil at a bulk density of 1.25 Mg m^{-3} and typical NH_4^+ concentrations for soil extracts.
NO_3^- (Nitrate)	5 (0.1–30)	Assumes 1 m^3 soil at a bulk density of 1.25 Mg m^{-3} and typical NO_3^- concentrations for soil extracts.

Nitrogen is present in various forms (Table 15.1)—primarily as dinitrogen gas, organic nitrogen (in plants, animals, and soil organic matter), and ammonium (NH_4^+) and nitrate ions. Microbially mediated processes transform nitrogen from one form to another. Certain bacteria can transform dinitrogen to ammonia by a process known as dinitrogen fixation

(Chapters 16 and 17). The processes of **mineralization**, **immobilization**, **nitrification**, and **denitrification** are responsible for moving the fixed nitrogen from one form to another in the soil and will be discussed in turn in this chapter. First, however, we need to review the various forms of nitrogen and their characteristics, because these important fundamentals influence the microbial transformations of nitrogen in soil.

Forms of nitrogen

The sizes of the nitrogen pools vary over several orders of magnitude (Table 15.1). Although we tend to ignore the relatively inert dinitrogen pool, probably because it is an invisible gas, it actually represents the largest pool of potentially biologically active nitrogen in terrestrial ecosystems. Soil organic nitrogen makes up the next largest pool of nitrogen and varies widely among soil types in relation to organic matter content. The variation in soil organic nitrogen is determined largely by the factors of soil formation, particularly temperature and precipitation. The amount of nitrogen in plant biomass is of intermediate size and varies as a function of vegetation type (forests versus grasslands), climate, and soil nitrogen availability. Soil inorganic nitrogen pools are usually small, generally just a few mg N kg^{-1} (milligrams of nitrogen per kilogram) in natural ecosystems and rarely exceeding 100 mg N kg^{-1} in the plow layer of recently fertilized agricultural soils.

Larger pools tend to be less reactive (i.e., they turn over more slowly) and smaller pools usually are more dynamic. For example, the atmospheric dinitrogen pool is the largest pool of nitrogen and has a mean residence time on the order of millions of years. Decades are required to turn over the total pool of organic nitrogen in soil, although turnover of specific organic nitrogen compounds varies from minutes to millennia. Nitrogen in plant biomass often turns over annually, whereas inorganic-nitrogen pools are so dynamic that they may turn over more than once a day.

Soil organic nitrogen

The composition of soil organic nitrogen is enigmatic, but modern spectroscopic methods suggest that most of the nitrogen is present in amide bonds—the type of bonds present in protein, with lesser amounts of amine and heterocyclic bonds. Naturally occurring organic-nitrogen compounds extracted from soils include high molecular weight biopolymers (e.g., proteins, microbial cell walls, nucleic acids), their smaller breakdown products (e.g., peptides and amino acids, hexosamines, nucleotides), and a variety of other monomers (e.g., quaternary ammonium compounds, vitamins, antibiotics, and metabolic intermediates) (Figs. 15.3 and 15.4).

Most of the organic nitrogen in soil is associated with soil minerals, either sorbed directly on particle surfaces or occluded in small pores within aggregates (>90%), some is present as partially decomposed particulate organic matter (1%–4%), and a small fraction is dissolved in the soil solution (0.1%–2%) (Table 15.1). The particulate organic nitrogen (PON) and dissolved organic N (DON) are thought to be more available to soil microorganisms, with DON often considered the most biologically active. The composition of DON is complex, about two-thirds is of high molecular weight (>1000 kDa) compounds too large to be taken up by microorganisms, with the remaining smaller molecules dominated by proteins, oligopeptides, amino acids, and quaternary ammonium compounds (Warren, 2014).

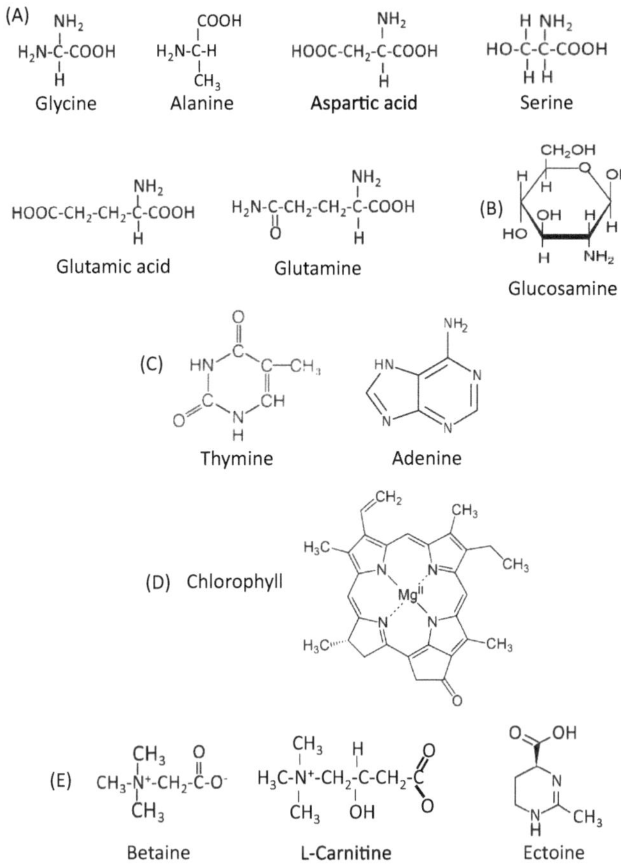

FIGURE 15.3 **Examples of important organic N compounds in soil.** (A) Common amino acids, (B) an amino sugar, (C) common nucleic acids, (D) chlorophyll, and (E) quaternary ammonium compounds.

Soil inorganic nitrogen

Unlike soil organic nitrogen, the important inorganic forms of nitrogen in soil ecosystems are well characterized, primarily because most inorganic-nitrogen compounds can be readily separated and measured. Inorganic-nitrogen pools in soil are usually small compared to organic nitrogen, but are nevertheless important because they serve as substrates to be oxidized, as alternate electron acceptors, or as intermediates or products of the many biological nitrogen transformations. Some key inorganic forms of nitrogen and their characteristics are shown in Table 15.2.

With this background behind us, it is time to delve more deeply into the microbial workings of the nitrogen cycle. Microbial processes can convert nitrogen from one form or another without changing the oxidation state of the nitrogen (e.g., mineralization, immobilization), by oxidizing a reduced form of nitrogen to gain energy (e.g., nitrification), or by reducing an oxidized form of nitrogen, often as an alternate electron acceptor to oxygen (e.g., denitrification). The remainder of this chapter is organized by these three types of nitrogen transformation.

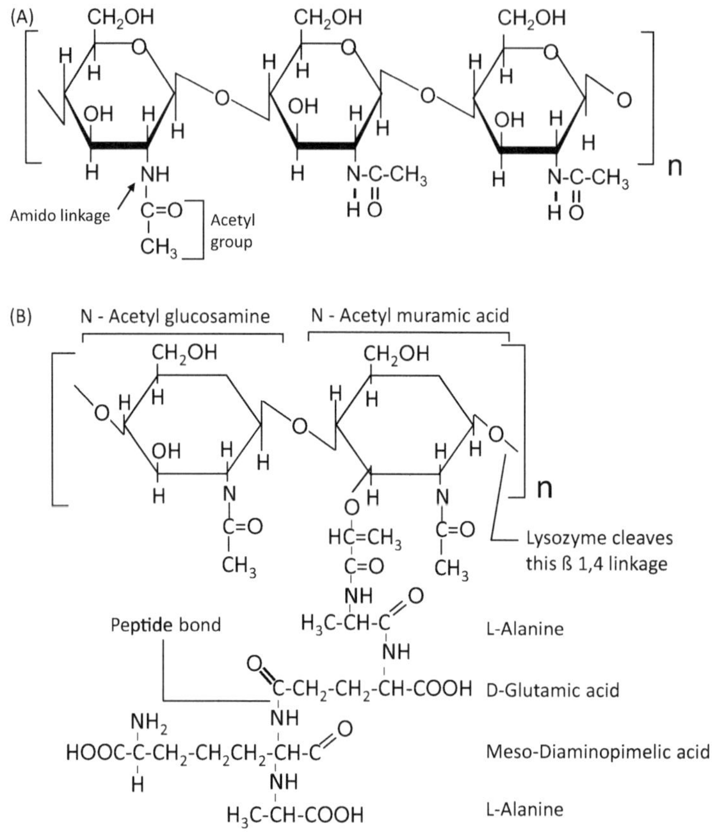

FIGURE 15.4 **Cell walls of common soil microorganisms are composed of polymers of hexosamines (e.g., amino sugars).** (A) Fungal cell walls contain chitin that consists of N-acetylglucosamine moieties connected by β−1,4 linkages, and (B) bacterial cell walls contain a peptidoglycan layer whose backbone is a polymer of N-acetylglucosamine and N-acetylmuramic acid connected by β−1,4 linkages, which has an oligopeptide side chain in Gram-negative bacteria.

Nitrogen transformations between organic and inorganic forms

The interconversion of nitrogen between organic and inorganic forms is typically called **mineralization** and **immobilization** by soil scientists and ecologists (Fig. 15.5). Nitrogen mineralization has several meanings. It is sometimes used in a generic sense for the production of inorganic nitrogen, both ammonium and nitrate, and sometimes more narrowly for the production of ammonium. The increase (or sometimes decrease) in inorganic nitrogen is most often called net nitrogen mineralization because it represents the difference of the concurrent ammonium production and consumption processes (Box 15.1). It is more precise to use **ammonification**, or gross nitrogen mineralization, to describe the biological transformation of organic nitrogen to ammonium.

TABLE 15.2 Important inorganic nitrogen compounds found or produced in soil.

Compound	Formula	Oxidation state	Form in soil	Major attributes
Ammonium	NH_4^+	−3	Fixed in clay lattice, dissolved, as gaseous ammonia (NH_3)	Cationic, rather immobile, volatilizes as NH_3 at high pH, assimilated by plants and microbes, substrate for autotrophic nitrification (NH_3 oxidation)
Hydroxylamine	NH_2OH	−1	Not detected	Intermediate in NH_3 oxidation
Dinitrogen	N_2	0	Gas	Largest pool of N, relatively insoluble, substrate for N_2 fixation, end-product of denitrification
Nitrous oxide	N_2O	+1	Gas, dissolved	Greenhouse gas and implicated in ozone destruction, very soluble, an intermediate in denitrification, by-product of nitrification
Nitric oxide	NO	+2	Gas	Chemically reactive, intermediate in denitrification, by-product of nitrification
Nitrite	NO_2^-	+3	Dissolved	Normally present at very low concentration, toxic, product of NH_3 oxidation, substrate for NO_2^- oxidation, an intermediate in denitrification
Nitrate	NO_3^-	+5	Dissolved	Anionic, mobile, readily leached, assimilated by plants and microbes, end product of nitrification, substrate for denitrification

Less confusion surrounds the term immobilization because it almost always describes the conversion of ammonium to organic nitrogen, primarily as a result of the **assimilation** of ammonium by the microbial biomass, a process that temporarily renders the nitrogen unavailable for other organisms. Less frequently, immobilization may refer to the assimilation of both ammonium and nitrate. The assimilation of nitrate by the microbial biomass is usually specified explicitly as nitrate immobilization. It is important to remember, however, that nitrate assimilation requires that nitrate be reduced to ammonium before the nitrogen can be incorporated into cellular constituents.

Ammonification (mineralization)

The conversion of organic-nitrogen compounds to ammonium is mediated by enzymes produced by microorganisms. Production of ammonium often involves several steps (Fig. 15.5). Extracellular enzymes first break down organic-nitrogen polymers into compounds that are low enough in molecular weight to pass across the cell membrane into the cell, where they may be further metabolized to generate energy, with the resulting production of ammonium, which is released into the soil solution.

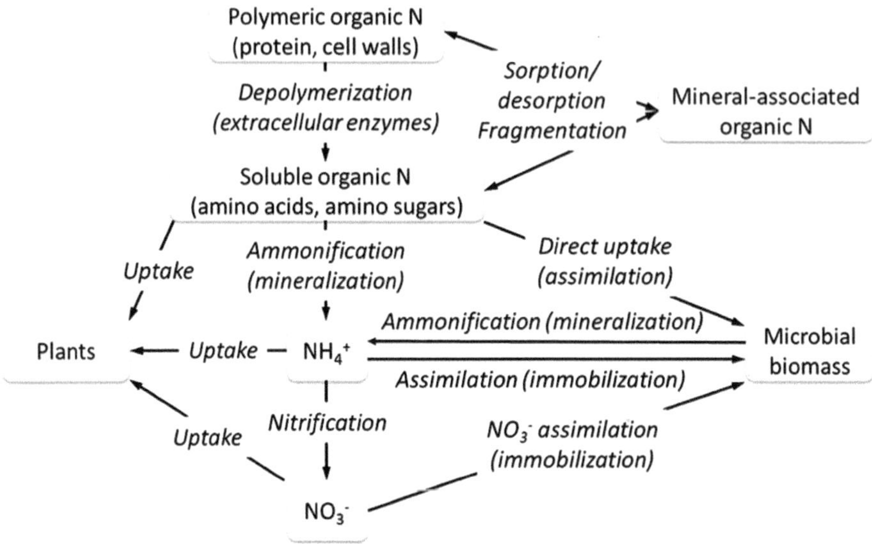

FIGURE 15.5 **Nitrogen mineralization-immobilization cycle in soil, including plant uptake and interactions of organic nitrogen and soil minerals.** Mineralization involves the steps of depolymerization followed by the release of ammonium, usually through deamination. Microbial immobilization includes the uptake and assimilation of small organic nitrogen compounds, ammonium, or nitrate. Microbes are in competition with plants for these same nitrogen compounds. *Source: Based on Schimel and Bennett (2004), Myrold and Bottomley (2008), Geisseler et al. (2010), Jilling et al. (2018), and Chacon et al. (2019).*

Extracellular enzymes important in nitrogen transformations

The major extracellular enzymes produced by microorganisms depolymerize proteins, aminopolysaccharides present in microbial cell walls, and nucleic acids (Fig. 15.6).

Proteins are broken down by a wide variety of extracellular proteolytic enzymes, also called proteinases, proteases, or peptidases, which are secreted into the environment (Nguyen et al., 2019). Peptidases are classified into families according to their active site. The relative abundance of secreted peptidases varies across microbial taxa. Serine peptidases are most abundant in both bacteria and fungi, with aspartic peptidases more common in fungi and cysteine and metallopeptidases more common in bacteria. All peptidases hydrolytically cleave peptide bonds; endopeptidases cleave internally to produce oligopeptides, and exopeptidases cleave off individual amino acids from either the amino- or carboxy-terminal end. It is generally assumed that microbes produce peptidases to acquire nitrogen, but peptidases may also be used to obtain carbon for energy generation (Geisseler et al., 2010).

Although microbial cell walls are thought to be relatively recalcitrant in soils, several common extracellular enzymes will degrade the polymers that make up the cell wall. Chitin (Figs. 15.4 and 15.6), which forms the cell wall of fungi and is also part of insect exoskeletons, is degraded by the combined activities of chitinase and chitobiase. Chitinase breaks chitin, a polymer of N-acetylglucosamine, into dimers (chitobiose), which are subsequently cleaved to two molecules of N-acetylglucosamine by chitobiase. This process is analogous to the enzymatic degradation of cellulose (Chapter 13). Several enzymes work to degrade the

BOX 15.1

Net versus gross rates of nitrogen cycle processes

From Fig. 15.5, we can see that several processes can both produce and consume ammonium. Here we focus on the ammonium pool and simplify with the input coming from ammonification of nitrogen from a single organic pool and output being assimilation into microbial biomass. Two different scenarios are shown.

The boxes represent the three different pools; the arrows represent the two different processes. The numbers above the arrows represent the gross rates of ammonium production and consumption; the number under the ammonium pool represents the net rate of ammonium production. As you can see, the net rate of ammonium production is the same in the two scenarios, despite a large difference in gross rates. Nitrogen is cycling about 5 times faster in one case than the other. An example of the ecological significance of gross and net rates is given in greater detail in Box 15.7.

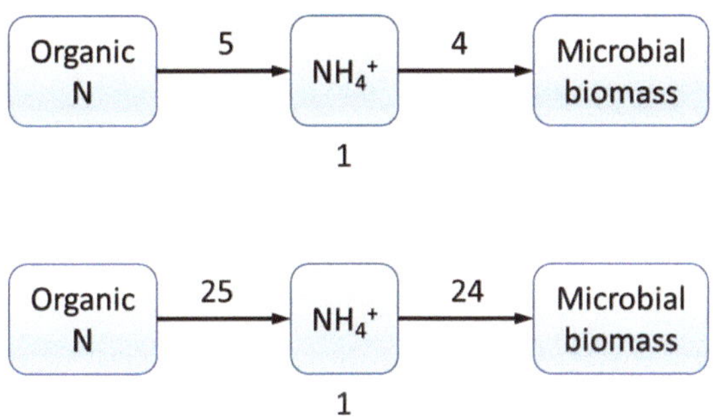

peptidoglycan portion of bacterial cell walls. Lysozyme is perhaps the most well-known. It breaks the β-1,4 linkage between N-acetylmuramic acid and N-acetylglucosamine. Individual hexosamines are the end products of the extracellular enzymes that degrade microbial cell walls.

Nucleic acids are degraded by ribonucleases (RNases) and deoxyribonucleases (DNases), which hydrolyze the ester bonds between the phosphate groups and pentose sugars of nucleic acids (Fig. 15.6). The known types of RNases and DNases are divided into exonucleases, which split off a single nucleotide from the end of the nucleic acid polymer, or endonucleases, which cleave within the nucleic acid polymer. Individual nucleotides are the ultimate product of the nucleases.

Many plants and microorganisms produce intracellular ureases, which hydrolyze urea into carbon dioxide and ammonia. Upon cell lysis, the ureases are released into soil where they maintain their function. Ureases function in the utilization of natural sources of urea (e.g., animal wastes) but perhaps most importantly in making the nitrogen in urea fertilizer available to plants.

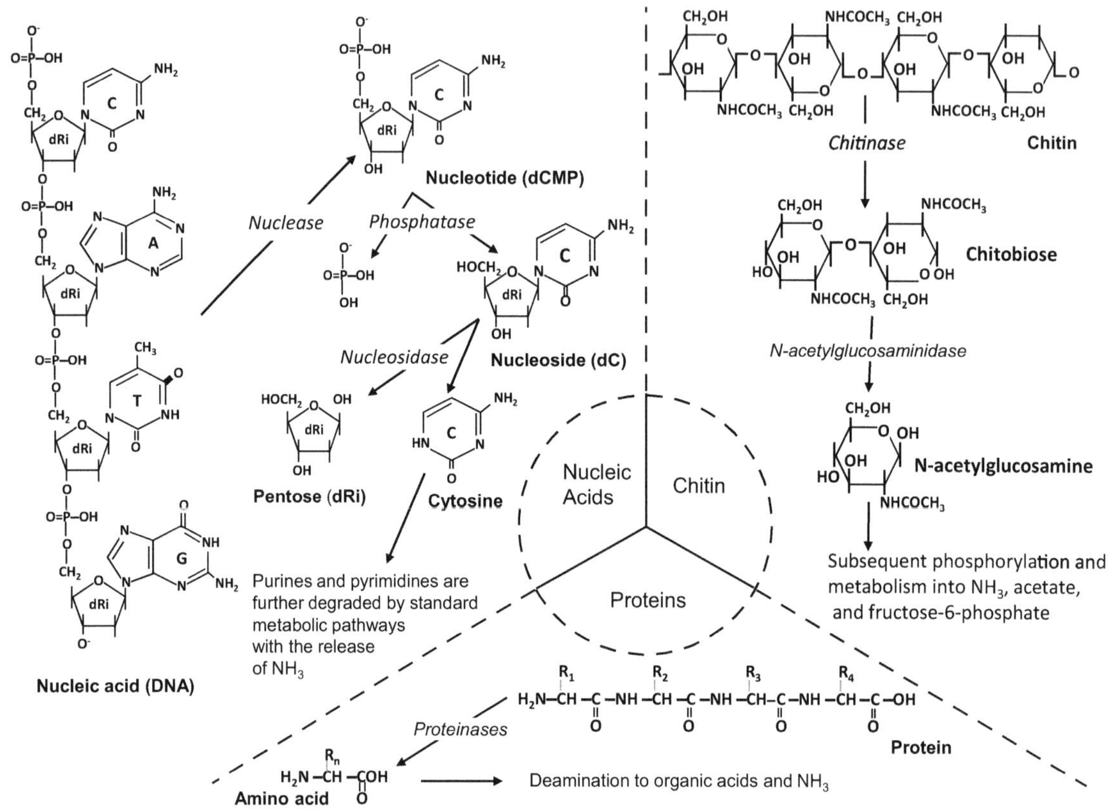

FIGURE 15.6 **Major routes for the depolymerization of organic nitrogen compounds through the action of extracellular enzymes.** *Source: Adapted from Myrold and Bottomley (2008). Used with permission.*

Considerable research has focused on the interactions between extracellular enzymes and soil mineral and organic constituents. These interactions are complex. For example, both the enzyme, which is a protein, and the substrate may be adsorbed onto clay surfaces. This may act to stabilize the enzyme or substrate and protect it from degradation; however, some minerals may destabilize and fragment proteins (Fig. 15.5). Stabilization on mineral surfaces provides one explanation for the presence of extracellular, or relic, DNA in soils. If the active conformation of an extracellular enzyme is altered by adsorption, this will likely inactivate the enzyme, but if the catalytic site is not affected, the enzyme may remain active. In the latter case, the protected, active enzyme may be an important catalyst as long as its substrate is accessible.

Intracellular enzymes important in nitrogen transformations

In most cases, the final production of ammonium occurs within microbial cells through the action of intracellular enzymes. Of course, some of these intracellular enzymes may become "extracellular" when microbial cells are lysed.

Two types of nitrogen are found in amino acids: the amine (NH_2-CR_3) and amide (NH_2-CR=O) functional groups. The amide-nitrogen group of asparagine and glutamine is cleaved from the amino acid in the form of ammonia by asparaginase and glutaminase. Amino nitrogen is also released as ammonia by amino-acid dehydrogenases and amino-acid oxidases in an oxidative process known as *deamination*. Dehydrogenases, such as glutamate dehydrogenase, use NAD^+ as a cofactor to accept electrons. Amino-acid oxidases are sometimes tethered to the cell wall so that deamination occurs outside the cell.

Amino sugars are metabolized in two steps. First, the amino sugar is phosphorylated by a *kinase* and then ammonia is released through a deamination reaction.

Degradation of nucleotides and the release of ammonium typically require several steps (Fig. 15.6). First nucleotides are hydrolyzed to produce nucleosides and phosphate (PO_4^{3-}). Following the dephosphorylation, the nucleosides are further hydrolyzed to purine or pyrimidine bases and pentose sugar moieties. Normal metabolic pathways then release ammonium during the catabolism of purines and pyrimidines, with urea as a prominent intermediate.

In most instances the microbial degradation of amino acids, amino sugars, and nucleic acids is driven by the need of heterotrophic microbes for energy and carbon. Thus, the ammonium released as a result of ammonification can be considered a by-product of catabolism.

Assimilation (immobilization)

Microbes, and other organisms, assimilate ammonium by two primary pathways (Fig. 15.7): glutamate dehydrogenase and glutamine synthetase-glutamate synthase (GOGAT). When ammonium is present in relatively high concentrations (>0.1 mM or about 0.5 mg N kg^{-1} soil), glutamate dehydrogenase, acting with $NADPH_2$ as a coenzyme, can add ammonium to α-ketoglutarate to form glutamate.

In most soils, ammonium is present at rather low concentrations, which results in low intracellular ammonium concentrations. Under these conditions, the second ammonium assimilation system is operable. The GOGAT pathway is quite complex. The first step requires ATP to add ammonium to glutamate to form glutamine. The second step transfers the ammonium from glutamine to α-ketoglutarate to form two glutamates. Once ammonium has been incorporated into glutamate, it can then be transferred to other carbon skeletons by various transaminase reactions to form additional amino acids.

Microorganisms (and plants—see Box 15.2) can also assimilate organic nitrogen and nitrate to meet their nitrogen needs. As noted in the previous section, the nitrogen in organic nitrogen compounds taken up by microbes can be released by mineralization as ammonium, which might be assimilated by other microbes, leading to what is known as the **mineralization-immobilization turnover** route. Alternatively, organic nitrogen can also be taken up and directly assimilated (e.g., amino acids incorporated into proteins) through the so-called **direct route**. Both routes can occur in the same soil, with the proportion likely depending upon the relative requirement of the microbial community for carbon or nitrogen (Geisseler et al., 2010).

The assimilation of nitrate by microorganisms (and plants) requires energy for the reduction of nitrate to ammonium and subsequent incorporation of ammonium into amino acids. Consequently, this process is regulated by nitrogen availability, and nitrate utilization is expected when energy is in excess relative to the concentrations of ammonium or

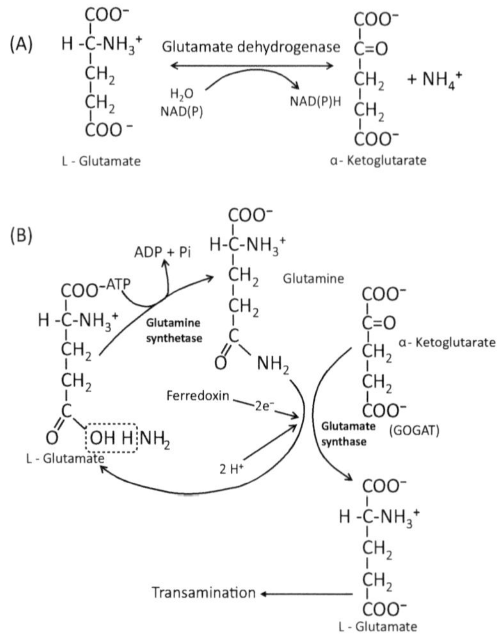

FIGURE 15.7 **Two main pathways of NH_4^+ assimilation.** (A) Glutamate dehydrogenase, which is a reversible reaction that works at higher NH_4^+ concentrations and (B) glutamine synthetase-glutamate synthase (GOGAT), which is induced and functions at low NH_4^+ concentrations and requires energy.

BOX 15.2

Exception to the rules 1: inorganic nitrogen is the only form of nitrogen important for plant uptake

Inorganic nitrogen fertilizers are the most widely used tool to manage crop nitrogen fertility, particularly in intensively managed agricultural and forested production systems. Even with organic nitrogen amendments, such as manure, it is commonly assumed that the organic nitrogen must first be converted into inorganic nitrogen before plants can utilize it. However, the ability of plants to take up organic nitrogen compounds, particularly amino acids, has been known for more than a century. Recently, there has been renewed interest in organic nitrogen in plant nutrition (Näsholm et al., 1998), particularly plants with roots colonized by certain types of mycorrhizal fungi (Chapter 12). Much of the most readily available organic nitrogen is likely found in soluble, small organic molecules, which make up the dissolved organic nitrogen (DON) pool. Although the size of the DON pool is typically smaller than the inorganic nitrogen in the soil solution of agricultural soils, DON can be the dominant form of soluble nitrogen in soils receiving low inputs of nitrogen. Studies in unpolluted forest ecosystems show that DON can be the major form of nitrogen lost to ground and surface waters (Perakis and Hedin, 2002). As a result of these recent findings, more attention is being paid to DON and its role in the nitrogen cycle.

organic-nitrogen compounds. For this reason, soil scientists believed assimilation of nitrate (sometimes called *nitrate immobilization*, a term which emphasizes that the nitrogen has been made unavailable to other organisms) by soil microorganisms to be minor. However, there is growing evidence that nitrate immobilization is an important process in some soils (see Box 15.7 in a later section).

Plants vary in their ability and preference for ammonium and nitrate uptake. When both ammonium and nitrate are equally available in soil solution, it is energetically more favorable for plants to use ammonium because nitrate must be reduced prior to use by the plant. In many cases, however, plants are not energy limited, so reducing power is available to convert nitrate to ammonium. This is particularly true for plants that reduce nitrate in leaf tissue, where this reduction is coupled to light energy and photosynthesis. The relative energy cost of ammonium versus nitrate metabolism is even more difficult to calculate in a heterogeneous medium like soil, because it may be more efficient for a plant to use nitrate than to put energy into growing a more extensive root system to access the less mobile ammonium.

Preference and competition for nitrogen

We know surprisingly little about the preferences for organic nitrogen, ammonium, and nitrate among microbial taxa. It is generally assumed that ammonium is preferred, and overall the abundance of pathways for ammonium assimilation was found to be higher than for nitrate assimilation in a study of soil metagenomes (Nelson et al. 2016). In reality, because the types of inorganic and organic nitrogen compounds in soil are limited, most organisms can probably use several of these sources, although preferences among taxa are likely. For example, Morrissey et al. (2018) found that some bacterial taxa assimilated more ^{15}N-labeled ammonium than other taxa.

Competition for nitrogen occurs not only among microorganisms but also between microorganisms and plants. It is difficult to generalize whether microorganisms or plants are winners in the competition for soil nitrogen, in part because most plants form symbiotic associations with mycorrhizal fungi, which blurs whether the competition is strictly between microorganisms and plants. A survey of studies on plant-microbial competition for nitrogen suggests that microorganisms have faster initial uptake and assimilation, winning in the short-term, but the ability of roots to grow allows them to garner resources in different locations and at different times, enabling them to be successful competitors in the end (Kuzyakov and Xu, 2013).

Net ammonium production or consumption

Several factors influence whether there is net production or consumption of ammonium by microorganisms in soil. The general principle is that net immobilization of ammonium occurs if the availability of nitrogen is limiting; otherwise net production occurs. In most soils, the growth and activity of heterotrophic microorganisms are limited primarily by the amount of available carbon, thus net nitrogen mineralization is the norm. This is actually not surprising, because plants, which require inorganic nitrogen, grow well in most soils.

When plant residues are returned to the soil or organic amendments are added, we can predict what affect this addition has on nitrogen availability. Adding a material high in carbon, such as sawdust, will likely immobilize inorganic nitrogen, whereas adding a material relatively high in nitrogen, such as manure, will provide available nitrogen. Most organic materials are about 45% carbon by mass; consequently, the ratio of carbon to nitrogen is determined largely by the concentration of nitrogen in the material. Decades of research have shown that when organic amendments with **C/N ratios** below 20/1 are added to soils, net ammonium production results; at wider C/N ratios, inorganic nitrogen from the soil is immobilized. This critical, or "break even," C/N ratio can be calculated from basic principles based on knowledge of the C/N ratio of the heterotrophic microorganisms and their **carbon use efficiency**, the amount of substrate carbon converted to microbial biomass (Box 15.3).

The C/N ratio of soil microorganisms ranges from 4 or 5 for bacteria to as high as 15 for fungi. Given that fungal biomass is often about twice that of bacteria in many soils, a typical C/N ratio for soil microbial biomass is about 8. Yield coefficients vary widely, depending on the type of organic matter (substrate quality), type of microorganism, and environmental conditions. Readily degradable organic compounds, such as simple sugars, may have yield coefficients as high as 0.6, whereas assimilation efficiencies for complex, recalcitrant compounds, such as lignin, may be less than 0.1. A reasonable average yield coefficient for plant-derived substrates is about 0.4. Environmental conditions that stress microorganisms would generally decrease the yield coefficient because more energy is required for cell maintenance than for growth. Another example of the effect of environmental factors is that yield coefficients are often much lower under anaerobic conditions, primarily because anaerobic metabolism normally produces less energy per mole of substrate (Chapter 3).

Net production of ammonium is influenced not only by environmental factors and the C/N ratio of substrates and microbes but also by other biotic factors. Most important of these is the role that soil animals play as predators of the primary decomposers, bacteria and fungi. About 30% of the yearly net nitrogen mineralization is directly caused by the activities of soil animals, such as protozoa and nematodes. When soil animals prey upon microorganisms, ammonium is often released as a waste product because the predators have a C/N ratio similar to their prey, which results in an excess of nitrogen due to a loss of carbon as carbon dioxide during metabolism. This phenomenon may be important in the rhizosphere where "grazing" by soil protozoa facilitates liberation of available nitrogen for plant growth (Chapter 8).

Fate of ammonium in soil

In addition to the mineralization-immobilization cycle, ammonium has several other fates in soil. It can be chemically held on cation exchange sites or become fixed in the lattice of clay minerals (ammonium fixation), such as illite and vermiculite. Ammonium may react chemically with organic compounds, such as quinones, or it may be volatilized at high pH. The loss of nitrogen through ammonia volatilization has potential negative consequences on fertilizer efficiency and the environment (Box 15.4). Major biological fates are plant uptake, microbial assimilation, or oxidation to nitrate by nitrifying microorganisms.

Oxidative processes

Some bacteria and archaea are chemolithotrophic, having the ability to oxidize reduced forms of inorganic nitrogen (ammonium or nitrite) in energy-yielding reactions. These microorganisms are also primarily autotrophs, fixing inorganic carbon, although some can

BOX 15.3

Calculating the critical C/N ratio that determines whether nitrogen is mineralized or immobilized

The critical C/N ratio for a decomposing organic substance can be calculated based on stoichiometric principles, the chemical composition of the substrate, knowledge of the physiological attributes of the decomposer(s), and how those attributes might be affected by environmental factors. For a pure culture of a bacterial species growing on a single organic N compound at a defined temperature, the critical C/N ratio will be well-defined. For a complex substrate, such as plant litter, which is being decomposed by a diverse microbial community in a variable environment, the critical C/N ratio is a rougher approximation. Nevertheless, a reasonable estimate can be obtained if we have an idea of the C/N ratio of the microbial community and the average efficiency of their conversion of substrate C into biomass C (their C use efficiency or growth yield—Y), as shown in this example.

Assumptions:

- All the organic C and N in decomposing plant litter are equally available to the microbial community.
- Fungi typically make up about two-thirds of the total microbial biomass; bacteria make up about one-third.
- Fungal cells commonly have a C/N ratio of about 10; bacterial cells commonly have a C/N ratio of about 4.
- The C use efficiency of soil microbial communities is highly variable, but a value of $Y = 0.4$ (i.e., 40% of C is assimilated into biomass) is representative of empirical measurements.

We can calculate that the decomposing organic substrate must have a C/N ratio of 20 or less in order to meet the nitrogen needs of the microbial decomposers by the following steps:

Step 1. Calculate the average microbial C/N ratio;

$C/N = (2/3)10 + (1/3)4 = 8$

Step 2. Divide the average microbial C/N ratio by the average microbial Y;

Critical substrate C/N ratio $= 8 \div 0.4 = 20$

The graph shows how the critical C/N ratio for a substrate varies with the organism's C/N ratio and its yield.

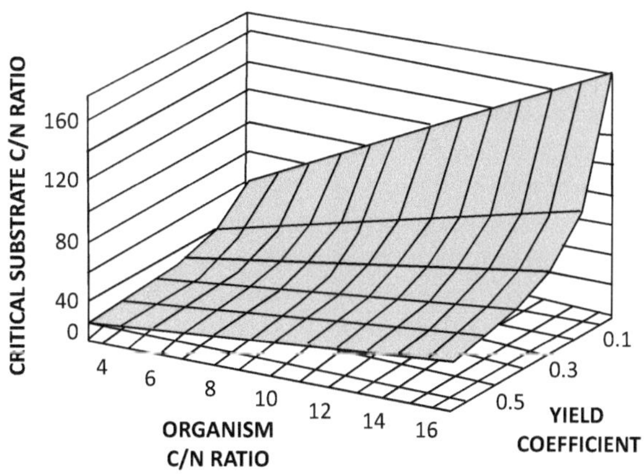

BOX 15.4

Ammonia volatilization

The ammonium ion (NH_4^+) in soil solution is in equilibrium both with cation exchange sites and with uncharged ammonia (NH_3), which can be lost to the atmosphere as a gas. Several factors influence the proportion of ammonia that is volatilized, including the cation exchange capacity of the soil, soil pH, soil moisture, soil temperature, and wind speed; some of these can be managed.

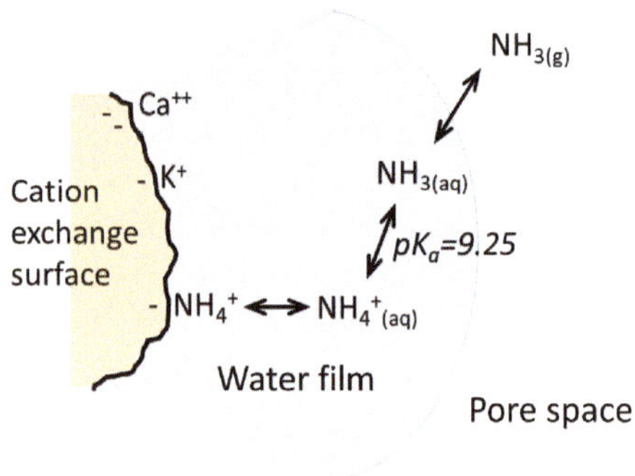

The higher the cation exchange capacity of a soil, the more ammonium can be held on those exchange sites. Similarly, the wetter the soil, the greater is the capacity for ammonium to be dissolved in the soil solution, and the more dilute the ammonium becomes, which has an influence on the ammonium-ammonia equilibrium. However, that equilibrium is more sensitive to soil solution pH, with lower pH favoring the ammonium form. At a pH of 7, less than 0.5% is present as dissolved ammonia, but this increases rapidly to more than 5% at pH 8. Microsite pH is especially important when ammonia-based fertilizer is applied, particularly urea, because hydrolysis of urea by soil ureases releases free NH_3 raising the pH by 2–3 units in the immediate vicinity of a urea granule:

$$H_2N{-}\underset{\displaystyle O}{\overset{}{\underset{\|}{C}}}{-}NH_2 + 3H_2O \xrightarrow{\text{Urease}} CO_2 + 2NH_4^+ + 2OH^-$$

The transfer from dissolved to gaseous ammonia (**volatilization**) increases with temperature and wind speed due to enhanced gas exchange with the above-ground atmosphere.

Volatilization of ammonia-based fertilizer represents nitrogen that is unavailable to the growing crop and may have detrimental effect to ecosystems downwind. Management practices that can help minimize volatilization losses include: application when soils are cool and moist, fertilizer placement at depth rather than on the surface, and the use of controlled-release fertilizers or addition of urease inhibitors (also known as nitrogen stabilizers) when using urea. Controlled-release fertilizer products include those coated with sulfur (which should additionally promote acidification localized around the fertilizer granule) and polymers that slowly biodegrade or control movement of water into the pellet. A common urease inhibitor is N-(n-butyl) thiophosphoric triamide (NBTP), which is marketed under several trade names. Reductions in ammonia volatilization through the use of NBTP or controlled-release fertilizers range from 30% to more than 50%.

BOX 15.5

Exception to the rules 2: anaerobic ammonia oxidation

The previous discussion of nitrification suggests that it is a strictly aerobic process. Although this is likely the predominant case in most soils, there is an exception. The story of anaerobic ammonia oxidation, or *anammox*, is a wonderful tale of discovery, which illustrates the importance of understanding the basic principles of microbial metabolism (Kraft et al., 2011).

About 25 years ago, it was postulated, based on thermodynamic considerations of redox reactions (Chapter 3), that a group of lithotrophic bacteria should exist that were capable of generating sufficient energy by oxidizing ammonium under anoxic conditions using nitrite as an electron acceptor.

$$NH_4^+ + NO_2^- \rightarrow N_2 + 2H_2O;$$
$$-358\ kJ\ mol^{-1}\ NH_4^+$$

Several years later, this process was observed in a wastewater treatment facility in the Netherlands and confirmed using ^{15}N-labeling experiments. Anammox bacteria are found in one order of the Planctomycetes phylum, a relatively unstudied bacterial group. Anammox bacteria are thought to be well adapted for habitats that have an oxic-anoxic interface. Anammox bacteria have been detected in soil, but the few studies of the anammox process in soil have found that it produces relatively little dinitrogen compared with denitrification. Nevertheless, the identification of a new group of bacteria with a novel metabolic pathway is an excellent example of how modern methods and fundamental principles can be combined to discover something new even about something that has been as well studied as the nitrogen cycle.

use a limited number of organic carbon compounds. For over 100 years, the microorganisms involved in the oxidation of reduced inorganic nitrogen (**nitrification**) were thought to be only aerobes. That changed in 1999 with the identification of bacteria that could oxidize ammonium anaerobically (Box 15.5). Anaerobic ammonia oxidation, known as **anammox**, couples the oxidation of ammonium with the reduction of nitrite.

Nitrification

Nitrification is the microbial production of nitrate from the oxidation of reduced nitrogen compounds using oxygen as the electron acceptor (Fig. 15.8). Since the 1890s, when the first nitrifying bacteria were isolated, nitrification was considered as a two-step, two-organism process of oxidizing ammonium to nitrate, in which the inorganic nitrogen served as the energy source for nitrifying bacteria that were chemolithoautotrophs. In the first step, ammonium (actually, ammonia at the enzyme level) was oxidized to nitrite by the ammonia-oxidizing bacteria of the "Nitroso-" genera (Table 15.3); in the second step, nitrite was oxidized to nitrate by the nitrite-oxidizing bacteria of the "Nitro-" genera. The recent discoveries of archaeal ammonia-oxidizers and of bacteria that are capable of complete ammonia oxidation (**comammox**) have expanded our perspective of the microorganisms involved in nitrification.

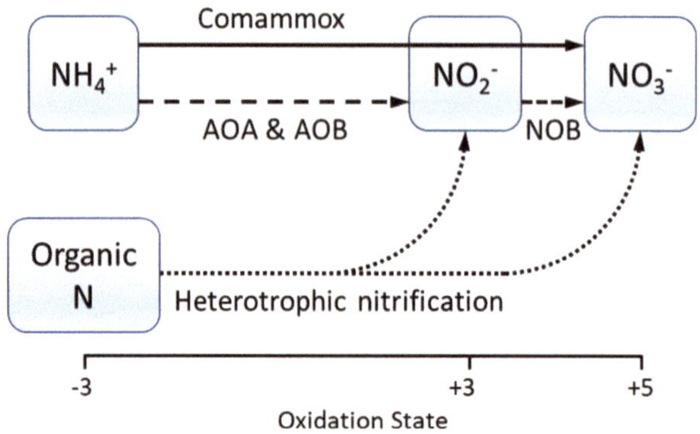

FIGURE 15.8 **Pathways of aerobic nitrification.** *Comammox*, complete ammonia oxidation; *AOA*, ammonia-oxidizing archaea; *AOB*, ammonia-oxidizing bacteria; *NOB*, nitrite-oxidizing bacteria.

TABLE 15.3 Chemolithoautotrophic nitrifying microorganisms found in soil.

Phylum or Class	Genus
NH_3 oxidizers	
Thaumarcheota	*Nitrososphaera*
Betaproteobacteria	*Nitrosotalea*
Gammaproteobacteria	*Nitrosomonas*
	Nitrosospira
	Nitrosococcus
NO_2^- oxidizers	
Nitrospira	*Nitrospira*
Alphaproteobacteria	*Nitrobacter*
Betaproteobacteria	*Nitrotoga*
Gammaproteobacteria	*Nitrococcus*
Comammox	
Nitrospira	*Nitrospira*

Based on Aakra et al. (2001), Alves et al. (2018), and Daims and Lucker (2016).

In addition to the oxidations by the chemolithoautotrophic nitrifying microorganisms, other microbes can produce nitrite and nitrate by enzymatic oxidation processes that are not linked to microbial growth. For example, the many genera of methane-oxidizing bacteria contain a membrane-bound methane monooxygenase enzyme that will oxidize ammonia as well as methane, which is an interesting linkage between carbon and nitrogen cycling. Perhaps more widespread is **heterotrophic nitrification** (discussed below), the oxidation of ammonium or organic-nitrogen compounds by a variety of heterotrophic bacteria and fungi.

Ammonia oxidation

Despite our expanded view of nitrification with the discovery of ammonia-oxidizing archaea and comammox bacteria, the capability to oxidize ammonia occurs in just four phylogenetic clades: the Thaumarcheota, Betaproteobacteria, Gammaproteobacteria, and

Nitrospira (Table 15.3). This sparse distribution is likely because of the relatively low amount of energy, and hence slow growth, obtained when ammonia is oxidized. The overall reaction for the conversion of ammonia to nitrite is:

$$NH_3 + 1.5O_2 \rightarrow NO_2^- + H^+ + H_2O$$

This oxidation is a $6e^-$ transfer that yields 275 kJ mol^{-1} NH_3. The first step in the reaction is the conversion of ammonia to hydroxylamine (NH_2OH) by the membrane-bound *ammonia monooxygenase* enzyme:

$$NH_3 + O_2 + 2H^+ + 2e^- \rightarrow NH_2OH + H_2O$$

This reaction is endergonic, requiring a small amount of energy. Net energy is obtained, and ATP is produced, when hydroxylamine is converted via nitric oxide to nitrite through still undefined steps. The enzymes and reactions appear to differ between ammonia-oxidizing bacteria and archaea (Lehtovirta-Morley, 2018).

Like many monooxygenase enzymes, ammonia monooxygenase has broad substrate specificity. It can oxidize methane but does so at much lower rates than methane-oxidizing bacteria. Ammonia monooxygenase has been shown to *cometabolize* several other small organic compounds, including some halogenated organics such as trichloroethylene, chlorinated ethanes, and chloroform. A practical characteristic of the broad substrate specificity of ammonia monooxygenase is that it will bind irreversibly to acetylene. Thus, acetylene is a useful inhibitor of ammonia oxidation in research settings. Several other inhibitors of ammonia oxidation have been developed and used in agriculture (Box 15.6).

Two other products of ammonia oxidation are nitrous oxide and acidity. Ammonia oxidizers contain a nitrite reductase, which is capable of reducing nitrite to nitrous oxide. Under aerobic conditions, the production of nitrous oxide by this mechanism is small, less than 1% of the ammonia oxidized. As oxygen availability decreases, however, relatively more nitrous oxide is produced as nitrite is used as the electron acceptor. In some habitats, nitrification may be a major source of gaseous nitrogen oxides. In the case of ammonia-oxidizing archaea, there is debate over the specifics of their nitrous oxide producing mechanism. It has been proposed that the nitrous oxide yield of ammonia-oxidizing archaea might be less than that of ammonia-oxidizing bacteria, which could have implications to agriculture. For example, there are inhibitors of bacterial ammonia oxidation that have been known for many years (Box 15.6) and are often ineffective on archaea.

Ammonia oxidation acidifies soils by releasing 1 mole of H^+ for every mole of ammonia oxidized. This presents a paradox, as most nitrifying bacteria and archaea generally grow best at neutral pH and their activity is often inhibited by low pH. Nevertheless, the production of acidity by ammonia oxidizers has been shown to be responsible for lowering the pH of natural and agricultural soils, and some ammonia-oxidizing archaea are acidophilic and function effectively at $pH < 4$.

Nitrite oxidation

Nitrite-oxidizing bacteria are phylogenetically more diverse than the ammonia oxidizers (Table 15.3), with *Nitrospira* and *Nitrobacter* spp. being most common in soil.

BOX 15.6

Agronomic use of chemical nitrification inhibitors

For several decades scientists have attempted to find specific inhibitors of ammonia oxidation with the ultimate goal of commercializing these compounds for use in agriculture. The initial motivation was to increase the efficiency of fertilizer nitrogen (typically ammonium compounds or urea) use by crop plants to maximize economic yield. More recently the goal has expanded to include minimizing environmental consequences of nitrate in excess of plant needs: leaching and N_2O emissions. Currently, four chemical nitrification inhibitors are in commercial use: nitrapyrin (2-chloro-6-(trichloromethyl)pyridine; N-Serve), AM (2-amino-4-chloro-6-methyl-pyrimidine), DCD (dicyandiamide), and DMPP (3,4-dimethyl pyrazole phosphate). Although their efficacy varies widely, a meta-analysis showed yield, economic, and environmental benefits: NO_3^- leaching decreased 48%, N_2O emissions decreased 44%, but NH_3 volatilization increased 20% (Qiao et al. 2015).

The oxidation of nitrite to nitrate is a one-step reaction, with the following stoichiometry:

$$NO_2^- + \frac{1}{2}O_2 \rightarrow NO_3^-$$

Nitrite is oxidized to nitrate by a membrane-bound *nitrite oxidoreductase*, which transfers oxygen from water and transfers a pair of electrons to the electron-transport chain for the production of ATP via electron transport phosphorylation:

$$NO_2^- + H_2O \rightarrow NO_3^- + 2H^+ + 2e^-$$

This reaction yields 77 kJ mol^{-1} NO_2^- utilized, about one-third that of ammonia oxidation.

Unlike ammonia oxidizers, which are strict autotrophs, several nitrite oxidizers are capable of heterotrophic growth under some limited conditions. Even anaerobic heterotrophic growth may be possible, with nitrite oxidoreductase reducing nitrate to nitrite. Heterotrophic growth by nitrite oxidizers is much slower than other heterotrophic bacteria and slower than when nitrite oxidizers grow autotrophically.

Comammox

The discovery in 2015 of complete ammonia-oxidizers in the Nitrospirae phylum confirmed an earlier prediction that such organisms should exist, as oxidation of ammonia all the way to nitrate generates 27% more energy than the partial oxidation to nitrite (Daims and Lucker, 2016). Comammox bacteria use ammonia monooxygenase and nitrite oxidoreductase enzymes as do their bacterial ammonia oxidizing and nitrite oxidizing counterparts. Recent surveys have found comammox bacteria present in a wide range of soils; however, their contribution to nitrate production is uncertain.

Heterotrophic nitrification

Several heterotrophic microorganisms oxidize either ammonium or organic nitrogen to nitrite or nitrate. Heterotrophic nitrifiers include both fungi (e.g., *Aspergillus*) and bacteria (e.g., *Alcaligenes*, *Arthrobacter*). A particularly interesting bacterium is *Paracoccus pantotrophus*, which is a heterotrophic nitrifier that can also denitrify under aerobic conditions. Unlike the autotrophic nitrifiers, heterotrophic nitrifiers gain no energy through this activity. In fact, it is uncertain what benefit heterotrophic nitrifiers gain by oxidizing organic nitrogen, although hydroxamic acids, which act as siderophores, compounds involved in iron acquisition, are one type of oxidized nitrogen product.

The relative importance of heterotrophic versus autotrophic nitrification is still debated. In pure cultures, the highest rates of nitrite or nitrate production are just one-tenth that of autotrophic nitrifiers, which would suggest that heterotrophic nitrifiers are of minor importance. The case is not as clear-cut in soils, however, when the relative rates of the two nitrification processes have been assessed with inhibitors (e.g., nitrapyrin and acetylene are thought to block only autotrophic nitrification) or the use of ^{15}N. Data suggest that heterotrophic nitrification may be favored in soils with a higher C/N ratio or a lower pH, conditions less favorable for most chemolithotrophic nitrifiers.

Factors affecting nitrification in the environment

The rate of nitrification in soil is a function of the interactions (e.g., competition, mutualism) among different groups of nitrifying microorganisms, which are influenced by several interacting soil and environmental factors. Conceptually, each group of nitrifiers occupies a niche that optimizes its fitness, with different groups active depending upon the existing conditions (Fig. 15.9).

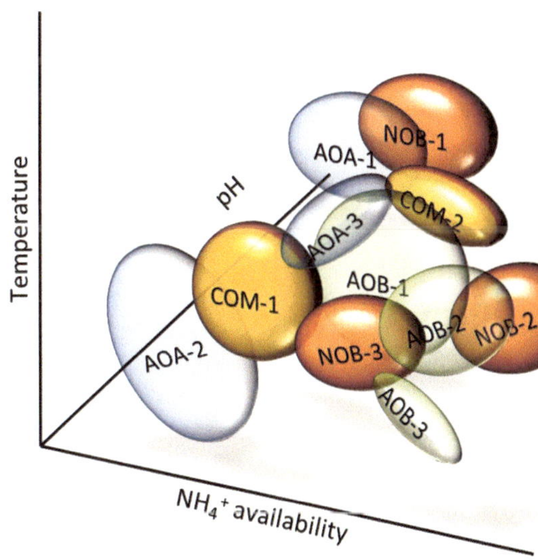

FIGURE 15.9 **Conceptual diagram of ecological niche space of different ecotypes of nitrifying microorganisms.** Regions of overlaps between ecotypes suggest potential for competition or cooperation. *AOA*-x, ammonia-oxidizing archaea; *AOB*-x, ammonia-oxidizing bacteria; *COM*-x, comammox bacteria; *NOB*-x, nitrite-oxidizing bacteria where "x" represents a particular ecotype. *Source: Adapted from Lu et al. (2020). Used with permission.*

Nitrifier populations. The number of copies of the *amoA* gene, coding for the ammonia monooxygenase of ammonia oxidizers, and of the *nxr* gene, coding for the nitrite oxidoreductase of nitrite oxidizers, is used as a proxy for the abundance of nitrifiers in soil. The sequence variation in these genes is used to characterize the composition of nitrifier communities.

Populations of nitrifiers vary widely among soils but some generalizations can be made. Ammonia-oxidizing archaea are the numerically dominant group of ammonia-oxidizers in most soils, where they can comprise up to 1% of the total archaeal and bacterial population. Ammonia-oxidizing bacteria are often 10% or less abundant than the ammonia-oxidizing archaea, and the few studies to date have found comammox bacteria to be 10% or less abundant than the ammonia-oxidizing bacteria. Nitrite-oxidizing bacteria are present in numbers roughly equivalent to that of the total ammonia-oxidizer population.

The overall size of the nitrifier populations varies primarily as a function of nitrogen availability. Unfertilized soils contain far fewer nitrifiers but, upon nitrogen fertilization, nitrifier populations, particularly ammonia-oxidizing bacteria, can increase 10- to 100-fold. A similar response is often seen when wildland soils are disturbed. In their natural state, many wildland soils have low concentrations of nitrate and small populations of nitrifiers (Box 15.7). If disturbance increases the availability of ammonium, nitrifier populations and rates of nitrification often increase gradually until they reach a new, higher steady state. Although nitrification activity might be expected to scale with the size of the nitrifier population, it is important to remember that not all nitrifiers may be metabolically active at a given time.

BOX 15.7

Exception to the rules 3: nitrification in soils of mature forests

Observations of nitrate pool sizes and net production of nitrate during incubations of undisturbed forest soils led to the suggestion that nitrate is a relatively unimportant pool of available nitrogen in these soils and that nitrification was not an important process. Various reasons were given for this, including low populations of autotrophic nitrifiers, limited substrate (ammonium) availability, and allelochemical inhibition. When mature forests were disturbed, for example, by clearcutting, a subsequent increase in soil and stream-water nitrate concentrations and net nitrification rates often resulted. This change was in part explained by a reduced amount of plant competition for ammonium. Collectively, these observations led to the dogma that as ecosystems mature, losses of nitrogen are reduced and nitrogen cycling becomes more conservative, primarily because nitrification is turned off.

Work using ^{15}N isotope dilution methods to measure the gross rates of ammonification and immobilization and nitrification and nitrate immobilization has revised our view of how nitrogen is conserved in mature forest ecosystems. An excellent example is the study by Davidson et al. (1992), who applied these methods to forest soils from an old-growth (more than 100 years) mixed-conifer

(Continued)

BOX 15.7 (*Cont.*)

forest and a 10-year-old mixed-conifer plantation in northern California and found the following:

Nitrogen cycling characteristic[a]	Young forest	Old-growth forest
Inorganic N (mg N m^{-2})[b]		
NH_4^+	340	210
NO_3^-	300	80
N mineralization		
(mg N m^{-2} d^{-1})[c]	6.8	2.6
Net	90	280
Gross		
Nitrification		
(mg N m^{-2} d^{-1})[c]	6.6	-0.4
Net	67	45
Gross		

[a]*Data are for top 9 cm of mineral soil.*
[b]*Mean of seven dates from November through September.*
[c]*Mean of three dates from November through April.*

In agreement with past studies, the researchers found higher inorganic-nitrogen concentrations and greater net nitrogen mineralization and net nitrification in soil from the young, recently disturbed stand. However, gross rates of inorganic-nitrogen production were 10–100 times greater than net rates, with rates of nitrification (gross nitrate production) being nearly as high in the soil from the old-growth forest as from the young stand. Thus, significant nitrification occurred in soils of both stands but immobilization rates of ammonium and nitrate were relatively higher compared to net production rates in the older stand. Probably a more significant point drawn from these data was how rapidly both the ammonium and nitrate pools turned over, with mean residence times on the order of days or hours and with greater turnover in the old-growth stands. This work agrees with the dogma that as forests age, soil nitrogen cycling becomes more conservative, not because nitrification is lessened but because nitrate immobilization is greater, and not because turnover slows down, because it may actually increase. Nitrogen seems to be conserved because of enhanced immobilization, which is likely fueled by greater carbon availability (Hart et al., 1994).

Substrate availability. Because nitrifiers gain their energy from ammonium or nitrite, the availability of these substrates is a primary determinant of the rate of nitrification as well as of the interactions among different groups of nitrifiers. Studies have shown that ammonium and nitrite oxidation follow Michaelis-Menten kinetics (Chapter 10). The saturation constants for these oxidations are in the same range as typical soil solution concentrations of ammonium and nitrite, which suggests that substrate availability is often limiting to growth and activity. Because different groups of nitrifiers vary in the ranges of their kinetic parameters, it is possible to predict which group should be more competitive, and thereby favored, under low and high substrate availability.

Many ammonia-oxidizing archaea have a greater affinity for ammonium than ammonia-oxidizing bacteria, which corresponds both to the higher populations and greater relative contribution of ammonia-oxidizing archaea observed in soils of low ammonium availability. Conversely, ammonia-oxidizing bacteria often increase in abundance and in their relative contribution to ammonia oxidation following the application of ammonium-based fertilizers. Comammox bacteria appear to have similar or greater affinity for ammonium as the ammonia-oxidizing archaea; however, it is not yet known how competitive they are in soils.

The two main genera of nitrite-oxidizing bacteria in soil also differ in their substrate affinity, with *Nitrospira* having a greater affinity for nitrite than *Nitrobacter*. In all but heavily fertilized soils, *Nitrospira* are numerically dominant and presumably responsible for most nitrite oxidation, although we don't have a convenient method to differentiate between the activities of different nitrite-oxidizers.

Soil pH. Because ammonia, and not ammonium, is the substrate for the ammonia monooxygenase enzyme, and the pK_a for the ammonium-ammonia equilibrium is 9.25, ammonia availability is strongly pH-dependent. At a pH of 6.2, ammonia is just 0.1% that of ammonium in soil solution. Thus, it is not surprising that nitrification rates are often low in soils below pH 4.5. Conversely, at high pH, nitrite can accumulate because of greater inhibition of nitrite oxidizers relative to ammonia oxidizers. These observations, along with the fact that most isolates of chemolithoautotrophic nitrifying bacteria grow best at neutral pH, support the generalization that ammonia-oxidizers and nitrite-oxidizers are neutrophilic. Nevertheless, high rates of nitrification or high concentrations of nitrate have been observed in many acid (pH <4.5) soils. Several explanations for this apparent paradox include acidophilic ammonia-oxidizing archaea, capability of some ammonia-oxidizers to use urea, heterotrophic nitrifiers, and neutral pH microsites (De Boer and Kowalchuk, 2001).

Ammonia-oxidizing archaea and bacteria, and nitrite-oxidizing bacteria have been isolated from low pH soils, although many prove to be neutrophilic or alkalophilic in pure culture. Different physiological mechanisms likely exist to allow for nitrifiers to function at low pH. Some ammonia-oxidizers may actively transport ammonium into their cells rather than depend upon the diffusion of ammonia and some are known to take up urea and hydrolyze it intracellularly to generate ammonia.

Perhaps the most common explanation given for nitrification in acid soils is the presence of heterotrophic nitrifiers. As mentioned previously, studies with either inhibitors specific for autotrophic nitrifiers or $^{15}NH_4^+$ have shown significant heterotrophic nitrification in some acid soils. Most isolated heterotrophic nitrifiers are not acidophilic, however.

In a habitat as diverse as soil, it would not be surprising to find microsites of higher pH. These might be associated with surfaces of minerals, organic matter, or even roots; however, conclusive evidence is lacking and empirical evidence is mixed. It is perhaps more likely that microsites are created by the activity of microorganisms themselves. For example, ammonia released during mineralization of organic nitrogen by heterotrophs or from urea hydrolysis may alter microsite pH. Such an effect would be difficult to distinguish from enhanced substrate availability, however.

Temperature. We assume that most nitrifiers in soils are mesophilic in their temperature optimum; however, ammonia-oxidizers and nitrite oxidizers have been found and isolated from both cold and hot environments. One recent observation is that the temperature optimum for ammonia oxidation in soil is lower for ammonia-oxidizing bacteria than for archaea, which suggests that their relative activity varies by season in temperate ecosystems (Taylor et al., 2017).

Miscellaneous soil and environmental factors. Clearly factors such as aeration, water potential, salinity, and availability of nutrients other than nitrogen all have the potential to affect the activity of nitrifiers. There is some evidence that nitrifiers may differ in their affinity for oxygen, which may allow for niche specialization, and production of nitrous oxide by nitrifiers increases as oxygen concentrations decrease. There have also been indications that phosphorus availability may limit nitrification rates in some soils.

The observation of low nitrate concentrations in soils of climax communities in natural ecosystems led to a theory of specific inhibition of nitrification by allelochemicals produced by the climax vegetation. This theory suggests that tannins and polyphenols are among the more important allelochemical agents. Subsequent research has demonstrated that this allelochemical theory is probably not the major reason for the low soil nitrate concentrations. More controlled studies using individual plant species, such as sorghum and signalgrass, have isolated individual organic compounds from root exudates that specifically inhibit nitrification. It is not known how effective such biological nitrification inhibitors may be in the field.

Fate of nitrate in the soil environment

Like ammonium, nitrate has many competing fates in the soil ecosystem (Fig. 15.2). Because it is an anion, nitrate is easily leached. Removal of nitrate from the soil by leaching has several consequences. Obviously, nitrate leaching represents a loss of available nitrogen from the plant-soil system. When nitrate is leached, it must be accompanied by an equivalent amount of cations to maintain charge balance. Thus, soils are also depleted of cations when nitrate is leached. The leaching of basic cations, such as K^+ and Ca^{2+}, reduces the base saturation of a soil and increases exchangeable acidity. Nitrate that leaches eventually enters ground and surface waters, where it may have potentially adverse environmental effects. High concentrations of nitrate in surface waters can lead to **eutrophication** (the sudden enrichment of natural waters with excess nutrients which can lead to the development of algal blooms and other vegetation). Current Federal regulations require that drinking water contain < 10 mg NO_3^--N L^{-1} (read as "mg nitrate nitrogen per liter"). High concentrations of nitrate may also be associated with methemoglobinemia (blue-baby syndrome), which is now quite rare. A further environmental hazard may be the production of carcinogenic nitrosamines from reactions between nitrite and secondary amines.

Biological fates for nitrate include its assimilation by plants and microorganisms and its use in the anaerobic, reductive processes discussed next.

Reductive processes

The reduction of oxidized nitrogen compounds occurs in biological nitrogen fixation, where dinitrogen is reduced to ammonia for the acquisition of nitrogen (Chapters 16 and 17), or when different oxygen-containing inorganic nitrogen compounds (nitrate, nitrite, nitric oxide, and nitrous oxide) are used as alternate electron acceptors in place of oxygen. The latter are often referred to as **dissimilatory** because a reduced nitrogen compound is released into the environment (Fig. 15.10).

Our discussion will focus on microbial processes but several chemical reactions can convert nitrite to various nitrogen gases, particularly nitric oxide, in a process known as **chemodenitrification**. Chemodenitrification reactions include: decomposition of nitrite to nitric

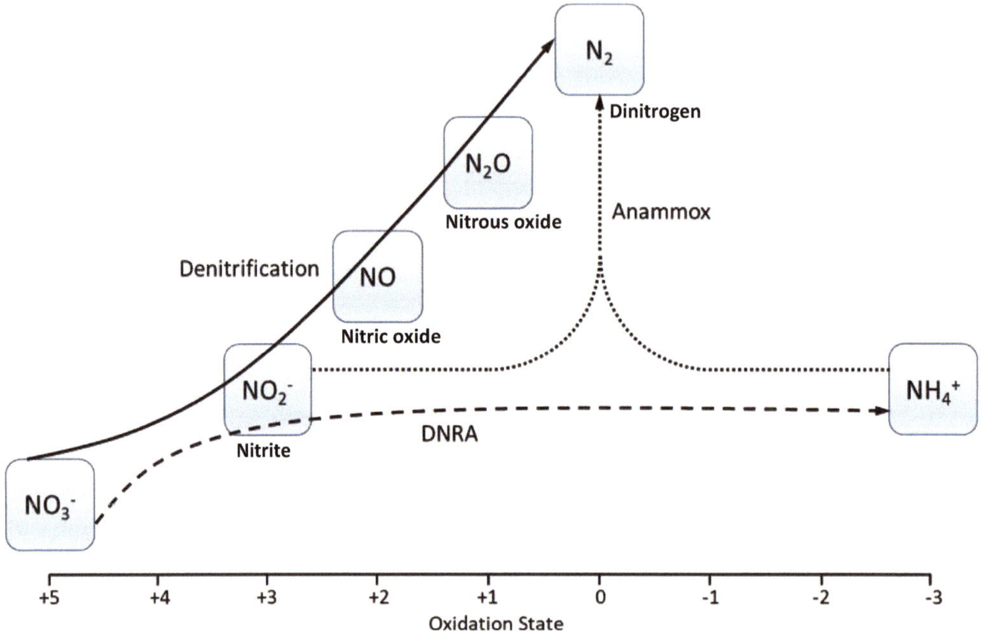

FIGURE 15.10 **Anaerobic dissimilatory nitrate reductive processes.** *DNRA*, dissimilatory nitrate reduction to ammonium; *Anammox*, anaerobic ammonia oxidation.

oxide in acidic soils of pH 5 or less; reduction of nitrite to nitric oxide or nitrous oxide by the oxidation of reduced metals, such as Fe^{2+}, which is favored at lower pH; and nitrosation reactions between nitrite and organic nitrogen compounds, which can release dinitrogen as well as nitric oxide and nitrous oxide (Heil et al., 2016). These chemodenitrification reactions are typically minor compared to biological dissimilatory processes of anammox (see Box 15.5), dissimilatory nitrate reduction to ammonium (DNRA), or denitrification.

Dissimilatory nitrate reduction to ammonium (DNRA)

Many soil bacteria are facultative anaerobes with the capability to reduce nitrate to nitrite when oxygen becomes depleted, a process known as **nitrate respiration**. In doing so, they gain energy via electron transport phosphorylation (161 kJ mol^{-1} NO_3^-). The reduction of nitrate to nitrite is catalyzed by the enzyme nitrate reductase. There are two forms of nitrate reductase: membrane-bound (encoded by the *nar* genes) or periplasmic (encoded by the *nap* genes), both of which contain molybdenum cofactors. Nitrate reductase is common to all microorganisms that use nitrate as an alternate electron acceptor.

The complete reduction of nitrate to ammonium is known as **dissimilatory nitrate reduction to ammonium**, or **DNRA**. Under anaerobic conditions, several genera of bacteria are capable of DNRA, including both obligate anaerobes (e.g., *Clostridium*, *Desulfovibrio*) and facultative bacteria (e.g., *Bacillus*, *Enterobacter*, *Erwinia*, *Pseudomonas*). The overall reaction for DNRA is:

$$NO_3^- + 4H_2 + 2H^+ \rightarrow NH_4^+ + 3H_2O$$

A total of $8e^-$ are transferred during this reduction, with an energy yield of 600 kJ mol^{-1} NO_3^-, or 150 kJ mol^{-1} $2e^-$ transferred. The first step in this reaction is the conversion of nitrate to nitrite; the second, definitive step, converts nitrite to ammonium using either a pentaheme cytochrome *c* form of nitrite reduction, encoded by the *nrf* genes, or more recently discovered octaheme enzymes (Kraft et al., 2011). Bacteria with a functioning electron transport chain can potentially gain energy via oxidative phosphorylation from both of these reductive steps; whereas fermenting bacteria can regenerate NAD and gain ATP by substrate level phosphorylation. Some DNRA bacteria may not gain any additional energy from the reduction of nitrite to ammonium, but presumably benefit from the detoxification of the nitrite intermediate or the regeneration of reducing equivalents through the oxidation of $NADH_2$. The latter process seems to be the most important, particularly when carbon availability is high. Generally, DNRA is favored in environments with high carbon-to-nitrate ratios, such as wetlands and rice paddies; however, DNRA has been measured in a wide range of soils, including croplands (Chen et al., 2015).

Denitrification

The canonical **denitrification** process refers to the stepwise reduction of nitrate to nitrite to nitric oxide to nitrous oxide to dinitrogen, which is coupled to energy production via oxidative phosphorylation. It is an example of anaerobic respiration, where an alternate electron acceptor other than oxygen is used. Other biological processes that reduce nitrate or nitrite to nitrous oxide or dinitrogen include **codenitrification** and **fungal denitrification** (Box 15.8).

BOX 15.8

Codenitrification and fungal denitrification

In the past few decades, more processes that can produce nitrous oxide and dinitrogen have been identified. **Codenitrification** refers to the cometabolic process that forms a hybrid N-N bond, where one nitrogen comes from an organic nitrogen compound, typically an amine (Spott et al., 2011). Thus, it is the biological analog of chemodenitrification. Although codenitrification has been found with archaea and bacteria, it was first observed with fungi and **fungal denitrification** (see later). Few measurements of codenitrification have been made in soil, but it has been shown to be elevated in urine spots, which are high in amino compounds.

A wide variety of fungi, primarily ascomycetes, are able to produce nitrous oxide (but not dinitrogen) from the reduction of nitrite (and sometimes nitrate). These fungi have a form of the NirK nitrite reductase but use a different form of nitric oxide reductase: heme-cytochrome P450 (P450Nor), which seems to operate via codenitrification. The contribution of fungi to nitrous oxide production in soils varies widely but it is sometimes higher than that of bacteria (Mothapo et al., 2015).

The overall stoichiometry of the denitrification reaction is:

$$2NO_3^- + 5H_2 + 2H^+ \rightarrow N_2 + 6H_2O$$

Denitrification gains slightly less energy per mol of nitrate than DNRA (560 kJ); however, it gains more per mol of $2e^-$ transferred (224 kJ). The higher thermodynamic yield per $2e^-$ is consistent with the observation that denitrification is likely to be the most important reductive process in soils where heterotrophic organisms are often limited by available carbon.

Denitrification occurs in archaea, bacteria, and fungi, although it is most widespread among bacteria. Denitrifiers comprise 0.1%–5% of the total bacterial population of soils spanning multiple phylogenetic lineages. Despite this diversity, soil denitrifiers are dominated by members of the genus *Pseudomonas*, with species of *Alcaligenes*, *Flavobacterium*, and *Bacillus* also common. Thus aerobic heterotrophs predominate, although autotrophic denitrifiers are also known. Furthermore, bacteria normally associated with other nitrogen transformations (e.g., *Azospirillum*, *Nitrosomonas*, and *Rhizobium*) denitrify under certain conditions. Because denitrification is considered an anaerobic process, it is interesting to note that some denitrifying bacteria are able to denitrify in the presence of oxygen.

Denitrification enzymes

The denitrification pathway involves four reductive steps, each with a corresponding enzyme (Fig. 15.11). As with other microorganisms that can reduce nitrate, denitrifiers may have either a membrane-bound (Nar) or periplasmic (Nap) nitrate reductase. Synthesis of nitrate reductases is inhibited by oxygen, as is the activity of preexisting enzymes.

The reduction of nitrite to nitric oxide is a defining characteristic of denitrifiers. This step is catalyzed by nitrite reductase (Nir). Two forms of Nir are known: one contains copper (encoded by *nirK*), and the other contains cytochromes *c* and d_1 (encoded by *nirS*). About

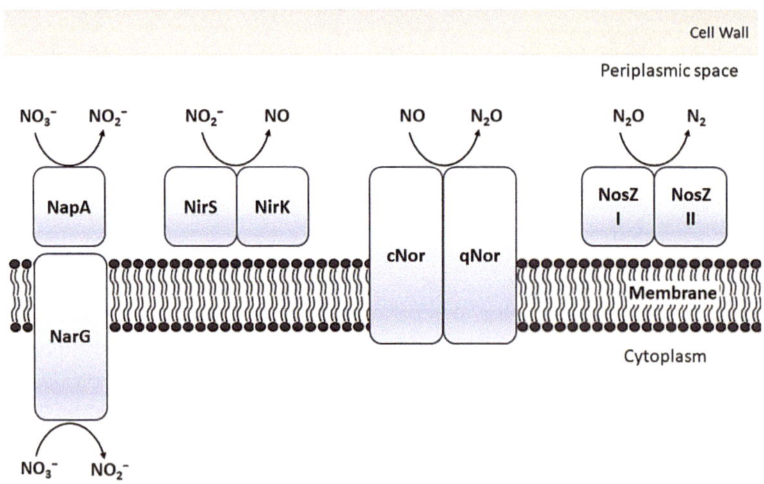

FIGURE 15.11 **Denitrification pathway showing the location of the four denitrification enzymes.** *NapA* and *NarG*, nitrate reductases; *NirS* and *NirK* , nitrite reductases; *cNor* and *qNor*, nitric oxide reductases; and *NosZ I* and *NosZ II*, nitrous oxide reductases. A given denitrifier will have just one type of each of the reductases, and some denitrifiers will not have all four reductases. *Source: Based on Kraft et al. (2011) and Hallin et al. (2018).*

two-thirds of the denitrifiers carry the *nirS* gene, including most of the *Pseudomonas* strains, *Alcaligenes*, *Paracoccus*, *Thiobacillus*, and *Azospirillum*. Copper is required for activity of NirK, which is less common than NirS but more widespread taxonomically. It is found in some *Pseudomonas* and *Alcaligenes* strains, *Bacillus*, *Rhizobium*, *Thiosphaera*, *Nitrosomonas*, and others. Researchers have not determined whether the two types of nitrite reductases are related to the performance of the denitrifiers in the environment, although differences in their distribution have been observed. Like Nar, Nir synthesis is repressed by oxygen; however, it is less clear if its activity is directly inhibited by oxygen. The presence of nitrate induces the expression of Nir.

Nitric oxide is an obligate intermediate in the denitrification pathway and it is converted to nitrous oxide by the activity of nitric oxide reductase (Nor). There are two types of nitric oxide reductases. Both are membrane bound: cNor is associated with cytochrome *c*; qNor is quinol dependent. Because it is membrane bound, the electron transport associated with Nor activity is presumably linked to ATP synthesis. An interesting point about nitric oxide reduction is that this is the step in which the N=N double bond is formed. Synthesis of Nor is repressed by oxygen and induced by nitrogen oxides.

The final enzyme of the denitrification pathway is nitrous oxide reductase (Nos, encoded by the *nosZ* gene), which reduces nitrous oxide to dinitrogen. As with the other enzymes in the denitrification pathway, nitrous oxide reductase exists in two forms, known as Clade I and Clade II. Both forms of nitrous oxide reductase are periplasmic and contain copper as a cofactor. Denitrifiers with Clade I Nos are mainly found in the Proteobacteria whereas those with Clade II Nos are more phylogenetically diverse. Although Nos is not associated with the cell membrane, energy production must be associated with this final reductive step because denitrifiers can grow with nitrous oxide as their sole electron acceptor. Indeed, there are many bacteria with Clade II Nos that do not have the other denitrifier enzymes and can reduce only nitrous oxide (Hallin et al., 2018). Most studies in soils have found the copy numbers of Clade II *nosZ* to be in greater abundance than those of Clade I.

Like other denitrification enzymes, synthesis of Nos is regulated by oxygen and nitrogen oxides. Its activity is more strongly inhibited by oxygen and more sensitive to low pH than the other enzymes. Thus, under high oxygen or low pH conditions, relatively more nitrous oxide than dinitrogen is produced.

Factors affecting denitrification in the environment

The regulation of denitrification can be thought of as a hierarchy from more to less important factors. The presence of denitrifiers is seldom a limitation. Denitrifiers make up a reasonably large fraction of the soil bacteria, probably because most of them normally exist as aerobic heterotrophs that switch to nitrate as an alternate electron acceptor when oxygen becomes unavailable. Variations in the diversity or composition of the denitrifier community may affect the rates or proportions of gaseous products emitted (Philippot et al., 2013).

Of the environmental factors that influence denitrification, the formation of anaerobic conditions is the most important controlling factor, followed in order by the availability of nitrate and carbon. Of course, there are always instances when other factors, such as temperature and soil pH, may be extreme enough to limit denitrification.

Soil aeration. Although aerobic denitrification can be done by a few denitrifiers, denitrification in soils and other natural habitats is predominantly an anaerobic process. Oxygen affects denitrification by regulating enzyme synthesis and by inhibiting enzyme activity. Enzyme synthesis is less sensitive to oxygen than is activity. Synthesis of Nar and Nir is derepressed when oxygen concentrations reach about one-tenth of atmospheric concentrations (2 kPa O_2 in the gas phase, which is in equilibrium with 29 μmol O_2 L^{-1} H_2O at 20°C). As oxygen concentrations decrease, inhibition of denitrifier enzyme activities is relieved sequentially, with Nar being the least oxygen sensitive and Nos being the most sensitive. The differential sensitivity of denitrifier enzyme activity explains why the ratio of nitrous oxide to dinitrogen increases as oxygen concentrations increase.

The oxygen concentration experienced by denitrifiers in soil is a complex function of the many interacting factors that control soil aeration. Aeration in soils occurs predominantly by diffusion, although there may be some transport by convection or even of oxygen dissolved in percolating water. Diffusion of oxygen is directly proportional to the concentration gradient of oxygen, which is largely a function of heterotrophic respiratory activity consuming oxygen in the soil and inversely proportional to the path-length for diffusion. The proportionality constant is called the diffusion coefficient (D). It varies over several orders of magnitude depending on soil texture, water content, and tortuosity. For example, D varies from 0.208 cm^2 s^{-1} in air to 2.6×10^{-5} cm^2 s^{-1} in water (i.e., oxygen diffuses through water about 10,000 times more slowly than through air). A detailed example will be covered later as a case study, but the general principle is that rates of denitrification are generally greatest in wet soils when more than 80% of the pore space is filled with water and where there is reasonably high respiratory activity (Box 15.9).

Once anaerobic conditions are established, denitrification rates are most often limited by either nitrate or carbon availability. Which of these two is more limiting depends upon their relative abundance, which is often related to soil type, plant community, or management practices.

Nitrate availability. In most natural soil systems, nitrate, the alternate electron acceptor, is more limiting than carbon even if heterotrophic microorganisms are carbon-limited under aerobic conditions. There are several reasons for this. First, the obligately aerobic heterotrophs, which make up the bulk of the microbial biomass, are no longer able to compete for carbon in the absence of oxygen. In effect, carbon is less limiting under anaerobic conditions. Second, in many wildland soils, such as those of forests, grasslands, and natural wetlands, net nitrogen mineralization and net production of nitrate are small. Habitats that are very anaerobic or experience long periods of anaerobiosis, such as some sediments and anaerobic digestors, are similarly limited in nitrate availability because nitrification is inhibited under anaerobic conditions. Finally, even in soils with relatively high rates of nitrification, there are many competing fates for the nitrate produced. Many of these fates, including leaching and DNRA, are enhanced under wet, anaerobic soil conditions.

Carbon availability. Researchers often find correlations between measures of carbon availability (e.g., respiration rates) and denitrification rates. This is evidence that denitrification rates are influenced by carbon availability; however, it is confounded because carbon utilization also influences oxygen supply. Nevertheless, some controlled studies have also shown a positive response of denitrification to additions of carbon under anaerobic conditions. It is likely that carbon limitation is greatest in soils with high nitrification rates or large nitrate pools, such as fertilized agricultural soils.

Miscellaneous soil and environmental factors affecting denitrification. Denitrification responds to temperature, as do most biological processes, increasing as temperature increases

BOX 15.9

Anaerobic microsites

An intriguing puzzle about denitrification is how this anaerobic process can occur in aerobic soils with nearly atmospheric concentrations of oxygen. Most of the evidence points to the establishment of anaerobic zones within an otherwise aerobic soil profile. The most studied and best documented of these anaerobic microsites occur within large soil aggregates.

Using several simplifying assumptions and typical values for diffusion and consumption parameters, it is possible to calculate how large a soil aggregate must be in order for an anaerobic microsite to develop (Currie, 1961; Smith, 1980). This calculation can be extended to estimate the anaerobic volume of soils using aggregate size distributions and to relate the anaerobic volume to denitrification rates.

The most direct demonstration of the existence of anaerobic microsites and their relationship to denitrification activity is the work of Sexstone et al. (1985b). Using oxygen microelectrodes, they measured oxygen profiles within saturated soil aggregates, detected anaerobic zones in their centers, and mapped oxygen contours.

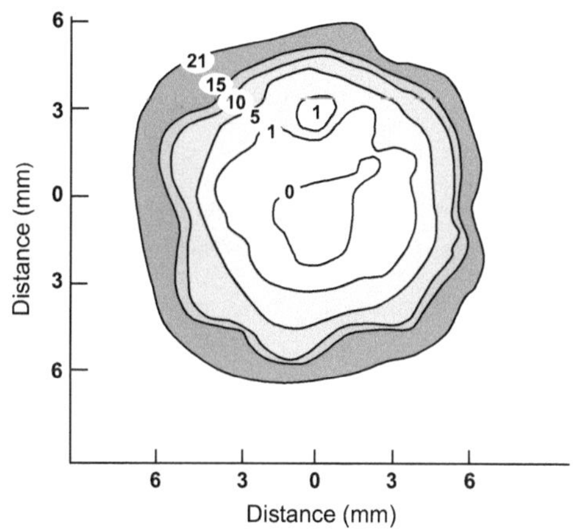

Oxygen contours in a saturated soil aggregate as measured with a microelectrode. Numbers on contour lines are volume percent oxygen. *Source: Adapted from Sexstone et al. (1985b).*

The measured anaerobic radii were highly correlated with those calculated with the oxygen consumption-diffusion model, and measurable denitrification rates were associated only with aggregates that had anaerobic zones. However, denitrification did not occur in all aggregates that had anaerobic zones, probably because factors other than aeration limited denitrification.

until a maximum is reached, above which activity declines rapidly. Mesophilic denitrifiers predominate in most soils, although activity has been measured near freezing and also under thermophilic conditions. Temperature is likely to have a more complex effect on denitrification than on some other soil processes because it also affects oxygen and nitrous oxide solubility, gas diffusion coefficients, and the oxygen consumption activity of other heterotrophs.

The response of denitrifiers to pH is similar to that of other soil heterotrophs, which usually function best near neutrality. Biological denitrification has been measured in some acidic soils, but rates are usually low and the measurements can be potentially confounded by chemodenitrification. Relatively little research has focused on denitrification in soils of high pH.

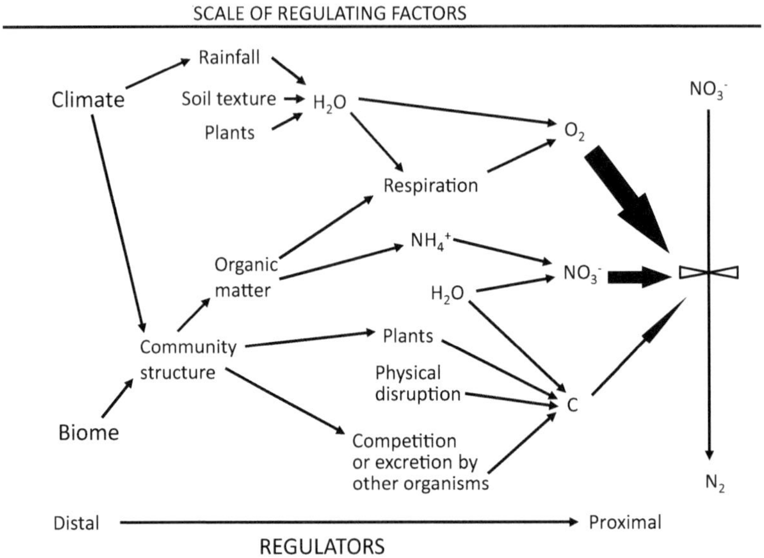

FIGURE 15.12 **Regulation of denitrification at various spatial scales.** The thickness of the right most *arrows* reflects the relative importance of oxygen, NO_3^-, and carbon as regulators of denitrification. *Source: Adapted from Tiedje (1988). Used with permission.*

Spatial scale and appropriate controlling factors. When considering the regulation of denitrification, it is useful to integrate the concept of spatial scale with the list of controlling factors (Fig. 15.12). Denitrifying bacteria are ultimately influenced by the concentrations of oxygen, nitrate, and carbon compounds just external to their cell surfaces. Consequently, measuring these concentrations is appropriate for physiological studies in the laboratory. Other properties are likely to be more useful and insightful when studying denitrification in soils, particularly as the spatial scale increases from the microbial cell through microcolonies, soil aggregates, soil columns, and field plots up to the landscape scale (Box 15.10). We are just beginning to understand how to scale-up processes such as denitrification.

Summary

The nitrogen cycle is both fascinating and frustrating in its complexity. A thorough knowledge of the cycle is fundamental, though, if one wishes to understand the functioning of natural ecosystems, to manage agricultural ecosystems for productivity and sustainability, and to ameliorate environmental problems.

Most nitrogen in soil is in organic forms. Organic nitrogen serves as a reservoir of N, slowly supplying the more dynamic and much smaller inorganic nitrogen pools. The conversion of organic nitrogen to inorganic nitrogen is called mineralization. A key step is the depolymerization of large organic molecules, such as proteins, into smaller molecules that can be taken up by microorganisms. The subsequent step of this process is the production of ammonium by ammonification, which is carried out by a wide variety of soil microorganisms and soil

BOX 15.10

Hot spots and hot moments of denitrification

Denitrification is notorious for being highly variable in time and space. Denitrification rates can vary more than 100-fold from one day to the next:

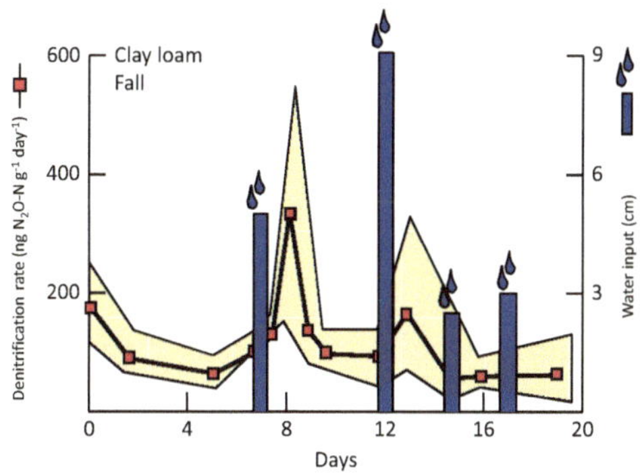

Temporal variation in daily denitrification rates of a clay loam soil in Michigan in autumn. The fine *solid lines* represent the upper and lower 95% confidence intervals and *vertical bars* represent water input. Note the response of denitrification to the addition of water. *Source: Adapted from Sexstone et al. (1985a).*

Such short-term changes are often associated with precipitation, irrigation, or sometimes nutrient additions (e.g., inorganic fertilizer or manures). Seasonal responses, largely to soil temperature or precipitation patterns, are also observed. For example, denitrification rates in the Pacific Northwest of the United States are highest in the fall and spring when both soil temperature and water content are relatively high, whereas low rates are found in the winter, because of low soil temperatures, and in the summer, because of very dry soil conditions.

Variations in soil denitrification rates span several spatial scales, although the field plot level has probably been studied the most. The most common observation when sufficient numbers of denitrification measurements are made is that most rates are low with just a few high or very high rates. This results in a skewed frequency distribution that is most often described as lognormal. This observation has often been attributed to the formation of "hot spots" of activity where optimal conditions of anaerobiosis, adequate nitrate, and available carbon concentrations coincide. The existence of such hot spots was perhaps best shown by the clever experiment of Parkin (1987). In this experiment, the denitrification activity of a soil core was monitored as it was subdivided into smaller and smaller units. Ultimately, over 85% of the activity of the core was found to be associated with one small piece of decaying vegetation.

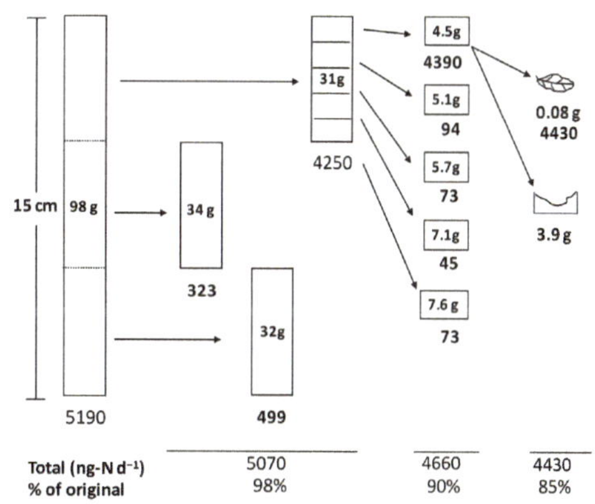

Evidence for "hot spots" as important locations for denitrification in soil. The numbers at the bottom of each soil section are denitrification rates (ng-N day^{-1}). Note that almost all the denitrification activity was associated with a single pigweed leaf. *Source: Adapted from* Parkin (1987).

animals. Ammonification is always counterbalanced by the opposite pattern of immobilizing ammonium into the soil biomass through assimilation. A major controlling factor determining whether net mineralization or immobilization of nitrogen occurs is the C/N ratio of the decomposing organic matter.

The mineralization process is continued further by the conversion of ammonium to nitrate by the nitrifiers, which we now know are much more diverse than we appreciated a decade ago. Typically, ammonia-oxidizing archaea and bacteria perform the first step of this two-step process, the transformation of ammonium to nitrite, and nitrite is further oxidized to nitrate by the nitrite-oxidizing bacteria. The recent discovery of bacteria that can completely oxidize ammonium to nitrate has broadened how nitrification can be done. The ammonia and nitrite oxidizers are chemoautotrophic bacteria that gain their energy from these inorganic oxidations.

Nitrate has many fates in the soil environment. It is readily taken up by plants and can also be immobilized by heterotrophic microorganisms. Because nitrate is relatively mobile, it can be readily leached, which represents not only a loss from the system but also a potential environmental problem. A final fate of nitrate is to be lost to the atmosphere through denitrification.

Denitrification occurs under anaerobic conditions, which can exist as microsites even in well-aerated soils. The reduction of nitrate to gaseous nitrous oxide and dinitrogen is accomplished by a wide range of bacteria, most of which normally function as aerobic heterotrophs. Denitrification is a relatively benign loss of nitrogen when dinitrogen is the dominant product; however, nitrous oxide production can be an environmental concern because it acts as a greenhouse gas and has been implicated in the destruction of ozone in the stratosphere.

The process of dinitrogen fixation closes the nitrogen cycle, which is ultimately the source of nearly all nitrogen that is transformed within the soil ecosystem. That process is covered in detail in Chapters 16 and 17.

Dedication

I dedicate this chapter to my longtime friend and collaborator, Peter Bottomley, for his thoughtful comments on a draft of this chapter, but especially for the delightful and insightful discussions through the years based on our mutual fascination with the nitrogen cycle.

Study questions

(Note: starred questions are more difficult.)

1. Draw your own diagram of the terrestrial nitrogen cycle including all major pools and transformations.
2. Describe the composition of soil organic nitrogen and how this relates to its biological availability.
3. Describe the steps involved in releasing ammonium from chitin.
4. A crop residue (45% C, 1.5% N) is chopped and mixed into soil. Assume that the soil microbial biomass has an average C/N ratio of 8 and efficiency (microbial

yield coefficient) of 0.6 in mineralizing this residue. Will net mineralization or immobilization of nitrogen occur? Show your calculations and explain your reasoning.

5. Describe how nitrate is formed and list its possible fates in the soil ecosystem.
6. Explain why the pH of agricultural soils decline when they are fertilized with ammoniacal N. Would this also happen if nitrate fertilizers were used?
7. Give an example of why net rates of nitrogen cycle processes may be misleading.
8. Contrast nitrification and denitrification with respect to the bacterial genera capable of carrying out each process and discuss how the respective substrates are utilized biochemically.
9. Denitrification rates are known to be highly variable in time and space. Provide a rationale for this based on what you know about the factors that regulate denitrification activity.

*10. The following data are based on an original paper that examined the effects of different grass species on nitrogen cycling (Oecologia 84:433–441). Use the data given in the table to fill in the missing values, calculate how much nitrogen is required for microbial growth in each grass system and estimate the annual net nitrogen mineralization rate for the Grass B system. Assume that roots are the only plant input, that roots are 45% carbon, that all the grass nitrogen is available, that lignin-carbon is unavailable, and that 80% of the nonlignin-carbon is metabolized in 1 year ($k = 1.6\ year^{-1}$). The microbial growth yield is 0.5, and the microbial C/N ratio is 8. Clearly state any other assumptions that you think are necessary. Show your work.

Root/soil property	Grass A	Grass B
Root biomass ($g\ m^{-2}$)	50	1200
Root C ($g\ m^{-2}$)		
Root N (%)	2.0	0.5
Root N ($g\ m^{-2}$)		
Root C/N	22.5	90
Root lignin-C (%)	5	15
Net N mineralization ($g\text{-}N\ m^{-2}\ year^{-1}$)	14	

*11. Production of allelochemicals that specifically inhibit autotrophic nitrification has been proposed as a mechanism through which plants conserve nitrogen within an ecosystem. Develop an experiment that would conclusively prove or disprove this theory.

*12. Discuss whether denitrification is likely to be enhanced in the rhizosphere compared to bulk soil. Develop a logical argument to support your view and describe an experiment that could be done to test your hypothesis.

References

Aakra, Å., Utåker, J.B., Pommerening-Röser, A., Koops, H.-P., Nes, I.F., 2001. Detailed phylogeny of ammonia-oxidizing bacteria determined by rDNA sequences and DNA homology values. Int. J. Syst. Evol. Microbiol. 51, 2021–2030.

Alves, R.J.E., Minh, B.Q., Urich, T., von Haeseler, A., Schleper, C., 2018. Unifying the global phylogeny and environmental distribution of ammonia-oxidising archaea based on amoA genes. Nat. Commun. 9, 1517, doi:10.1038/s41467-018-03861-1.

Anderson, J.M., Spencer, T., 1991. Carbon, nutrient and water balances of tropical rainforests subject to disturbance. MAB Digest No. 7. UNESCO, Paris, France.

Battye, W., Aneja, V.P., Schlesinger, W.H., 2017. Is nitrogen the next carbon? Earth's Future 5, 894–904, doi10.1002/2017ef000592.

Chacon, S.S., Reardon, P.N., Burgess, C.J., Purvine, S., Chu, R.K., Clauss, T.R., et al., 2019. Mineral Surfaces as Agents of Environmental Proteolysis: Mechanisms and Controls. Environ. Sci. Technol. 53, 3018–3026, doi:10.1021/acs.est.8b05583.

Chen, Z.M., Ding, W.X., Xu, Y.H., Muller, C., Rutting, T., Yu, H.Y., et al., 2015. Importance of heterotrophic nitrification and dissimilatory nitrate reduction to ammonium in a cropland soil: Evidences from a N-15 tracing study to literature synthesis. Soil Biol. Biochem. 91, 65–75, doi:10.1016/j.soilbio.2015.08.026.

Currie, J.A., 1961. Gaseous diffusion in the aeration of aggregated soils. Soil Sci. 92, 40–45.

Daims, H., Lucker, S., Wagner, M., 2016. A new perspective on microbes formerly known as nitrite-oxidizing bacteria. Trends Microbiol. 24, 699–712, doi:10.1016/j.tim.2016.05.004.

Davidson, E.A., Hart, S.C., Firestone, M.K., 1992. Internal cycling of nitrate in soils of a mature coniferous forest. Ecology 73, 1148–1156.

De Boer, W., Kowalchuk, G.A., 2001. Nitrification in acid soils: micro organisms and mechanisms. Soil Biol. Biochem. 33, 853–866.

Fowler, D., Coyle, M., Skiba, U., Sutton, M.A., Cape, J.N., Reis, S., et al., 2013. The global nitrogen cycle in the twenty-first century. Philos. Trans. R. Soc. B Biol. Sci. 368, 20130164, doi:doi:10.1098/rstb.2013.0164.

Geisseler, D., Horwath, W.R., Joergensen, R.G., Ludwig, B., 2010. Pathways of nitrogen utilization by soil microorganisms—A review. Soil Biol. Biochem. 42, 2058–2067, doi:10.1016/j.soilbio.2010.08.021.

Hallin, S., Philippot, L., Loffler, F.E., Sanford, R.A., Jones, C.M., 2018. Genomics and ecology of novel N_2O-reducing microorganisms. Trends Microbiol. 26, 43–55, doi:10.1016/j.tim.2017.07.003.

Hart, S.C., Nason, G.E., Myrold, D.D., Perry, D.A., 1994. Dynamics of gross nitrogen transformations in an old-growth forest: The carbon connection. Ecology 75, 880–891.

Heil, J., Vereecken, H., Bruggemann, N., 2016. A review of chemical reactions of nitrification intermediates and their role in nitrogen cycling and nitrogen trace gas formation in soil. Eur. J. Soil Sci. 67, 23–39, doi:10.1111/ejss.12306.

Houlton, B.Z., Morford, S.L., Dahlgren, R.A., 2018. Convergent evidence for widespread rock nitrogen sources in Earth's surface environment. Science 360, 58–62, doi:10.1126/science.aan4399.

Jilling, A., Keiluweit, M., Contosta, A.R., Frey, S., Schimel, J., Schnecker, J., et al., 2018. Minerals in the rhizosphere: overlooked mediators of soil nitrogen availability to plants and microbes. Biogeochemistry 139, 103–122, doi:10.1007/s10533-018-0459-5.

Kraft, B., Strous, M., Tegetmeyer, H.E., 2011. Microbial nitrate respiration—Genes, enzymes and environmental distribution. J. Biotechnol. 155, 104–117, doi:org/10.1016/j.jbiotec.2010.12.025.

Kuzyakov, Y., Xu, X.L., 2013. Competition between roots and microorganisms for nitrogen: mechanisms and ecological relevance. New Phytol. 198, 656–669, doi:10.1111/nph.12235.

Lehtovirta-Morley, L.E., 2018. Ammonia oxidation: Ecology, physiology, biochemistry and why they must all come together. FEMS Microbiol. Lett. 365, fny058. doi:10.1093/femsle/fny058.

Lu, X., Taylor, A.E., Myrold, D.D., Neufeld, J.D., 2020. Expanding the perspectives of soil nitrification to include ammonia-oxidizing archaea and comammox bacteria. Soil Sci. Soc. Am. J. 84, 287–302, doi: 10.1002/saj2.20029.

Morrissey, E.M., Mau, R.L., Schwartz, E., Koch, B.J., Hayer, M., Hungate, B.A., 2018. Taxonomic patterns in the nitrogen assimilation of soil prokaryotes. Environ. Microbiol. 20, 1112–1119, doi:10.1111/1462-2920.14051.

Mothapo, N., Chen, H.H., Cubeta, M.A., Grossman, J.M., Fuller, F., Shi, W., 2015. Phylogenetic, taxonomic and functional diversity of fungal denitrifiers and associated N_2O production efficacy. Soil Biol. Biochem. 83, 160–175, doi:10.1016/j.soilbio.2015.02.001.

Myrold, D.D., Bottomley, P.J., 2008. Nitrogen mineralization and immobilization, in: Schepers, J.S., Raun, W.R., (Eds.), Nitrogen in Agricultural Systems. American Society of Agronomy, Madison, WI, pp. 157–172. doi: 10.2134/agronmonogr49.c5.

Näsholm, T., Ekblad, A., Nordin, A., Geisler, R., Högberg, M., Högberg, P., 1998. Boreal forest plants take up organic nitrogen. Nature 392, 914–916.

Nelson, M.B., Martiny, A.C., Martiny, J.B.H., 2016. Global biogeography of microbial nitrogen-cycling traits in soil. Proc. Nat. Acad. Sci. USA 113, 8033–8040, doi:10.1073/pnas.1601070113.

Nguyen, T.T.H., Myrold, D.D., Mueller, R.S., 2019. Distributions of extracellular peptidases across prokaryotic genomes reflect phylogeny and habitat. Front. Microbiol. 10, 413. doi:10.3389/fmicb.2019.00413.

Olson, R.A., Kurtz, L.T., 1982. Crop nitrogen requirements, utilization, and fertilization, in: Stevenson, F.J. (Ed.), Nitrogen in Agricultural Soils. American Society of Agronomy, Madison, WI, pp. 567–604.

Parkin, T.B., 1987. Soil microsites as a source of denitrification variability. Soil Sci. Soc. Am. J. 51, 1194–1199.

Perakis, S.S., Hedin, L.O., 2002. Nitrogen loss from unpolluted South American forests mainly via dissolved organic compounds. Nature 415, 416–419.

Philippot, L., Spor, A., Henault, C., Bru, D., Bizouard, F., Jones, C.M., et al., 2013. Loss in microbial diversity affects nitrogen cycling in soil. ISME J. 7, 1609–1619, doi:10.1038/ismej.2013.34.

Post, W.M., Pastor, J., Zinke, P.J., Strangenberger, A.G., 1985. Global patterns of soil nitrogen storage. Nature 317, 613–616.

Qiao, C.L., Liu, L.L., Hu, S.J., Compton, J.E., Greaver, T.L., Li, Q.L., 2015. How inhibiting nitrification affects nitrogen cycle and reduces environmental impacts of anthropogenic nitrogen input. Global Change Biol. 21, 1249–1257, doi:10.1111/gcb.12802.

Schimel, J.P., Bennett, J., 2004. Nitrogen mineralization: Challenges of a changing paradigm. Ecology 85, 591–602.

Sexstone, A.J., Parkin, T.B., Tiedje, J.M., 1985a. Temporal response of soil denitrification to rainfall and irrigation. Soil Sci. Soc. Am. J. 49, 99–103.

Sexstone, A.J., Revsbech, N.P., Parkin, T.B., Tiedje, J.M., 1985b. Direct measurement of oxygen profiles and denitrification rates in soil aggregates. Soil Sci. Soc. Am. J. 49, 645–651.

Smith, K.A., 1980. A model of the extent of anaerobic zones in aggregated soils, and its potential application to estimates of denitrification. J. Soil Sci. 31, 263–277.

Spott, O., Russow, R., Stange, C.F., 2011. Formation of hybrid N_2O and hybrid N_2 due to codenitrification: First review of a barely considered process of microbially mediated N-nitrosation. Soil Biol. Biochem. 43, 1995–2011, doi:10.1016/j.soilbio.2011.06.014.

Taylor, A.E., Giguere, A.T., Zoebelein, C.M., Myrold, D.D., Bottomley, P.J., 2017. Modeling of soil nitrification responses to temperature reveals thermodynamic differences between ammonia-oxidizing activity of archaea and bacteria. ISME J. 11, 896–908, doi:10.1038/ismej.2016.179.

Tiedje, J.M., 1988. Ecology of denitrification and dissimilatory nitrate reduction to ammonium, in: Zehnder, A.J.B. (Ed.), Biology of Anaerobic Microorganisms. John Wiley and Sons, New York, pp. 179–244.

Waring, R.H., Schlesinger, W.H., 1985. Forest Ecosystems—Concepts and Management. Academic Press, Inc., Orlando, FL.

Warren, C.R., 2014. Organic N molecules in the soil solution: what is known, what is unknown and the path forwards. Plant Soil 375, 1–19, doi:10.1007/s11104-013-1939-y.

Further reading

Bothe, H., Ferguson, S., Newton, W.E., 2007. Biology of the Nitrogen Cycle. Elsevier Science, Boston, MA, p. 427.

Gorman, H.S., 2013. The Story of N: A Social History of the Nitrogen Cycle and the Challenge of Sustainability. Rutgers University Press, New Brunswick, NJ, p. 241.

Revsbech, N.P., Sørensen, J., 1990. Denitrification in Soil and Sediment. Plenum Press, New York, NY, p. 349.

Schepers, J.S., Raun, W.R. (Eds.), 2008. Nitrogen in Agricultural Systems. In: Agronomy Monographs Series, vol. 49. ASA, CSSA, SSSA, Madison, WI.

Ward, B.B., Arp, D.J., Klotz, M.G., 2011. Nitrification. American Society for Microbiology Press, Washington, DC, p. 454.

CHAPTER

16

Biological dinitrogen (N_2) fixation: introduction and nonsymbiotic

David A. Zuberer

Department of Soil and Crop Sciences, Texas A&M University, College Station, TX, United States

"In nature there is an enormous reserve of organic matter poor in nitrogen. One should ask how all this organic carbon could be circulated in nature without the existence of organisms capable of fixing free nitrogen. These organisms...could be nothing else but microbes."—**Sergei Winogradsky, 1895**

Biological **dinitrogen fixation,** the reduction of atmospheric dinitrogen (N_2) to two molecules of ammonia (NH_3) is, after photosynthesis, the second most important biological process on earth. In the absence of modern fertilizers or animal wastes, natural ecosystems rely on the biological conversion of atmospheric dinitrogen, to forms available for plant and microbial growth, by a variety of prokaryotic microbes. Dinitrogen fixation is mediated exclusively by prokaryotes including many genera of bacteria, archaea (Chapter 5), and cyanobacteria (Chapter 7) and the actinomycete (actinobacteria) *Frankia* (Chapter 17). The N_2-fixing microbes can exist as independent, **free-living** organisms or in associations of differing degrees of complexity with other microbes, plants, and animals. These range from loose associations (e.g., the **associative symbioses**) to complex symbiotic associations (Fig. 16.1) in which there is exquisite molecular communication between the bacterium and the host plant, as well as shared physiological functions (see the Chapter 17 for detailed information on symbiotic N_2-fixing plant/microbe associations). Central to all these systems is an N_2-fixing prokaryote containing the enzyme complex **nitrogenase**, which is responsible for the conversion of dinitrogen to ammonia. Only certain prokaryotes fix dinitrogen! Thus, when you see statements about "nitrogen-fixing" plants, keep in mind that they do so by virtue of their prokaryotic partner, not on their own. Organisms or symbiotic associations that can use atmospheric dinitrogen as their sole source of nitrogen for growth are referred to as **diazotrophs** ("diazo" for dinitrogen, "troph" for feeding; the term coined by Burns and Hardy, 1975).

Principles and Applications of Soil Microbiology. http://dx.doi.org/10.1016/B978-0-12-820202-9.00016-2
Copyright © 2021 Elsevier Inc. All rights reserved.

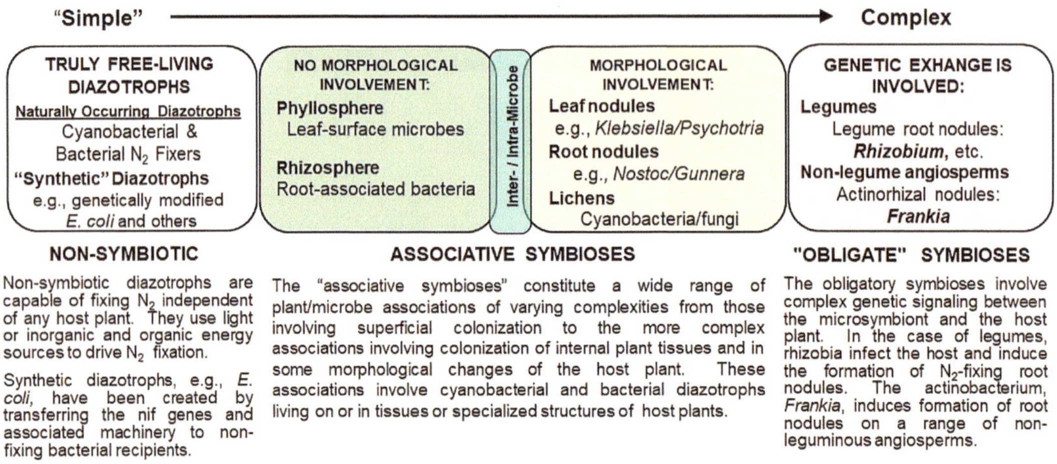

FIGURE 16.1 **Biological nitrogen fixation is mediated exclusively by prokaryotic free-living diazotrophs and a wide variety of plant-microbe associations of varying complexities.** The most elaborate of these are the root-nodule symbioses between rhizobia and legumes and between the actinobacterium, *Frankia*, and a variety of nonleguminous trees and shrubs. Between the free-living diazotrophs and the root-nodule symbioses, there is a broad array of associations involving plants and diazotrophs. These are the so-called "associative symbioses" which are characterized, more or less, by loose associations between the host plant and diazotroph. The term "diazotroph" was coined by Burns and Hardy, 1975, and is used to describe any microbe or symbiotic association that can fix dinitrogen.

Historical background

The agricultural importance of legumes was recognized very early in the history of humans, reaching a milestone in 1888 with the isolation and description of the first root-nodule bacterium *Rhizobium*, determined (though much later) to have the ability to fix N_2. *Rhizobium* and its relatives fix only small amounts of dinitrogen in pure culture, and significant rates of dinitrogen fixation only occur when the bacteria are in root nodules (Chapter 17). In 1893, Winogradsky isolated the first free-living bacterium, an anaerobe, capable of significant N_2-fixation. He named this organism, *Clostridium pasteurianum*, in recognition of Louis Pasteur. Winogradsky, though a Russian, did much of his scientific work at the Pasteur Institute in Paris. Somewhat later (1901), the Dutch microbiologist, Martinus Beijerinck, isolated an aerobic free-living, N_2-fixing bacterium that he named *Azotobacter chroococcum* (derived from the French word "azote" meaning "lifeless" in recognition of the inertness of dinitrogen).

Thus by the late 19th century, the foundation was laid for the extensive studies of N_2 fixation that would follow. These early discoveries began what has become more than 130 years of research on the fascinating and important subject of biological N_2 fixation.

Significance of biological N_2 fixation

It is a fundamental principle of agriculture and plant physiology that plants require relatively high levels of nitrogen to produce abundant biomass (yield). All forms of life require nitrogen to synthesize proteins, nucleic acids, and other important biochemicals, and nitrogen is often the limiting nutrient for plant and microbial growth in soil. In natural systems, nitrogen for plant growth comes from the soil, some from rainfall or other atmospheric deposition, or through biological N_2 fixation.

Industrial fixation of nitrogen (Haber-Bosch process) amounts to about 95 Tg year^{-1} (teragram = 10^{12} g or 1,000,000 MT) (Smil, 1999) and requires substantial inputs of energy, usually in the form of natural gas (Box 16.1). However, this is less than the global contribution of fixed N acquired through biological N_2 fixation. Accurate estimates of the magnitude of biological N_2 fixation are hard to derive but reported values range from 100 to 290 Tg year^{-1} (195 Tg year^{-1} mean value, Cleveland et al., 1999). Fowler et al. (2013) suggest that about one third of anthropogenic nitrogen input to the terrestrial ecosystem comes through biological N_2 fixation (58 Tg ± 50%) while the rest is from fertilizer inputs (120 Tg ± 10%). Note the wide range of values for biological N_2 fixation indicating the difficulty in obtaining accurate values. Most of the biological input is through symbiotic N_2 fixation (Chapter 17) but nonsymbiotic and associative N_2 fixation are reportedly of significance in crops such as sugarcane and sorghum, which have a C_4 photosynthetic pathway, and may be grown in ecosystems where N for plant growth is a limiting factor. Because of its cultivation in flooded soils, rice derives significant benefit from the activities of free-living diazotrophic bacteria and cyanobacteria that find favorable habitats (e.g., root exudates and low oxygen concentrations) in the flooded soils.

Biological N_2 fixation remains an effective alternative to the use of expensive ammonium-based fertilizer nitrogen but it must be emphasized that the high-yielding agricultural systems of the United States and elsewhere are difficult to sustain relying only on biological N_2 fixation and the use of animal manures. The growing world population will continue to dictate production systems that will require judicious use of synthetic and, where possible, organic fertilizers. Legumes, other N_2-fixing systems (e.g., the *Azolla-Anabaena* symbiosis), and perhaps genetically modified plants yet to be developed will continue to play an important part in sustainable agricultural production. Dinitrogen-fixing plants are also of high value in restoring disturbed or impoverished soils. They serve as excellent cover crops, green manures, and forage crops for livestock production. Use of N_2-fixing crops also has the potential to reduce the contamination of groundwater with nitrate and release of nitrogenous "greenhouse gases." Further aspects of the utility of symbiotic nitrogen fixation in agriculture and other ecosystems are discussed in Chapter 17.

Nitrogenase enzyme complex

The process of biological N_2 fixation is restricted, so far as known, to prokaryotic microbes including many genera of soil bacteria, cyanobacteria (formerly the "blue-green algae"), and a few actinobacteria (the "actinomycetes", most notably *Frankia*). These microbes can fix N_2 as **free-living** forms (i.e., not plant associated), in loose associations with higher plants called **associative symbioses** (e.g., many grass-bacteria associations) (Fig. 16.1), and in truly complex symbiotic partnerships with higher plants (the **root-nodule symbioses** involving rhizobia and *Frankia*).

A trait shared by all these microbes is that they produce the enzyme complex called **nitrogenase** which mediates biological N_2 fixation. It is most appropriately called the "nitrogenase complex" because it is actually composed of two protein components, each composed of multiple subunits, and it also houses several metallic clusters involved in catalyzing the

BOX 16.1

The energy costs of N_2 fixation

Dinitrogen is described as the most stable diatomic molecule known. The two atoms of nitrogen in the diatomic molecule are joined by a very stable triple bond. This triple bond requires lots of energy (945 kJ, or 226 kcal mol^{-1}) to be broken and therein lies one of the major challenges of fixing dinitrogen chemically or biologically. Dinitrogen fixation is "energy expensive" because it requires much energy to break the triple bond of N_2 and also to provide the hydrogen necessary to reduce N_2 to $2NH_3$. The chemical fixation of nitrogen is accomplished most widely through the high-pressure catalytic method called the Haber-Bosch process named after the German scientists who developed the process in 1914. The process served as a source of ammonium and nitrates for manufacture of explosives during World War I. The reaction for chemical fixation of dinitrogen can be written as follows:

$$N_2 + 3H_2 \xrightarrow[\text{(catalysts)}]{K_2O, Al_2O_3} 2NH_3(aq.)$$

$$\Delta G = -53\,kJ\,(12.7\,kcal)$$

To obtain high yields from this reaction in reasonable time, high pressures (approximately 200 atm. or 20 MPa) are applied to the reaction vessel and the temperature is raised to 400°C–500°C. Natural gas (methane) is usually employed as the feedstock for the hydrogen needed to reduce the N_2 and is also required to heat the reaction vessel. To put these energy requirements in terms more easily understood, it takes about 31,000 cu. ft. (cubic feet) of natural gas, 5.50 barrels of oil, or 2 t (ton) of coal to fix 1 t (2,000 lbs.) of ammonia (Dixon and Wheeler, 1986). About 40% of this is used to provide the necessary heat and the remainder provides the H_2 needed for the reaction. Because we rely on fossil fuel (e.g., natural gas) in the production of ammonium-based fertilizers, fertilizer costs can be high as they are inextricably linked to fossil fuel prices and the whims of that market. For example, an oil embargo against the U.S. in the early 1970's brought about long lines at gas pumps and quadrupled fertilizer prices. The high fertilizer prices were due also, in part, to short supplies at a time of peak demand. It is estimated that fertilizer nitrogen production consumes 1 to 2% of the world's total energy on an annual basis. It should be noted that the efficiency of ammonia production has increased markedly. For example, a nitrogen plant built today typically uses 30% less energy per tonne of ammonia produced than one constructed 40 years ago. (IFA, 2014).

Just as the chemical fixation of dinitrogen is energy intensive, so too is its fixation in biological systems. The principal differences lie in the sources of reductant and energy and the fact that biological N_2 fixation takes place at ambient pressures and temperatures; quite a feat when one considers the rigors of the process described earlier. Energy for biological N_2 fixation comes mainly from the oxidation of organic carbon sources such as glucose or from light in the case of photosynthetic diazotrophs.

conversion of N_2 to NH_3. Recent advances in understanding the workings of the various nitrogenase enzymes are nicely detailed by Newton (2015).

The nomenclature of the nitrogenase complex is as follows (Evans and Burris, 1992):

- the overall complex is known as **nitrogenase**
- the molybdenum-iron (MoFe) protein is **dinitrogenase** (the "type species" substrate is dinitrogen; enzymes conventionally are named relative to their substrate).
- the iron (Fe) protein is designated **dinitrogenase reductase**; the consensus is that its function is the reduction of dinitrogenase.

Characteristics of the two proteins are summarized in Table 16.1. A diagrammatic representation of the nitrogenase complex is shown in Fig. 16.2.

What sets the nitrogenase complex apart is that it:

- consists of two proteins; the MoFe protein (dinitrogenase) and the Fe protein (dinitrogenase reductase);
- is destroyed by oxygen;
- contains iron and molybdenum or vanadium;
- needs Mg^{2+} ions to be active;
- converts ATP to ADP when functioning;
- is inhibited by ADP;
- reduces dinitrogen and several other small triply bonded molecules; and
- reduces H^+ to H_2 even when dinitrogen is present.

The overall reaction for biological N_2 fixation using nitrogenase is shown in Fig. 16.2. Two MgATP are required for each electron transferred from dinitrogenase reductase to dinitrogenase; thus the reaction, as written, shows a requirement of 16 molecules of ATP (112 kcal). Under natural conditions, however, probably 20–30 MgATP are needed as the process is less efficient than when observed under optimum laboratory conditions (Burris and Roberts, 1993). A consensus on the general model for the mechanism of nitrogenase has evolved over quite a few years. The mechanism can be summarized as follows:

- dinitrogenase reductase (the Fe protein) accepts electrons from a low-redox donor such as reduced ferredoxin (Fd_{red}) or flavodoxin and it binds two MgATP (Fig. 16.2);

TABLE 16.1 Characteristics of dinitrogenase (the MoFe protein) and dinitrogenase reductase (the Fe protein).

Dinitrogenase:	Dinitrogenase reductase:
• Mol. wt.: 218,000–227,000	• Mol. wt.: 55,000–66,700
• Has 4 subunits:	• Has 2 subunits: 27,500–34,000
2-α approximately 50,700 mol. wt.	• No molybdenum
2-β approximately 59,500 mol. wt.	• 4 iron atoms per molecule
• 2 molybdenum atoms per molecule	• 4 labile sulfur atoms per molecule
• 21–35 iron atoms per molecule (depending on species)	• Half-life in air: 0.5–0.75 s
• 18–24 labile sulfur atoms per molecule (depending on species)	
• Half-life in air: up to 10 min	

Postgate (1998).

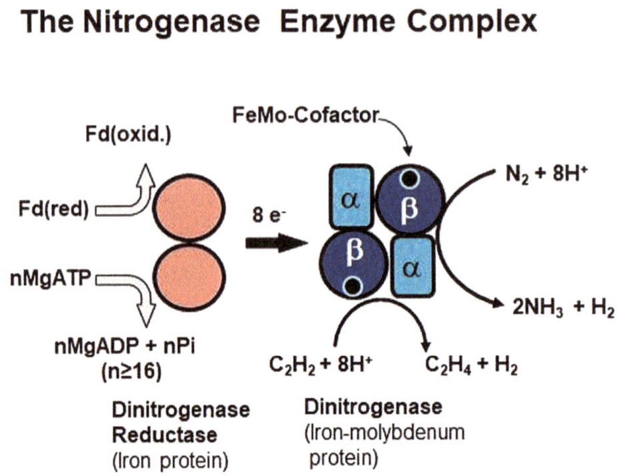

Overall reaction for nitrogen fixation by nitrogenase:

$$N_2 + 8H^+ + 8e^- + 16\ \text{Mg-ATP} \xrightarrow{N_2\text{ase}} 2NH_3 + 16\ \text{Mg-ADP} + 16\ \text{Pi} + H_2$$

FIGURE 16.2 **The nitrogenase enzyme complex consists of two protein components, dinitrogenase reductase (the Fe protein) and dinitrogenase.** The dinitrogenase reductase gathers electrons from low-redox carriers such as ferredoxin and flavodoxin and passes them on to dinitrogenase (the MoFe protein). Nitrogen is bound to the FeMo-cofactor on the dinitrogenase protein where it is reduced. See text for further details of the dinitrogen reduction process.

- it transfers electrons, one at a time, to dinitrogenase (the MoFe protein);
- dinitrogenase reductase and dinitrogenase form a complex, the electron is transferred, and two MgATP are hydrolyzed to two MgADP + Pi (phosphate);
- dinitrogenase reductase and dinitrogenase dissociate and the process is then repeated;
- when dinitrogenase has collected enough electrons, it binds a molecule of dinitrogen at the FeMo-cofactor, reduces it and releases NH_4^+; and
- the MoFe protein (dinitrogenase) then accepts additional electrons from dinitrogenase reductase (the Fe protein) to repeat the cycle.

In each cycle of N_2 fixation, dinitrogenase and dinitrogenase reductase bind together, MgATP is hydrolyzed, and an electron is transferred. The dissociation of the MoFe protein–Fe protein complex is the rate-limiting step of the process. In fact, the nitrogenase complex is "remarkably slow —it takes 1.25 sec for a molecule of enzyme to form two of NH_3. The two proteins have to come together and separate 8 times to reduce one dinitrogen molecule" (Postgate, 1998). A consequence of the slowness of nitrogenase is that N_2-fixing bacteria must synthesize a lot of the protein. Nitrogenase can commonly account for 10% of the cell's proteins and levels up to 40% have been recorded (Postgate, 1994, pg.144).

Returning to the reaction in Fig. 16.2 for biological N_2 fixation, we can observe that for every $8e^-$ transferred via the nitrogenase complex, $2e^-$ are consumed in the formation of H_2. The production of H_2 that accompanies the fixation of N_2 is obligatory. One H_2 (requiring 4 MgATP) is released for each N_2 reduced to $2NH_3$ (requiring 12 MgATP). Thus, 25% of the

TABLE 16.2 Physiological requirements for dinitrogen fixation by free-living diazotrophic microbes.

Trace elements
Molybdenum or vanadium, iron for nitrogenase and or dinitrogenase
Magnesium—for production of MgATP
ATP
A minimum of 16 ATP per N_2 fixed. Probably 20–30 under natural conditions. However, *Azotobacter* may use as little as 6 ATP per N_2 fixed.
Needed for nitrogenase activity and for nitrogenase synthesis.
The high ATP requirement means an abundant supply of energy-yielding substrates must be readily available for vigorous N_2 fixation.
Minimize H_2 evolution
Up to 35% of the ATP diverted to nitrogenase may be consumed in H_2 evolution.
Acceptable temperature
Most diazotrophs are mesophiles.
Many will not grow on media at 37°C.
Nitrogenase activity falls off rapidly at about 40°C (except in thermophiles).
Oxygen excluded from the enzyme complex
Nitrogenase is destroyed by oxygen
Source of low-redox reductant
This is restricted to the naturally occurring ferredoxins and flavodoxins.

energy from MgATP is "lost" in the production of H_2. Interestingly, some diazotrophs contain an uptake hydrogenase that allows them to oxidize some of the H_2 and to generate a reduced electron carrier or MgATP. This can then be used in the N_2 fixation reaction thereby recapturing some of the energy lost.

The physiological requirements for a free-living diazotroph to fix N_2 are summarized in Table 16.2. These requirements stem in large part from the unique properties and requirements of the nitrogenase complex including its exceptional sensitivity to molecular oxygen, the metal content of the complex (iron and molybdenum or vanadium), and the need for adequate supplies of reducing power and MgATP.

Substrates for nitrogenase

The principal substrate for nitrogenase is dinitrogen (N≡N). Note that the two atoms of nitrogen are joined by a strong triple bond. In addition to reducing protons to H_2, nitrogenase also reduces several other small, triply bonded molecules (Table 16.3).

Of particular interest among this list of substrates is the gas acetylene. In the late 1960s, it was discovered that nitrogenase can reduce acetylene (HC≡CH) to ethylene ($H_2C = CH_2$) (Dilworth, 1966; Burris, 1974). The significance of this finding lays in the fact that acetylene and ethylene can be very easily measured using a gas chromatograph to separate the two gases. Thus was born the simple, sensitive, and rapid **acetylene reduction** (AR) **assay** for nitrogenase activity. The development of this assay led to an explosion in research on N_2 fixation because it removed the need to have an expensive mass spectrometer or to deal with the tedious and insensitive Kjeldahl analyses to measure nitrogen in soil or biological

TABLE 16.3 Substrates for nitrogenase: the triple-bond connection.

Name	Formula	Major products
Proton	H^+	H_2
Dinitrogen	$N\equiv N$	$NH_3 + H_2$
Nitrous oxide	$N\equiv N^+\text{-}O^-$	$N_2 + H_2O$
Azide	$[N\equiv N^+\text{-}N]^-$	$N_2 + NH_3 + N_2H_4$
Acetylene	$HC\equiv CH$	$H_2C = CH_2$
Cyanide	$[C\equiv N]^-$	$CH_4 + NH_3$
Carbon monoxide	$C\equiv O$	Recent research indicates that nitrogenase reduces CO and an assortment of carbon-containing compounds including CO_2, HCOOH, ethane, propane, propylene, and butylene. Seefeldt et al. (2013)

materials. However, the technique is not without limitations, which are discussed later in this chapter.

Alternative nitrogenases

Prior to the early 1980s researchers thought that there was only one type of nitrogenase, and that molybdenum was essential for N_2 fixation. However, in the early 1980s Dr. Paul Bishop and his colleagues in North Carolina isolated a second nitrogenase (nitrogenase 2) from *Azotobacter chroococcum* that was produced only under conditions of molybdenum starvation (Bishop and Premakumar, 1992). With this discovery it became clear that molybdenum was not required for N_2 fixation in all bacteria and could be supplanted with vanadium (V). Nitrogenase 2 has in common with nitrogenase 1 the fact that it consists of two proteins, a VFe protein and an Fe protein, produces H_2, and is sensitive to O_2. Nitrogen binds to the VFe protein suggesting that metals other than Mo can participate in binding the nitrogen to the protein for reduction. Subsequently, a third nitrogenase was described as nitrogenase 3. This complex does not appear to contain either molybdenum or vanadium and is synthesized under starvation for both metals.

All the "nitrogenases" are complexes of a dinitrogenase reductase and a dinitrogenase component. Nitrogenases 2 and 3 appear to evolve more H_2 than nitrogenase 1, and nitrogenases 2 and 3 reduce acetylene to ethane (C_2H_6) rather than ethylene. In the late 1990s, yet another *putative nitrogenase* was described (Ribbe et al., 1997, Gadkari, 2004). It was reported to occur in *Streptomyces thermoautotrophicus*, a unique, thermophilic (it can grow at 65°C), chemoautotrophic actinomycete that reportedly uses carbon monoxide (CO) as an energy source, oxidizing it to CO_2, or it can also use H_2. It fixed N_2 and was unique in that its "nitrogenase" was not inhibited by CO, as are the others, it did *not* reduce acetylene and was *not* poisoned by O_2. The "nitrogenase" of this streptomycete reportedly consisted of three proteins: dinitrogenase, a superoxide oxidoreductase, and a carbon monoxide dehydrogenase that were responsible for nitrogen fixation. The manganese-containing enzyme is in fact a superoxide dismutase that supplied electrons to the molybdoprotein that is the "dinitrogenase." Because the reduction of N_2 depended obligately on O_2^-, the N_2-fixing system of *S. thermoautotrophicus* was designated as a superoxide-dependent nitrogenase (sdn) (Gadkari, 2004). It was astonishing that not only was this "nitrogenase" reportedly not poisoned by O_2, quite strikingly, O_2 was *required* in the reaction mechanism of the enzyme complex.

However, perhaps as an example of how science advances, in 2016, MacKellar and coworkers reported seemingly definitively that *S. thermoautotrophicus* does not fix nitrogen. The research was carried out in seven laboratories including in the United States, Europe, and South America. Working with the original isolate of the organism as well as two additional isolates, the international team was unable to demonstrate that any of the cultures could incorporate ^{15}N-labeled dinitrogen ($^{15}N_2$) into cell constituents. The authors proposed "*reclassification of the species containing strains UBT1 (the original culture), H1, and P1-2 as a non-Streptomycete, non-diazotrophic, facultative chemolithoautotroph and conclude that the existence of the previously proposed oxygen tolerant nitrogenase is extremely unlikely*" (MacKellar et al., 2016). The organism reportedly shares >99% nucleotide identity of its genes with *Hydrogenibacillus schegelii* (another carbon-monoxide oxidizing bacterium). Thus, as of this writing, the story of N_2−fixation by *S. thermoautotrophicus* appears in want of further verification, though the results of the international team cast significant doubt about the ability to fix N_2 using a radically different type of oxygen-insensitive nitrogenase.

The finding that dinitrogen can be fixed in the absence of molybdenum is interesting because prior to the discovery of the alternative nitrogenases, microbiologists and biochemists believed that the element was an obligate requirement for N_2 fixation. It now seems that the predominance of bacteria containing nitrogenase 1 (the Mo-containing enzyme) may have been due to the fact that researchers habitually used Mo-containing media to isolate N_2-fixing bacteria. The other nitrogenases may be quite common in nature though this is not yet known. The study of alternative nitrogenase systems is young and many fundamental questions remain.

Genetics of nitrogenase

As one might surmise from the complexities of the system described earlier, the genetics of the process are equally complex. Much of our knowledge of the genetics of the nitrogenase system has come from the intensive study of the bacterium *Klebsiella pneumoniae* (subsequently renamed *Klebsiella oxytoca*), a member of the family Enterobacteriaceae and a close relative of *Escherichia coli*. Forty plus years of research on this bacterium and, more recently, other diazotrophic bacteria has shown that the nitrogenase complex and supporting systems are under the control of 20 or more genes (Fig. 16.3; Table 16.4).

The functions of the nitrogen fixation (*nif*) genes are summarized in Table 16.4, starting with the gene (J) on the left-hand side of Fig. 16.3.

Genomic sequencing analysis showed that nearly all diazotrophs contain a minimal gene set consisting of six conserved genes; *nifH*, *nifD*, *nifK*, *nifE*, *nifN*, and *nifB*, consistent with the minimal *nif* gene numbers required to assemble FeMo-co in vitro (Dos Santos et al., 2012). Clearly, the complexity of this system does not make it an easy target for genetic manipulation and transfer to higher organisms. Still, some hold out the prospect that we may someday see genetically engineered crops like corn, wheat, and other cereals that fix their own nitrogen. However, this remains one of the more elusive goals facing dinitrogen-fixation researchers (Pankievicz et al., 2019; Burén and Rubio, 2018; Vicente and Dean, 2017; de Bruijn, 2015; Oldroyd and Dixon, 2014).

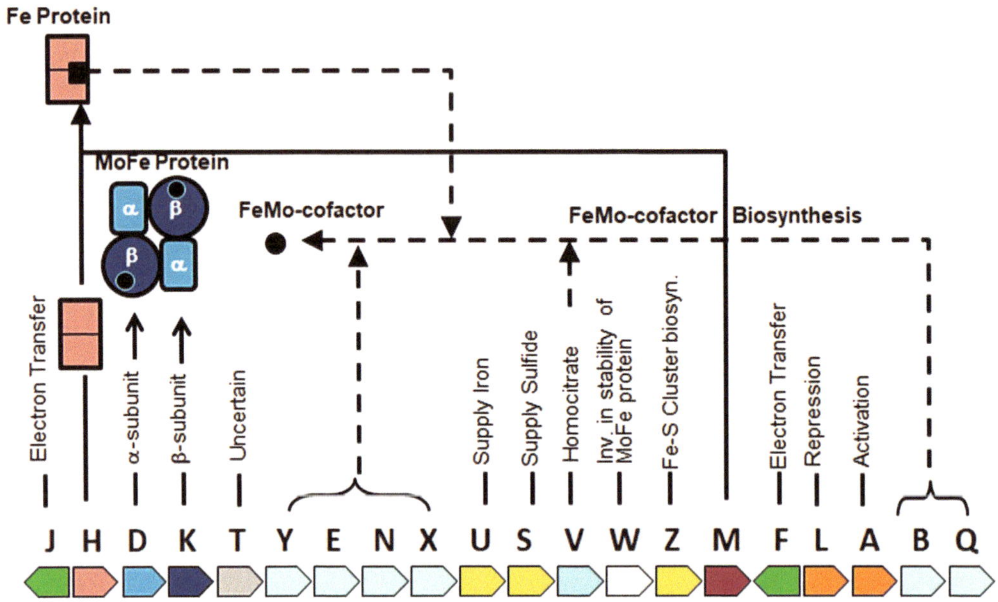

FIGURE 16.3 **Organization of the nitrogen fixation (*nif*) genes in *Klebsiella pneumoniae* with the corresponding known or putative functions of the gene products.** Individual genes with the same color have related functions. Brief descriptions of the function(s) of each gene are given in Table 16.4. *Source: Redrawn from Fisher and Newton (2002). Used with permission. Color coding is after Figure 3 of Oldroyd and Dixon (2014). Genes are color-coded according to their functions: medium blue, α subunit of the MoFe protein; dark blue, β subunit of the MoFe protein; dark red, Fe protein maturation (nifM); light blue, FeMo-co biosynthesis (nifY, nifE, nifN, nifX, nifV, nifB, nifQ); yellow, Fe–S cluster biosynthesis (nifU, nifS, nifZ); green, electron transport (nifJ, nifF); orange, transcriptional regulation (nifL, nifA); white, (nifW) involved in stability of the MoFe protein (Rubio and Ludden, 2005; Burén et al., 2020); grey, unknown function (nifT).*

TABLE 16.4 The *nif* gene products and their role (known or proposed) in nitrogen fixation.

Gene	Identity and/or role of gene product
nifJ	Pyruvate: flavodoxin (ferredoxin) oxidoreductase. Electron donor to Fe protein in *K. pneumoniae*.
nifH	Fe protein (dinitrogenase reductase). Obligate electron donor to MoFe protein during nitrogenase turnover. NifH is also required for FeMo-co biosynthesis and apo-MoFe protein maturation.
nifD	α subunit of MoFe protein (dinitrogenase). Forms an α2β2 tetramer with the β subunit. The site of substrate reduction, FeMo-co, is within the α subunit of MoFe protein.
nifK	β subunit of MoFe protein (dinitrogenase). Forms an α2β2 tetramer with the α subunit. P-clusters are present at each αβ subunit-interface.
nifT	Unknown.
nifY/ *nafY*	Chaperone for the apo-MoFe protein. NafY is also a FeMo-co carrier and is proposed to aid in the insertion of FeMo-co into apo-MoFe protein.
nifE	Scaffold for NifB-co maturation into FeMo-co. Forms an $\alpha_2\beta_2$ tetramer with NifN.
nifN	Scaffold for NifB-co maturation into FeMo-co. Forms an $\alpha_2\beta_2$ tetramer with NifE.
nifX	Binds NifB-co and VK-cluster. Interacts with apo-NifEN. Proposed carrier of FeMo-co precursors.
nifU	Supplies [Fe–S] clusters precursors for nitrogenase proteins and cofactors.

(*Continued*)

TABLE 16.4 The *nif* gene products and their role (known or proposed) in nitrogen fixation. (*Cont.*)

Gene	Identity and/or role of gene product
nifS	Mobilizes S for [Fe–S] cluster synthesis on NifU.
nifV	Homocitrate synthase. Involved in FeMo-co synthesis.
nifW	Associates with apo-MoFe protein preceding P-cluster maturation. Involved in stability of MoFe protein.
nifZ	Involved in P-cluster maturation in combination with the Fe protein.
nifM	Essential for the maturation of the Fe protein. Proposed peptidyl-prolyl cis–trans isomerases that would act over the Pro^{258} residue of NifH.
nifF	Flavodoxin. Physiological electron donor to the Fe protein.
nifL	Sensor protein. Acts as NifA antiactivator. Negative regulatory element.
nifA	Transcriptional activator. Forms a two-component regulatory system with NifL. Positive regulatory element.
nifB	Essential for FeMo-co synthesis. Catalyzes NifB-co formation, the first committed step in FeMo-co biosynthesis.
fdxN	2x[4Fe-4S] ferredoxin. Required for NifB in the formation of NifB-co.
nifQ	Processes Mo prior to its incorporation into FeMo-co.

Derived from Burén et al. (2020) https://pubs.acs.org/doi/10.1021/acs.chemrev.9b00489. Further permissions related to the material excerpted should be directed to the ACS. See also Rubio and Ludden (2002).

Free-living nitrogen-fixing bacteria

As described earlier, biological N_2 fixation is restricted to prokaryotic organisms such as the bacteria and archaea (remember the cyanobacteria and the actinomycetes (actinobacteria) are in fact specialized bacteria). Of the 10,000, or so, bacterial genera named, only about 100 contain bonafide diazotrophic species (Table 16.5). While this may appear to be a somewhat limited representation, it should be stressed that these species represent all sorts of physiological types and occupy all sorts of ecological niches. An extensive listing of diazotrophic species was published by J.P.W. Young (1992) and others have been added in more recent surveys.

It should also be noted that recent studies probing soil genomes for the presence of *nif* genes (notably *nifH* and *nifD*) suggest that the list of N_2-fixing species will be expanded in the future (Dos Santos et al., 2012). Some representative genera of free-living diazotrophs are listed in Table 16.5. These organisms encompass groups such as heterotrophic and chemoautotrophic bacteria, photoautotrophs (bacteria and cyanobacteria), and photoorganotrophs with respect to carbon metabolism (see Chapter 3). Note also that they are well represented by aerobes, **microaerophiles** (grow at low O_2 concentrations), facultative anaerobes, and obligate anaerobes. Such metabolic diversity enables some type of diazotroph to colonize almost any imaginable sort of habitat. Indeed, diazotrophs are widespread in nature as free-living microbes and in a large number of associations with plants and animals. What this great metabolic diversity means is that, in all sorts of environments (terrestrial and aquatic), diazotrophs can make contributions to the supply of "fixed" nitrogen (the term "reactive nitrogen" is now commonly used to refer to "fixed" forms such as NH_4^+ and NO_3^-) for growth of

TABLE 16.5 Some genera of microbes that include dinitrogen-fixing species or strains. The list is not intended to be all-inclusive.

	Genus or type	Species (examples only)
AEROBES	*Azotobacter*	*A. chroococcum*[a], *A. vinelandii*[a]
	Azotococcus	*A. agilis*[a]
	Azomonas	*A. macrocytogenes*[a]
	Beijerinckia	*B. indica*[a], *B. fluminis*[a]
	Derxia	*D. gummosa*[a]
	Pseudomonas	*P. stutzeri*, *P. saccharophila*
	Azoarcus	*A. communis*, *A. indigens*
	Gluconacetobacter	*G. diazotrophicus*, *G. johannae*
FACULTATIVE	*Klebsiella*	*K. pneumoniae*, *K. oxytoca*
(aerobic when not fixing nitrogen)	*Bacillus*	*B. polymyxa*, *B. macerans*
	Enterobacter	*E. agglomerans* (*Erwinia herbicola*)
	Citrobacter	*C. freundii*
	Escherichia	*E. intermedia*
	Propionibacterium	*P. shermanii*, *P. petersonii*
	Paenibacillus	*P. polymyxa*, *P. macerans*, *P. azotofixans*
MICROAEROPHILES	*Xanthobacter*	*X. flavus*[a], *X. autotrophicus*
(normal aerobes when not fixing nitrogen)	*Thiobacillus*	*T. ferrooxidans*
	Azospirillum	*A. lipoferum*[a], *A. braziliensis*[a]
	Aquaspirillum	*A. perigrinum*[a], *A. fascicilus*[a]
	Methylosinus	*M. trichosporum*
	Herbaspirillum	*H. seropedicae*
	Burkholderia	*B. brasilense*
STRICT ANAEROBES	*Clostridium*	*C. pasteurianum*[a], *C. butyricum*
	Desulfovibrio	*D. vulgaris*, *D. desulficuricans*
	Methanosarcina	*M. barkeri*
	Methanococcus	*M. thermolithotrophicus*
PHOTOTROPHS	*Anabaena*	*A. cylindrica*, *A. inaequalis*
(aerobic) cyanobacteria	*Nostoc*	*N. muscorum*
	Calothrix	
	(Seven other genera of heterocystous cyanobacteria)	
	Gloeothece	*G. alpicola*
PHOTOTROPHS	*Plectonema*	*P. boryanum*
(microaerophiles) cyanobacteria	*Lyngbya*	*L. aestuarii*
	Oscillatoria	
	Spirulina	

(Continued)

TABLE 16.5 Some genera of microbes that include dinitrogen-fixing species or strains. The list is not intended to be all-inclusive. (*Cont.*)

	Genus or type	Species (examples only)
PHOTOTROPHS	*Rhodospirillum*	*R. rubrum*
(facultative) bacteria	*Rhodopseudomonas*	*R. palustris*
PHOTOTROPHS	*Chromatium*	*C. vinosum*
(anaerobes) bacteria	*Chlorobium*	*C. limicola*
	Thiopedia	
	Ectothiospira	*E. shapovnikovii*

[a] *Signifies that all reported strains of the species fix dinitrogen. Adapted from Postgate (1998). See also Young (1992).*

nonfixing microbes and higher forms of life. For example, as will be discussed further, diazotrophs colonizing the roots and rhizospheres (see Chapters 2 and 11) of many plant species make small quantities of fixed nitrogen available to the plants.

Factors affecting dinitrogen fixation by free-living diazotrophs

In the preceding sections, the complexities and uniqueness of the nitrogenase enzyme complex have been described. In this section, the factors that must be successfully integrated for a nonsymbiotic diazotroph to fix N_2 will be discussed and then, in a later section, the integration of these factors with respect to the functioning of nonsymbiotic diazotrophs in association with higher plants (the so-called **associative symbioses**) will be described.

Sources of energy for diazotrophs

Except for the phototrophic bacteria and cyanobacteria, all diazotrophs require an organic or inorganic (in the case of chemoautotrophs) energy source. A requirement for abundant energy sources is dictated by the high energy demands of nitrogenase. Remember, a minimum of 16 ATP's is required to make two molecules of NH_3 from N_2. A wide variety of carbon sources, ranging from methane (CH_4) to complex carbohydrates, can be used by one diazotroph or another. It is not usually the type of carbon source that limits N_2 fixation; rather it is the lack of an ample supply in many habitats that limits fixation by the nonsymbiotic bacteria. Recall also, the soil is not an "organic soup" abundantly supplied with readily available C sources and also that diazotrophs must compete with all the other soil microbes for the same carbon. In pure cultures this is not a problem.

Regarding utilization of carbon sources for energy, it is interesting to look at the efficiency of N_2 fixation in terms of the amount of C consumed per nitrogen fixed (Table 16.6). Note that the efficiencies of N_2 fixation in terms of milligrams of N fixed per gram of carbon source are low, averaging about 15 mg N g^{-1} of C source. Also, the assimilation of NH_4^+ is about twice as efficient as N_2 fixation and that fixation by anaerobes is generally much less efficient

TABLE 16.6 Carbon and energy source requirements for heterotrophs grown in chemostats limited by the carbon and energy source.

Organism	Carbon and energy source	Efficiency of N incorporation during growth (mg N g^{-1} C and energy source used)		Carbon and energy expenditure for N_2 fixation	
		N_2	NH_4^+	g C g^{-1} N_2	lb C / 100 lb of N_2 Fixed
Anaerobic growth by fermentation					
Clostridium pasteurianum	Sucrose	11	22	46	4,600 (2,054)[a]
Klebsiella pneumoniae	Glucose	8	19	72	7,200 (3,214)
Aerobic growth					
Klebsiella Pneumoniae (O_2-limited)	Glucose	15	35	38	3,800 (1,696)
Azospirillum Brasilense (9 μM O_2)	Malate	26	48	19	1900 (848)
Azotobacter Vinelandii (2–10 μM O_2)	Sucrose	16	63	47	4,700 (2,098)
Azotobacter Vinelandii (180 μM O_2)	Sucrose	7	38	117	11,700 (5,223)

[a]*The 100-lb figure is chosen as representative of a typical amount of fertilizer N that might be applied to a crop. Numbers in ( ) are values in kg.*
Adapted from Hill (1992). See also Giller and Day (1985).

than by aerobes. An exception to this generalization, however, is shown by the fact that the efficiency of fixation by *Azotobacter* at higher levels of O_2 is much reduced. An efficiency of 15 mg g^{-1} of carbon source is equivalent to about 13 kg (30 lb.) of N fixed per ton of carbon source. The need for large amounts of available energy sources for N_2 fixation is clear.

In nature, the carbon sources for N_2 fixation by heterotrophic bacteria normally come from the remains of crop residues and roots decomposing in the soil as well as carbon released from active roots (root exudates, lysates, sloughed cells, etc.). Addition of readily available carbon sources to soil generally stimulates N_2 fixation for short periods of time.

Effects of combined nitrogen on N_2 fixation

Since N_2 fixation is so costly to a cell, it should not be surprising that the nitrogenase complex is under strong regulation by the level of combined N (i.e., ammonium, nitrate, organic N) in the medium or the environment. The formation of the nitrogenase complex is repressed by ammonium at fairly low levels. For example, Alexander and Zuberer (1989a) showed that as little as 4.2 μg mL^{-1} of ammonium N in a nutrient solution eliminated acetylene reduction associated with corn roots at an O_2 concentration of 2% (0.1 atm or 2 kPa) around the roots. Nitrate and organic N also prevent the synthesis of nitrogenase. By repressing the formation of nitrogenase at very low levels of combined N, the organisms avoid the high expense of synthesizing and operating an enzyme system that is not needed under conditions of N sufficiency. Additionally, there is a "switch off" mechanism through which ammonia shuts down nitrogenase activity.

Effects of O_2 on nitrogenase activity

An interesting characteristic of nitrogenase, and one that poses considerable problems for most free-living diazotrophs, is its extreme sensitivity to molecular oxygen (O_2) (Tables 16.1 and 16.2). In most bacteria, the enzyme complex is irreversibly "poisoned" by exposure to O_2. This sensitivity to O_2 has led to the development of some unique strategies among the diazotrophs to protect the enzyme system. Most unusual among these is the ability of some *Azotobacter* species to protect their enzymes through **respiratory** or **conformational protection**. Box 16.2 illustrates the many adaptations that allow for N_2 fixation to occur in environments with widely divergent aeration conditions. It is apparent from the discussion earlier that O_2 plays a significant role in governing the occurrence of N_2 fixation in various environments. It is not surprising that the higher rates of nitrogen fixation reported in the literature are generally associated with seasonally or chronically wet soils where the O_2 levels tend to be lower due to saturation. It is for these reasons that paddy rice is one of the few agricultural crops that derive significant benefit from the activities of nonsymbiotic diazotrophs and N_2-fixing cyanobacteria.

BOX 16.2

How microbes "solve" the oxygen problems for nitrogenase

Avoidance—Anaerobes and facultative anaerobes fix dinitrogen only in the absence of oxygen with the exception of *Klebsiella pneumoniae* that can tolerate very low levels of oxygen. In fact, oxygen is one of the factors that regulate the synthesis of nitrogenase in this bacterium.

Microaerophily—Most aerobic diazotrophic bacteria fix dinitrogen maximally at low partial pressures of oxygen thereby lessening the exposure of nitrogenase to oxygen. For example, it has been reported that *Azospirillum* fixes dinitrogen most rapidly at 0.007 atm oxygen (Okon et al., 1977).

Respiratory protection—Respiration functions in all aerobes to divert oxygen away from nitrogenase to some extent. In certain *Azotobacter* species, an exaggerated form of respiratory protection is observed. In fact, this bacterium exhibits one of the highest respiration rates of all lifeforms. The high respiration rate serves to scavenge oxygen and to keep it away from nitrogenase. As a consequence, the bacterium must consume large amounts of substrate to scavenge oxygen and growth is very inefficient in terms of carbon consumed under these conditions.

Production of specialized cells—Many diazotrophic cyanobacteria produce specialized, thick-walled cells called **heterocysts** in which the nitrogenase is compartmentalized. These cells do not evolve oxygen in photosynthesis and the thick wall excludes external oxygen. The diazotrophic actinomycete, *Frankia* produces **vesicles** in which nitrogenase is protected from oxygen. The vesicle wall becomes thicker as the oxygen concentration increases in the medium.

(*Continued*)

BOX 16.2 (*Cont.*)

Slime—Aerobic diazotrophs grown on nitrogen-free agar media frequently produce large gummy colonies due to the production of **extracellular polysaccharides**. The gum serves as a diffusion barrier to the free flow of oxygen into the colony such that cells in the interior of the colony are not as exposed to oxygen. The efficacy of this mechanism is questionable.

Conformational protection—Some *Azotobacter* species produce a protein that binds to the nitrogenase and protects it from damage by oxygen. In the presence of oxygen, and when respiration cannot keep up with the incoming oxygen supply, the protein binds to the nitrogenase complex, and alters its conformation (shape) to protect it from oxygen. Upon exposure to oxygen, the organism stops fixing N_2 abruptly but, when conditions of oxygenation return to a more favorable state, the organism resumes N_2 fixation. The enzyme is not destroyed.

Temporal or spatial separation of dinitrogen fixation and oxygen-evolving processes—Non-heterocystous cyanobacteria solve the oxygen problem by fixing N_2 primarily during the dark phase of growth (i.e., temporal separation) when oxygen evolution is not occurring, and respiration also serves to scavenge oxygen away from nitrogenase. Others form aggregates of cells and some of these cells then function in a microaerophilic environment with spatial separation.

Root nodule symbioses—In the root nodules of leguminous plants (Chapter 17), the O_2 concentration is maintained at a level favorable for N_2 fixation by the presence of leghemoglobin which controls the level of O_2 reaching the rhizobia. Some actinorhizal plants also produce hemoglobin in their nodules or rely on thickened cell walls or lipid layers to exclude O_2.

After Postgate (1998), used with permission.

Other environmental factors

Energy supply (available carbon), presence or absence of sufficient combined N, and the abundance of molecular oxygen exert primary control over the synthesis and level of nitrogenase activity expressed by bacteria in a given environment. Other environmental factors are of lesser significance but cannot be ignored. Nitrogenase is active over a relatively narrow temperature range. At the lower limits (5°C–10°C), the nitrogenase activity is quite low whereas at the upper limits, 37°C–40°C, nitrogenase activity falls off rapidly because of the sensitivity of the enzyme to heat. Most diazotrophs are mesophiles, but there are a few exceptions. For example, some thermophilic cyanobacteria grow in hot springs. *Methanococcus thermolithotrophicus* (a methanogen) fixes N_2 at 64°C.

Numerous other factors can affect the growth and survival of diazotrophs and thus directly or indirectly influence N_2 fixation. Among these are: adequate supplies of phosphorus (N_2 fixation requires high levels of P), other inorganic nutrients (especially trace metals), and the acidity or alkalinity of the environment. Biotic factors are also involved (Table 16.7).

TABLE 16.7 Biotic and abiotic factors affecting N_2 fixation by free-living soil bacteria.

Factor	Biotic or abiotic	Effect on N_2 fixation
Carbon/energy source	Biotic/abiotic	Lack of an abundant supply available organic C is considered the principal limiting factor for N_2 fixation by free-living soil diazotrophs. (Recall the need for large quantities of ATP for nitrogenase activity.)
Oxygen	Abiotic	Nitrogenase is in most cases irreversibly damaged by exposure to O_2. N_2 fixation by aerobes is usually most vigorous when the O_2 level is much reduced and it must generally be absent for the anaerobes and facultative anaerobes.
Combined nitrogen (NH_4^+, NO_3^-, etc.)	Abiotic	Nitrogenase is strongly controlled by the presence of combined nitrogen [i.e., ammonium, nitrate, and organic N compounds (amino acids, etc.)] in soil.
Competition	Biotic	N_2-fixing bacteria must compete with all other soil microbes for carbon supplies, etc. It is generally agreed that diazotrophs constitute 1%–10% of the cultivable bacterial population.
Others	Biotic	Like all other soil bacteria, diazotrophs are subject to predation by protozoa and lysis by bacterial viruses (bacteriophages).
	Abiotic	pH, temperature, trace element availability, etc.

N_2 fixation by plant-associated bacteria: the associative symbioses

Nature of the association

There are numerous examples of associations in which an otherwise free-living diazotroph associates with a plant. These more or less casual associations generally exhibit no morphological modification of the "host" or overt genetic interaction between the plant and the bacterium such as those observed in the root nodule symbioses described in the next chapter. Instead, they tend to result from the bacterium establishing itself in an environment where carbon and energy sources are being made available, albeit somewhat inadvertently, by the plant. Diazotrophic bacteria have been found on leaf surfaces (the so-called phyllosphere) and in and around the roots (in the rhizosphere) of a wide range of plant species.

Diazotrophic bacteria have been observed in the intercellular spaces of the outer cell layers of roots, in the moribund cells of the outer layers and in vascular elements of some plants. The bacteria residing in the rhizosphere and at the root surface (see Figs. 16.4 and 16.5) utilize root exudates, secretions, lysates, and sloughed cells as carbon and energy for growth and N_2 fixation.

Some endophytic bacteria, such as *Gluconacetobacter diazotrophicus* (formerly *Acetobacter*), may occupy the internal root tissues including the vascular elements of the plant (sugarcane in this instance). This and other so-called **endophytic** (see Chapter 11 for discussion of endophytic bacteria) N_2-fixing bacteria are the subjects of continuing interests in N_2-fixing nonleguminous plants. As a consequence of their activities, some N_2 is fixed and is eventually made available for plant growth. Bacteria that take up residence in the vascular elements gain the obvious advantage of being "first in line" for substrates and thereby reduce the problem of competition from nondiazotrophs (Fig. 16.4). It is widely accepted that diazotrophic bacteria can be active colonizers of plant roots.

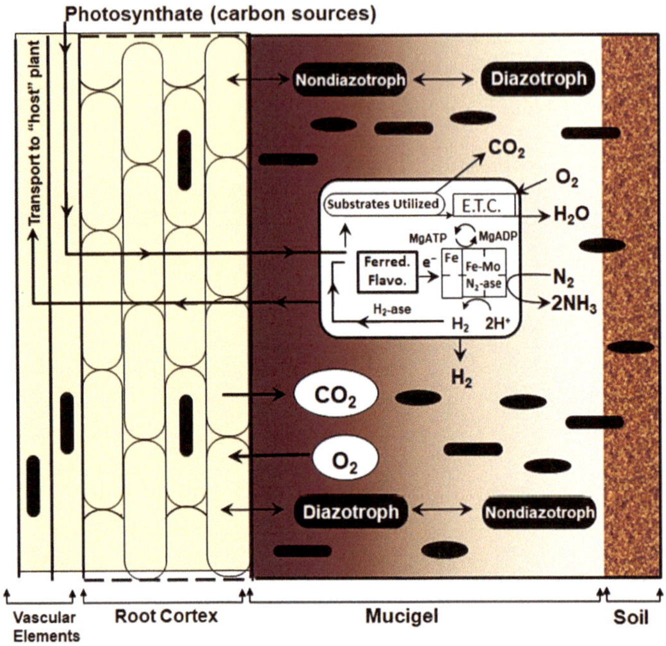

FIGURE 16.4 **A model for root and rhizoplane-rhizosphere colonization by diazotrophic bacteria.** The central reactions of the diazotroph are in the enlarged "cell" at center. Diazotrophs and nondiazotrophs (indicated by *black bars*) colonize the rhizosphere soil, the mucigel layer, the rhizoplane (root surface), and the outer cell layers of the root. In some instances, certain bacteria (e.g., the endophyte *Gluconacetobacter diazotrophicus*) can colonize the vascular elements and other internal tissues. The bacteria grow at the expense of carbon sources (photosynthate, exudates, secretions, lysates, and sloughed cells) provided by the plant and forming a gradient of substrates external to the root *(shaded area)*. Diazotrophs must compete for substrates with nondiazotrophs. Dinitrogen is fixed by bacteria in all of the locations shown and eventually becomes available to the plant or other microbes when the cells die and become mineralized. Interactions between diazotrophs and nondiazotrophs may be beneficial (e.g., a nondiazotroph lowering the oxygen concentration or immobilizing ammonium or nitrate) or detrimental (e.g., nondiazotrophs consuming carbon that otherwise might have been available to support N_2 fixation). Abbreviations: *Ferred.*, ferredoxin; *Flavo.*, flavodoxin; *E.T.C.*, electron-transport chain.

Knowledge that diazotrophic bacteria are part of the root and rhizosphere bacterial population has prompted interest in using the bacteria as an alternate source of N for crop growth. In the mid to late 1970s, considerable research was launched as a result of reports of diazotrophic bacteria fixing large amounts of N (up to 90 kg ha^{-1} $year^{-1}$) in association with the roots of tropical grasses such as *Paspalum notatum* cv. *batatais* and *Digitaria* species. The association between *Paspalum notatum* and *Azotobacter paspali* is interesting because of the apparent specificity involved. The bacterium associates only with the tetraploid cultivar "batatais" of the grass. These and other studies stimulated much more research on the use of free-living diazotrophs as inoculants for crop plants. Further aspects of the use of these bacteria as inoculants are discussed in Box 16.3.

Today, it is widely accepted that rates of N_2 fixation by root-associated bacteria are probably in the range of 5–25 kg N ha^{-1} $year^{-1}$ (about 4–22 lbs. ac^{-1} $year^{-1}$) with most estimates

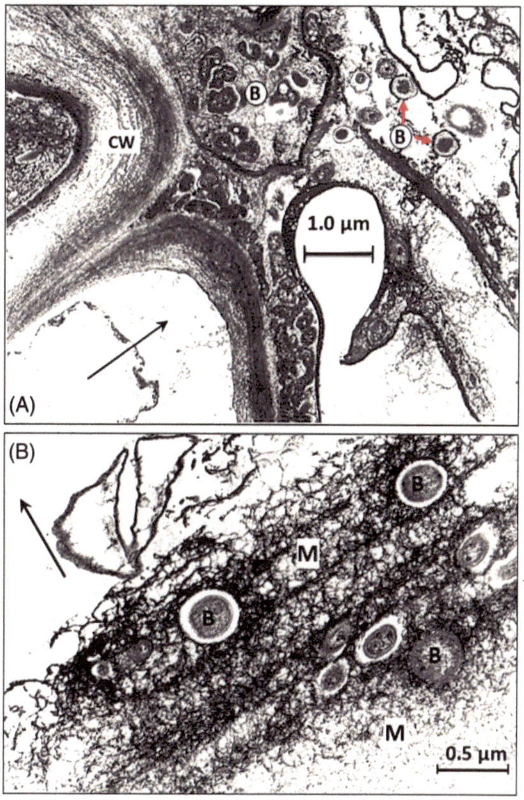

FIGURE 16.5 **Photomicrographs of ultra-thin sections of mangrove roots viewed with a transmission electron microscope.** (A) Junction between two root cells (note the multilayered wall material at left) and bacterial cells in the mucilage (M) and sloughed cell debris at the root surface. Numerous bacterial cells can be seen in this material. The *arrow* points away from the central axis of the root. (B) Mucilage at the roots surface containing embedded bacterial cells. Note the fibrillar nature of this material; in this instance, it appears to have been laid down in discrete layers, perhaps due to periods of increased exudation of carbon. For reference orientation, these photos represent the immediate exterior of the intact root that could be encompassed in a zone such as that marked with the box in Fig. 16.6 and "blown up" graphically in Fig. 16.4. *Marker bars* are 1.0 and 0.5 µm respectively in (A) and (B). *CW*, Cell wall.

being at the lower end of this range. Compare these rates with those reported for a range of N_2-fixing plant-microbe associations in Table 16.8. These low rates of N_2 fixation by associative bacteria are probably insignificant for production of the major cereal crops; however, they are significant for rangeland and prairie grasses where the only other inputs of N are from atmospheric deposition and possibly from the excretion products of animals. They may also prove yet to be significant in the culture of low-maintenance grasses for amenity planting and roadside vegetation, and they may be important for growth of tropical plants other than grasses.

Why are the rates of N_2 fixation by root- and rhizosphere-colonizing bacteria so low in dry-land systems? The answer lies in the requirements of the typical nitrogenase system. The enzyme complex is irreversibly harmed by exposure to oxygen and it is stringently regulated

BOX 16.3

Does inoculation with free-living diazotrophs benefit crop production?

Among the most widely studied of these **associative symbioses** is the association between *Azospirillum* and roots of numerous grass species, including most of the important cereal crops. The subject has been thoroughly reviewed by Okon et al. (2015). The idea of exploiting free-living diazotrophs for crop production is not new. Throughout the 1950s and 60s, the Russians treated vast acreages of wheat with *Azotobacter* in a formulation called "azotobacterin" in hopes of increasing yields without applications of fertilizer nitrogen. This practice of "bacterization" met with little success. A small percentage of the experiments showed increased growth due to inoculation and, when increases were observed, they were on the order of 10% or just barely detectable.

Okon and Labandera (1994) summarized the results of 20 years of field inoculation experiments with the bacterium *Azospirillum*. They concluded that many of the experiments (60%–70%) led to statistically significant yield responses on the order of 5%–30%. They further concluded that successful experiments "appeared to be those in which researchers have paid special attention to the optimal numbers of cells of *Azospirillum* in the inoculant, using inoculation methods where the optimal number of cells remained viable and available to colonize the roots." More recently, Okon et al. (2015) reported that in most grain and forage crops the contribution of nitrogen fixation by *Azospirillum* has been no more than 10 kg N ha^{-1} $year^{-1}$.

Although some results are encouraging for the use of bacterial inoculants to increase or maintain crop yields, it must be pointed out that the scientific literature on the subject tends to be slanted by the traditional bias against publishing papers reporting negative results. Thus, experiments where no inoculation responses are observed tend not to be published. Another important point to be learned from the large body of work dealing with inoculation of plants with *Azospirillum* is that it is now widely accepted that increases in plant growth due to inoculation with the bacterium are likely not due to increased N_2 fixation. Rather, they may be attributed to the production of plant growth-promoting hormones (e.g., auxins, gibberellins, and cytokinins) or to a variety of plant physiological responses. For example, inoculated plants are reported to be more effective in taking up inorganic nutrients, and it has been reported that the bacteria applied as inoculants elicit changes in the permeability of the membranes of root cells. In this context, *Azospirillum* is not only a nitrogen-fixing associate of the roots of many plant species but also has the attributes of other so-called **plant-growth-promoting rhizobacteria (PGPR**; Chapters 11 and 23). Today, researchers continue to look for microbes that exert beneficial effects on plants. Whether we will see the development of highly effective inoculants for agricultural crops is still a "work in progress" and it should be viewed with guarded optimism.

TABLE 16.8 Range of rates of BNF measured under field conditions by different diazotrophic organisms and their associations.

Diazotrophic bacteria and their associations	N_2 fixed (kg N $ha^{-1}year^{-1}$)
Free-living, N_2-fixing bacteria	
Soil, decaying wood and litter, biological soil crusts	1–10
Associative N_2-fixing bacteria	
Azolla-cyanobacteria	33
Endophytic, root associations	0–15
Endophytic, sugarcane association	25
Legume-rhizobial symbioses	
Grain and oilseed crops (peas, chickpeas, lentils, soybeans, etc.)	20–180
Forage crops (alfalfa, clover, vetch, etc.)	110–230
Trees (*Acacia, Gliricidia, Leucaena, Sesbania*, etc.)	50–300
Actinorhizal-Frankia symbioses	
Alder	50–300
Casuarina	75
Ceanothus	10–100

Bottomley and Myrold (2015). With permission. See also Stevenson (1982).

by the level of combined nitrogen (e.g., ammonium and nitrate) in the vicinity of the cell. Remember also that large amounts of energy and therefore large amounts of substrates are needed to reduce dinitrogen. Next, consider the conditions that might be encountered around the roots of a plant growing in a normally well-aerated (i.e., not waterlogged) soil. In the associative grass-bacteria systems, the diazotrophic bacteria live on or in the root or in the rhizosphere soil immediately surrounding the root. They are encouraged there by the liberation of substrates from the "host" root (refer to Fig. 16.4).

The oxygen concentration around roots can vary considerably but, in general, it is high enough that some limitation on N_2 fixation would result from the effect of molecular oxygen on nitrogenase. A secondary effect of the oxygen is that the diazotrophs would respire more carbon to protect the nitrogenase and reduce the efficiency of N_2 fixation. On the other hand, the formation of microsites with low oxygen concentrations results from the consumption of oxygen by root and microbial respiration so that some diazotrophs probably benefit from the reduced oxygen concentrations. Alexander and Zuberer (1989a) showed that nitrogenase activity associated with roots of corn and sorghum was greatest at oxygen concentrations ranging from 1.3% to 2.1% (1.3–2.1 kPa) and markedly lowered as the oxygen concentration approached 5%. They demonstrated that plants whose roots were exposed to $^{15}N_2$ at the low oxygen concentration contained more than 200 times as much ^{15}N as plants whose roots were exposed to air (Alexander and Zuberer, 1989b). Thus, we see that inhibition of nitrogenase activity by oxygen is reduced as the oxygen content of the environment decreases but that even at values approaching 5% the activity was nearly shut down.

Since nitrogenase activity is also closely regulated by combined nitrogen, the bacteria will also only fix dinitrogen when it is absolutely necessary. Thus, nitrogen levels in the soil solution around roots of fertilized crop plants may be enough to cause limitations on root-associated N_2 fixation. Dinitrogen fixation would be expected to occur only in areas where plant and microbial uptake (by nondiazotrophs) depleted the available supply of soil nitrogen to levels sufficient to derepress nitrogenase synthesis. This may not be unlikely in grasslands where dense rooting and microbial colonization of roots consistently keep the supply of available nitrogen at very low levels, but the prospects for high rates of N_2 fixation are limited in well fertilized soils.

Perhaps the most significant constraint on root- and rhizosphere-associated diazotrophs is lack of an adequate supply of readily available carbon sources coupled with the extremely high competition for any carbon liberated from the root. The rhizosphere is not populated exclusively with diazotrophic bacteria. In fact, they usually comprise about 1%–10% of the recoverable rhizosphere population. Competition among these organisms for the limited supply of available carbon can place severe restrictions on N_2 fixation in the root zone. However, if the diazotroph can become established inside the root as an endophyte, the competition for substrates is lessened, particularly for cells in the vascular elements of the plant (see Box 16.4 on N_2 fixation in sugarcane). Many studies have shown that N_2 fixation in the rhizosphere is carbon limited. For example, roots exhibiting little or no acetylene reduction by associative bacteria showed very high rates shortly after the addition of available carbon sources (sugars or organic acids) to the root systems. Many studies have also shown marked stimulation of N_2 fixation upon addition of readily available carbon sources to soils. The addition of the ready supply of carbon and the resulting immobilization of combined soil N make conditions favorable for high rates of N_2 fixation. Even incorporation of straw into soil can support some biological N_2 fixation.

To put the limitation of carbon supply on N_2 fixation into perspective, consider a statement from Giller and Day (1985) concerning estimation of maximal possible rates of N_2 fixation for a given amount of root carbon (see also Jones et al., 2003):

"If one assumes that:

- the efficiency of conversion of root carbon is 10 g C g N^{-1} fixed,
- dinitrogen-fixing bacteria comprise 10% of the total rhizosphere population (a high estimate) and acquire organic carbon in proportion to their numbers,
- all the carbon translocated below ground is equally available for use by all bacteria,

then:

- if 1500 kg C ha^{-1} are translocated below ground as estimated for a wheat crop, the maximum potential N_2 fixation is 15 kg N ha^{-1}."

Recall that most estimates of N_2 fixation in grasslands range from 5 to 25 kg N ha^{-1}. It is clear that conditions are conducive for N_2 fixation in the rhizosphere, at least part of the time, during the growing season. It is also likely that the integration of biotic and abiotic factors (Table 16.7) provides favorable microsites for N_2 fixation to occur at relatively high rates for short periods or at low rates for longer periods.

The factors (biotic and abiotic) affecting N_2 fixation by bacteria associated with belowground plant parts are summarized in Fig. 16.6 and Table 16.7. Remember that the bacteria

BOX 16.4

Dinitrogen-fixing rice and sugarcane

Within the last 4 decades, a great deal of effort has been expended in looking for N_2-fixing plant-bacteria associations capable of providing agriculturally significant amounts of fixed N for crop production. The optimistic reports of the early 1970s suggesting high rates of N_2 fixation associated with roots of grasses were soon tempered by numerous reports that failed to corroborate the earlier findings. However, from this large body of work, scientists have been able to learn much about the nature of the associations between roots of grasses and a large variety of free-living, N_2-fixing bacteria (diazotrophs). What they learned is that there appear to be some systems where free-living diazotrophs can be of significant benefit to their "host" plants.

Two crops that derive meaningful benefit from their diazotrophic associates are rice grown under flooded conditions and sugarcane. Interestingly, both crops have been observed to be able to produce appreciable yields in the absence of added fertilizers. The ability of plants to grow well without exogenous fertilizers raises the suspicion that these plants might somehow be benefiting from N_2-fixing bacteria.

Consider the case for N_2 fixation in wetland (paddy) rice culture. Rice is the major staple crop for two thirds of the world's people. For centuries in Asian countries, rice has been cultured without the use of fertilizers other than small inputs of human and animal manures, yet the crop bears good yields. Scientists suspected for a long time that N_2-fixing bacteria might somehow be involved in providing fixed N for the rice plant. As it turns out, because of the tolerance of rice plants for growth in flooded, often anoxic soils, they derive benefit from a variety of N_2-fixing microbes including free-living heterotrophic bacteria, N_2-fixing cyanobacteria, phototrophic bacteria, and a symbiotic partnership between the floating aquatic fern, *Azolla*, and the heterocystous cyanobacterium, *Anabaena*. The flooded conditions of the rice paddy overcome some of the environmental limitations on N_2-fixing bacteria and cyanobacteria. First, the floodwater is a much better environment for growth of algae and cyanobacteria than is an upland "dry" soil. Second, bacteria in the mud and associated with roots of rice plants find the low-oxygen environment protective of their nitrogenase enzyme. Thus, the oxygen poisoning of nitrogenase is at least partially alleviated by the naturally reduced oxygen status of the muds and floodwater. Third, before the rice plants grow tall enough to shade the paddy surface, considerable light is available for growth of the diazotrophic photosynthetic bacteria and cyanobacteria. It has been estimated that rice may gain from 30 to 50 kg N ha^{-1} per growing season through the combined activities of these free-living microbes. The *Azolla-Anabaena* symbiosis (discussed further in the next chapter) can provide from 150 to 300 kg N ha^{-1} per growing season. Researchers have also reported a unique association of *Rhizobium leguminosarum* bv *trifollii* and rice plants (Yanni and Dazzo, 2015). They reported very significant increases in rice growth and yield resulting from inoculation and that these benefits "more likely involve rhizobial modulation of the plant's root architecture for more efficient acquisition of certain soil nutrients

(*Continued*)

BOX 16.4 (*Cont.*)

[e.g. N, phosphorus (P), potassium (K), magnesium (Mg), calcium (Ca), zinc (Zn), sodium (Na) and molybdenum (Mo)] rather than biological N_2 fixation."

While sugarcane is not grown in flooded soils, it too has been observed to yield well with seemingly little input of fertilizer nitrogen. During the 1980s Dr. Johanna Döbereiner and others working in Brazil discovered that some cultivars of sugarcane contained a diazotrophic bacterium that they named *Acetobacter diazotrophicus* (later renamed *Gluconacetobacter diazotrophicus*). The organism can be isolated from the internal tissues of roots, stems, and leaves. It is unique in that it will grow in a 30% sucrose solution, fixes N_2 optimally at pH 5.5 (and less well at values approaching pH 3) and, apparently has an N_2-fixing capability more tolerant of oxygen than other free-living bacteria. Researchers in Brazil and elsewhere have since confirmed the unique association between this bacterium and sugarcane. The bacterium has been observed in the intercellular spaces as well as in the xylem elements of sugarcane stems and it may achieve numbers on the order of 1–10,000,000 cells g^{-1} of these tissues and 1–10,000 cells mL^{-1} of fluid in the intercellular spaces (James et al., 1994; Dong et al., 1994). Occupying such a unique location, the organism is poised to intercept a large supply of sugar to drive N_2 fixation and is free of the competition from other bacteria that organisms in the rhizosphere must deal with! Early estimates suggested that some sugarcane cultivars might derive as much as 100−150 kg N ha^{-1} from this unique association. However, more recent research indicates that sugarcane derives up to 40 kg N ha^{-1} $year^{-1}$ from these associations and the benefits vary among cultivars of the crop (Urquiaga et al., 2012, Antunes et al., 2019). Moreover, there is some question as to whether yield increases are due to nitrogen fixation or other factors contributed by microbial inoculants (Schultz et al., 2014).

Considerable world-wide research efforts are ongoing in search of ways to enhance crop production through the application of beneficial microbes. Rice and sugarcane are two major crops that have been proven to benefit from a variety of plant-associated microbial processes. See Dent and Cocking (2017) for some perspectives on extending nitrogen fixation to cereals and other nonleguminous plants.

depend on the "host" plant for provision of carbon sources, thus anything that affects the photosynthesis of the plant indirectly affects the bacteria associated with the roots or other belowground structures such as nodules (see also Chapter 17). Also important are the combined effects of the soil variables (moisture and redox conditions, aeration, pH, temperature, and nitrogen content) on plant and microbial growth. By studying the simple models presented in Figs. 16.4 and 16.6, we can develop an appreciation of how all factors must come together at appropriate levels to allow for expression of N_2 fixation by bacteria in the roots or rhizospheres of nonleguminous plants like the grasses. We can also gain an appreciation of the many factors that are "taken care of" by the host plants in the exquisite root-nodule symbioses discussed in the next chapter.

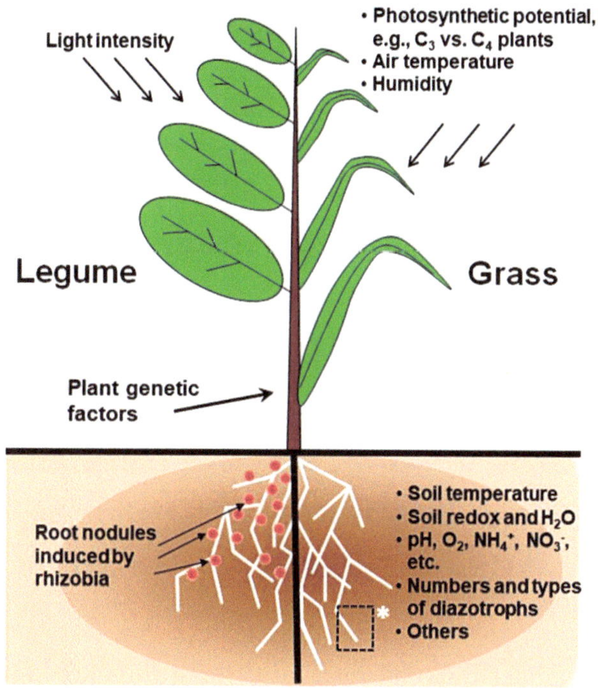

FIGURE 16.6 **Main factors affecting nitrogenase activity belowground include: light intensity, potential of the photosynthetic apparatus, for example, C_3 versus C_4 plants, air temperature (19°C min.), air water deficit (affects stomatal activity and water balance), soil temperature, soil redox and water content (18% water content considered an approximate minimum), other soil factors (e.g., pH, O_2, and N content), plant genetic factors, and the nature and numbers of the diazotrophs associated with roots.** * The *boxed area* is represented in Fig. 16.4; the rhizosphere model.

Methods for measuring dinitrogen fixation

The agronomic and ecological importance of biological N_2 fixation has driven intensive study of the process for the past century and into the next. Early studies of biological N_2 fixation were limited by the lack of techniques sensitive enough to measure nitrogen gains in plants or to measure the activity of the enzyme complex itself. As developments in electronics have evolved rapidly so too has the development of sophisticated instrumentation for measurement of biological processes including N_2 fixation. This section is not intended to give a detailed chronicling of the development of techniques for measuring N_2 fixation. Rather, the methods routinely used for measurement of N_2 fixation or nitrogenase activity are introduced. Detailed reviews of the subject are available elsewhere (Weaver and Danso, 1994 or Sprent and Sprent, 1990).

Measurement of dinitrogen fixation by the nitrogen difference method

It is possible to estimate N_2 fixation by comparing the yields and nitrogen content of plants grown with and without N_2-fixing bacteria. One example is comparing yields of nodulated and nonnodulated legumes grown under similar conditions of soil N. The nitrogen in the nonfixing plant is a measure of the N acquired from soil and is subtracted from the nitrogen of the fixing plant and the difference is attributed to N_2 fixation. Weaver and Danso (1994) indicate that the sensitivity of the method is such that an increase of about 20 kg N ha^{-1} would be required for detection of significant differences. The method is limited to the study of systems where N_2 fixation rates are high.

Measurement of nitrogen fixation by the acetylene reduction assay

The ability of the nitrogenase complex to reduce acetylene (C_2H_2) to ethylene (C_2H_4) has already been mentioned and forms the basis for the **acetylene reduction** (AR) **assay** (see Burris, 1974, for a description of the development of the method). The sensitivity of this method based on gas chromatography allowed detection of very low levels of nitrogenase activity. In fact, it was the sensitivity of the AR assay that facilitated attempts to measure N_2 fixation by grasses.

The reaction for the reduction of N_2 to ammonia is as follows:

$$N_2 + 8H^+ + 8e^- \rightarrow 2NH_3 + H_2.$$

The conversion of acetylene to ethylene occurs by the following reaction:

$$C_2H_2 + 2H^+ + 2e^- \rightarrow C_2H_4.$$

The reduction of N_2 (including reductant used in H_2 evolution) requires eight electrons, whereas the reduction of acetylene requires two electrons. Thus, reduction of four molecules of acetylene is equivalent to reduction of one molecule of N_2, that is, the ratio of acetylene reduced to nitrogen fixed is 4:1. In practice, this ratio has been shown to vary anywhere from 1.5 to 25:1, depending on the system being measured. Most investigators assume a ratio of 3 or 4:1.

In the AR assay, the system to be measured (e.g., whole plants, isolated roots, soil cores, and bacterial cultures) is exposed to an atmosphere containing 11% C_2H_2 and incubated under appropriate conditions. The concentration of acetylene is such that the nitrogenase complex is saturated with this substrate and N_2, the normal substrate (Fig. 15.2), is no longer reduced. Samples of the gas phase are periodically removed and injected into a gas chromatograph for quantification of ethylene (C_2H_4) production from acetylene.

The AR assay is not without limitations and careful attention must be paid to these in order to avoid gross over-estimations of N_2 fixation. Giller (1987) summarized the pitfalls of the AR assay for measurement of associative N_2 fixation. A few of these include:

- microbial production of ethylene, which is not from acetylene,
- oxidation of ethylene by other soil bacteria,
- nitrogen starvation during prolonged incubations due to inhibition of nitrogenase by acetylene, and
- failure to calibrate the AR versus N_2 fixation ratio appropriately.

Measurement of dinitrogen fixation by stable isotope (^{15}N) methods

The most definitive measurements of biological N_2 fixation make use of the stable, heavy isotope of nitrogen, ^{15}N. Incorporation of $^{15}N_2$ into microbial cells or plant tissues provides direct, quantitative proof of dinitrogen fixation. Chalk et al. (2017) provide a 75-year retrospective of the use of $^{15}N_2$ as a tracer of biological N_2 fixation. ^{15}N methods require access to a mass spectrometer and while the availability of these instruments is not as major a limitation as it once was, ^{15}N methods can still be costly, both because of the high costs of the ^{15}N-labeled N sources and the cost of analysis. Chalk and Craswell (2018) recently described the role of ^{15}N methodologies in quantifying biological N_2 fixation and the dynamics of N_2 fixation in agroecosystems in a thorough review of the subject.

The principal methods involved in studies using ^{15}N to measure N_2 fixation are:

- measurement of the direct incorporation of $^{15}N_2$ (labeled N_2) into plant or microbial cells,
- isotope dilution methods in which the content of ^{15}N in plant tissue is measured and the ratio of ^{15}N to ^{14}N is calculated, and
- natural abundance of ^{15}N in soils or plants.

The first method is straightforward. Samples are exposed to an atmosphere of about 10% $^{15}N_2$, usually in a balance of argon or helium (with sufficient oxygen and carbon dioxide) to eliminate competition from $^{14}N_2$. Control plants not exposed to $^{15}N_2$ are included to measure background ^{15}N. Following incubation, the samples (cells or ground plant materials) are digested in acid with heating and the ^{15}N content of the materials is determined using a mass spectrometer. Detection of ^{15}N in tissues or cells provides definitive proof of N_2 fixation and allows very accurate quantification of the amount of N_2 fixed. However, care must be taken to ensure that the $^{15}N_2$ is free of contamination by other labeled N compounds (Chalk and Ji-Zheng He, 2017; MacKellar et al., 2016; Dabundo et al., 2014).

The **isotope dilution** and **natural abundance** methods rely on the fact that in nature the ratio of $^{15}N/^{14}N$ is remarkably constant. ^{15}N and ^{14}N constitute about 0.3663% and 99.6337% of the N in the atmosphere, respectively. A substance containing more ^{15}N is said to be ^{15}N-enriched. Such materials can be exploited to measure N_2 fixation in the field. In practice, the soil is fertilized to a low level with the ^{15}N-enriched fertilizer and the plants are grown to maturity or a suitable stage for harvest. The ^{15}N content in the tissues is analyzed and the nitrogen derived from the atmosphere can be calculated. The "fixing" plant will derive part of its N from the atmosphere and will contain more ^{14}N than the soil N thus the ^{15}N in the fixing plant is "diluted" with the ^{14}N from atmospheric N_2. The ^{15}N natural abundance method is based on the same principle as the isotope dilution method. Natural materials contain nitrogen that is naturally enriched with ^{15}N because of isotope discrimination during biological transformations of nitrogen in soil. Plants containing less ^{15}N than the natural abundance therefore can be suspected of supporting N_2 fixation. Weaver and Danso (1994) presented the assumptions and formulae for calculating the amount of N fixed and the rationale for the calculations.

Summary

In this chapter you have been introduced to some of the fundamentals of the process of biological N_2 fixation and the nature of the enzyme complex, nitrogenase, which mediates the conversion of N_2 to $2NH_3$. The process is unique in that only prokaryotic microbes, so far as is known, possess this enzyme system. You have learned that just as it is expensive to fix dinitrogen industrially, it is also expensive biologically for the microbes that fix N_2. Therefore, high rates of biological N_2 fixation occur only where there is a ready supply of available energy (light for phototrophs; organic C for organotrophs). For this reason, the organotrophic N_2-fixing microbes are usually found in association with plants or in soils receiving organic C additions.

You have learned that despite the cumbersome nature of the nitrogenase enzyme complex, about two thirds of the global supply of fixed N comes through biological N_2 fixation; mostly through the legume-rhizobia symbiosis you will learn about more thoroughly in the next chapter. While the contributions of fixed N from the nonsymbiotic diazotrophs are generally low (5–25 kg N ha^{-1} $year^{-1}$), they are significant in areas where no other source of N is available. The associative symbioses between grasses and free-living diazotrophs continue to receive attention not only for their interesting biology but also with the objective of enhancing such associations for practical use in sustaining agricultural and natural ecosystems.

It is difficult to foresee what breakthroughs will spring from the continued investigations of nitrogenase and the microbes that possess it. Imagine if chemists, biochemists, and microbiologists do someday figure out how to mimic the action of nitrogenase in huge tanks to fix dinitrogen without the high costs of the fossil fuel currently needed. Imagine also if molecular biologists and plant breeders someday figure out how to get N_2-fixing root nodules on nonleguminous plants such as wheat, rice, and corn or to establish a functional N_2-fixing enzyme system in the leaves or roots of these crops. With modern genetic modification techniques, we may yet see the development of cereal crops that can fix their own nitrogen (what some have called the "Holy Grail" of N_2 fixation); a feat that would have a major impact in supplying food for a growing world population and a lessening of the negative consequences of nitrogen fertilizers.

Then, a whole new set of questions would arise. For example, can these plants provide the carbon necessary to support N_2-fixing nodules or enzymes and still produce abundant grain? These are dreams perhaps, but it is reasonably certain that dinitrogen fixation has a strong future in agricultural research and production, and in the maintenance of our relatively unmanaged ecosystems such as forests and grasslands.

Study questions

1. Discuss the significance of biological dinitrogen fixation in terms of its relative contribution to the world's supply of "fixed" nitrogen.
2. Write the biochemical reaction for the process of biological N_2 fixation.
3. What kinds of microorganisms can carry out biological N_2 fixation? Why are these microbes called "diazotrophs"?
4. Describe, in general terms, the nitrogenase enzyme complex.
5. What are some unusual properties of the nitrogenase enzyme complex?
6. What are some problems, inherent in the nitrogenase system, which limit our abilities to use nonsymbiotic N_2-fixing bacteria to improve the yields of agriculturally important crop plants?
7. Compare and contrast the processes of nonsymbiotic and symbiotic N_2 fixation.

8. What is an "associative symbiosis"? How do rates of dinitrogen fixation compare in associative symbioses and root nodule symbioses between legumes and *Rhizobium*?
9. Why is it not a simple matter to genetically engineer higher plants to "fix" their own dinitrogen for growth?
10. Why do rice and sugarcane benefit from the activities of nonsymbiotic N_2-fixing bacteria more so than other nonleguminous plants?
11. Given that a free-living diazotroph can fix about 15 mg of N g^{-1} of carbon source, how much carbon would have to be supplied to fix 100 lbs. (about 45 kg) of nitrogen? Give your answer in pounds and kilograms. Show your calculations.

References

Alexander, D.B., Zuberer, D.A., 1989a. Impact of soil environmental factors on rates of N_2 fixation associated with intact maize and sorghum plants, in: Skinner, F.A., Boddey, R.M., Fendrik, I. (Eds.), Nitrogen Fixation with Non-Legumes. Kluwer Academic Press, Dordrecht, The Netherlands, pp. 273–285.

Alexander, D.B., Zuberer, D.A., 1989b. $^{15}N_2$ fixation by bacteria associated with maize roots at a low partial O_2 pressure. Appl. Environ. Microbiol. 55, 1748–1753.

Antunes, J.E.L., Freitas, A.D.S., Oliveira, L.M.S., Lyra, M.C.C.P., Fonseca, M.A.C., Santos, C.E.R.S., Oliveira, J.P., Araújo, A.S.F., Figueiredo, M.V.B., 2019. Sugarcane inoculated with endophytic diazotrophic bacteria: Effects on yield, biological nitrogen fixation and industrial characteristics. An. Acad. Bras Cienc. 91, e20180990, doi 10.1590/0001-3765201920180990.

Bishop, P.E., Premakumar, R., 1992. Alternative nitrogen fixation systems, in: Stacey, G., Burris, R.H., Evans, H.J. (Eds.), Biological Nitrogen Fixation. Chapman & Hall, NY, pp. 736–762.

Bottomley, P.J., Myrold, D.D., 2015. Biological N inputs, in: Paul, E.A. (Ed.), Soil Microbiology, Ecology and Biochemistry. Academic Press, NY, pp. 447–470.

Burén, S., Rubio, L.M., 2018. State of the art in eukaryotic nitrogenase engineering. FEMS Microbiol. Lett. 365 (2), 1–9.

Burén, S., Jiménez-Vicente, E., Echavarri-Erasun, C., Rubio, L.M., 2020. Biosynthesis of nitrogenase cofactors. Chem. Rev. 120 (12), 4921–4968, DOI: 10.1021/acs.chemrev.9b00489.

Burns, R.C., Hardy, R.W.F., 1975. Nitrogen Fixation in Bacteria and Higher Plants. Springer-Verlag, NY.

Burris, R.H., 1974. Methodology, in: Quispel, A. (Ed.), The Biology of Nitrogen Fixation. American Elsevier Publishing Company, NY, pp. 10–33.

Burris, R.H., Roberts, G.P., 1993. Biological nitrogen fixation. Ann. Rev. Nutr. 13, 317–335.

Chalk, P.M., Craswell, E.T., 2018. An overview of the role and significance of ^{15}N methodologies in quantifying biological N_2 fixation (BNF) and BNF dynamics in agro-ecosystems. Symbiosis 75, 1–16, https://doi.org/10.1007/s13199-017-0526-z.

Chalk, P.M., He, J.-Z., Peoples, M.B., Chen, D. 2017. $^{15}N_2$ as a tracer of biological N_2 fixation: A 75-year retrospective. Soil Biol. Biochem. 106, 36–50.

Cleveland, C.C., Townsend, A.R., Schimel, D.S., Fisher, H., Howarth, R.W., Hedin, L.O., Perakis, S.S., Latty, E.F., Von Fischer, J.C., Elseroad, A., Wasson, M.F., 1999. Global patterns of terrestrial N fixation. Global Biogeochem. Cycles 13 (2), 623–645.

de Bruijn, F.J., 2015. The quest for biological nitrogen fixation in cereals: A perspective and prospective. In: de Bruijn, F.J. (Ed.), Biological Nitrogen Fixation. John Wiley & Sons, Inc, NY, doi:10.1002/9781119053095.ch108.

Dabundo, R., Lehmann, M.F., Treibergs, L., Tobias, C.R., Altabet, M.A., Moisander, P.H., Granger, J., 2014. The contamination of commercial $^{15}N_2$ gas stocks with ^{15}N–labeled nitrate and ammonium and consequences for nitrogen fixation measurements. PLoS One. 9(10), e110335. https://doi.org/10.1371/journal.pone.0110335.

Dent, D., Cocking, E., 2017. Establishing symbiotic nitrogen fixation in cereals and other non-legume crops: The greener nitrogen revolution. Agric. Food Secur. 6, 7, 10.1186/s40066-016-0084-2.

Dilworth, M.J., 1966. Acetylene reduction by nitrogen-fixing preparations from *Clostridium pasteurianum*. Biochim. Biophys. Acta. 127 (2), 285–294.

Dixon, R.O.D., Wheeler, C.T., 1986. Nitrogen Fixation in Plants. Chapman & Hall, NY.

Dong, Z., Canny, M.J., McCully, M.E., Roberedo, M.R., Cabadilla, C.F., Ortega, E., Rodés, R., 1994. A nitrogen-fixing endophyte of sugarcane stems. A new role for the apoplast. Plant Physiol. 105, 1139–1147.

Dos Santos, P.C., Fang, Z., Mason, S.W., et al., 2012. Distribution of nitrogen fixation and nitrogenase-like sequences amongst microbial genomes. BMC Genom. 13, 162, https://doi.org/10.1186/1471-2164-13-162.

Evans, H.J., Burris, R.H., 1992. Highlights in biological nitrogen fixation during the last 50 years, in: Stacey, G., Burris, R.H., Evans, H.J. (Eds.), Biological Nitrogen Fixation. Chapman & Hall, NY, pp. 1–42.

Fisher, K., Newton, W.E., 2002. Nitrogen fixation: A general overview, in: Leighton, G.J. (Ed.), Nitrogen Fixation at the Millennium. Elsevier Science B.V., pp. 1–34.

Fowler, D., Coyle, M., Skiba, U., Sutton, M.A., Cape, J.N., Reis, S., Sheppard, L.J., Jenkins, A., Grizzetti, B., Galloway, J.N., Vitousek, P., Leach, A., Bouwman, A.F., Butterbach-Bahl, K., Dentener, F., Stevenson, D., Amann, M., Voss, M., 2013. The global nitrogen cycle in the twenty-first century. Philos. Trans. R. Soc. Lond. B Biol. Sci. 368, 20130164, http://dx.doi.org/10.1098/rstb.2013.0164.

Gadkari, D., 2004. Super-oxide Dependent Nitrogenase. In: Smith, B.E., Richards, R.L., Newton, W.E. (Eds.), Catalysts for Nitrogen Fixation: Nitrogenases, Relevant Chemical Models, and Commercial Processes, Kluwer Academic Publishers, pp. 309–332.

Giller, K.E., 1987. Use and abuse of the acetylene reduction assay for measurement of "associative" nitrogen fixation. Soil Biol. Biochem. 19, 783–784.

Giller, K.E., Day, J.M., 1985. Nitrogen fixation in the rhizosphere: Significance in natural and agricultural systems, in: Fitter, A.H. (Ed.), Ecological Interactions in Soil. Blackwell Scientific Publications, Oxford, England, pp. 127–147.

Hill, S., 1992. Physiology of nitrogen fixation in free-living heterotrophs, in: Stacey, G., Burris, R.H., Evans, H.J. (Eds.), Biological Nitrogen Fixation. Chapman & Hall, NY, 87–131.

IFA, 2014. IFA Fertilizer Facts "Ammonia Production: Moving Towards Maximum Efficiency and Lower GHG Emissions." International Fertilizer Association (IFA), Paris, France.

James, E.K., Reis, V.M., Olivares, F.L., Baldani, J.I., Döbereiner, J., 1994. Infection of sugarcane by the nitrogen-fixing bacterium *Acetobacter diazotrophicus*. J. Exp. Bot. 45, 756–766.

Jones, D., Farrar, J., Giller, K., 2003. Associative nitrogen fixation and root exudation—What is theoretically possible in the rhizosphere? Symbiosis 35, 1–3.

Newton, W.E., 2015. Recent advances in understanding nitrogenases and how they work, in: de Bruijn, F.J. (Ed.), Biological Nitrogen Fixation, vol. 1, first ed., John Wiley & Sons, Inc., NY.

MacKellar, D., Lieber, L., Norman, J.S., Bolger, A., Tobin, C., Murray, J.W., Oksaksin, M., Chang, R.L., Ford, T.J., Nguyen, P.Q., Woodward, J., Permingeat, H.R., Joshi, N.S., Silver, P.A., Usadel, B., Rutherford, A.W., Friesen, M.L., Prell, J., 2016. *Streptomyces thermoautotrophicus* does not fix nitrogen. Nat. Sci. Rep. 6, 20086, doi: 10.1038/srep20086.

Okon, Y., Houchins, J.P., Albrecht, S.L., Burris, R.H., 1977. Growth of *Spirillum lipoferum* at constant partial pressures of oxygen, and the properties of its nitrogenase in cell-free extracts. J. Gen. Microbiol. 98, 87–93. doi: 10.1099/00221287-98-1-87.

Okon, Y., Labandera, C., 1994. Agronomic application of *Azospirillum*: An evaluation of 20 years worldwide field inoculation. Soil Biol. Biochem. 26, 1591–1601.

Okon, Y., Labandera-Gonzales, C., Lage, M., Lage, P., 2015. Agronomic applications of *Azospirillum* and other PGPR, in: de Bruijn, F.J. (Ed.), Biological Nitrogen Fixation, vol. 2, first ed., John Wiley & Sons, Inc., NY.

Oldroyd, G.E.D., Dixon, R., 2014. Biotechnological solutions to the nitrogen problem. Curr. Opin. Biotechnol. 26, 19–24.

Pankievicz, V.C.S., Irving, T.B., Maia, L.G.S., Ané, J.M., 2019. Are we there yet? The long walk towards the development of efficient symbiotic associations between nitrogen-fixing bacteria and non-leguminous crops. BMC Biol. 17, 99.

Postgate, J.R., 1994. The inertness of nitrogen, in: The Outer Reaches of Life. Cambridge University Press, Cambridge, England, pp. 137–153.

Postgate, J.R., 1998. Nitrogen Fixation, third ed. Cambridge University Press, Cambridge, England.

Ribbe, M., Gadkari, D., Meyer, O., 1997. N_2 fixation by *Streptomyces thermoautotrophicus* involves a molybdenum-dinitrogenase and a manganese-superoxide oxidoreductase that couples N_2 reduction to the oxidation of superoxide produced from O_2 by a molybdenum CO dehydrogenase. J. Biol. Chem. 272, 26627–26633.

Rubio, L.M., Ludden, P.W., 2002. The gene products of the *nif* regulon, in: Jeffery Leighton, G. (Ed.), Nitrogen Fixation at the Millennium. Elsevier Science B.V., pp. 101–136.

Rubio, L.M., Ludden, P.W., 2005. Maturation of nitrogenase: A biochemical puzzle. J. Bacteriol. 187, 405–414.

Schultz, N., Alves da Silva, J., Silva Sousa, J., Monteiro, R.C., Pedula Oliveira, R., Almeida Chaves, V., Pereira, W., Flores da Silva, M., Baldani, J.I., Boddey, R.M., Reis, V.M., Urquiaga, S., 2014. Inoculation of sugarcane with diazotrophic bacteria. Rev. Bras. de Ciên. do Solo 38 (2), 407–414, https://doi.org/10.1590/S0100-06832014000200005.

Seefeldt, L.C., Yang, Z.Y., Duval, S., Dean, D.R., 2013. Nitrogenase reduction of carbon-containing compounds. Biochim. Biophys. Acta 1827 (8-9), 1102–1111, doi:10.1016/j.bbabio.2013.04.003.

Smil, V., 1999. Nitrogen in crop production: An account of global flows. Global Biogeochem. Cycles 13, 647–662.

Sprent, J.I., Sprent, P., 1990. Nitrogen Fixing Organisms. Pure and Applied Aspects. Chapman & Hall, NY.

Stevenson, F.J., 1982. Origin and distribution of nitrogen in soil, in: Stevenson, F.J. (Ed.), Nitrogen in Agricultural Soils, Agronomy No. 22, American Society of Agronomy, Madison, WI, pp. 1–42.

Urquiaga, S., Xavier, R.P., de Morais, R.F., et al., 2012. Evidence from field nitrogen balance and ^{15}N natural abundance data for the contribution of biological N_2 fixation to Brazilian sugarcane varieties. Plant Soil 356, 5–21, https://doi.org/10.1007/s11104-011-1016-3.

Vicente, E.J., Dean R.F., 2017. Keeping the nitrogen-fixation dream alive. PNAS 114 (12), 3009–3011.

Weaver, R.W., Danso, S.K.A., 1994. Dinitrogen fixation. In: Weaver, R.W., Angle, S., Bottomley, P., Bedzdicek, D., Smith, S., Tabatabai, A., Wollum, A. (Eds.), Methods of Soil Analysis, Part 2. Microbiological and Biochemical Properties. Soil Science Society of America, Book Series, No. 5. Madison, WI, pp. 1019–1045.

Yanni, Y.G., Dazzo, F.B., 2015. Occurrence and ecophysiology of the natural endophytic Rhizobium–rice association and translational assessment of its biofertilizer performance within the Egypt Nile Delta, in: de Bruijn, F.J. (Ed.), Biological Nitrogen Fixation, John Wiley & Sons, Inc., NY. doi:10.1002/9781119053095.ch111.

Young, J.P.W., 1992. Phylogenetic classification of nitrogen-fixing organisms, in: Stacey, G., Burris, R.H., Evans, H.J. (Eds.), Biological Nitrogen Fixation. Chapman & Hall, NY, pp. 43–86.

Further reading

de Bruijn, F.J. (Ed.), 2015. Biological Nitrogen Fixation, vols. 1 and 2, John Wiley & Sons, Inc., NY.

Döbereiner, J., Pedrosa, F.O., 1987. Nitrogen-fixing bacteria in nonleguminous crop plants. Science and Technology Publishing, Madison, WI.

Elmerich, C., Zimmer, W., Vielle, C., 1992. Associative nitrogen-fixing bacteria in: Stacey, G., Burris, R.H., Evans, H.J. (Eds.), Biological Nitrogen Fixation. Chapman & Hall, NY, pp. 212–258.

Ormeño-Orrillo, E., Hungria, M., Martinez-Romero, E., 2013. Dinitrogen-fixing prokaryotes. In: Rosenberg, E., DeLong, E.F., Lory, S., Stackebrandt, E., Thompson, F. (Eds.), The Prokaryotes. Springer, Berlin, Heidelberg.

Marchal, K., Vanderleyden, J., 2000. The "oxygen paradox" of dinitrogen-fixing bacteria. Biol. Fertil. Soils 30, 363–373.

Santi, C., Bogusz, D., Franche, C., 2013. Biological nitrogen fixation in non-legume plants. Ann. Bot. 111, 743–767.

Schlesinger, W.H., 1997. Biogeochemistry, second ed. An Analysis of Global Change. Academic Press, NY.

Zuberer, D.A., 1990. Soil and rhizosphere aspects of N_2-fixing plant-microbe associations, in: Lynch, J.M. (ed.), The Rhizosphere. John Wiley & Sons, NY, pp. 317–353.

CHAPTER

17

Biological dinitrogen fixation: symbiotic

Janice E. Thies
Cornell University, Ithaca, NY, United States

"Both plant and bacteria have co-evolved for thousands of years to achieve a delicately poised, precisely balanced, and amazingly intricate symbiotic relationship, the expression of which allows nitrogen to be taken from the atmosphere and used directly for plant growth and seed production."
— ***E.L. Schmidt, University of Minnesota***

A **mutualistic symbiosis** is an association between two organisms from which each derives benefit. It is usually a longer-term relationship and in the case of symbiotic dinitrogen (N_2) fixation often involves a special structure to house the microbial partner. Each N_2-fixing symbiotic association involves an N_2-fixing prokaryotic organism, the microsymbiont (e.g., *Rhizobium, Klebsiella, Frankia,* or *Anabaena*) and a eukaryotic, usually photosynthetic, host (e.g., leguminous or non-leguminous plant, water fern, or liverwort). Together, microorganism and host convert inert N_2 gas from the atmosphere into ammonia, benefiting the host, while the microsymbiont has a protective environment that is supplied with photosynthate and other nutrients for growth and N_2 fixation. Symbiotic N_2 fixation is the single greatest natural contributor to the global cycle of nitrogen. Important symbiotic associations in which dinitrogen is fixed include:

- leguminous plants and their associated rhizobia
- actinorhizal plants in symbiosis with *Frankia*
- the water fern *Azolla* and its microsymbiont *Anabaena* (*Nostoc*)
- the lichen symbiosis involving cyanobacteria

Estimates of the contribution of these different symbioses to the global nitrogen economy vary, but Herridge et al. (2008) projected recently that 50–70 Tg N year^{-1} is fixed in global agricultural systems. Crop and pasture legumes alone are grown on more than 250 million ha worldwide, where pasture and fodder legumes are estimated to fix about 12–25 Tg N year^{-1}, with pulses and oilseeds contributing over 21 Tg N year^{-1}. Rates of symbiotic N_2 fixation in legumes vary with host, microsymbiont, and environment, but rates as high as 600 kg N fixed

Principles and Applications of Soil Microbiology. http://dx.doi.org/10.1016/B978-0-12-820202-9.00017-4
Copyright © 2021 Elsevier Inc. All rights reserved.

ha^{-1} $year^{-1}$ have been reported in temperate clover pastures, and different grain legumes fix from 165 to 450 kg N ha^{-1} (Unkovich and Pate, 2000).

The direct availability of the N_2 fixed to the host allows it to grow in environments that are low in nitrogen and to minimize N losses through denitrification, volatilization, and leaching. This enhances the sustainability of agricultural and natural systems that are legume based. Greatest dependence on N_2 fixation is usually associated with small-holder farms where access to inputs may be limited; organic producers who use green manures rather than chemical fertilizers; and environments where rainfall is unreliable and extensive use of fertilizer is risky. With the world's population projected to increase to 9.8 billion by 2050, and with much of that increase to occur in developing countries where fertilizer use is most limited, a marked increase in overall N_2 fixation will be needed to maintain food security in many regions. In contrast, where fertilizer nitrogen is still relatively inexpensive, agriculture is intensive, or manures are land applied, N_2 fixation in agriculture will be important for reasons of soil quality and agricultural sustainability.

The symbiosis between legumes and rhizobia

Legumes have been used in crop rotations since the time of the Romans; Theophrastus (370–285 BC) stated "...beans are not a burdensome crop to the ground, they even seem to manure it....wherefore the people of Macedonia and Thessaly turn over the ground when it is in flower." However, it was not until detailed nitrogen balance studies became possible that legumes were shown to accumulate nitrogen from sources other than soil and fertilizer. Hellriegel and Wilfarth, (1886) associated the ability of legumes to convert dinitrogen from the atmosphere into compounds that could be used by the plants with the presence of swellings or **nodules** on the legume root (Fig. 17.1). They related this to the presence within the nodule of particular bacteria, now collectively known as **rhizobia**. It was a series of short but important steps to the isolation of rhizobia from nodules by Beijerinck in 1888 and then to the completion of **Koch's postulates** (Chapter 1), which demonstrated their ability to reinfect the legume and to fix dinitrogen in symbiosis.

Groupings of rhizobia and their separation into species

Each rhizobial **strain** or **nodule isolate** has a finite host range, nodulating certain legumes but not others. In earlier studies, this led to the concept of **cross-inoculation groups**, with legumes clustered according to the different rhizobia with which they formed nodules. Thus, rhizobia isolated from alfalfa (*Medicago sativa*) would also nodulate hosts in the genera *Melilotus* and *Trigonella*, and *vice versa*, but none of the *Medicago* or *Melilotus* rhizobia could nodulate *Trifolium* spp. (clovers). Host specificity is still important for characterizing rhizobia, but more recently other traits, particularly genome-sequence analyses, have assumed greater significance in their classification. There are a number of reasons for this, including many anomalous results. One study identified more than 500 examples of strains that were either **promiscuous** and nodulated legumes from other cross-inoculation groups, or else failed to nodulate all legumes in their particular cross-inoculation group. In addition, nodulation genes of rhizobia may be plasmid-borne (see Box 17.3). Strains that lose the symbiotic plasmid also lose

FIGURE 17.1 **Variations in nodule morphology.** (A) Cowpea, (B) Peanut, (C) *Sesbania*, (D) *Medicago*, and (E) Soybean. *Source: Part (A): Photo: D. Zuberer, used with permission; Part (B): Photo: Leah Ellman-Stortz, Texas A&M Univ. with permission; Part (C): E.K. James with permission; Part (D): Wikimedia Commons; Part (E): Photo, Mariangela Hungria, with permission.*

the ability to form nodules and yet are otherwise essentially identical to nodulating strains. Another anomaly is the bacterial strain NGR234, originally isolated from *Lablab purpureus* (the hyacinth bean) that nodulates at least 232 species of legumes in 112 genera, as well as the non-legume, *Parasponia andersonii* (Stanley and Cervantes, 1991). Even now, less than 15% of the roughly 19,000 species of legumes have been evaluated for nodulation. Lastly, when taxonomic methods based on differences in cell DNA or RNA were used, the groupings that resulted were often different from those obtained using host range. Some of the traits used in identifying rhizobia and in distinguishing among the different types are shown in Table 17.1. The minimal standards for describing new genera and species of rhizobia are given in de Lajudie et al. (2019) (Box 17.1). A key criterion for belonging to the rhizobia remains the capacity of a strain to form nodules on a given legume.

TABLE 17.1 Characteristics used in the phenotypic and phylogenetic characterization of rhizobia.

Phenotypic traits[a]

- Range of substrates usable as sources of energy (sugars, sugar alcohols, complex carbohydrates, etc.)
- Range of substrates usable as sources of nitrogen (amino acids, urea, nitrate, etc.)
- Resistance to specific antibiotics
- Electrophoretic mobility of different cell enzymes
- Tolerance to different stresses (salt, temperature, pH)
- Fatty acid methyl ester (FAME) analysis of cell membranes

Phylogenetic traits [b] (refer also to Chapter 4)

- 16S rRNA sequence analysis
- Pattern of banding of DNA restriction fragments (RFLPs, Chapter 4)
- Degree of hybridization with specific DNA probes
- *nif* and *nod* gene sequence analysis

[a] *Phenotypic traits are those that can be determined from cultured cells.*
[b] *Phylogenetic traits are those determined using cell DNA or RNA.*

BOX 17.1

Minimum standards for the description of new genera and species of rhizobia (de Lajudie et al., 2019)

- Provide a genome sequence of the proposed type strain
- Provide evidence that the strain differs from other species based on genome sequence comparisons
- Document genetic variation within the species with sequence data from several clearly different strains
- Describe phenotypic features and document their variation with data from a relevant set of representative strains
- Provide information on nodulation phenotypes, with relevant gene sequences
- Assign a name that conforms to the International Code of Nomenclature of Prokaryotes
- Validate the name by publication either directly in the International Journal of Systematic and Evolutionary Microbiology or in a validation list when published elsewhere
- Deposit the type strain in two international culture collections in separate countries

Identification and classification of the rhizobia continues to evolve. The group now includes two classes of bacteria, Alphaproteobacteria and Betaproteobacteria, 19 genera, and more than 240 species (Table 17.2). A significant gap between the fast-growing rhizobia, and the slow-growing *Bradyrhizobium* is recognized by most rhizobiologists. The so-called "fast-growing" rhizobia form colonies in 3–5 days on agar media and produce acidic metabolites, whereas the "slow-growing" rhizobia produce visible colonies in 7–10 days and produce alkaline metabolites. Surprisingly, some of the latter have been shown to grow as **hydrogen autotrophs** and produce **bacteriochlorophyll**. In addition, species of methyl-oxidizing bacteria in the betaproteobacteria have also been shown to nodulate legumes such as *Sesbania, Crotalaria,* and *Lotononis* (Moulin et al., 2001).

TABLE 17.2 Recognized families, genera, and representative species of rhizobia nodulating legumes. Legume genera in the *square brackets* refer to better-known hosts nodulated by each species of root-nodule bacteria. Common names are included for well-known legume genera. In several examples, different species of root-nodule bacteria can nodulate the same legume.[a]

- Alphaproteobacteria
  - Bradyrhizobiaceae[b]
    - *Bradyrhizobium*
      - *B. japonicum* [*Glycine*, soybean]
  - Brucellaceae
    - *Ochrobactrum*
      - *O. lupine* [*Lupinus*, lupin]
  - Hyphomicrobiaceae
    - *Devosia*
      - *D. neptuniae* [*Neptunia natans*]
  - Methylobacteriaceae
    - *Methylobacterium*
      - *M. nodulans* [*Crotalaria*]
    - *Microvirga*
      - *M. vignae* [*Vigna*, cowpea]
  - Phyllobacteriaceae
    - *Aminobacter*
    - *Mesorhizobium*
      - *M. ciceri* [*Cicer*, chickpea]
      - *M. loti* [*Lotus*]
    - *Phyllobacterium*
      - *P. loti* [*Lotus*]
  - Rhizobiaceae
    - *Allorhizobium*
      - *A. undicola* [*Neptunia*]
    - *Sinorhizobium* (syn. *Ensifer*)
      - *S. meliloti* [*Melilotus*, sweetclover; *Medicago*, alfalfa]
    - *Neorhizobium*
      - *N. galegae* [*Galega*]
    - *Pararhizobium*
    - *Rhizobium*
      - *R. leguminosarum*
        - *bv trifolii* [*Trifolium*, clover]
        - *bv viciae* [*Pisum*, peas; *Vicia*, field beans; *Lathyrus*; and *Lens*, lentil]
        - *bv phaseoli* [*Phaseolus*]
    - *Shinella*
      - *S. kummerowiae* [*Kummerowia*]
  - Xanthobacteriaceae
    - *Azorhizobium*
      - *A. caulinodans* [*Sesbania*]
- Betaproteobacteria
  - Burkholderiaceae
    - *Cupriavidus*
    - *Paraburkholderia*
      - *P. symbiotica* [*Mimosa*]
    - *Trinickia*

Updates available at: https://sites.google.com/view/taxonomyagrorhizo (see also de Lajudie et al., 2019).

[a] *Other genus and species names exist in the literature. Some predate the present names, others have not been accepted as valid.*

[b] *Strains which have not yet been recognized as belonging to any named species are usually identified by the host from which they were isolated—for example,* Rhizobium *spp.* (Acacia) *or* Bradyrhizobium *spp.* (Lupinus).

The infection process

Nodule initiation and development

Rhizobia can infect their hosts and induce root- or stem-nodules using several different mechanisms, including:

- penetration of root hairs and formation of infection threads, as found in plants such as clovers and beans;
- entry via wounds or sites of lateral root emergence, as found in peanuts and the pasture legume *Stylosanthes*; and
- penetration of root primordia found on the stems of plants such as *Sesbania* (Fig. 17.1C).

The processes of attachment, root hair curling, and infection thread invagination leading to nodule formation are illustrated in Box 17.2.

Visible changes during root hair infection. Infection begins with attachment of rhizobia to immature, emerging root hairs of a compatible host (Box 17.2A). This becomes evident within minutes of inoculation, with attached rhizobia capping the root-hair tip and often oriented end-on to their host. Deformation and curling of the root hair follow, with the root hair surface at the point of infection hydrolyzed to permit penetration of the rhizobia. Rhizobia then move down the root hair toward the root cortex. Rhizobia do not gain direct intracellular access to their host. During infection, and as they move down the root hair, they remain enclosed within a plant-derived **infection thread** (Box 17.2C). In some of the more primitive legumes, for example *Chamaecrista* (partridge pea), rhizobia never escape this containment. Root hair penetration and infection-thread formation (Box 17.2) are paralleled by other changes in the host including host-cell proliferation in the root cortex adjacent to the infected root hairs. In indeterminate nodules, this includes formation of a nodule primordium, with rhizobia then released into modified cells of the root cortex, where again they are enclosed within a plant-derived **peribacteroid membrane,** thus comprising the **symbiosomes** (Fig. 17.2). These membranes protect the bacteria from the defense responses of the host. Indeterminate nodules (see section on nodule development and function below) have a persistent meristem and will continue to grow and fix dinitrogen throughout the season, whereas dinitrogen fixation in determinate nodules begins to decrease following flowering and nodules do not persist past a single season.

Molecular changes during root hair infection. The visible changes evident during root-hair infection are paralleled by molecular interchanges that prepare both host and rhizobia for symbiosis. Studies examining the genes required for nodulation demonstrated that some genes occurred in all rhizobia and were essentially interchangeable, such as the "common" nodulation genes (*nodABC*), while others were only found in organisms that nodulated a specific legume(s). It was also shown that only one gene (*nodD)* was still expressed in the absence of a suitable host. A search for host factors needed to trigger expression of the remaining nodulation genes followed, with **flavonoids** or isoflavonoids (complex phenolic compounds exuded from legume roots) shown to be responsible (Table 17.3). Further, it was shown that legume species vary in the flavonoids they produced and that rhizobia respond differently to specific flavonoids. Thus, luteolin was identified as the principal flavonoid required for *nod*-gene expression in *Sinorhizobium meliloti* (syn. *Ensifer meliloti*), whereas naringenin and genistein were

BOX 17.2

Events leading to nodulation and nitrogen fixation in legumes

- Attachment of rhizobia to the root begins within 1 minute of inoculation (A)
- Number of attached rhizobia increases with time up to several hours
- Root hair curling begins within 5 hours (B) (Limpens and Bissleing, 2003)
- Infection threads become visible in the root hair within 3–4 days of inoculation (C,D)
- Nodules become visible within 5–12 days
- N_2 fixation is often evident in 15-day-old plants

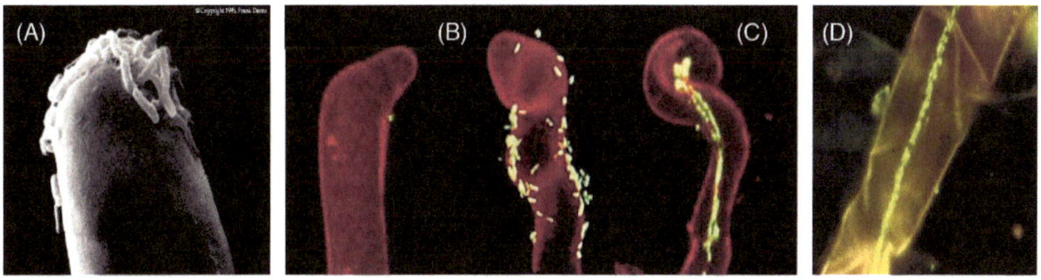

Courtesy of Frank Dazzo, with permission. Limpens and Bisseling, 2003, with permission. Courtesy of B. Ben Bohlool, with permission.

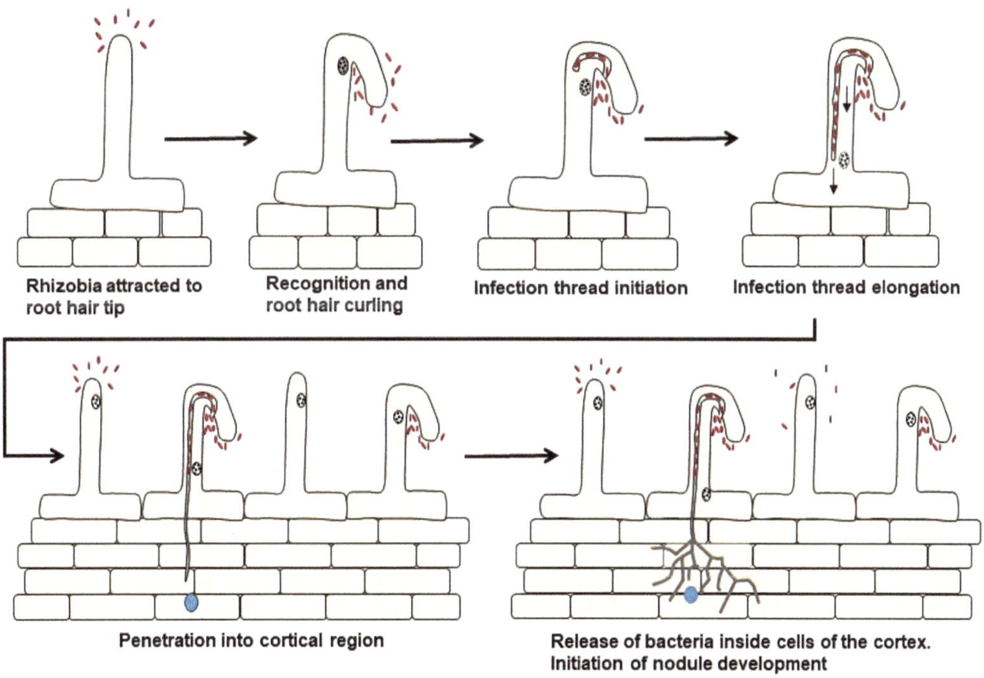

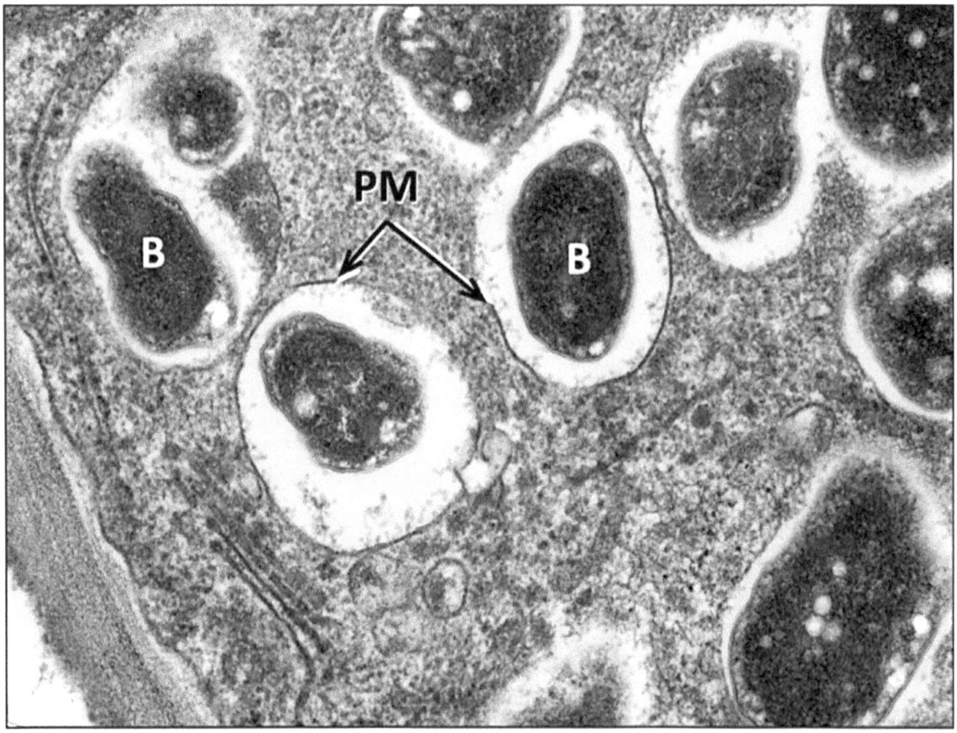

FIGURE 17.2 **Transmission electron micrograph of a cross section through a soybean (*Glycine max* Essex) root nodule.** The micrograph shows part of an infected cell with single *Bradyrhizobium japonicum* bacteroids (*B*) within their symbiosomes surrounded by a peribacteroid membrane (*PM*) of host origin. *Source: Photo by Louisa Howard, Dartmouth Electron Microscope Facility, Public Domain.*

required for *nod*-gene expression in *Bradyrhizobium japonicum*. In fact, a given flavonoid may be a *nod*-gene activator for some rhizobial species, but a repressor for others. In addition, for some symbioses, plant-derived compounds other than flavonoids can act as *nod* gene inducers, including betaines (*S. meliloti*), xanthones (*B. japonicum*), jasmonates (*R. leguminosarum* and *B. japonicum*), or phenolic compounds (*Sinorhizobium.* sp. NGR234) (Table 17.3). These latter are generally less specific and act at higher concentrations (μM or mM) than flavonoids, many of which are effective at nM concentrations (Janczarek et al., 2015).

The *nodD* gene product, which is produced constitutively in rhizobia, is a transcriptional regulator that is responsible for recognizing specific flavonoids and inducing transcription of the nodulation genes. Study of the different nodulation genes and their functions in the rhizobia led to the identification of a series of substances, termed **lipo-chitooligosaccharides** (LCOs) or **nod factors**. These molecules all have the same core structure, coded for by the *nodABC* genes, but differ according to species of rhizobia in the side chains each carries, strongly affecting host range (Fig. 17.3). Strains of rhizobia may produce several nod factors differing slightly in composition. They act as very powerful plant morphogens; at concentrations

TABLE 17.3 Signals and other plant-derived and bacterial compounds involved in legume-rhizobia interactions.

Legume-derived signal molecules
Nod gene inducers
Flavonoids and isoflavonoids
luteolin, genistein, daidzein, narigenin
Non-flavonoid inducers of nod genes
betaines, xanthones, aldonic acids, jasmonates, phenolic compounds
Plant proteins enabling root hair infection
nod factor receptors, lectins, trifoliins, remorins
Other symbiotic modulators
Hydrogen peroxide (H_2O_2), nitric oxide (NO)
Rhizobium-derived signal molecules
Nod factors—lipochitooligosaccharides (LCOs)
Low molecular weight compounds—acyl homoserine lactones (AHLs)
Surface polysaccharides
Exopolysaccharides (EPS), lipopolysaccharides (LPS), K antigen polysaccharide, glucomannan, cellulose
Rhizobial proteins
NodO, rhicadhesin, cellulase, glycanases, adhesins

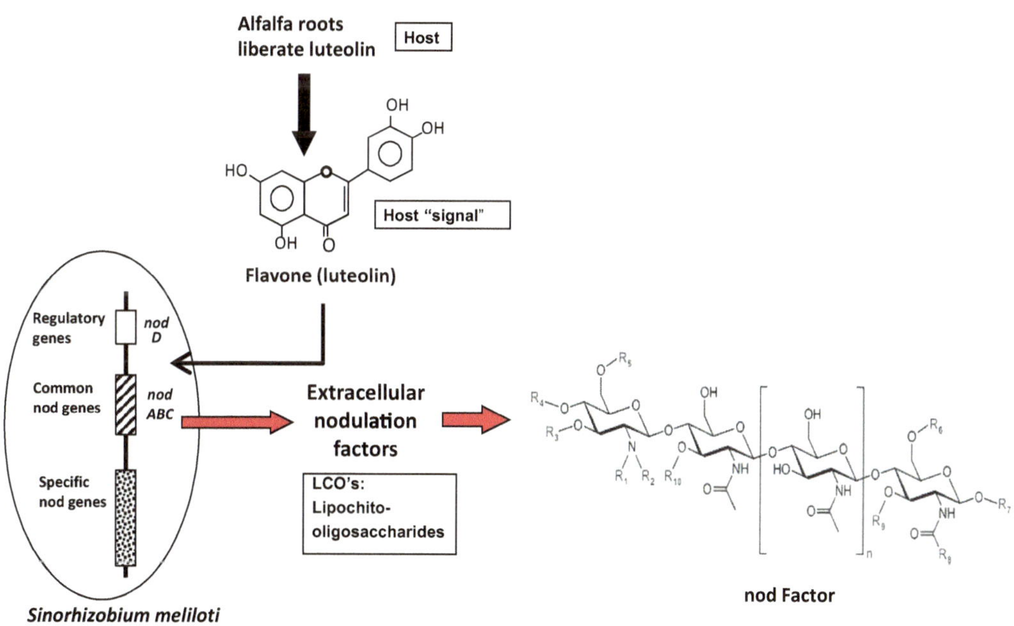

FIGURE 17.3 Early signaling interactions between the host legume, alfalfa, and its microsymbiont, *Sinorhizobium meliloti*. The flavonoid luteolin produced by alfalfa induces the transcription of the common nodulation genes (*nodABC*) to produce **nod factor**. The basic core structure of nod factor is the same in all rhizobia. Components R1 to R10 vary with strain and affect host specificity.

as low as 10^{-13} molar, they can induce many of the root hair deformation and cortical cell changes in the root that are typical of nodule formation. Nod factors are recognized by compatible host plants and are bound by plant LysM-type receptors on the root hair epidermis, which triggers the symbiosis (SYM) signaling pathway in the host leading to induction of nodule formation. Nod-factor binding thus stimulates many changes in the plant including deformation of the root hair, localized depolarization of the plasma membrane, calcium spiking, preinfection thread formation, localized cortical cell divisions, and induction of a variety of host genes (Poole et al., 2018).

Lectins are plant proteins and glycoproteins that can reversibly bind specific carbohydrates. They are expressed on the surface of root hairs and allow binding of the rhizobia to the root hair via specific cell surface components (Fig. 17.4). Many proteins and cell surface components of the rhizobia become involved at this point in establishing the symbiosis. These include complex polysaccharides located on or in the outer membrane, such as **lipopolysaccharide** (LPS), K antigen polysaccharide, **extracellular polysaccharide** (EPS), glucomannan (neutral polysaccharide), and cellulose. EPS is of particular significance in establishing and stabilizing the symbiosis. EPS interacts with lectins on the root hair and facilitates rhizobial attachment to the host. With indeterminate nodules, EPS is also important in initiating and elongating infection threads, releasing bacteria from infection threads, and in bacteroid development. LPS is important in suppressing host defenses and in differentiation of rhizobia into bacteroids and stably maintaining them in the symbiosome. Glucomannan and cellulose function in attachment of rhizobia to root hairs (Janczarek et al., 2015).

Other rhizobial compounds, such as **N-acyl homoserine lactones** (AHLs), hopanoids, luminochrome, and proteins, such as rhicadhesin, cellulase, and glycanases also participate in the early interactions between the microsymbiont and its host as the symbiosis becomes established (Table 17.3). The AHLs, in particular, are involved in **quorum sensing** (Chapter 10) in rhizobia, which is the density-dependent promotion of the transcription of genes whose products are important for the initial interactions of rhizobia with their hosts and for facilitating symbiosome development and dinitrogen fixation. Other rhizobial proteins, such as NodO, form cation-selective channels in membranes and promote cation movement across the root-hair membrane. Glycanases serve to modify EPS, and rhizobial adhesins play a role in attachment (Janczarek et al., 2015; Poole et al., 2018). Lastly, hydrogen peroxide and nitric oxide are also important in signaling during early symbiotic interactions and in nodule functioning (Hichri et al., 2016). Hydrogen peroxide controls differentiation of rhizobia into bacteroids, and nitric oxide both acts as a developmental signal and is responsible for optimal establishment of the symbiosis.

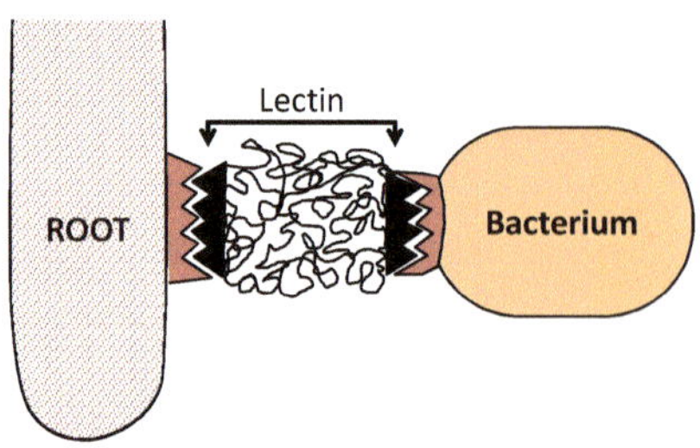

FIGURE 17.4 **Lectins serve to bind compatible rhizobia to legume root hairs prior to infection thread formation.**

The nodulation (*nod*, *nol*, and *noe*) and dinitrogen fixation (*nif*, *fix*) genes in *Rhizobium* species are located on large plasmids (pSym), whereas those of *Bradyrhizobium* and *Mesorhizobium* are chromosomally integrated but are otherwise analogous to those found in *Rhizobium* (Box 17.3). Even in *Rhizobium*, some genes contributing to nodule formation (e.g., genes regulating LPS and EPS composition) are chromosomally located.

Nodule development and function

As the infection thread penetrates the root cortex, host root-cell division and enlargement results in the formation of a visible nodule. Root nodules differ in appearance and structure, a trait that is determined by the host legume (Fig. 17.1). *Determinate* nodules such as occur in soybean, cowpea, and *Phaseolus* (beans) are spherical and do not show any clear developmental gradient. They develop from cell divisions in the middle or outer cortex of the root and lack a persistent meristem (Fig. 17.5A, C). In contrast, the *indeterminate* nodules of peas, medics, and clovers are elongated with a pronounced meristematic region, increasing in length over the course of the growing season (Fig. 17.5B, D).

Examining an indeterminate alfalfa nodule under the microscope reveals four distinct zones, as shown in Fig. 17.5D:

- a meristematic region in which host cells are undergoing active division but are not yet infected by rhizobia (D-I)
- a region in which many plant cells are infected but in which the bacteria have not undergone changes in size and shape, and N_2 fixation is limited (D-II)
- a region of active N_2 fixation, often red or pink in color due to the presence of **leghemoglobin** (Chapter 16), in which host cells contain many rhizobia that may be misshapen. Such bacteria are referred to as **bacteroids** (D-III), these are surrounded by a peribacteroid membrane, thus comprising the symbiosomes (Fig 17.2). See Box 17.4 for an interesting sidelight on a modern application of soy leghemoglobin.
- a region of nodule **senescence** in which the symbiosis is breaking down. Bacteroids may undergo lysis, and the degradation of leghemoglobin results in a green or brown coloration (D-IV)

Bacteroids differ between the two types of nodules. Determinate nodule symbiosomes house one to two bacteria similar in size to free-living cells, whereas indeterminate nodule symbiosomes house a single swollen bacteroid cell with varying morphology. The swollen shape of bacteroids of indeterminate nodules is due to endoduplication of the genome without associated cell division, a phenomenon that is thought to be caused by production of hundreds of cysteine-rich peptides. The bacteroids of indeterminate nodules have highly permeable membranes and are unable to revert to an undifferentiated, non-swollen state. Thus, they are considered to be "terminally differentiated." Indeterminate bacteroids can also become dependent on the host for various co-factors, such as homocitrate, which is required for nitrogenase function. In fact, Terpolilli et al. (2012) suggested that these indeterminate bacteroids should be considered as a type of facultative plant organelle and suggested the term "ammoniaplasts" to describe them. In contrast, bacteroids from many determinate nodules, where the cells do not terminally differentiate, do not exhibit endoduplication of the genome, have less permeable membranes, and are able to be re-isolated.

BOX 17.3

The discovery of a symbiosis island in *Mesorhizobium loti* and a proposed mechanism for its transfer between symbiotic and non-symbiotic rhizobia in the soil

When *Lotus corniculatus* (bird's-foot trefoil) was first introduced to a New Zealand tussock grassland, it required inoculation with *Mesorhizobium* strain ICMP3153 in order for these legumes to nodulate and fix nitrogen. Up to 19% of the nodules from field-grown plants 7 years after the stand was established contained mesorhizobia that varied in colony morphology, when only a single strain was expected to be re-isolated (Sullivan et al., 1995). When these isolates were examined genetically by Sullivan and Ronson at Otago University, NZ, they found that they all contained the same genetic element containing the nodulation and nitrogen fixation genes but were otherwise quite diverse chromosomally. The *nod* and *nif* genes are known to be chromosomally integrated in *Mesorhizobium*, thus horizontal gene transfer was at first thought not to be the mechanism for the diverse isolates obtained, as has been shown for *Rhizobium* sp. whose *nif* and *nod* genes are located on large, transferable plasmids, depicted below.

In a series of landmark papers, Sullivan and Ronson went on to show that the *nif* and *nod* genes in *Mesorhizobium* strain ICMP3153 were located on a transmissible island they termed a symbiotic island (Sullivan and Ronson, 1998). They went on to show that non-symbiotic mesorhizobia could be isolated from the soil and were able to be transformed with the mobile genetic element in the laboratory, enabling them to nodulate *Lotus*. The symbiosis island was shown to be a 500-kb chromosomally integrated transferable element that excises from and integrates into a phenylalanine tRNA gene, reconstructing the gene at one end of the island and leaving a 17 bp direct repeat of the phe-tRNA gene at the opposite end. A phage-related integrase at one end is necessary for excision of the island as a circle. The discovery of the symbiosis island provided the first direct evidence that lateral transfer of symbiotic DNA in the environment was not limited to plasmids and that such transfer was responsible for the diverse mesorhizobia that arose from the initial inoculation of *L. corniculatus* with *Mesorhizobium* strain ICMP3153. The symbiosis island of *Mesorhizobium* is now known to belong to a widespread group of mobile genetic elements called integrative and conjugative elements (ICEs) depicted below.

Genome organization in the rhizobia

The genomes of rhizobia tend to be large (~5–10 Mb) and consist of a chromosome and additional plasmids of varying size. The symbiosis genes (*nod, nif,* and *fix*) are either located on the chromosome or are plasmid-encoded.

In the *Rhizobium* and *Sinorhizobium* (syn. *Ensifer*) genera and in the *Burkholderia* and *Cupriavidus* genera, the symbiosis genes are carried on large plasmids. The symbiosis plasmid pSymA in *Sinorhizobium meliloti* 1021, for example, is 1.35 Mb. pSymB does not contain symbiosis genes but does

(*Continued*)

BOX 17.3 (*Cont.*)

contain several essential genes including that for arginine tRNA and several genes needed for cell division. Symbiosis genes in *Bradyrhizobium*, *Mesorhizobium*, and *Azorhizobium* are chromosomally integrated as described above.

The symbiosis islands of *Mesorhizobium loti* and *Azorhizobium* are actually integrative and conjugative elements (ICEs). These elements can excise from the chromosome to form a plasmid-like structure that can be transferred through conjugation to a nonsymbiotic recipient cell. In the recipient, the ICE integrates into the host genome at conserved sites, such as the tRNA genes as shown for *M. loti*. Some symbiosis ICEs may not be contiguous on the chromosome, but when excised recombine into a single element prior to transfer. It is unknown whether the chromosomally located symbiosis regions of *Bradyrhizobium* spp. are transferable (after Poole et al., 2018).

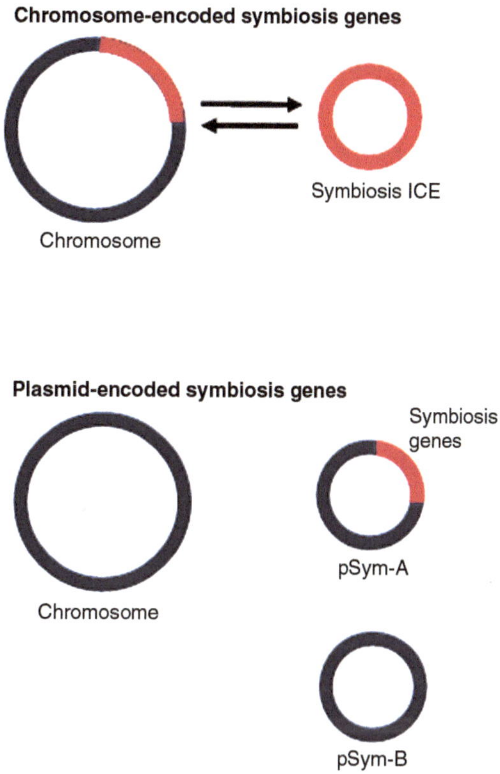

Nodules that have pink or red interiors are usually active in N_2 fixation and are said to be *effective*. Nodules with white or greenish-brown interiors indicate that the symbiosis is either *ineffective* or that the nodule is senescing (Box 17.5). Ineffective nodules tend to be smaller and scattered throughout the root system, whereas effective nodules tend to form near the crown of the root system when inoculants are applied but may be more scattered where native rhizobia nodulate the host. A typical nodule may remain active for only 50–60 days, meaning that plants in the field may have more than one flush of nodules formed during the growing season. Relatively few infections result in root-nodule formation. Successful infections may be visible 5–6 days after inoculation with rhizobia, with active N_2 fixation beginning 8–15 days thereafter (Box 17.2).

Transport of fixed nitrogen

In indeterminate nodules such as those produced on alfalfa, bacteroids produce ammonia, which is exported to the host cell and is there converted via glutamine, glutamate, and

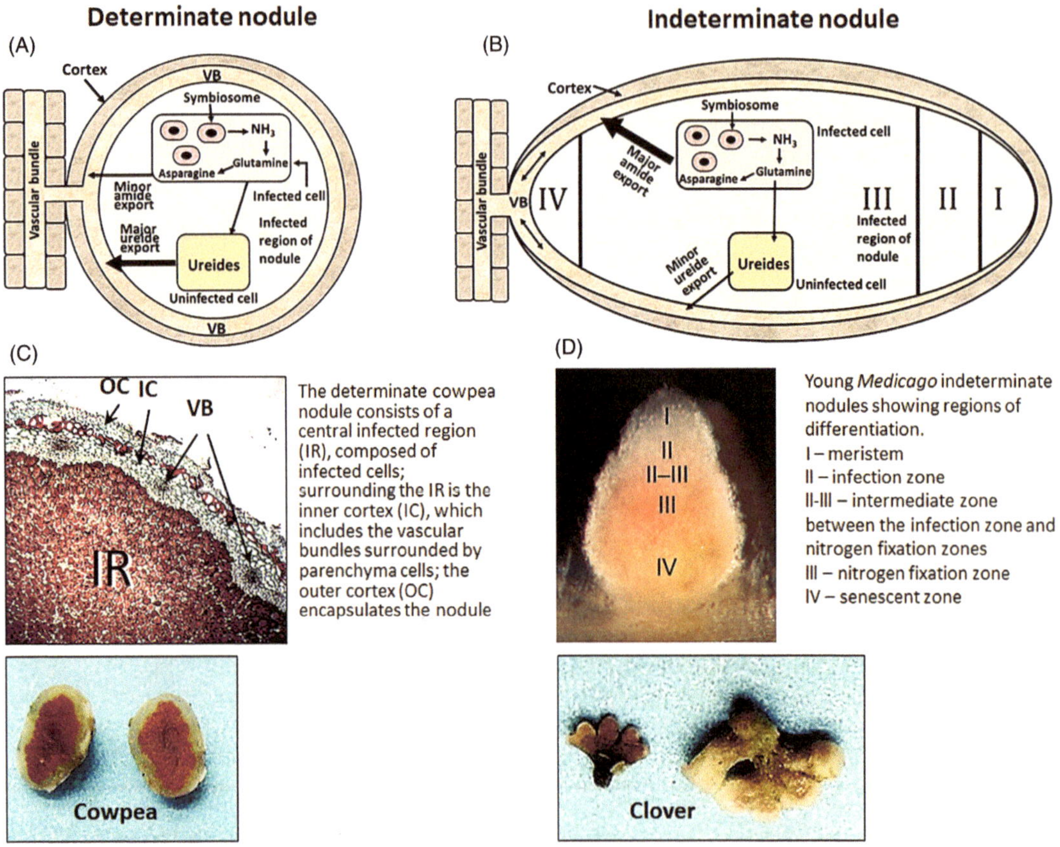

FIGURE 17.5 *Top panel*: The anatomical and metabolic differences between (A) determinate and (B) indeterminate legume nodules. Determinate nodules are generally from tropical origins and export fixed N_2 mostly as ureides. Indeterminate nodules generally originate from temperate regions and export fixed N_2 largely as amides (asparagine and glutamine). *Middle panel*: (C) Longitudinal cross section of the determinant soybean nodule and (D) the indeterminant *Medicago truncatula* nodule. The determinant soybean nodule consists of a central infected region (IR), composed of infected cells dispersed between uninfected cells; surrounding the IR is the inner cortex (IC), which includes the vascular bundles (VB) surrounded by parenchyma cells; the outer cortex (OC), encapsulates the nodule. The meristem of the indeterminant *M. truncatula* nodule remains active throughout nodule development, allowing continual growth. As a result, the nodule is divided into sequential zones of development, including the meristem (I); invasion zone (II), where rhizobia infect divided cortical cells; an intermediate zone (II–III), composed of infected cells containing differentiating bacteria; a nitrogen fixation zone (III) and a zone of senescence (IV) close to the root. *Bottom panel*: Root nodules of cowpea (determinate) and clover (indeterminant). *Source: Top panel: (redrawn from Kleinert et al., 2017); Middle panel: (Wikimedia ; Creative Commons Attribution-Share Alike 3.0 Unported (https://creativecommons.org/licenses/by-sa/3.0/deed.en) license; Bottom panel: Photo credit: D. Zuberer, used with permission.*

aspartate to asparagine. Asparagine is then exported to the shoot (Fig. 17.5B). The overall equation for this series of reactions is:

$$2\ \text{ammonia} + 3\ \text{ATP} + \text{oxaloacetate} \rightarrow \text{L-asparagine} + \text{AMP} + 2\text{ADP} + 2\text{P}_i + \text{NAD} + \text{Pp}_i \qquad (17.1)$$

Determinate nodules export a very different end product. Glutamate and aspartate are produced but are then used to synthesize purines such as xanthine. These are converted to the

BOX 17.4

Soybean leghemoglobin provides that "meaty" mouth feel for vegetable-derived meat products

Hemoglobin in our bloodstream is responsible for carrying oxygen throughout our bodies. Likewise, leghemoglobin is a protein produced by legumes that acts in the root nodule to scavenge oxygen and protect nitrogenase from deactivation. Leghemoglobin, just as the hemoglobin in our blood, carries a heme or iron (Fe) molecule at its center, which causes the red color of blood and the red interior of nodules that are active in nitrogen fixation. Heme also gives meat its flavor.

The leghemoglobin used in today's vegan meat-alternative products is not extracted from soybean root nodules. Instead, the gene coding for leghemoglobin was isolated and genetically engineered into a yeast. The engineered yeast is grown in batch fermentation cultures to produce the leghemoglobin for the recently marketed meat-substitute products. The question arose as to whether the leghemoglobin produced by the yeast was safe for human consumption. The FDA recently released its decision that leghemoglobin is generally recognized as safe (GRAS) and thus not subject to regulation. In fact, leghemoglobin was approved in 2019 for use more generally as a food coloring agent.

Researchers and scientists all over the world are working to develop plant-based versions of pork, chicken, seafood, and dairy products. Pressure-cooked wheat proteins are often the main ingredient, which provide the firmness and "chewiness" needed to simulate meat. Protein extracted from potatoes is also part of the blend. Then comes the leghemoglobin or heme. Heme is what creates the flavor in meat and what replicates that meat flavor in vegetable-derived meat products. Tiny amounts of carbohydrates are added as binding agents. Coconut oil is used to provide the mouth feel of fat and create the "sizzle" of the burger during cooking.

(A) A solution of leghemoglobin illustrating the deep red color due to the heme moiety of the leghemoglobin molecule. (B) Sliced soybean root nodule showing the central region containing the N_2-fixing bacteria surrounded by leghemoglobin in the cells of the nodule. *Source: Part (A): Photo courtesy of Impossible Foods Inc./©Aubrie Pick Photography used with permission; Part (B): Photo courtesy of M. Hungria used with permission.*

BOX 17.5

Infectiveness and effectiveness

- ***Infectiveness*** is the ability of a rhizobial strain to induce nodule formation with a particular legume.
- ***Effectiveness*** is the ability of nodules formed to fix dinitrogen. There are varying degrees of effectiveness with the symbiosis. Some rhizobia fix N_2 vigorously while others are less active on the same hosts and some produce nodules with no N_2 fixation activity (ineffective).

ureides, allantoin and allantoic acid (Fig. 17.5A). The level of these substances in nodules and xylem sap is sometimes used to estimate dinitrogen fixation (Peoples et al., 2009).

The main factor controlling dinitrogen fixation in nodules is oxygen concentration, with low concentrations (10–21.5 nM) needed for maintaining nitrogenase activity. Oxygen regulation in the nodule is achieved by the legume-derived protein **leghemoglobin**. In addition to leghemoglobin, homocitrate produced by host plants is used to control dinitrogen fixation in bacteroids through its interaction with the FeMo cofactor in nitrogenase. The *nifV* gene in rhizobia is needed to synthesize homocitrate. Most rhizobia lack this gene and thus cannot fix dinitrogen when living as saprophytes in the soil (Terpolilli et al., 2012).

Host specificity

Given the signaling that occurs between host and root-nodule bacteria during nodule formation, it is not surprising that specificity can occur at several stages of nodule formation. The range of this specificity is described as follows.

Some legumes never nodulate

Among the subfamilies of the Leguminosae, 97% of the Papilionoideae bear nodules, but less than 20% of Caesalpinioideae are nodulated. Because of such differences in the frequency of nodulation, and because some legumes can be nodulated by rhizobia from several different genera, it has been suggested that ability to nodulate with rhizobia might have arisen on more than one occasion in the evolution of legumes. In one often-cited example, the genus *Cassia* is not nodulated, whereas the very-closely related *Chamaecrista* (partridge pea) bears nodules ranging in structure from primitive to quite highly developed.

Nodulation mutants within normally nodulated plant species

A number of different nodulation mutants have been identified in legumes, particularly soybean and pea. In some cases, these plants will nodulate only with specific rhizobia; in others they are never nodulated. Non-nodulating soybean, common bean, and pea lines are often used as control plants in inoculation studies.

Cross-inoculation group specificity

No rhizobium nodulates all legumes, although some rhizobia are *promiscuous* and will nodulate hosts from a number of different species and genera of legumes. The example of strain NGR234 has already been mentioned. *Rhizobium tropici* strain UMR1899 is also promiscuous, nodulating species of *Leucaena, Dalea, Phaseolus, Macroptilium, Onobrychis,* and *Coronilla.* Other strains may only nodulate hosts belonging to one genus of legume (e.g., *Trifolium*), and may not even nodulate all species of that genus. As a consequence, more than 100 different strains of rhizobia are needed to satisfy the inoculation requirements of currently important legume species.

Infectiveness sub-groups

According to the old cross-inoculation group concept, rhizobia isolated from one host in a cross-inoculation group should nodulate all members of that group. However, particular species, or even cultivars, of legumes can require different rhizobia. A well-documented example is the pea landrace "Afghanistan" that is highly specific in *Rhizobium* requirements. It nodulates freely with pea rhizobia from the center of origin of *Pisum sativum* in the Middle East but not with rhizobia recovered from peas in other areas. Similarly, when soybean varieties such as "Bossier" developed in the United States were introduced into Africa around 1980, they failed to nodulate with indigenous soil rhizobia and had to be inoculated. In contrast, varieties such as "Orba" and "Malaysian" from Asia nodulated with the indigenous strains and rarely responded to inoculation. Such symbiotic specificities can be a problem when introducing and evaluating new plant germplasm. For this reason, collectors of legume germplasm should also collect nodule or soil samples from which the appropriate rhizobia can be isolated. Host–rhizobial interactions also influence levels of N_2 fixation. Peanut and cowpea, for example, are nodulated by and fix dinitrogen with a range of soil rhizobia, whereas *Centrosema* (butterfly pea) and *Desmodium* (tick trefoil) species will often nodulate with these strains but fix little dinitrogen. A consequence is that when cowpeas or peanuts are introduced into a new area, they will often grow well without inoculation, whereas *Centrosema* and *Desmodium* species may have many nodules but grow poorly.

Environmental factors affecting symbiosis in legumes

Environmental factors influence all aspects of nodulation and symbiotic N_2 fixation, in some cases reducing rhizobial survival and diversity in soil while in others affecting nodulation, N_2 fixation, or even growth of the host. Critical factors include acidity, temperature, mineral nutrition, salinity and alkalinity, heavy metals, toxins, waterlogging, and drought.

Acidity

In Latin America alone there are more than 800 million hectares of Oxisols and Ultisols having a pH less than 5.0. Dinitrogen fixation in these soils can be markedly reduced due to:

- hydrogen ion concentration per se;
- toxic levels of aluminum and manganese; and
- induced deficiencies of calcium, phosphorus, and molybdenum.

Soil acidity can directly limit rhizobial growth and persistence in soil, with isolates of *Sinorhizobium meliloti* being particularly acid sensitive. Fast-growing rhizobia are generally more sensitive than are the slow-growing bradyrhizobia, but relatively low pH-tolerant strains exist in many species. Surprisingly, not all strains recovered from acid soils are acid tolerant, suggesting that microsites of more favorable pH can occur within the soil.

Failure to nodulate is common in acid soils, in part because fewer rhizobia persist, but also because acid pH affects attachment of the rhizobia to their host. For many strains, problems in nodulation can be expected once soil pH falls below pH 5.2. Acid soils in the United States are often limed to near neutrality. Other ways to limit the negative effects of soil pH include the use of acid-tolerant inoculant strains and host cultivars and the pelleting (coating) of inoculated seed with ground rock phosphate, limestone, or biochar. Amending acid soils with biochar may also enhance the ability of indigenous rhizobia to nodulate legumes (Box 17.6). In Australia, the use of relatively acid-tolerant *Sinorhizobium meliloti* strains such as WSM419,

BOX 17.6

Biochar increases nodulation and nitrogen fixation in common beans in greenhouse and field trials in Kenya

Crop production in tropical soils is often constrained by acidity and related low nutrient availability, especially of phosphorus. Phosphorus is essential for nodulation and nitrogen fixation in legumes, and both processes are impaired in these soils (Figure A below), even if there are sufficient populations of rhizobia present in the soil to nodulate the legume. Biochar is the solid co-product of the thermal condensation of plant biomass under low oxygen conditions (pyrolysis) and, used as a soil amendment, can help alleviate these constraints (Chapter 14).

In field trials conducted in western Kenya on an acidic Acrisol (A), common bean (*Phaseolus vulgaris* L.) grows poorly and does not nodulate in the absence of management inputs. Inoculation alone also fails to induce nodulation. Furrow application of 10 t ha^{-1} biochar prepared from sugarcane bagasse increased plant growth over no inputs and dramatically promoted nodulation and nitrogen fixation resulting in improved crop growth.

In greenhouse experiments, Güereña et al. (2015), tested the ability of seven biomass feedstocks prepared at two pyrolysis temperatures (350°C and 550°C) with four post-pyrolysis treatments to improve nodulation and nitrogen fixation in common beans in a Kenyan humic Acrisol. On average, regardless of pyrolysis temperature or post-pyrolysis treatment (removal of volatile matter, ash or pH adjustment), biochar incorporated into this soil at 15 t ha^{-1} increased shoot biomass by 262%, nodule biomass by 3575%, and N derived from the atmosphere (Ndfa) by 2126% over the unamended control (B). Many biochars are alkaline and the ash component contains phosphorus and other sedimentary minerals. However, neither the phosphorus content of the biochar, nor its alkalinity, could explain observed increases in nodulation and nitrogen fixation. Instead, the significant increase in shoot phosphorus observed was associated with increased mycorrhizal infection of roots in soils amended with biochar. Increased P uptake via mycorrhizal colonization enabled indigenous rhizobia to nodulate and fix nitrogen in this system.

(*Continued*)

BOX 17.6 (*Cont.*)

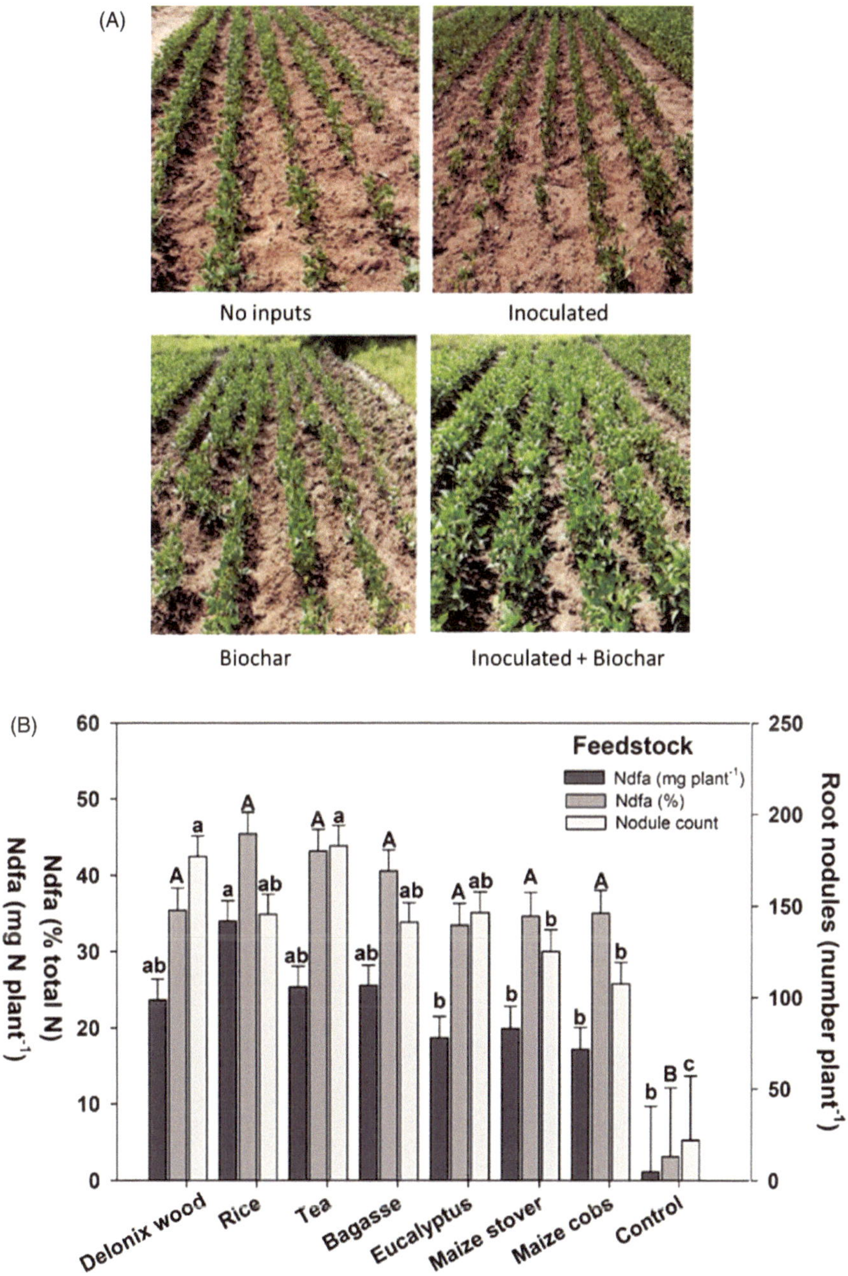

Source: Photo credit: Part (A) J.E. Thies, used with permission; Part (B): D. Güereña, used with permission.

together with *Medicago* species collected from acid soils in Sardinia, has permitted extension of the area sown to annual medics by more than 350,000 ha since 1985 (Ewing and Howieson, 1989).

Plant species vary in tolerance to aluminum and manganese but are generally more affected by these ions than are rhizobia. Some rhizobia tolerate 100 μM aluminum and 300 μM manganese. In contrast, reduced root growth of alfalfa (*Medicago sativa*) occurs at only 8 μM aluminum, and nodulation in cowpea can be inhibited at 25 μM aluminum.

Temperature

Rhizobia are mesophiles and most do not grow below 10°C or above 37°C. Exceptions are the rhizobia from some Arctic legumes and bradyrhizobia collected from the hot, dry Sahel savannah of Africa. Temperature during the shipment and storage of rhizobial inoculants, after seed inoculation, and at planting is particularly critical. Exposure to high temperatures at these times can lead to the loss of the symbiotic plasmid in *Rhizobium*, or it can reduce cell numbers below the levels needed for good nodulation.

Temperature also influences nodule growth, N_2 fixation, and the period of time over which nodules remain active. The optimum temperature for many legumes is around 25°C, and exposure to temperatures in excess of 40°C for even short periods can cause irreparable loss of nodule function. In contrast, minimum tillage in the American Midwest can result in cool-wet soil conditions in the early spring and delay nodulation.

Mineral nutrition

Although well-nourished legumes generally nodulate and fix dinitrogen better than those that are nutrient limited, several elements have specific functions in nodulation and symbiotic N_2 fixation (Table 17.4). Adequate levels of these elements are essential for effective N_2

TABLE 17.4 Elements having specific functions in the nodulation or nitrogen fixation of legumes.

Molybdenum	FeMoCo protein of nitrogenase (Chapter 16)
Phosphorus	Energy transformations in the nodule
Iron	Fe- and FeMoCo proteins of nitrogenase
	Leghemoglobin
	FeS centers of nitrogenase (Chapter 16)
	Unspecified function in nodule development
Calcium	Attachment of rhizobia to root hairs
	Cell wall integrity in *Rhizobium*
Sulfur	FeS centers of nitrogenase
	FeMoCo proteins of nitrogenase
	Asparagine and arginine synthesis, N export from nodules
Cobalt	Nodule coenzyme function
Nickel	Hydrogenase function

fixation and failure to supply them will result in the generalized yellow chlorosis typical of nitrogen deficiency. Several of these elements warrant specific mention.

Phosphorus

Dinitrogen-fixing leguminous plants usually require more phosphorus than similar plants supplied with fertilizer nitrogen. Nodules are an important phosphorus sink and commonly have the highest concentration of that element in the plant. The high energy cost of N_2 fixation, with its need for ATP, along with the phosphorus required to build and maintain functioning nodules, leads to this elevated requirement. Under P-deficient conditions, phosphorus fertilization will usually result in enhanced nodule number and mass and greater N_2 fixation per plant. Problems of phosphorus fertilization are particularly severe in Africa, where the abandonment of traditional slash-and-burn methods and the limited supply and high cost of fertilizers for small-holder farmers have resulted in significant phosphorus loss from soils. Complicating this issue, global phosphorus reserves are only projected to last 50–80 years. Strain and cultivar differences in phosphorus use efficiency have been demonstrated and will need to be emphasized more in the future.

Field-grown legumes form tripartite symbioses with both *Rhizobium* or *Bradyrhizobium* and arbuscular-mycorrhizal fungi (Chapter 12). This has an added energy cost to the host but can result in striking growth improvement due to enhanced phosphorus uptake (Box 17.6).

Sulfur

Sulfur is frequently overlooked as a potentially critical nutrient for sustained dinitrogen fixation. Recall that nitrogenase contains two key subunits, the iron protein (dinitrogenase reductase) and the iron-molybdenum protein (dinitrogenase) (Chapter 16), both of which contain sulfur. In both determinate and indeterminate nodules, fixed nitrogen is exported in forms requiring sulfur, such as glutathione. Where sulfur is insufficient in site soils, sulfur applications can increase nitrogen fixation two- to sixfold.

Molybdenum

Molybdenum is a component of the nitrogenase enzyme complex, with only 100–500 g ha^{-1} of sodium or ammonium molybdate needed to meet this requirement. How to supply this small amount can be a problem, especially under acid soil conditions where adsorption reduces the availability of soil molybdenum to the plant. Molybdenum salts have sometimes been included in the rhizobial inoculants, but this commonly kills the rhizobia and is not recommended. Liming the soil can also enhance molybdenum availability. An alternate approach adopted in Brazil is to fertilize plants being grown for seed with molybdenum applied to the foliage. This provides enough molybdenum for the next growing cycle.

Iron

Iron is a component of leghemoglobin that functions in the regulation of oxygen supply to bacteroids (Chapter 16). It also occurs in both the Fe- and FeMo- proteins of the nitrogenase complex and is essential for early nodule development. Plants that are iron-deficient can develop hundreds of nodule initials but few mature nodules. Both host and microsymbiont can differ in efficiency of iron utilization. In the case of the bacteria, this is because some strains produce iron-sequestering **siderophores** (Chapter 19), and so compete more effectively for iron in the rhizosphere. For example, in peanut, bradyrhizobial strains with the ability to produce catechol can significantly enhance the growth and iron nutrition of iron-inefficient cultivars.

Salinity and alkalinity

The effects of saline or alkaline conditions are likely to be greater on the host or symbiosis than on the microsymbiont. Alkaline soil conditions limit the availability of iron, zinc, manganese, and boron in the soil, thereby reducing plant growth and N_2 fixation. Foliar fertilization with micronutrients is often an effective remedy. Legumes as a group are also markedly sensitive to salt, with some species affected by concentrations as low as 80 mM. In contrast, strains of rhizobia from *Medicago* and *Acacia* often tolerate 500 mM salt. Cells of rhizobia exposed to high salt concentrations may accumulate compatible solutes such as glutamic acid, trehalose, glycine betaine, and proline that help to maintain turgor in the cell and limit the damage caused by salts.

Legume inoculation

When a new legume species is introduced into a region, the soil is unlikely to contain appropriate rhizobia, and inoculation is usually needed for adequate nodulation and N_2 fixation. Yield increases following this initial inoculation can exceed 50%, with clear differences evident between inoculated and uninoculated plants (Fig. 17.6). Inoculation in subsequent years is often not needed. In fact, where a legume has a long history of cultivation in an area, most soils will contain appropriate rhizobia, and even uninoculated plants will be heavily nodulated. Strains used in any subsequent inoculation usually give rise to only a small fraction of the nodules formed. Less than 50 rhizobia g^{-1} soil is often adequate to nodulate the host and lead to a lack of yield response to inoculation (Thies et al., 1991). Unfortunately,

FIGURE 17.6 **Response to inoculation of soybean.** Plants in uninoculated or ineffectively nodulated treatments are smaller and chlorotic and have symptoms of nitrogen deficiency; those inoculated with effective strains of *Bradyrhizobium* are green and vigorous. *Photo source: R. Stewart Smith, used with permission*

many indigenous soil rhizobia are less than fully effective in N_2 fixation. Response to inoculation in these areas may be limited, thus adoption of new inoculant technologies by farmers/growers in such areas is usually much less than for regions where legumes are of relatively recent introduction.

Need for inoculation

A simple, three-treatment experiment will establish the need for inoculation:

- uninoculated control plots
- plots inoculated with a strain of rhizobium effective on the host legume being evaluated
- plots inoculated with the same strain, but also supplied with fertilizer nitrogen

Extensive nodulation of the uninoculated plants means that the soil already contains indigenous rhizobia able to nodulate the host. The contrast between the uninoculated plants and those supplied with nitrogen will then be a measure of the effectiveness in N_2 fixation of the indigenous rhizobia. If the uninoculated plants are green and vigorous, inoculation is probably not necessary. Absence of nodulation in the uninoculated plants, coupled with heavy nodulation of plants receiving inoculation, indicates the need for inoculation. The difference in plant growth in the three treatments is then an indicator of the efficiency in N_2 fixation of the inoculant strain. Excellent plant growth in all three treatments indicates that either the native rhizobia are highly effective, and thus inoculation is not necessary or that the site is high in available nitrogen. Poor growth in all treatments would imply that a factor other than nitrogen is limiting plant growth.

Methods of inoculation

When inoculation is necessary, there are a number of approaches that can be used. The goals of inoculation are to:

- supply the number of rhizobia needed for good nodulation and effective N_2 fixation
- ensure that the highly effective rhizobial strains used in inoculants form most of the nodules produced
- favor inoculant strain persistence in soil and its domination in nodules over subsequent years

Prior to the 20th century, inoculation was often achieved by taking soil from a productive site and applying it to a new crop or revegetation area. This often resulted in the transfer of undesirable soil organisms as well as those that were beneficial. Inoculant rhizobia are now commercially produced and packaged, with different formulations available to meet different seeding requirements. Four inoculation procedures are common.

Seed inoculation

The inoculant is mixed with milk or some other slightly-adhesive material, and the seed is uniformly covered with this suspension. The seed is dried in the shade and sown the same day. This procedure requires that the inoculant strain be packaged in a relatively fine carrier material (commonly finely ground and limed peat) or liquid that will adhere to the seed.

Seed pelleting

A stronger adhesive, such as gum arabic or methyl cellulose, is used with the inoculated seed and then rolled in ground limestone, rock phosphate, or biochar. Pelleting combats unfavorable soil conditions such as low pH or high temperature and is used for aerial sowing. The aim is to provide a microenvironment around the seed favorable to rhizobial survival. Preinoculation of legume seed, advocated for and sold by commercial vendors as a time savings, is not recommended because rhizobial numbers on the seed can decline dramatically during storage.

Soil inoculation with a granular peat or liquid

The inoculant is banded into the seed furrow so that it can make contact with the emerging root. The inoculant is packaged in a coarse peat or a liquid that is dribbled into the row alongside or below the seed. Granular soil inoculation is less time consuming, allows higher rates of inoculation when soil conditions are unfavorable, and permits separation of seed and inoculants, when the seed has been treated with fungicides.

Inoculation in the planter seed box

The inoculant is mixed directly with seeds in the planter box. This simple method is sometimes used as insurance when soils already contain rhizobia. However, inoculant and seed tend to separate providing uneven coverage, and the practice is not recommended.

Strain selection and testing

Where inoculation is required, it must be carried out using an inoculant preparation having a rhizobial strain (or strains) with the following characteristics:

- able to form highly effective nodules with all commonly used varieties of the legume species for which it is recommended
- competitive in nodule formation and persistent in the soil
- able to tolerate soil environmental stresses such as low pH and high temperature
- have good growth in simple, inexpensive culture media
- be genetically stable, and not subject to mutation
- able to survive well on the seed prior to seed germination
- have the ability to persist in soil between crops

Less than one strain in 100 is likely to meet these criteria, so rhizobia for inclusion in inoculants must undergo extensive testing. The initial step is usually a growth chamber or greenhouse evaluation of numerous strains obtained from other research laboratories or from nodules collected in the field. Marked variation in nodulation and N_2 fixation will usually be evident.

Strains that perform well in the initial testing must be further evaluated using other cultivars or species of legume for which the inoculant strain may be recommended. Host variety-strain interactions are common, and the goal is to identify strains that are broadly effective. As the number of strains being considered is reduced, testing under field conditions is recommended. Field trials should be conducted at sites varying in soil type and numbers of indigenous rhizobia. The latter helps to provide a measure of strain competitiveness over time and, in succeeding seasons, can be used to determine inoculant rhizobia persistence in the soil. Finally, environmental and cultural factors that could influence strain performance, for example acid or temperature tolerance, need to be considered. Designation of an organism as being of inoculant quality should not be done lightly. Once added to the soil, they can be difficult to displace.

Inoculants and inoculation

The industry that manufactures rhizobia for seed and soil inoculation currently generates gross revenues between US$60–75 million per year for the four leading firms in the Americas and Europe (Phillips, 2004). Inoculants range from simple tube cultures sufficient for small quantities of exotic seed, to large-scale, fermenter-grown cultures mixed with peat or another carrier material. The latter are used in the commercial inoculation of the large areas planted to soybean, bean, peanut, and clover. The large-scale production of inoculants is a relatively simple process that should result in at least 10^9 highly effective rhizobia per gram of product.

Once the inoculant culture has been produced, different carrier materials can be used to keep these bacteria viable until used by the farmer. Most allow rhizobial survival for periods of at least 6 months. Characteristics of a good inoculant carrier are:

- high water-holding capacity
- nontoxic to rhizobia
- available, inexpensive, and easily produced
- sterilizable by autoclaving or, preferably, radiation
- adheres well to seed
- good buffering capacity

The most commonly used carrier material is peat but, as peat is not universally available and is a limited resource, alternatives including compost, sugarcane press mud (derived from the milling of sugar cane), coal, polyacrylamide, vegetable oils, clays and clay granules, biochar, liquid preparations, and soil have all been used successfully. No listing of physical or chemical properties can fully explain why some materials make suitable inoculant carriers and others do not. Previously sterilized carriers generally support a higher population of rhizobia and have longer shelf-lives than nonsterilized materials.

Inoculant quality and inoculation failure

The number of rhizobia desired per seed varies with environmental conditions and is affected by seed size. In countries where inoculant quality is regulated by law, such as Canada and Australia, the usual standard is for 10^4 rhizobia per seed for small-seeded legumes including clover and 10^6 rhizobia per seed for larger-seeded species such as bean and soybean. Inoculants from many countries do not meet these standards. Factors contributing to the poor quality of many inoculants include:

- inadequate testing in the selection of the inoculant strain
- mutation in the inoculant strain after repeated subculture or storage at high temperature
- use of inappropriate culture media
- contamination of the rhizobial culture
- use of carrier materials that do not support suitable populations of rhizobia
- poor storage and transport conditions

Inoculants can fail for other reasons including high levels of soil nitrogen, contact with acidic fertilizers or with pesticides, use of inappropriate gums or stickers, and use of the wrong inoculant culture. Catroux et al. (2001), in a review of trends in inoculant production and use, concluded: "We enter the era of biotechnology knowing more and more about the mechanisms of N_2 fixation at the gene level, but except for some manufacturers in developed countries ...still lacking good quality and reliable inoculants (p. 27)." Only by continued research in strain selection and evaluation, and the imposition of quality standards, can inoculation success be improved.

Frankia and the actinorhizal symbiosis

Frankia (genus) is an actinobacterium that forms N_2-fixing (actinorhizal) nodules (Fig. 17.7) with a range of angiosperms (Table 17.5). A number of the hosts involved are important in agroforestry, in the ecology and nitrogen economy of marginal soils, or in mine-spoil reclamation or sand dune stabilization. Reported rates of N_2 fixation are highly variable but in *Coriaria arborea* have averaged 90 kg^{-1} N ha^{-1} annually over a 20-year period (Silvester, 1975). Isolation of *Frankia* from root nodules was not achieved until 1978 and from some hosts, such as *Datisca*, is still difficult. Increasing numbers of isolates are now available, and information on their morphological, genetic, and specificity differences is accumulating.

FIGURE 17.7 **Nodules of *Alnus* showing a typical multi-lobed structure that is derived from modified lateral roots.** *Source: Cwmhiraeth / CC BY-SA (https://creativecommons.org/licenses/by-sa/3.0).*

TABLE 17.5 The monophyletic genus and current species of *Frankia* and the actinorhizal hosts they nodulate.

Actinobacteria		Actinorhizal Hosts
Frankiaceae		
Cluster 1a	*Frankia alni, F. torreyi*	(*Betulaceae* (*Alnus*), *Casuarinaceae, Myricaceae*)
1b	*F. canadensis*	
1c	*F. casuarineae*	
Cluster 2	*F. coriariae*	(*Coriariaceae, Datiscaceae, Rosaeae, Rhamnaceae*)
Cluster 3	*F. elaeagni, F. discariae, F. irregularis*	(*Elaeagnaceae, Myricaceae Gymnostoma*)
Cluster 4	*F. inefficax, F. asymbiotica, F. saprophytica*	(Ineffective on *Alnus* or asymbiotic)

Derived from Gtari et al., 2019.

Frankia

Members of the genus *Frankia* are Gram-positive, filamentous organisms characterized by multilocular **sporangia** and N_2-fixing **vesicles** (Fig. 17.8). Few isolates produce spores within the nodule, and those doing so are generally less effective in N_2 fixation. Vesicle production occurs under conditions of nitrogen limitation with mature vesicles in the nodule having a pronounced, multilayered, hopanoid (bacterial sterol lipid) envelope that protects the nitrogenase enzyme from oxygen deactivation. *Frankia* vesicles are functionally similar to heterocysts in the nitrogen-fixing cyanobacteria (see Fig. 17.10 in later section). In the symbioses in which vesicles are not formed, such as with *Casuarina*, lignification of infected and adjacent cells provides an oxygen diffusion barrier.

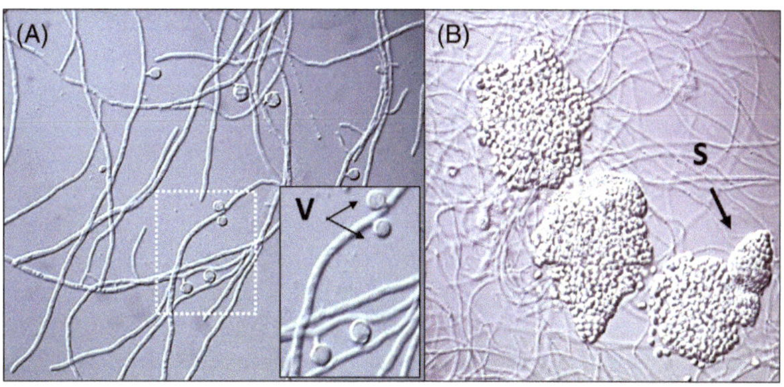

FIGURE 17.8 Hyphae, vesicles (A), and sporangium (B) of the actinomycete *Frankia*. *Source: Part (A): https://commons.wikimedia.org/wiki/File:Hypha_and_vesicle_of_Franikia.jpg Kkucho / CC BY-SA (https://creativecommons.org/licenses/by-sa/3.0); Part (B): https://commons.wikimedia.org/wiki/File:Sporangium_of_Frankia.jpg Kkucho / CC BY-SA (https://creativecommons.org/licenses/by-sa/3.0).*

Sequence analysis of 16S rRNA (Chapter 4) from cultured and uncultured nodule endophytes has permitted 11 species within four clusters to be identified within the monophyletic genus *Frankia* (Gtari et al., 2019, Table 17.5).

Specificity and symbiosis

Knowledge of the *Frankia*-actinorhizal symbiosis lags behind that for the nodulated legumes. However, as legumes and actinorhizal species share common ancestors (Rosid I Clade), contrasts between the two symbioses have allowed significant recent progress (Pawlowski and Demchenko, 2012). Commonalities with nodulation by rhizobia include:

- flavonoids from seed washes of *Alnus rubra* affect nodulation
- root-hair deformation assays suggest involvement of nod-factors in nodulation; however, in at least some strains, these appear to be heat stable and chitinase resistant, suggesting a structure quite different from that found in the rhizobia
- regulation of nodulation has been shown in *Discaria* and *Alnus*
- nodule-specific gene products have been identified including expression of a hemoglobin conserved between actinorhizal *Casuarina* and legumes

Strains of *Frankia* form root nodules on eight families of dicotyledonous plants, including *Betulaceae, Casuarinaceae,* and *Myricaceae* in the Fagales; *Elaeagnaceae, Rhamnaceae,* and *Rosaceae* in the Rosales; and *Coriariaceae and Datiscaceae* in the Cucurbitales. With the exception of the *Datiscaceae*, all actinorhizal species are woody plants. Infection is via root hairs in genera such as *Casuarina* (Australian pine) and *Myrica* (Myrtle) in the Fagales but proceeds intracellularly in others such as *Discaria, Dryas* (Mountain avens), and *Ceanothus* (Buckthorn) in the Rosales. The infection mechanism in the Curcurbitales remains unknown as no microsymbionts in this group have yet been cultured. Nodules are perennial modified lateral roots with lobes up to 5 cm in length (Fig. 17.7). Unlike rhizobia bacteroids, *Frankia* in nodules formed on *Datisca glomerata* (Durango root) assimilates the dinitrogen fixed and undertakes arginine biosynthesis (Berry et al., 2011), thereby demonstrating metabolic independence from the host. Berry et al. (2011) suggest a mechanism for nitrogen transfer to the host whereby arginine is first exported from *Frankia* to infected host cells and then to uninfected host cells, where degradation and ammonia re-assimilation may take place. Alternatively, arginine could be catabolized to glutamine and ammonia and then to glutamine for possible transport in the xylem. Precise mechanisms for transport of fixed dinitrogen from *Frankia* to its hosts remain poorly described.

Cross-inoculation studies with *Frankia* have been complicated by difficulties in isolating the microsymbionts from some actinorhizal hosts. Though many more tests are needed, it appears that microsymbionts differ quite markedly in promiscuity. *Alnus* and *Casuarina* are specific in their *Frankia* requirements, whereas isolates from *Myrica, Gymnostoma, Eleagnus,* and *Atriplex* are much more promiscuous. Simonet et al. (1999) found seven phylogenetically distinct *Frankia* groups in the nodules of Australian *Casuarina* and *Allocasuarina* species, yet they recovered only one group from *Casuarina* plants grown outside Australia. These isolates were more easily cultured and survived better in soil apart from their host. It appears from this that saprophytic competence is as important for *Frankia* as for the rhizobia.

Cyanobacterial symbioses

Cyanobacteria have the ability to both photosynthesize and fix dinitrogen, making them ideal pioneering species (Chapter 7). Some cyanobacteria also form symbioses; their hosts include lichenous fungi, liverworts such as *Blasia*, the aquatic fern *Azolla*, and the angiosperm *Gunnera*. Of these, *Azolla* is the most important commercially, whereas lichens containing cyanobacteria and cyanobacteria themselves are important components of soil biological crusts in dryland regions; both are considered here.

Anabaena (Nostoc) and the *Azolla* symbiosis

In many rice growing areas, *Azolla* is maintained and propagated in slow-flowing creeks or overwintered in protected beds, then introduced into flooded paddies between plantings of rice. After some period for growth, the fern can be incorporated into drained soil before the transplanting of rice seedlings or left to be shaded out as the rice canopy develops. The low C/N ratio of the fern ensures rapid mineralization after incorporation, with yields in the subsequent rice crop enhanced by up to 1000 kg ha^{-1}. Under favorable conditions, rates of N_2 fixation can reach 2 kg ha^{-1} day^{-1}; in the field, rates of 100 kg ha^{-1} year are more likely.

Dinitrogen fixation in *Azolla* has traditionally been attributed to a heterocystous cyanobacterium, *Anabaena azollae,* growing within cavities in the dorsal leaf lobe (Fig 17.9). The dorsal lobe of the *Azolla* frond contains mucilaginous ellipsoidal cavities within which are located 2000–5000 cyanobacterial cells surrounded by an envelope of plant origin. Unlike other cyanobacterial symbioses, *Anabaena* cells are persistent in *Azolla* and pass to the progeny during fragmentation of the host thallus.

Dinitrogen fixation in *Anabaena* species occurs predominantly in specialized larger, thick-walled cells termed **heterocysts** (Fig. 17.10). When such organisms are grown in culture away from the host, only 6%–10% of the cells in the filament are heterocysts, though this frequency increases under nitrogen deprivation. In the mature *Azolla* frond, heterocysts occur at a frequency of 20%–30% irrespective of nitrogen level.

Although the yield benefits from the use of *Azolla* can be appreciable, the intensification of agriculture, increased labor costs, and the availability in some areas of relatively cheap sources of fertilizer nitrogen, have resulted in reduced use of green manures such as *Azolla*. This has dramatically reduced the land area devoted to *Azolla*.

FIGURE 17.9 **Location of the symbiotic cyanobacterium *Anabaena azollae* within the leaf frond of the water fern *Azolla*.** The leaf frond has been cleared, with the filaments of the microsymbiont (M) evident as a darker region toward the center of the frond. *Photo source: CIMMYT Archives, used with permission.*

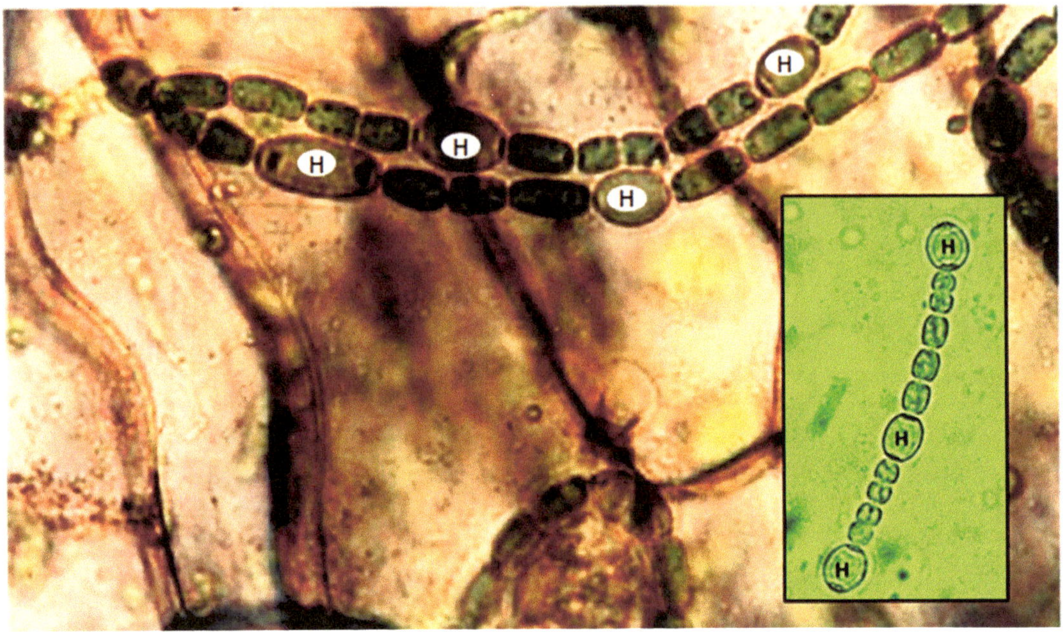

FIGURE 17.10 **In the *Azolla* frond, much of the N_2 fixation by *Anabaena* occurs within larger, modified cells known as heterocysts.** The frequency of such cells within the frond may be 3–4 times greater than is found in free-living filaments of *Anabaena. Photo source: CIMMYT Archives, used with permission. Inset, heterocysts of the cyanobiont of Azolla (Photo: D. Zuberer, used with permission).*

Soil biological crusts

Cyanobacteria, both free-living and in the lichen symbiosis, are important components of soil biological crusts (biocrusts, cryptobiotic crusts) that are formed in the upper few centimeters of soil in drylands around the world (Chapter 7). Exudates of cyanobacteria growing between sand grains help to glue these particles together, stabilizing the soil surface. Biocrusts contribute significantly to primary production and soil fertility, particularly in desert and tundra areas, where they can constitute up to 70% of the soil cover. Models estimate that soil crust communities contribute more than 4% of the dinitrogen fixed globally, with some model estimates as high as 50%, but these estimates vary widely and remain highly uncertain given variable modeling approaches and limited knowledge of the extent and structure of these communities in many dryland ecosystems (Ferrenberg et al., 2017). Clearly, their contribution to global nitrogen cycling is significant and may be seriously threatened by changing climate and land use trends. Biocrusts are extremely fragile and easily damaged by compression. They are also sensitive to warming, drying trends and thus are now under threat in most drylands globally.

The lichen component of biocrusts represents a mutualism between either an alga or cyanobacterium (the **phycobiont**) and a fungus (the **mycobiont**). The phycobiont captures carbon from the atmosphere, while the fungal partner may secrete organic acids to mineralize nutrients and absorb them from the surfaces on which they grow. For those lichens that have a cyanobacterial partner, dinitrogen may also be fixed. Lichens colonize the surfaces of the soil, rocks, and higher plants, growing on the bark of trees both living and dead (e.g., in forest canopies), but deriving little nutritive value from them. In the tundra, lichens may represent a key food source for grazing animals. Lichens are also extremely sensitive to air pollution and have been used to assess its extent in peri-urban environments.

Summary

This chapter emphasizes the potential for N_2 fixation in legumes, as well as some of the problems associated with using this symbiosis. It also introduces the *Frankia*-actinorhizal and *Anabaena-Azolla* symbioses. The greater exploitation of all three symbioses will require a multidisciplinary approach and better understanding of microbial ecology.

Symbiotic dinitrogen fixation currently accounts for almost 50% of the nitrogen used in agriculture. With world population increases in the period to 2050 concentrated in developing countries, that contribution will have to increase. Many of the research skills needed to resolve specific problems in N_2 fixation are already in place. They include methods to breed for enhanced nitrogen fixation, fertilization practices appropriate to the inoculated legume, limiting the effects of environmental stress, and better integration of pesticide and inoculation methodologies. Although inoculant production in many regions of the world has considerable room for improvement, quality inoculant preparations are within the reach of most countries. Consistent support for applied research in N_2 fixation, for technology transfer and training, and for the better understanding of the nitrogen needs of both modern and traditional systems should improve the benefits coming from this important technology in both intensive and extensive production systems.

Acknowledgment

Portions of this chapter are derived from the work of Dr. Peter Graham (deceased) who authored a similar chapter for the Second Edition of the textbook. The current author and editors are grateful for his contributions to the earlier editions of the text as well as the Third Edition.

Study questions

1. As a Peace Corps volunteer in Nepal you are assigned the task of helping to develop an inoculant industry. What steps might be necessary in completing such an undertaking?
2. Explain why the presence of inorganic nitrogen in the soil inhibits biological nitrogen fixation.
3. Describe how to manage nitrogen additions to a mixed grass/legume sward in order to stimulate nitrogen fixation in the legume, while also obtaining equal growth and representation of both plant species.
4. What are the primary determinants of a legume's response to inoculation? Consider host, microsymbiont, and environmental factors in your answer.

5. You are a consultant to a legume introduction project in Latin America and advised your counterpart to use "stickers" in the inoculation of their seeds. They chose Coca Cola for simple seed inoculation and carpenter's glue for pelleted seeds. In both cases inoculation failed. Why?
6. Protecting nitrogenase from inactivation by oxygen is a key constraint to biological nitrogen fixation. Compare and contrast the mechanisms used to control the interaction between oxygen and nitrogenase in the legume/rhizobia, actinorhizal/*Frankia*, and the *Azolla*/*Anabaena* symbioses.
7. Name the two organisms involved in the lichen mutualism and then name the biological role of each partner in maintaining the existence of both partners in a given environment.
8. Legumes abound in most situations. Find one such plant, carefully dig it up, and describe its nodulation. How many nodules does it have, how are they distributed, and what type of nodules are they? Describe the internal appearance of one such nodule. How would you determine whether the organism contained in the nodule was *Rhizobium* or *Bradyrhizobium*?

References

Berry, A.M., Mendoza-Herrera, A., Guo, Y.-Y., Hayashi, J., Persson, T., Barabote, R., Demchenko, K., Zhang, S., Pawlowski, K., 2011. New perspectives on nodule nitrogen assimilation in actinorhizal symbioses. Funct. Plant Biol. 38, 645–652.

Catroux, G., Hartmann, A., Revellin, C., 2001. Trends in rhizobial inoculant production and use. Plant Soil 230, 21–30, https://doi.org/10.1023/A:1004777115628.

de Lajudie, P.M., Andrews, M., Ardley, J., Eardly, B., Jumas-Bilak, E., Kuzmanovic, N., Lassalle, F., Lindström, K., Mhamdi, R., Martinez-Romero, E., Moulin, L., Mousavi, S.A., Nesme, X., Peix, A., Pulawska, J., Steenkamp, E., Stępkowski, T., Tian, C.-F., Vinuesa, P., Wei, G., Willems, A., Zilli, J., Young, P., 2019. Minimal standards for the description of new genera and species of rhizobia and agrobacteria. Int. J. Syst. Evol. Microbiol. 69, 1852–1863, doi 10.1099/ijsem.0.003426.

Ewing, M.A., Howieson, J.G., 1989. The development of *Medicago polymorpha* L. as an important pasture species for southern Australia (pp. 197–198). In: Proceedings XVI International Grassland Congress, Nice, France.

Ferrenberg, S., Tucker, C.L., Reed, S.C., 2017. Biological soil crusts: Diminutive communities of potential global importance. Front. Ecol. Environ. 15 (3), 160–167, doi:10.1002/fee.1469.

Güereña, D., Lehmann, J., Thies, J.E., Enders, A., Karanja, N., Neufeldt, H., 2015. Partitioning the contributions of biochar properties to enhanced biological nitrogen fixation in common bean (*Phaseolus vulgaris*). Biol. Fertility Soils 51 (4), 479–491.

Gtari, M., Nouioui, I., Sarkar, I., Ghodhbane-Gtari, F., Tisa, L.S., Sen, A., Klenk, H.-P., 2019. An update on the taxonomy of the genus *Frankia* Brunchorst, 1886, 174AL. Antonie van Leeuwenhoek 112, 5–21.

Hellriegel, H., Wilfarth, H., 1886. Untersuchungen uber die stickstoffnahrung der gramineen und leguminosen. Beilageheft Ztsch. Ver. Rubenzucker-Industrie Deutschen Reichs, pp. 234.

Herridge, D.F., Peoples, M.B., Boddey, R.M., 2008. Global inputs of biological nitrogen fixation in agricultural systems. Plant Soil 311, 1–18.

Hichri, I., Meilhoc, E., Boscari, A., Bruand, C., Frendo, P., and Brouquisse, R., 2016. Nitric oxide: Jack-of-all-trades of the nitrogen-fixing symbiosis? In: Wendehenne, D. (Ed.), Advances in Botanical Research, vol. 77, Academic Press, NY.

Janczarek, M., Rachwał, K., Marzec, A., Grządziel, J., Palusińska-Szysz, M., 2015. Signal molecules and cell-surface components involved in early stages of the legume-rhizobium interactions. Appl. Soil Ecol. 85, 94–113.

Kleinert, A., le Roux, M., Kang, Y., Valentine, A.J., 2017. Oxygen and the regulation of N_2 fixation in legume nodules under P scarcity. In: Sulieman, S., Tran, L-S.P. (Eds.), Legume Nitrogen Fixation in Soils with Low P Availability. Springer International Publishing, Switzerland.

Limpens, E., Bisseling, T., 2003. Signaling in symbiosis. Curr. Opin. Plant Biol. 6, 343–350.

Moulin, L., Munive, A., Dreyfus, B., Boivin-Masson, C., 2001. Nodulation of legumes by members of the β subclass of proteobacteria. Nature 411, 948–950.
Pawlowski, K., Demchenko, K.N., 2012. The diversity of actinorhizal symbiosis. Protoplasma 249, 967–979.
Peoples, M.B., Unkovich, M.J., Herridge, D.F., 2009. Measuring symbiotic nitrogen fixation in legumes. In: Emmerich, D.W., Krishnan, H.B. (Eds.), Nitrogen Fixation in Crop Production. American Society of Agronomy, Madison, WI.
Phillips, P.W.B., 2004. An economic assessment of the global inoculant industry. Crop Manage. 3, 0, doi:10.1094/CM-2004-0301-08-RV.
Poole, P., Ramachandran, V., Terpolilli, J., 2018. Rhizobia: from saprophytes to endosymbionts. Nat. Rev. Microbiol. 16, 291–303.
Silvester, W.B., 1975. Ecological and economic significance of the non-legume symbioses. In: Newton, W.E., Nyman (Eds.), First International Symposium on Nitrogen Fixation, Washington State University Press, WA.
Simonet, P., Navarro, E., Rouvier, C., Reddell, P., Zimpfer, J., Dommergues, Y., Boudin, R., Combarro, P., Hamden, J., Domenach, A.M., Goubiere, F., Prin, Y., Dawson, J.O., Normand, P., 1999. Coevolution between *Frankia* populations and host plants in the family Casuarinaceae, and consequent patterns of global dispersal. Environ. Microbiol. 1, 525–533.
Stanley, J., Cervantes, E., 1991. Biology and genetics of the broad host range *Rhizobium* sp. NGR234. J. Appl. Bacteriol. 70, 9–19.
Sullivan, J.T., Patrick, H.N., Lowther, W.L., Scott, D.B., Ronson, C.W., 1995. Nodulating strains of *Rhizobium loti* arise through chromosomal symbiotic gene transfer in the environment. Proc. Natl. Acad. Sci. USA 92, 8985–8989.
Sullivan, J.T., Ronson, C.W., 1998. Evolution of rhizobia by acquisition of a 500-kb symbiosis island that integrates into a phe-tRNA gene. Proc. Natl. Acad. Sci. USA 95, 5145–5149.
Terpolilli, J.J., Hood, G.A., Poole, P.S., 2012. What determines the efficiency of N_2-fixing Rhizobium-legume symbioses? Adv. Microb. Physiol. 60, 325–389.
Thies, J.E., Singleton, P.W., Bohlool, B.B., 1991. Influence of the size of indigenous rhizobial populations on establishment and symbiotic performance of introduced rhizobia on field-grown legumes. Appl. Environ. Microbiol. 57, 19–28.
Unkovich, M.J., Pate, J.S., 2000. An appraisal of recent field measurements of symbiotic N_2 fixation by annual legumes. Field Crops Res. 65, 211–228.

Further reading

Lhuissier, F.G.P., de Ruijter, N.C.A., Sieberer, B.J., Esseling, J.J., Emons, A.M.C., 2001. Time course of cell biological events evoked in legume root hairs by *Rhizobium* nod factors. State of the art. Ann. Bot. 87, 289–302.
Rai, A.N., Soderback, E., Bergman, B., 2000. Cyanobacterium-plant symbioses. New Phytol. 147, 449–481.
Santi, C., Bogusz, D., Franche, C., 2013. Biological nitrogen fixation in non-legume plants. Ann. Bot. 111, 743–767.
Somasegaran, P., Hoben, H., 1994. Handbook for Rhizobia. Springer, Verlag, NY.
Syiem, M.B., Singh, A.K., Rai, A.N., 2017. N_2-fixing cyanobacterial systems as biofertilizer. In: Singh, J., Seneviratne, G. (Eds.), Agro-Environmental Sustainability, Volume 1: Managing Crop Health. Springer International Publishing, Switzerland.

CHAPTER

18

Microbial transformations of sulfur in soil

V.V.S.R. Gupta[a], James J. Germida[b]

[a]CSIRO Agriculture and Food, Locked Bag 2, Glen Osmond, SA, Australia; [b]Department of Soil Science, University of Saskatchewan, Saskatoon, SK, Canada

... sulfur is a devilish substance.... discharged from the bowels of the earth, by volcanoes or evil-smelling hot springs.....surely the effluent of Hell itself—**J.R. Postgate**

This chapter introduces the basic principles of the sulfur cycle and the microorganisms that drive it. This information will give one an appreciation of how important it is to understand the biogeochemistry of sulfur and other elements is soil and other ecosystems.

Sulfur (S) is an essential element for the growth and activity of all living organisms. It is 1 of the 10 major bio-elements required by organisms in relatively high concentrations (i.e., $>10^{-4}$ M). Sulfur is required for the synthesis of the amino acids cysteine, cystine, and methionine, and plays an active role in plants, animals, and microorganisms as an important constituent of vitamins, hormones, structural components, and for metabolic processes (e.g., coenzyme A). For example, the **disulfide bond** formed between cysteine residues helps stabilize the tertiary structure of proteins (Box 18.1). Several sulfur compounds are also important for plant protection against pests, diseases, and for food quality Box 18.1.

Sulfur is an important source of metabolic energy for many bacteria. For example, certain chemoautotrophic (e.g., chemolithotrophic) bacteria obtain energy for cell growth and division by oxidizing reduced inorganic sulfur compounds. In fact, some exotic ecosystems, such as hot sulfur springs and hydrothermal-vent communities, are driven by energy generated in the oxidation of sulfur, which they use to fix carbon (Box 18.2). These environments also support diverse communities of microorganisms (Box 18.3).

Although sulfur is considered to be a macronutrient, its availability to biota and plants can be limiting in some ecosystems; and then it must be applied as a supplement or fertilizer. However, in other instances, too much sulfur may be present and it can be considered a pollutant. For example, burning high–sulfur–content (e.g., bituminous and lignite)

Principles and Applications of Soil Microbiology. http://dx.doi.org/10.1016/B978-0-12-820202-9.00018-6
Copyright © 2021 Elsevier Inc. All rights reserved.

BOX 18.1

Disulfide bonds stabilize protein structure

A disulfide bond is a covalent linkage between two sulfur atoms (-S-S-) which plays an important role in protein folding and stability. This type of linkage is commonly found in polypeptides and protein molecules (i.e., tertiary and/or quaternary structures of proteins) and may occur as an interchain or intrachain bond between sulfur-containing amino acids. This type of bond is not common to all proteins but is usually critical to those that do possess it.

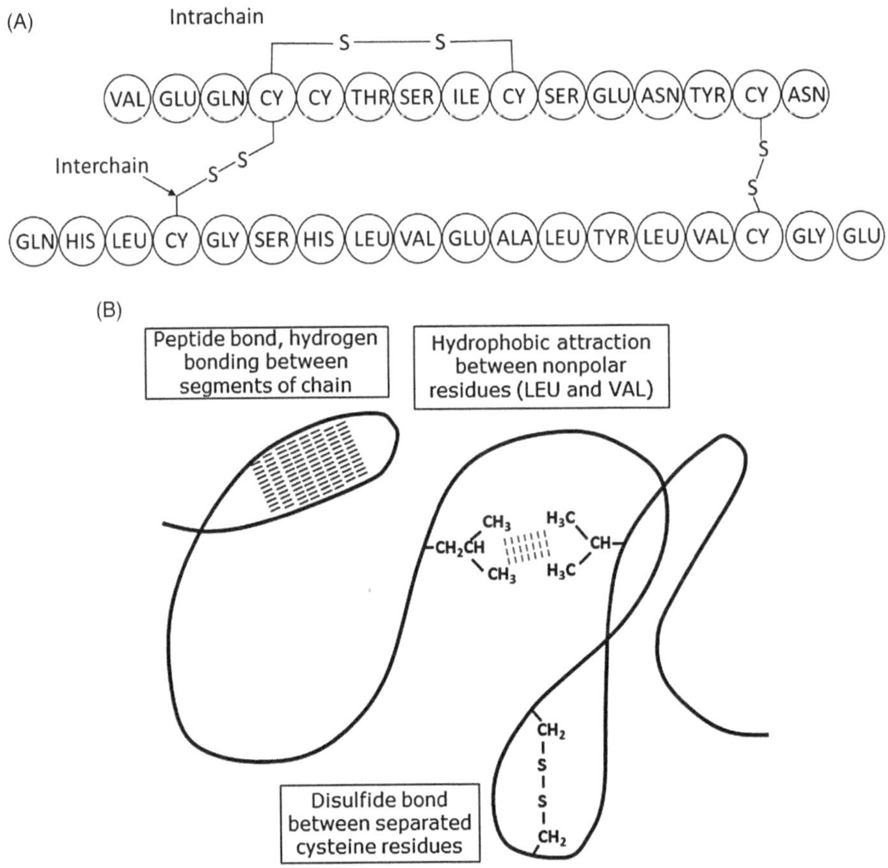

(A) Amino acid sequence of two polypeptide chains illustrating how intrachain and interchain disulfide covalent bonds form between cysteine residues. (B) Diagrammatic representation of how disulfide bonds and other stabilizing forces confer a unique spatial conformation on proteins. Not all types of forces or their frequency of occurrence are shown.

BOX 18.2

Life without light: the importance of chemosynthesis

All life requires energy. Furthermore, it is established that the sun's energy trapped through photosynthesis is the driving force for primary productivity in all ecosystems. About 50 years ago, scientists made one of the most startling discoveries of 20th-century biology. They found bizarre, exotic, and complex biological communities living in the deep ocean where hydrothermal fluids rise through the earth's crust. The communities at these deep sea-thermal vents consist of giant tubeworms and masses of large clams. Life here is sustained by primary productivity based on microbial sulfur oxidation (Jannasch and Mottl, 1985).

Hydrogen sulfide (H_2S) is the most abundant compound in the deep sea-thermal vent fluid. This compound is extremely toxic to higher animals, but can be used by some sulfur-oxidizing bacteria as an energy substrate. Considerable energy is released when hydrogen sulfide is oxidized. Through a process called chemosynthesis, analogous to photosynthesis, vent bacteria employ chemical rather than light energy to fix inorganic carbon to make organic compounds. More surprisingly, the giant tubeworms and clams living at the vent are in a symbiotic relationship with these sulfur-oxidizing bacteria that live in their tissue. Also, the distinct habitats in the hot core of the chimney, cooler middle layers and outer surfaces, in terms of pH, redox potential and metals, support diverse and novel bacterial and archaeal communities. This is truly a remarkable example of how life adapts to extreme environments and is found where least expected. In fact, it now appears that microbial life exists in all environments with an oxidizable energy source and favorable conditions for microbial life.

NOAA Okeanos Explorer Program, Galapagos Rift Expedition 2011—Flickr NOAA Photo Library. Photo of one of the largest concentrations of *Riftia pachyptila* observed, with anemones and mussels colonizing in close proximity. From the 2011 NOAA Galapagos Rift Expedition. The original NOAA image has been modified by increasing brightness. The bacteria found in symbiosis with tube worms at hydrothermal vents use chemosynthesis to fix carbon to make organic compounds. Shown in the sketch is the breathing organ, which contains the symbiotic association. The bacteria use the hydrosulfide ion (HS-) and oxygen (O_2) to generate chemical energy (chemosynthesis). The hydrosulfide ion serves as the electron donor; the energy is released when HS- and O_2 are combined, driving a series of reactions called the Calvin cycle. In the Calvin cycle, organic compounds are formed as the carbon dioxide (CO_2) present in both vent water and sea water is fixed. In this illustration, *R* stands for the auxiliary compounds, including CO_2-fixing enzymes, involved in the cycle. Note that sulfuric acid is produced as a by-product. The process is analogous to photosynthesis, in which green plants use light energy to fix carbon dioxide and form organic compounds. *Source: Adapted from Tunnicliffe (1992).*

BOX 18.3

Biodiversity of sulfur bacteria at hydrothermal sea vents

Oceanic (deep-sea) hydrothermal vents, the distinctive chimney-like structures (e.g., black smokers) are S-rich environments supporting diverse communities of bacteria and archaea. They are formed when sulfates and metal sulfides precipitate from mixing of hot anaerobic fluids with cold, oxic seawater. Based on maximal available energy of representative chemolithotrophic metabolisms, it is hypothesized that S-trophic chemolithotrophy would be always predominant in any type of hydrothermal vent and at any temperature range.

Earlier studies of these environments concentrated on bacteria obtained through culturing of hyperthermophiles and mesophiles including S-reducing bacteria. Culture independent techniques and fluorescent in situ hybridization (FISH) using 16S rRNA specific probes to study microbial diversity have identified the extensive diversity and novel species of bacterial and archaeal communities in these vents (Harmsen et al., 1997).

A diverse group of archaeal 16S rRNA gene phylotypes representing three distinct phyla and unclassified archaeal groups have been recorded from deep-sea hydrothermal environments. For example, novel bacteria identified from a mid-Atlantic ridge vent included members of the genus *Desulfurobacterium*, while other sequences were related to ß-Proteobacteria, the group containing the genus *Thiobacillus*. Some sequences belong to the ε-Proteobacteria such as *Sulfurimonas*, which were also detected at hot springs at Yellowstone National Park. Using gene probes for dissimilatory bisulfate reductase genes (primers highly effective with *Desulfovibrio*, *Desulfobulbus*, and *Desulfobacter* spp.), bacteria were also detected at sea vents. These studies provide clues to the evolutionary origins of terrestrial sulfur bacteria.

A study investigating a diffusely venting chimney spire from the East Pacific Rise (9°N), found archaea representing families of hyperthermophilic, anaerobic archaea with optimal growth temperatures between 75°C and 106°C (Kormas et al., 2006). Some of them were species related to the genera *Thermococcus*, *Pyrococcus*, and *Palaeococcus* (Thermococcales), the sulfate-reducing and iron-reducing genera *Archaeoglobus*, *Geoglobus*, and *Ferroglobus* (Archaeoglobales), the methanogenic genus *Methanocaldococcus* (Methanococcales), the sulfur-reducing or fermentative genera within the Desulfurococcales, and the sulfur-reducing, heterotrophic genera within the Thermoproteales.

coals produces gaseous sulfur emissions that can become components of acid precipitation. Microbial oxidation of reduced sulfur minerals associated with coal seams, and the subsequent leaching of the resulting **oxyanions** (e.g., SO_4^{2-}) gives rise to acid mine drainage which can pollute streams or acidify reclaimed soils. Finally, sulfur is also important for many industrial processes, and can be used, for example, in the production of chemicals, concrete, and asphalt. As you read this chapter it will become evident that sulfur is an essential, versatile, and economically and ecologically important element.

Sulfur cycle in agroecosystems and terrestrial environments

The sulfur cycle bears many similarities but is more complex than the nitrogen cycle. Both of these elements exist in a number of oxidation states and undergo similar types of chemical reactions and biological transformations, including volatilization. The majority of sulfur is found in the **lithosphere** (Table 18.1).

Most nitrogen on earth is also in the lithosphere; however, dinitrogen in the atmosphere is the major pool of biologically available nitrogen through dinitrogen fixation (Chapter 16). Only a small portion of the sulfur pool is found in the atmosphere, and most sulfur that cycles through the atmosphere is because of human activities. In fact, since the Industrial Revolution, increased burning of fossil fuels has almost doubled the rate of sulfur entering the atmosphere to approximately 1.5×10^{11} kg S year^{-1}. The volatilization of sulfur as hydrogen sulfide, carbonyl sulfide, and dimethyl sulfide, for example, from marine algae, marsh lands, mud flats, plants, and soils also contributes to the global circulation of sulfur through the atmosphere. Biological sources of sulfur volatilization total about 4.05×10^{10} kg S year^{-1}.

The nature and quantities of the various sulfur pools in surface soils are the basis for sulfur cycling in terrestrial environments. These sulfur pools are influenced by pedogenic factors such as climate, regional vegetation, and local topography. For example, the total sulfur content of soils ranges from 0.002% to 10%, with the highest levels being found in tidal flats and in saline, acid sulfate, and organic soils. The impact of pedogenic factors on the total sulfur

TABLE 18.1 Estimated quantities of sulfur in major global sulfur pools.

Pool	Mass (kg S)
Atmosphere	3.6×10^{9}
Hydrosphere	1.3×10^{18}
Oceans	1.3×10^{18}
Marine organisms	2.4×10^{10}
Fresh waters	3.0×10^{12}
Ice	6.0×10^{12}
Lithosphere	24.1×10^{18}
Igneous rocks	5.0×10^{18}
Metamorphic rocks	11.4×10^{18}
Sedimentary rocks	7.7×10^{18}
Evaporites	5.1×10^{18}
Shales	2.0×10^{18}
Limestones	0.1×10^{18}
Sandstones	0.3×10^{18}
Soils	2.6×10^{16}
Soil organic matter	1.0×10^{13}
Biosphere	7.6×10^{12}

From Pierzynski et al. (2000).

TABLE 18.2 Amounts and distribution of sulfur in some world soils.

Type of soil	Location	Total S (mg kg^{-1})
Agricultural	Alberta, Canada	80–1,593
	British Columbia, Canada	214–438
	Saskatchewan, Canada	88–760
	New South Wales, Australia	38–545
	Queensland, Australia	11–725
	New South Wales, Australia	128–133
	South Australia, Australia	182–279
	Western Australia, Australia	50–770
	Victoria, Australia	83–224
	Iowa, United States	78–452
	Delaware, United States	1200–1,700
	Iowa, United States	2300–3,700
	Georgia, United States	2900–3,700
	West Indies	110–510
	New Zealand	240–1,960
	Nigeria	25–177
	Argentina	164–287
	Brazil	43–398
Forest	British Columbia, Canada	162–2,328
	Germany	74–328
	Illinois, U.S.	112–555
	New Hampshire, United States	452–1,563
	New York, United States	452–1,563
Tidal marsh	Carolinas, United States	3,000–35,000
Surface	Iowa, United States	55-618
Organic	England	7,405
Acid	Scotland	300–800
Calcareous	Scotland	460–1,790
Volcanic	Hawaii, United States	180–2,200
	Tochigi, Japan	820–1270

Adapted from Pierzynski et al. (2000), Paul (2007), Zhao et al. (2015), Tanikawa et al. (2014), Wyngaard and Cabrera (2015).

concentrations in the surface is noted when comparing values for soils from diverse geographic areas, as shown in Table 18.2.

Nature and forms of organic and inorganic sulfur in soil

Organic sulfur constitutes more than 90% of the total sulfur present in most surface soils. There is a diverse array of organic sulfur compounds in soils and the precise nature of all the organic sulfur compounds in soil cannot be clearly identified. Thus, organic sulfur is grouped into two broad categories, organic sulfates and carbon-bonded sulfur. Examples of these organic compounds are given in Fig. 18.1.

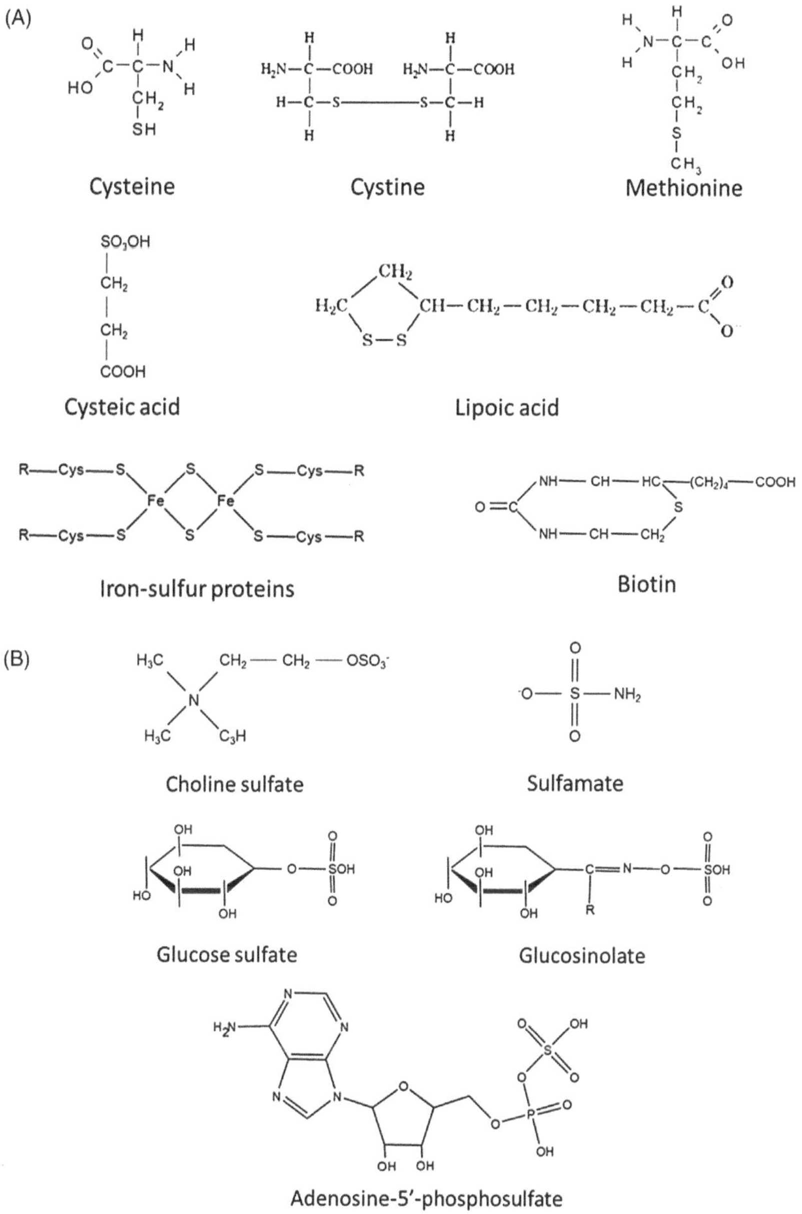

FIGURE 18.1 **Examples of some of the organic sulfur compounds found in soils.** (A) Amino acids and other compounds (cysteic and lipoic acids) containing carbon-bonded sulfur. Note that lipoic acid also has a disulfide bond. (B) Compounds that possess ester sulfate bonds (e.g., C-O-S and C-N-S). This class of compounds is sometimes used by microorganisms to store sulfur, although some bacteria are able to store elemental sulfur.

Organic sulfates (R-O-S) include sulfate esters (C-O-S), sulfamates (C-N-S), and sulfated thioglycosides (N-O-S). Organic sulfates constitute 30%–75% of total organic sulfur in soil. Carbon-bonded sulfur (C-S) includes the sulfur present in amino acids, proteins, polypeptides, heterocyclic compounds (e.g., biotin and thiamine), sulfinates, sulfones, sulfonates, and sulfoxides (Table 18.3). A large portion of carbon-bonded sulfur present in soil has yet to be identified; however; in some cases, the carbon-bonded sulfur of amino acids may constitute up to 30% of the organic sulfur in soil.

Inorganic forms of sulfur account for less than 25% of the total sulfur in most agricultural soils. Sulfur exists in a number of forms with a wide range of oxidation states (−6 to +2) (Table 18.3). Sulfide, elemental sulfur, sulfite, thiosulfate, tetrathionate, and sulfate are the

TABLE 18.3 Important forms of sulfur and their oxidation states.

Compound	Formula	Oxidation state(s) of sulfur
Inorganic		
Sulfide	S^{2-}	–2
Polysulfide	S_n^{2-}	–2, 0
Sulfur[a]	S_8^0	0
Hyposulfite (dithionite)	$S_2O_4^{2-}$	+3
Sulfite	SO_3^{2-}	+4
Disulfite	$S_2O_5^{2-}$	+5
Sulfur dioxide	SO_2	+2
Sulfur trioxide	SO_3	+6
Thiosulfate[b]	$S_2O_3^{2-}$	–1, +5
Dithionate	$S_2O_6^{2-}$	+6
Trithionate	$S_3O_6^{2-}$	–2, +6
Tetrathionate	$S_4O_6^{2-}$	–2, +6
Pentathionate	$S_5O_6^{2-}$	–2, +6
Sulfate	SO_4^{2-}	+6
Organic[c]		
Di- and poly-sulfides	S_n^{2-}	0
Thiophenes	C_4H_4S	0
Sulfonium compounds	H_3S^+	+1
Sulfoxides[c]	R-S(=O)-R′	+2
Sulfones	R-S(=O_2)-R′	+4
Sulfonates	RSO_3^-	+5
Ester sulfates	R-C-O-SO_3^-	+6

[a] *Occurs in an octagonal ring in crystalline form.*
[b] *For compounds with covalent bonds, including organic compounds, the outer S has a valence of −1; inner S has a valence of +5.*
[c] *R and R′ are a grouping of C and hydrogen atoms.*
From Vairavamurthy et al. (1993), Vairavamurthy (1998), Lomans et al. (2002).

main forms of inorganic sulfur in agricultural soils. Sulfate is the most common form of inorganic sulfur found in well-aerated agricultural soils, whereas sulfides account for less than 1% of total sulfur and measurable quantities of thiosulfate and tetrathionate are usually detected only in soils treated with sulfur fertilizer or those receiving pollutants.

Microbial transformations of sulfur in soil

The sulfur cycle—emphasizing soil and plant sulfur transformations in agroecosystems—is illustrated in Fig. 18.2.

As indicated earlier, the major forms of sulfur in soil include elemental sulfur (S^0), sulfides (S^{2-}), sulfates (S^{6+}), and organic sulfur compounds. Most of this sulfur enters the soil as soluble inorganic forms produced during the weathering of minerals, from fertilizers and atmospheric deposition, or as soluble organic and inorganic forms from the decomposition of organic matter. Losses of sulfur occur through leaching, surface runoff, and crop removal. Volatilization losses of sulfur can also occur especially under water-logged conditions.

Soil microorganisms drive the sulfur cycle. Hence, sulfur undergoes many microbially mediated transformations in soil, including:

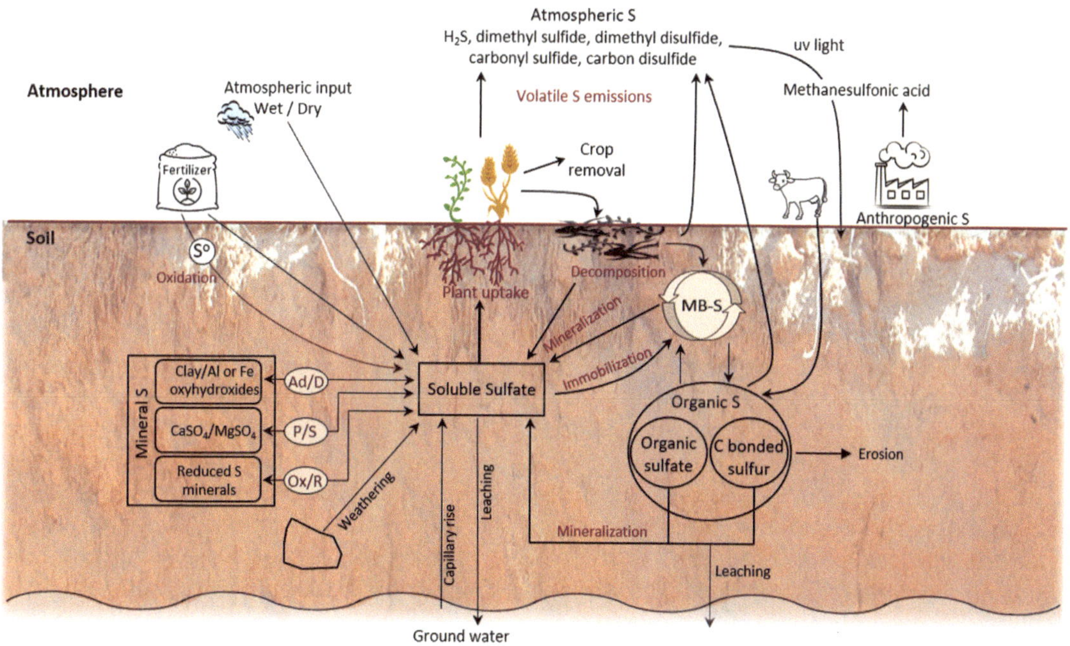

FIGURE 18.2 **Conceptual sulfur cycle in agroecosystems.** *Ad/D*, adsorption/desorption; *P/S*, precipitation/solubilization; *Ox/R*, oxidation/reduction; *MBS*, microbial biomass sulfur. The relative flux of sulfur between pools is reflected in the thickness of the *arrows* depicting microbial conversion.

- oxidation and reduction reactions,
- mineralization and immobilization reactions, and
- volatilization reactions.

The soil microbial biomass is the key driving force behind all sulfur transformations. This biomass acts as both a source and sink for inorganic sulfate, whereas microbial activity regulates both the fluxes of sulfur between different pools (inorganic sulfate, labile organic sulfur, resistant organic sulfur) and the losses of sulfur from these pools (e.g., conversion of complex organic sulfur compounds into mobile forms that may be lost by leaching). Most of the sulfur in soil (75%–90%) is found in organic complexes. These complexes are either stable, passive fractions that turn over very slowly or active, dynamic fractions that are readily transformed or metabolized. Actually, the continuum of sulfur organic complexes in soil ranges from very old, stable (e.g., organic sulfur found in stable soil organic matter) fractions to very young, short-lived (e.g., organic sulfur found in the amino acid cysteine) fractions. The microbial biomass is the engine for the conversion of passive fractions into active fractions, and vice versa. This is illustrated in the following diagram, where the relative flux of sulfur between pools is reflected in size of the arrows depicting microbial conversion:

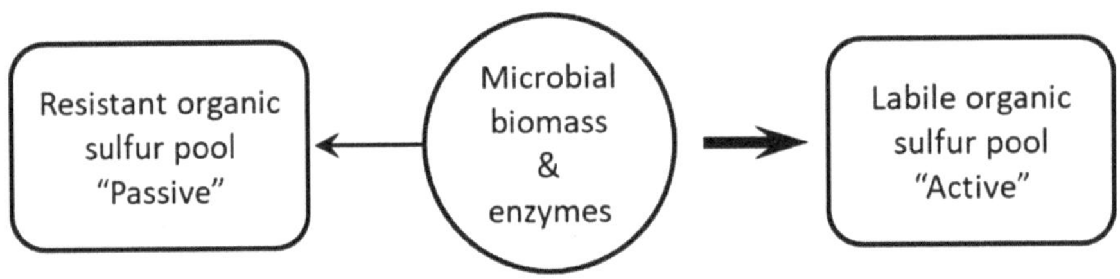

Understanding the sources, sinks, and transformations of sulfur in soil is crucial to ensure adequate supplies of sulfur for the biota of various ecosystems and to protect the environment from the detrimental effects of too much sulfur. All the main biological reactions in the sulfur cycle are closely related to the carbon cycle and the amount of carbon involved varies depending upon the type of organisms undertaking the process. Because microorganisms are so important for the conversions between the active and passive organic matter pools, any factor that disturbs or otherwise has an impact on the microbial biomass influences sulfur cycling. For example, crop rotations and soil cultivation typically increase sulfur cycling in soil. When a soil is cultivated, it is mixed and churned and broken up into smaller pieces. This increases soil aeration and exposes soil particles and organic matter that was previously "hidden/protected" from the soil microflora. Fresh organic matter, containing different ratios of carbon, nitrogen, phosphorus, and sulfur, is mixed into the soil. Some microbial biomass is activated due to the flush of available carbon and nutrients, and some of the biomass might be killed. For example, fungal hyphae help to hold soil aggregates together; as cultivation breaks up these aggregates, the hyphae are broken, resulting in dead biomass. This dead biomass is now available to be mineralized. As new microbial biomass is formed during decomposition of the newly exposed or added "active" organic material, nutrient elements (carbon, nitrogen, phosphorus, sulfur, and micronutrients) are cycled back and forth between active to passive states as organic and inorganic forms of the element. The overall consequence is increased nutrient cycling.

Similarly, interactions among different microbial groups, such as predation and parasitism, tend to increase the turnover rate of microbial biomass sulfur and hence affect sulfur fluxes. One can think of this in the following way. Bacteria and fungi store sulfur in the organic and inorganic sulfur compounds that comprise their cells. The proportion of ester sulfate sulfur is generally higher in fungi (14%–45%) compared to that in bacteria (6%–23%). When predators, such as soil amoebae (Chapter 8), eat bacterial cells or pieces of fungal hyphae, that biomass material (labile or active sulfur pool) is broken down into smaller, nonmetabolized sulfur containing organic fragments resistant to further metabolism by the amoebae. These fragments are then excreted into the soil, along with any inorganic sulfur or metabolizable organic sulfur not needed by the amoebae for growth. The excess inorganic sulfur or metabolizable organic sulfur is now available for plants and other organisms to use. This is an example of the opposing processes of **mineralization** and **immobilization** of nutrients.

$$\text{Organic S} \underset{\text{Immobilization}}{\overset{\text{Mineralization}}{\rightleftharpoons}} \text{Inorganic S}$$

Biological mineralization and immobilization are processes that occur concurrently and exhibit a strong relationship with the soluble sulfate pool in soil. Immobilization occurs as a result of the microbial assimilation of nutrients that are then rendered unavailable for further plant or microbial uptake until the cell dies and is remineralized. Immobilization of sulfur may also involve precipitation as metal sulfide, especially pyrite, as in salt marshes. Because these transformations are mediated by microorganisms, soil factors that influence the growth and activity of microorganisms (e.g., pH, temperature, moisture, and availability of carbon) will also affect the rate of sulfur transformations. To estimate or predict the available sulfur status of soils, it is necessary to understand the factors that influence these processes.

Immobilization (assimilation) of sulfur

Microbial assimilation and conversion of inorganic sulfate into organic sulfur through the ***assimilatory sulfate reduction pathway*** leads to temporary immobilization of sulfur from plant or microbial availability. This process involves **ATP sulfurylase** and two energy-rich sulfate nucleotides, **APS** (adenosine 5′-phosphosulfate) and **PAPS** (3′-phosphoadenosine-5′-phosphosulfate). The overall reaction of SO_4^{2-}−S uptake from soil and incorporation into amino acids in plants is as shown in Fig. 18.3.

Most of the sulfur accumulated by microorganisms is in the form of amino acids in proteins; however, microorganisms also accumulate sulfate esters, sulfonates, vitamins, and cofactors. Some microorganisms, such as fungi, accumulate especially large amounts of sulfate esters (Fig. 18.1). This is important because organic sulfates (e.g., sulfate esters and thioglucosides) are considered to be the most labile form of organic sulfur in soil and may comprise up to 30%–70% of the total sulfur in surface soils. The relative proportion of fungal biomass to bacterial biomass in soil (approximately 2:1) underscores the potential importance of microorganisms accumulating ester-sulfur compounds.

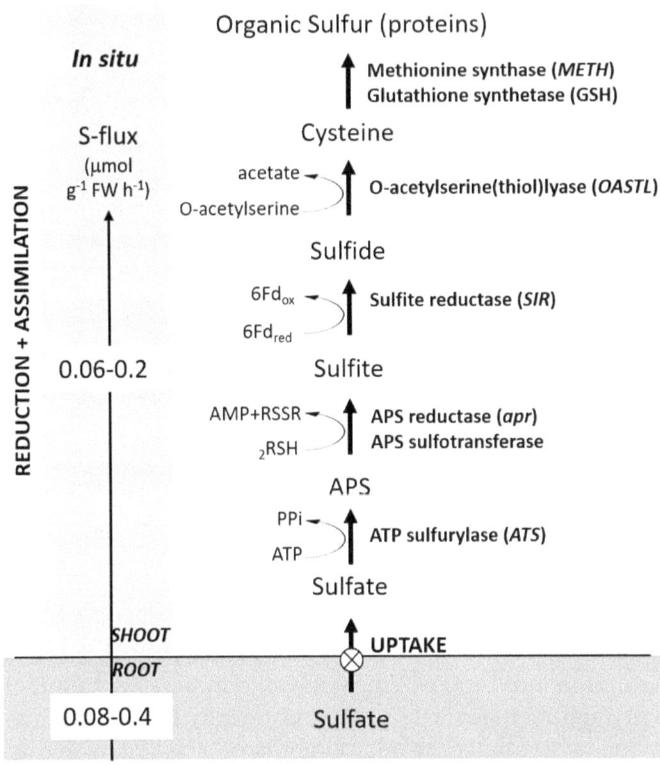

FIGURE 18.3 **Assimilatory reduction pathway of sulfate and formation of organic sulfur (proteins) with enzymes and genes.** (*RSH*, thioredoxin; *APS*, adenosine 5′-phosphosulfate; Fd_{red}, Fd_{ox}, reduced and oxidized ferredoxin; *RSH, RSSR*, reduced and oxidized glutathione). Left side: Ranges for rate (µmol g^{-1} fresh weight h^{-1}) of root sulfate uptake and its reduction and assimilation in the shoots of different plant species grown with optimal sulfur supply. *Source: Modified from De Kok et al. (2002), Gonzalez-Ballester and Grossman (2009).*

Factors affecting the immobilization of sulfur in soil

Typically, the addition of inorganic SO_4^{2-}–S to soil leads to its quick incorporation into the organic sulfur fractions via microbial assimilation. The rate and magnitude of this immobilization is increased in the presence of an energy source, such as metabolizable organic matter including plant residues or addition of easily degradable carbon sources like glucose. Later, much of this accumulated sulfur is found in the fulvic acid fractions, especially as organic sulfates.

Mineralization of sulfur

Mineralization of organic sulfur in soil is largely mediated by microbial activity. The various known pathways of sulfur mineralization are summarized in Fig. 18.4. Carbon-bonded sulfur is mineralized either through oxidative (aerobic) decomposition or desulfurization (anaerobic) processes, whereas various **sulfatases** (enzymes that remove sulfate groups) are

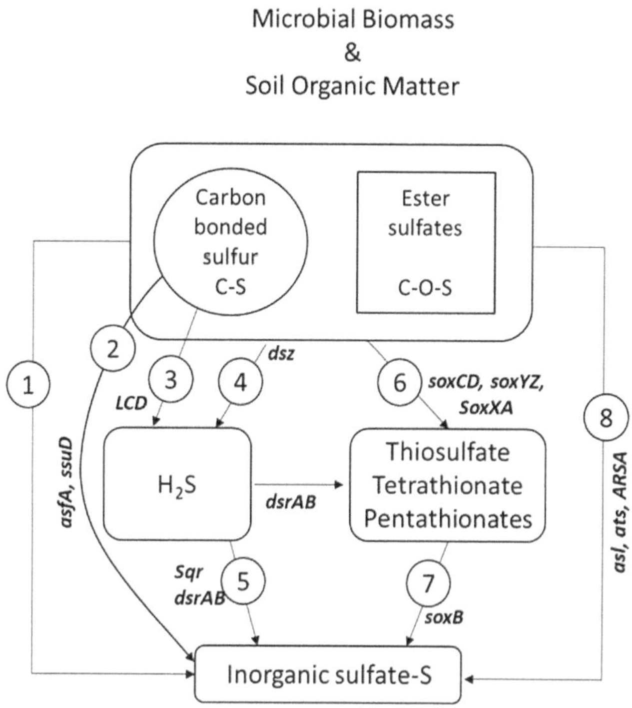

FIGURE 18.4 **Known pathways and examples of associated functional genes (e.g., *soxCD*) for mineralization of organic sulfur compounds in soil.** (1) Biological mineralization (direct) during oxidation of carbon as an energy source; (2) mineralization through desulfonation of aromatic and aliphatic sulfonates; (3) hydrolysis of cysteine by cysteine desulfhydrolase; (4) anaerobic mineralization (desulfurization) of organic matter; (5) biological oxidation of hydrogen sulfide to sulfate through elemental sulfur and sulfite by sulfide:quinone oxidoreductase; (6) incomplete oxidation of organic sulfur into inorganic sulfur compounds; (7) oxidation of tetrathionate, thiosulfate through sulfide; and (8) biochemical (indirect) mineralization of sulfate esters through hydrolysis by sulfatases.

involved in the mineralization of sulfate esters from bacteria, archaea, and fungi (Box 18.4). The mineralization process may be cell mediated, involving viable microbial cells, or enzymatic, involving enzymes such as sulfatases. In the case of cellular mineralization, elements such as nitrogen and sulfur in direct association with carbon are mineralized as microorganisms oxidize the organic carbon compounds to obtain energy. Heterotrophic soil microorganisms decompose organic sulfur compounds in order to grow; as the carbon-sulfur bond is broken, the sulfur is released, usually as sulfide. Because this process involves actively growing microorganisms, their requirement for sulfur may meet or even exceed the sulfur supplied by the substrate. Thus, net mineralization of sulfur by this process may not be reflected by increases in the sulfate-sulfur pool in soil. In the case of enzymatic mineralization, those elements that exist as sulfate esters are hydrolyzed by intracellular or extracellular enzymes. This process occurs mainly outside the cell and may be regulated by end-product inhibition, that is, the sulfate level. Cellular mineralization is controlled by the microbial need for carbon and energy sources, whereas enzymatic mineralization is controlled by factors influencing enzyme synthesis, activity, and kinetics.

BOX 18.4

Enzymes drive mineralization of sulfate esters in soil

Sulfatases (technically referred to as sulfohydrolases EC 3.1.5) are enzymes of the esterase class that hydrolyze sulfuric acid esters, where the linkage with sulfate is in the form of R-O-S and R represents a diverse group of organic moieties. This reaction can be shown as follows:

$$R\text{-}O\text{-}SO_3^- + H_2O \rightarrow R\text{-}OH + SO_4^{2-} + H^+$$

Arylsulfatases, encoded by genes such as *aslA*, *ats* and ARSA, are by far the most studied enzymes involved with sulfur cycling in soil. This is because organic sulfates are abundant in soil (30%–75% of total sulfur), and hence arylsulfatases play an important role in the mineralization of organic soil sulfur to plant-available sulfur. For example, ARSA is a protein-encoding gene that provides instructions for making the enzyme Arylsulfatase A. The enzyme may be extracellular or associated with cell debris, and its activity is easily determined based on colorimetric assays. Variations in the activity and kinetics of arylsulfatase in different soils indicate that the sources of the enzyme vary depending upon land-use management and soil physico-chemical properties such as organic matter content and sulfate sulfur concentrations.

Factors affecting the mineralization of sulfur in soil

Mineralization is generally measured as net mineralization, that is, the amount of SO_4^{2-}–S accumulated during the time period under study, or the difference between gross mineralization and microbial assimilation. Thus, for higher net mineralization to occur, the mineralization-assimilation process has to be driven toward mineralization. A break-even point for mineralization and immobilization can be calculated based on the carbon-to-sulfur ratio of the substrate, the decomposing organisms, and the yield coefficient. For example, if we consider the decomposition of crop residues, net mineralization will generally occur with a carbon-to-sulfur ratio of 200 or less, whereas net sulfur mineralization activity is the driving force for mineralization and immobilization, these processes are significantly influenced by all factors affecting microbial metabolism:

- energy and nutrient supply,
- carbon-to-sulfur ratio,
- abundance of organic sulfur,
- water availability,
- pH,
- temperature, and
- redox potential.

In addition, actively growing plants may significantly increase sulfur mineralization in soils. Plants supply energy sources to the rhizosphere in the form of root exudates that increase microbial growth and activity, thus increasing sulfur mineralization. However, the re-assimilation of inorganic sulfates released by the growing microorganisms may result in no increase in the sulfur pool available to plants and may even result in reduction of that pool when the microbial demand exceeds the rate of sulfur mineralization. Different plants excrete different types and amounts of root exudates and require different amounts of sulfur for growth. Hence, crop rotations can have a significant impact on sulfur cycling in soil. Additionally, the composition of the microbial community, for example, the bacteria:fungi ratio, could also affect the balance between mineralization and immobilization.

Detecting sulfur mineralization genes in soil

Unlike the cultivation-based methods, cultivation independent approaches allow the measurement of the abundance (quantitative PCR method) and genetic composition (sequencing-based method) of groups of microorganisms involved in specific biological processes. Mineralization of organic sulfur is achieved by a multi-enzyme complex that is regulated by a diverse set of genes in bacteria and fungi (Fig. 18.4). For example, the enzymes involved in the mobilization of organosulfonates, e.g., aromatic sulfonate sulfur, are regulated by genes such as *asfA* and *ssuD*(Fig. 18.4). The *asfA* gene-based analysis indicated that bacteria belonging to Gamma- and Beta-proteobacteria were the dominant desulfonating group (~30%) in the rhizosphere and hyphosphere (associated with arbuscular mycorrhizal hyphae) environments both in agricultural and non-agricultural soils (Fig. 18.5). It is suggested that this group of bacteria has the potential to improve S supply to plants when sulfate supply is limiting in soils as sulfate levels can alter the community composition (Schmalenberger et al., 2010).

Sulfatases are highly conserved enzymes and the genetic capacity for enzymatic hydrolysis of ester sulfates occurs in bacteria, fungi, and archaea, encoded by genes such as *asl*, *ats*, and ARSA (Fig. 18.4). Plant rhizosphere soils generally show higher arylsulfatase activity and abundance of microbes harboring sulfatase genes. Genomic analysis of rhizosphere soils suggested that plant species affect the abundance and composition of this functional bacterial community through their rhizodeposits in turn influencing overall sulfur dynamics (Cregut et al., 2009).

Microbial oxidation of inorganic sulfur compounds

The element sulfur can exist in a number of oxidation states ranging from −2 to +6 in the forms of tetrathionate ($S_4O_6^{2-}$), thiosulfate ($S_2O_3^{2-}$), sulfite (SO_3^{2-}), sulfide (S^{2-}), and elemental sulfur (S^0) (Table 18.3). Sulfur oxidation is carried out by microorganisms with diverse metabolic capabilities including: (1) chemoautotrophs (lithotrophs); (2) chemoheterotrophs (chemolithoorganotrophs and chemoorganoheterotrophs); (3) mixotrophs (capable of autotrophic and heterotrophic growth); (4) heterotrophs which do not gain energy but derive benefits; and (5) phototrophic sulfur bacteria.

Chemoautotrophic and chemoheterotrophic sulfur oxidation

The abiotic oxidation of reduced sulfur compounds can occur to a limited extent in soils, but microbial reactions clearly dominate the process. For example, the biological oxidation of elemental sulfur in soils apparently takes place via the following sequence of reactions (i.e., those most common to heterotrophs), although some of the products may result from abiotic side reactions:

$$\underset{\text{Elemental sulfur}}{S^0} \rightarrow \underset{\text{Thiosulfate}}{S_2O_3^{2-}} \rightarrow \underset{\text{Tetrathionate}}{S_4O_6^{2-}} \rightarrow \underset{\text{Sulfate}}{SO_4^{2-}}$$

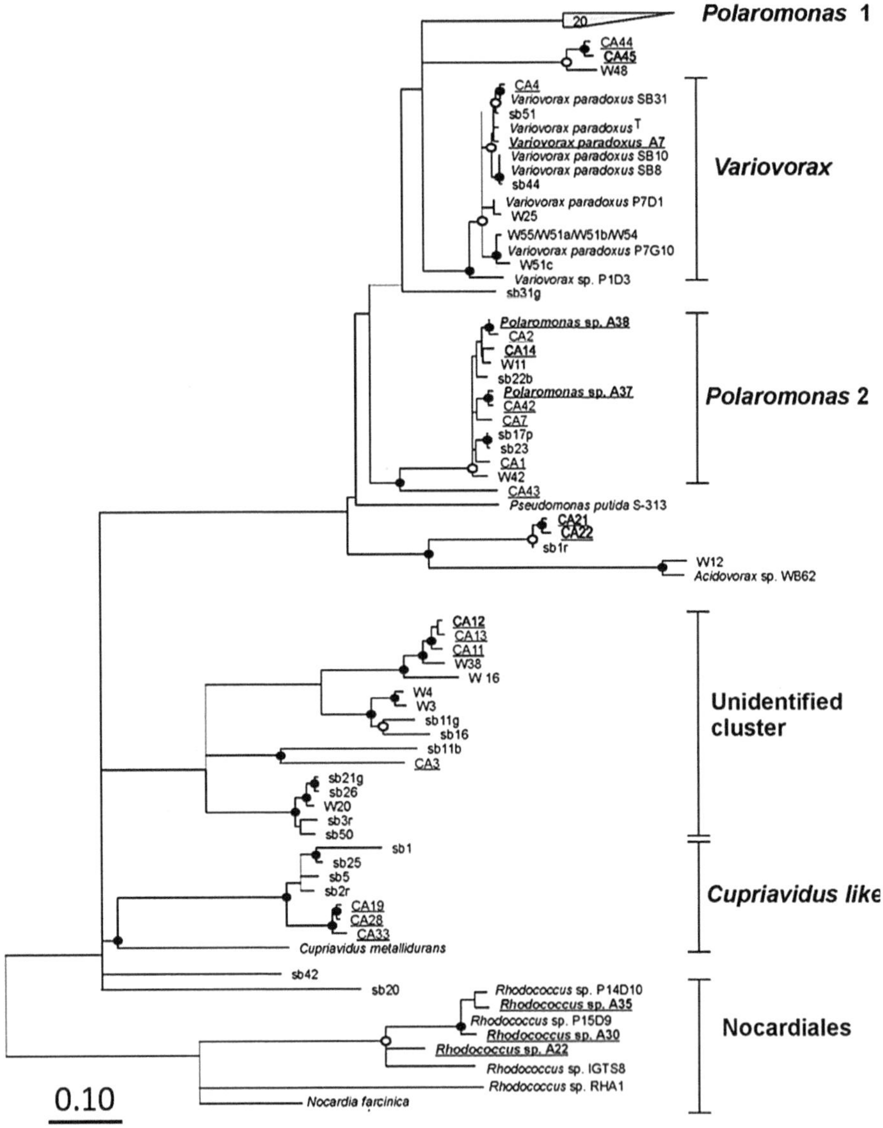

FIGURE 18.5 Phylogenetic (randomized accelerated maximum likelihood) tree of truncated *asfA* sequences including cultivated and molecular isolates from rhizospheres of wheat, spring barley, and Agrostis grassland, AMF hyphosphere/hyphoplane samples and Damma glacier forefield. The *asfA* gene was amplified with primers *asfAF*1all (50-YTSTCVGGCATGGAGTTYT-30) and *asfBtoA* (Schmalenberger and Kertesz, 2007). *Source: Adapted from Schmalenberger et al., 2010.*

Sulfur-using bacteria

Many different microorganisms are important for the oxidation, reduction, and cycling of sulfur in soil and other ecosystems (Table 18.4). In the case of sulfur oxidation, the microorganisms can be divided into the following groups:

- chemoautotrophs (lithotrophs), including members of the genera *Thiobacillus* and *Acidithiobacillus* (Box 18.5),

TABLE 18.4 Sulfur-using bacteria occurring in soil and aquatic habitats.

Group	Sulfur conversion	Habitat requirements	Habitat example	Examples of Genera
Heterotrophs that use oxidized S species as electron acceptors	$SO_4^{2-} \rightarrow HS^-$ $S_2O_3^{2-} \rightarrow HS^-$ or S^0 $S^0 \rightarrow HS^-$ $SO_3^- \rightarrow HS^-$	Anaerobic; organic substrates available; light not required	Anoxic sediments and soils	*Desulfomonas; Desulfovibrio; Desulfotomaculum; Desulfuromonas; Campylobacter*
Obligate and facultative autotrophs that use reduced S as an energy source	$HS^- \rightarrow S^0$ $S^0 \rightarrow SO_4^{2-}$ $S_2O_3^{2-} \rightarrow SO_4^{2-}$	H_2S-O_2 interface; light not required	Mud; hot springs; mine drainage; soils	*Acidithiobacillus, Thiobacillus, Thiomicrospira, Achromatium, Beggiatoa*
Phototrophs that use reduced S as an electron donor	$HS^- \rightarrow S^0$ $S^0 \rightarrow SO_4^{2-}$	Anoxic; H_2S; light	Shallow water; anoxic sediments; metalimnion or hypolimnion; anoxic water	*Chlorobium; Chromatium; Ectothiorhodospira; Thiopedia; Rhodopseudomonas*
Heterotrophs that use organic S compounds as energy sources or that hydrolyze esters	Organic S $\rightarrow HS^-$ Organic S $\rightarrow$ volatile Organic S-ester $SO_4 \rightarrow SO_4^{2-}$	Source of organic S compounds	Sediments; soils; water column	Members from all major phyla
Microorganisms that use SO_4^{2-} or H_2S in biosynthesis	$SO_4^{2-} \rightarrow$ Protein $HS^- \rightarrow$ protein $SO_4^{2-} \rightarrow$ DMSP[a]	Nonspecific	Sediments; soils; water column	Members from all major phyla
Heterotrophs that use aliphatic and aromatic sulfonates as energy sources	C-$SO_3^- \rightarrow SO_3^{2-}$	Aerobic and anaerobic	Soils, shallow water sediments	*Variovorax, Pseudomonas, Azotobacter, Cupriavidus*
Chemolithotrophs that use reduced S compounds or elemental S	$S^0 \rightarrow S_2O_3^{2-}$ $S_2O_3^{2-} \rightarrow SO_4^{2-}$	Aerobic condition, source of reduced S including elemental S	Soils	α-proteobacteria (*Paracoccus, Bradyrhizobium, Methylobacterium*) β-proteobacteria (*Janthinobacterium, Pandoraea, Thiobacillus*) γ-proteobacteria (*Halothiobacillus*) Firmicutes (*Bacillus*) Actinobacteria (*Rubrobacter, Mycobacterium*)

[a] *Dimethylsulfoniopropionate; Cook and Kelly (1992), Zhao et al. (2016), Schmalenberger and Kertesz (2007).*

- photoautotrophs, including species of purple and green sulfur bacteria, and
- chemoheterotrophs (organotrophs), including a wide range of bacteria and fungi.

The chemoautotrophs and heterotrophs are largely responsible for oxidizing sulfur in most aerobic, agricultural soils.

Many chemoautotrophic bacteria (e.g., thiobacilli) are capable of oxidizing reduced inorganic sulfur compounds. The biochemistry of sulfur oxidation by thiobacilli growing in vitro has been extensively reviewed (Postgate and Kelly, 1982; Pronk et al., 1990; (Kelly, 1999);

BOX 18.5

Taxonomic changes in the classification of the genus *Thiobacillus*

New knowledge from using 16S rRNA sequencing and DNA-DNA hybridization led to the recent reclassification of many *Thiobacillus* species. *Acidithiobacillus thiooxidans, A. albertensis,* and *A. ferrooxidans* were formerly *Thiobacillus* spp., as were *Halothiobacillus halophilus, H. hydrothermalis, Guyparkeria hydrothermalis,* and *Thermithiobacillus tepidarius.* Bacteria formerly called *Thiobacillus* spp. are now classified into different subclasses of *Proteobacteria.*

Current name	Old name	Characteristics
Acidithiobacillus spp.		Acidophilic
A. thiooxidans	*T. thiooxidans*	
A. albertensis	*T. albertis*	
A. ferrooxidans	*T. ferrooxidans*	
A. caldus	*T. caldus*	
A. ferrivorans		
A. ferrivorans		
A. ferridurns		
A. ferriphilus		
Halothiobacillus spp.		Halophilic
H. tepidarius	*T. tepidarius*	
H. neapolitanus	*T. neapolitanus*	
H. halphilus	*T. halophylus*	
Guyparkeria spp.		
G. halophila	*H. halophila*	Halophilic
G. hydrothermalis	*H. hydrothermalis*	
Thermithiobacillus spp.	*T. tepidarius*	Warm-water dwelling
Thiobacillus spp.	*T. thioparus*	
	T. denitrificans	
	T. aquaesulis	

Adapted from Kelly and Wood (2000), Boden (2017).

Ghosh and Dam, 2009). For acidophilic thiobacilli, the most common sequence of reactions involved in sulfur oxidation is:

$$\underset{\text{Elemental sulfur}}{S^0} \rightarrow \underset{\text{Sulfite}}{SO_3^{2-}} \rightarrow \underset{\text{Sulfate}}{SO_4^{2-}}$$

A great variety of thiobacilli can be isolated from natural habitats. They include obligate acidophilic chemoautotrophs, facultative chemoautotrophs (thiobacilli that grow autotrophically with reduced inorganic sulfur compounds as energy sources, but are also capable of heterotrophic growth), and **mixotrophs**, which can use mixtures of inorganic and organic compounds simultaneously. The thiobacilli differ in their physiological characteristics and in the reduced inorganic sulfur compounds used as energy sources (Table 18.5). The majority of these thiobacilli are obligate aerobes, although some, like *Thiobacillus denitrificans*, can grow anaerobically by using nitrate as a terminal electron acceptor. Other species of thiobacilli use electron donors such as ferrous iron (*A. ferrooxidans*) and thiosulfate (*T. thioparus*) in addition to sulfur.

TABLE 18.5 Characteristics of species of sulfur chemolithotrophs.

Species	Electron donor	Electron acceptor	Facultative heterotroph	Facultative anaerobe	pH optimum
Acidithiobacillus thiooxidnas	H_2S, $S_2O_3^{2-}$, S^0	O_2	–	–	2.2
Acidithiobacillus ferrooxidans	S^0, $S_2O_3^{2-}$, Fe^{2+}	O_2	+	–	3.0
Acidithiobacillus albertensis	S^0, $S_4O_6^{2-}$	O_2	+	–	2.7–3.0
Acidiphilium acidophilus	ND[a]	O_2	+	–	3.0
Halothiobacillus neapolitanus	S_2, $S_2O_3^{2-}$	O_2	–	–	6.6
Thiobacillus thioparus	S^0, $S_2O_3^{2-}$, NCS^-	O_2, NO_2^-	–	+	6.9
Thiobacillus kabobis	S^0, $S_2O_3^{2-}$	O_2	–	–	3.0
Thiobacillus denitrificans	S^0, $S_2O_3^{2-}$, $S_4O_6^{2-}$	O_2, NO_3^-	–	+	7.0
Thiobacillus novellas	$S_2O_3^{2-}$	O_2	+	–	8.4
Thiobacillus organoparus	ND[a]	O_2	+	–	3.0
Sulfurimonas denitrificans	H_2S, $S_2O_3^{2-}$, S^0	O_2 and NO_3^-	–	+	7.0
Thermithiobacillus tepidarius	$S_2O_3^{2-}$	O_2	–	–	7.0
Thiomonas intermedius	S^0, $S_2O_3^{2-}$	O_2	+	–	5.5–6.0
Thiomonas perometabolis	$S_2O_3^{2-}$, S^0	O_2	+	–	5.5–6.0
Thiomonas delicata	$S_2O_3^{2-}$, $S_4O_6^{2-}$, Fe^{2+}	O_2, NO_3^-	+	–	5.5–6.0
Paracoccus versutus	$S_2O_3^{2-}$	O_2, Organic C	+	+	8.2

[a]ND = no data

Germida and Janzen (1993), Konopka et al. (1986), Reynolds et al (1981), Kuenen and Beudeker (1982), Katayama-Fujimura et al. (1984), Kelly and Wood (2000), Takai et al. (2006), Vardanyan and Vardanyan (2014).

Although thiobacilli can oxidize sulfur to plant-available sulfate in some soils, this process is also evidently mediated by many different heterotrophic soil microorganisms. Bacteria, such as *Arthrobacter, Bacillus, Micrococcus*, and *Pseudomonas*, some actinobacteria, and a wide range of fungi are also capable of oxidizing elemental and reduced forms of sulfur. Many of these sulfur-oxidizing heterotrophs have been isolated from soil and may:

- oxidize sulfur, producing mainly thiosulfate,
- oxidize sulfur, producing sulfate, and
- oxidize thiosulfate to sulfate.

The pathway by which heterotrophic microorganisms produce these sulfur oxyanions has not been established, although several studies suggest that it is enzymatic in fungi. Apparently, no energy is derived by the organisms through these oxidations, and the transformations are incidental to the major metabolic pathways. Because the heterotrophic organisms are generally more numerous in soils than chemoautotrophs, mixed populations of heterotrophs probably play the dominant role in sulfur oxidation in many aerobic, neutral, and alkaline agricultural soils.

The opposing view that thiobacilli play the dominant role in sulfur oxidation in soils is largely based on the observation that these bacteria achieve rates of sulfur oxidation in culture far in excess of those achieved by heterotrophs growing under the same conditions. Most thiobacilli are facultative or obligate chemoautotrophs, which means that they are able to oxidize sulfur independently of the supply of available organic carbon. Marked increases in numbers of thiobacilli may follow the addition of reduced forms of sulfur to some soils, supporting the concept that populations of thiobacilli are important oxidizers of the added sulfur. However, no consistent correlation has been found between sulfur-oxidation rates and the incidence of thiobacilli, except that rates of sulfur-oxidation are generally low in soils that lack these organisms and are accelerated in soil inoculated with thiobacilli. It is probable that in many soils the initial oxidizers of reduced sulfur compounds are heterotrophic organisms until the pH is reduced sufficiently (pH $\leq$5) to permit oxidation by chemolithotrophs. In addition, there is good evidence that consortia of heterotrophs and autotrophs work together to bring about the oxidation of sulfur in agricultural soils.

Other sulfur bacteria

Other bacteria may also oxidize sulfur compounds. The gliding sulfur oxidizers include those bacteria that have a gliding motion on the substrate; their cells are arranged in **trichomes**. The most important members of this group in relation to sulfur-oxidation in soils are species of *Beggiatoa*, bacteria that participate in sulfide oxidation in the root zone of rice. All strains of *Beggiatoa* deposit sulfur intracellularly in the presence of hydrogen sulfide. Phototrophic bacteria, such as *Chromatium* and *Chlorobium*, also play an important role in sulfide oxidation in rice paddy soil, but not in aerobic agricultural soils. A number of nonfilamentous, chemolithotrophic sulfur-oxidizing bacteria, such as *Thermothrix, Sulfovorum* and *Thiomicrospira*, have also been isolated from special habitats, but the importance of these bacteria in sulfur oxidation in soils has yet to be determined. Additionally, aerobic chemolithoautotrophic sulfur-oxidizing archaea such as *Sulfolobus* and *Sulfurococcus* have been found under autotrophic and mixotrophic conditions. The activity of different groups of sulfur-oxidizing bacteria may be predicted based on the relative turnover rates of inorganic sulfur compounds and organic substrates (Kuenen and Beudeker, 1982; Box 18.6).

Phototrophic sulfur-oxidizing bacteria that include green sulfur bacteria (e.g., *Chlorobium* and *Pelodictyon*) and purple sulfur bacteria (e.g., *Chromatium* and *Thiococcus*) utilize light as an energy source and hydrogen sulfide (H_2S) as substrate. Whereas, the purple non-sulfur bacteria, for example, *Rhodovulum*, use hydrogen or sulfide as an electron source and grow as chemoorganotrophs.

BOX 18.6

Predicting the occurrence of sulfur-oxidizing microorganisms in different habitats

Consider a hot sulfur spring like those found in Yellowstone National Park, United States. The molar ratio of inorganic sulfur compounds to organic substrates in the water is very large (i.e., lots of dissolved, reduced sulfur compounds in the water and very little organic matter), hence we predict that obligate autotrophic (e.g., chemolithotrophic) sulfur-oxidizing bacteria would be abundant. We also know these bacteria are thermophiles because they are living in water at a temperature greater than 60°C. Alternatively, in a soil that contains 1% organic matter and 0.01% thiosulfate, or elemental sulfur, the ratio of inorganic sulfur compounds to organic substrates would be very small, and thus heterotrophic sulfur-oxidizing organisms including bacteria and fungi would be most abundant.

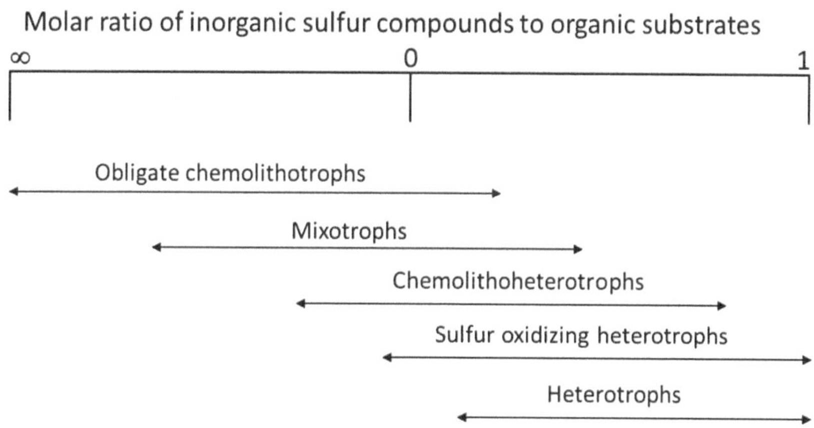

Biogenesis and oxidation of metal sulfides

Metal sulfides may be formed through biotic or abiotic reactions. In both cases the metal sulfide results from the interaction between a metal ion and a sulfide ion:

$$M^{2+} + S^{2-} \rightarrow MS \downarrow$$

Many sulfate-reducing bacteria, such as *Desulfovibrio* spp. or *Desulfotomaculum* spp. and some archaea are involved in the biogenesis of sulfides of antimony, cobalt, cadmium, iron, lead, nickel, and zinc. The extent of metal-sulfide genesis depends on many factors, the most important of which is the relative toxicity of the metal ion. In nature, this toxicity is probably reduced when the metal ions are adsorbed on clays or complexed with organic matter. The formation of metal sulfides during the mineralization of organic sulfur compounds is also possible, although little is known about this phenomenon.

The oxidation of metal sulfides in soil involves both chemical and microbial processes and, as a result, is a more complex process than is sulfur oxidation. Chalcocite (Cu_2S), chalcopyrite ($CuFeS_2$), galena (PbS), pyrite (FeS_2), and nickel sulfide (NiS) are just a few examples of metal sulfides that are subject to microbial transformations. For example, the biological oxidation of pyrite follows a series of oxidation steps described in the following equations. These biotic oxidations are responsible for the formation of acid mine drainage and acid soil formation in surface mine spoils. First, ferrous sulfate is formed as the result of an abiotic oxidation step:

$$2FeS_2 + 2H_2O + 7O_2 \rightarrow 2Fe_2SO_4 + 2H_2SO_4$$

This reaction is then followed by the bacterial oxidation of ferrous sulfate, generally by *A. ferrooxidans*:

$$4FeSO_4 + O_2 + 2H_2SO_4 \rightarrow 2Fe_2(SO_4)_3 + 2H_2O$$

This reaction occurs chemically but can be accelerated 10^6–10^8 times by thiobacilli. Subsequently, ferric sulfate is reduced and pyrite oxidized by a strictly chemical reaction:

$$Fe_2(SO_4)_3 + FeS_2 \rightarrow 3FeSO_4 + 2S^0$$
$$2S + 6Fe_2(SO_4)_3 + 8H_2O \rightarrow 12FeSO_4 + 8H_2SO_4$$

The elemental sulfur produced is finally oxidized by *A. thiooxidans* and *A. ferrooxidans*, and the acidity produced helps the whole process to continue:

$$2S^0 + 3O_2 + 2H_2O \rightarrow 2H_2SO_4$$

Note the net production of 10 molecules of H_2SO_4 during the process.

Although several sulfur-oxidizing thiobacilli and heterotrophs can be isolated from acid sulfate soils in which pyrite is being oxidized, only *A. ferrooxidans* appears to play an important role in the process. The biological oxidation of sulfides and other reduced sulfur compounds can have severe consequences for the environment (Box 18.7). For example, acid mine drainage contaminates several thousand miles of streams in the Appalachian coal mining region of the United States. Alternatively, they can be used in a beneficial manner in the bioremediation of soils contaminated with metals (Box 18.8) or in the bioextraction of precious metals from low-grade ores not suitable for smelting.

BOX 18.7

Problems associated with sulfur oxidation

Oxidation of reduced sulfur compounds leads to the formation of acidic products. As a consequence, sulfur oxidation can have detrimental effects on the environment. This is especially true in the case of sulfide minerals such as pyrite (FeS_2). Some of the more serious problems associated with sulfur oxidation include:

- formation of acid mine drainage,
- formation of acid sulfate soils,
- corrosion of concrete structures, and
- corrosion of metals.

BOX 18.8

Metal bioremediation using sulfur bacteria

Bioremediation refers to clean-up of a polluted environment via transformation of toxic heavy metals into less toxic forms by microbes or their enzymes. Many industrial sites are contaminated by metals, including: arsenic, cadmium, copper, chromium, mercury, lead, and zinc. These are typically toxic to animals, including humans, ecosystem functions, and can bioaccumulate in plants. Hence, it is important to remediate these soils. Remediation of polluted soil or sludge by adding chemicals like strong acids to leach heavy metals and applying electrokinetics may be possible but the cost of these processes is often inhibitory. Using sulfur-oxidizing microorganisms is an alternative. This process exploits the same reaction that produces acid mine drainage. *Acidithiobacillus thiooxidans* produces a high concentration of sulfuric acid, dissolving heavy metals in soil:

$$2S + 3O_2 + 2H_2O \rightarrow 2H_2SO_4$$

Alternately, where iron is present in the soil, *Acidithiobacillus ferrooxidans* can also produce sulfuric acid:

$$4FeS_2 + 15O_2 + 2H_2O \rightarrow 2Fe_2(SO_4)_3 + 2H_2SO_4$$

These reactions can drop the pH to as low as 1.5, providing an environment where leaching easily occurs. Many heavy metals can be dissolved up to 80%. After leaching, metals can be recovered by sulfate-reducing bacteria. The soil or sludge can be moved to an anaerobic bioreactor, where sulfate-reducing bacteria like *Desulfovibrio* spp. can remove both acidity and metals. The rate and efficiency of these processes depend on the soil type, inoculum density, and the initial soil pH. Thus, consortia of sulfur bacteria can be used to detoxify, extract, or sequester heavy metal pollutants from soil, sludge, and mine waste.

Detecting sulfur oxidation genes in soil

The oxidation of sulfur consists of several different enzymatic systems and pathways (Box 18.9). Sulfur oxidation pathways are categorized into (1) sulfur oxidation enzymes (e.g., sulfur dioxygenase, HDR-like complex), (2) thiosulfate oxidation enzymes (sulfur oxidizing enzyme (sox) system, tetrathionate intermediate thiosulfate oxidation (S4I) pathway, and

BOX 18.9

Enzymes and genes involved in sulfur oxidation

The oxidation of thiosulfate (most abundant reduced inorganic sulfur species) is considered to play a key role in the biogeochemical sulfur cycle. The Sox pathway (A) capable of oxidizing thiosulfate, sulfide, sulfite, and elemental sulfur to sulfate consists of four components: SoxXA, SoxYZ, SoxB, and SoxCD, **(B)** the branched thiosulfate oxidation pathway mostly attributed in lithotrophic bacteria takes place either inside or outside the cells with an initial reductive cleavage of thiosulfate to sulfide and sulfite, and occurs in the absence of SoxCD but with the participation of a reverse-acting dissimilatory sulfite reductase enzyme (DsrAB).

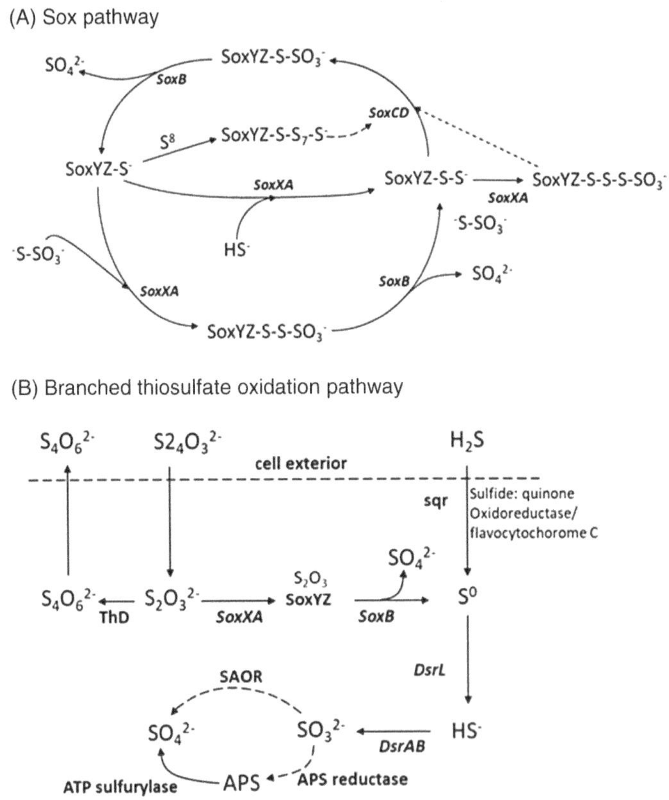

Enzymes and genes (in bold) involved in thiosulfate oxidation in bacteria. (A) Sox pathway, (B) branched thiosulfate oxidation pathway. *APS*, adenosine phosphosulfate; *ATP*, adenosine triphosphate; *SAOR*, sulfite:acceptor dehydrogenase; *ThD*, thiosulfate dehydrogenase. *Source: Modified from to Ghosh and Dam, 2009, Zhao, 2015.*

thiosulfate dehydrogenase), and (3) sulfide and sulfite oxidation enzymes (e.g., sulfide:quinone oxidoreductase). Analysis of functional gene markers, such as *sqr*, *apr*, *dsr*, and *sox*, related to the individual transformation processes allows the quantification of the abundance and diversity of S-oxidizing bacteria and archaea. By using culture-independent methods, bacteria and archaea involved in sulfur oxidation have been found to be taxonomically and phylogenetically diverse and found in most soil and sediment ecosystems.

The *soxB* gene, which codes the SoxB component of the periplasmic thiosulfate-oxidizing Sox enzyme complex involved in the oxidation of thiosulfate to sulfate, is considered as an indicator gene for sulfur oxidation ability of soils and sediments. Comparative analysis of the SoxB- and the 16S rRNA-based phylogenetic tree indicated multiple events of lateral *soxB* gene transfer among distant phylogenetic members of sulfur-oxidizing bacteria. These types of culture-independent studies allow detailed understanding of the biogeography of sulfur-oxidizing microbial communities and the identification of key members that drive sulfur oxidation under specific environmental and soil conditions. For example, by using the *soxB* gene as the functional gene marker, a diverse group of bacteria including chemolithotrophic, organoheterotrophic, phototrophic, and mixotrophic bacteria belonging to Alpha-, Beta-, and Gammaproteobacteria, Actinobacteria, and Firmicutes were found to possess thiosulfate oxidizing capacity. In Australian soils, *soxB*-based quantitative PCR and sequencing analyses showed a close link between the abundance and diversity of chemolithotrophic sulfur-oxidizing bacteria and the oxidation of elemental sulfur in agricultural soils, suggesting that sulfur oxidation rate can be better predicted by considering both soil microbiological and chemical properties (Zhao et al. 2016).

Microbial reduction of inorganic sulfur compounds

Bacterial sulfate reduction

The reduction of sulfate to hydrogen sulfide is a process that is mediated mainly by anaerobic, sulfate-reducing bacteria and archaea. This process may be significant in anaerobic, waterlogged soils, but is usually not important in well-aerated agricultural soils, except in anaerobic microsites. Nevertheless, sulfate reduction is a major component of the sulfur cycle in soils exposed to waterlogging or periodic flooding, especially where readily decomposable plant residues are present.

Microorganisms reduce oxidized sulfur compounds by either an assimilatory or dissimilatory process. Some use *assimilatory sulfate reduction* to meet their sulfur requirements. In *dissimilatory sulfate reduction*, bacteria use sulfate as a terminal electron acceptor, and large quantities of hydrogen sulfide (H_2S) are released. This process is analogous to the denitrification process discussed in Chapter 15. Like most denitrification, dissimilatory sulfate reduction is a strictly anaerobic process. In this case it is carried out by bacteria such as *Desulfovibrio* spp., *Desulfomonas* spp., and *Desulfotomaculum* spp. (Table 18.6). These bacteria utilize end products of other fermentations such as lactate, malate, and ethanol as electron donors.

Factors influencing sulfate reduction

When a soil is flooded, electron acceptors become reduced in an ordered sequence; first oxygen, followed by nitrate, nitrite, manganic, and ferric compounds, and finally sulfate and

TABLE 18.6 Dissimilatory sulfate-reducing bacteria and archaea.

Category	Description
Bacteria	
Genera	*Desulfobacter, Desulfobulbus, Desulfococcus, Desulfonema, Desulfosarcina, Desulfotomaculum, Desulfovibrio, Desulfurella, Desulfomicrobium, Desulfatibacillum, Desulfobacterium, Desulfohalobium, Desulfotalea, Syntrophobacter, Thermodesulfovibrio*
General characteristics	Strict anaerobes Grow at mildly acid to mildly alkaline pH Generally mesophilic, but some species thermophilic
Substrates	Most sulfate reducers will also reduce sulfite and thiosulfate Some species reduce elemental S Organic matter utilization varies with genus and species As a group, capable of completely oxidizing fatty acids from C1–C18, lactate, pyruvate, low molecular weight, alcohols, and some aromatic compounds
Habitats	Anaerobic sediments of freshwater, brackish water, and marine environments, thermal regions, water-logged soils, and animal intestines
Archaea	
Genera	*Archaeoglobus fulgidus, A. profundus, Caldivirga maquilingenes*
General characteristics	Generally thermophilic, optimum temperature 82°C–85°C Grow at mildly acidic (pH 4.5) to mildly alkaline (pH 7.5) Hyper saline systems (optimum 1.5%–1.8% NaCl)
Substrates	Grow chemolithoautotrophically on H_2, CO_2, thiosulfate, sulfate, or sulfite as energy source but not on S^0 Grow chemo-organoheterophically on formate, formamide, D- and L-lactate, acetate, pyruvate, glucose, starch, peptone, gelatin, casein, and meat- or yeast extract as carbon sources
Habitats	Hydrothermally heated shallow marine sediments, submarine hydrothermal systems, crater of an active Polynesian submarine volcano

From Trudinger (1986), Thauer and Kunow (1995), Pereira et al., 2011.

carbon dioxide (Chapter 3). Although the reduction of one compound does not have to be completed before another is reduced, oxygen and nitrate must be removed before the reduction of ferric and sulfate ions can occur. Because of this reaction sequence, sufficient ferrous ions generally are available to react with any hydrogen sulfide produced, and as a result, free hydrogen sulfide is rarely liberated from soils. Sulfate reduction increases with the period of soil submergence and following the addition of organic matter. Sufficient organic substrates to stimulate the process are also liberated from seeds and from roots into the rhizosphere, with the result that in paddy soils, blackening caused by ferrous sulfide deposits often occurs in the root region. There is evidence, however, that rice roots can aerate the soil sufficiently such that ferric iron is observed on the root surface. In general, the rate of sulfate reduction increases with decreasing redox potential, with the optimum being a function of soil pH, around −300 mV at pH 7. Sulfate-reducing bacteria are active in soil, sediments, polluted water, oil-bearing strata, and shales. Their activity may be beneficial or detrimental to the surrounding environment and have serious economic consequences (Box 18.10).

BOX 18.10

Consequences of sulfate reduction

Sulfate reducing bacteria can thrive under a variety of environmental conditions including in fresh water, thermophilic, and halophilic systems. The activity of sulfate-reducing bacteria and the problems they pose can be seen in many examples from our daily lives:

- Microbially influenced corrosion due to sulfate-reducing bacteria is a major cause of corrosion of underground storage tanks and iron pipes, causing ~$30–40 billion of damage annually in the United States alone.

  [https://www.mswmag.com/online_exclusives/2019/04/the-cost-of-sewer-pipe-corrosion_sc_00125].
- Turf managers find that anaerobic sulfate-reducing bacteria can produce a black layer under golf course greens and even biofouling by using the organic matter in soil and root exudates to reduce soil sulfates to ferrous sulfides.
- It has been posited that the water in the canals of Venice is polluted with hydrogen sulfide produced by sulfate-reducing bacteria and, as a consequence, the gondolas of Venice turn black regardless of their original color. However, an opposing view is that in 1562, the gondole were ordered to be painted black by a local magistrate except for those carrying important people.
- Not all activity by sulfate-reducing bacteria is bad, though. Sulfate-reducing bacteria can also be used in metal recovery and bioremediation of acid mines and heavy-metal contaminated soils. Sources of heavy-metal contamination include burning of fossil fuels, industrial processes, and sewage effluent. Sulfate-reducing bacteria such as *Desulfovibrio* and *Desulfomicrobium* play a vital role in the control and treatment of mine waste, generating alkalinity and neutralizing the acidic waste. Leachate produced by sulfur-oxidizing bacteria can be transferred to bioreactors, where sulfur-reducing bacteria can remove acidity and immobilize toxic metals, producing an effluent with low levels of heavy metals that can be returned to the environment.

Detecting sulfur reduction genes in soil

The depletion of oxygen under the anaerobic soil conditions such as those seen in the surface 1-cm layer in flooded rice fields and rhizosphere soils can promote significant reduction of inorganic sulfates. The final step in anaerobic sulfate reduction, reduction of sulfite to sulfide, is catalyzed by dissimilatory (bi)sulfite reductase which is encoded by the *dsrAB* gene. A diverse group of prokaryotes including genera from five bacterial and two archaeal phyla have been found to harbor this functional gene marker. Analysis of *dsrAB* genes amplified from nucleic acids from rice paddy soils in southern China has indicated a high degree of diversity in different fertilization regimes and identified novel sulfate-reducing bacteria (Liu et al., 2009). Also, using functional gene-specific quantitative PCR methods, the highest sulfate-reducing prokaryotic abundance was observed in or near oxygenated areas and rhizosphere soils.

Volatilization of inorganic and organic sulfur compounds from soil

A number of sulfur gases are released from soils, marshes, peats, and sediments, or from anthropogenic sources. These gases may be inorganic or organic and play an important role in the cycling of sulfur through the atmosphere. Many different fungi and heterotrophic bacteria are responsible for the formation of these volatile compounds during the metabolism of organic sulfur compounds (Table 18.7; see Box 18.11).

Environmental aspects of sulfur pollutants

Acid sulfate soils

Acid sulfate soils contain metal sulfides, mainly in the form of pyrites, which may be oxidized to yield free and adsorbed sulfates. They are characterized by yellow mottling due to the formation of jarosites [$AFe_3(SO_4)_2(OH)_6$, where A = K^+ , NH_4^+ , Na^+ or H_3O^+], and have a pH typically below 4. Although these soils cover large areas of the tropics, they tend to be of only local importance in temperate regions. The acidification of these soils results from the abiotic and microbial oxidation of ferrous ions in pyrite which serve as oxidants for the metal sulfides and for most of the intermediate sulfur compounds that are further oxidized

TABLE 18.7 Biochemical origin of volatile sulfides produced in soils by microbial degradation of organic matter under aerobic and anaerobic conditions.

Volatile S compound	Formula	Biochemical precursors
Hydrogen sulfide	H_2S	Proteins, polypeptides, cystine, cysteine, glutathione
Methyl mercaptan (Methanethiol)	CH_3SH	Methionine, methionine sulfoxide, methionine sulfone, S-methylcysteine, methoxylated aromatic compounds, dimethylsulfoniopropionate
Dimethyl sulfide	CH_3SCH_3	Methionine, methionine sulfoxide, methionine sulfone, S-methylcysteine, homocysteine, S-methyl-methionine, dimethylsulfoniopropionate
Dimethyl disulfide	CH_3SSCH_3	Methionine, methionine sulfoxide, methionine sulfone, S-methylcysteine
Carbon disulfide	CS_2	Cysteine, cystine, homocysteine, lanthionine, djenkolic acid
Carbonyl sulfide	COS	Lanthionine, djenkolic acid

From Andreae and Jaeschke (1992), Lomans et al. (2002).

BOX 18.11

Methane-eating microorganisms in a sulfur-rich environment

Methane is one of the most potent greenhouse gases and it is estimated that 3%–4% of the global carbon flux to the seafloor is converted by microorganisms to methane (10 trillion tons). However, very little of the methane escapes to the atmosphere. Until recently, it was not understood what happens to the methane. It was hypothesized that bacteria assimilated methane, but most known methane-eating (methanotrophic) bacteria are aerobic, which obviously cannot survive deep in the ocean floor. However, it is known that sulfur-reducing bacteria can function under these conditions. Recently, scientists recognized that sulfate disappeared at the same place in the sediment as the methane (Boetius et al. 2000; Orphan et al. 2001; Jørgensen et al., 2019). Investigations using molecular and biochemical techniques revealed that anaerobic methanotrophic archaea (ANME) lived in association with sulfate-reducing bacteria (SRB) like *Desulfosarcina* (Deltaproteobacteria) and *Desulfotomaculum* (Gammaproteobacteria) as syntrophic aggregates in the sediment layer. The consortia carry out the following reaction in which sulfate serves as the electron acceptor with the chemical species dissolved in the aqueous phase:

$$SO_4^{2-} + CH_4 \rightarrow HCO_3^- + HS^- + H_2O$$

This biochemical symbiosis results in a highly efficient method of transferring intermediates, where the archaeal partner activates and metabolizes methane leading to an intermediate that is scavenged as electron donor by the sulfate-reducing bacterial partner. Recent results indicated that methane pressure and sulfide concentration influence the growth of different ANME-SRB consortia. The occurrence of anaerobic methane oxidation associated with sulfate reducers in low-saline environments was also shown using molecular and isotopic techniques. This is a good example of how novel associations of microorganisms, for example, bacteria with archaea, are discovered in unique environments.

to sulfuric acid. Problems in producing crops on these soils occur because of aluminum and manganese toxicity, rather than to the direct effects of acidity. Acid sulfate soils can also form in coastal environments with an endless supply of sulfate from seawater, where microbial decomposition of organic matter in water-logged soils and sediments which are rich in metal cations, creates the reducing conditions thereby producing metal sulfides. Acid sulfate soils can be reclaimed by:

- controlling the water table through adequate drainage,
- adding lime,

- planting crops tolerant of aluminum, manganese, and iron, and
- improving soil fertility generally.

Deposition of atmospheric sulfur in soils

Soils subject to atmospheric pollution receive sulfur from the atmosphere largely in the form of dilute sulfuric acid. Thus, sulfate is the main sulfur ion entering soils from the atmosphere; smaller quantities of sulfite and bisulfite may also contaminate these soils. Atmospheric pollution deposits, consisting largely of soot, may also be locally important sources of reduced sulfur compounds, particularly in areas adjacent to industrial plants, such as coking and steel works. Because sulfate is the major sulfur ion entering soil from atmospheric pollution, we expect that the major sulfur transformations that occur will involve sulfur mineralization and sulfur reduction rather than sulfur oxidation. However, when reduced sulfur compounds in the atmosphere are deposited on soils, they are rapidly oxidized.

Concrete corrosion by sulfur-oxidizing bacteria

Concrete sewer pipes are often corroded and sulfur-oxidizing bacteria seem to be the culprits. Concrete has a pH of around 12, so it was surprising to find that these bacteria were responsible, since sulfur-oxidizing bacteria are acidophilic. Microbial succession aids the corrosion process. It appears that the concrete is neutralized by carbonation and acidic H_2S followed by colonization by sulfate-producing bacteria. The first colonizers are moderately acidophilic *Thiobacillus* species, followed by *Acidithiobacillus thiooxidans*. Some studies have shown that this organism often occurs in relationship with acidophilic heterotrophic microorganisms, which degrade the organic compounds excreted by the sulfur-oxidizing bacteria. This helps to maintain the acidic environment. Some co-existing organisms may oxidize H_2S producing thiosulfate, which acts as a source for *A. thiooxidans*. Corroded concrete typically has a pH of 6.2–1.5.

The inner surface of sewer pipes is corroded in only certain areas of the pipe (Fig. 18.6). The anaerobic region covered by sewage water is typically not corroded, while the region near the top of the pipe is most corroded. In the sewer, H_2S is volatilized to the sewer atmosphere and dissolves in the condensate on the sewer's crown. Sulfur-oxidizing bacteria are capable of oxidizing the elemental sulfur. Several mm of concrete can typically be corroded each year. The actual corrosion occurs when the sulfuric acid produced by sulfide oxidation reacts with the calcium hydroxide binder in the concrete to form calcium sulfate.

$$H_2SO_4 + Ca(OH)_2 \rightarrow CaSO_4 + 2H_2O$$

In newer sewer systems, attempts are made to reduce the amount of anaerobic activity to control sulfuric acid production (Vincke et al., 2001).

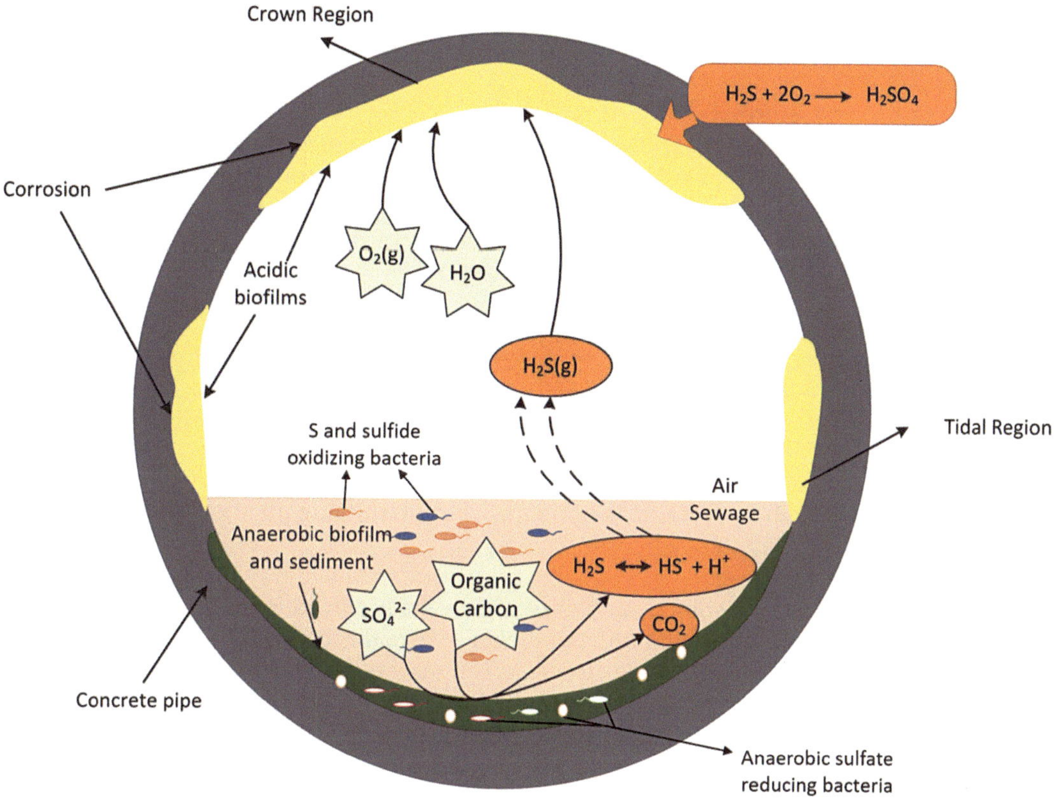

FIGURE 18.6 A conceptual diagram summarizing the major processes that lead to acid formation in the anaerobic biofilms and the onset of corrosion in a concrete sewer pipe. The activities of sulfate-reducing and sulfur-oxidizing bacteria in concrete sewer pipes can lead to serious corrosion problems. Hydrogen sulfide (H_2S) produced in the anoxic effluents diffuses to the top of the pipe where the actions of sulfur (sulfide)-oxidizing bacteria (e.g., *Acidithiobacillus thiooxidans*, *A. ferrooxidans*, *Acidiphilium* spp.) convert it to sulfuric acid, which can corrode a few millimeters of the concrete per year. Eventually, the roof of the sewer pipe can collapse. *Source: Modified from Li et al. (2017).*

Summary

Sulfur is an essential element for all living organisms and is the basis for primary productivity in some exotic communities. It exists in a number of oxidation states as inorganic and organic compounds that undergo a number of biotic and abiotic transformations. These transformations can be beneficial or detrimental to ecosystems depending on the forms and fluxes of sulfur. The cycling of sulfur through aquatic, terrestrial, and atmospheric ecosystems is similar to that of other elements, such as carbon and nitrogen, and is influenced by natural and anthropogenic processes.

We currently have a basic understanding of the forms and amounts of sulfur in terrestrial ecosystems and the processes controlling the supply of sulfur to plants. The exchange of sulfur gases between the soil-plant system and the atmosphere is less well documented and understood. Through the use of the latest *omics* tools (metagenomics, transcriptomics and metabolomics) it is now possible to describe, in-depth, the networks of members involved in different sulfur transformation processes. Understanding how key processes in the sulfur cycle respond to environmental factors, (e.g., construction of models of mineralization and volatilization processes that include temperature, moisture, substrate, and microbial composition response functions) will help us predict accurately the impact of human-induced or natural changes on sulfur fluxes in all components of the biosphere.

Study questions

1. What is meant by the term chemosynthesis, and how does this relate to the sulfur cycle?
2. Discuss how the oxidation of reduced sulfur compounds drives primary productivity at deep sea thermal vents.
3. Sketch a model that helps predict the types of sulfur-oxidizing microorganisms that will be found in certain habitats.
4. What is a sulfur mixotroph?
5. Sketch the sulfur cycle, listing examples of organisms involved in each phase of the cycle, and then discuss how soil microorganisms drive this cycle.
6. List and discuss some of the similarities between the sulfur and nitrogen cycles.
7. What is a disulfide bond and why is it important?
8. Why might crop rotations influence sulfur cycling in soil?
9. What techniques can be used to study and classify novel sulfur bacteria?
10. Compare and contrast cellular mineralization and enzymatic mineralization processes.
11. What types of microorganisms could be used to mine metals?
12. Discuss how and why a group of microorganisms might act together as a consortium to oxidize sulfur in soil.
13. Discuss how and what molecular techniques can be used to investigate the diversity and functional capacity of S cycling microorganisms.
14. How can concrete corrosion caused by sulfur-oxidizing bacteria be prevented?
15. In what type of soil would you expect to find active sulfate-reducing bacteria?
16. Why is the enzyme arylsulfatase important in the sulfur cycle in soil?
17. How can sulfur bacteria prevent the release of greenhouse gases from oceanic sources?

References

Andreae, M.O., Jaeschke, W.A., 1992. Exchange of sulfur between biosphere and atmosphere over temperate and tropical regions. In: Howarth, R.W., Stewart, J.W.B., Ivanov, M.V. (Eds.), Sulphur Cycling on the Continents: Wetlands, Terrestrial Ecosystems, and Associated Water Bodies. SCOPE 48. John Wiley and Sons, Chichester, England, pp. 27–61.

Boden, R., 2017. Reclassification of *Halothiobacillus hydrothermalis* and *Halothiobacillus halophilus* to *Guyparkeria* gen. nov. in the Thioalkalibacteraceae fam. nov., with emended descriptions of the genus *Halothiobacillus* and family Halothiobacillaceae. Int. J. Syst. Evol. Microbiol. 67, 3919–3928.

Boetius, A., Ravenschlag, K., Schubert, C.J., Rickert, D., Widdel, F., Gieseke, A., Amann, R., Jørgensen, B.B., Witte, U., Pfannkuche, O., 2000. A marine microbial consortium apparently mediating anaerobic oxidation of methane. Nature 407, 623–626.

Cook, R.B., Kelly, C.A., 1992. Sulphur cycling and fluxes in temperate dimictic lakes. In: Howarth, R.W., Stewart, J.W.B., Ivanov, M.V. (Eds.), Sulphur Cycling on the Continents: Wetlands, Terrestrial Ecosystems, and Associated Water Bodies. SCOPE 48. John Wiley and Sons, Chichester, England, pp. 145–188.

Cregut, M., Piutti, S., Vong, P.-C., Slezack-Deschaumes, S., Crovisier, I., Benizri, E., 2009. Density, structure, and diversity of the cultivable arylsulfatase-producing bacterial community in the rhizosphere of field-grown rape and barley. Soil Biol. Biochem. 41, 704–710.

De Kok, L.J., Castro, A., Durenkamp, M., Stuiver, C.E., Westerman, S., Yang, L., Stulen, I., 2002. Sulphur in plant physiology. In: Proceedings No. 500. International Fertiliser Society, York, UK, pp. 1–26.

Germida, J.J., Janzen, H.H., 1993. Factors affecting the oxidation of elemental sulfur in soils. Fert. Res. 35, 101–114.

Ghosh, W., Dam, B., 2009. Biochemistry and molecular biology of lithotrophic sulfur oxidation by taxonomically and ecologically diverse bacteria and archaea. FEMS Microbiol. Rev. 33, 999–1043.

Gonzalez-Ballester, D., Grossman, A.R., 2009. Sulfur: from acquisition to assimilation. In: Harris, E.H., Stern, D.B., Witman, G.B. (Eds.), The Chlamydomonas Source Book. Academic Press, pp. 159–188.

Harmsen, H.J.M., Prieur, D., Jeanthon, C., 1997. Group-specific 16S-rRNA-targeted oligonucleotide probes to identify thermophilic bacteria in marine hydrothermal vents. Appl. Environ. Microbiol. 63, 4061–4068.

Jannasch, H.W., Mottl, M.J., 1985. Geomicrobiology of deep-sea hydrothermal vents. Science 229, 717–725.

Jørgensen, B.B., Findlay, A.J., Pellerin, A., 2019. The biogeochemical sulfur cycle of marine sediments. Front. Microbiol. 10, 849, doi: 10.3389/fmicb.2019.00849.

Katayama-Fujimura, Y., Kawashima, I., Tsuzaki, N., Kuraishim, H., 1984. Physiological Characteristics of the Facultatively Chemolithotrophic *Thiobacillus* Species *Thiobacillus delicatus* nom. rev., emend., *Thiobacillus perometabolis*, and *Thiobacillus intermedius*. Int. J. Syst. Bacteriol. 34, 139–144.

Kelly, D.P., 1999. Thermodynamic aspects of energy conservation by chemolithotrophic sulfur bacteria in relation to the sulfur oxidation pathways. Arch. Microbiol. 171, 219–229.

Kelly, D.P., Wood, A.P., 2000. Reclassification of some species of *Thiobacillus* to the newly designated genera *Acidithiobacillus* gen. nov., *Halobacillus* gen. nov. and *Thermithiobacillus* gen. nov. Int. J. Syst. Evol. Microbiol. 50, 516.

Konopka, A.E., Miller, R.H., Sommers, L.E., 1986. Microbiology of the sulfur cycle. In: Tabatabai, M.A. (Ed.), Sulfur in agriculture. In: Agronomy Monographs, vol. 27. American Society of Agronomy, Madison, Wis, pp. 23–56.

Kormas, K.A., Tivey, M.K., Von Damm, K., Teske, A., 2006. Bacterial and archaeal phylotypes associated with distinct mineralogical layers of a white smoker spire from a deep-sea hydrothermal vent site (9°N East Pacific Rise). Environ. Microbiol. 8, 909–920.

Kuenen, J.G., Beudeker, R.F., 1982. Microbiology of thiobacilli and other sulphur-oxidizing autotrophs, mixotrophs and heterotrophs. Philos. Trans. R. Soc. London Ser. B. 298, 473–497.

Li, X., Kappler, U., Jiang, G., Bond, P.L., 2017. The ecology of acidophilic microorganisms in the corroding concrete sewer environment. Front. Microbiol. 8, 683, doi: 10.3389/fmicb.2017.00683.

Liu, X.Z., Zhang, L.M., Prosser, L. J.I., He, J.Z., 2009. Abundance and community structure of sulfate reducing prokaryotes in a paddy soil of southern China under different fertilization regimes. Soil Biol. Biochem. 41, 687–694.

Lomans, B.P., van der Drift, C., Pol, A., Op den Camp, H.J.M., 2002. Microbial cycling of volatile organic sulfur compounds. Cell. Mol. Life Sci. 59, 575–588.

Orphan, V.J., House, C.H., Hinrichs, K.U., McKeegan, K.D., DeLong, E.F., 2001. Methane-consuming archaea revealed by directly coupled isotopic and phylogenetic analysis. Science 293, 484–487.

Paul, E.A., 2007. Soil Microbiology, Ecology and Biochemistry. Academic Press, Burlington, MA.

Pereira, I.A.C., Ramos, A.R., Grein, F., Marques, M.C., da Silva, S.M., Venceslau, S.S., 2011. A comparative genomic analysis of energy metabolism in sulfate reducing bacteria and archaea. Front. Microbiol. 2, 1–22, doi: 10.3389/fmicb.2011.00069.

Pierzynski, M., Sims, J.T., Vance, G.F., 2000. Soils and Environmental Quality, second ed. Boca Raton, Fla, CRC Press.

Postgate, J.R., Kelly, D.P. (Eds.), 1982. Sulphur Bacteria. The Royal Society. Cambridge, England.

Pronk, J.T., Meulenberg, R., Hazeu, W., Bos, P., Kuenen, J.G., 1990. Oxidation of reduced inorganic sulphur compounds by acidophilic thiobacilli. FEMS Microbiol. Rev. 75, 293–306.

Reynolds, S.D., Laishley, M.E.J., Costerton, J.W., 1981. Physiological and ultrastructural characterization of a new acidophilic *Thiobacillus* species (*T. kabobis*). Can. J. Microbiol. 59, 151–161.

Schmalenberger, A., Kertesz, M.A., 2007. Desulfurization of aromatic sulfonates by rhizosphere bacteria - high diversity of the *asf*A gene. Environ. Microbiol. 9, 535–545.

Schmalenberger, A., Telford, A., Kertesz, M.A., 2010. Sulfate treatment affects desulfonating bacterial community structures in Agrostis rhizospheres as revealed by functional gene analysis based on *asf*A. Eur. J. Soil Biol. 46, 248–254.

Takai, K., Suzuki, M., Nakagawa, S., Miyazaki, M., Suzuki, Y., Inagaki, F., Horikoshi, K., 2006. *Sulfurimonas paralvinellae* sp. nov., a novel mesophilic, hydrogen- and sulfur-oxidizing chemolithoautotroph within the Epsilonproteobacteria isolated from a deep-sea hydrothermal vent polychaete nest, reclassification of *Thiomicrospira denitrificans* as *Sulfurimonas denitrificans* comb. nov. and emended description of the genus *Sulfurimonas*. Int. J. Syst. Evol. Microbiol. 56, 1725–1733, doi:10.1099/ijs.0.64255-0.

Tanikawa, T., Noguchi, K., Nakanishi, K., Shigenaga, H., Nagakura, J., Sakai, H., Akama, A., Takahashi, M., 2014. Sequential transformation rates of soil organic sulfur fractions in two-step mineralization process. Biol. Fertil. Soils 50, 225–237.

Thauer, R., Kunow, J., 1995. Sulfate reducing archaea. In: Barton, L.L. (Ed.), Sulfate-Reducing Bacteria. In: Biotechnology Handbooks, vol. 8. Springer Science Business Media, New York, pp. 33–45.

Trudinger, P.A., 1986. Chemistry of the sulfur cycle. In: Tabatabai, M.A., (Ed.), Sulfur in Agriculture. In: Agronomy Monographs, vol. 27. American Society of Agronomy, Madison, Wis, pp. 1–22.

Tunnicliffe, V., 1992. Hydrothermal-vent communities of the deep sea. Am. Scientist 80, 336–349.

Vairavamurthy, A., Manowitz, B., Luther, III, G.W., Jeon, Y., 1993. Oxidation state of sulfur in thiosulfate and implications for anaerobic energy metabolism. Geochim. Cosmochim. Acta 57, 1619–1623.

Vairavamurthy, A., 1998. Using X-ray absorption to probe sulfur oxidation states in complex molecules. Spectrochim. Acta Part A 54, 2009–2017.

Vardanyan, N.S., Vardanyan, A.K., 2014. New sulphur oxidizing bacteria isolated from bioleaching pulp of zinc and copper concentrates. Univ. J. Microbiol. Res. 2 (2), 27–31.

Vincke, E., Boon, N., Verstraete, W., 2001. Analysis of the microbial communities on corroded concrete sewer pipes-A case study. Appl. Microbiol. Biotechnol. 57, 776–785.

Wyngaard, N., Cabrera, M.L., 2015. Measuring and estimating sulfur mineralization potential in soils amended with poultry litter or inorganic fertilizer. Biol. Fertil. Soils 51, 545–552.

Zhao, C., 2015. Elemental sulphur oxidation in Australian cropping soils, Ph.D. thesis submitted to the University of Adelaide, South Australia.

Zhao, C., Degryse, F., Gupta, V.V.S.R., McLaughlin, M.J., 2015. Elemental sulfur oxidation in Australian cropping soils. Soil Sci. Soc. Am. J. 79 (1), 89–96.

Zhao, C., Gupta, V.V.S.R., Degryse, F., McLaughlin, M.J., 2016. Abundance and diversity of sulphur-oxidising bacteria and their role in oxidising elemental sulphur in cropping soils. Biol. Fertil. Soils 53, 159–169.

Further reading

Brady, N.C., Weil, R.R., 1996. The Nature and Properties of Soils, eleventh ed. Prentice Hall, Upper Saddle River, NJ.

Ehrlich, H.L., 1996. Geomicrobiology. Marcel Dekker, New York.

Freney, J.R., 1986. Forms and reactions of organic sulfur compounds in soils. In: Tabatabai, M.A. (Ed.), Sulfur in Agriculture. In: Agronomy Monographs, vol. 27. American Society of Agronomy, Madison, WI, pp. 207–232.

Germida, J.J., Wainright, M., Gupta, V.V.S.R., 1992. Biochemistry of sulfur cycling in soil. In: Bollag, J.-M., Stotzky, G. (Eds.), Soil Biochemistry, vol. 7, Marcel Dekker, New York, pp. 1–53.

Howarth, R.W., Stewart, J.W.B., Ivanov, M.V. (Eds.), 1992. Sulphur cycling on the continents: Wetlands, terrestrial ecosystems, and associated water bodies. SCOPE 48. John Wiley and Sons, Chichester, England.

Meyer, B., Imhoff, J.F., Kuever, J., 2007. Molecular analysis of the distribution and phylogeny of the soxB gene among sulfur-oxidizing bacteria—evolution of the Sox sulfur oxidation enzyme system. Environ. Microbiol. 9, 2957–2977.

Postgate, J.R., 2000. Microbes and Man, fourth ed. Cambridge University Press, Cambridge, England.

Schoenau, J.J., Germida, J.J., 1992. Sulphur cycling in upland agriculture systems. In: Howarth, R.W., Stewart, J.W.B., Ivanov, M.V. (Eds.), Sulphur Cycling on the Continents: Wetlands, Terrestrial Ecosystems, and Associated Water Bodies. John Wiley and Sons, Chichester, England, pp. 261–306.

Schlegel, H.G., Bowien, B. (Eds.), 1989. Autotrophic Bacteria. Science Tech Publishers, Madison, WI.

Schlesinger, W.H., 1997. Biogeochemistry: An Analysis of Global Change, second ed. Academic Press, San Diego, Calif.

Tabatabai, M.A. (Ed.), 1986. Sulfur in Agriculture. Agronomy Monographs, vol. 27, American Society of Agronomy, Madison, Wis.

CHAPTER

19

Phosphorus and selected metals and metalloids

Shiping Deng
Department of Plant and Soil Science, Oklahoma State University, Stillwater, OK, United States

"There is no life without phosphorus… It's in your DNA."
— ***The Sustainable Phosphorus Alliance***

In addition to C, N, and S, several other elements are required by living organisms at macro- and micro-quantities. As with C, N, and S, cycling of these elements is primarily driven by microbial transformations. The mechanisms by which soil microorganisms transform these elements include:

- **Mineralization** and **immobilization**: reactions that convert organic to inorganic forms (mineralization) or vice versa (immobilization).
- **Reduction** and **oxidation**: reactions that involve the transfer of electrons from (oxidation) or to (reduction) the element of interest.
- **Solubilization** and **bioavailability**: reactions that convert elements to forms with increased aqueous solubility or membrane permeability, which often leads to increased bioavailability to living organisms.
- **Biomethylation** of elements, such as mercury, increases mobility of the element and is often done as a detoxification mechanism.

In this chapter, microbial transformations of phosphorus and selected metals and metalloids will be discussed. Phosphorus (P) is an essential macronutrient, while iron (Fe), manganese (Mn), and selenium (Se) are essential micronutrients for living organisms. Mercury (Hg) and arsenic (As) have no known essential function and are toxic to living systems. Of these metals and metalloids, As, Hg, and Se often receive attention as environmental pollutants. Discussion of these elements serves to illustrate the range of elemental cycling reactions performed by soil microorganisms.

Phosphorus

Carbon has long been recognized as the backbone of life. Perhaps it is time to recognize that phosphorus is the spirit of life. Phosphorus was discovered in 1669 by Hennig Brandt,

Principles and Applications of Soil Microbiology. http://dx.doi.org/10.1016/B978-0-12-820202-9.00019-8
Copyright © 2021 Elsevier Inc. All rights reserved.

BOX 19.1

Discovery of phosphorus: from Pee to P

Hennig Brandt was not looking for P when he boiled over 1000 gallons of urine in 1669. The alchemist hoped to isolate the philosopher's stone, which was believed capable of transforming other metals into gold and providing eternal life. Brandt's method yielded a chemiluminescent product he described as "cold fire." Phosphorus was also referred to as the "Devil's Element" due to the eerie glow, tendency to combust spontaneously, and being the 13th known element at the time. The element was named "phosphoros," Greek for bringer of light. Brandt thus became the first person credited with scientific discovery of an element. He subsequently sold the secrets of his methods to others who carried on the Magnus Opus (Great Work) of discovering the philosopher's stone. Brandt's efforts were captured by the artist Joseph Wright.

Information sources: The Alchymist, In Search of the Philosopher's Stone, Discovers Phosphorus, and prays for the successful Conclusion of his operation, as was the custom of the Ancient Chymical Astrologers, by Joseph Wright of Derby, now in Derby Museum and Art Gallery, Derby, U.K. *Source: The Alchemyst by Joseph Wright (1771).*

a German merchant, pharmacist, and alchemist (Box 19.1). Phosphorus is an essential macronutrient of living organisms and a structural component of many important cellular compounds such as DNA (deoxyribonucleic acid), phospholipids in membranes, and ATP (adenosine triphosphate, a key element in cellular energy transfer). Phosphorus is not only an essential nutrient for life, it is embedded in our DNA. Without phosphorus, there is no life.

Global phosphorus issues are receiving increasing attention. Phosphorus is often a limiting nutrient in food production, whereas phosphorus pollution and harmful algal blooms (HABs) are causing significant concerns (Box 19.2). Global phosphorus issues are, in part, instigated by the concern that phosphorus reserves are diminishing.

BOX 19.2

Harmful algal blooms (HABs)

The main driver behind HAB proliferation is phosphorus pollution. Phosphorus is usually the limiting nutrient for algal and cyanobacterial growth in aquatic systems. Algae can acquire carbon through photosynthesis and are often able to obtain available forms of nitrogen in eutrophied surface waters. Unlike algae, some cyanobacteria can both photosynthesize and obtain nitrogen from the atmosphere by biological dinitrogen fixation. Thus, cyanobacteria more easily respond to excess phosphorus in surface waters. The occurrence of HABs in the United States has considerable economic impacts on affected communities, requiring targeted efforts to reduce nutrient loading and the occurrence of HABs. In fact, several US states have limited or banned use of phosphorus in lawn fertilizers due to these issues. Although phosphorus mining in the United States was valued at $1.8 billion in 2018 (USGS, 2019), the estimated annual cost of dealing with HABs approaches $2.2 billion (Sim, 2016).

This aerial view showing algal blooms around the Great Lakes was captured on July 28, 2015 by the NASA Operational Land Imager. The algal blooms are visible as swirls of green in western Lake Erie.

Source: NASA (The National Aeronautics and Space Administration in the United States).

Global phosphorus reserve

Phosphorus is the 11th-most abundant element on Earth, with an estimated global reserve of 10^{14} kg. The largest reserve is in oceanic sediment (99.98%) and is not subject to significant turnover. Oceans are sinks that accumulate terrestrial phosphorus released from soil and rock by weathering and biota. Of the terrestrial phosphorus, 85% is in soil, 13% in rocks, and 2% in biota. Clearly, a very small portion of global phosphorus reserves serves as a vital component of life. Phosphorus nutrition and fertilizers are obtained through mining phosphate rock. Since phosphate rock forms slowly, in geological time, it is essentially non-renewable and non-substitutable. The global extractable phosphorus supply is diminishing over time. In 2018, world mining production was 0.107 billion kg P, and estimated global reserves of economically extractable phosphorus were 28 billion kg P (USGS, 2019). To date, the world phosphate supply has been able to meet increasing demand (Table 19.1). Yet, global distribution of economically extractable phosphorus is concentrated in only a few countries (Fig. 19.1) making supply vulnerable to trade and policy issues. Various projections suggest that global P reserves will last anywhere from 50 to 400 years (Smil, 2000; IFDC, 2010). Despite differences

TABLE 19.1 World phosphate supply, demand, and balance 2015–20 ($\times 10^6$ kg P).

Year	2015	2016	2017	2018	2019	2020
H_3PO_4 theoretical capacity	25,060	25,480	26,602	27,056	27,510	28,226
H_3PO_4–based fertilizer supply	18,216	18,485	18,918	19,491	19,931	20,195
H_3PO_4–based fertilizer demand	16,425	16,746	17,273	17,661	18,033	18,388
Non-H_3PO_4 fertilizer demand	1,535	1,559	1,578	1,593	1,612	1,626
Potential H_3PO_4 balance	1,791	1,739	1,645	1,830	1,898	1,808

FAO (2017). Supply refers to the maximum achievable production, while demand is the purchase of fertilizer within a given timeframe. The non-H_3PO_4 fertilizer demand refers to industrial and other demand that includes P_2O_5 through super phosphate and rock phosphate.

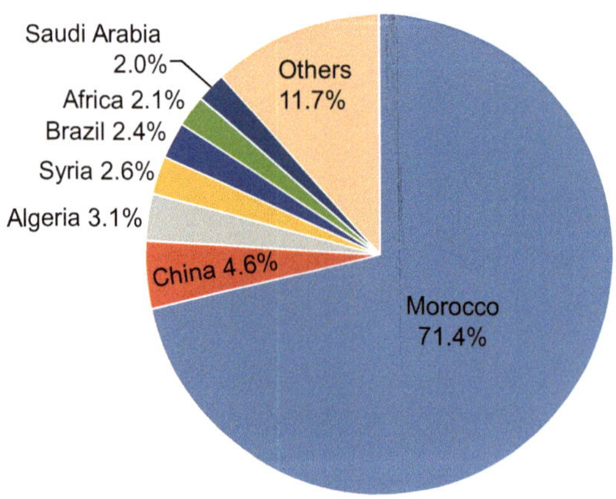

FIGURE 19.1 Estimates of global economically extractable phosphorus. The top seven countries account for 88.3%, with the vast majority found in Morocco (USGS, 2019).

in the projections, one thing is clear: phosphorus reserves are finite. According to the USGS (2019), world resources of phosphate rock are more than 300 billion tons (equivalent to 1.19×10^{14} kg P). Even though there are no imminent shortages of phosphate rock globally, concerns about the long-term availability of economically extractable phosphorus make understanding the cycling of phosphorus important for both increasing phosphorus use efficiency and reducing phosphorus pollution in the environment.

Phosphorus in soil

As discussed earlier, 85% of terrestrial phosphorus is in soil. The amount of total phosphorus found in surface soils can vary greatly, ranging from <100 μg P g^{-1} (200 kg P ha^{-1}) to >1,000 μg P g^{-1} (2,000 kg P ha^{-1}). The C/P ratio in strongly weathered soil can be as high as 200. In most cultivated agricultural soils, this ratio is around 100. The primary mineral form of phosphorus is rock phosphate, or apatite (Table 19.2, Box 19.3). Phosphorus-bearing minerals have low solubility as evidenced by their solubility product values (Table 19.2). Highly weathered soils may have very little remaining apatite, while relatively unweathered and

TABLE 19.2 Examples of inorganic phosphorus minerals in soil and their solubilities with respect to dissolution of the cation and PO_4^{3-}.

Name	Formula	Solubility product (Log K_{sp})[a]
Fluorapatite	$Ca_5(PO_4)_3F$	–59
Hydroxyapatite	$Ca_5(PO_4)_3OH$	–57
Tricalcium phosphate	$Ca_3(PO_4)_2$	–29
Variscite	$AlPO_4 \cdot 2H_2O$	–21
Strengite	$FePO_4 \cdot 2H_2O$	–26

[a] *Note: These values represent very low water solubility. For example, the K_{sp} for $Ca_3(PO_4)_2$ is 1.0×10^{-29} (or log -29). The molar solubility is 6.2×10^{-7} M which is equivalent to 1.9×10^{-4} g/L or 0.19 ppm.*
Solubility products calculated from equilibrium data (Lindsay, 1979).

BOX 19.3

Phosphate rock

Rock phosphate minerals are called apatite (Greek for "deceiving"). As a group of minerals, its compositions are complex and not fully known. The general formula for apatite is $M_{10}(PO_4)_6X_2$, where M is commonly calcium but could also be Al or Fe. The X can be F^-, OH^-, Cl^-, or CO_3^{2-}, resulting in fluor-, hydroxyl-, chloro-, and carbonate apatites. Fluorapatite [$Ca_{10}(PO_4)_6F_2$] is most common and is practically insoluble in water (log Ksp = −59.6 at 25°C). Ground phosphate rock has been directly applied as fertilizer to acidic soils or made more water-soluble by treating with acid and/or heat. Rock phosphate is used for manufacture of fertilizers, phosphoric acid, super phosphates, phosphorus, and phosphorus-containing compounds.

alkaline soils formed from basic parent materials may have high apatite contents. Weathering of apatite results in the release of orthophosphate, which is referred to as phosphoric acid by the International Union of Pure and Applied Chemistry (IUPAC). Pure phosphoric acid (H_3PO_4) is non-toxic and solid at room temperature and pressure. Phosphoric acid is a weak acid. In aqueous solution it forms conjugate bases, dissociating to dihydrogen phosphate ($H_2PO_4^-$) at pH 2.14; monohydrogen phosphate (HPO_4^{2-}) at pH 7.2; and phosphate (PO_4^{3-}) at pH 12.4. Therefore, orthophosphate solutions would be comprised of essentially 100% $H_2PO_4^-$ at around pH 4.7, and as HPO_4^{2-} at around pH 9.8. Although orthophosphate contributes a very small portion, usually < 1%, of the total phosphorus in soil, it is the predominant form of bioavailable phosphorus in the soil solution.

Concentrations of solution phosphorus commonly range from 0.1 to 1 mg L^{-1} in soil. Of this, more than half may be soluble organic compounds originating from dead cells or colloidal organic compounds. Orthophosphate is also released into the soil solution following addition of soluble phosphate fertilizers or mineralization of organic phosphorus. Phosphorus concentrations in several common fertilizers are shown in Table 19.3. Orthophosphate release into the soil solution can be followed by precipitation as iron, manganese, and aluminum phosphates in acid soils, calcium phosphates in alkaline soils (stable-inorganic phosphorus), or by adsorption to iron and aluminum oxides (labile-inorganic phosphorus). These processes result in low orthophosphate concentrations in the soil solution. The optimum availability of orthophosphate occurs at about pH 6.5 when precipitation as aluminum and calcium phosphates is lowest. Therefore, liming acidic soils or acidifying alkaline soils would result in increased orthophosphate concentrations in the soil solution.

Many organic forms of phosphorus are found in soils. As an essential element for life, the return of plant and animal remains or waste products to soil is accompanied by the introduction of organic phosphorus. When microorganisms grow, reproduce, and go through life cycles, they also produce organic phosphorus compounds through biosynthesis. Organic phosphorus usually represents 30%–50% of total phosphorus in most soils. The chemical nature of the soil organic phosphorus fraction is not fully understood. Commonly found organic phosphorus compounds in soil are inositol phosphates, phospholipids, and nucleotides (Fig. 19.2). As key players in cellular energy transfer, adenosine phosphates (ATP, ADP, and AMP) are present and detected in soil (structures are shown in Fig. 3.1). Of these organic phosphorus compounds, inositol polyphosphates (also termed phytate or phytic acids), are typically present in the greatest quantity, comprising 10%–50% of the total organic soil phosphorus. Although phytate-bound phosphorus is a major storage form of phosphorus in plants, most of the inositol phosphates and other polymeric forms of organic phosphorus in soil are thought to be of microbial origin. Phospholipids and nucleic acids account for 1%–5% of the organic phosphorus. Other identifiable organic phosphorus compounds are typically present in trace amounts.

TABLE 19.3 Percentage of total phosphorous and water-soluble phosphorus in several common fertilizers.

Phosphorus fertilizer source	Total P (%)	Water soluble P (%)
Superphosphate (OSP)	9.2	7.8
Triple superphosphate (TSP)	19.6	16.7
Monoammonium phosphate (MAP)	21.4	17.5
Ammonium polyphosphate (APP)	14.8	14.8
Rock phosphate	14.8	0.0

University of Minnesota Cooperative Extension Service. https://extension.umn.edu/phosphorus-and-potassium/understanding-phosphorus-fertilizers.

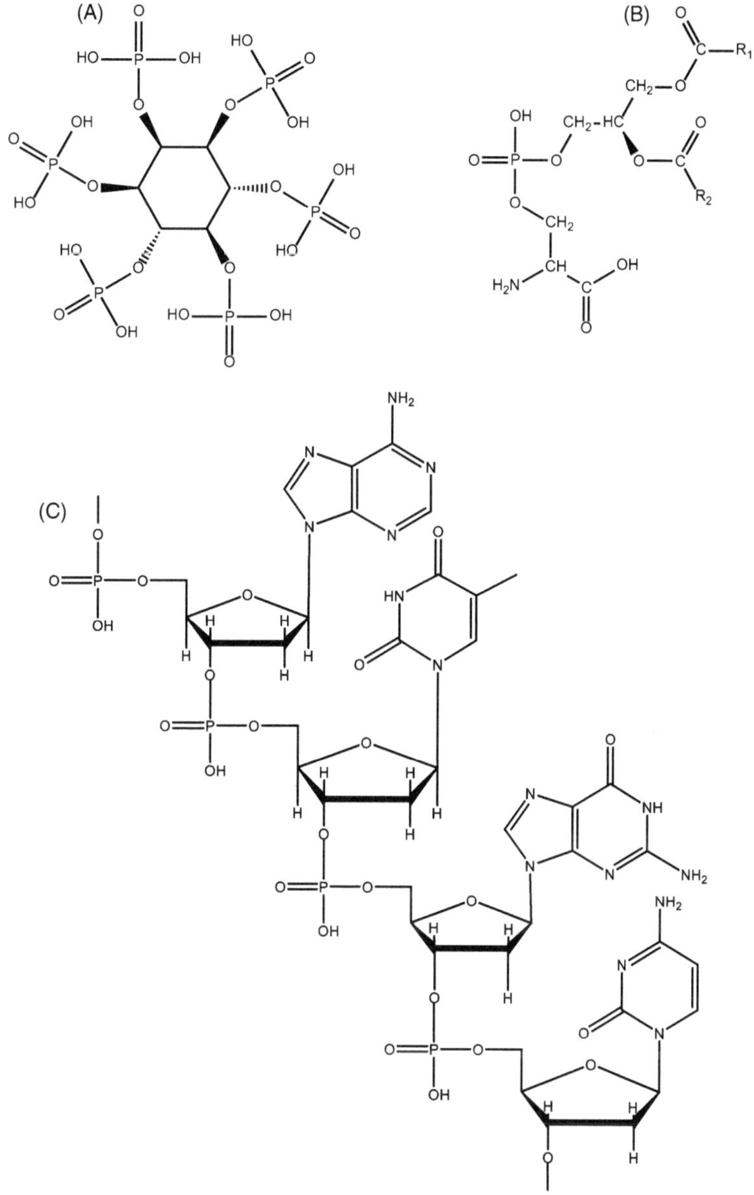

FIGURE 19.2 Chemical structures of representative forms of organic phosphorus in soil. (A) Inositol hexaphosphate (phytate or phytic acid); (B) Phospholipid—with R1 and R2 representing two fatty acids, each containing 14–20 carbons, with varying degrees of saturation; and (C) Polynucleotide—deoxyribonucleic acid (DNA). As shown in this DNA fragment, each nucleotide consists of a nitrogenous base (adenine—A, thymine—T, guanine—G, or cytosine—C), a five-carbon sugar (deoxyribose), and a phosphate.

Another pool of organic phosphorus is the soil microbial biomass. This fraction is an actively cycling pool and is part of the labile soil organic phosphorus. A method of measuring microbial biomass phosphorus is presented in Box 19.4. Microbial biomass phosphorus in soil commonly ranges from 5 to 75 μg biomass P g^{-1} soil (equivalent to 2.6–38.5 kg biomass P ha^{-1}). Brookes et al. (1984) reported that the biomass-phosphorus fraction ranged from 2%

BOX 19.4

Measurement of microbial biomass phosphorus content in soil

The overall approach is analogous to the one described in Box 13.2 for determining microbial biomass based on C, but the current method is based on quantifying inorganic orthophosphate (P*i*) in fumigated and non-fumigated fresh soils. Microbial biomass P is calculated from chloroform ($CHCl_3$)-released P*i* divided by an extraction efficiency factor (k_p). A k_p value of 0.4 indicates that 40% of microbial biomass P is released by $CHCl_3$ fumigation and detected by the method. Briefly, soil is fumigated with $CHCl_3$ for 24 h and P*i* in the soil is extracted using 0.5 M $NaHCO_3$. Chloroform fumigation lyses microbial cells, releasing cellular components to the soil. A portion of the released organic P is converted to inorganic P*i* during the 24 h of $CHCl_3$ fumigation. The k_p value is determined by the recovery of known amounts of microbial phosphorus from microorganisms added to soil. Therefore, k_p values may vary depending on microbial composition and diversity in soil. Because P*i* is easily adsorbed to soil, a correction for phosphorus fixation is necessary. A known amount of H_2PO_4, typically 25 µg P g^{-1} soil, is added to a nonfumigated soil sample, followed by $NaHCO_3$ extraction and P*i* quantification. The difference between the P*i* in this extract and the nonfumigated control represents the amount recovered from the added P*i*. More details of this method can be found in Brookes et al. (1982). Microbial biomass P (MBP) is calculated as follows using a k_p value of 0.4 and spiking 25 µg P*i* g^{-1} soil to the nonfumigated soil:

$$\frac{\mu g\ \mathrm{MBP}}{g\ \mathrm{soil}} = \left(\frac{25}{c-a}\right) \times \left(\frac{b-a}{0.4}\right)$$

where: a = µg P*i* extracted from the nonfumigated control

b = µg P*i* extracted from the fumigated soil

c = µg P*i* extracted with 0.5 M $NaHCO_3$ from a nonfumigated soil that was spiked with 25 µg P*i* g^{-1} soil

to 5% of the total organic phosphorus in arable soils. In forest soils, microbial biomass-phosphorus ranged from 16% to 73% of total soil phosphorus, commonly contributing 25%–35% of total soil phosphorus. Microbial biomass-phosphorus contents range from 1.5% to 10% of the magnitude of organic carbon in forest soils, and the percentages decrease with increasing decomposition. Cultivation of soil promotes decomposition of organic matter, reducing both total organic phosphorus and the proportion of total organic phosphorus that is present as microbial biomass. Therefore, tillage depletes the labile organic phosphorus pool more rapidly than the stable organic phosphorus fraction. The microbial biomass-phosphorus concentrations in soils are similar to microbial biomass-nitrogen concentrations. There is no known storage form for excess N in microbial cells, but phosphorus storage can account for 10%–20% dry weight in yeast cells (Kornberg, 1995) and >10% dry weight in some bacteria (Deinema et al., 1985). Since microorganisms can take up phosphorus efficiently, high phosphorus availability causes enhanced P uptake, while P limitation promotes homeostasis (the maintenance and regulation of intracellular processes with little influence from surrounding environmental changes).

Phosphorus cycling

Soil phosphorus pools and cycling are intimately connected to the mining of rock phosphate (Fig. 19.3) and transporting of terrestrial phosphorus to the ocean sediments. A model of the terrestrial phosphorus cycle, illustrated in Fig. 19.4, shows the predominant pools and major transformation processes. In contrast to nitrogen, the phosphorus cycle is sedimentary with no major gaseous forms of phosphorus produced. The cycling of phosphorus is affected by both biological and chemical reactions. In this model, the solution phosphorus pool serves as the central point in the overall cycle.

In the biological phosphorus subcycle, soil solution phosphorus can be assimilated by plants and soil microorganisms, leading to the formation of orthophosphate and biota phosphorus, such as phospholipids, nucleotides (DNA, RNA), ATP, phytate, and biomass phosphorus. These processes result in immobilization of phosphorus, which is the reverse process

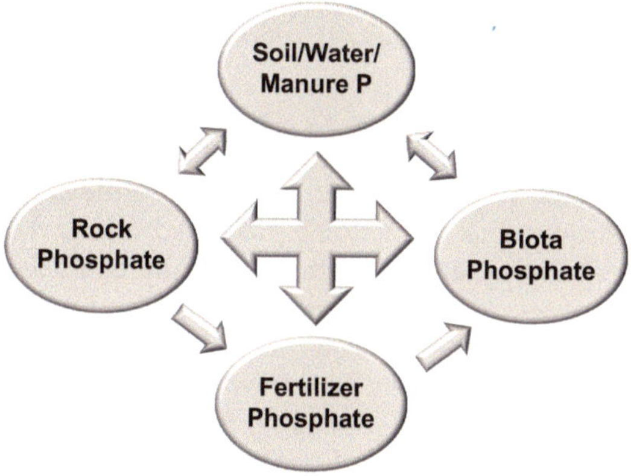

FIGURE 19.3 **Relationships of major terrestrial phosphorus pools.**

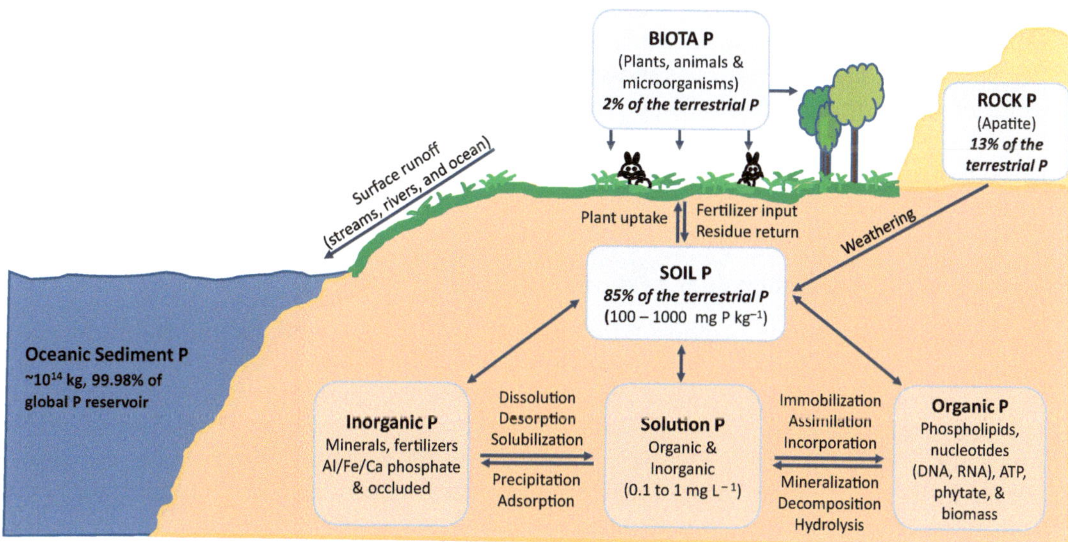

FIGURE 19.4 **The phosphorus cycle: predominant pools and major transformation processes.** Plants take up primarily phosphate from the solution P portion of the soil P pool.

of mineralization. As plant residues and animal remains and wastes are returned to soil, the organic phosphorus may be directly incorporated into persistent soil organic matter (incorporation), mineralized to orthophosphate, or immobilized (assimilated) into the microbial biomass. Biomass phosphorus is subject to incorporation into soil organic matter as well as mineralization and immobilization reactions. The turnover or cycling of biomass contributes significantly to the labile organic phosphorus pool. Crop removal and erosion are two processes that lead to the loss of organic phosphorus.

In the chemical phosphorus subcycle, orthophosphate is released into the soil solution by chemical and biochemical weathering processes from rock phosphate, minerals, inorganic phosphorus fertilizers, and Al/Fe/Ca/Mn phosphates occluded by secondary minerals. Dissolution and desorption of phosphate minerals releases orthophosphate into the soil solution. Solubilization occurs through organic acids produced by plants and microbes. Orthophosphate can be adsorbed to aluminum and iron oxides (labile inorganic phosphorus) or precipitated as aluminum, iron, or calcium phosphates (secondary minerals). Through time, phosphorus finds its way into streams, lakes, rivers, and eventually the oceans where it ends up in sediments.

There is growing evidence that microbes also produce reduced P species, phosphite (HPO_3^{2-}) and phosphine gas (PH_3), in soil under anaerobic conditions. Some microorganisms have been demonstrated to produce phosphine gas under highly anaerobic conditions. Likewise, studies have shown phosphine to be produced under wetland environments such as rice paddy fields. In some aquatic systems, phosphite can account for 10%–30% of dissolved P (Figueroa and Coates, 2017). Although high levels of naturally occurring reduced P species are unlikely to exist in most aerobic, agricultural soils, phosphite is sometimes added to these soils as a fungicide, fertilizer, and/or biostimulant. Phosphite can positively influence plants by disease suppression or providing phosphorus nutrition through oxidation to phosphate by microorganisms. Added phosphite appears to oxidize slowly in soil with a reported half-life of weeks to months. Nevertheless, the ability to oxidize phosphite appears to be widespread among microorganisms, possibly because of reduced P being the dominant form on early Earth. Microbial oxidation of phosphite to phosphate by assimilatory phosphite oxidation has been known since the 1950s. More recently, a bacterium (*Desulfotignum phosphitoxidans* strain FiPS-3) was shown to possess dissimilatory phosphite oxidation capabilities (Schink and Friedrick, 2000). This bacterium was isolated from brackish sediments and is capable of growth by coupling phosphite oxidation to the reduction of either sulfate or carbon dioxide. So far, this is the only bacterium known to carry out dissimilatory phosphite oxidation.

Bioavailability and microbial participation

Limited bioavailability of phosphorus hinders crop growth and agricultural production. Although 85% of terrestrial phosphorus is in soil, phosphorus is second only to nitrogen among mineral nutrients that most limit plant growth. Microorganisms in soil contribute to solution phosphorus by solubilization of labile phosphorus and mineralization of organic phosphorus. Bioavailability of phosphorus in soil is governed by multiple processes outlined earlier in the phosphorus cycle. Three key processes, mineralization, immobilization, and solubilization, are further discussed later.

Mineralization and immobilization of phosphorus

Phosphorus mineralization is an enzymatic process, employing phosphatases that catalyze a variety of hydrolytic mineralization reactions leading to the release of phosphate from organic phosphorus compounds to the soil solution. **Phosphatases** in soil are primarily synthesized and released by microorganisms but can also originate from plants and animals. Examples of enzymes catalyzing hydrolytic mineralization reactions in the soil solution include:

- **Phosphomonoesterases** hydrolyze the phosphate from monoester forms of phosphorus, such as phosphate groups present in nucleotides and phospholipids (Box 19.5).

BOX 19.5

Assay of phosphomonoesterase (phosphatase) activity in soil

This procedure measures the potential of a soil to hydrolyze phosphomonoester bonds in organic phosphorus compounds and release orthophosphate (P*i*). The assay involves incubating soil with an organic phosphate analog (*p*-nitrophenyl-phosphate) as substrate in a buffer solution, followed by quantification of enzymatically released *p*-nitrophenol (Tabatabai, 1994). The soil is treated with toluene to inhibit microbial activity and the buffer solution maintains the reaction pH. Activity of acid phosphatase is measured when pH of the buffer is in the acidic range (pH 6.5 is commonly used). Activity of alkaline phosphatase is measured when pH of the buffer is in the alkaline range (pH 11 is commonly used). After incubation, NaOH is added to the reaction mixture to stop the reaction and adjust to an alkaline pH where *p*-nitrophenol turns yellow. Colorimetric determination of *p*-nitrophenol concentration permits calculation of the rate of enzyme activity, which is expressed as μg *p*-nitrophenol released/g soil/hour. The reaction is:

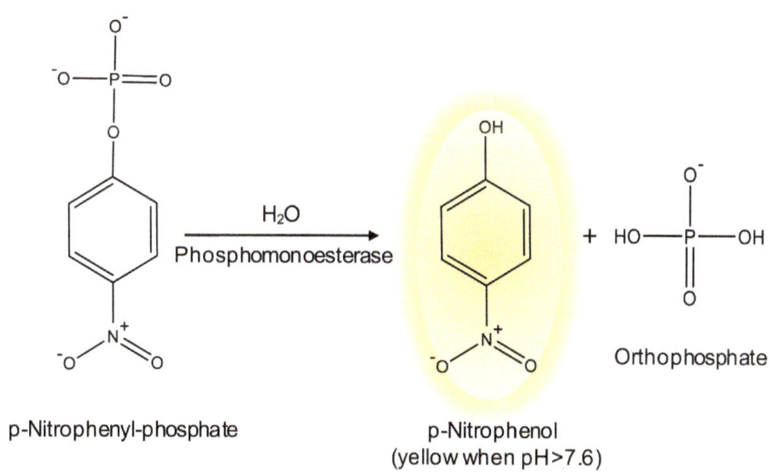

BOX 19.6

Phytase

Three phytase enzymes are recognized by the International Union of Biochemistry and Molecular Biology: 3-phytase or myo-inositol-hexakisphosphate 3-phosphohydrolase (EC 3.1.3.8), 4-phytase or myo-inositol-hexakisphosphate 4-phosphohydrolase (EC 3.1.3.26), and 5-phytase or myo-inositol-hexakisphosphate 5-phosphohydrolase (EC 3.1.3.72) (IUBMB, 2020). Studies on phytase and its substrate, phytate, have been carried out for almost 100 years. Phytase activity was first detected in rice bran. However, research on phytase progressed slowly until the 1960s when efforts were launched to commercially produce phytase. A soil fungus, *Aspergillus niger* NRRL 3135, was identified to be the most efficient producer of phytase after screening over 2000 organisms (Shieh and Ware, 1968). This strain produces two phytases, one has a pH optimum of 2.5 and the other enzyme has a maximum rate at pH 5.5. Phytase has been approved as a feed additive in many countries and has been marketed as a food additive in the United States since 1996.

- **Phosphodiesterases** hydrolyze phosphate from diester forms of phosphorus, e.g., cyclic nucleotide phosphodiesterases in nucleic acids.
- **Phytases** hydrolyze phosphate from inositol phosphates, including those found in plant seeds (Box 19.6).

Once phosphorus is mineralized, orthophosphate ions can be taken up by plants, immobilized back into microbial cells, or form insoluble inorganic complexes. Orthophosphate can also be incorporated into organically bound forms in the organism. For example, orthophosphate is enzymatically added to ADP (adenosine diphosphate) to form ATP. The extent of microbial immobilization is affected by the carbon to phosphorus (C/P) ratio of the organic materials being decomposed and the amount of available phosphorus in the soil solution.

The C/P ratio of an added residue determines whether phosphorus in the soil is mineralized or immobilized. If the added residue is high in carbon and low in phosphorus (a high C/P ratio), microorganisms involved in decomposition will use phosphorus from the soil solution, which leads to net immobilization of soil phosphorus. Conversely, if the plant residue is rich in phosphorus and has a low C/P ratio, residue phosphorus will be released from the decomposition process to the soil solution, resulting in net mineralization of phosphorus. Generally, a C/P ratio <200 results in net mineralization and a C/P ratio >300 results in net immobilization. For ratios between 200 and 300, there is little net change in solution phosphorus concentrations. These processes are similar to mineralization and immobilization of nitrogen and sulfur and are also influenced by soil and environmental variables (pH, temperature, aeration, and soil moisture) that affect microbial activities as well as the form and solubility of soil phosphorus. The element that is most limiting controls the overall decomposition rate of a residue. As easily decomposable organic carbon levels decline, a portion of the microbial community will die and P will be mineralized, becoming available for uptake by living organisms.

Solubilization of inorganic phosphorus

Although organic phosphorus may be 30%–50% of total phosphorus in soil, inorganic forms of phosphorus are often predominant. As described earlier, inorganic phosphorus minerals such as aluminum, manganese, and iron phosphates in acidic soil and calcium phosphates in alkaline soils have low solubility in water. Orthophosphate in soil solution is supplied to the roots primarily by diffusion. Thus, chemical equilibria between orthophosphate, adsorbed orthophosphate, and inorganic phosphate minerals are important in supplying phosphorus to plants and microorganisms. As solution phosphorus is assimilated by living organisms, it is replenished from phosphorus minerals.

Some microorganisms are capable of solubilizing and converting insoluble soil phosphorus and rock phosphate to bioavailable forms. Microbial solubilization of phosphorus has long been recognized. Bacteria capable of phosphate solubilization are referred to as **phosphobacteria**, but the solubilizing activities are also observed among other microorganisms. Attempts have been made to isolate **phosphorus-solubilizing microorganisms** (PSM) for use as soil and/or seed inoculants. Examples of PSM (including bacteria, fungi, actinomycetes, and algae) are shown in Table 19.4. Phosphorus solubilizing *Pseudomonas*, *Agrobacterium*, and

TABLE 19.4 Examples of phosphorus-solubilizing microorganisms (PSM) enhancing crop production.

PSM	Test crop	Result
Aspergillus niger	Wheat	Increased plant growth
Serratia sp.	Wheat	Increased plant growth
Aspergillus awamori S29	Mung bean	Increased total P content and plant biomass
Burkholderia gladioli	Sweetleaf	Increased plant growth
Pseudomonas aeruginosa	Chinese cabbage	Increased biomass production and shoot length
Pseudomonas putida	Moss	Increased plant growth
Azotobacter chroococcum, *Saccharomyces cerevisiae*, and *Bacillus megaterium*	*Moringa oleifera*	Increased shoot and root lengths and weights, and content of vitamin C and protein in leaves
Burkholderia gladioli	Oil palm	Increased growth and phosphate uptake
Aspergillus niger and *Penicillium aculeatum*	Chinese cabbage	Increased plant growth
Bacillus thuringiensis	Rice	Increased shoot length
Pseudomonas striata and *Glomus fasciculatum*	Soybean-wheat	Improved root development and increased grain yield
Burkholderia cepacia	Maize	Increased plant growth
Azotobacter chroococcum and *Bacillus subtilis*	Wheat	Improved productivity
Pseudomonas favisporus TG1R2	Soybeans	Increased dry biomass
Rhizobium tropici CIAT899	Beans	Increased nodule number and mass, shoot dry weight, and root development

Adapted from Alori et al. (2017).

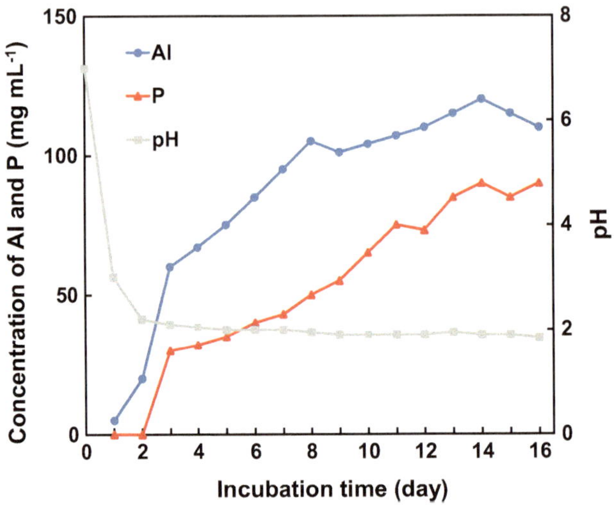

FIGURE 19.5 Changes in aluminum and phosphorus concentrations and pH during the solubilization of $AlPO_4$ by the fungus *Aspergillus niger*. *Source: Adapted from Illmer et al. (1995).*

Bacillus are commonly isolated. Several strains of *Burkholderia* have also been found to be effective. For decades, Russian farmers used phosphobacterin, *B. megaterium* var. *phosphaticum*, as a bacterial inoculant. Unfortunately, the effectiveness of PSM is not consistently supported in the scientific literature. Limited success may be due, in part, to limited knowledge of the precise mechanisms by which the organisms solubilize rock phosphate and the conditions that promote solubilizing activity. It is generally accepted that mineral phosphate solubilization by PSM strains may occur due to a variety of mechanisms including lowering pH via the release of organic acids (Fig. 19.5) and producing siderophores that chelate Fe^{3+} from iron phosphate minerals. A wide range of low molecular weight acids is produced by microorganisms and plants, including inorganic acids such as carbonic acid (H_2CO_3) and organic acids such as citrate, oxalate, and gluconate. Carbonic acid can promote dissolution of calcium and magnesium phosphate compounds. The hydroxyl and carboxyl groups of the acids chelate the cations bound to phosphate (i.e., aluminum, iron, calcium, and magnesium phosphates), thereby resulting in the release of orthophosphate into the soil solution. One group of organisms that may be especially important in this regard is the mycorrhizal fungi, which form symbioses with plant roots and enhance the uptake of phosphorus and other nutrients (Chapter 12). Mycorrhizal fungi play a key role in phosphorus acquisition in the environment through forming a symbiotic relationship with the roots of nearly all vascular plants. Mycorrhizal symbioses are recognized for their importance in plant nutrition, particularly in phosphorus uptake. Under water-logged conditions, hydrogen sulfide, produced by sulfate-reducing bacteria or other processes, can also displace metal cations from insoluble phosphates, leading to the release of phosphate. The incorporation of readily mineralizable carbon sources, such as manure, can enhance phosphorus solubilization through increased biological activity. Further, the added organic carbon can complex aluminum in acidic soils, thereby reducing aluminum phosphate precipitation.

Environmental aspects

Nutrients or pollutants?

Some nutrients essential to living organisms can also become environmental pollutants. Phosphorus exemplifies this dual nature. As discussed earlier in this chapter, phosphorus is an essential nutrient for life. It is second only to nitrogen in limiting agricultural production, being deficient in nearly 67% of crop-production land worldwide (Batjes, 1997). Yet phosphorus is one of the most widespread environmental pollutants threatening water quality. How can an element essential for life be a major pollutant in the environment? To fully understand the dilemma, we need to revisit the fundamentals of phosphorus storage, forms, and transformation processes that were discussed earlier in this chapter. Building on this base, the focus will be on reactions and processes involving phosphorus that are relevant to environmental health, and the role of soil microorganisms in governing these processes.

Phosphorus pollution of the environment is largely related to its mobility, which is highly related to its chemical forms. Organic phosphorus is more mobile than inorganic P in soil. The high mobility of organic P in soils was noted as early as 1902 by Dyer and supported by numerous studies thereafter. A major portion of total P in soil leachates is in organic form, and the percentage of dissolved organic P (DOP) increases with soil depth, with up to 88% of total dissolved P being DOP (Parham et al., 2002). As discussed earlier in the chapter, phytate is typically the predominant organic phosphorus form in soil. Phytate possesses a plane of symmetry with six phosphate groups (Fig. 19.2A). Phytate-bound phosphorus is a major storage form of phosphorus in plants, constituting between 60% and 90% of total phosphorus in grains and oilseeds. The release of phosphorus from phytate occurs via the enzyme phytase (Fig. 19.6). Unfortunately, phytate-phosphorus is essentially unavailable to monogastric animals (non-ruminants), including humans, chickens, and pigs, because they neither produce phytase nor sustain microorganisms in their digestive track that produce phytase. For this reason, animal feed is sometimes supplemented with inorganic phosphorus. Further, phytate is an effective chelating agent that binds metals and proteins in animal feedstuffs. The chelation of cations, such as Ca^{2+}, Zn^{2+}, Mg^{2+}, Fe^{3+}, Mn^{2+}, Cu^{2+}, and Mo^{2+}, renders them less available to monogastric animal nutrition. Because of this, phytate is considered an antinutrient for

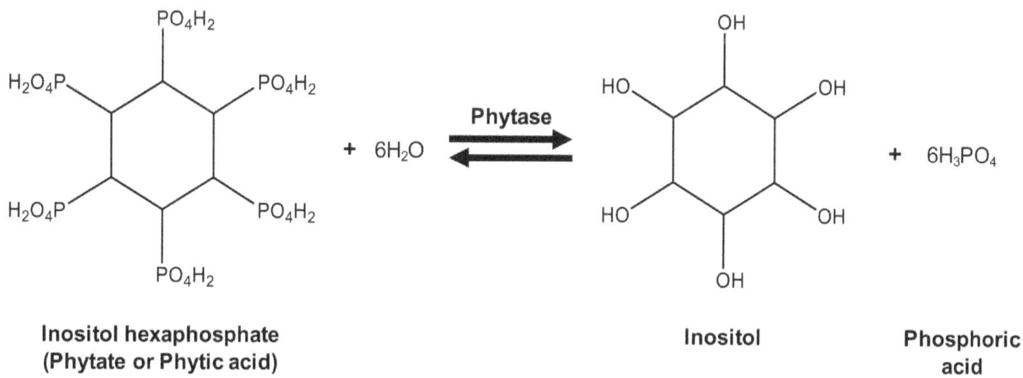

FIGURE 19.6 **Hydrolysis of phytic acid to inositol and phosphoric acid by phytase.**

monogastric animals, and inorganic phosphate and trace elements such as Zn^{2+}, Mg^{2+}, Fe^{3+}, and Cu^{2+} are supplemented in the feed to satisfy their nutritional needs. Phytate, excess inorganic phosphorus, and trace elements excreted in animal waste are ultimately released to the environment because land application of animal manure is the most economical route for waste disposal.

Soil microorganisms hydrolyze phytate and release phosphoric acid to the environment. The environmental concern is the release of phytate-phosphorus and subsequent transport to nearby aquatic systems. If a large amount of phosphorus is applied to soil, it may move to surface waters with runoff water carrying dissolved phosphorus and eroded soil particles with bound phosphorus. Phosphorus runoff and loading into freshwater ecosystems can lead to eutrophication, resulting in oxygen depletion, fish kills, algal blooms, and biodiversity losses (Stokstad, 2008). Phosphorus has long been recognized as the most limiting nutrient for primary productivity in many ecosystems, particularly aquatic systems (Schindler, 1974; Box 19.7). In the 1960s, David Schindler identified detergent phosphates as a primary pollutant in lakes through long-term monitoring of lakes and streams. Certain cyanobacteria are capable of photosynthesis and dinitrogen fixation to obtain carbon and nitrogen nutrition from the atmosphere (Chapter 7). This leaves phosphorus as the most limiting nutrient to support cyanobacterial growth in the aquatic environment. When excess phosphorus enters a body of water, it promotes HABs (Box 19.2) that are often noticeable as a green "scum" or film on the water surface. As the algae die, they settle to the bottom where bacterial decomposition of the nutrient-rich algae occurs. The decomposition of the algae requires oxygen consumption, which may lead to oxygen depletion. If the process continues, the lake slowly becomes anaerobic, negatively impacting many forms of aquatic life. Nonpoint sources of phosphorus, such as from agricultural activities, are often associated with increased phosphorus in surface water. For example, runoff from cropland and soils receiving animal manures can lead to significant phosphorus inputs into surface waters. In industrialized countries, nonpoint sources are the major input of phosphorus. In non-industrialized countries with

BOX 19.7

Phosphate as a pollutant killing lakes

In the 1960s, David Schindler identified detergent phosphates as a pollutant killing lakes. He was the founding director of the Experimental Lakes Area (ELA) in northwestern Ontario, Canada, where ecosystem-scale experiments with a variety of pollutants and long-term monitoring of lakes and streams were initiated in 1968. An article by Stokstad (2008) detailed some of the effort and illustrated this with an aerial view of Lake 226 during the eutrophication experiment of the 1970s. Schindler fertilized half of the lake with carbon and nitrogen and the other half with carbon, nitrogen, and phosphates. Within 2 months of fertilization, the half of the lake fertilized with phosphate was covered by an algal bloom, while no increases in algae or species changes were observed in the half of the lake that was fertilized with carbon and nitrogen only. This demonstrated the vital role of phosphorus in governing productivity of aquatic systems (Schindler, 1974).

limited sewage-treatment capacity, raw sewage discharge into surface waters (a point source) is a significant source of phosphorus pollution.

Remediation, prevention, and sustainability

The natural ability of soils to adsorb phosphorus from solution can be used to remove phosphorus from contaminated media and wastewater and may therefore be an economical remediation strategy. Natural and constructed wetlands serve as a sink for the removal of phosphorus from wastewaters before their release to a river or lake. Attempts to decrease the amount of phosphorus being excreted from animal production systems have focused on supplementing diets only to required phosphorus levels, adding microbial-derived exogenous phytase enzymes to increase the phosphorus digestibility of grains, and using genetically engineered low-phytate maize as the feeding material. Thus, phytate and phytase are key components of efforts to reduce phosphorus pollution. Unfortunately, these approaches have had limited success in reducing the amount of manure-phosphorus entering the environment. Continued inputs of fertilizer and manure phosphorus in excess of crop requirements have led to widespread build-up of soil phosphorus levels that are of environmental rather than agronomic concern (Sharpley et al., 1994).

One of the targeted efforts to mitigate phosphorus pollution in several US states centers on establishment of a **Phosphorus Index**, a risk assessment tool to quantify the potential for P runoff from a field (Box 19.8). With increasing interest in practicing no-till agricultural

BOX 19.8

What's the Phosphorus (P) Index?

The P Index is a risk assessment tool used to quantify the potential for P runoff from a field to surface water so that conservation management practices can be developed to lower the risk of eutrophication in nearby water bodies. The P Index was developed by Iowa State University, the National Soil Tilth Lab, and the Natural Resources Conservation Service. The Index is based on soil and water quality research conducted in Iowa and other Midwestern states and incorporates P source and transport factors. Source factors include soil test P, total soil P, and the rate, method, and timing of P application from commercial fertilizer, manure, and other organic sources. Transport factors influencing P runoff potential include soil erosion, water connectivity, precipitation, drainage, slope of the site, texture of the soil, and management practices. The risk of P movement is mathematically determined to be either very low, low, medium, high, or very high based on erosion, runoff, and drainage factors. A high P Index suggests that field application of organic fertilizer, such as manure, needs to be based on P content in the soil and the fertilizer. When the P Index is low, organic fertilizers can be applied based on the need for N. Further, if the P Index is high and the factors that are responsible for the higher risk of P loss are identified, the risk of P runoff can be reduced through effective management practices. For example, if the P Index is high because of high soil erosion, implementing soil conservation best management practices may lower the risk and allow safe manure application.

production, there is an emerging opportunity to reduce phosphorus pollution while strengthening nutrient retention and productivity of the soil through continuous no-tillage.

Iron, manganese, mercury, selenium, and arsenic

Elements that serve as nutrients to living organisms can also become toxic depending on their forms and concentrations. As an element's bioavailable concentration increases in the environment, organisms may take up more of the element into their cells, thereby potentially exceeding required levels and causing adverse physiological reactions (toxicity). For macronutrients, the sufficiency range is broad, and toxicity occurs rarely. For micronutrients, the difference between deficient and toxic levels may be quite narrow, making toxicity a concern. The upper tolerance level that triggers physiological responses varies among organisms and the nutrient involved. Some elements are of little nutritional value to living organisms and are often regarded as pollutants. For example, iron (Fe), manganese (Mn), and selenium (Se) are important nutrients and beneficial for microbial growth at certain concentrations, while mercury (Hg) and arsenic (As) are not required for growth and are usually considered pollutants regardless of their concentration.

Transformations of these elements are often carried out by both microorganisms and abiotic reactions, changing their solubility, bioavailability, and mobility in the environment. Biologically mediated redox reactions are often linked to energy transduction in an organism. For instance, chemoautotrophic bacteria oxidize reduced inorganic compounds to extract electrons for use in ATP production during oxidative phosphorylation (Chapter 3). Alternatively, elemental reduction occurs during energy transduction under anaerobic conditions with the element serving as an alternative to oxygen as the terminal electron acceptor. Microbially mediated redox reactions of selected metals and metalloids are listed in Table 19.5. Iron and manganese are two of the most commonly transformed metals in the soil environment. Oxidation

TABLE 19.5 Microbial oxidation-reduction reactions of selected metals and metalloids.

Element	Common oxidation states	Reaction oxidized ↔ reduced	Example of ecosystem function	Example of involved bacteria
Arsenic (As)	As^{5+}, As^{3+}	$As^{5+} + 2e^- \leftrightarrow As^{3+}$	Reduction leads to increased mobility	*Pseudomonas, Alcaligenes, Micrococcus, Thiobacillus*
Iron (Fe)	Fe^{3+}, Fe^{2+}	$2Fe^{3+} + 2e^- \leftrightarrow 2Fe^{2+}$	Reduction leads increased solubility	*Pseudomonas, Geobacter, Thiobacillus, Desulfovibrio, Acidithiobacillus*
Mercury (Hg)	Hg^{2+}, Hg^0	$Hg^{2+} + 2e^- \leftrightarrow Hg^0$	Reduction leads to detoxification	*Pseudomonas, Streptomyces, Bacillus*
Manganese (Mn)	Mn^{4+}, Mn^{2+}	$Mn^{4+} + 2e^- \leftrightarrow Mn^{2+}$	Oxidation leads to detoxification	*Pseudomonas, Bacillus, Geobacter, Arthrobacter*
Selenium (Se)	Se^{6+}, Se^{4+}, Se^0, Se^{2-}	$SeO_4^{2-} + 2e^- \leftrightarrow SeO_3^{2-}$ $Se^0 + 2e^- \leftrightarrow Se^{2-}$	Reduction leads to reduced mobility	*Thiobacillus, Desulfovibrio, Bacillus, Clostridium, Micrococcus*

and reduction of iron and manganese can affect plant bioavailability and soil properties and can be indicative of imperfect drainage (i.e., mottling and manganese concretions; Chapter 2). Because these elements are also subject to oxidation and reduction by chemical means, it is difficult to assess the extent to which these transformations are biotic versus abiotic. Mercury, selenium, and arsenic are environmentally significant elements that are also affected by another interesting microbial transformation, biomethylation. Biomethylation results in the formation of volatile organic forms of these elements, which is a detoxification mechanism for the organisms involved. Transformations of iron, manganese, mercury, selenium, and arsenic are briefly discussed in the following sections to further illustrate a select number of ways microorganisms impact metal and metalloid cycling in soil. It should be noted that microorganisms can transform many other metals beyond those discussed—even radionuclides like uranium.

Iron

Iron is an essential nutrient for life. It is a component of enzymes such as peroxidase, catalase, cytochrome oxidase, and nitrogenase. As discussed in Chapter 16, nitrogenase is the enzyme complex responsible for biological dinitrogen fixation. Iron participates in oxidation-reduction reactions, such as ferredoxin participation in electron transfer. Dominant forms in soil solutions are Fe^{2+}, Fe^{3+}, $Fe(OH)_2^{+}$, and $Fe(OH)^{2+}$. Although iron is among the most abundant elements in soil, the bioavailability of common soil forms is typically very low in aerobic soil environments, causing Fe deficiencies for living organisms. However, some soils contain Fe in forms soluble enough to be toxic to plants. Total iron contents range from less than 0.05% (0.5 g kg^{-1}) in coarse-textured soils to more than 10% (100 g kg^{-1}) in highly weathered Oxisols found in the tropics.

The iron cycle (Fig. 19.7) is characterized by the oxidation and reduction of iron compounds in soils and sediments. The solubilization of iron from inorganic compounds by microorganisms will be discussed later in this chapter in a separate section on microbial enhancement of

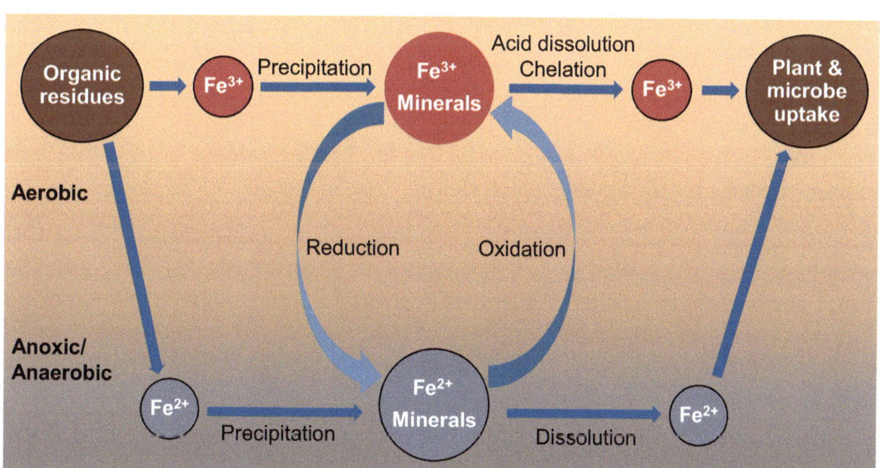

FIGURE 19.7 **The iron cycle.** Transformations include oxidation and reduction of iron minerals, precipitation of Fe^{2+} and Fe^{3+} minerals, dissolution of iron minerals to plant-available forms, and mineralization of organically bound iron.

iron bioavailability. Chemical (abiotic) oxidation of Fe^{2+} occurs rapidly under aerobic conditions at pH > 3 and is the major pathway of iron oxidation in most soil environments. Under acidic conditions, ferrous iron (Fe^{2+}) can be oxidized to ferric iron (Fe^{3+}) by chemoautotrophic bacteria such as *Acidithiobacillus ferrooxidans*. A representative reaction for ferrous iron oxidation by *A. ferrooxidans* is the oxidation of ferrous sulfate ($FeSO_4$) to ferric sulfate and ferric hydroxide.

$$12FeSO_4 + 3O_2 + 6H_2O \rightarrow 4Fe_2(SO_4)_3 + 4Fe(OH)_3$$

Microorganisms also carry out dissimilatory iron reduction (Lovley, 1993). Most organisms that reduce Fe^{3+} also reduce Mn^{4+}. Dissimilatory iron reduction is often coupled to the oxidation of fermentation products from other organisms. Complex organic matter, sugars, and amino acids can be partially oxidized by fermentation, resulting in the production of organic acids, alcohols, hydrogen, and methane. Bacteria from several genera can oxidize these fermentation products with Fe^{3+} serving as the terminal electron acceptor. The reduction of Fe^{3+} by *Geobacter metallireducens* coupled with the oxidation of acetate occurs as follows:

$$CH_3COO^- + 8Fe^{3+} + 4H_2O \rightarrow 2HCO_3^- + 8\,Fe^{2+} + 9H^+$$

Some organisms can couple the complete oxidation of monoaromatic compounds with Fe^{3+} reduction. *Geobacter metallireducens* oxidizes environmental contaminants, such as toluene and phenol, to carbon dioxide in the presence of Fe^{3+}. These organisms may play important roles in the remediation of aquifers contaminated by aromatic hydrocarbons. There are also reports of bacteria that couple the oxidation of H_2 to H^+ with the reduction of Fe^{3+}.

Dissimilatory Fe^{3+} reduction primarily takes place in anoxic environments lacking sufficient oxygen to act as terminal electron acceptor. Examples of its functions in the environment include:

- Fe^{3+} is an important electron acceptor in organic matter decomposition in aquatic sediments and saturated soils. This can lead to gleying of the soil (i.e., the characteristic gray to gray-blue colors associated with removal of iron oxide coatings or the production of certain reduced iron compounds that indicate drainage problems).
- Fe^{3+} reduction by some obligate anaerobic bacteria may be coupled with the use of a variety of organic pollutants as carbon/energy sources that result in bioremediation of these contaminants.
- Fe^{3+} reduction in iron-phosphate minerals can lead to the release of phosphate for subsequent uptake by plants and microbes.

The bioavailability of Fe^{3+} in soil solutions is low and its concentrations increase/decrease 1000-fold with each unit change in pH. For example, assume that $Fe(OH)_3$ is the solid phase controlling bioavailable Fe^{3+} in the soil solution. At pH 7, the calculated concentration of Fe^{3+} in soil solution would be approximately 10^{-17} M (Lindsay, 1979). Since solution Fe^{3+} decreases as pH increases, the low solubility of Fe^{3+} in alkaline soils can result in iron-deficient conditions for plants and microorganisms.

Some microorganisms have the ability to enhance iron bioavailability by producing Fe^{3+}-complexing compounds known as siderophores when iron is limiting (Box 19.9). These low-molecular-weight, iron-transporting compounds have high Fe^{3+} affinity and are able to solubilize Fe^{3+} for uptake by the organism. Fungi and some bacteria produce siderophores with hydroxymate active sites, whereas other bacterial siderophores are catechol derivatives (Fig. 19.8). The organism has siderophore receptor proteins on its outer membrane that bind

BOX 19.9

Siderophores

Siderophores (Greek for "iron carrier") are low-molecular-weight, high-affinity iron-chelating compounds that are produced by organisms to solubilize Fe^{3+} for uptake. They are produced by aerobic and facultative microorganisms in oxic environments to scavenge iron from insoluble iron minerals. Solubility of Fe^{3+} is increased by forming complexes with siderophores. The iron-siderophores bind with siderophore receptors in the outer membrane of the organism. Iron is released and transported into the cell following reduction from Fe^{3+} to Fe^{2+}, a form with low affinity for siderophores. Evidence suggests that iron-siderophores can also be taken up by active transport mechanisms across cell membranes, followed by iron reduction and the releasing of Fe^{2+}. The exact mechanism for how iron enters the cell using these transporters is not well understood. Over time, siderophore receptors have become "hijacked" by lethal agents, including bacteriophage, bacteriocins (killer proteins), and antibiotics. These agents can also bind to siderophore receptors, facilitating a negative outcome for the bacterial cells. For example, the receptor for a siderophore of *Yersinia enterocolitica* also binds the toxin pesticin. Siderophores are usually classified by the ligands used to chelate the ferric iron. The major groups of siderophores include the catecholates, hydroxamates, and carboxylates. A model complex based on catechol is shown in Fig. 19.8.

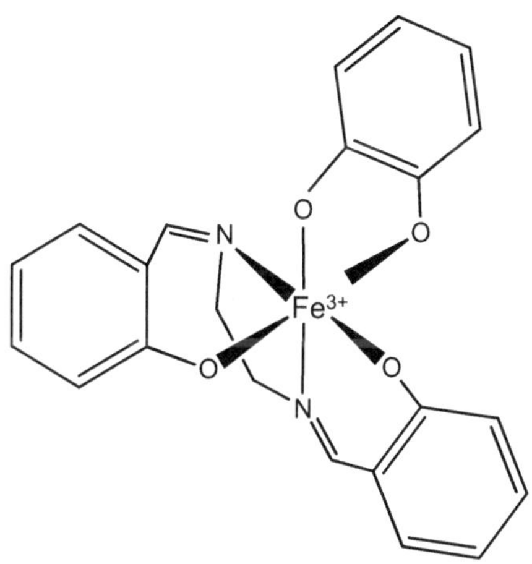

FIGURE 19.8 **Structure of a model catecholate-Fe^{3+} siderophore complex.** *Source: Modified from Yamahar et al. (2002).*

the siderophore-Fe^{3+} complex. Iron is released from the siderophore via reduction of Fe^{3+} to Fe^{2+}. The ferrous ion is then transported into the cell and the siderophore is released to complex more Fe^{3+}. In some cases, the siderophore complex enters the cell before reduction of Fe^{3+} to Fe^{2+}. The chelation of Fe^{3+} from iron phosphate minerals also results in the liberation of soluble orthophosphate.

Some pseudomonads produce a yellow-green fluorescent siderophore called *pseudobactin*, which binds iron tightly, preventing its use by other organisms. By this mechanism, the pseudomonad can control some fungal root pathogens in iron-limited environments.

Manganese

Manganese is essential for a variety of transformations in microorganisms and plants being a co-factor of many enzymes and required for activity of decarboxylase, dehydrogenase, and oxidase enzymes. Manganese also plays important roles in photosynthesis, nitrogen metabolism, and nitrogen assimilation.

Total manganese levels in soil span a wide concentration range, from < 20 mg to > 3000 mg kg^{-1} soil. The mobile and plant available forms are variable but comprise a small percentage of total Mn. It can exist in 11 oxidation states ranging from −3 to +7. The most common ones are +2, +4, and +7. Manganese and its compounds that are commonly found in soil include dissolved, exchangeable, solid (precipitated as sparingly soluble hydrous oxides), and organic forms. The form that is most common in soil solution and most available for plant growth is the reduced Mn^{2+} ion. Mn^{2+} is often complexed by organic compounds and its concentration in soil solution decreases by about 100-fold with each unit of pH increase. Some Mn is present within the structural lattice of crystalline minerals as either a primary constituent or impurity resulting from isomorphous substitution. This Mn is essentially unavailable except in terms of geological time as a consequence of natural mineral weathering.

Manganese cycling is similar to that of iron. Chemical and microbial oxidation results in the formation of relatively insoluble manganese oxides (MnO_2). Microbial oxidation of manganese can occur in soil at pH > 5. The reaction occurs as follows, with the rate increasing with increasing alkalinity up to about pH 8:

$$Mn^{2+} + 2H_2O \rightarrow MnO_2 + 4H^+ + 2e^-$$

The MnO_2 formed is very insoluble. Manganese concretions, or nodules, are often found in soils with cyclical oxidation and reduction processes, such as those with fluctuating water tables. Several bacteria enzymatically oxidize Mn^{2+} with no energy gain, although few are true chemoautotrophs. Some of the bacteria are mixotrophs; they derive some energy from the oxidation of Mn^{2+} but obtain carbon from organic sources. Other organisms oxidize Mn^{2+} to MnO_2, coupled with H_2O_2 reduction, via the catalase enzyme:

$$Mn^{2+} + H_2O_2 + 2OH^- \rightarrow MnO_2 + 2H_2O$$

This reaction does not provide energy for the organism but serves to remove potentially harmful H_2O_2. Manganese can also be oxidized nonenzymatically when microorganisms alter the pH of the microenvironment to 8 or above, where chemical oxidation occurs readily, or by producing compounds that chemically oxidize Mn^{2+}. These reactions may serve to protect the cell from high concentrations of Mn^{2+} at low pHs.

Microorganisms also carry out dissimilatory manganese reduction (Lovley, 1993). Most organisms that reduce Mn^{4+} also reduce Fe^{3+}. Dissimilatory Mn^{4+} reduction is similar to dissimilatory Fe^{3+} reduction and takes place primarily in anoxic environments with the oxidized metal acting as terminal electron acceptor. Mn^{4+} reduction can lead to the release of soluble Mn^{2+} into the soil solution. Oxidation of Mn^{2+}can result in the formation of manganese concretions in the soil, an indicator of fluctuating water tables in the soil profile.

Mercury

Mercury has no known biological function in humans. All forms of mercury are toxic, and each exhibits different health effects. Exposure of humans to mercury can cause nausea, vomiting, abdominal pain, damage to the central nervous system, and birth defects, depending on level and duration of the exposure. Anthropogenic activity is a major source of mercury emissions and soil contamination. The main sources include mining and smelting, burning of fossil fuels, industrial production of sodium hydroxide and chloride, and waste incineration, with trace amounts in fertilizers used in agriculture. At one time, mercury compounds were common in pesticides. Mercury abundance in the Earth's crust is generally low, ranging from 0.02 to 0.06 mg kg^{-1}, but it is concentrated naturally in some mineral deposits (e.g., cinnabar). Total mercury concentrations average about 0.3 mg kg^{-1} soil (Li et al., 2017). In UK soils, mercury concentrations were found to be markedly higher in urban than rural soils, with a mean value of 0.13 mg kg^{-1} in rural soil compared to a mean of 0.35 mg kg^{-1} in urban soils (Environment Agency, 2007). Mercury can be present in many forms, including elemental (Hg^0), inorganic (Hg^{2+}), and organic [methylmercury, CH_3Hg^+, or dimethylmercury, $(CH_3)_2Hg$]. Over 95% of the mercury in air is present as elemental mercury vapor.

Mercury undergoes several transformations by microorganisms in the environment (Fig. 19.9). Many organisms reduce mercury to volatile Hg^0 as a detoxification mechanism.

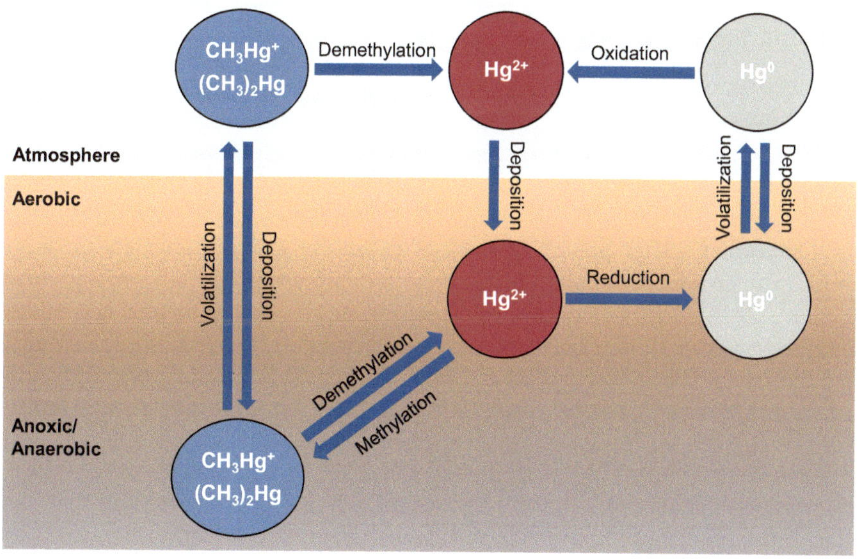

FIGURE 19.9 The mercury cycle. Transformations of mercury including reduction of ionic Hg^{2+} to elemental Hg^0, methylation of Hg^{2+} under anoxic conditions, and demethylation under aerobic conditions. Methylmercury and elemental mercury are volatile and can be lost to the atmosphere. Mercury enters the food chain as methylmercury, which is toxic and can accumulate in fatty tissues of animals. Mercury is increasingly concentrated as methylmercury moves up in the food chain (biomagnification).

Reduction can occur during aerobic growth and is not linked to energy transduction pathways. Several aerobic and facultative bacteria (e.g., *Bacillus, Pseudomonas, Corynebacterium, Micrococcus*, and *Vibrio*) catalyze the reduction of Hg^{2+} to Hg^0. In mercury-contaminated environments, Hg^{2+}-resistant communities of bacteria are generally present, and Hg^{2+} reduction is fairly rapid. Bacterial reduction of Hg^{2+} to Hg^0 may be a useful remediation strategy for contaminated sediments.

The methylation of mercury occurs under anoxic conditions, such as those found in wetland soils and lake sediments, resulting in increased solubility and volatility, as well as movement into the food chain. Methylmercury is lipophilic and can be concentrated in fish in contaminated waters. In the 1950s, industrial discharges of methylmercury into Minamata Bay in Japan resulted in the contamination of fish, the poisoning of thousands of people, and several hundred deaths.

Evidence indicates that methylation is mediated by sulfate-reducing bacteria. The methylation reaction occurs via the transfer of methyl groups from methylcobalamin or methylated vitamin B_{12} to Hg^{2+}.

$$Hg^{2+} + B_{12}\text{-}CH_3 \rightarrow CH_3Hg^{+} + \text{reduced-}B_{12}$$

Methylcobalamin is produced by sulfate-reducing bacteria using fermentation products from other heterotrophs. Methylation generally stops when fermentation ceases, although sulfate reduction may continue. As methylmercury migrates into more aerobic zones, demethylation can occur, followed by reduction of Hg^{2+}, resulting in the formation of elemental mercury. Even though elemental mercury has a very low vapor pressure, it is volatile and can escape to the atmosphere.

Selenium

Selenium is essential for animal nutrition in trace quantities but is toxic at high concentrations. Infirmities such as the white-muscle disease of cattle result from selenium-deficient feeds produced on selenium-deficient soils. On the other hand, chronically high levels of selenium in animal diets can lead to embryonic deformities, poor growth, and reproductive problems (Ohlendorf, 1989). Selenium toxicity occurs in grazing animals in rangeland soils that are high in selenium; blind staggers and even death are potential consequences of Se toxicity. Selenium does not appear to be essential for plant metabolism, and can adversely affect plants. Irrigation water in dry regions may contain enough dissolved selenium to damage sensitive crops, even if the original levels of Se in the soil were not very high. In most soils, the total concentration of Se is between 0.01 and 2 mg kg^{-1}. However, in seleniferous soils, concentrations as high as 38 mg kg^{-1} have been reported.

The biochemistry of Se is similar to that of sulfur. Common species of selenium in aerobic environments are selenate (SeO_4^{2-}; Se^{6+}) and selenite (SeO_3^{2-}, Se^{4+}). Selenide (Se^{2-}) is present as H_2Se and insoluble metal precipitates in reduced environments. Both selenate and selenite are soluble and toxic to animal life in high concentrations.

A diagram of the selenium cycle is provided in Fig. 19.10. Several bacterial genera, including *Pseudomonas, Clostridium*, and *Flavobacterium*, use oxidized selenium (Se^{6+}, Se^{4+}, and Se^0) as terminal electron acceptors. The reduction of oxidized selenium to insoluble elemental Se^0 is

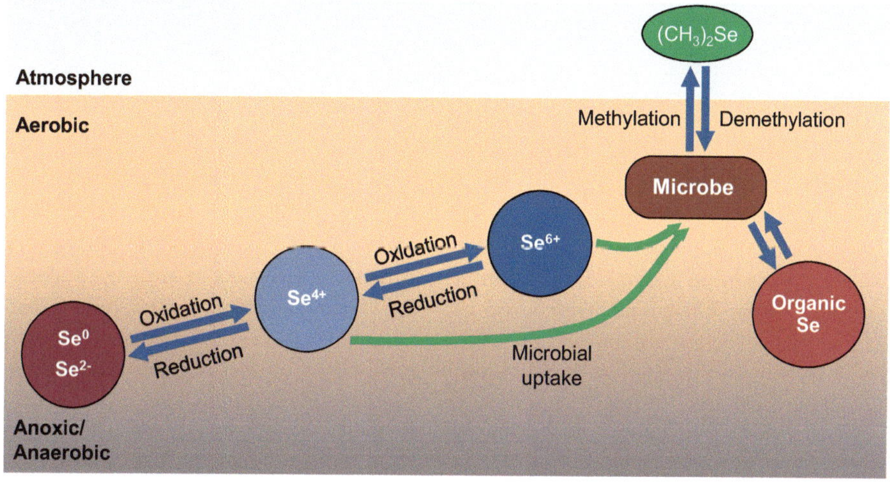

FIGURE 19.10 **Selenium cycling in soils and sediments.** Transformations of selenium include oxidation, reduction, methylation, and demethylation. Oxidized forms (Se^{6+} and Se^{4+}) are mobile in soil and are available for microbial uptake. Methylated forms of selenium are volatile, and microbial methylation has been used as a mechanism to remove selenium from contaminated soils. *Source: Adapted from Winkel et al. (2015).*

a principal means for immobilizing soluble selenate and selenite from contaminated environments (Losi and Frankenberger, 1997).

There is evidence that dissimilatory selenium reduction in some bacteria may be catalyzed by the nitrate reductase enzyme; in some cases, NO_3^- inhibits SeO_4^{2-} reduction, while in one isolate (*Enterobacter cloacae* SLD1a-1), NO_3^- and SeO_4^{2-} reduction occur simultaneously. The use of bacteria in bioreactors to reduce oxyanions of selenium to insoluble Se^0 is a promising method to remediate selenium-contaminated waters, such as agricultural wastewaters.

Selenium can also be biomethylated by bacteria and fungi to forms such as dimethylselenide [$(CH_3)_2Se$] by a mechanism similar to that described for mercury. Volatilization of dimethylselenide from contaminated waters and soils removes selenium from the immediate environment. This methylation reaction has potential use for remediation of seleniferous environments (Losi and Frankenberger, 1997; Box 19.10).

Arsenic

The poisonous nature of arsenic and its potential health effects have long been recognized (Box 19.11). Arsenic concentrations in soil vary from about 0.2 to 40 mg As kg^{-1} (averaging 5 mg kg^{-1}), with background concentrations typically less than 15 mg As kg^{-1}. In addition to naturally occurring sources, arsenic may be introduced into the environment as an industrial waste. The wood preservative industry is a major user of As in the manufacture of rot-resistant CCA (copper, chromium, arsenic)-treated wood. Prior to 2003, approximately 90% of industrial As was used as a wood preservative in the United States. However, CCA-treated wood has since been banned in most residential construction in the United States and restricted in the European Union and several other countries. Arsenic is also used in applications such as paints, dyes, drugs, soaps, and semiconductors. Mining operations and fly ash

BOX 19.10

Selenium contamination in a California reservoir

Kesterson Reservoir is comprised of a series of 12 ponds covering approximately 500 ha in the southern part of the Kesterson National Wildlife Refuge in California (Ohlendorf, 1989). The ponds were used as collection and evaporation basins for agricultural drainage water from the San Joaquin Valley. In the 1980s, the drainage waters were found to be high in selenium and salinity. Selenium concentrations approached 300 μg L^{-1} (the EPA drinking water standard for selenium is 10 μg L^{-1}), resulting in the transport of approximately 9000 kg of selenium into the reservoir between 1981 and 1986. Bioaccumulation and biomagnification of selenium in the food chain occurred, resulting in abnormal rates of mortality and deformity in waterfowl. These events resulted in the closure of the reservoir to drainage waters. Since the closure, many of the low-lying areas have been filled with soil, and management plans have been implemented to dissipate selenium through plant bioaccumulation and volatilization by microbial biomethylation. In one study, 68%–88% of the total selenium in research plots at Kesterson was dissipated over a 100-month period (Flury et al., 1997). Thus, microbially-mediated removal of selenium from contaminated sites may prove more cost effective than soil removal and disposal.

BOX 19.11

Arsenic: occurrence, health effects, and drinking water regulation

Arsenic is the 20th most abundant element in the Earth's crust. It is naturally occurring, found in rocks, soil, water (primarily groundwater), air, and biota. Arsenic toxicity is well recognized. Arsenic poisoning is thought to be responsible for the deaths of well-known historical figures such as Claudius, Pope Pius III, Pope Clemente XIV, Charles Francis Hall, and Napoleon (US EPA, 2015). Arsenic has been linked to cancer of the bladder, lungs, skin, kidney, nasal passages, liver, and prostate, as well as non-cancerous cardiovascular, pulmonary, immunological, and neurological effects. Therefore, arsenic was one of the first regulated drinking water contaminants. In 1942, The US Public Health Service established an arsenic drinking water standard of 50 μg L^{-1}. Following extensive research and much debate on cost, The US Environmental Protection Agency established a health-based, non-enforceable Maximum Contaminant Level Goal (MCLG) for arsenic of zero and an enforceable Maximum Contaminant Level (MCL) for arsenic of 10 μg L^{-1}. The Final Arsenic Rule was published in the Federal Register on January 22, 2001 (66 FR 6976).

from coal-fired power plants can result in As contamination. Arsenical pesticides were used for decades before the introduction of organochlorine compounds, such as dichlorodiphenyltrichloroethane (commonly known as DDT). Pesticides containing As were applied heavily in orchards and in cattle operations, resulting in significant soil contamination. Herbicides such as monosodium methanearsonate (MSMA) have been widely applied in cotton and turf grass production. Additionally, phosphate fertilizers typically contain trace amounts of As.

In aerobic soils, inorganic As is present as As^{5+} ($H_2AsO_4^-$ or $HAsO_4^{2-}$; arsenate). Under moderately reduced conditions, As^{3+} ($HAsO_2$; arsenite) is the dominant form. Both forms are very soluble in water and are highly bioavailable. However, As^{5+} is adsorbed by iron oxides and hydroxides in soil; therefore, its bioavailability tends to be higher in sandy soils where the frequency of oxide-coated minerals surfaces is lower. The solubility and mobility of arsenic depends upon its oxidation state and changes with pH. The mobility of As^{5+} is low in acidic environments while As^{3+} is more mobile. As^{5+} is more prevalent in well-oxygenated aerobic environments, while As^{3+} is more likely to occur in anoxic environments.

An overview of the arsenic cycle in terrestrial systems is shown in Fig. 19.11. Solution arsenic levels in soil are governed by weathering of soil and mineral rock and by sorption and desorption to and from iron and manganese oxides. Although arsenic is highly toxic to many forms of life, some microorganisms have evolved to tolerate or even thrive under relatively high concentrations of this metalloid, using it as a source of energy or terminal electron acceptor for growth or converting it to less toxic forms. Microbial oxidation and reduction

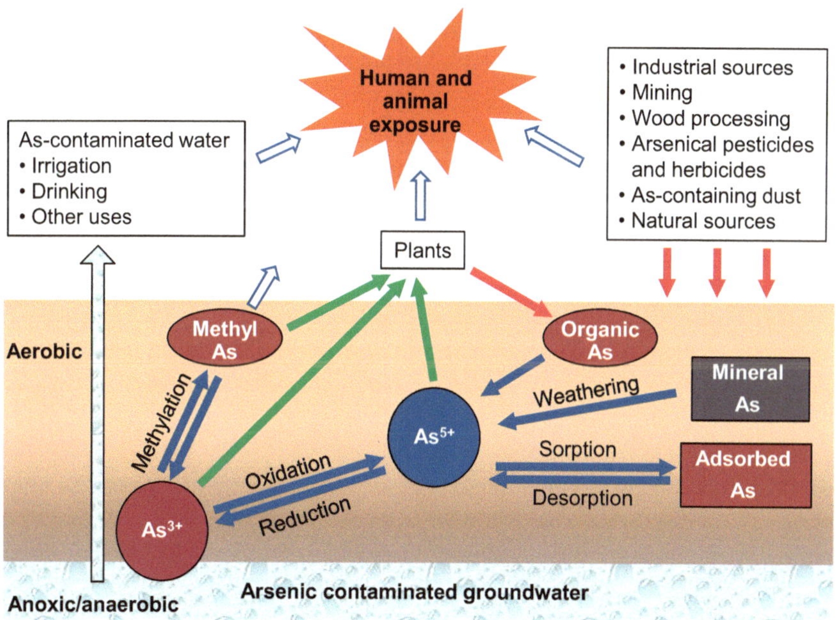

FIGURE 19.11 **Arsenic cycling in the environment.** Transformations of arsenic include oxidation, reduction, methylation, and demethylation. Arsenic commonly occurs as As^{5+}, adsorbed to iron oxides, in aerobic soils and can be reduced to As^{3+}, a more mobile form, in anoxic environments. A variety of microorganisms can produce different forms of methylated arsenic, which can be taken up by plants and/or potentially removed from the soil via volatilization.

result in conversions between As^{3+} and As^{5+}, while methylation leads to the formation of volatile methyl-As. Oxidation of As^{3+} to As^{5+} by chemoautotrophic bacteria results in the release of energy that could be available to support microbial life, with the electrons from this transformation transferred to a suitable electron acceptor such as nitrate or oxygen. Conversely, microbial reduction of As^{5+} to As^{3+} under anaerobic conditions may be coupled to the oxidation of organic and some inorganic substances to yield energy to support microbial growth, with arsenate serving as an electron acceptor (arsenate respiration). Reduction of As^{5+} to As^{3+} results in a more toxic product that is generally considered more mobile in the environment. Methylated-As is volatile, and biomethylation has been suggested as a potential method for arsenic bioremediation. As with mercury and selenium, biomethylation is an important reaction with arsenic. Bacterial methylation produces mono- and di-methylated species, while fungi usually produce tri-methylated species. Methylation occurs under aerobic or anaerobic conditions and is likely a detoxification mechanism.

Microbial mobilization of arsenic from contaminated soils and sediments can cause pollution of downstream areas. The anaerobe *Sulfurospirillum barnesii* reduces As^{5+} (arsenate) while oxidizing lactate, resulting in potential mobilization of the As^{3+} form. Solubilization from iron arsenates by dissimilatory arsenate reduction may be partially responsible for significant efflux of arsenic from anoxic sediments in the environment. A widespread problem with arsenic that has been related to microbial activities is the contamination of well water (Box 19.12).

BOX 19.12

Arsenic contamination of well water in Bangladesh

To meet water needs of an increasing population, the government of Bangladesh, with the assistance of United Nations Children's Fund (UNICEF) and other aid organizations, began installing millions of tube wells starting in the 1970s. This program greatly decreased the incidence of waterborne diseases. However, many of these wells were placed into groundwater containing significant levels of arsenic. Over 50,000 tube wells have been tested for arsenic contamination since the mid-1970s, with nearly 30% found to contain arsenic at concentrations greater than 50 $\mu g\ L^{-1}$, the maximum permissible concentration in drinking water in Bangladesh and many other countries. With approximately 4 million tube wells in place, some have estimated that over half of the population has been exposed to high arsenic concentrations in their water. The arsenic in the water has resulted in a number of chronic health problems in large areas of Bangladesh over the past several decades.

The mechanisms underlying this arsenic mobilization are not completely agreed upon. One mechanism that is generally agreed on is that microbial reduction of FeOOH in the aquifer sediments has resulted in the release of geological arsenic that is adsorbed to the iron oxyhydroxide. The reduction of FeOOH is coupled to the oxidation of sedimentary deposits of organic matter in the aquifer beds. Another hypothesis, fueled by observations of microbial arsenate reduction, is that As^{5+} is directly reduced by dissimilatory arsenate reduction coupled to oxidation of the organic matter. In either case, it is apparent that microbial activities have had a severe impact on human health in Bangladesh and surrounding areas.

Microorganisms as catalysts for mineral formation

In this chapter, we have considered the microbial transformations of select metals and metalloids. A related topic is how microbial cells facilitate the formation of minerals in soils and sediments. These processes are often termed microbial **biomineralization** (Heim, 2011). Here, only a short introduction to the topic is feasible.

Several mechanisms of microbial biomineralization are recognized. Microorganisms may produce minerals from an oversaturated solution through nucleation and growth of crystals on the outside surface of cells through chemically reactive mineral binding sites on the cell walls and external sheaths of bacterial cells. These binding sites are predominantly negatively charged and include carboxyl, phosphate, and amino groups on the cell wall surfaces. These properties afford the ability of microbial cell surfaces to act as nucleation points for mineral formation. The reactions occur predominantly on metabolically inactive cell surfaces, since actively metabolizing microorganisms produce protons which compete with the metals. Metal ions in solution interact with the charged cell-surface functional groups. This is essentially an electrostatic cation exchange mechanism resulting in the stoichiometric sorption of metals. The metals, which are bound in the matrix of the cell wall, then act as nucleation sites for the subsequent deposition of metal precipitates. The type of mineral formed depends on the types of negative counter-ions, such as carbonate, phosphate, sulfate, sulfide, or silicate, in the system.

Microbial mineral formation may also result from metabolic activities inside or outside of microbial cells. Microbial activity can change solution chemistry such as pH that leads to oversaturation and mineral precipitation. Enzymes produced by microorganisms catalyze redox reactions that may alter the form and solubility of metals and facilitate mineral formation. In soil, iron/manganese oxides are deposited by bacteria that actively oxidize their soluble, reduced forms to generate energy for growth. In sulfidic mine areas, thiobacilli catalyze the formation of iron oxides in aerobic environments, whereas sulfate-reducing bacteria can facilitate the formation of iron sulfides in anoxic soil and sediments. In the laboratory, Thiel et al. 2019 demonstrated pyrite (FeS_2) formation mediated at ambient temperature through a microbially catalyzed redox process. Bacterial cells can also facilitate the formation of clay-like silicates, carbonate minerals, and a variety of elemental deposits, such as those of copper and gold.

Although knowledge on microbial biomineralization is limited, its importance in many environments is widely recognized. Microbial biomineralization plays a major role in global elemental cycling. For example, microbial pyrite precipitation leads to the transfer of dissolved iron and sulfur into marine sediments. Knowledge of Fe^{3+} minerals and the role of Fe^{2+}-oxidizing bacteria in the deposition of ancient banded iron formations has been used in the reconstruction of early biogeochemical systems on Earth (Posth et al., 2014). Furthermore, microbiological precipitation of phosphate minerals contributes to the incorporation of phosphorus into sediments, and microbial biomineralization drives the precipitation of carbonate minerals that serve as a sink for atmospheric carbon dioxide.

Biomineralization also has considerable commercial potential. Some biominerals have properties of nanoparticles that are valuable in developing biomaterials for industrial and medical applications (Gadd, 2010; Uebe and Schüler, 2016). In the petroleum industry, microbial biomineralization may extend oil production by sealing off small pores to prevent water entering oil reservoirs. In another example relevant to construction and a wide variety of applications, formation of stable biomineral coatings on iron objects could result in reduced metal corrosion (Kooli et al., 2018).

Summary

Phosphorus, iron, manganese, and selenium are essential nutrients, while mercury and arsenic are not known to have essential functions in living systems. Although beneficial at low concentrations, some nutrients can be directly toxic to living organisms or result in other adverse effects at elevated concentrations. For example, phosphorus, arsenic, mercury, and selenium are common environmental pollutants. Whether it may be nutrition or pollution, microorganisms play a vital role in mediating reactions and governing transformation processes of these elements, such as mineralization/immobilization, reduction/oxidation, solubilization, and methylation.

The phosphorus cycle includes mineralization and immobilization reactions mediated by the microbial biomass pool, and the solubilization of inorganic phosphorus-containing minerals by organic acids, inorganic acids, and chelating agents produced by soil organisms. Mineralization of organic phosphorus is a significant source of phosphorus for plant growth, particularly in nonagricultural ecosystems. For other nutrients, transformations are often associated with redox reactions used for energy generation with the resulting changes in redox state altering the solubility and mobility of compounds in soil. For example, reduced forms of iron and manganese (Fe^{2+} and Mn^{2+}) can be oxidized by chemoautotrophic bacteria, providing a source of electrons for energy transduction, while their oxidized forms (Fe^{3+} and Mn^{4+}) can serve as terminal electron acceptors during dissimilatory reduction processes. Reduction increases solubility of iron and manganese and contributes to the characteristic gray or bluish colors in soils with poor drainage. In other cases, redox reactions or other transformations are used as detoxification mechanisms. For instance, the ionic form of mercury can be reduced to volatile elemental mercury (Hg^0), resulting in mercury transport to the atmosphere. Ionic Hg^{2+} can be methylated in sediments by sulfate-reducing bacteria and other microorganisms, producing methylmercury that is lipophilic and can accumulate in the food chain. Transformation reactions involving selenium and arsenic are similar to those of sulfur and mercury, undergoing oxidation, reduction, and methylation. In contrast with mercury, methylation of selenium and arsenic results in decreased toxicity to humans and other higher organisms. Microorganisms also play key roles in the formation of soil and sediment minerals, serving as nucleation sites for the formation of a variety of minerals. Therefore, microbially mediated transformations of elements in soil are vital to plant and microbial nutrition and play critical roles in understanding and addressing environmental pollution.

Study questions

1. What are the major forms of phosphorus in soil?
2. Why would tillage of soil result in a more rapid decrease in microbial biomass phosphorus relative to stable organic phosphorus?

3. Describe the importance of phosphorus mineralization to plant nutrition in undisturbed soil ecosystems. Why would this process be of primary importance to plant growth?
4. If a waste material was added to a soil that had a C/N/P ratio of 500/20/1, would phosphorus likely be mineralized from the waste or immobilized from the soil solution? Why?
5. Why is a soil pH of approximately 6.5 important for plant and microbial growth?
6. You have measured total organic phosphorus and biomass phosphorus for two soils, one tilled and the other a long-term, no-till cropping system. For biomass determinations, you measured inorganic phosphorus (P*i*) in a non-fumigated sample (1), a fumigated sample (2), and a non-fumigated sample extracted with 0.5 M $NaHCO_3$ containing KH_2PO_4 to provide 25 μg P*i* g^{-1}. The data are as follows:

Sample	Soil 1	Soil 2
	(μg P*i* g^{-1} soil)	
Non-fumigated	5.8	4.3
Fumigated	20.9	7.9
Non-fumigated + KH_2PO_4	25.2	23.0
Total organic P	262.4	212.6

   a. Determine the microbial biomass phosphorus for each soil.
   b. Indicate which soil has been in cultivation and which has been in continuous no-till for several years. Justify your answer.
7. Why is phytate-P of little nutritional value to monogastric animals?
8. Why are we concerned about potential eutrophication in aquatic environments when large quantities of animal waste are applied to land near a body of water?
9. Under what conditions might microbial reduction of Fe^{3+} compounds occur? If NO_3^- is present, will Fe^{3+} reduction be important? Explain.
10. How might a microorganism facilitate the solubilization of both iron and phosphorus simultaneously?
11. Both mercury and selenium can be biomethylated by microorganisms. Why is the methylation of mercury a potential problem while selenium methylation is not?

References

Alori, E.T., Glick, B.R., Babalola, O.O., 2017. Microbial phosphorus solubilization and its potential for use in sustainable agriculture. Front. Microbiol. 8, 971. doi: 10.3389/fmicb.2017.00971.

Batjes, N.H., 1997. A world dataset of derived soil properties by FAO–UNESCO soil unit for global modelling. Soil Use Manage. 13, 9–16.

Brookes, P.C., Powlson, D.S., Jenkinson, D.S., 1982. Measurement of microbial biomass phosphorus in soil. Soil Biol. Biochem. 14, 319–329.

Brookes, P.C., Powlson, D.S., Jenkinson, D.S., 1984. Phosphorus in soil microbial biomass. Soil Biol. Biochem. 16, 169–175.

Deinema, M.H., van Loosdrecht, M., Scholten, A., 1985. Some physical characteristics of *Acinetobacter* spp. accumulating large amounts of phosphate. Water Sci. Technol. 17, 119–125.

Environment Agency, 2007. UK Soil and Herbage Pollutant Survey. Report No. 7: Environmental concentrations of heavy metals in UK soil and herbage. Bristol: Environment Agency.

FAO, 2017. World fertilizer and trends outlook to 2020. Food and Agriculture Organization of the United Nations, Rome.

Figueroa, I.A., Coates, J.D., 2017. Microbial phosphite oxidation and its potential role in the global phosphorus and carbon cycles. Adv. Appl. Microbiol. 98, 93–117.

Flury, M., Frankenberger, Jr., W.T., Jury, W.A., 1997. Long-term depletion of selenium from Kesterson dewatered sediments. Sci. Total Environ. 198, 259–270.

Gadd, G.M., 2010. Metals, minerals and microbes: geomicrobiology and bioremediation. Microbiology 156, 609–643.

Heim, C., 2011. Microbial biomineralization. In: Reitner, J., Thiel, V. (Eds.), Encyclopedia of Geobiology. Encyclopedia of Earth Sciences Series. Springer, Dordrecht.

IFDC, 2010. World Phosphate Rock Reserves and Resources. International Fertilizer Development Center Publications, Muscle Shoals, AL.

Illmer, P., Barbato, A., Schinner, F., 1995. Solubilization of hardly-soluble $AlPO_4$ with P-solubilizing microorganisms. Soil Biol. Biochem. 27, 265–270.

IUBMB (International Union of Biochemistry and Molecular Biology), 2020. Enzyme nomenclature. Available from: https://iubmb.org/biochemical-nomenclature/ (accessed 05.2020.).

Kooli, W.M., Comensoli, L., Maillard, J., Albini, M., Gelb, A., Junier, P., Joseph, E., 2018. Bacterial iron reduction and biogenic mineral formation for the stabilization of corroded iron objects. Sci. Rep. 8, 764, https://doi.org/10.1038/s41598-017-19020-3.

Kornberg, A., 1995. Inorganic polyphosphate: toward making a forgotten polymer unforgettable. J. Bacteriol. 177, 491–496.

Li, R., Wu, H., Ding, J., Fu, W., Gan, L., Li, Y., 2017. Mercury pollution in vegetables, grains and soils from areas surrounding coal-fired power plants. Sci. Rep. 7, 46545, https://doi.org/10.1038/srep46545.

Lindsay, W.L., 1979. Chemical Equilibria in Soils. John Wiley & Sons, New York.

Losi, M.E., Frankenberger, Jr., W.T., 1997. Bioremediation of selenium in soil and water. Soil Sci. 162, 692–702.

Lovley, D.R., 1993. Dissimilatory metal reduction. Ann. Rev. Microbial. 47, 263–290.

Ohlendorf, H.M., 1989. Bioaccumulation and effects of selenium in wildlife. In: Jacobs, L.W. (Ed.), Selenium in Agriculture and the Environment, SSSA Special Publication vol. 23., Soil Science Society of America, Madison, Wis, pp. 133–177.

Parham, J.A., Deng, S.P., Raun, W.R., Johnson, G.V., 2002. Long-term cattle manure application in soil I. Effect on soil phosphorus levels, microbial biomass C, and dehydrogenase and phosphatase activities. Biol. Fert. Soils 35, 328–337.

Posth, N.R., Canfield, D.E., Kappler, A., 2014. Biogenic Fe(III) minerals: From formation to diagenesis and preservation in the rock record. Earth-Sci. Rev. 135, 103–121.

Schindler, D.W., 1974. Eutrophication and recovery in experimental lakes: Implications for lake management. Science 184, 897–899.

Schink, B., Friedrick, M., 2000. Phosphite oxidation by sulphate reduction. Nature 406, 37.

Sharpley, A.N., Chapra, S.C., Wedepohl, R., Sims, J.T., Daniel, T.C., Reddy, K.R., 1994. Managing agricultural phosphorus for protection of surface waters: Issues and options. J. Environ. Qual. 23, 437–451.

Shieh, T.R., Ware, J.H., 1968. Survey of microorganisms for the production of extracellular phytase. Appl. Micriobiol. 16, 1348–1351.

Sim, R., 2016. The essential element: How phosphorus scarcity endangers the world. Harvard Int. Rev. 10, 2016.

Smil, V., 2000. Phosphorus in the environment: Natural flows and human interferences. Annu. Rev. Energy Environ. 25, 53–88.

Stokstad, E., 2008. Canada's experimental lakes. Science 322, 1316–1319.

Tabatabai, M.A., 1994. Soil enzymes. In: Weaver, R.W. et al., (Ed.), Methods of Soil Analysis, Part 2. Microbiological and Biochemical Properties. Soil Science Society of America, Madison, Wis, pp. 775–833.

Thiel, J., Byrne, J.M., Kappler, A., Schink, B., Pester, M., 2019. Pyrite formation from FeS and H_2S is mediated through microbial redox activity. Proc. Natl. Acad. Sci. USA 116, 6897–6902.

Uebe, R., Schüler, D., 2016. Magnetosome biogenesis in magnetotactic bacteria. Nat. Rev. Microbiol. 14, 621–637.

US EPA (Environmental Protection Agency), 2015. The arsenic rule: Background and rule provisions. Available from: https://www.epa.gov/sites/production/files/2015-09/documents/train1-background.pdf (accessed 24.02.2020.).

USGS (The United States Geological Survey), 2019. Mineral commodity summaries. Available from: https://pubs.er.usgs.gov/publication/70202434 (accessed 05.04.2020.).
Winkel, L.H.E., Vriens, B., Jones, G.D., Schneider, L.S., Pilon-Smits, E., Bañuelos, G.S., 2015. Selenium cycling across soil-plant-atmosphere interfaces: a critical review. Nutrients 7, 4199–4239.
Yamahar, R., Ogo, S., Masuda, H., Watanabe, Y., 2002. (Catecholato)iron(III) complexes: structural and functional models for the catechol-bound iron(III) form of catechol dioxygenases. J. Inorg. Biochem. 88, 284–294.

Further reading

Douglas, S., Beveridge, T.J., 1998. Mineral formation by bacteria in natural microbial communities. FEMS Microbial. Ecol. 26, 79–88.
Frankenberger, W.T., Benson, S., 1994. Selenium in the Environment. Marcel Dekker, Inc, New York.
Kucey, R.M.N., Janzen, H.H., Leggett, M.E., 1989. Microbially mediated increases in plant-available phosphorus. Adv. Agron. 42, 199–228.
Lovley, D.R., 1995. Microbial reduction of iron, manganese, and other metals. Adv. Agron. 54, 175–231.
Sigel, A., Sigel, H., 1997. Mercury and its Effects on Environment and Biology. Marcel Dekker, New York.
Smith, E., Naidu, R., Alston, A.M., 1998. Arsenic in the soil environment: A review. Adv. Agron. 64, 149–195.
Whitelaw, M.A., 2000. Growth promotion of plants inoculated with phosphate-solubilizing fungi. Adv. Agron. 69, 99–151.
WHO (World Health Organisation), 1988. Phosphine and Selected Metal Phosphides. Environmental Health Criteria, 73. Joint sponsorship of UNEP, ILO and WHO, Geneva.

CHAPTER

20

Global gases

Wendy H. Yang[a], *Steven J. Hall*[b], *Gavin McNicol*[c]

[a]Departments of Plant Biology and Geology, University of Illinois at Urbana-Champaign, Urbana, IL, United States; [b]Department of Ecology and Evolutionary Biology, Iowa State University, Ames, IA, United States; [c]Department of Earth and Environmental Sciences, University of Illinois at Chicago, Chicago, IL, United States

It is widely accepted that microorganisms have played a key part in determining the atmospheric concentrations of greenhouse gases, including carbon dioxide, methane and nitrous oxide (which have the greatest impact on radiative forcing), throughout much of Earth's history. What is more open to debate is the part that they will play in the coming decades and centuries, the climate feedbacks that will be important, and how humankind might harness microbial processes to manage climate change.

— ***(Singh et al., 2010)***

Soil microorganisms play critical roles in the cycling of all the major elements, including (but not limited to) carbon, oxygen, nitrogen, sulfur, phosphorus, and iron. Several of these elemental cycles have gaseous components such as carbon dioxide, dinitrogen, and sulfur dioxide. Thus, soil microorganisms play a role in controlling the composition, chemistry, and physics of the atmosphere. They can affect climate, ozone depletion, smog production, the rate at which organic chemicals are "scrubbed" from the atmosphere, and nutrient inputs to downwind ecosystems. Despite the substantial atmospheric effects of gases produced by soil microorganisms, most of these gases are present only at very low concentrations in the atmosphere. They are **trace gases**.

Defining trace gases

Atmospheric composition

If it were not for the presence of life on Earth, our atmosphere would be largely carbon dioxide (CO_2) with smaller amounts of oxygen (O_2), nitrogen (N_2), and carbon monoxide (CO), and traces of nitrous oxide (N_2O) and methane (CH_4). Instead, our atmosphere is dominated by nitrogen (78%) and oxygen (21%). All the rest of the gases put together comprise only about 1% of the atmosphere (Fig. 20.1). However, almost all of this 1% is the inert gas argon

Principles and Applications of Soil Microbiology. http://dx.doi.org/10.1016/B978-0-12-820202-9.00020-4
Copyright © 2021 Elsevier Inc. All rights reserved.

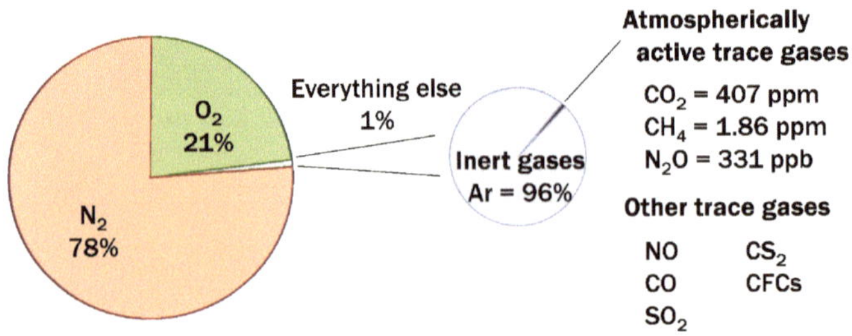

FIGURE 20.1 **Composition of the atmosphere in 2018.** (*CFCs*, chlorofluorocarbons; *ppm*, part per million; *ppb*, part per billion). *Source: Adapted by Wendy Yang from Schimel and Holland (2005).*

(Ar). The rest of the atmosphere (0.04%) comprises a collection of hundreds, if not thousands, of chemically and radiatively active gases that play important roles in atmospheric processes. We call them trace gases because they are present in such small concentrations.

The concentrations of trace gases are controlled by both natural and **anthropogenic** factors. Atmospheric concentrations of carbon dioxide, nitrous oxide, and methane have risen at unprecedented rates since the turn of the 20th century (Table 20.1) due to anthropogenic emissions. However, the rates of these increases have varied over time. It is still not fully understood why, for example, methane concentrations stabilized during the 2000s. Although overall increases in atmospheric trace gas concentrations are due to human activity, soil microbes

TABLE 20.1 Properties of selected atmospheric trace gases.

	Abundance		Annual trend	100 Year global	
Compound	**2018**[a]	**1750**[b]	**2017–18**[a]	**Lifetime**[c]	**Warming potential**[d]
CO_2 (*ppm*)	407	278	2.4 ppm $year^{-1}$	–[e]	1
CH_4 (*ppb*)	1859	722	8.2 ppb $year^{-1}$	12.4 year	34
N_2O (*ppb*)	331	270	1.1 ppb $year^{-1}$	121 year	298
CFC-12 (*ppt*)	506	0	−2.8 ppt $year^{-1}$	100 year	10,200
NO_x ($NO + NO_2$) (*ppt*)	–[f]	unknown	29% $year^{-1}$ in Asia and −7% $year^{-1}$ in North America and Europe over 1996–2006[b]	hours	–[g]
CO (*ppb*)	94[h]	unknown	1% $year^{-1}$ over 2002–10[b]	2–3 months	–[g]

[a] *Data from Dlugokencky et al. (2019).*
[b] *Data from Hartmann et al. (2013).*
[c] *Data from Myhre et al. (2013). The lifetimes for CH_4 and N_2O are perturbation lifetimes, which account for feedbacks from increasing atmospheric concentrations of the given trace gas and are used to calculate warming potential.*
[d] *Data from Myhre et al. (2013). The warming potentials for CO_2, CH_4, and N_2O include climate-carbon feedback.*
[e] *No single lifetime for CO_2 can be given, according to Myhre et al. (2013).*
[f] *NO_x concentrations vary locally due to the rapid removal of these gases from the atmosphere and poor atmospheric mixing. Therefore, global NO_x concentrations are not reported.*
[g] *CO and NO_x are indirect greenhouse gases, serving as precursors for the formation of tropospheric ozone.*
[h] *Data from Ciais et al. (2013); for the year 2011.*

strongly regulate the magnitude of soil trace gas emissions and contribute in important ways to the global budgets of each trace gas, via both gas production and consumption. The interactions of soil processes and climate may influence the variability of the rates of increase.

Importance of global gases

If trace gases comprise such a small proportion of the total atmosphere, why do we care about them? Trace gases:

- influence Earth's radiation balance and thereby influence global climate,
- regulate the ability of the atmosphere to cleanse itself of pollutants, and
- provide nutrients to the **biosphere,** particularly nitrogen containing gases.

Long-lived gases (e.g., CO_2, N_2O, and CH_4) play multiple roles in the atmospheric system. The sun's radiation is dominated by the short wavelengths of visible light, but these are absorbed and reradiated at longer wavelengths, largely in the infrared region. This infrared radiation can be absorbed by carbon dioxide, nitrous oxide, and methane in the lower atmosphere, which warms the Earth's surface. This is the **greenhouse effect**. While greenhouse gases in the atmosphere are essential to maintaining temperatures conducive for life on Earth, the increasing atmospheric accumulation of greenhouse gases from human activities is causing climate change that now threatens life on Earth. Nitrous oxide and methane further impact atmospheric chemistry when they are eventually transported to the *stratosphere* (the region 20–50 km above Earth's surface), where they contribute to ozone (O_3) destruction. Thus, these gases contribute to the Antarctic "ozone hole." The ozone layer present in the stratosphere blocks much of incident ultraviolet radiation, which is harmful to living organisms and also influences Earth's radiation balance. With the production of chlorofluorocarbons phased out by international treaties, nitrous oxide is now the largest contributor to the destruction of stratospheric ozone.

Short-lived gases such as carbon monoxide and nitrogen oxides (NO_x, which includes nitrogen dioxide, NO_2, and nitric oxide, NO) play an important role in *tropospheric* (near Earth surface) chemistry as air pollutants and their precursors. In the stratosphere, NO_x and other nitrogen compounds primarily destroy ozone but, in the troposphere, NO_x can react with radicals produced from the oxidation of carbon monoxide, methane, and other volatile organic compounds (VOCs) to form the secondary pollutant, ozone. Tropospheric ozone is an important pollutant, causing crop and ecosystem damage and respiratory illnesses. In less polluted regions, soil nitric oxide production can strongly influence atmospheric chemistry by pushing atmospheric NO_x concentrations above the threshold (5–10 ppt, parts per trillion) needed for net ozone production. Soil microbes regulate atmospheric photochemistry in the troposphere through the production and consumption of these short-lived gases.

Gases released from soils

A large number of gases are exchanged between the atmosphere and terrestrial ecosystems. A subset of these is produced in soils (Fig. 20.2), accounting for roughly 50% of carbon dioxide, 30% of methane, 60% of nitrous oxide, and 30% of nitric oxide released to the

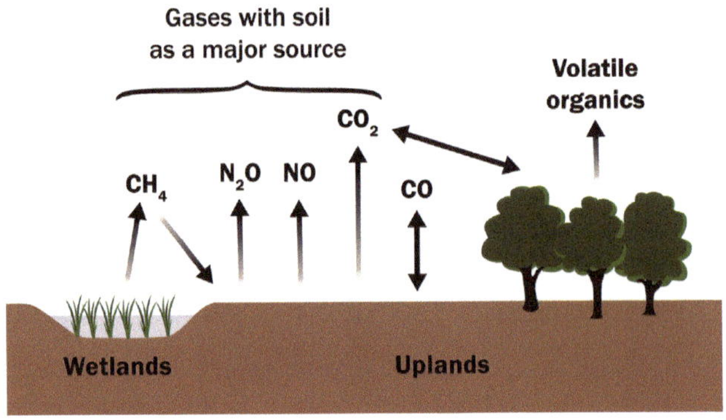

FIGURE 20.2 **Sources of gases contributed to the atmosphere, showing the ecosystems that are the dominant sources for each gas.** *Source: Adapted by Sada Egenriether from Schimel and Holland (2005).*

atmosphere on an annual basis. Several other gases or groups of compounds, such as carbon monoxide and volatile organic compounds, are known to be emitted from soils, but soils tend to be minor sources of these gases at the global scale. Our focus in this chapter is on carbon dioxide, methane, nitrous oxide, and nitric oxide, for which soils act as important sources at the global scale. However, as soil carbon dioxide emissions are roughly balanced by plant growth, the net increase in atmospheric concentrations over recent decades has been largely driven by fossil fuel combustion. In addition, soil microbes can also consume methane, nitrous oxide, and nitric oxide to attenuate soil emissions and mitigate some portion of emissions resulting from human activities.

While trace gas fluxes from terrestrial ecosystems are often large from the perspective of global atmospheric chemistry, they may be small from the perspective of biogeochemical cycling within an individual ecosystem, with the exception of carbon dioxide and sometimes methane. For example, many ecosystems have net nitrogen mineralization rates as high as 100 kg-N ha^{-1} $year^{-1}$, but nitrous oxide fluxes are rarely greater than several kg-N ha^{-1} $year^{-1}$. The production of trace gases can also be extremely variable in space and time; for instance, nitrous oxide production from denitrification often increases sharply immediately after rainfall (Chapter 15). These factors can make measuring and modeling trace gas fluxes a challenge.

We have a good understanding of the basic microbial physiology and processes affecting most trace gas fluxes. Advances in molecular approaches to characterize microbial communities and measurement approaches to quantify process rates enable continual refinement of this understanding. At the same time, we continue to face the great challenge of integrating our understanding of microbial ecology with our understanding of soil processes and then extrapolating it to regional and global scales. To do this, we need to quantify how the specific microbial and environmental controls on soil trace gas dynamics vary in space and time. Then we can estimate large-scale fluxes of trace gases that can be linked to global atmospheric chemistry and climate models, while remaining solidly grounded in the basic soil processes. An important component of developing this large-scale perspective is understanding how land-use practices and other human activities affect trace gas dynamics.

The focus of this chapter is, therefore, to take the more detailed information on specific soil processes or transformations presented in Chapters 13–19 and discuss how these processes

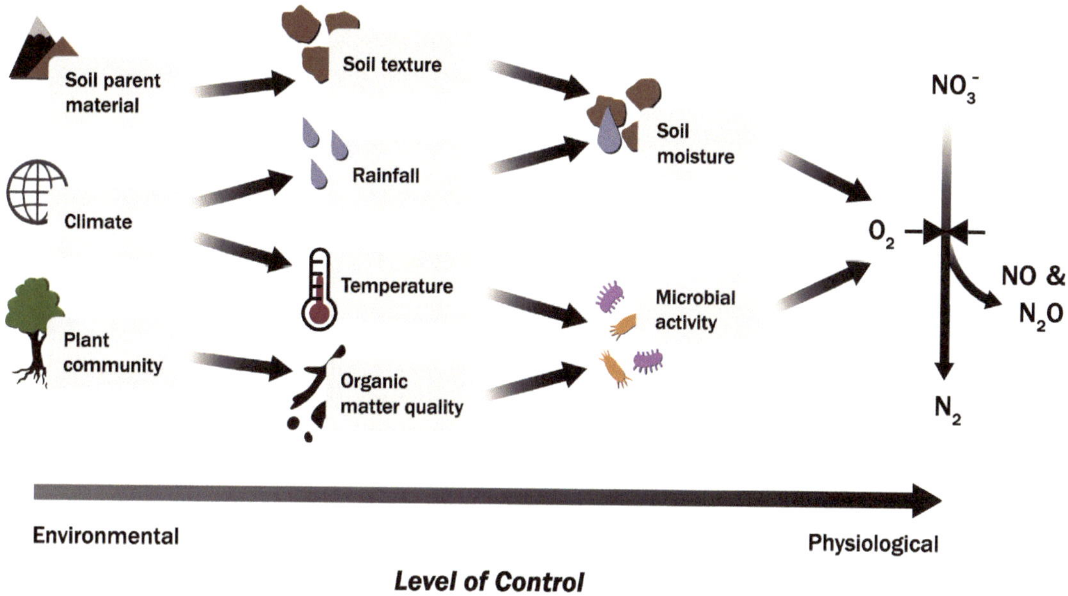

FIGURE 20.3 **Relationship between physiological controls on nitrogen gas production and the large-scale environmental factors that mediate them.** This figure is similar to the upper portion of Fig. 15.12. *Source: Adapted by Sada Egenriether from Schimel and Holland (2005).*

control atmospheric trace gases at a global scale. The essence of this approach is to understand the links between the immediate, physiological controls on a process and the large-scale environmental factors that mediate those controls. For example, oxygen availability controls denitrification rates at the cellular level, but soil moisture and respiration control oxygen diffusion and availability (Fig. 20.3). At a larger scale, rainfall and soil texture control soil moisture, while temperature and the quality and availability of organic matter control respiration. At an even larger scale, the fundamental state factors of climate, relief, plant community, and soil parent material control rainfall, soil texture, and organic matter quality (Jenny, 1994). Therefore, we can use a basic understanding of microbial physiology and soil processes to work our way to large-scale, mechanistic models of trace-gas dynamics.

General controls on trace gas production

Temperature, moisture, redox potential, and substrate availability are probably the four factors that exercise the greatest control on trace gas dynamics across ecosystems.

Temperature

All biological reactions are temperature sensitive, and in many cases reaction rates approximately double for every 10°C increase in temperature. The proportional change in rates following a 10°C temperature increase is referred to as a "Q_{10} value." However, this concept is increasingly debated in the scientific community, as responses of trace gas production to temperature change remain difficult to predict under field conditions. This is because other ecosystem factors—especially soil moisture—strongly interact with temperature. That is, if some other environmental factor limits microbial activity (e.g., low available soil water), trace gas production may show little response to temperature change. Overall, however, across seasons and climatic zones, temperature plays a major role in controlling the rates of trace gas production.

Moisture

Moisture is important for several reasons. First, all living organisms require water. At low soil moisture contents, physiological stress from low water potential can limit microbial activity. Second, water films on soil particles become thinner as soil dries, causing them to become discontinuous and thereby restricting the movement of nutrients to microbes. Third, soil water restricts the diffusion of gases into and out of soil. At high soil moisture, limited diffusion of atmospheric oxygen into soil can create *sub-oxic* (low oxygen) or *anoxic* (no oxygen) conditions conducive to anaerobic processes such as denitrification, methanogenesis, and anaerobic methane oxidation. In contrast, at intermediate soil moisture, rapid oxygen diffusion into soil favors aerobic processes such as aerobic respiration, nitrification, and aerobic methanotrophy (oxygen-dependent microbial consumption of methane). Soil aggregation (clumping of soil particles) can create spatial heterogeneity in soil moisture and oxygen that can allow aerobic and anaerobic processes to occur simultaneously in unsaturated soils (Box 15.9).

Redox potential

At the level of microbial physiology, redox potential determines the relative thermodynamic favorability of different reduction-oxidation (redox) reactions that produce or consume trace gases in soil. Microbial metabolisms that use oxygen as a terminal electron acceptor yield more energy than anaerobic metabolisms that use other electron acceptors (such as nitrate, oxidized iron, or sulfate; Chapter 3). Therefore, aerobic processes utilizing oxygen are more likely to occur in well-aerated soils. When oxygen becomes depleted, alternative electron acceptors are thought to be used sequentially by different members of microbial communities according to their thermodynamic favorability. This often leads to characteristic spatial and temporal patterns of trace gas production. For example, when a soil is flooded and oxygen becomes limiting, nitrogen gases are quickly produced by denitrification, one of the most energetically favorable anaerobic metabolic processes. In contrast, methane production, a metabolism that yields relatively little energy, only occurs at significant rates after other electron acceptors have been exhausted. While redox potential plays a critical role in determining the potential for different microbial metabolisms to occur, it is challenging to apply this conceptual framework to predict trace gas dynamics at the ecosystem scale because the redox potential of the soil environment cannot be measured accurately nor at the scales experienced by soil microbes. The simultaneous occurrence of aerobic and anaerobic processes in unsaturated soils underscores this challenge of scaling trace gas dynamics from the microbial scale to the ecosystem scale.

Substrate availability

Substrate availability, particularly organic matter quantity, quality, and accessibility, is a key control on trace gas dynamics. The availability of substrates directly controls the dynamics of carbon-based gases, such as carbon dioxide and methane, and the activity of heterotrophic bacteria, such as denitrifiers (Chapter 15). Similarly, the availability of inorganic nitrogen is a direct control on nitrogen gas dynamics.

Substrate quality refers to how easily organic matter can be decomposed, both in terms of speed and the number of different enzymes needed to convert a given substrate to carbon dioxide. Rapidly decomposable carbon sources are referred to as *labile* carbon. High-quality substrates contain relatively high amounts of nitrogen and low amounts of lignin. This reflects the importance of nitrogen for synthesis of microbial biomass and enzymes, including extracellular enzymes which are responsible for initiating organic matter decomposition (Chapters 3 and 13). It also reflects the biochemical challenge of lignin degradation, which yields little energy for microbial decomposers but is necessary to access higher-quality carbon sources such as cellulose, which are often protected by lignin. The composition of the plant community influences the quantity and quality of organic matter inputs into the soil and thus also impacts carbon and nitrogen availability for microbial decomposers. For example, grasslands and some deciduous forests produce relatively nutrient-rich, lignin-poor litter, whereas coniferous forests generally produce nutrient-poor, lignin-rich litter. Nutrient cycling and trace gas dynamics in these systems differ, in part, because of the differences in plant litter chemistry.

The capacity of microbes to access nutrients or organic matter is influenced by physical and chemical properties of soils (e.g., mineralogy and structure). For example, the majority of carbon and nitrogen in many soils may be bound to soil minerals or physically isolated within soil aggregates, such that microbes (and extracellular enzymes) cannot effectively access these substrates.

Carbon-based gases

Carbon dioxide

Soil carbon dioxide emissions derive from two distinct but closely related processes—the respiration of living roots (*autotrophic respiration*) and the respiratory decomposition of organic matter by microorganisms (*heterotrophic respiration*) (Chapter 13). These processes are intimately linked because many microbes live in close association with plant roots and use organic compounds supplied by the roots as sources of carbon and energy to produce carbon dioxide. On average, carbon dioxide emissions from a given soil are about equally split between heterotrophic and autotrophic respiration, but this ratio can vary as a consequence of numerous biological and environmental factors. Quantifying the rates and controls of heterotrophic respiration is especially important because it is the dominant pathway of soil carbon loss in oxic soils (methane production is also important in anoxic wetland soils, as discussed later). When plant growth remains constant, an increase in heterotrophic respiration will lead to net losses of soil carbon as carbon dioxide, whereas a decrease in heterotrophic respiration will lead to accumulation of carbon in soil organic matter (Chapter 14).

Carbon dioxide budget

Carbon dioxide is the most abundant trace gas in the atmosphere (~407 ppm, parts per million or μL per liter), and the fluxes of carbon dioxide between the biosphere and the atmosphere are the greatest of any trace gas. Approximately 60 Pg CO_2-C year^{-1} (Pg = 10^{15} grams) is released from the heterotrophic decomposition of plant litter and soil organic matter. This accounts for roughly 25% of the non-anthropogenic biological CO_2 flux to the atmosphere, while plant respiration accounts for another 25%; marine sources account for most of the remaining 50%. In context, human inputs from fossil fuel combustion, which is primarily responsible for the increase in atmospheric carbon dioxide concentrations observed since the Industrial Revolution, contribute about 10 Pg CO_2-C year^{-1}. Although much more carbon dioxide is released from the biosphere via plant and soil heterotrophic respiration, these sources are approximately balanced by plant uptake of carbon dioxide during photosynthesis on an annual scale. Because these fluxes are so large, even small changes in the relative rates of heterotrophic respiration and photosynthesis can have a significant impact on atmospheric carbon dioxide. Major areas of current research are focused on understanding how heterotrophic respiration and photosynthesis may respond differentially to global change and how the resulting "feedbacks" may further contribute to changes in atmospheric concentrations of carbon dioxide and other greenhouse gases. For example, warmer temperatures melt frozen soils and increase microbial activity, thereby increasing rates of soil carbon dioxide production (Schädel et al., 2016).

Globally, soils likely hold between 1,500 and 3,000 Pg of carbon, considerably more carbon than contained in plant biomass and the atmosphere combined (Chapter 13). The precise amount of soil carbon remains uncertain due to challenges in measuring carbon in deep soils (i.e., below 1 m depth) and the disproportionate impact of some ecosystem types. For example, peatland ecosystems occupy ~3% of the land surface (mainly in cool north temperate regions, but also in the tropics), but account for roughly 21% of total soil carbon storage worldwide (Leifeld and Menichetti, 2018; Yu et al., 2010).

Carbon dioxide production

Carbon dioxide is produced in soil by microbial metabolism during organic matter decomposition. Thus, the major controls on its production are essentially those that influence general heterotrophic microbial activity. Organic matter within a given soil, and the carbon that it contains, is characterized by large differences in *turnover times*. Most heterotrophic respiration derives from recently senesced plant biomass—mainly roots and leaves, termed *litter*—which decomposes over timescales of months to years. A significant portion of heterotrophic respiration also derives from the rapid decomposition (on the order of hours to days) of sugars and other organic compounds released by roots (root exudates; Chapter 11). Decomposition of fresh litter and exudates is strongly influenced by temperature and moisture, and generally proceeds fastest under warm and moist (but not saturated) conditions. The availability of nutrients, especially nitrogen, also greatly affects rates of litter decomposition. Litter with greater nitrogen availability, as reflected by lower C/N ratios, typically decomposes more rapidly than litter low in N. Together, climate and litter nitrogen predict most of the variation in leaf and root litter decomposition rates at the global scale.

In contrast to litter and exudates, most of the carbon mass within a given soil decomposes much more slowly, over timescales of decades or even centuries to millennia. Why does this soil carbon persist over such long timescales? One reason is that some of this carbon is in chemically complex compounds, making it less susceptible to decomposition (Chapter 13). Perhaps more importantly, various physical and chemical mechanisms can protect soil carbon from microbial decomposition. Assemblages of soil particles, termed aggregates (Chapter 2), can physically shield carbon from microbial enzymes. In addition, reactive minerals and metals can bind with carbon compounds, chemically removing them from the soil solution that microbes can access. The limited availability of oxygen in soil and/or the presence of strongly acidic (pH <5) or cold conditions can also partially constrain microbial decomposition. Spatial isolation of carbon compounds in soil pores too small to access, or remote from established microbial biofilms, may also hinder decomposition. Together, these mechanisms can explain the long-term persistence of carbon compounds that would otherwise be readily decomposed to carbon dioxide by heterotrophic respiration.

From a global perspective, climate is among the most important factors that control soil carbon storage. Decomposition is slow in colder and water-saturated environments, leading to the buildup of thick layers of decaying organic material, known as peat. Decomposition is also slow in hot, dry environments, but carbon inputs to soils under these conditions are constrained by low primary production. In hot, moist climates, decomposition is usually rapid and carbon dioxide production rates are usually high, but soil carbon inputs from plant growth are also high, leading to significant soil carbon storage on balance. Furthermore, rapid weathering processes in warm, moist soils generate reactive minerals that protect carbon from decomposition. Even small changes in climate may have large impacts on soil carbon storage across these diverse ecosystem types. For example, warming and drying can greatly accelerate the decomposition of carbon that was previously protected in frozen and flooded permafrost or peatland soils. In the tropics, where warming may lead to increased organic matter decomposition relative to plant carbon inputs, soil carbon stocks may also significantly decrease. The net release of carbon dioxide from soil carbon stocks contributes to increases in atmospheric carbon dioxide, thereby causing further climate warming.

Within a given climate zone, the impact of substrate quality on trace gas production becomes relatively more important. As mentioned previously, substrate quality is a somewhat ill-defined term. Various measures of litter quality have been suggested, including C/N ratio, lignin content, lignin/N ratio, and the lignin-cellulose index. The most appropriate choice may differ depending on the particular ecosystem or process under consideration and other additional measures may provide better estimates. For example, microbial activity may be more limited by phosphorus than nitrogen on highly weathered soils such as found in the wet tropics. In these cases, litter-quality metrics including phosphorus may be more predictive of decomposition rates than metrics including nitrogen, which are more applicable in nitrogen-limited ecosystems often found in boreal and temperate biomes.

Methane

The microbial metabolism of methane in soils is more complex than that of carbon dioxide. Soils can both produce and consume methane, even simultaneously, under the proper environmental conditions (Fig. 20.4). Soil microbial sources and, to a lesser extent, soil microbial sinks affect atmospheric methane concentrations.

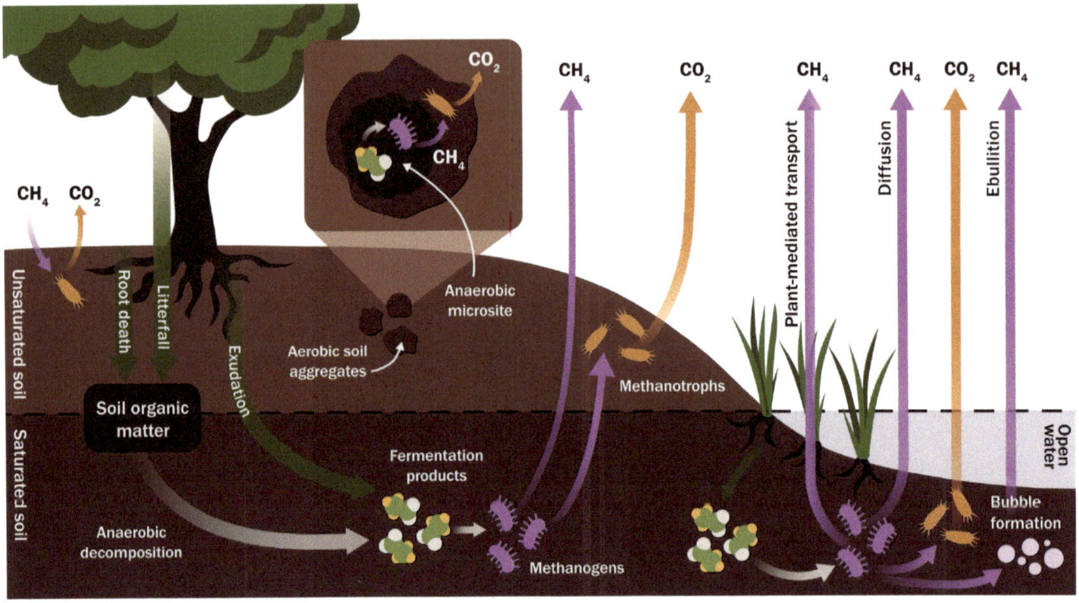

FIGURE 20.4 **Methane production, consumption, and transport processes in upland to wetland ecosystems.** *Source: Adapted by Sada Egenriether from Schimel and Holland (2005).*

Methane budget

Methane is the second most abundant trace gas in the atmosphere (~1.86 ppm) and accounts for over 23% of climate change since the Industrial Revolution (Saunois et al., 2016). Global methane sources exceed global sinks, causing an annual growth in atmospheric methane of around 6 Tg year^{-1} for the period 2000–2009 (Table 20.2). This imbalance is driven primarily by human enhancement of otherwise naturally occurring microbiological sources, including flooded soils for rice cultivation, animal livestock, and waste in landfills.

The biosphere releases approximately 553 Tg methane to the atmosphere annually, with natural wetland soils as the largest natural source of methane (Table 20.2). However, global wetland methane budget estimates produced by different models vary widely, spanning 80 Tg CH_4 year^{-1}. This large uncertainty stems from difficulty in mapping the global extent of natural wetlands, especially in the tropics, together with complex controls on methane transport to the atmosphere. In wetlands, methane oxidizers can consume up to 90% of the methane diffusing upward from the anoxic zone (Fig. 20.4) before it reaches the atmosphere. Methane transport through plants and in gas bubbles can bypass methane oxidizers to rapidly release methane to the atmosphere. Although any changes in methane oxidation have large potential impacts on the net gain of atmospheric methane from wetlands, it is often an "invisible" process from the atmospheric perspective. In contrast, unsaturated soils in upland ecosystems consume methane from the atmosphere (approximately 28 Tg year^{-1}) and account for roughly 6% of the total sink for atmospheric methane.

TABLE 20.2 Atmospheric budgets for methane and nitrous oxide. Fluxes are in Tg CH_4 $year^{-1}$ or Tg N_2O-N $year^{-1}$.

Sources		Sinks		Atmospheric increase
Type	**Flux**	**Type**	**Flux**	
Methane (2000–09) [a]				
Natural	**234**	**Natural**	**546**	
Natural wetlands	112	Atmospheric oxidation	514	
Freshwaters	75	Soil	32	
Geological	32			
Wild animals	6.1			
Termites	5.5			
Wildfires	1.8			
Open ocean	1.2			
Permafrost soils	0.6			
Anthropogenic	**319**			
Fossil fuels	112			
Enteric agriculture	103			
Landfills and waste	57			
Biomass burning	30			
Rice cultivation	29			
Total	**552**	**Total**	**546**	**6**
Nitrous oxide (2007–16) [b]				
Natural	**9.7**	**Natural**	**13.5**	
Natural soils	5.6	Atmospheric oxidation	13.5	
Oceans	3.4	Surface sink	0.01	
Lightning	0.4			
Natural waters	0.3			
Anthropogenic	**7.3**			
Agriculture	3.8			
Fossil fuel	1			
Nitrogen deposition	0.9			
Deforestation pulse	0.8			
Biomass burning	0.6			
Coastal waters	0.5			
Climate and CO_2 effects	0.5			
Waste and wastewater	0.3			
Deforestation reduction	−1.1			
Total	**17**	**Total**	**13.5**	**4.3**

[a] *Data from Saunois et al. (2016).*
[b] *Data from Tian et al. (2020).*

Methane production (methanogenesis)

Methanogenesis (methane production) is an ancient metabolic pathway performed exclusively by a group of anaerobic Archaea known as methanogens. Geological evidence suggests methanogens evolved more than 2.5 billion years ago, before oxygen became abundant in Earth's atmosphere. Methanogens persist ubiquitously in soils, even in dry desert environments. However, their activity, which yields little energy, is limited to low redox environments created by depletion of more energetically favorable electron acceptors.

Soil methane production rates are highest when soils are saturated (since oxygen diffuses 10,000x slower in water than air) or rich in labile organic matter that can fuel biological oxygen demand by heterotrophic respiration. This leads to anoxic conditions and the low redox conditions necessary for methanogenesis. Thus, at the ecosystem scale, the persistent presence of and depth to a water table control the soil volume in which methane production can occur. However, even in unsaturated soil or soil low in organic matter, anoxic conditions can develop at the very small (μm) scale where the methanogens live (see Box 15.9 and earlier section on General Controls on Trace Gas Production). This explains how methane production can occur in oxic soils with high measured redox potential, albeit at lower rates than in anoxic soils.

Methane production is sensitive to a complex series of anaerobic decomposition steps, even though methanogens themselves use only the simplest of substrates. Methanogens in soil produce methane by two primary pathways:

$$CO_2 + 4H_2 \rightarrow CH_4 + 2H_2O \text{ (carbon dioxide reduction)}$$

$$CH_3COOH \rightarrow CO_2 + CH_4 \text{ (acetate fermentation)}$$

Both acetate and hydrogen are by-products of anaerobic fermentation (Chapter 3); thus, methanogenesis is dependent on a complex food chain. Most plant detritus is made up of poorly soluble *polymers*, such as cellulose, lignin, lipids, and proteins. This material must first be broken down into soluble units or *monomers*, usually by extracellular enzymes, before it can be metabolized by soil microbes. The monomers must then be fermented to produce the methanogenic substrates, primarily hydrogen and acetate. Plant roots also exude carbon compounds that microbes can anaerobically decompose to produce acetate. Methane production therefore depends on anaerobic decomposition and fermentation to supply substrates rather than on organic matter quantity or quality directly.

Future temperature increases and precipitation changes due to climate change will alter soil methane production, which may have feedback effects on climate change. However, these feedbacks are difficult to predict. Rates of methanogenesis will increase at higher temperatures, as is generally true for soil microbial processes, but the magnitude of the increase is uncertain. The temperature sensitivity of methane production was previously overestimated ($Q_{10} > 5$) due to concurrent seasonal changes in temperature and substrate availability, a feature of most terrestrial ecosystems. Warmer climates could lead to longer growing seasons, increasing plant productivity, and therefore greater methane emissions due to increased organic matter supply. Also, because methane production requires anoxic conditions, changes in precipitation patterns that alter the extent of flooded land would greatly change soil methane emissions. Soil drying from reduced precipitation would both slow methane production and increase aerobic methane oxidation, thus reducing methane emissions (i.e., net methane efflux). Increased intensity and frequency of precipitation already occurring in some regions of the world could increase *wetland extent* (the land surface area covered by wetlands), thereby increasing global methane emissions to the atmosphere. As permafrost melts, wetland extent and methane production may also increase, though there is no evidence of a permafrost thaw methane feedback to date.

Methane consumption

Methane is consumed in soils primarily by groups of bacteria known as methanotrophs. These bacteria oxidize methane to carbon dioxide, using methane as a primary source of carbon and energy. Methane oxidation occurs under oxic conditions via this overall reaction pathway:

$$CH_4 + 2O_2 \rightarrow CO_2 + 2H_2O \text{ (aerobic methane oxidation)}$$

In addition to aerobic methanotrophy, methane can also be consumed in soil via *anaerobic methane oxidation (AMO)* pathways that utilize alternative electron acceptors in the absence of oxygen. Anaerobic methane oxidation was first discovered in marine sediments, where syntrophic microbial communities couple methane oxidation to the reduction of sulfate. In wetland and rice paddy soils, methane oxidation can be coupled to the reduction of nitrite, nitrate, and metals such as iron. It is still uncertain how much AMO mitigates the large CH_4 source from wetland soils.

One of two different groups of methanotrophs dominate in soils depending on environmental conditions that determine soil methane concentrations. *High-affinity (Type I) methanotrophs* produce enzymes that allow them to consume methane at low concentrations (< 0.0002% soil gas volume)—that is, the enzymes have high affinities for methane. Type I methanotrophs are important in oxidizing atmospheric methane in unsaturated upland soils where abundant oxygen limits methanogenesis and promotes methane oxidation. In these environments, such as are found in forest, grassland, and agricultural ecosystems, methanotrophs primarily live on atmospheric methane (1.8 ppm) and are strongly substrate limited. Thus, methane oxidation rates are controlled largely by atmospheric methane diffusion into the soil profile. For this reason, upland soils comprise only about 6% of all sinks (32 Tg CH_4 $year^{-1}$) in the global methane budget even though upland soils cover a large proportion of the Earth's land surface. In contrast, *low-affinity (Type II) methanotrophs* produce enzymes with low affinities for methane and thus can only oxidize methane at high concentrations. Type II methanotrophs are most abundant near oxic-anoxic interfaces, where relatively high concentrations of oxygen from the upper unsaturated soil layers and methane from the lower saturated soil layers co-occur. They can reduce emissions of methane produced in the saturated, anoxic zone by as much as 90%. Therefore, although Type II methanotrophs do not consume atmospheric methane, they are important in mitigating global methane sources to the atmosphere.

Global change factors, such as land-use change and nitrogen deposition, can influence soil methane consumption rates. For example, conversion of native lands to row crop agriculture can decrease methane consumption via reductions in the diversity of methanotrophic communities (Levine et al., 2011). In addition, high soil ammonium concentrations resulting from fertilization or atmospheric nitrogen deposition can directly inhibit methane consumption. The mechanism of ammonium inhibition results from the molecular similarity of methane and ammonia. Ammonia can compete with methane for binding at the enzyme's active site, thus inhibiting methane oxidation. Additionally, methanotrophs oxidize ammonia to nitrite, which is toxic to them if it accumulates. Predicting global change effects on soil methane consumption, therefore, requires consideration of controls on methanotrophy beyond regulation of methane oxidation itself.

Controls on soil-atmosphere methane fluxes

The balance of methane production (methanogenesis) and consumption (methane oxidation) controls net soil-atmosphere methane fluxes. Positive net fluxes indicate that the soil is a source of methane to atmosphere, and negative net fluxes indicate that the soil is a sink for atmospheric methane. To predict the net flux, we need to understand the individual controlling processes and their interaction with gas transport pathways (Fig. 20.4).

Water table depth is one of the most critical factors in controlling the overall flux of methane from a wetland system. The depth of the water table controls the volume of anoxic soil and thereby the rate of methane production. In a similar manner, it also controls the amount of oxic soil above the anoxic, methanogenic zone and the likelihood that methane is oxidized before it reaches the atmosphere. While anaerobic methane oxidation can also consume methane below the water table, its relative importance compared to aerobic methanotrophy is yet to be determined broadly.

Vegetation is also a critical control on wetland methane fluxes because it affects both the rate of methane production and the fate of methane (Fig. 20.4). Methane production is sensitive to the supply of easily decomposable (labile) carbon. Often the major source of such carbon is plant roots, either as exudates or from recently dead roots. Because of the importance of plant carbon in driving methane production, strong connections occur between primary productivity in an ecosystem and the magnitude of methane flux out of that system. However, plants also control methane export from the soil. Many wetland plants have gas exchange channels, in specialized tissue called aerenchyma, within their stems. Aerenchyma allows oxygen to diffuse downward to the roots, but also allow methane to diffuse upward through the plant, bypassing the gauntlet of methanotrophs. In many wetland systems, 90% or more of the methane that makes it out of the soil is transported through plants. Trees growing in wetter soils can also act as conduits for methane, via both aerenchyma and other pathways, but their global importance remains uncertain (Covey and Megonigal, 2019). Gas transport processes can also play an important role in determining wetland methane fluxes in unvegetated, open-water areas of wetlands. Methane bypasses methanotrophs when gas bubbles containing high concentrations of methane are released from sediments through the water column to the atmosphere via a gas transport process called **ebullition**.

Although we understand the basic controls on methane flux, developing and applying models across a range of ecosystems has been challenging. The difficulty may result from the complexity of processes controlling methane flux. First, models must account for methane production, so they must consider the production and breakdown of organic matter and account for its fermentation to methane. Then a model must account for the fate of methane—whether it escapes from the soil or is oxidized by methanotrophs—which requires an understanding of plant transport and ebullition dynamics. Non-linear relationships and interactions among controlling variables challenge our ability to represent the basic controls on these processes, such as the temperature sensitivity of methane production. Developing effective integrated models of methane dynamics remains an important area for research linking soil biology with atmospheric chemistry.

An example of the complex interactions among trace gases is described in Box 20.1.

BOX 20.1

What are the net effects of wetland drainage and restoration on total global warming potential?

Over the last 200 years, wetlands around the world have been drained for conversion to dryland agriculture, such as growing grain crops. More recently, greater value has been placed on other ecosystem services provided by wetlands, such as water storage and nitrate removal via denitrification. This has led to wetland restoration in some of these agricultural lands via re-flooding.

Consider the climatic impacts of draining and re-flooding a wetland due to altered ecosystem exchanges of carbon dioxide and methane with the atmosphere. Drainage lowers the water table, decreasing the anoxic zone for methane production and increasing the oxic zone for methane oxidation. This reduction in methane fluxes would reduce the global warming potential. In contrast, increasing the oxic zone also leads to rapid decomposition of accumulated organic matter. This increase in carbon dioxide emissions would increase global warming potential. Re-flooding can reverse high carbon dioxide emissions and rebuild the peat, but this comes with the cost of renewed methane emissions (Hemes et al., 2018). The relative magnitude of these opposing processes dictates the overall net effect on global warming potential, with a possible warming or cooling effect on the climate.

The concept of *global warming potential* is used to balance the ecosystem exchanges of multiple greenhouse gases, including carbon dioxide, methane, and nitrous oxide. Methane traps more heat than carbon dioxide on a per molecule basis, meaning that it has a higher global warming potential. However, this difference in global warming potential effectively becomes smaller over time because methane has a relatively short atmospheric lifetime (9 years). The time horizon considered is crucial in understanding how wetland drainage changes affect total global warming potential. Over years to decades, the high global warming potential of methane emitted from re-flooded wetlands can offset the negative carbon dioxide emissions associated with slowed organic matter decomposition. Thus, in the short term, the global warming potential of re-flooded wetlands is often greater than that of drained wetlands, where methane emissions are low. However, over centuries to millennia, the global warming potential of methane becomes much smaller and is typically more than offset by the long-term uptake of atmospheric carbon dioxide in plant biomass and its storage in soil organic matter. A further consideration is that agricultural fertilizer application to drained wetlands will increase soil emissions of nitrous oxide, which is even more effective than methane at trapping heat. In contrast, nitrous oxide emissions from consistently flooded wetlands are typically small. Therefore, we should consider the long-term effects of land use changes on all potentially important trace gases.

Nitrogen-based gases

Nitrous oxide (N_2O) and nitric oxide (NO) are two important nitrogen gases that are produced by soil microorganisms. Although these gases can be produced by many soil microbial and chemical processes, nitrification and denitrification are predominantly responsible for their production. The physiology of these processes is discussed thoroughly in Chapter 15.

Nitrous oxide and nitric oxide budgets

Balancing nitrous oxide sources and sinks to develop a global nitrous oxide budget (Table 20.2) has been a major challenge for researchers. Soils from natural and agricultural ecosystems account for more than half of global nitrous oxide emissions. The remaining nitrous oxide is mostly produced in the oceans, by fossil fuel combustion and industrial processes, and from biomass burning (e.g., forest fires). Nitrous oxide is removed from the atmosphere mostly by photochemical reactions in stratosphere, which is a slow process: the average residence time of an N_2O molecule is more than 100 years. Soil microbes can remove small amounts of atmospheric nitrous oxide under some circumstances, but this is usually counterbalanced by much greater rates of nitrous oxide production.

A relatively consistent annual increase in atmospheric N_2O accumulation (~4 Tg year^{-1} of N_2O-N) has occurred over the last several decades. This observed increase agrees with the difference between our global estimates of sources (~17 Tg N year^{-1}) and sinks of nitrous oxide (13 Tg N year^{-1}; Table 20.2). Thus, at the global scale, our "bottom-up" estimates of nitrous oxide emissions from multiple sources and removal pathways approximately match "top down" observations of atmospheric nitrous oxide. However, at smaller scales these estimates can disagree considerably. One widely used approach for estimating bottom-up emissions involves *emissions factors*, for which a given fraction of nitrogen inputs (e.g., 1% of synthetic fertilizer) or ecosystem nitrogen pools (e.g., 1% crop residues) is predicted to be released from soil as nitrous oxide. The United States Corn Belt is a region of intensive agriculture that receives large inputs of reactive nitrogen from fertilizer and biological nitrogen fixation. Top-down measurements in this region show much greater atmospheric nitrous oxide concentrations than can be accounted for by soil nitrous oxide emissions estimated from emissions factors. Small streams that drain from agricultural watersheds can produce very high nitrous oxide emissions from denitrification and could thus account for at least part of this missing source (Griffis et al., 2017). However, long-term measurements have also shown that high nitrous oxide emissions from agricultural soils also contribute to regional underestimates by default emissions factors (Gillette et al., 2018). Improving our capacity to estimate nitrous oxide emissions at the site and regional scales remains an active area of research that is critical to inform policy and management for decreasing the emissions of this potent greenhouse gas.

In contrast to nitrous oxide, nitric oxide is a very short-lived compound in the atmosphere (hours–weeks), and its budget is even harder to estimate than for nitrous oxide. Based on extrapolation from field measurements of soil nitric oxide emissions, global nitric oxide emissions from soil average 21 Tg NO-N year^{-1} but with large uncertainty spanning 4–10 Tg NO-N year^{-1} (Davidson and Kingerlee, 1997).

Nitric oxide and nitrous oxide production

In an ecological context, the overall controls on the production of nitric oxide and nitrous oxide by nitrification and denitrification have best been considered within the "leaky-pipe" model of Firestone and Davidson (1989) (Fig. 20.5). This conceptual model links overall nitrogen dynamics with nitric oxide and nitrous oxide production. The basis of this conceptual

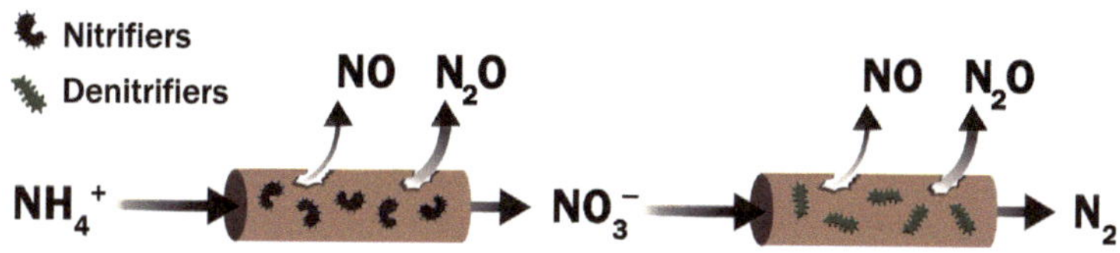

FIGURE 20.5 **"Leaky pipe" model of the control of nitric oxide and nitrous oxide production from nitrification and denitrification.** *Source: Adapted by Sada Egenriether from Firestone and Davidson (1989).*

model is that nitric oxide and nitrous oxide are side products of nitrification and denitrification. Thus, the production of these gases is dependent on both the total rate of the processes (*flow through the pipe*) and the proportion of the nitrogen that is released as either nitric oxide or nitrous oxide (*the size of the holes in the pipe*). By considering these factors separately, we can build effective models for how soil processes control the emissions of these globally important gases.

Production from nitrification

Flow through the pipe. As it is true for many soil microbes, the main limitation on nitrifier activity is substrate availability, ammonium in this case. Nitrification rates often correlate well with nitrogen mineralization rates and ammonium availability. If ammonium availability is great enough, it can even allow nitrifiers to overcome other factors that generally limit nitrification, such as low pH. This explains why the very acidic forest soils of New England exhibit rapid nitrification after clear-cutting, which is almost unmeasurable in the native forest; by eliminating plant uptake, ammonium is made available for nitrifiers.

One of the major controls on ammonium availability is plant activity (Fig. 20.6). Plant uptake can limit ammonium supply, but plants also release large amounts of organic matter into

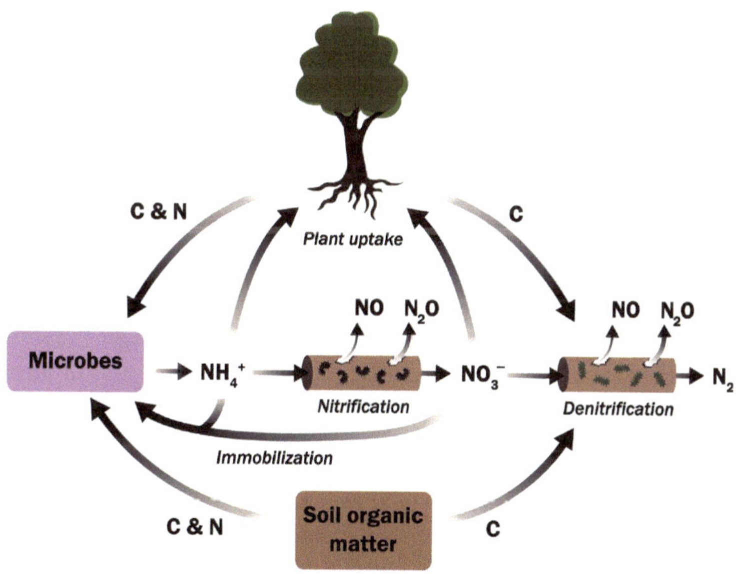

FIGURE 20.6 **Biotically mediated interactions of carbon and nitrogen cycling in controlling production of nitric oxide and nitrous oxide in soils.** *Source: Adapted by Sada Egenriether from Schimel and Holland (2005).*

the soil as leachates, root exudates, or litter. If this material is nitrogen rich, it can stimulate mineralization and nitrification. If the material is nitrogen poor, it stimulates microbial immobilization, reducing the availability of ammonium to nitrifiers.

Water exerts a major secondary control on nitrification. First, nitrification is an aerobic process, and wet soils slow oxygen diffusion into the soil. Increasing soil moisture, therefore, limits the oxygen supply to nitrifiers and reduces nitrification rates. Second, as soils dry, water films around soil particles become thinner and the water film may become discontinuous. This makes substrate diffusion more difficult and reduces substrate supply to limit nitrification rates.

Size of the holes in the pipe. Nitric and nitrous oxide are side products of nitrification, rarely accounting for more than about 1% of the total ammonium nitrified. However, the fraction of nitrified nitrogen released as nitric oxide and nitrous oxide is variable and controlled by a number of factors. Oxygen is a major control of the proportion of those gases produced, with the proportion of the nitrified nitrogen evolved as nitric oxide and nitrous oxide (the size of the holes in the pipe) increasing at lower oxygen levels. However, the overall rate of nitrification (the flow through the pipe) also decreases with decreasing oxygen concentrations. Hence, while low oxygen causes a higher proportion of gaseous products from nitrification, the overall lower rate of nitrification limits rates of nitric oxide and nitrous oxide production.

Production from denitrification

Flow through the pipe. Some of the controls on denitrification rates are the same as those on nitrification; others are complementary. Denitrification is predominantly an anaerobic, heterotrophic process in which nitrate or nitrite is the terminal electron acceptor. Because denitrification is an anaerobic process, soil moisture is an important control on denitrification rates. Similar to methane production, denitrification can occur in anoxic microsites in an otherwise aerated soil. Producing these microsites requires that the pores within aggregates are water-filled to limit oxygen diffusion or that microsites are rich in labile organic matter to stimulate microbial oxygen consumption in excess of oxygen inputs by diffusion. When anoxic microsites are present, nitrate availability becomes a major control on denitrification such that nitrification rates can be important in controlling denitrification rates by supplying nitrate. Plant uptake of nitrate and microbial immobilization can also influence nitrate supply to denitrifiers (Fig. 20.6). The availability of organic carbon, the electron donor in denitrification, can also control denitrification rates but is usually secondary to water and nitrate. Thus, at a large scale, water availability and plant activity are major controls on denitrification. Water controls oxygen diffusion, while plants control both nitrogen availability and carbon supply to the denitrifiers (Chapter 15).

Size of the holes in the pipe. Both nitric oxide and nitrous oxide are obligate intermediates in denitrification, but the factors controlling the proportion of nitrous oxide evolved during denitrification are better understood than those for nitric oxide. Nitric oxide is usually a minor product because it can be toxic to microbes if it is allowed to accumulate; therefore, microbes usually reduce nitric oxide rapidly to limit its accumulation and resulting emission from soil. In contrast, nitrous oxide can range from being none to all of the final product. Three factors that appear to be particularly important in controlling the proportion of nitrous oxide produced in denitrification are oxygen, pH, and the ratio of nitrate to available carbon. The parallel factors for nitric oxide are less-well understood.

Although denitrification rates generally decline as oxygen increases, the proportion of nitrogen evolved as nitrous oxide increases as oxygen concentrations rise. Thus, as with nitrification, a tradeoff exists between maximizing the denitrification rate and maximizing the proportion of product as nitric and nitrous oxide. Also similar to nitrification, the oxygen concentration that maximizes nitrous oxide production from denitrification appears to be about 1%.

Low pH generally inhibits the reduction of nitrous oxide to dinitrogen. Thus, at low pH, nitrous oxide can be the dominant product of denitrification. However, highly acidic soils often have low nitrogen availability and, therefore, low nitrification and denitrification rates. As a result, the highest rates of nitrous oxide production from denitrification generally occur in moist soils that cycle nitrogen rapidly, regardless of pH.

When nitrate, an electron acceptor, is present in excess relative to organic carbon, an electron donor, denitrifiers are "wasteful" of nitrate and generally forego the final reduction step to dinitrogen, thereby producing nitrous oxide as a major product. When nitrate is present in limited amounts, however, denitrifiers are more likely to reduce nitrous oxide to dinitrogen. Abiotic reactions termed "chemodenitrification" can also produce nitric and nitrous oxide from substrates (e.g., nitrite) generated by microbial processes, although their overall importance remains poorly understood.

Other soil microbes that are not denitrifiers can also reduce nitrous oxide to dinitrogen (Jones et al., 2014; Sanford et al., 2012). These microbes may not produce nitrous oxide to contribute to the flow through the denitrification pipe in the "leaky pipe" model; however, by consuming nitrous oxide, they do contribute to the size of the holes in the pipe (i.e., how much nitrous oxide is released from the soil to the atmosphere). It is still unknown when and where non-denitrifiers play an important role in mitigating soil emissions of nitrous oxide. Research efforts in this area are complicated by the difficulty in measuring rates of nitrous oxide reduction to dinitrogen due to the high atmospheric background concentration of dinitrogen (78%).

Nitrification versus denitrification as sources of nitrogen gases

Because nitric oxide can be consumed by both biotic and abiotic reactions, it is generally only released from the soil in large quantities when relatively dry soil conditions limit biotic activity. Thus, in most cases, the major source of nitric oxide is nitrification. Nitrous oxide, however, is less reactive in soil and can escape from relatively wet soil. Therefore, nitrous oxide can be produced in large quantities by either nitrification or denitrification, although the largest nitrous oxide fluxes are probably from denitrification because the "holes-in-the pipe" for denitrification are "larger" than for nitrification.

Soil moisture can roughly predict nitric oxide and nitrous oxide production from nitrification and denitrification. Under extremely dry conditions, microbial activity and gas emissions are limited. Under moist soil conditions, nitrification increases and nitric oxide is produced. In wetter soils, both nitrification and denitrification produce nitric oxide and nitrous oxide. When water-filled pore space increases above approximately 60%, nitrification slows, most nitric oxide is consumed before it can diffuse out of the soil, and nitrous oxide is the main nitrogen gas produced from denitrification. In completely water-logged soils (at 100% water-filled pore space), nitrification is limited and denitrification can occur rapidly until available nitrate is consumed. Under these conditions, dinitrogen is likely to be the main denitrification product and relatively little nitric or nitrous oxide will be produced. However, water-filled pore space may poorly predict soil nitric oxide and nitrous oxide emissions from processes other than nitrification and denitrification. For example, chemodenitrification might sustain significant nitric oxide emissions even under saturated soil conditions (Hall et al., 2018).

We can combine our understanding of plant-soil carbon and nitrogen dynamics to predict both which ecosystems are likely to produce large quantities of nitric oxide and nitrous oxide and which processes are likely to be responsible. For example, the Great Plains of the western United States are semiarid and dominated by grasslands, in which decomposition and nitrogen cycling are fairly rapid. Thus, nitrification is relatively rapid, plants use nitrate for a substantial proportion of their nitrogen uptake, and anoxic microsites are typically scarce. Nitrification generally dominates nitrogen gas emissions and the ratio of nitric oxide to nitrous oxide is fairly high, except shortly after a rainstorm. When a rainstorm temporarily saturates the pore spaces within soil aggregates, anoxic sites become common, and a pulse of denitrification occurs, with nitrous oxide dominating nitric oxide as a product. Another example from a tropical system is presented in Box 20.2.

BOX 20.2

Effects of converting old-growth tropical forest to pasture

Tropical forests are being rapidly cleared for conversion to pastures and farmland for food and bioenergy crops. In addition to the tremendous amount of carbon dioxide, carbon monoxide, nitrous oxide, and other gases released during burning, subsequent changes in ecosystem characteristics alter trace gas fluxes. Common results may include decreased methane consumption, increased methane production, and increased nitric oxide and nitrous oxide production. What factors contribute to and control these changes in gas fluxes?

- *Changes in nitrogen cycling.* Initially, the ash produced during burning and left on the soil surface contains a lot of nitrite, which is converted to nitric oxide chemically. This chemodenitrification can increase nitric oxide fluxes by more than 10-fold for a short period. Later conversion to pasture or farmland can accelerate nitrogen turnover in decomposing organic matter and thereby enhance nitric and nitrous oxide emissions. Both nitrous oxide and nitric oxide emissions are enhanced for years following conversion to pasture. The increased ammonium availability from this accelerated nitrogen cycling can also inhibit methane oxidation and thereby reduce methane consumption.
- *Altered soil microclimate.* Clearing the forest and litter layer removes their buffering effect on soil temperature and moisture. This is especially true after burning, because the black ash layer is particularly good at absorbing sunlight. This can cause drying of the surface soil, making it a less favorable zone for microbial activity. Desiccation can be a particular problem for methane-oxidizing bacteria because it may reduce their numbers in the upper soil layers, thereby decreasing methane consumption rates. As a new plant canopy develops, some of these effects can be reduced.
- *Altered soil structure.* Forest clearing and conversion can cause soil compaction. This results from hauling timber during clearing and trampling by pasture animals. The compaction reduces gas diffusion into the soil and stimulates anaerobic processes when soils are moist, enhancing denitrification and methane production, while simultaneously inhibiting nitrification and methane oxidation. Compaction also decreases oxygen diffusion into the soil, thereby increasing the proportion of products from nitrification and denitrification evolved as nitric oxide and nitrous oxide.

Summary

Soil microbes are key players in producing and consuming atmospheric trace gases. Questions remain about the atmospheric budgets of some trace gases (e.g., methane, nitrous oxide, nitric oxide, and carbon monoxide) and the physiology of some key processes (e.g., methane oxidation, nitric oxide production, and nitrous oxide reduction). However, even for those gases for which the atmospheric budgets and microbial physiologies are fairly well understood (e.g., carbon dioxide, methane production), the challenges of linking the divergent scales from process and site to global dynamics remain. Models of the whole Earth system require understanding trace gas fluxes and how they will change under altered climate and land-use practices. This in turn requires a firm mechanistic understanding of the processes controlling trace gas production and consumption and how these processes operate at different scales from the microbial to the global.

Acknowledgments

Portions of this chapter, with permission, are derived from the work of Drs. Josh Schimel and Elisabeth A. Holland who authored a similar chapter for the Second Edition of the textbook. The current authors thank them for the use of their material and for graciously supporting production of the Third Edition.

Study questions

1. What is the difference between methane production and methane flux? Under what conditions would you use one term or the other?
2. Some trace gases, such as methane, nitric oxide and nitrous oxide, are both produced and consumed in soil. Why is it problematic to deduce trace gas production and consumption rates from measurements of net soil-atmosphere trace gas exchange (e.g., trace gas emissions or uptake)?
3. Many atmospheric models assign each ecosystem a specific, fixed flux for each trace gas. What are some of the problems with this approach? How would you improve it?
4. Given that there will be relatively large uncertainties in any global gas budget, what are the scientific reasons for constructing them?
5. Reforestation is one of the proposed solutions for slowing the accumulation of carbon dioxide in the atmosphere. What are the possible implications of a global reforestation program for the other trace gases?
6. The carbon and nitrogen cycles are coupled in many ways. What is one way in which nitrogen pools or fluxes can influence the production or consumption of carbon-based gases? What is one way in which carbon pools or fluxes can influence the production or consumption of nitrogen-based gases?
7. For which trace gases would soil emissions increase with increasing precipitation? With increasing temperatures? Why?

References

Ciais, P., Sabine, C., Bala, G., Bopp, L., Brovkin, V., Canadell, J., Chhabra, A., DeFries, R., Galloway, J., Heimann, M., Jones, C., Le Quere, C., Myneni, R.B., Piao, S.L., Thornton, P., 2013. Carbon and other biogeochemical cycles. In: Stocker, T.F., Qin, D., Plattner, G.K., Tignor, M.M.B., Allen, S.K., Boschung, J., Nauels, A., Xia, Y., Bex, V., Midgley, P.M. (Eds.), Climate Change 2013: The Physical Science Basis. Contribution of Working Group I to the Fifth Assessment Report of the Intergovernmental Panel on Climate Change. Cambridge University Press, Cambridge, United Kingdom and New York, NY, USA.

Covey, K.R., Megonigal, J.P., 2019. Methane production and emissions in trees and forests. New Phytol. 222, 35–51.

Davidson, E.A., Kingerlee, W., 1997. A global inventory of nitric oxide emissions from soils. Nutr. Cycling Agroecosyst. 48, 37–50.

Dlugokencky, E.J., Hall, B.D., Montzka, S.A., Dutton, G., Muhle, J., Elkins, J.W., 2019. 2019: Long-lived greenhouse gases [in "State of the Climate in 2018"]. Bull. Am. Meteorol. Soc. 100, S48–S51.

Firestone, M.K., Davidson, E.A., 1989. Microbiological basis of NO and N_2O production and consumption in soil. Andreae, M.O., Schimel, D.S. (Eds.), Exchange of Trace Gases between Terrestrial Ecosystems and the Atmosphere. John Wiley and Sons Ltd, Chichester, UK.

Gillette, K., Malone, R.W., Kaspar, T.C., Ma, L., Parkin, T.B., Jaynes, D.B., Fang, Q.X., Hatfield, J.L., Feyereisen, G.W., Kersebaum, K.C., 2018. N loss to drain flow and N_2O emissions from a corn-soybean rotation with winter rye. Sci. Total Environ. 618, 982–997.

Griffis, T.J., Chen, Z., Baker, J.M., Wood, J.D., Millet, D.B., Lee, X., Venterea, R.T., Turner, P.A., 2017. Nitrous oxide emissions are enhanced in a warmer and wetter world. Proc. Nat. Acad. Sci. 114, 12081–12085.

Hall, S.J., Reyes, L., Huang, W., Homyak, P.M., 2018. Wet spots as hotspots: Moisture responses of nitric and nitrous oxide emissions from poorly drained agricultural soils. J. Geophys. Res. Biogeosci. 123, 3589–3602.

Hartmann, D.L., Tank, A., Rusticucci, M., Alexander, L.V., Bronnimann, S., Charabi, Y.A.R., Dentener, F.J., Dlugokencky, E.J., Easterling, D.R., Kaplan, A., Soden, B.J., Thorne, P.W., Wild, M., Zhai, P.M., 2013. Observations: Atmosphere and surface. In: Stocker, T.F., Qin, D., Plattner, G.K., Tignor, M.M.B., Allen, S.K., Boschung, J., Nauels, A., Xia, Y., Bex, V., Midgley, P.M. (Eds.), Climate Change 2013: The Physical Science Basis. Contribution of Working Group I to the Fifth Assessment Report of the Intergovernmental Panel on Climate Change. Cambridge University Press, Cambridge, United Kingdom and New York, NY, USA.

Hemes, K.S., Chamberlain, S.D., Eichelmann, E., Knox, S.H., Baldocchi, D.D., 2018. A biogeochemical compromise: The high methane cost of sequestering carbon in restored wetlands. Geophys. Res. Lett. 45, 6081–6091.

Jenny, H., 1994. Factors of Soil Formation: A System of Quantitative Pedology. Courier Corporation, New York.

Jones, C.M., Spor, A., Brennan, F.P., Breuil, M.C., Bru, D., Lemanceau, P., Griffiths, B., Hallin, S., Philippot, L., 2014. Recently identified microbial guild mediates soil N_2O sink capacity. Nat. Clim. Change 4, 801–805.

Leifeld, J., Menichetti, L., 2018. The underappreciated potential of peatlands in global climate change mitigation strategies. Nat. Commun. 9, 1071. https://doi.org/10.1038/s41467-018-03406-6.

Levine, U.Y., Teal, T.K., Robertson, G.P., Schmidt, T.M., 2011. Agriculture's impact on microbial diversity and associated fluxes of carbon dioxide and methane. ISME J. 5, 1683–1691.

Myhre, G., Shindell, D., Breon, F.M., Collins, W., Fuglestvedt, J., Huang, J.P., Koch, D., Lamarque, J.F., Lee, D., Mendoza, B., Nakajima, T., Robock, A., Stephens, G., Takemura, T., Zhang, H., 2013. Anthropogenic and natural radiative forcing. In: Stocker, T.F., Qin, D., Plattner, G.K., Tignor, M.M.B., Allen, S.K., Boschung, J., Nauels, A., Xia, Y., Bex, V., Midgley, P.M. (Eds.), Climate Change 2013: The Physical Science Basis. Contribution of Working Group I to the Fifth Assessment Report of the Intergovernmental Panel on Climate Change. Cambridge University Press, Cambridge, United Kingdom and New York, NY, USA.

Sanford, R.A., Wagner, D.D., Wu, Q.Z., Chee-Sanford, J.C., Thomas, S.H., Cruz-Garcia, C., Rodriguez, G., Massol-Deya, A., Krishnani, K.K., Ritalahti, K.M., Nissen, S., Konstantinidis, K.T., Loffler, F.E., 2012. Unexpected nondenitrifier nitrous oxide reductase gene diversity and abundance in soils. Proc. Natl. Acad. Sci. USA 109, 19709–19714.

Saunois, M., Bousquet, P., Poulter, B., Peregon, A., Ciais, P., Canadell, J.G., Dlugokencky, E.J., Etiope, G., Bastviken, D., Houweling, S., Janssens-Maenhout, G., Tubiello, F.N., Castaldi, S., Jackson, R.B., Alexe, M., Arora, V.K., Beerling, D.J., Bergamaschi, P., Blake, D.R., Brailsford, G., Brovkin, V., Bruhwiler, L., Crevoisier, C., Crill, P., Covey, K., Curry, C., Frankenberg, C., Gedney, N., Höglund-Isaksson, L., Ishizawa, M., Ito, A., Joos, F., Kim, H.S., Kleinen, T., Krummel, P., Lamarque, J.F., Langenfelds, R., Locatelli, R., Machida, T., Maksyutov, S., McDonald, K.C., Marshall, J., Melton, J.R., Morino, I., Naik, V., O'Doherty, S., Parmentier, F.J.W., Patra, P.K., Peng, C., Peng, S., Peters, G.P., Pison, I., Prigent, C., Prinn, R., Ramonet, M., Riley, W.J., Saito, M., Santini, M., Schroeder, R., Simpson, I.J., Spahni,

R., Steele, P., Takizawa, A., Thornton, B.F., Tian, H., Tohjima, Y., Viovy, N., Voulgarakis, A., Van Weele, M., Van Der Werf, G.R., Weiss, R., Wiedinmyer, C., Wilton, D.J., Wiltshire, A., Worthy, D., Wunch, D., Xu, X., Yoshida, Y., Zhang, B., Zhang, Z., Zhu, Q, 2016. The global methane budget 2000-2012. Earth Syst. Sci. Data 8 (2), 697–751. doi: 10.5194/essd-8-697-2016.

Schädel, C., Bader, M.K.F., Schuur, E.A.G., Biasi, C., Bracho, R., Čapek, P., De Baets, S., Diáková, K., Ernakovich, J., Estop-Aragones, C., Graham, D.E., Hartley, I.P., Iversen, C.M., Kane, E., Knoblauch, C., Lupascu, M., Martikainen, P.J., Natali, S.M., Norby, R.J., O'Donnell, Jonathan A., Chowdhury, T.R., Šantrůčková, H., Shaver, G., Sloan, Victoria L., Treat, C.C., Turetsky, M.R., Waldrop, M.P., Wickland, K.P., 2016. Potential carbon emissions dominated by carbon dioxide from thawed permafrost soils. Nat. Clim. Change 6, 950–953.

Schimel, J., Holland, E.A., 2005. Global gases. In: Sylvia, D.M., Fuhrmann, J.J., Hartel, P.G., Zuberer, D.A. (Eds.), Principles and Applications of Soil Microbiology. Pearson, New Jersery, pp. 491–509.

Singh, B.K., Bardgett, R.D., Smith, P., Reay, D.S., 2010. Microorganisms and climate change: terrestrial feedbacks and mitigation options. Nat. Rev. Microbiol. 8, 779–790.

Tian, H., Xu, R., Canadell, J.G., Thompson, R.L., Winiwarter, W., Suntharalingam, P., Davidson, E.A., Ciais, P., Jackson, R.B., Janssens-Maenhout, G., Prather, M.J., Regnier, P., Pan, N., Pan, S., Peters, G.P., Shi, H., Tubiello, F.N., Zaehle, S., Zhou, F., Arneth, A., Battaglia, G., Berthet, S., Bopp, L., Bouwman, A.F., Buitenhuis, E.T., Chang, J., Chipperfield, M.P., Dangal, S.R. S., Dlugokencky, E., Elkins, J., Eyre, B.D., Fu, B., Hall, B., Ito, A., Joos, F., Krummel, P.B., Landolfi, A., Laruelle, G.G., Lauerwald, R., Li, W., Lienert, S., Maavara, T., MacLeod, M., Millet, D.B., Olin, S., Patra, P.K., Prinn, R.G., Raymond, P.A., Ruiz, D.J., Werf, G.R. v. d., Vuichard, N., Wang, J., Weiss, R., Wells, K.C., Wilson, C., Yang, J., Yao, Y., 2020. A comprehensive quantification of global nitrous oxide sources and sinks. Nature. 586(7828), 248–256.

Yu, Z.C., Loisel, J., Brosseau, D.P., Beilman, D.W., Hunt, S.J., 2010. Global peatland dynamics since the Last Glacial Maximum. Geophys. Res. Lett. 37, Article L13402.

Further reading

Butterbach-Bahl, K., Baggs, E.M., Dannenmann, M., Kiese, R., Zechmeister-Boltenstern, S., 2013. Nitrous oxide emissions from soils: how well do we understand the processes and their controls? Philos. Trans. R. Soc. B Biol. Sci. 368, 20130122.

Conrad, R., 2009. The global methane cycle: recent advances in understanding the microbial processes involved. Environ. Microbiol. Rep. 1, 285–292.

Conrad, R., 2020. Importance of hydrogenotrophic, aceticlastic and methylotrophic methanogenesis for methane production in terrestrial, aquatic and other anoxic environments: A mini review. Pedosphere 30, 25–39.

Lehmann, J., Kleber, M., 2015. The contentious nature of soil organic matter. Nature 528, 60–68.

Schlesinger, W.H., Bernhardt, E., 2020. Biogeochemistry: An Analysis of Global Change. Academic Press, San Diego, Calif.

Schmidt, M.W.I., Torn, M.S., Abiven, S., Dittmar, T., Guggenberger, G., Janssens, I.A., Kleber, M., Koegel-Knabner, I., Lehmann, J., Manning, D.A.C., Nannipieri, P., Rasse, D.P., Weiner, S., Trumbore, S.E., 2011. Persistence of soil organic matter as an ecosystem property. Nature 478, 49–56.

CHAPTER

21

Biodegradation of environmental pollutants

William J. Hickey

Department of Soil Science, University of Wisconsin, Madison, WI, United States

The microbes will have the last word.—**Louis Pasteur**

Biodegradation is more than a process; it is a lifestyle. The vast majority of microbes carry out some form of biodegradation in order to grow and have been honing these abilities for more than 3 billion years. Not surprisingly, microbes have become highly adept in this capacity, and it is from this deep evolutionary history and driving force for survival upon which the variety of microbial biodegradation processes acting on organic pollutants is derived. As interest in pollutant biodegradation centers on the applicability for removal of compounds from the environment, characteristics of the environment that modulate the potential for biodegradation are as important as the biological systems. Thus, the goal of this chapter is to provide a holistic view of pollutant biodegradation that examines both the biological processes as well as their environmental controls. To efficiently explore these two dimensions of pollutant biodegradation, the chapter will focus on select groups of chemicals that are among the pervasive environmental pollutants.

Biological processes

Types of biodegradation

Two general types of processes can contribute to biodegradation, those that support microbial growth and those that do not. This chapter will focus primarily on growth supporting processes, as these are generally most important in the application of biodegradation for bioremediation. There are instances where processes that do not support growth have a role in pollutant degradation, and these will be addressed later in this chapter.

Principles and Applications of Soil Microbiology. http://dx.doi.org/10.1016/B978-0-12-820202-9.00021-6
Copyright © 2021 Elsevier Inc. All rights reserved.

The essential feature that defines biodegradation supporting microbial growth is that the compound degraded is utilized as a source of nutrition and/or for generation of energy (Fig. 21.1). Nutrition is the assimilation of elements needed to form molecules and build cells, and carbon is the element that must be obtained in the greatest quantities. Thus, for heterotrophic bacteria, biodegradation of organic compounds is driven in part by a need to obtain carbon for biosynthesis. Biodegradation of organic compounds plays a dual function, as energy bound in these materials can be captured and used to power biosynthesis (Fig. 21.1). In this role, the organic compounds are used as **electron donors** in **respiration** (Chapter 3). This collection of processes is generally simplified and referred to as "use as a carbon and energy source," which is the dominant process underlying biodegradation in aerobic environments.

To utilize pollutants as carbon and energy sources, the general strategy microbes employ is to decompose these compounds stepwise in a series of enzymatic reactions that are collectively referred to as **degradation pathways**. The ultimate goal of these pathways is to funnel the pollutant compound into a cell's central metabolism (e.g., TCA cycle; Chapter 3) from which its components are then directed to either biosynthesis (carbon) or to energy generation (e.g., NADH). While central metabolism pathways are shared by many organisms, the enzymes that constitute degradation pathways are not, and it is this **genotype** (possession of genes that collectively encode biodegradation pathways) that defines a bacterium as a degrader of a given chemical.

A second type of growth support occurs when a compound is used in respiration as the **terminal electron acceptor** and serves in energy generation alone. In this case, carbon needed for biosynthesis is obtained from another source, either another organic compound or carbon dioxide. This type of growth support is limited to halogenated organic compounds and is a type of **anaerobic respiration** termed **organohalide respiration**. The organohalide compound serves the same role as oxygen in aerobic respiration, nitrate in denitrification, sulfate/sulfite in sulfate reduction, or carbon dioxide in methanogenesis.

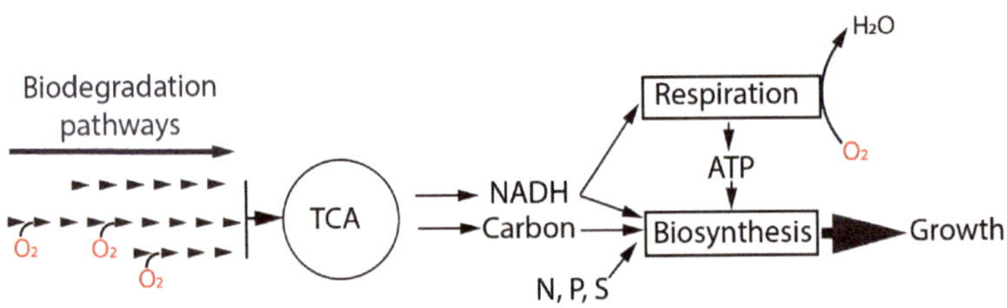

FIGURE 21.1 **Overview of biodegradation processes.** Biodegradation pathways are shown schematically on the left as a series of arrowheads, which represent enzymatic reactions converting the initial compound into a series of metabolites. The variation in arrowheads indicates that pathways vary in the number of steps needed to reach the TCA cycle. The involvement of oxygen (O_2) in aerobic metabolism (degradation pathways and respiration) is shown in *red*. In the pathways, oxygen utilization reflects the activity of oxygenases, which occur in some, but not all, pathways. Thus, while oxygenases are important enzymes in aerobic biodegradation, they are not required for metabolism of all compounds. The occurrence of multiple oxygenases in a pathway illustrates different roles for these enzymes, such as the initial hydroxylation of aromatic rings followed by cleavage of the ring (Box 21.1). The termination of the pathways at the vertical line indicates that, although there are numerous biodegradation pathways, these all culminate in only a few "central intermediates," which are the thresholds for entry into the TCA cycle. The right-hand side illustrates the flow of carbon to biosynthesis and the flow of reducing equivalents to NADH, the latter serving as the electron donor in respiration as well as reducing agent in some biosynthetic reactions.

In organohalide respiration, electrons and hydrogen are transferred to a halogenated compound, which displaces a halogen. The replacement of halogens with hydrogen represents a reduction of the carbon atoms (in the same way oxygen in O_2 is reduced to H_2O in aerobic respiration), thus the process is referred to as **reductive dehalogenation** (Fig. 21.2). The core organic compound is not degraded, and the end product of organohalide respiration is either a compound with fewer halogens or one that is completely dehalogenated. Organohalide respiration can occur with halogenated aliphatic compounds (haloalkanes and haloalkenes), halogenated aromatics (e.g., chlorophenols, chlorobenzenes), and halogenated polyaromatics (e.g., polychlorinated biphenyls, PCBs). For environmental remediation, organohalide respiration is useful because the nonhalogenated structures are either non-toxic (as in the case of haloethenes converted to ethane, Fig. 21.2) or may be rendered more susceptible to degradation by other processes (e.g., conversion of highly chlorinated PCBs to modified PCBs having few chlorine substituents, Fig. 21.2). Organohalide respiration is the most effective biodegradation process for highly halogenated compounds, but is limited to anaerobic environments and specific types of anaerobic bacteria (see section on Anaerobic repiration, below).

Aerobic biodegradation

The prevalence of oxygen (O_2) is the defining characteristic of surface environments exposed to air, and microbes utilize oxygen in two ways in their biodegradation of environmental pollutants. First, for pollutants degraded as a source of carbon and energy, O_2 serves as the terminal electron acceptor. Second, the most effective degradation of some compounds involves enzymes, termed **oxygenases**, which require oxygen and utilize it directly in the reaction by inserting one or both oxygen atoms into the substrate. Oxygenases are a large category of enzymes present in aerobic microbes as well as in insects, plants, and animals (Box 21.1). But microbes, mainly bacteria, employ oxygenases in unique ways to enable their growth on a wide range of compounds.

One of the main categories of environmental pollutants for which aerobic biodegradation is particularly effective, because of the activity of oxygenases, is hydrocarbons. Oxygenases are key because, without these enzymes, hydrocarbons are relatively inert. Hydrocarbons that occur most frequently as environmental pollutants are oil and oil-derived fuels (gasoline, diesel) and materials derived from coal. The coal materials include coal tar, a waste product of the past practice of coal gasification, and creosote, a substance applied to wood as an antimicrobial preservative. All of these materials are liquids comprised of mixtures of chemicals that include several classes (Box 21.2). Most compounds in these mixtures are composed solely of carbon and hydrogen, although some rings contain an atom of oxygen, nitrogen or sulfur (NSO compounds). Thus, hydrocarbons are a rich source of carbon and energy, but are deficient in the other essential macronutrients (primarily nitrogen, but also phosphorus and sulfur) needed to support growth. In cases of hydrocarbon spills, the amount of carbon introduced into the environment far exceeds the levels of these other macronutrients needed to support growth, and thus a common practice to enhance aerobic biodegradation of hydrocarbons is supplementation with O_2 and nutrients.

Hydrocarbons are natural compounds, and the microbial capacity (metabolic machinery) to degrade them is ubiquitous in the environment. This is not to say, however, that all hydrocarbons are equal with respect to microbial biodegradation potential, which becomes apparent

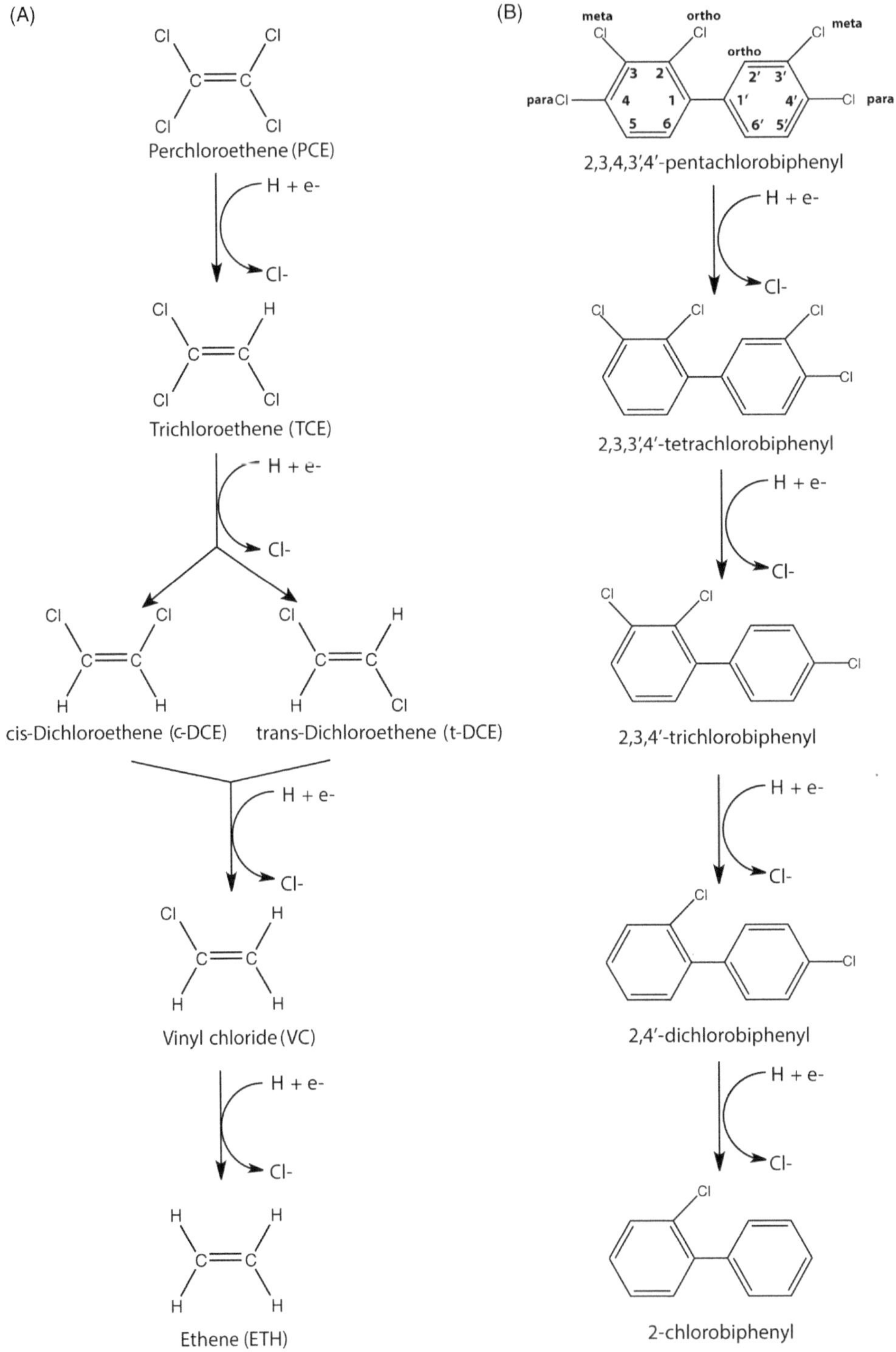

(A)
Perchloroethene (PCE)
H + e-
Cl-
Trichloroethene (TCE)
H + e-
Cl-
cis-Dichloroethene (c-DCE)
trans-Dichloroethene (t-DCE)
H + e-
Cl-
Vinyl chloride (VC)
H + e-
Cl-
Ethene (ETH)
(B)
meta
ortho
meta
ortho
para
para
2,3,4,3′,4′-pentachlorobiphenyl
H + e-
Cl-
2,3,3′,4′-tetrachlorobiphenyl
H + e-
Cl-
2,3,4′-trichlorobiphenyl
H + e-
Cl-
2,4′-dichlorobiphenyl
H + e-
Cl-
2-chlorobiphenyl

FIGURE 21.2 Reductive dechlorination patterns in the organohalide respiration of a chloroaliphatic compound (A) and aromatic compounds (B). For both cases, each stepwise displacement of a halogen releases energy, which can be captured to fuel cellular processes in the same manner as oxygen in aerobic respiration (Fig. 21.1). For the chloroaliphatic compound, the end product of organohalide respiration is fully dehalogenated. In contrast, for PCBs, the process terminates with compounds that are not fully dehalogenated but instead possess fewer halogen atoms than the parent compound. PCBs in the environment occur as mixtures varying in the number and location of the halogen substituents, and PCB dehalogenation follows a pattern of preferential removal of halogens at the "*meta*" and "*para*" positions. Thus, the end products are PCBs depleted in chlorine at these sites and enriched in PCBs containing chlorines in the "*ortho*" positions. In panel B, the top structure (2,3,4,3′, 4′-pentachlorobiphenyl) includes a key for the naming of carbon atoms as *ortho, meta, para,* as well as for the convention for carbon numbering (numbers inside rings).

when microbes are presented with hydrocarbon mixtures. In the case of oil spills, hydrocarbon classes are preferentially degraded in the order (first to last) *n*-alkanes ("normal" alkanes; lacking branches), branched alkanes, polyaromatic compounds, and lastly NSO compounds. There is also variation within hydrocarbon classes in biodegradation potential. For example, biodegradation rates of *n*-alkanes decrease with increasing length of the carbon chain, which is in part attributable to the progressive decrease in aqueous solubility that accompanies increasing chain length. Thus, the composition of hydrocarbon mixtures changes over time such that they are depleted in those fractions most susceptible to biodegradation and enriched in the compounds that are relatively resistant to microbial decomposition. Physical-chemical processes (e.g., volatilization, photooxidation, sorption) also contribute to the changes in the hydrocarbon mixture, and the resulting aged material is referred to as "weathered residual."

BOX 21.1

Examples of oxygenase transformations of hydrocarbons

There are two general types of oxygenases–monooxygenases (top portion of figure found below), which incorporate one oxygen atom from O_2 into a substrate, and dioxygenases (bottom), which incorporate both oxygen atoms of O_2 into the substrate. Monooxygenases act on both aliphatic and aromatic structures. Dioxygenases act only on aromatic compounds, and different types of dioxygenases carry out two different types of reactions–ring hydroxylation and ring cleavage. The ring-hydroxylating dioxygenases (RHDs) are a large family of enzymes, and each subfamily is associated with transformation of a particular class of aromatic compounds (de Gannes and Hickey., 2017). Thus, the genes that encode RHDs are frequently targeted in molecular approaches to assess composition of microbial communities at contaminated sites (de Gannes and Hickey., 2017). The ring cleavage reaction gives an example of the variety existing in biodegradation pathways, in that a single compound (catechol) can be metabolized through two routes—the "*ortho*" and "*meta*" pathways—to arrive at the same place (TCA cycle). The color-coding of bonds in catechol indicates the targets of dioxygenase activity in *ortho* versus *meta* cleavage.

(*Continued*)

BOX 21.1 (Cont.)

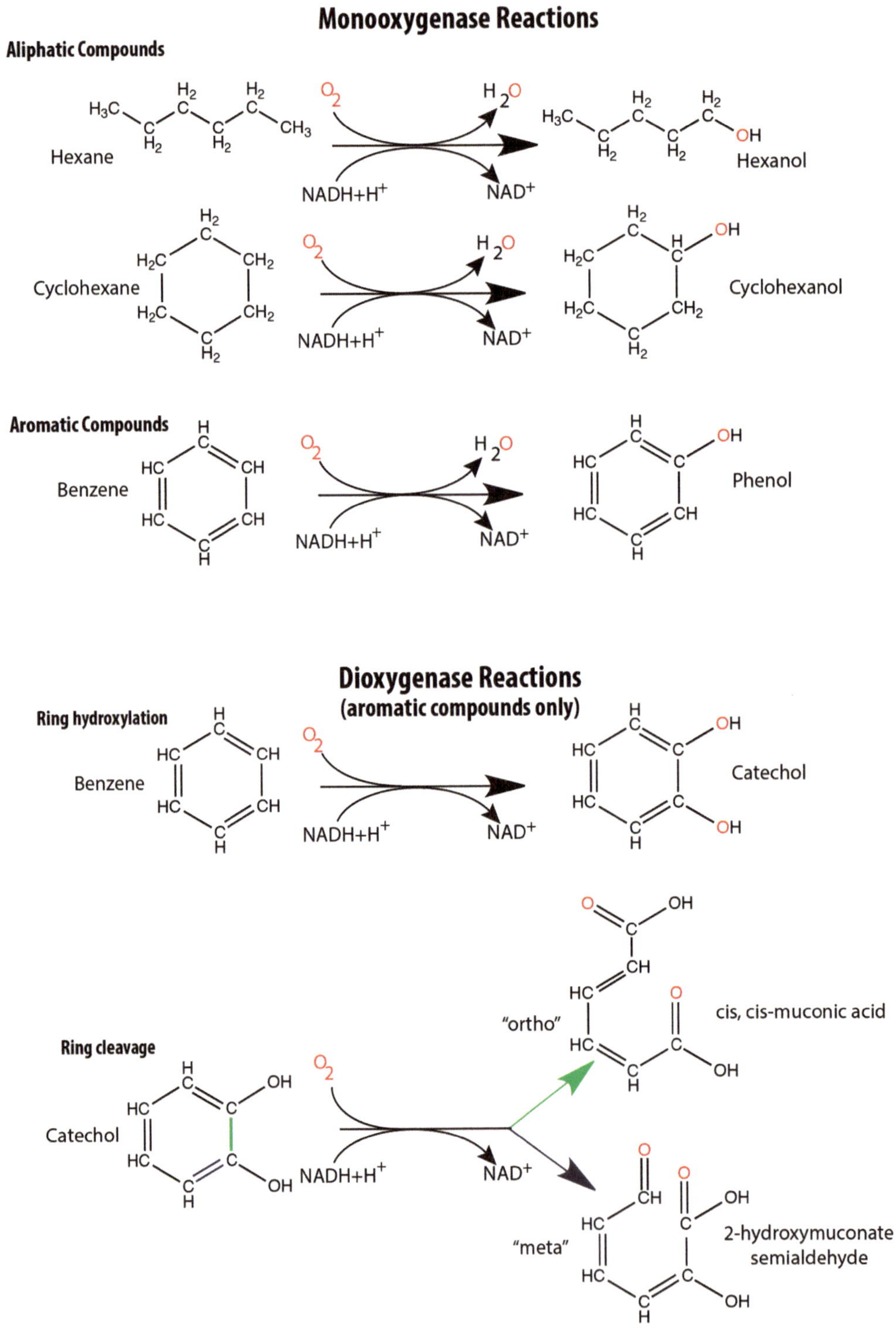
Monooxygenase Reactions
Aliphatic Compounds
Hexane
O2
H2O
NADH+H+
NAD+
Hexanol
Cyclohexane
Cyclohexanol
Aromatic Compounds
Benzene
Phenol
Dioxygenase Reactions
(aromatic compounds only)
Ring hydroxylation
Benzene
Catechol
Ring cleavage
Catechol
"ortho"
cis, cis-muconic acid
"meta"
2-hydroxymuconate semialdehyde

BOX 21.2

Illustration of the major classes of compounds in hydrocarbon pollutants

Some common aliphatic and aromatic pollutants are shown. For the aliphatic and monoaromatic hydrocarbon pollutants, a general range in molecule size for each class is given by the number of constituent carbon atoms. Additionally, in the case of the polycyclic aromatics, the range in the number of fused rings is indicated. The four examples of monoaromatics given represent the fraction of fuels (e.g., gasoline) that are generally the main targets of bioremediation. These are benzene, toluene, ethylbenzene, and xylene (BTEX).

Aliphatic Compounds

Chains (5 to 18 Carbons)

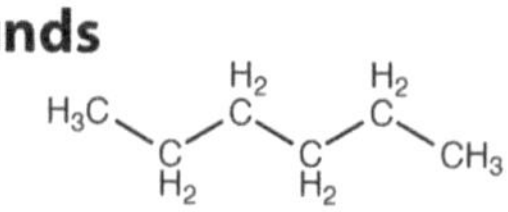

Hexane

Rings (3 to 6 Carbons)

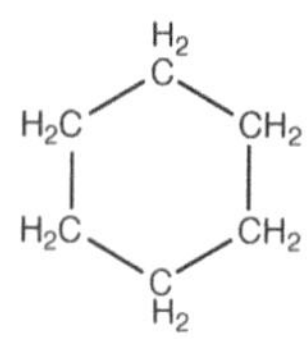

Cyclohexane

Aromatic Compounds

Monoaromatics (6 to 8 Carbons)

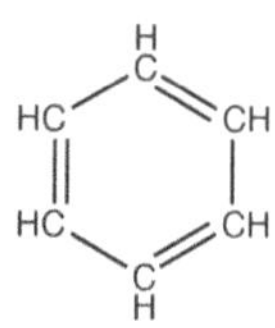

Benzene

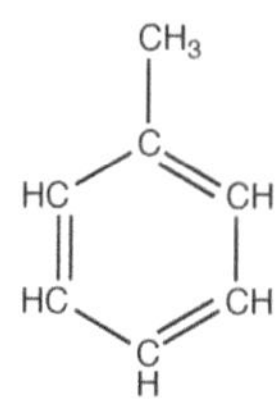

Toluene

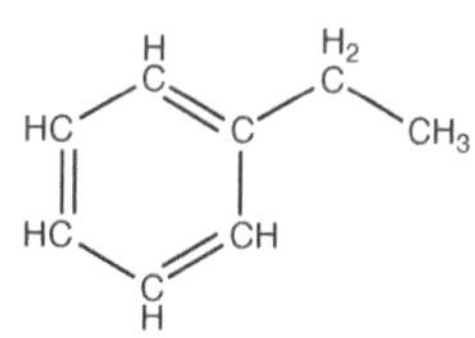

Ethylbenzene

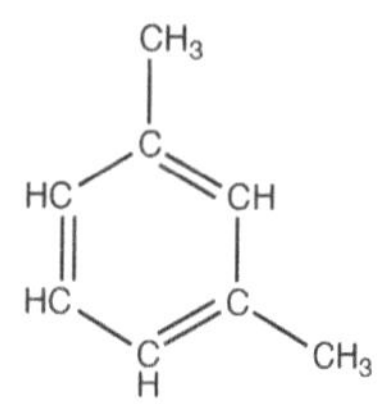

Xylene

Polycyclic aromatics (10 to 22 Carbons, 2 to 5 fused rings)

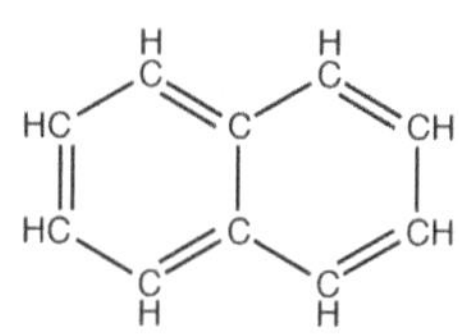

Naphthalene

In contrast to most hydrocarbons, **xenobiotic compounds** (from *xeno* = "foreign", *bios* = "life") constitute a large class of chemicals created by synthetic chemical processes and are not natural components of the biosphere. Xenobiotic compounds can be composed of a variety of structures, but one of the most important is the presence of multiple halogen atoms. Because these structures lack close analogs in the biosphere, biodegradation mechanisms are limited or lacking and the compounds can persist in the environment. The most problematic of these compounds were internationally recognized (Stockholm Convention, 2001) and designated "**Persistent Organic Pollutants**" (POPs). These compounds pose significant environmental health hazards for humans and wildlife as their persistence can lead to **bioaccumulation** in fatty tissues and **biomagnification** in food chains. It is important to note that not all xenobiotics display the characteristics of POPs, and many xenobiotics can be degraded more quickly in the environment than are some natural organics. Xenobiotics that are relatively susceptible to biodegradation have chemical structures that have analogs in the biosphere and lack some features that tend to impede biodegradation, such as branching.

Biodegradation of xenobiotics (as well as natural compounds) is also facilitated by the association of genes encoding key catabolic functions with mobile genetic elements (MGEs), which serve as vectors for gene transfer between bacteria. There is a wide variety of MGEs (e.g., plasmids, transposons; Chapter 4), and genome sequencing has revealed that the association of biodegradation genes with MGEs is more common than not. Because MGEs can transfer biodegradation genes both into and out of bacterial genomes, biodegradation genotypes are fluid and vary at the sub-species (strain) level. For example, one strain of a particular bacterial species may possess a given set of biodegradation genes, which are absent in a second strain of the same species. Conversely, because some MGEs cross boundaries of genera and species, the same biodegradation genes can be possessed by different bacterial genera or species. Thus, biodegradation capabilities do not track with specific bacterial taxa. Association with MGEs is likely a key factor that has contributed to the broad distribution of biodegradation genotypes throughout microbial communities and is likely important in contributing to their ability to respond to the introduction of pollutants into the environment.

Anaerobic biodegradation

Biodegradation of compounds that are used as carbon and energy sources aerobically can also occur in anaerobic environments with these processes coupled to nitrate, ferric iron, or sulfite/sulfate as terminal electron acceptors. A key difference is that the anaerobic biodegradation pathways cannot make use of oxygenases, which are key to effective degradation of certain classes of chemicals (e.g., hydrocarbons). Thus, anaerobic biodegradation pathways of these chemicals utilize a different set of enzymatic transformations that are generally less effective (slower) than those used by aerobes.

In contrast, organohalide compounds, especially those containing multiple halogen atoms, are poorly degraded by aerobes but are excellent substrates for anaerobes capable of organohalide respiration. There are two types of **organohalide respiring bacteria** (OHRB)—those which rely solely on this process for growth (obligate OHRB), and those which may also grow anaerobically by using other natural electron acceptors, such as sulfur oxides (facultative OHRB). Obligate and facultative OHRB differ in the range of compounds that may be used for electron donors—obligate OHRB are limited to hydrogen (H_2), whereas facultative OHRB can utilize H_2

as well as a broad range of other donors such as alcohols and organic acids (e.g., acetate, lactate). Obligate OHRB belong to the phyla Chloroflexi (genera *Dehalococcoides* and *Dehalogenimonas*) and Firmicutes (genus *Dehalobacter*), while facultative OHRB occur in the phyla Proteobacteria (genera *Sulfurospirillum* and *Geobacter*) and Firmicutes (genus *Desulfitobacterium*).

For both OHRB groups, key enzymes are reductive dehalogenases (RDases) that mediate dehalogenation. Genomes of obligate OHRB harbor 17–38 putative reductive dehalogenase genes, while facultative OHRB possess only one to five such genes (Türkowsky et al., 2018). While there is a great diversity of putative dehalogenases, especially in the obligate OHRB, relatively few of these have been characterized (hence, the designation "putative"). The RDases that have been studied are those primarily connected to metabolism of PCE, TCE, or other chloroaliphatics and are distinguished by the types of organohalides on which they act (Fig. 21.3). Strains of *Dehalococcoides* are so far unique in possessing RDases capable of effecting complete dehalogenation of vinyl chloride (VC) to ethene, which is the last step needed to the complete dehalogenation of halogenated alkenes (Figs. 21.2 and 21.3). Complete dehalogenation is important for remediation of halogenated alkenes because the intermediate, VC, is more toxic than the parent compounds. Thus, of all OHRB, the obligate OHRB (*Dehalococcoides*) are of particular importance in the anaerobic bioremediation of haloalkenes.

While the dehalogenation reactions mediated by OHRB are the focus of organohalide biodegradation, the activity of these organisms is dependent on food-web interactions with

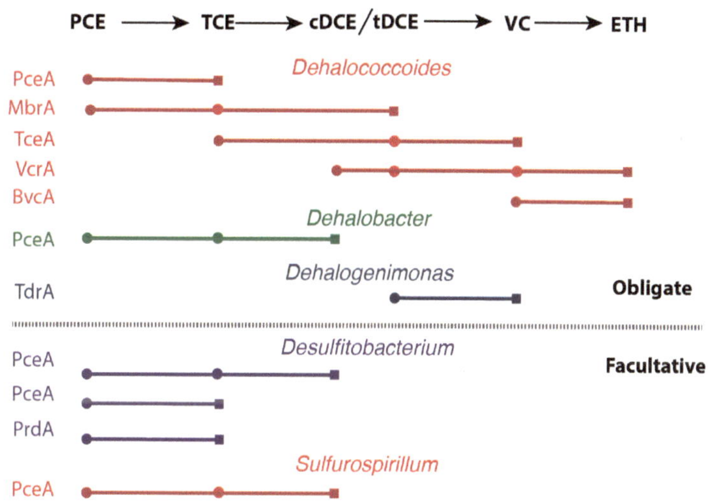

FIGURE 21.3 **Examples of variation in substrate ranges of RDases of genera active in organohalide respiration with chloroalkenes.** *Circles* and *squares* on each line indicate RDase substrates and end products, respectively, and correspond to the compounds in the PCE degradation pathway at the top (see Fig. 21.2 for structures and key to abbreviations). Names of each RDase are listed along the left; the designation "PceA" has been applied to several different RDases characterized in different bacteria. Data are shown for three genera of obligate OHRB (top section) and two genera of facultative OHRB (bottom section). The importance of obligate OHRB in mediating the complete dechlorination of chloroalkenes is demonstrated by activity encoded by the RDases VcrA and BvcA, an ability that is so far unique to obligate OHRB. Variation between RDases in production of *cis*-DCE versus *trans*-DCE is also apparent, which accounts in part for the mixtures of *cis*-DCE and *trans*-DCE that are detected in environments polluted with PCE or TCE. Plots are based on data summarized in Fincker and Spormann (2017).

other bacteria. An essential function of the food web is carrying out fermentation and acetogenesis, which supplies OHRB with electron donors (H_2 and/or acetate) that drive respiration. In applied bioremediation processes utilizing OHRB, a common biostimulation approach is introduction of organic supplements that can be metabolized in a manner that yields H_2 and/or acetate needed by the OHRB. Food–web interactions also provide corronoid cofactors (e.g., vitamin B_{12}), that are required by RDases for their activity, but which obligate OHRB (*Dehalococcoides*) cannot produce.

Ecological factors

The physiological processes described above establish the potential of microbes to biodegrade compounds. But, it is the environment that ultimately controls the type of biodegradation processes that might occur, as well as the rates and extent. The oxygen level of an environment is generally the single most important factor affecting the types of biodegradation that may occur, as its presence or absence demarcates the vastly different worlds of aerobic versus anaerobic processes, which were discussed in the preceding sections. Thus, the following review of ecological controls on biodegradation will focus on factors other than oxygen levels that impact the extent and rates of biodegradation. For this discussion, ecological factors are those that arise from interactions of: pollutants with the environment, microbes with the environment, or microbes with pollutants.

Physical distribution

In the environment, chemicals will be distributed (partitioned) across three phases—liquid, solid, and gas. Of these, occurrence in the liquid (aqueous) phase is of particular importance to biodegradation. For the aerobic processes discussed earlier, all of the enzymatic transformations occur within the cytoplasm. In the case of organohalide respiration, the RDases of OHRB are embedded in the cytoplasmic membrane with active sites facing outward, and dehalogenation reactions thereby occur within the periplasm (Wang et al., 2018; Chapter 5). Thus, substrates for all of these enzymes must enter the cell, and their uptake occurs from the aqueous phase. As such, the extent to which a compound partitions into water is a major determinant of its potential bioavailability to degrader organisms.

A variety of physicochemical characteristics affect the partitioning behavior of a compound, but partitioning can be broadly predicted based on polarity. Compounds that are charged (ionic) are polar and typically have greater aqueous solubility than noncharged (neutral, nonionic) compounds that are nonpolar. Nonpolar compounds have a strong propensity to partition into sorbents in the environment that provide physicochemical environments more compatible with these compounds than is water. The extent to which compounds exist in the aqueous phase versus associated with solid phase sorbents is a fundamental characteristic affecting a compound's behavior. Thus, **partition coefficients** have been developed which are useful to assess movement of a compound between aqueous and organic phases, and one of the most widely used is the octanol-water partition coefficient ($\log K_{ow}$, Box 21.3).

Low bioavailability (strong sorption) of contaminants can be a key factor limiting their bioremediation in soils and sediments. Materials that serve as sorbents are minerals (e.g., expanding clays called "smectites") and soil organic matter (SOM). The tendency of hydrophobic organic compounds to partition into SOM is well established and analogous

BOX 21.3

Partitioning coefficients are descriptors of chemical behavior

The movement of a chemical between different physical or biological compartments is a fundamental process and is essential to an understanding of contaminant degradation. Natural systems are complex, and an organic compound's movement is governed by many factors. Nevertheless, a basic component is the transit between dissolution in water and sorption by organic matrices.

The octanol:water partition coefficient (K_{ow}) is widely used to gauge the behavior of a chemical in systems containing aqueous and organic phases. In this assay, a chemical is added to a vessel containing water (polar solvent) and *n*-octanol, which is a non-polar, organic liquid immiscible with water. After vigorous mixing, the water and *n*-octanol phases are allowed to separate and the concentration of the chemical in each phase measured at equilibrium. These measurements of a test compound "X" are then expressed through the ratio:

$$K_{ow} = [X]_{n-\text{octanol}} / [X]_{\text{water}}$$

The greater the hydrophobicity of a compound, the greater its equilibrium concentration in *n*-octanol. Because values of this ratio range over several orders of magnitude, the ratio is usually expressed in $\log_{10}$ form, log K_{ow}, which typically ranges from −3 ($K_{ow} = 10^{-3}$, highly hydrophilic) to 10 ($K_{ow} = 10^{10}$, very hydrophobic). Examples of logK_{ow} are given in the table below.

In environmental sciences, log K_{ow} is used either directly to assess behavior of organic compounds in the environment or indirectly to derive other partitioning coefficients that are more specific to the process in question. For example, in soils and sediments, sorption by organic matter is generally the dominant physical process affecting the fate of organic chemicals and is described by the organic carbon-water partition coefficient, K_{oc}. However, because direct determination of K_{oc} is more complicated and laborious than is measurement of K_{ow}, K_{oc} is often estimated from regression analyses with K_{ow}. For example (European Commission, 1996):

$$\log K_{oc} = 0.81 \log K_{ow} + 0.1$$

In biological sciences, logK_{ow} is referred to as logP. One of the uses of logP values is to gauge the potential toxicity of organic compounds to microbes. As logP values increase, compounds become progressively more hydrophobic. Consequently, they have increasing propensity to partition into biological structures that offer a hydrophobic environment. A biologically important location where such environments exist is the interior space of the lipid bilayers that constitute cellular membranes. High levels of compounds entering cellular membranes disrupt the bilayer structure (increase membrane fluidity), which causes numerous problems including impairment of membrane-based functions (e.g., respiratory electron transfer) and cellular leakage.

In general, compounds with logP in the range of 1–4 have the greatest potential to exert membrane toxicity. These compounds possess both comparatively strong hydrophobicity and significant water solubility, and a sufficiency of the latter is needed to

(*Continued*)

BOX 21.3 (Cont.)

create a relatively large pool of the compound available for uptake. Compounds with logP > 4 have low water solubility, and thus microbes (and their membranes) are not exposed to levels sufficient to create toxicity. Compounds with logP < 1 are highly soluble in water but weakly hydrophobic.

Example $\log K_{ow}$ (logP) values	
Hydrocarbons[a]	
Benzene	2.13
Tolune	2.73
Ethylbenzene	3.15
p-Xylene	3.15
Naphthalene	3.30
n-Hexane	4.00
Phenanthrene	4.52
Chloroethenes[b]	
Trichloroethene	2.29
Tetrachloroethene	3.40

[a] *Sangster (1989).*
[b] *USEPA (2001).*

to the partitioning behavior of organic compounds into organic liquids (e.g., K_{OW}) (Box 21.3). Thus, sorption strength of a compound is predicted to increase with increasing $\log K_{ow}$ values. While these trends generally hold true, the sorption processes in soils and sediments are physically different and far more complex than partitioning between water and organic liquids.

One of the many complexities of organic compound sorption in soils and sediments is the heterogeneity of SOM, which is composed of two distinct types of sorbent materials (Semple et al., 2013; Ren et al., 2018). One type, termed "soft carbon" (also called "amorphous organic matter," AOM), is derived from the microbial decomposition of organic material deposited in soil or sediment (i.e., humic substances; Chapter 13). The other type, "hard carbon" consists of kerogen and combustion products of organic materials (also called "black carbon"), such as soot and charcoal. The soft- and hard-carbon fractions of SOM differ greatly in physicochemical characteristics, which affect their interactions with chemicals (sorbates).

A key distinction between the soft versus hard carbon types is that soft carbon is believed to take up chemicals by absorption (dissolution of sorbate within the sorbent) while hard carbon does so by adsorption (adherence of sorbate on the sorbent surface). Also, adsorption to hard carbon occurs on particle external surfaces and as well as within nanopores (e.g., diameter $< 2 \times 10^{-3}$ µm) that permeate particles. Desorption of chemicals absorbed in soft carbon can be relatively fast; thus, chemicals absorbed by soft carbon can quickly re-enter the pool bioavailable to degraders. In contrast, desorption of chemicals from hard carbon nanopores is very slow, such that adsorption in these sites is regarded as effectively irreversible. Thus, the extent to which chemicals are rendered non-bioavailable is in many cases considered to be controlled by the extent of their adsorption to hard carbon.

Population density and growth of degraders

Along with the bioavailability of a compound, the population density of microbes capable of its degradation has a critical impact on the rate and level of decomposition. In general, native population levels (i.e., in non-contaminated environments) of pollutant degraders are low. In these cases, the introduction of the pollutant compound, such as oil, creates a niche (a unique carbon and energy source) that oil-degraders can exploit to grow. It is only through population growth and establishment of a relatively high density of degraders that biodegradation significantly reduces contaminant levels. The relation of population expansion to reduction of contaminant levels is illustrated in Box 21.4, which gives an example typical of

BOX 21.4

Relationship among population densities, carbon demand, and apparent changes in substrate level

Panel A. This example assumes a starting population density of bacteria that utilize the compound as a carbon source of 10^4 cells g^{-1} soil. Each cell is assumed to be composed of 30 fg (femtogram) carbon (Fukuda et al., 1998), and the efficiency of carbon incorporation into cells is assumed to be 60% (60 g cellular C/100 g substrate C). Four different masses of C are present initially; the corresponding total masses of the substrate would be twice the values presented, assuming the substrate is 50% C. In each case, several population doublings occur before substrate degradation becomes apparent. If this were an actual environmental analysis, a measurement error of roughly 10% would also be realistic and is indicated by the *grey* box. Thus, clear indications of substrate degradation would not be apparent until the degraders establish a threshold population density indicated by the symbol plotted in white for each concentration. Once the threshold population is established, subsequent doublings exert a carbon demand that consumes a significant fraction of the remaining substrate carbon.

Panel B. Threshold population densities (*white symbols*, panel A) plotted against the relevant substrate level. The different symbol shapes plotted correspond to the different initial substrate concentrations in Panel A. The threshold population density decreases with decreasing substrate level because smaller amounts of substrate degradation are needed to reach the threshold level. In the example illustrated in Panel A, the substrate needs to be reduced by 10% to reach the threshold level. For a substrate initially present at 500 μg, this requires removal of 50 μg, while a substrate initially present at 10 μg reaches the threshold after degradation of only 1 μg.

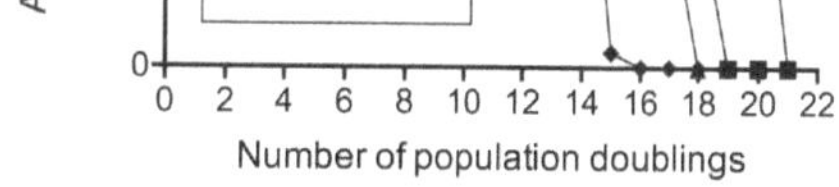

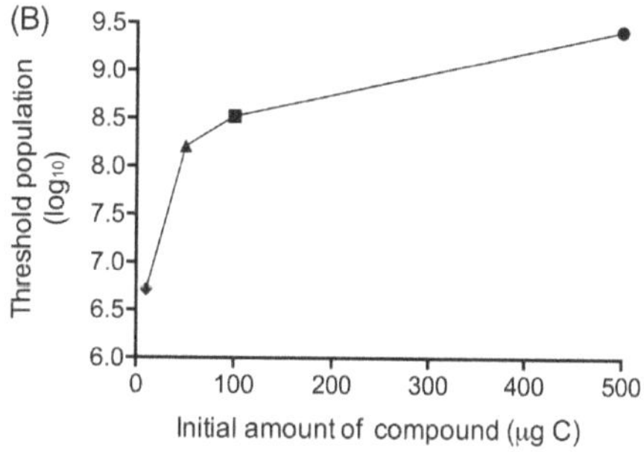

aerobic biodegradation wherein the compound is used as a carbon and energy source. This example illustrates that, although microbial degradation of the compound is constantly ongoing, there is initially no apparent change in the amount of chemical. Although the number of degraders is increasing during this "lag" period, the pollutant degrader population remains relatively low and total carbon consumption is small. It is only after the degraders attain a critical amount of biomass (the "threshold population") that sufficient carbon demand exists to cause significant degradation of the compound.

In order for a microbe to utilize a compound as a carbon and energy source, supporting nutrients are needed at levels that enable production of new cells and population expansion. As noted in Chapter 3, carbon constitutes about half of the dry mass of microbial cells and, consequently, is the nutrient that must be obtained in the greatest amount. After carbon, nitrogen (N) and phosphorus (P) are the two elements needed in the largest amounts. The amount of N and P that cells contain is often expressed as a molar ratio relative to carbon—the ratios C/N and C/P, respectively. The magnitude of these ratios vary between types of microbes (e.g., bacteria versus fungi) and, for any given type of microbe, the ratios can differ in response to a variety of biological and environmental conditions. However, as a broad guidance, commonly used values for soil microbial biomass are C/N = 8 and C/P = 60 (Cleveland and Liptzin, 2007). These ratios mean one unit of N is needed for every 8 units of carbon that are assimilated into biomass, and one unit of P is needed for every 60 units of carbon assimilated.

Using the data introduced in Box 21.4, we can explore the relationship between C utilization and N demand. The total amount of biomass generated from the hypothetical pollutant range from 350 to 7 μg biomass-C (Table 21.1). Using a C/N = 8, the total N demand ranges from 43.75 to 0.88 μg N, respectively (Table 21.1). The levels of mineral N in soils vary widely according to soil type, climate, time of year, and other factors. Thus, whether or not native levels of mineral N are sufficient to support biodegradation of a given amount of compound must be assessed for each site. For example, oil spills introduce large amounts of carbon, and native levels of N (and P) may be insufficient to support extensive biodegradation. In such cases, natural rates of replenishment for these nutrients may be deemed insufficient to sustain biodegradation at rates needed to attain remediation within a desired time period. Materials delivering N and P may then be added to the affected site in an effort to overcome these rate-limiting factors; this practice is a type of bioremediation termed "biostimulation" (Chapter 22).

TABLE 21.1 Relationships between amount of carbon supplied to microorganisms, the resulting amount of carbon in microbial biomass, and the corresponding nitrogen demand, assuming a biomass C/N = 8.

Initial compound-C supplied	Resulting biomass-C[a]	Corresponding biomass-N demand
	μg	
500	350	43.75
100	70	8.75
50	35	4.38
10	7	0.88

[a] *Assumes a 70% conversion of compound-C to biomass-C.*

While carbon is the main element that microbes extract from most organic compounds to support growth, some chemicals may also be utilized to provide other nutrient elements in addition to, or instead of, carbon. The pesticide atrazine is an example of a xenobiotic compound that can support growth by serving as a C source, a combined C and N source, or an N source alone (Topp, 2001; Udiković-Kolić et al., 2012). There are two parts to the atrazine degradation pathway. The "upper" pathway includes the conversion of atrazine to cyanuric acid, which is then decomposed to ammonium and carbon dioxide in the "lower" pathway (Fig. 21.4). Atrazine degraders may grow on this compound by employing both pathways and completely degrade the chemical, or only the upper pathway that results in cyanuric acid as the product.

Bacteria that utilize atrazine solely as an N source process the compound through both the upper and lower pathways but assimilate only the N contained in the *s*-triazine ring. These organisms are unable to catabolize the alkylamine side chains released in the upper pathway and thereby are unable to access the nitrogen contained in these structures. An important consequence of this metabolic approach is that exogenous N sources (e.g., ammonium, nitrate, urea) inhibit degradation. Inhibition occurs because production of the lower pathway enzymes is negatively controlled by N availability, meaning exogenous N sources repress the production of the enzymes needed for cyanuric acid metabolism (Govantes et al., 2009). This inhibitory effect of exogenous N on cyanuric acid metabolism is mediated by networks that regulate cellular N metabolism, and the mechanism by which the lower pathway genes are controlled is similar to that operative with the bacterial *nif* genes that encode nitrogenase (Chapter 16).

Bacteria that use atrazine as both a C and N source can do so in one of two ways. One approach is to assimilate C and N that is contained in the alkylamine side chains alone

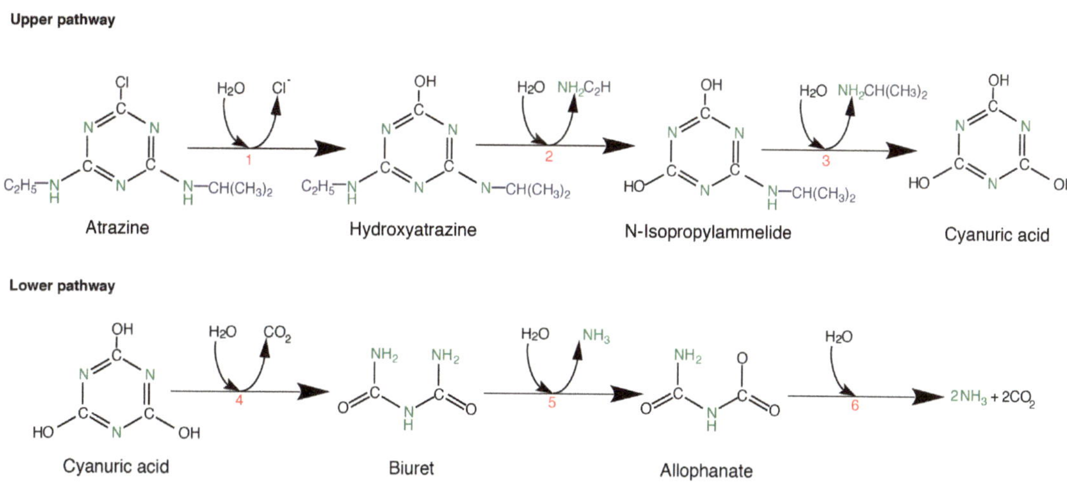

FIGURE 21.4 **Atrazine biodegradation illustrating the "upper" and "lower" pathways.** Parts of atrazine that can be used in carbon nutrition are colored *blue*, those useful in nitrogen nutrition are *green*, and parts colored *black* have no nutritional value. Note that the carbon atoms of the *s*-triazine ring are *black* to indicate they are already at the maximum oxidation state for carbon (+4) and not biologically useful as a carbon or energy source; these carbon atoms are ultimately released as carbon dioxide. The numbers beneath each of the transformations refer to the following: *1*. Dechlorination via a hydrolytic dehalogenease; *2*. Hydrolytic release of ethylamine side chain; *3*. Hydrolytic release of isopropylamine side chain; *4–6*, Hydrolase-mediated cleavage of *s*-triazine ring.

(Fig. 21.4). In this case, atrazine is processed only through the upper pathway and cynauric acid is released as a byproduct. In the other strategy, atrazine is metabolized through both the upper and lower pathways; for these organisms, N may be assimilated from the s-triazine ring and/or the alkylamine side chains (i.e., ethylamine, isopropylamine). A key characteristic common to both approaches is that N can be obtained from catabolism of the alkylamine side chain, a process which is not impacted by the N regulatory network that controls the lower pathway. Thus, availability of exogenous N does not inhibit atrazine degradation by these organisms as it does for bacteria utilizing atrazine as a sole N source alone.

Another group of xenobiotics that may be degraded as a source of nutrients other than, or in addition to, carbon are organophosphorus compounds (also called organophosphates). This is a large class of xenobiotic compounds that, as the name indicates, all contain organic structures attached to a central phosphate group and may also contain nitrogen and/or sulfur. The most widely used organophosphate is glyphosate (Box 21.5), which can be degraded

BOX 21.5

Glyphosate: From bacterial metabolism to genetically engineered crops

Glyphosate is more commonly known by the trade name of the widely used commercial herbicide "Roundup", in which it is the active ingredient. As a pesticide, glyphosate works by blocking production of aromatic amino acids in bacteria and plants by inhibiting an enzyme named 5-enolpyruvylshikimate-3-phosphate synthase (EPSPS). This activity is highly effective against weeds that impact production of crops such as corn, soybeans, cotton, and many others. However, because the activity is nonspecific and affects all susceptible organisms, its agricultural use was limited to periods either before crop growth began or after growth was finished. Thus, it could not be used for weed control when it would be most useful, that is after crops had emerged (post-emergence). Glyphosate-resistant bacteria were known, and the physiological basis of this trait was traced not to enhanced biodegradation of the compound but rather to possession of EPSPS variants that were insensitive to glyphosate. These variants were termed class II EPSPS, which maintained high levels of activity in the presence of glyphosate. A glyphosate-resistant soil bacterium, *Agrobacterium* sp. strain CP-4, was identified as possessing a class II EPSPS, and the gene responsible for resistance was genetically cloned and introduced into soybeans. The resulting transgenic plants produced the bacterial class II EPSPS and were not harmed by exposure to glyphosate (Padgette et al., 1995). Thus, "Roundup Ready" soybeans were born, which enabled use of glyphosate for post-emergence weed control, thereby greatly enhancing its agricultural utility. The line of "Roundup Ready" plants was subsequently expanded to include alfalfa, canola, corn, cotton, sugar beets, and wheat. Cultivation of these genetically engineered resistant crops (GERCs) accounts for 56% of the global glyphosate usage (Givens et al., 2009). Moreover, glyphosate is now considered the most widely applied agrochemical in human history (Gilbert, 2013), a status largely attributable to the widespread adoption of GERCs.

by bacteria using it as a carbon source (Nourouzi et al. 2011) and/or as a sole source of N or P (Singh and Walker, 2006).

Glyphosate biodegradation may occur via two pathways (Fig. 21.5) named for the first intermediate formed, either sarcosine or AMPA (aminophosphonic acid). A key reaction in the biodegradation of glyphosate (and all organophosphates) is the breaking of the

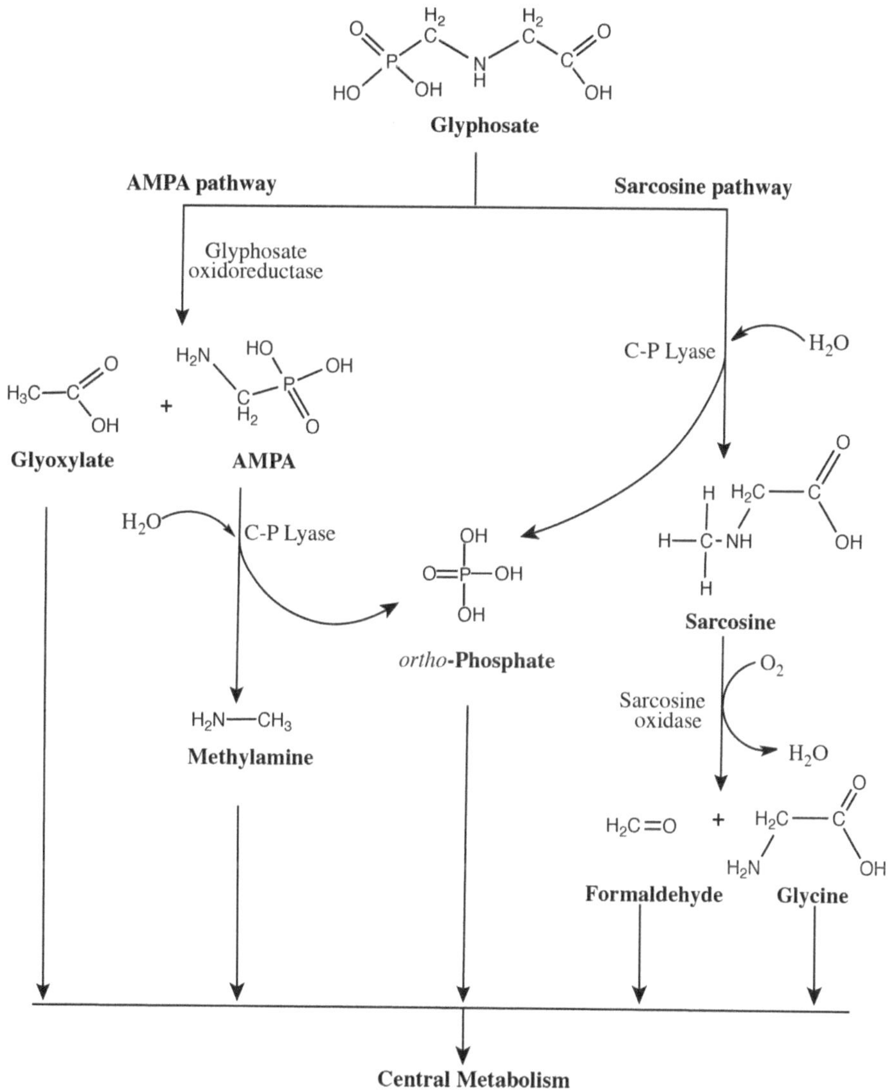

FIGURE 21.5 **Biodegradation of glyphosate via the sarcosine (right) and AMPA (left) pathways illustrating the central role of C-P lyase.** As C-P lyase production is expressed only when the cell is experiencing phosphorus starvation, this activity will frequently be absent. Without C-P lyase, the initial step of the sarcosine pathway cannot occur and degradation is thus directed along the AMPA pathway. The enzyme initiating the AMPA pathway, glyphosate oxidoreductase, can function under either aerobic or anaerobic conditions (Borggaard and Gimsing, 2008). However, catabolism along the AMPA pathway can also become limited by C-P lyase, as this enzyme is needed to breakdown AMPA. When C-P lyase is expressed, either the AMPA or sarcosine pathways may be functional, and the simultaneous operation of both pathways within a given bacterium or bacterial community can occur (Wang et al., 2016).

carbon-phosphorus bond (C-P) via an enzyme called a C-P lyase (Fig. 21.5), which eliminates the biological activity (toxicity) of organophosphates. This reaction may occur either directly on glyphosate or on the intermediate, AMPA (Fig. 21.5). The C-P lyase is an example of a large class of enzymes called **hydrolases** (Table 3.2), which utilize water as a reactant in their catalysis in a manner analogous to the way that oxygenases use O_2 (see earlier section).

Bacterial production of C-P lyase is controlled by the Pho regulon, which is a global network that monitors cellular P status and controls expression of genes connected to phosphate metabolism (McGrath et al., 2013). Thus, genes encoding the C-P lyase active in glyphosate degradation are expressed only under conditions of P starvation. Cells typically experience limitations of C and N before encountering P shortages. Consequently, glyphosate biodegradation occurs predominantly through the AMPA pathway. But, because a C-P lyase catalyzes AMPA decomposition, degradation stalls at this point and AMPA tends to persist in the environment. In this case, the partial biodegradation of glyphosate is driven by acquisition of the carbon available in glyoxylate (Fig. 21.5). Thus, the degradation of glyphosate and atrazine is similar in that, while each has the potential for catabolism to be driven by a nutritional need other than carbon, the systems needed to do so are tightly controlled by networks that monitor the cell's status for each of these elements.

Pollutant concentration

Biodegradation rates typically vary as a function of a compound's concentration. Rates increase with increasing levels of a contaminant, reach a maximum, and then decline with further increases in concentration. Declining growth levels with increasing concentrations reflect toxicity, most commonly resulting from disruption of cellular membranes. This effect is termed "membrane toxicity" (Sikkema et al., 1994) and leads to disruption of membrane-based functions (e.g., respiratory electron transfer) and cellular leakage.

Membrane toxicity is particularly well documented for hydrocarbons, many of which can be either very good in supporting microbial growth or highly toxic depending upon their concentration. Hydrocarbons tend to partition out of water and into membranes because the internal space of their lipid bilayers is nonpolar, as are hydrocarbons (Box 21.3). Not all hydrocarbons have the same potential to exert membrane toxicity; those with the greatest potential to do so exhibit a balance of fairly strong hydrophobicity and relatively high aqueous solubility. The extent to which a compound exhibits this balance of hydrophobicity and aqueous solubility can be gauged by the value of the partition coefficient logP, and the potential for membrane toxicity is greatest for compounds having logP values between 1 and 4 (Box 21.3).

Membrane toxicity can be environmentally relevant in cases where there are highly contaminated "hot spots" of certain pollutants. For example, in cases of groundwater contamination with fuels (gasoline, diesel) or chloroethenes, a separate liquid phase, termed a non-aqueous phase liquid (NAPL), often exists in hot spots. The NAPL is a point source from which its constituent compounds continuously dissolve into the aqueous phase, creating a plume with a concentration gradient of the pollutant. For fuels, the compounds composing the "BTEX" fraction (Box 21.2) are of particular concern, because these present risks to human health (particularly benzene) and have logP values that confer a high potential for membrane toxicity (Box 21.3). Chloroethenes, such as tetrachloroethene and trichloroethene, present similar risks to human health and also have logP values that qualify them as membrane toxicants (Box 21.3).

Thus, for remediation of groundwater contaminated by fuels or chloroethenes, removal of the NAPL source is the first step. For hydrocarbons, this can be done fairly easily because the NAPLs have a specific gravity less than that of water (i.e., < 1 g cm^{-3}) and are therefore called "light" NAPL (LNAPL). These float on the groundwater surface and can be removed by a physical remediation process called siphoning. However, chloroethene NAPLs have a specific gravity greater than one ("dense" NAPL, or DNAPL) and thus sink through water, accumulating at the bottom of the aquifer as well as on top of retentive zones (e.g., rocks, clay lenses) within the aquifer. Typically, neither the DNAPL spots dispersed throughout an aquifer, nor the accumulations at its bottom, can be directly removed by a physical remediation technique such as siphoning. Therefore, because DNAPL zones are not effectively removed by either biodegradation (because of membrane toxicity) or by physical processes, these can be sources of groundwater pollution that persist for years or decades.

At low concentrations, a variety of processes may contribute to slow biodegradation rates. One possibility is that the low levels do not effectively induce enzyme systems (pathways) that are essential for effective transformation. Alternatively, the low availability of the compound may not provide sufficient energy to support the growth of degrader populations. Also, there are always low levels of natural organic compounds present (e.g., sugars, organic acids), and preferential degradation of these may impede that of a contaminant compound present at low levels. This type of inhibition (preferential degradation) is mediated by "global" regulatory systems, such as carbon catabolite repression (CCR), that coordinate cellular metabolism in the presence of multiple potential carbon sources (Rojo, 2009). As discussed earlier, cellular regulatory networks control the catabolism of organic pollutants as sources of nitrogen and phosphorus, and CCR regulates the use of pollutant compounds as carbon sources.

Slow biodegradation of low levels of pollutants can be a significant problem for bioremediation because the levels to which contaminant concentrations must be reduced to achieve mandated cleanup levels are generally lower than these threshold concentrations. For example, for the herbicide atrazine, attainment levels (concentrations considered "safe" and which must be reached in clean-up operations) in groundwater are 0.1 $\mu g\ L^{-1}$, but in a well-mixed laboratory system, one atrazine-degrading bacterium studied was only able to reduce atrazine levels to 10 $\mu g\ L^{-1}$ (Kundu et al., 2019). Chlorophenoxyalkanoic acids such as 2,4-dichlorophenoxyacetic acid (2,4-D) and mecoprop (MCPP) are also widely used herbicides and frequent groundwater contaminants. Above threshold levels, these may be quickly degraded in days to weeks, but below threshold concentrations their biodegradation slows dramatically (Box 21.6). Thus, at sub-threshold concentrations, compounds such as 2,4-D and MCPP can persist for months to years at levels that may be considered to pose human health risks.

Pollutant interactions

In addition to the effects of natural organic matter on pollutant biodegradation, interactions may also occur between contaminant compounds that affect biodegradation processes. These interactions may have either positive or negative effects on the degradation of a particular compound. Interactions more commonly occur between compounds that are potential substrates for the same enzymes. Thus, these substrate interactions are most common within families of chemicals, that is, compounds that are structurally related.

One of the most common types of substrate interactions is termed **cometabolism**, which concerns interactions between a compound that supports growth and one or more other

BOX 21.6

Effect of low substrate levels on biodegradation rates

Biodegradation rates typically decrease as the concentrations of chemicals drop below threshold levels. Data in the table below are from a study examining the biodegradation of 2,4-D and MCPP ("A" in figure) in groundwater microcosms. The rate of biodegradation is expressed as a half-life, which, as the name indicates, is the time period during which half of the compound degrades. Thus, in the first half-life period, a compound decreases in concentration from 100% to 50%. The second half-life period thus begins with a total mass of the compound that is 50% of the original amount and concludes with 25%. As degradation continues, the concentration of the chemical asymptotically approaches zero ("B" in figure).

In the tabular data presented below, biodegradation of both compounds was rapid at levels ≥25 μg L^{-1}. At 100 μg L^{-1}, 2,4-D biodegradation is particularly rapid with a half-life of 0.09 d; 50% of this compound was degraded in just over 2 hours. At sub-threshold concentrations, slow biodegradation rates occur in part because the amount of compound present is insufficient to sustain growth of the degrader population.

It is also noteworthy that, although 2,4-D and MCPP are both chlorophenoxyalkanoic acids and structurally very similar, MCPP biodegradation is typically slower than that for 2,4-D (Paszko et al., 2016). Thus, what may appear to be minor structural variations between compounds in the same chemical family can have significant impacts on biological activity.

Biodegradation rates of 2,4-D and MCPP as a function of concentration[a]

Concentration (μg L^{-1})[b]	2,4-D	MCPP
	Half-life (d)	
100	0.09	2.73
25	2.33	7.74
1	81.5	213

[a] Calculated from data presented in Torang et al. (2003).
[b] μg L^{-1} = ppb (parts per billion).

(A)

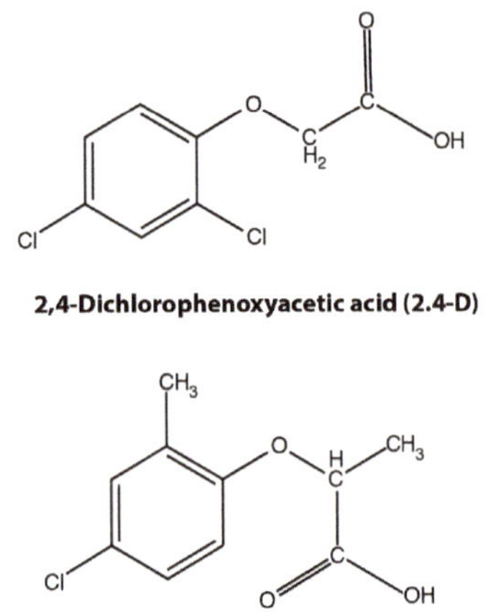

2,4-Dichlorophenoxyacetic acid (2.4-D)

Methylchlorophenoxypropionic acid (MCPP, Mecoprop)

(B)

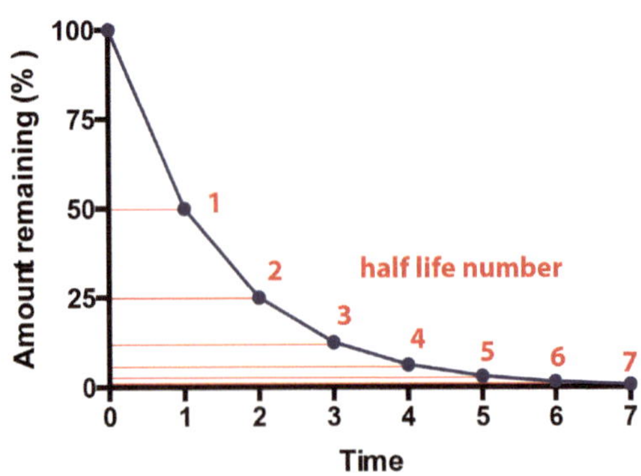

compounds that do not support growth. This process can occur when environmental contaminants are composed of mixtures, such as hydrocarbon pollutants (i.e., oil, fuels, coal-derived materials). In aerobic biodegradation, the oxygenases that are central to hydrocarbon biodegradation can also cometabolize compounds that are not their intended substrates; this characteristic of enzymes is often referred to as "broad spectrum" or "relaxed substrate specificity."

A class of chemicals for which cometabolism is common is polycyclic aromatic hydrocarbons (PAHs), which occur in oil, fuels, and coal-derived materials. The PAH family consists of 86 compounds having benzene fused to other ring structures, either additional benzene rings or pyrrole rings, and range from having two rings to compounds with five rings or more. Bacterial catabolism of PAHs is generally initiated by a ring-hydroxylating dioxygenase (PAH-RHD) that inserts two adjacent hydroxyl groups into the PAH (Fig. 21.6). For PAHs

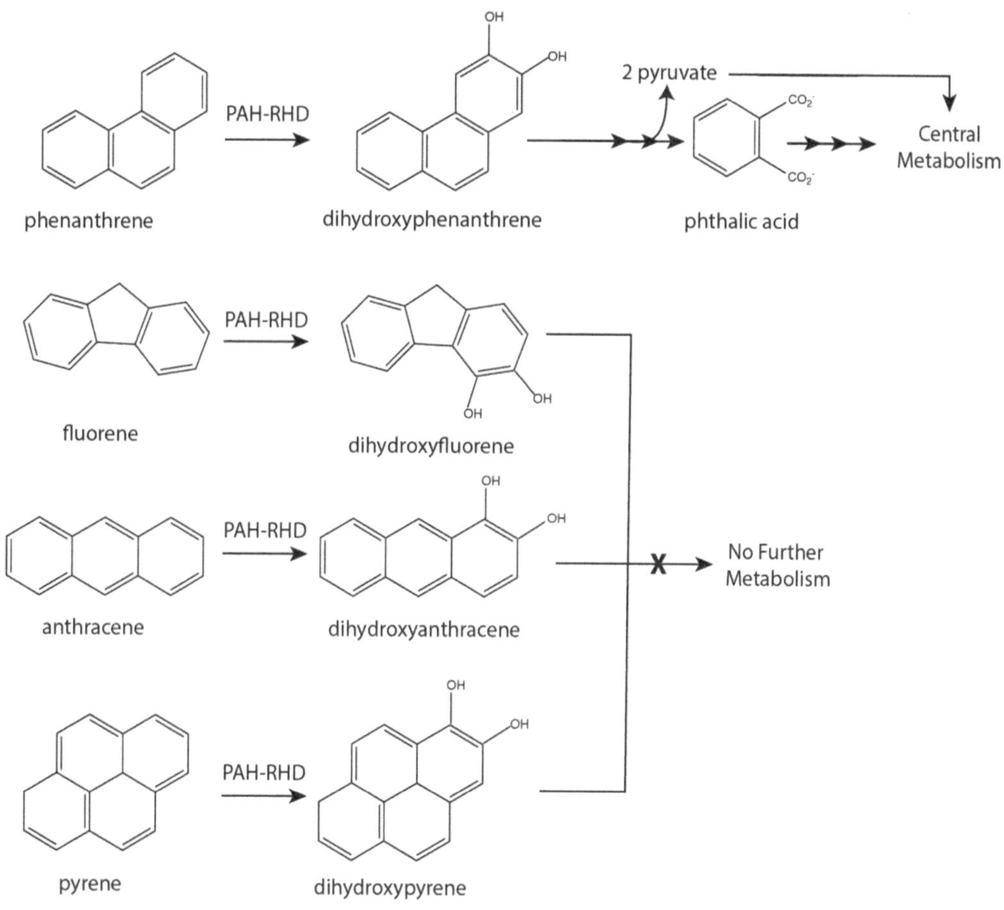

FIGURE 21.6 **Growth and cometabolism in PAH degradation.** In this example, phenanthrene is the only one of the four PAHs shown that is utilized for growth. Because of relaxed substrate specificity, the PAH-RHD that initiates metabolism of phenanthrene can also catalyze hydroxylation of a range of other PAHs. However, the enzymes that further degrade the dihydroxyphenanthrene produced by PAH-RHD activity on phenanthrene, thereby linking it to the growth-supporting machinery in central metabolism, exhibit more stringent substrate specificities and will not transform the PAH-RHD products derived from the other PAHs. For phenanthrene, the series of reactions needed to link dihydroxyphenanthrene to central metabolism are shown in simplified form. For PAHs that are not growth substrates (i.e., that are cometabolized), the dihydroxylated PAHs are dead-end products released from the cell.

that serve as growth substrates, subsequent reactions result in ring cleavage and the generation of metabolites that can enter central metabolism. The PAH-RHDs characteristically exhibit relaxed substrate specificities and will act on a range of PAHs, some of which may not undergo further catabolism and are thereby "dead-end" metabolites for that particular organism. Moreover, because the same enzyme acts on both the growth supporting PAHs and the cometabolized PAHs, there is competition between the two substrates that may reduce catabolism of the former and thereby reduce growth.

Although substrate competition during cometabolism may cause some negative effects at the individual level, the net effect of cometabolism at the community level is generally positive because it enhances degradation of pollutant mixtures, that is, the limitations of any one organism can be offset by complementary activities of other community members.

In anaerobic biodegradation, a key type of substrate interaction occurs in organohalide respiration of chloroalkenes. As discussed earlier, this process involves stepwise removal of halogens (Fig. 21.2). Thus, a system originally containing only PCE evolves into a mixture of the parent compound and the dechlorination products TCE, *cis*-DCE/*trans*-DCE, and VC (Fig. 21.7). The more highly chlorinated compounds can inhibit transformation of those with fewer chlorines and result in the accumulation of less chlorinated metabolites. Moreover, sites contaminated with chloroalkenes are also often polluted with chloroalkanes (e.g., 1,1,1-trichloroethane; 1,1,1-TCA) and/or chloromethanes (e.g., trichloromethane; chloroform, CF), and both 1,1,1-TCA and CF can inhibit dechlorination of VC. The metabolic

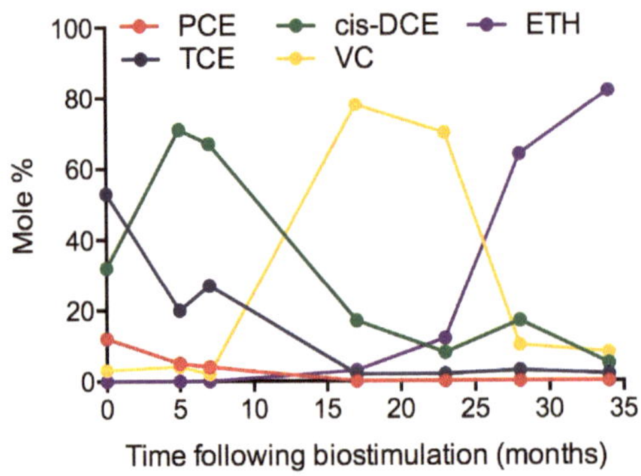

FIGURE 21.7 Variation in chloroethene and ethene concentrations in contaminated groundwater undergoing biostimulation. The plot begins at "time 0" when biostimulation started with injection of a vegetable oil emulsion, a substrate that supports anaerobic formation of acetate and H_2 (electron donors). The amounts of each chloroethene and ethene are presented as "mole %", which is calculated by converting the measured concentration of each compound (e.g., μg/L) to moles, summing the molar amounts of all compounds, and then expressing the molar amount of each compound as a percentage of the total moles. Using the mole % approach facilitates monitoring of organohalide respiration as, for example, one mole of PCE is transformed to one mole of TCE. Five months after biostimulation began, enhanced activity was documented by the increase in *cis*-DCE levels. Over the next 10 months, levels of TCE and *cis*-DCE decreased with a concomitant accumulation in VC, which remained high for several months before significant conversion of VC to ethene began. As noted in the text, a number of factors can negatively affect the transformation rates of *cis*-DCE and VC during organohalide respiration, which could in part account for the transient accumulations of these compounds during the 3 years of observation. See Fig. 21.2 for a legend to abbreviated compound names. The plots are based on data presented in Leeson et al. (2004) for bioremediation applied at Travis Air Force Base, CA.

interactions in these environments are complex, and inhibition can occur at various levels, including for a specific enzyme (RDase), for individual cells (alteration in metabolism), and/or for the larger community (change in type or level of interaction between cells). Thus, deciphering the underlying mechanism resulting in accumulation of intermediates (stalling) in actual contaminated environments is often difficult. However, for bioremediation of chloroalkenes like PCE, dissipating the dechlorination products is essential because they can have greater toxicity (e.g., VC) than the parent compound. Driving the dechlorination process to completion can be approached by biostimulation to enhance overall electron donor supply and/or by bioaugmentation with obligate OHRB cultures that possess RDases particularly effective in dehalogenation of DCE and VC (i.e., *Dehalococcoides* that possess VcrA or BvrA, Fig. 21.3).

Summary

Microbes have tremendous capacity to degrade pollutants, an activity that intersects with their natural roles as catalysts of biogeochemical cycles. Decades of study have yielded general principles governing biodegradation and provided insights into the mechanisms underlying some of these processes. That knowledge has proven essential for guiding the application of microbial biodegradation systems for environmental bioremediation, and it is clear that the biological and environmental dimensions are equally important in determining the outcome of bioremediation. While there are still many unanswered questions about the biological mechanisms and environmental controls involved, probing the latter is a far greater challenge because technologies do not yet exist with capabilities that enable researchers to directly address key questions. Perhaps most important among these issues is determining the exact physical and chemical nature of sorbents that impact bioavailability.

Study questions

1. Explain the roles of pollutants in supporting biodegradation by serving as electron donors or electron acceptors. What characteristics of a compound make it suitable for use in each of these roles? Give examples. What characteristics of the environment may dictate if the example compounds can serve in the proposed roles?
2. The concentration of a pollutant may affect its rate of biodegradation. Explain how both a low level and high level of a compound may result in its persistence.
3. For which types of compounds are oxygenases effective catalysts in biodegradation? For which types of compounds are oxygenases relatively ineffective?
4. Explain how the potential for dechlorination of haloalkenes is affected by the number of halogen atoms present on the compound.

References

Borggaard, O.K., Gimsing, A.L., 2008. Fate of glyphosate in soil and the possibility of leaching to ground and surface waters: A review. Pest. Manag. Sci. 64, 441–456.

Cleveland, C.C., Liptzin, D., 2007. C:N:P stoichiometry in soil: Is there a "Redfield ratio" for the microbial biomass? Biogeochem. 85, 235–252.

de Gannes, V., Hickey, W.J., 2017. Genetic adaptations of bacteria for metabolism of polycyclic aromatic hydrocarbons. In: Cravo-Laureau, C., Cagon, C., Lauga, B., Duran, R. (Eds.), Microbial Ecotoxicology. Springer International Publishing, Cham, Switzerland, pp. 133–164.

European Commission, 1996. Technical guidance document in support of commission directive 93/67/EEC on risk assessment for new notified substances and commission regulation (EC) No. 1488/94 on risk assessment for existing substances. Office of Publications of the European Communities, Luxembourg.

Fincker, M., Spormann, A.M., 2017. Biochemistry of catabolic reductive dehalogenation. Annu. Rev. Biochem. 86, 357–386.

Fukuda, R., Ogawa, H., Nagata, T., Koike, I., 1998. Direct determination of carbon and nitrogen contents of natural bacterial assemblages in marine environments. Appl. Environ. Microbiol. 64, 3352–3358.

Gilbert, N., 2013. A hard look at GM crops. Nature 497, 24–26.

Givens, W.A., Shaw, D.R., Johnson, W.G., Weller, S.C., Young, B.G., Wilson, R.G., Owen, M.D.K., Jordan, D., 2009. A grower survey of herbicide use patterns in glyphosate resistant cropping systems. Weed Technol. 23, 156–161.

Govantes, F., García-González, V., Porrúa, O., Platero, A.I., Jiménez-Fernández, A., Santero, E., 2009. Regulation of the atrazine-degradative genes in *Pseudomonas* sp. strain ADP. FEMS Microbiol. Lett 310, 1–8.

Kundu, K., Marozava, S., Ehrl, B., Merl-Pham, J., Elsner, M., 2019. Defining lower limits of biodegradation: Atrazine degradation regulated by mass transfer and maintenance demand in *Arthrobacter aurescens* TC1. ISME J. 13, 2236–2251.

Leeson, A., Beevar, E., Henry, B., Fortenberry, J.,Coyle, C., 2004. Principles and practices of enhanced anaerobic bioremediation of chlorinated solvents. Technical Report TR-2250-ENV. Naval Facilities Engineering Command. Port Hueneme, CA.

McGrath, J.W., Chin, J.P., Quinn, J.P., 2013. Organophosphonates revealed: New insights into the microbial metabolism of ancient molecules. Nat. Rev. Microbiol. 11, 412–419.

Nourouzi, M.M., Chuah, T.G., Choong, T.S.Y., Lim, C.J., 2011. Glyphosate utilization as the source of carbon: Isolation and identification of new bacteria. E-J. Chem. 8, 1582–1587.

Padgette, S.R., Kolacz, K.H., Delannay, X., Re, D.B., LaVallee, B.J., Tinius, C.N., Rhodes, W.K., Otero, Y.I., Barry, G.F., Eichholz, D.A., Peshke, V.M., Nida, D.L., Taylor, N.B., Kishore, G.M., 1995. Development, identification and characterization of a glyphosate-tolerant soybean line. Crop Sci. 35, 1451–1461.

Paszko, T., Nski, P.M., Materska, M., Bojanowska, M., Kostecka, M., Jackowska, I., 2016. Adsorption and degradation of phenoxyalkanoic acid herbicides in soils: A review. Environ. Toxicol. Chem. 35, 271–286.

Ren, X., Zeng, G., Tang, L., Wang, J., Wan, J., Liu, Y., Yu, J., Yi, H., Ye, S., Deng, R., 2018. Sorption, transport and biodegradation – An insight into bioavailability of persistent organic pollutants in soil. Sci. Total Environ. (610–611), 1154–1163.

Rojo, F., 2009. Degradation of alkanes by bacteria. Environ. Microbiol. 11, 2477–2490.

Sangster, J., 1989. Octanol-water partition coefficients of simple organic compounds. J. Phys. Chem. Ref. Data 18, 1111–1227.

Semple, K.T., Riding, M.J., McAllister, L.E., Bending, G.D., 2013. Impact of black carbon on the bioaccessibility of organic contaminants in soil. J. Hazard. Mat. 261, 808–816.

Sikkema, J., de Bontt, J.A.M., Poolman, B., 1994. Interactions of cyclic hydrocarbons with biological membranes. J. Biol. Chem. 269, 8022–8028.

Singh, B.K., Walker, A., 2006. Microbial degradation of organophosphorus compounds. FEMS Microbiol. Rev. 30, 428–471.

Topp, E., 2001. A comparison of three atrazine-degrading bacteria for soil bioremediation. Biol. Fertil. Soils 33, 529–534.

Torang, L., Nyholm, N., Albrechtson, H.-J., 2003. Shifts in biodegradation kinetics of the herbicides MCPP and 2,4-D at low concentrations in aerobic aquifer materials. Environ. Sci. Technol. 37, 3095–3103.

Türkowsky, D., Jehmlich, N., Diekert, G., Adrian, L., von Bergen, M., Goris, T., 2018. An integrative overview of genomic, transcriptomic and proteomic analyses in organohalide respiration research. FEMS Microbiol. Ecol. 94, 1–21.

Udiković-Kolić, N., Scott, C., Martin-Laurent, F., 2012. Evolution of atrazine-degrading capabilities in the environment. Appl. Microbiol. Biotechnol. 96, 1175–1189.

USEPA, 2001. Sources, emission and exposure for trichloroethylene (TCE) and related chemicals. EPA/600/R-00/099. National Center for Environmental Assessment, Office of Research and Development. USEPA, Washington, DC.

Wang, S., Qiu, L., Liu, X., Xu, G., Siegert, M., Lu, Q., Juneaue, P., Yu, L., Liang, D., He, Z., Qiu, R., 2018. Electron transport chains in organohalide-respiring bacteria and bioremediation implications. Biotechnol. Adv. 36, 1194–1206.

Wang, S., Seiwert, B., Kästner, M., Miltner, A., Schäffer, A., Reemtsma, T., Yang, Q., Nowak, K.M., 2016. (Bio)degradation of glyphosate in water-sediment microcosms – A stable isotope co-labeling approach. Water Res. 99, 91–100.

CHAPTER

22

Bioremediation of contaminated soils

Robyn A. Barbato, C. Mike Reynolds

US Army Cold Regions Research and Engineering Laboratory, Hanover, NH, United States

Microbes are the best chemists in the world.—***Werner Stumm***

Earlier chapters described the fundamentals of soil microbiology and the soil environment, the nature of soil microorganisms, their interactions with soil, and how these factors influence major cycles in soil. In this chapter, we present strategies to exploit the pathways described in Chapter 21 to remediate contaminated soils. The use of organisms to clean up contaminated environments is termed **bioremediation**. There are several definitions of bioremediation, and many of these explicitly include the concept of using microorganisms to degrade contaminants. The US Environmental Protection Agency has defined bioremediation as "a treatment process that uses naturally occurring microorganisms (yeast, fungi, or bacteria) to break down, or degrade, hazardous substances into less toxic or nontoxic substances." Although the use of bioremediation is an evolving discipline with new approaches being developed continuously, there are central themes to bioremediation that, once understood and applied, will improve the likelihood of success. The growing use of bioremediation is testimony to its efficacy and increased acceptance by the regulatory community, stakeholders, and the public. Importantly, bioremediation is an interdisciplinary endeavor that encompasses many fields, a partial list of which includes soil microbiology, soil chemistry, soil physics, geology, microbial ecology, geo-hydrology, risk assessment and toxicology, statistics, contaminant chemistry, analytical chemistry, engineering, molecular biology, and plant physiology.

The topic of bioremediation is vast. A PubMed search (https://www.ncbi.nlm.nih.gov/pubmed) using the keywords "soil" and "bioremediation" yielded over 14,000 publications from 1998 to mid-2020. Soil bioremediation represents a continuum of efforts from the molecular level to large-scale field applications, and the development of effective bioremediation techniques is continuously improving. Research advances extend from an improved understanding of bioremediation mechanisms at the genetic level to development of techniques to alter groundwater and soil conditions hundreds of feet underground or across large areas of land. Characteristics common to bioremediation are that it:

- addresses a wide range of chemicals
- includes processes driven by many microorganisms that are often acting in concert and sequentially

Principles and Applications of Soil Microbiology. http://dx.doi.org/10.1016/B978-0-12-820202-9.00022-8
Copyright © 2021 Elsevier Inc. All rights reserved.

- is implemented under a variety of soil conditions, in both surface and subsurface soils
- is interdisciplinary

As our knowledge improves, so do the techniques and protocols used for both bioremediation and monitoring its efficacy.

In addition to providing an overview of bioremediation practices, the objectives of this chapter are to describe how an understanding of soil microbiology can be used to develop effective bioremediation systems, show how knowledge of microbial degradation pathways can be used to monitor bioremediation, and describe emerging strategies for assessing bioremediation success. A prerequisite to the successful implementation of a bioremediation strategy is to translate critical conditions identified in small-scale laboratory studies to a field scale, the oft-cited "flask to field" paradigm. Central to any successful bioremediation application involving microbes is the realization that soil conditions, contaminant degradation pathways, hydrology, and microbial ecology must be understood and integrated, particularly under dynamic field conditions.

Sources of contamination

Contaminants in soil are chemicals that have the potential to cause harm. Typically introduced by human activities, these chemicals may enter the soil environment from errors made during intentional applications, such as excessive pesticide application rates, or from accidental releases of chemicals not intended for application to soil, such as spills or leaks of solvents or petroleum products. Disposal practices that were acceptable in the past, due to a lack of detection and/or understanding of the harmful effects of a particular chemical, have also resulted in soils considered contaminated today. Naturally occurring compounds in soils, such as metals (e.g., arsenic) and petroleum seeps, also may be problematic and require bioremediation.

Common soil contaminants

There are literally thousands of chemicals used in industry and agriculture for human benefit. A portion of these chemicals become soil contaminants, resulting in diverse and complex mixtures entering the environment (Gerhardt et al., 2009; Alexander, 1999; Baker and Herson, 1994). A brief summary of some of the more common pollutants and classes of contaminants is provided as follows:

- *BTEX*—Specific components of petroleum and petroleum-related compounds are often referred to as BTEX, an acronym for benzene, toluene, ethylbenzene, and xylene. BTEX compounds are both relatively mobile in soil and **carcinogenic**. Fortunately, microbial degradation of petroleum compounds is well characterized and can often be readily implemented under both oxic (aerobic) and anoxic (anaerobic) conditions. In most cases, the solution lies in adjusting in situ conditions to promote BTEX degradation.
- *MTBE*—Methyl tertiary-butyl ether (MTBE), a chemical compound known as an oxygenate, was used beginning in 1979 as a gasoline additive to increase the octane

rating but was later replaced by ethanol. Interestingly, MTBE replaced lead because it was thought to not have the harmful health effects of lead. However, it was found that MTBE is harmful to human health, is poorly retained by soil (e.g., by soil organic carbon), and can migrate to groundwater faster than BTEX compounds. Although previously considered too highly recalcitrant for bioremediation, both aerobic and anaerobic pathways for MTBE transformation have been identified.

- *PAHs*—Polycyclic aromatic hydrocarbons (PAHs) are present in many petroleum-based compounds such as gasoline, crude oil, and coal. They are a class of compounds with characteristic aromatic rings in their molecular structures. Less mobile than BTEX compounds, they tend to remain near the soil surface. Many PAHs are carcinogenic and are also more recalcitrant than BTEX.
- *PCBs*—Polychlorinated biphenyls (PCBs) commonly occur in mixtures of compounds called congeners, which consist of different arrangements of chlorine on the underlying biphenyl structure comprised of aromatic rings. The resulting mixes have desirable properties for industrial uses and are classified as aroclor mixtures. Many of these congeners are carcinogens. Due to their chemical structure, PCBs are quite recalcitrant in the environment.
- *Chlorinated solvents*—Chlorinated solvents, such as trichloroethylene (TCE) and tetrachloroethylene (PCE), have long been used as cleaning agents. These solvents are often groundwater contaminants and pose a serious health threat to those down gradient of the contamination plumes. These compounds can also exist as vapors in buildings situated above contaminant plumes. Many chlorinated solvents are carcinogenic. The ease of biodegradation of chlorinated solvents varies and is typically related to the redox potential of soil through oxic cometabolism or reductive dechlorination. Under oxic conditions, PCE/TCE can be degraded to daughter products cis-1,2-dichloroethene, vinyl chloride, and/or ethylene through cometabolism when other compounds such as phenol are present (Fig. 21.3). More commonly, under anoxic conditions, PCE/TCE can also be transformed to the daughter products. Careful consideration should be paid if the final degradation product is vinyl chloride because it is more toxic than TCE.
- *Energetics and explosives*—Energetics and explosives, such as 2,4,6-trinitrotoluene (TNT) and hexahydro-1,3,5-trinitro-1,3,5-triazine (RDX) contaminate both surface soils and groundwater. Although typically confined to military lands, transport of these compounds from these facilities by groundwater flow to adjacent areas occupied by the public is a growing concern.
- *Metals*—Metals, such as mercury, chromium, arsenic, selenium, and lead have long been contaminants of soils and man-made structures (e.g., lead paint). Remediation of soils contaminated with heavy metals poses unique challenges because, although their chemical speciation (oxidation state) may change as a result of biological processes (Chapter 19), they are not subject to degradation per se.
- *Radionuclides*—Radionuclides, such as plutonium and uranium associated with nuclear power plants, also pose novel remediation challenges due to special considerations in both their handling and treatment.
- *Inorganic compounds*—Nutrients such as nitrogen and phosphorus may be the most pervasive soil and water contaminants known. Due to agricultural and industrial runoff, these compounds can contaminate vast expanses, such as the "dead zone" in the Gulf of

Mexico that stretches west from the Mississippi River Delta toward the Texas border. The dead zone results from excessive nutrients (eutrophication), from both agricultural lands and other sources within the Mississippi River watershed, entering the gulf and causing excessive growth of algae (algal blooms). As the algae die, other microbes oxidize the carbon in the algal biomass, thereby causing dissolved oxygen in the water to become depleted faster than it is replenished. This results in low oxygen, or hypoxic, conditions. Hypoxia can result in the death of marine life and severely impact commercial fishing.

- *Emerging perfluorinated chemicals*—The extent of contamination by perfluorinated compounds was only recently recognized. This group of chemicals include perfluorocarbons (PFCs), per- and poly-fluoroalkyl substances (PFASs), perfluorooctanoic acid (PFOA), and perfluorooctane sulfonate (PFOS). There are more than 4,700 PFASs. Manufactured for their hydrophobic properties, they are used in firefighting foam and the production of Teflon. Though present in small quantities, they persist in the environment, are resistant to biodegradation, and cause human health problems such as neurodevelopmental effects, suppressed immune system effects, and cancer.

Human and environmental risks

Contaminated soils can present a human health risk by several routes. Contaminants released at the Earth's surface can be transported vertically and laterally into surface and ground waters and eventually become ingested by private and municipal water users. Contaminants adsorbed onto soil can be inhaled or ingested by humans directly. They can also be taken up by plants, accumulated in animal tissues, and eventually detected in the foods ingested by humans. Dermal exposure to soils contaminated with toxic chemicals may lead to absorption through the skin. In each route described, deleterious effects of the contaminant depend on both the amount absorbed or ingested and the inherent toxicity of the contaminant.

Contamination of soil can also be harmful to the environment. **Biomagnification** of a contaminant, in which a contaminant becomes more concentrated in tissues as it moves up the food chain, can cause ecological harm at various trophic levels. Soil contamination can also limit land value by preventing its optimal use and cause corrosion to buildings and structures.

Magnitude of the problem

Number of sites

It is difficult, and probably impossible, to obtain an accurate and inclusive listing of sites where soil is contaminated, for reasons ranging from historical (the identification of compounds as pollutants has varied over time) to the high costs associated with sample collection and analysis needed to identify polluted sites. From 2001 to 2015, the United States generated from 20.3 to 29.1 million tons of hazardous waste per year as determined through the Resource Conservation Recovery Act originally enacted by Congress in 1976 (US EPA, 2018). The magnitude of the problem can be appreciated by looking at petroleum-contaminated sites alone, although it is important to note that many such sites are also contaminated with other chemicals. Petroleum-contaminated sites possibly number in the tens of thousands. It is estimated that more than 100,000 of the estimated 1.4 million buried gasoline and fuel tanks in the United States are leaking. According to US Environmental Protection Agency estimates, service station tanks are leaking over 40 million L of gasoline per year (Riser-Roberts, 1992). Some estimates place the volume of used motor oil lost from at-home, do-it-yourself projects, and automobile mechanics as higher than that lost by leaking tanks. Although difficult to quantify, the Integrated Waste Management Board in California estimated that, of the 125 million L of oil purchased statewide by the public, 64% could not be accounted for by known means, such as collection points (Integrated Waste Management Board, 2000).

High-resolution analytical chemistry techniques coupled with detailed ecotoxicity studies have resulted in the identification of new contaminants that are harmful. A current example is the perfluorinated solvents mentioned previously. It is only recently that analytical tools such as tandem mass spectrometry could effectively detect these compounds. Ecotoxicity studies are detailing the hazardous effects of these chemicals on aquatic life. Using zebrafish as a vertebrate model, Jantzen et al. (2016) found that perfluorinated solvents caused morphological and behavioral changes to the zebrafish including decreased body length and altered locomotive ability.

Volume of contaminated soil and groundwater

At any given site, the volume of soil that is contaminated can vary widely. Past spills and disposal practices have contaminated thousands of cubic meters of soil at a single site. Examples of such sites include disposal lagoons and seepage pits from past manufacturing practices. At other sites, concentrated areas of contamination, generally referred to as hot spots, can be several cubic yards in size, and there may be hundreds of such hot spots distributed within several acres. Characterizing the nature and distribution of contamination (often from several pollutants) at such sites is a significant task, but it is an important first step toward selecting an appropriate and cost-effective remediation strategy.

Cost of treatment

Accounting for chemicals, either those emerging or already on the market, is not trivial. Triggered by the Frank R. Lautenberg Chemical Safety for the 21 Century Act in 2016, the Environmental Protection Agency updated its Toxic Substances Control Act (TSCA) Inventory to include a total of over 86,000 chemicals manufactured, processed, and imported to the United States through 2018 (see https://www.epa.gov/tsca-inventory for more information). Though comprehensive, the TSCA Inventory does not include the new chemicals which appear on the market each year. New production and handling practices coupled with improved guidelines and awareness have significantly reduced the number of accidental releases and improper disposal practices, as well as enhanced the identification of point-sources (known sources of pollution such as chemical plants). Increasingly, new products include the concept of life-cycle costs, in which the environmental costs associated with production, persistence, and ultimate fate of chemicals are factored into their development and use. Although reliable estimates of the current value of bioremediation are unavailable, past estimates were on the order of tens of billions of dollars per year in the United States alone (Cunningham and Mather, 1995). Properties whose use or land value is negatively affected by pollutants are commonly referred to as "**brownfields**." Bioremediation on these sites has been shown to improve property values between 5% and 15% (Sullivan, 2017). Because of the diversity and magnitude of contaminants in soils, there is a critical need for cost-effective and feasible remediation techniques.

What is bioremediation?

Bioremediation is a strategy or process that uses microorganisms, plants, or more exactly, the enzymes produced by microorganisms, plants, or both, to transform or degrade contaminants in soil, groundwater, and other environmental matrices. Depending on the approach used, bioremediation may result in varying degrees of transformation or degradation of contaminants:

- **Biotransformation** is a general term that refers to the modification of a parent compound to daughter compounds.
- **Mineralization,** or complete biodegradation, is the complete conversion of an organic contaminant to its inorganic constituents, generally carbon dioxide and water and possibly other constituents such as chloride.
- **Cometabolism** refers to the transformation of a contaminant without the contaminant providing carbon or energy for the degrading microorganisms, sometimes called "fortuitous degradation".

General approaches to bioremediation

Most approaches to bioremediation seek to reduce the limitations of "natural" bioremediation. **Natural attenuation** is a bioremediation approach in which indigenous microorganisms and/or plants degrade contaminants in the environment without any manipulation to the organisms present or the environment. As one might anticipate, this approach can have significant shortcomings which limit its usefulness. These shortcomings can be overcome by any of several methods, all of which are based on either or both of the following general approaches:

- **Biostimulation** is the addition of limiting nutrients, such as nitrogen and phosphorus, or other amendments to stimulate indigenous microorganisms in soil. Adding small amounts of the contaminant or a non-toxic chemical with similar structure can also act as a stimulant by encouraging production of degradative enzymes. This practice is sometimes appropriate for enhancing degradation of residual levels of a contaminant or in other instances in which contaminant levels are low.
- **Bioaugmentation** is the inoculation of a contaminated site with known degradative microorganisms in order to facilitate biodegradation. Bacteria are the organisms most commonly used for bioaugmentation. A single species or a group (**consortium**) of microorganisms can be employed for biodegradation. Accordingly, the inoculant may contain wild-type or genetically engineered microorganisms as either a single microbial species or a consortium of several species. In any case, organisms are selected for their high potential to degrade the contaminant of concern and their ability to survive under field conditions. When implementing this process, promoting their ability to sustain

growth requires careful consideration. As such, bioaugmentation is often combined with biostimulation.

These terms and approaches are all used in the context of converting contaminants to less toxic intermediates or mineralizing them to their inorganic forms. Riser-Roberts (1992), Skipper and Turco (1995), Das (2014), and Sharma et al. (2018) provide additional reviews on bioremediation.

Unless deposited directly into water bodies, contaminants usually first enter the upper unsaturated layer of soil. The eventual location of the contaminants, either above or below the water table, depends in part on the nature of the contaminant (its tendency to adsorb to soil constituents), climate (leaching events), topography, and soil drainage class. The unsaturated region above ground water is known as the **vadose zone**. This zone is typically **oxic** (aerobic) due to the presence of air-filled pores. Conversely, the *saturated zone* below the water table is often **anoxic** (anaerobic) due to microbial respiration and restricted oxygen diffusion. In addition to its effect on oxygen availability, the degree of soil wetness affects the availability of other electron acceptors (e.g., Fe^{3+}, SO_4^{2-}, NO_3^-), microbial activity, microbial community structure, contaminant distribution, mixing, and a range of other properties that influence bioremediation.

Approaches to bioremediation include in situ and ex situ approaches (Table 22.1). In situ approaches treat the contaminated material in place, that is, without excavation and removal. Natural attenuation is the simplest in situ approach; more advanced, and generally more effective, technologies include bioventing, phytoremediation, biosparging, and bioslurping. Ex situ approaches involve excavation and treatment of the contaminated material elsewhere and include the use of biopiles, bioreactors, and land farming.

TABLE 22.1 List and description of in situ and ex situ bioremediation approaches.

***In situ* approaches:**

Passive or intrinsic bioremediation is the "natural" bioremediation of a contaminated site by indigenous (naturally occurring) microorganisms. Many contaminants are degraded by these microorganisms, although the rate of degradation is often too slow for practical benefit. A challenge with passive bioremediation is that it is difficult to monitor and, therefore, also difficult to predict migration and decay. In general, one or several factors are sub-optimal for biodegradation to occur. Most of the following techniques have been developed to reduce such limitations.

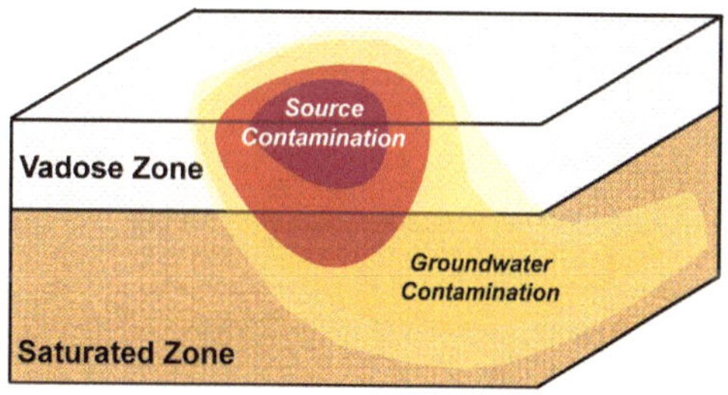

Monitored natural attenuation (MNA) is the "natural" remediation of a contaminated site by indigenous microorganisms and possibly abiotic processes. It is similar to passive or intrinsic bioremediation but includes an agreed-upon monitoring plan to confirm that remediation processes are occurring. For some of the contaminants present in groundwater, notably BTEX and TCE, many of the degradative processes are well characterized and can be measured, and groundwater systems are relatively well mixed compared to surface soils. For example, if anaerobic respiration is using BTEX as its carbon source, N and Fe may show a speciation difference concomitant with lower BTEX concentrations, i.e., oxidized forms (NO_3^- or Fe^{3+}) versus reduced forms. A gradient of speciation of electron acceptors, from oxidized to reduced forms, that coincides with lower BTEX concentrations may be an indication of anaerobic respiration using BTEX as the carbon source and alternative electron acceptors.

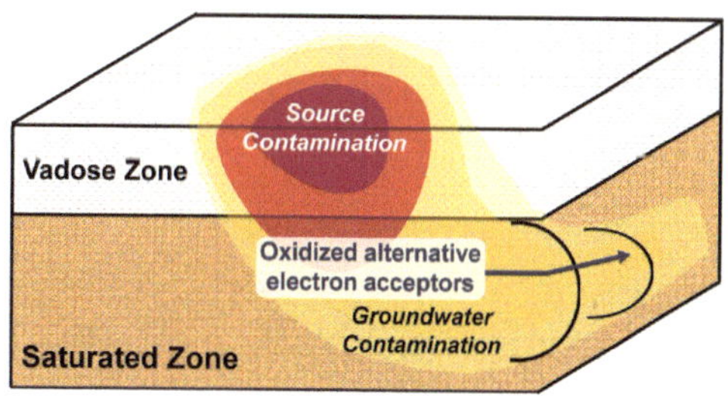

(Continued)

TABLE 22.1 List and description of in situ and ex situ bioremediation approaches. (*Cont.*)

Phytoremediation is an umbrella term describing the use of plants to remove, contain, or transform contaminants. Although not always explicitly stated, microbes can contribute to many of these strategies.

Phytoextraction: Certain plants take up significant amounts of nutrients. This capability can be exploited to remove excess nutrients from soils. Some plants accumulate compounds, such as metals, to a degree greater than the concentration in the soil solution. This is termed hyperaccumulation and is another example of phytoextraction. Plants with accumulated metals can be harvested and disposed of or, in some cases, recovery of the metals may be feasible. Microbial activity affects the availability of many metals and nutrients to plants.

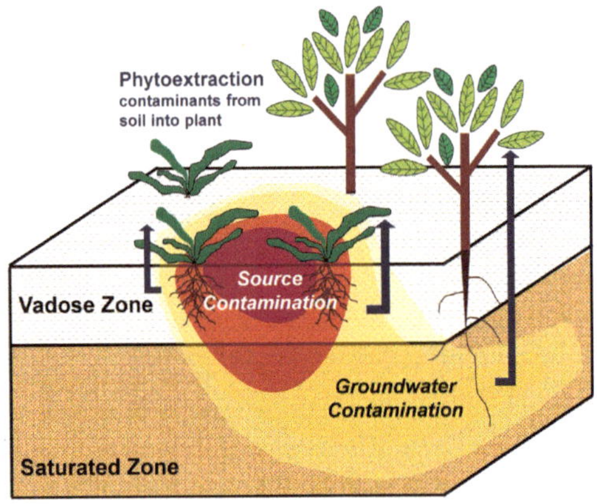

Phytodegradation: Some plants are also capable of taking up and degrading relatively water-soluble organic contaminants, such as TCE. In some plants, such as some hybrid poplars, TCE can be degraded by enzyme systems in the plant.

Phytovolatilization: Some contaminants that are phytoextracted may be volatilized from plant tissue, perhaps in concert with transpiration.

Hydraulic Control: Some plants, such as poplar trees, can transpire sufficient water to influence flow of shallow groundwater. This can be beneficial by limiting groundwater transport of contaminants and can be coupled with phytoextraction and phytodegradation.

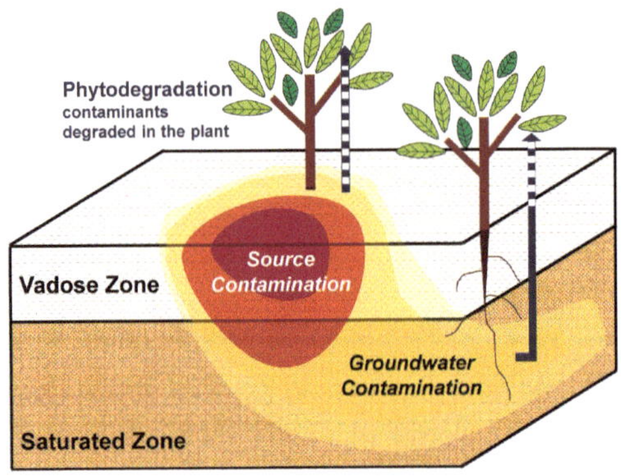

Rhizodegradation: Carbon exudations and secretions from roots stimulate microorganisms in the rhizosphere (zone of soil next to the roots). The enhanced microbial activity in the rhizosphere in turn can enhance degradation of contaminants through cometabolism and by providing limiting nutrients.

Phytostabilization: Plants, in concert with microorganisms, also influence the turnover and net accumulation of organic matter into the soil, a process referred to as mineralization-immobilization turnover (MIT). Some contaminants or their transformation products can be chemically bound or incorporated into soil organic matter, a process known as humification, or physically trapped in the soil humic or mineral fractions, a process known as sequestration.

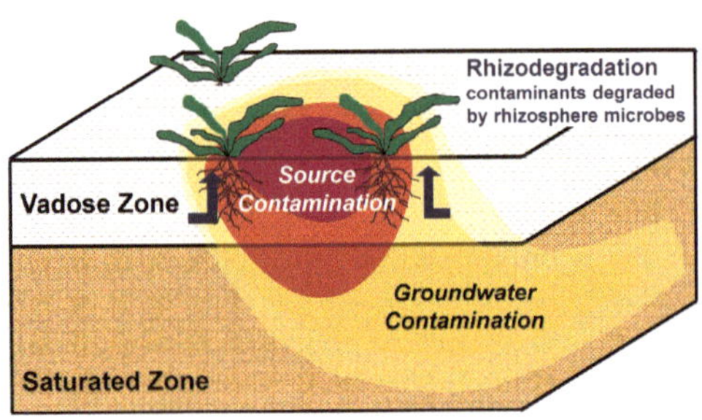

(*Continued*)

TABLE 22.1 List and description of in situ and ex situ bioremediation approaches. (*Cont.*)

Bioventing is a form of biostimulation in which gaseous stimulants, such as air or oxygen (to promote aerobic processes) and methane (as a carbon and energy source to promote cometabolism), are added to vadose zone soils, generally by pumping them into wells in the soil.	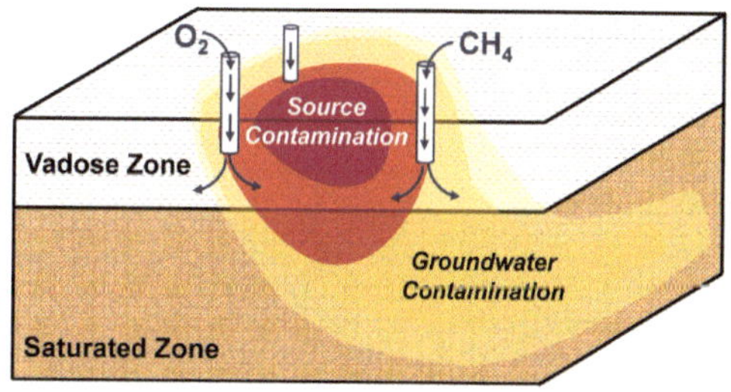
Biosparging is bioventing in the saturated zone. Biosparging can be used to add biostimulants, improve aeration within the groundwater, and also promote oxic conditions and in turn aerobic processes in the overlying unsaturated soil.	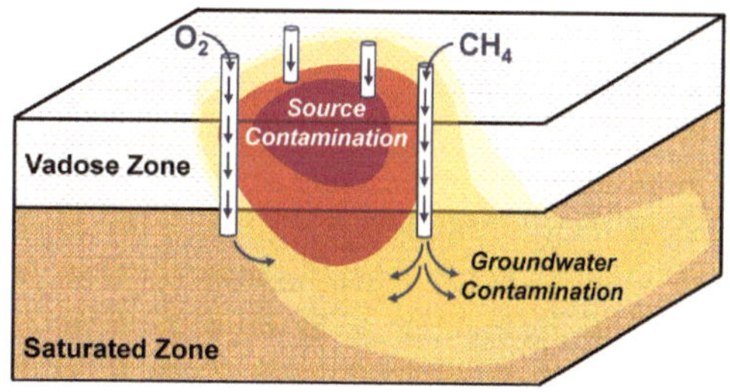
Biobarriers are permeable "walls" formed by placing biologically active material in a trench in the flow-path of shallow groundwater. Conditions in the permeable barrier enhance the degradation of contaminants. Biobarriers have been created by placing readily oxidized organics in the trench (e.g., fresh wood chips) so that the conditions are sufficiently reduced to degrade halogenated compounds in shallow groundwater.	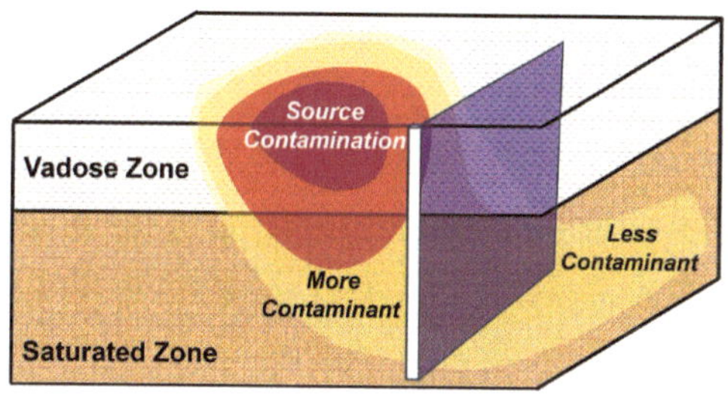

(*Continued*)

TABLE 22.1 List and description of in situ and ex situ bioremediation approaches. (*Cont.*)

Ex situ approaches:

Landfarming is the spreading and mixing of contaminants, contaminated soils, or wastes onto a surface, such as non-contaminated soil. The area is underlain with a barrier of some sort, such as a natural or constructed clay layer, to prevent leachates from contaminating the groundwater. The soil is plowed or disked to provide mixing, aeration, and moisture. If there are regions in the material in which the concentration of the contaminant is too high for easy biodegradation (i.e., "hot spots"), plowing or disking redistributes the contaminant and thereby reduces its average concentration. Finally, if coupled with biostimulation or bioaugmentation, plowing or disking gives a more uniform distribution of fertilizer and microbial inoculant, respectively.

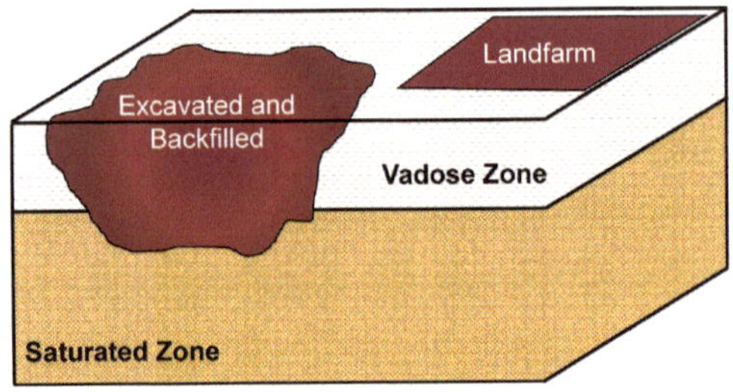

Composting is the use of aerobic, thermophilic microorganisms in constructed piles or windrows, composed of a mixture of soil, organic residues, and bulking agents, to degrade contaminants. The bulking agents help promote good aeration within the piles. The piles are physically mixed and moistened periodically to promote microbial activity and enzyme-contaminant contact.

Pile bioventing relies on air injected into stockpiled soils to stimulate aerobic degradation. It can be considered a combination of landfarming and composting. It takes less space than landfarming, and pumping air into the pile supports aerobic degradative processes without physical mixing.

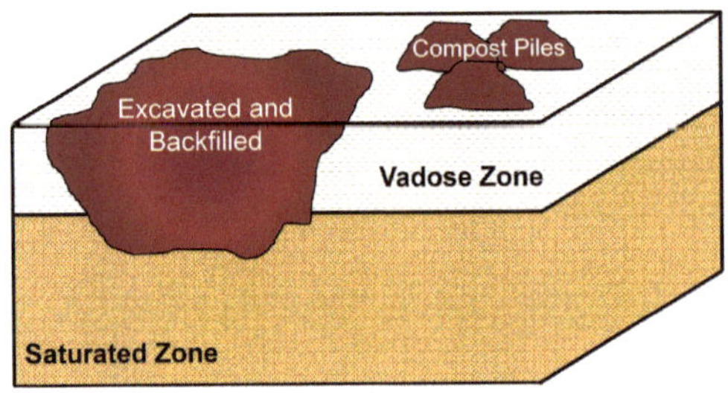

Bioreactors are large tanks or vessels that can hold excavated contaminated soil, water, nutrients, substrates and, if necessary, an inoculum of microorganisms primed to degrade the contaminant. Conditions can be controlled and optimized in bioreactors, but the volumes that can be contained are relatively small.

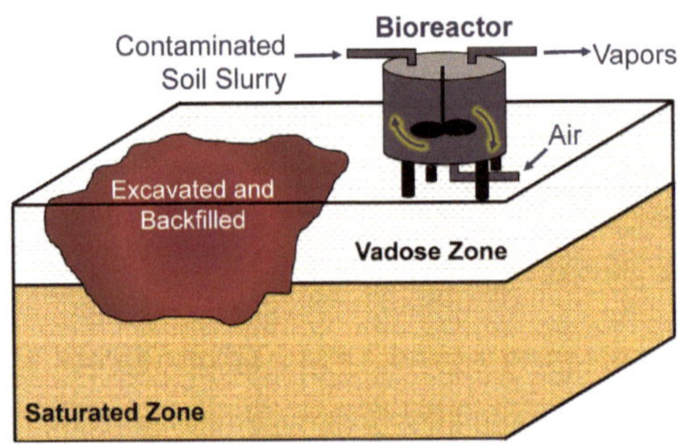

Advantages and disadvantages of bioremediation

Bioremediation has several potential advantages compared to non-biological treatment technologies such as excavation and incineration, excavation and placement in a landfill, or chemical oxidation by adding a strong oxidant (e.g., permanganate). It is essentially a natural process and, as a result, generally has a low environmental impact. Bioremediation also tends to have high public acceptance because it is an environmentally friendly or "green" approach. Bioremediation performed in situ can be less expensive because it avoids the costs associated with excavation and hauling of contaminated materials necessary for ex situ approaches such as those involving land farming and bioreactors. Additionally, at some sites, there are simply no other feasible alternatives to in situ bioremediation due to location, cost, or available resources. As an example, in situ bioremediation is commonly used in remote regions of

Alaska affected by oil spills and heavy metals, where the cost of removal often exceeds the cost of in situ treatment.

Conversely, bioremediation is often neither as rapid nor as predictable as some of the more aggressive alternative treatments. For some sites, the need to rapidly treat the soil takes precedence over cost. The certainty of technologies such as incineration can be particularly attractive when future liabilities must be minimized. In general, there is a compromise between time and cost for remediation technologies (Fig. 22.1). The concepts shown in Fig. 22.1 are general, and the specific conditions at any site must be considered when choosing a bioremediation approach. The increasing knowledge gained from both research and the field application of various systems has lessened the uncertainties, such as the length of time needed for successful treatment to occur, associated with the less costly bioremediation technologies.

For many bioremediation applications, the concentration of the contaminant will decrease progressively with continued treatment, but it will not be reduced to zero. Instead, the concentration becomes asymptotic to zero and a residual quantity remains (Fig. 22.2). There are two major reasons for this phenomenon. First, when contaminant levels are very low, the net energy gains to microbes from degrading a compound are unfavorable to support continued biodegradation. The second reason centers on the **bioavailability** of the contaminant,

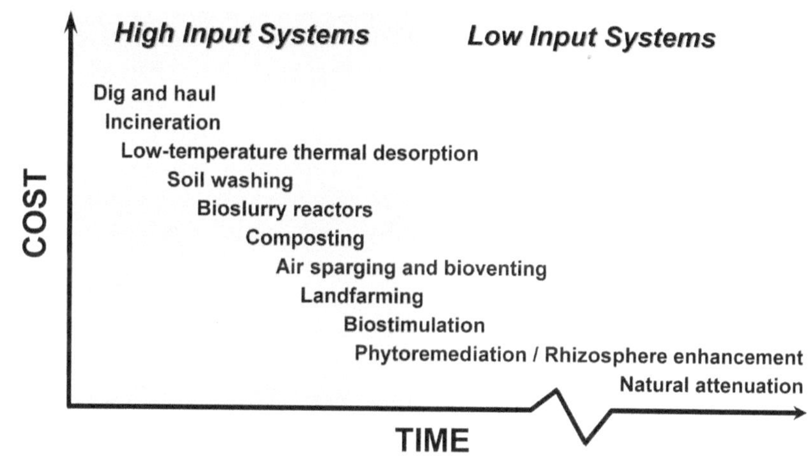

FIGURE 22.1 The choice of bioremediation approach is often a balance between more costly methods that achieve results in a relatively short time versus less expensive approaches that need more time for desirable results.

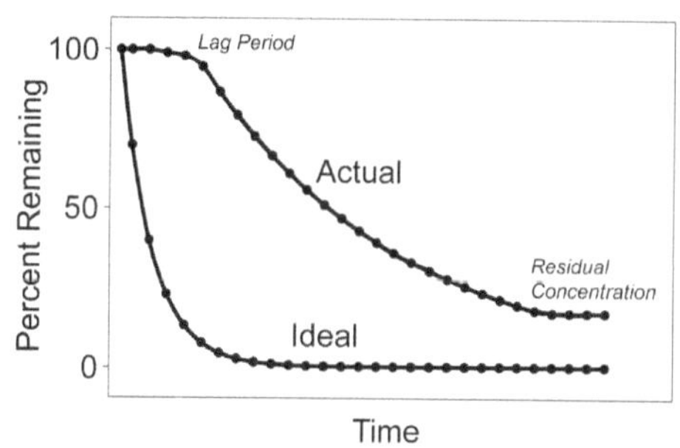

FIGURE 22.2 With increasing duration of bioremediation, the concentration of a contaminant will decrease progressively to a residual concentration, but it will not be reduced to zero.

that is, the ability of a contaminant to be accessed and transformed by microorganisms. Very low levels of contaminants can be detected analytically using methods that generally involve extracting the contaminant from the soil with a strong organic solvent that solubilizes the contaminant. Therefore, the more aggressive organic solvents can extract contaminants that are, at least in the short term, unavailable to microorganisms. Importantly, bioavailability is not a static value but is dependent on many criteria. For example, contaminants can be sorbed to soil surfaces, held within soil aggregates, or they may become coated or otherwise protected from enzymatic attack, thereby making them less bioavailable. Yet many contaminants can, with changing conditions, *desorb* (i.e., de-adsorb) from the soil or diffuse out of aggregates into soil pores. Adding *surfactants* or changing soil solution conditions are means to help solubilize contaminants making them available to microorganisms.

It is worth emphasizing here that a contaminant can be reduced to a concentration of near zero bioavailability but still be detected and measured using chemical extraction techniques described earlier. Currently, remediation researchers, users, and the regulatory community wrestle with the significance of this common phenomenon, that is, "If a contaminant is not bioavailable, is there a need to further reduce its concentration?" Conversely, "If a biologically unavailable contaminant can still be chemically extracted from a soil, is the soil 'clean' enough?" Research has shown that bioavailability is relative, in that the amount of a contaminant that is bioavailable can vary among organisms. For example, snails (*Helix aspersa, Cantareus aspersus*) are able to access, that is, make bioavailable to themselves, cadmium that was considered non-available to other organisms (Scheifler et al., 2003; Pauget et al., 2012). Thus, when estimating the bioavailability of a contaminant, we can use the efficacy of a particular extractant (e.g., organic solvent) as a standard, but we also need to consider the potential modifying influences of varying extraction efficacies among organisms.

Sites to be remediated can contain mixtures of contaminants, both inorganic and organic. At such sites, one contaminant might be readily biodegradable whereas another might be highly toxic to biodegrading microorganisms; one compound might require oxic conditions for aerobic biodegradation while another might require anoxic conditions for anaerobic biodegradation; one compound might require neutral to alkaline conditions for biodegradation but another might require acidic conditions. At such sites, designing a bioremediation strategy that sequentially attacks the targeted compounds requires a thorough understanding of the biodegradation pathways specific to the classes of compounds present at the site (see Chapter 21), the environmental conditions necessary to support the biodegradation activities, and the availability of technologies to manipulate the environment.

Practical considerations for selecting bioremediation

Prior to selecting bioremediation as the treatment strategy, certain criteria must be considered. In brief, the fundamental goal of all bioremediation strategies is to have the contaminant, the proper microorganisms, and the correct soil conditions present simultaneously for a period of time sufficient for the desired process to progress to a satisfactory endpoint.

Known contaminant and degradation pathways

The chances for successful use of bioremediation are greatly improved if the metabolic pathways for the transformation of the contaminant, and the conditions that favor those pathways, are known (Chapter 21). It is particularly important to confirm that daughter products are less harmful than parent products. For instance, vinyl chloride, a daughter product of TCE, is more toxic than its parent chemical. The organisms in the soil must have the necessary catabolic activity to degrade the contaminant at a reasonable rate and to bring the concentration of the contaminant to a level that meets regulatory standards. Much of the current research and implementation trials in bioremediation are focused on adjusting the conditions in contaminated soil to favor the appropriate microbial processes.

The *Principle of Microbial Infallibility* states that for every naturally occurring organic compound, there is a microbe or enzyme system capable of its degradation (Gale, 1952). In the past, a common assumption was that, because of their great metabolic flexibility, microbes could degrade all synthetic compounds as well as naturally occurring compounds. Although the vast majority of synthetic compounds can be degraded, a few are extremely recalcitrant and do not degrade at acceptable rates.

As new pathways and enzymes are characterized, some compounds formerly thought to be recalcitrant have been shown to be degradable. One technique to stimulate degradation is through forced molecular evolution whereby a consortium of microorganisms in a bioreactor is exposed to increasing concentrations of a recalcitrant chemical. Through genetic adaptations and horizontal gene transfer, the consortium may gradually develop the metabolic machinery necessary to use the contaminant as an energy source. A second technique to examine the potential for the biodegradation of recalcitrant contaminants is the addition of chemical analogs (termed **analog enrichment**) to microbial consortia. Chemical analogs are chemicals that have a similar structure to that of the recalcitrant compound under examination and are, therefore, capable of inducing the production of enzymes that may also degrade the structurally similar contaminant via cometabolism. There is convincing evidence that analog enrichment occurs in nature and can be enhanced by plants through their production of secondary metabolites (Singer et al., 2003). These secondary metabolites often have structural similarities to xenobiotics and, therefore, enzymes produced by the resident microorganisms degrading them may also be able to degrade the corresponding xenobiotics. As an example, PAH compounds produced and released by some plants appear to serve as *cometabolites* that selectively favor the growth and activity of bacteria that degrade PAH contaminants (Yergeau et al., 2018). Similar processes occur for PCBs (Leigh et al. 2002) and daughter products of TCE (Fraraccio et al., 2017).

Modifying soil conditions

Due to the natural consequences of the laws of thermodynamics, compounds in nature ultimately tend to degrade but, unless the conditions are favorable, the rates can be so slow that they are impractical for bioremediation purposes. A key for using bioremediation is to alter the environmental conditions so as to improve degradation rates and, from a practical perspective, to do this in a cost-effective way. A useful way to approach the problem is to apply Liebig's Law. *Liebig's Law of the Minimum* states that biological growth is limited by and proportional to the availability of the most limiting nutrient. The addition of the limiting nutrient will result in an increased yield until another nutrient becomes limiting. To apply this concept, one needs to know the nature of the desired microbial pathway and the conditions that favor the catabolic process. Knowing this, a bioremediation plan that addresses one or more of the dominant factors that limit catabolism can be developed. Questions to be addressed are "What are the *most* limiting factors at this specific site and for this specific contaminant?" and "Are there conditions that can be altered inexpensively and that will significantly improve the desired microbial processes?" Environmental conditions such as temperature, soil water potential, redox potential, and pH all influence metabolic pathways. Two situations commonly limit the use of bioremediation: (1) the contaminant and proper microorganisms are not in proximity, and/or (2) the local soil environment conditions do not favor the specific and necessary enzymatic processes for degradation of the contaminant. Overcoming these limitations typically requires some form of mixing, either by physically mixing the soil or by infusing some or all of the necessary ingredients (microorganisms, supplementary substrates, other environment-modifying substances) through the soil to the contaminant. This is often a costly and difficult limitation to overcome. If we know the degradation pathway for a contaminant and the necessary conditions to support this pathway, the questions become "Can the limitation be addressed at a reasonable cost?" and "How does the cost of reducing the limitation compare to the cost of the alternative remediation strategies?" Increasingly, bioremediation is viewed from the perspective of how best to exploit natural processes in soil rather than attempting to alter soil conditions by energy intensive and costly approaches such as excavation, incineration, or extracting contaminated groundwater by pumping.

Temperature

Although large masses of soil can be heated, it is relatively costly on a large scale. Bioreactors, or large vessels in which excavated soil can be treated, can be operated at controlled temperatures, but they treat relatively small volumes of soil per batch. Bioreactors have been used with success in situations where the throughput of the bioreactor matched the needs of the remediation plan, such as for soils contaminated with petroleum hydrocarbons (Safdari et al., 2018). Soil heating has also been used in cold regions to improve microbial activity and in turn the degradation rate of the contaminant, though scaling up for large chemical spills are generally cost prohibitive (Errington et al., 2018).

Water

For surface soils, irrigation often can be employed on a practical scale. Where irrigation is not feasible, insufficient water can be a significant limitation. Water is usually not an issue for groundwater systems unless the water has been displaced by pure contaminant, which may happen with dense substances such as TCE and PCE which can form non-aqueous phase liquids (NAPLs; Chapter 21). Where pure product exists, it is often pumped and recovered. The remaining contaminated soil can then be treated by bioventing (Table 22.1).

Nutrients and substrates

Nutrients are often needed to stimulate microbial activity and are relatively inexpensive to apply. For the degradation of petroleum compounds, which are carbon-rich relative to other nutrients, nitrogen is frequently most limiting followed by phosphorus. In landfarming systems designed to treat petroleum-contaminated soil, nitrogen can be surface applied and incorporated in soil using common agricultural practices such as tilling. When applied in appropriate amounts, nitrogen fertilizer results in enhanced degradation of the petroleum hydrocarbons.

Despite notable successes, the axiom "One size does not fit all" is nevertheless true for employing fertilizers for biostimulation. Nitrogen and other nutrients are usually added as fertilizer salts and, if added in excess, can have several detrimental osmotic and environmental effects. For example, petroleum contaminants are primarily carbon. Adding nitrogen fertilizers to reach preselected C/N ratios without consideration of their potential to elevate dissolved salt concentrations in the soil solution can cause osmotic stress to the microorganisms (Walworth et al., 1997). Additionally, in aerobic subsurface aquifers, nitrogen added as ammonium can be oxidized to nitrate by the process of nitrification. Excess nitrate in groundwater is itself polluting and can cause blue baby syndrome (methemoglobinemia) when consumed in drinking water and can eventually result in algal blooms when it migrates to surface waters. Balancing nitrate additions with microbial demand for nitrate as an electron acceptor for anaerobic metabolism is also a necessary consideration.

Redox potential

The redox potential of the soil matrix is an important consideration in many reductive pathways, especially for those involving dehalogenation, as is the availability of appropriate electron acceptors. Increasingly, bioremediation systems are being developed that supply readily metabolized carbon to drive the system to a sufficiently chemically reduced state to promote the continued biodegradation of chlorinated solvents.

Monitoring

A critical part of bioremediation is monitoring its performance. Monitoring may appear trivial but actually can be, and usually is, quite difficult because it is time and labor intensive. It often requires both analysis of the contaminant itself and an assessment of the state of associated degradative processes. As bioremediation applications move toward enhancing and monitoring natural attenuation processes, monitoring performance becomes critical.

A common and reasonable monitoring approach is to measure changes in concentration of the contaminant periodically, with the intent of observing decreases in the contaminant concentration. This approach is viable if the contaminant is uniformly distributed, the degradation occurs relatively evenly across the site, and the time scale for sampling is appropriately matched to the rate of bioremediation. Unfortunately, a uniform distribution of the contaminant and uniform degradation rates across a site are atypical (e.g., due to hot spots), but they can sometimes be obtained by mixing the system (e.g., in a bioreactor). In addition, the duration and frequency of sampling employed needs to correspond meaningfully to the rate of degradation. For example, if enhanced natural attenuation is used, a decrease in contaminant concentration may not be statistically detectable for many years or decades due to spatial variability of the contaminant in the soil. For these and similar cases, less frequent monitoring conducted over a longer period of time may be more appropriate than annual monitoring for the initial several years.

Relative to surface or vadose zone soils, soluble contaminants in groundwater may appear to be well-mixed. However, even in groundwater, contaminants may become strongly adsorbed within clay lenses or be present as NAPLs. These then serve as point sources of a contaminant, continually releasing small but significant amounts of the contaminant into the surrounding groundwater. Contaminant distribution in surface soils is far less uniform because of the lack of a network of water-filled pores that promote diffusion. Because the contaminant distribution is heterogeneous and the variability associated with measuring contaminant concentrations is great, measuring significant differences in contaminant concentrations over time or space is extremely difficult. A number of alternatives have been used. Taking more samples can improve the measurement estimates (i.e., by reducing statistical variance), but the added samples and analyses are expensive and there is no guarantee that the variability will be significantly reduced. Mixing the system to reduce the variability can be effective, but mixing is disruptive and extremely costly. For instance, mixing may result in volatilization of contaminants, and the measurements obtained may therefore not be reflective of the true extent of biodegradation.

Because of the difficulties in measuring statistically significant changes in the concentrations of contaminants themselves, alternative monitoring techniques have been developed. Although the concentrations of petroleum in surface soils can be quite heterogeneous, often the initial *composition* of the petroleum is relatively uniform. Petroleum contains a broad mixture of hydrocarbon compounds in addition to BTEX. A monitoring approach that is useful in non-mixed heterogeneous systems is based on changes in the relative amounts of these contaminants rather than their absolute or overall concentration. Such approaches combine an understanding of microbial degradation pathways with known differences in biodegradation rates for various compounds in the petroleum mixture. For example, many petroleum products contain PAHs, compounds that are relatively recalcitrant. Hopane, or 17a(H)21b(H)-Hopane (a C30 pentacyclic triterpane), is the most abundant biomarker in many crude oils and is highly recalcitrant. During bioremediation, hopane is largely conserved as other compounds in the petroleum mixture are biodegraded. By comparing the ratios of other petroleum compounds to hopane, the degree of bioremediation can be estimated. This approach circumvents the monitoring problems associated with heterogeneously distributed contaminant concentrations by examining changes in contaminant composition rather than overall concentration.

Characterizing microbially driven changes that occur concomitantly with contaminant degradation has also been an effective monitoring technique. Once again, this approach is based on an understanding of the degradation pathways for the contaminant. As an example, BTEX plumes in groundwater are a common contaminant issue. As BTEX is oxidized in groundwater by aerobic degradation pathways, respiratory depletion of dissolved O_2 causes the groundwater system to become anoxic. Because anaerobic respiration pathways occur in many bacteria through the use of alternative electron acceptors, arrays of sampling wells can be monitored for both BTEX compounds and changes in the concentrations of alternative electron acceptors. For example, hydrocarbons metabolized by anaerobic microbial respiration often coincide first with the electron acceptor nitrate being reduced to ammonium and nitrogen gas. Comparable electron transfers occur under progressively anaerobic conditions with manganese, iron, sulfate, and finally carbon dioxide (Jorgensen, 1989). By measuring microbially driven phenomena, such as changes in the concentrations of potential electron acceptors and their reduced counterparts, in addition to contaminant loss, multiple lines of evidence are developed to support the argument that active biodegradation is occurring on site or in situ. Multiple lines of evidence that are consistent with a known metabolic pathway, such as hydrocarbon degradation, are considerably more convincing than evidence developed from contaminant chemistry alone.

Although there can be exceptions, the biological potential to bioremediate many contaminants often exists in a given soil. However, the contaminant may be degraded at suboptimal rates and often not to completion (i.e., to CO_2). Because one of the primary goals in practical bioremediation is to adjust or modify soil conditions to favor microorganisms, measuring some response of the microbial community in addition to changes in contaminant levels can be useful in demonstrating that bioremediation is occurring. Particularly in the past, estimations of the growth responses of microbial populations were made by viable counting methods (i.e., "colony counts," Chapter 5), sometimes using media selective for degradative populations. General microbial activity can also be estimated from respiration studies that measure the efflux of metabolically produced gases from the soils. Finally, degradation processes can be monitored by measurements of the activities of enzymes, or genes encoding for the enzymes, involved in either general metabolism or degradative pathways. However, many of these approaches, particularly those that are culture-based, have limitations and often introduce measurement bias.

Recent advances in measuring bioremediation efficacy

Traditionally, methods focusing on the measurement of chemical constituent, that is, either the contaminant itself or metabolic byproducts, have been the primary means of monitoring bioremediation efficacy. However, recent advances allow for microbial community function and structure in soils to be characterized in relation to the bioremediation process. These technologies are based on methods that extract cellular components directly from the soil and use techniques, based on genetic or molecular phenotypic analyses, to characterize the soil microbial status relative to control treatments. For instance, the quantitative polymerase chain reaction (qPCR) has been used to detect genes encoding for alkane monooxygenase in soils contaminated with petroleum hydrocarbons (Barbato et al., 2017). Recent advances in metagenomics (Chapters 4 and 10) can elucidate the microbial community's response to exposure to a contaminant. This is particularly helpful because it avoids the severe bias associated with culture-based methods; however, genetic approaches can also introduce bias (Chapter 4). Finally, new techniques combining stable isotope probing (SIP) and microbial sequencing enable assessment of the microbial community that is actively degrading the contaminants (Xie et al., 2011). SIP involves introducing contaminant molecules containing stable isotopes of elements that can subsequently be tracked by mass spectrometric techniques; the presence of these isotopes in sequenced nucleic acids such as 16S rRNA genes permits identification and possibly relative quantification of the microbes actively degrading the contaminant. A drawback of SIP is that the labeled contaminant required for such techniques must be added to the microcosm, thereby disrupting the natural system. Finally, methods to quantify the expression of contaminant degradative genes in situ (e.g., mRNA-based approaches) are being developed, which, if successful, could lead to measurement of degradative gene expression in a soil.

Case studies of bioremediation

Although bioremediation is rapidly changing as we improve our understanding of microbial degradation pathways in soils, applied bioremediation is no longer in its infancy and there are numerous examples of successful implementation. Due to the many variables involved at any site, it is inappropriate to automatically recommend one bioremediation strategy over another. As these case studies demonstrate, combining more than one approach is often a useful strategy.

A case study involving biostimulation

This case study marks one of the first major successes in the use of bioremediation. In March 1989, the Exxon Valdez oil tanker hit a reef in Prince William Sound, Alaska, United States, releasing 40.8 million L of crude oil into the near pristine sound within 5 h. Over 1,500 km of the rocky intertidal shoreline in the sound and the Gulf of Alaska were contaminated to varying degrees by crude oil. Because oil is inherently high in carbon and low in nitrogen and phosphorus, a portion of the shoreline was selected for biostimulation. After several potential fertilizer candidates were evaluated, a microemulsion, Inipol EAP22 (henceforth, Inipol), was selected. Inipol, an oleophilic fertilizer, is a stable water-in-oil formulation that yields a N/P/K ratio of 7.3/0.8/0. The nitrogen source is urea, and the phosphorus source is trilaureth (4)-phosphate, a lipophilic molecule. At room temperature, Inipol has the consistency and appearance of honey, and it must be heated to 90°C before it can be applied by spraying. Inipol was applied as a coating to the shore at a rate of approximately 300 mL m^{-2} (Hinton, 1995). As the microemulsion mixed with the weathered crude oil, the crude oil destabilized Inipol to release its urea-nitrogen. In addition, a surface-active organic material, oleic acid, in Inipol served as a readily degradable carbon and energy source to increase the activity and number of indigenous hydrocarbon-degrading bacteria. When the oleic acid was depleted, the increased biomass of hydrocarbon-degrading bacteria supported enhanced biodegradation of the petroleum. Visual observations and chemical assays showed dramatic evidence that biostimulation contributed to the remediation of the site. Although passive bioremediation also occurred in the absence of the fertilizer, the accelerated rate of biodegradation observed with Inipol was critical to a successful bioremediation effort. In 2011 and 2012, oxygen in the form of hydrogen peroxide, nitrogen in the form of sodium nitrate or lithium nitrate, and phosphorus in the form of sodium tripolyphosphate were injected in four beaches over the course of 2 years, with an outcome of enhancing the natural degradation rate by 12- to 60-fold (Boufadel et al., 2016). Nevertheless, nearly 30 years later, petroleum hydrocarbons are still present on the shoreline, with long-term impacts to the surrounding ecosystem and wildlife such as orcas, seabirds, and subtidal organisms (Barron et al., 2020).

Two case studies combining biostimulation and bioaugmentation

At its St. Gabriel (Louisiana, US) plant site, Ciba-Geigy Corp. had 19,000 m^3 of atrazine-contaminated soil contained in an 8-ha Biological Cleanup Unit (BCU). The BCU was separated into six cells for monitoring and sampling purposes. The bioremediation strategy employed was to apply fertilizer and landfarm the site by plowing or disking the BCU 4 times a week combined with weekly soil samplings to measure the disappearance of atrazine. The fertilizer (898 kg of fertilizer ha^{-1}; N/P/K, 13/13/13) was added to the contaminated soil for biostimulation. For bioaugmentation, 2,000 L of a concentrated preparation of a proprietary, atrazine-degrading consortium of pseudomonads (*Pseudomonas* spp.) was applied to the 8 ha BCU. Initial *s*-triazine concentrations were approximately 100 mg kg^{-1} of soil, but after 20 weeks, triazine concentrations were below the target level of 10 mg kg^{-1} of soil. High variability was observed in the BCU Cell #3 due to "hot spots" caused by the incomplete mixing and breakdown of clay agglomerates. However, after some additional time, even these hot spots fell below the cleanup standard of 10 mg kg^{-1} (Finklea and Fontenot, 1995). Ciba-Geigy spent approximately $1,050,000 to bioremediate the soil versus a projected cost of $5,300,000 to excavate and to dispose of the atrazine-contaminated soil in an approved hazardous waste disposal facility. Thus, bioremediation saved an estimated $4,250,000 at this site. Landfarming remains a current, though less popular, practice in bioremediation and is often combined with bioaugmentation at specific sites (Maila and Cloete, 2004). Recently, landfarming was combined with bioaugmentation to clean up a petroleum contaminated site in Italy (Guarino et al., 2017).

At the Kelly Air Force Base in San Antonio, Texas, the groundwater was contaminated with chlorethenes ranging from tetrachloroethene (PCE; four chlorines) to cis-1,2-dichloroethene (cDCE; two chlorines). The concentrations of PCE were approximately 1 mg L^{-1} in the groundwater below the site. In a pilot scale study, biostimulation and bioaugmentation were used to reductively dehalogenate PCE to ethene (Major et al., 2002). For biostimulation, methanol and acetate were injected into the groundwater via the injection well for 230 d of the pilot study, starting at day 89. Then, at day 176, an anaerobic culture enriched from PCE/TCE-contaminated soil and groundwater from Southern Ontario, termed KB-1, was injected into the groundwater as anoxic conditions were maintained. The KB-1 consortium contained bacteria related to *Dehalococcoides ethenogenes*, which is known to reductively dechlorinate TCE to ethene. Complete dechlorination was targeted because vinyl chloride, a daughter product of PCE degradation, is more toxic than its parent compounds. Results from the pilot study showed that biostimulation alone was insufficient to achieve complete dehalogenation, resulting in a 90% reduction in PCE with the dominant daughter product being cDCE. However, the combination of biostimulation and bioaugmentation led to the successful dechlorination of PCE to ethene, as evidenced by the presence of bacterial species related to *D. ethenogenes* in the groundwater where complete reductive dehalogenation was observed. The final concentrations of the parent and daughter products were below 5 μg L^{-1}. This study demonstrates the utility of coupling bioaugmentation and biostimulation to bioremediate PCE in groundwater.

A case study involving phytoremediation

Phytoremediation is the use of plants to remove, contain, or transform contaminants. This approach to bioremediation is relatively new, although thoroughly documented case studies from the field are increasingly reported in the scientific literature. *Indirect plant effects* include cases in which root exudates released into the rhizosphere stimulate resident microbial populations which fortuitously degrade various compounds by cometabolism, whereas *direct plant effects* include degradation or sequestration by the plant itself following uptake by roots. Common contaminants shown to be susceptible to phytoremediation include pesticides (Lee et al., 2012; Arslan et al., 2017), trichloroethylene (TCE, Doty et al., 2017), munitions (Rylott et al., 2011), and heavy metals (Alahabadi et al., 2017; Busby et al., 2018). Plants that accumulate metals are termed hyperaccumulators.

One example where phytoremediation of a nonagricultural contaminant was effectively implemented is at the Middlefield-Ellis-Whisman (MEW) Superfund Study Area in the Moffett Field/NASA-Ames Research Park in Mountain View, California. Though there are co-contaminants at this site due to manufacturing, industrial, and military activities, this example highlights effective phytoremediation of TCE (Doty et al., 2017). Remedial investigations of the extent of contamination and initial treatment occurred as early as the 1980's. A remedial investigation is often the first step in assessing the extent of a contaminant at a site through site characterization, sample collection, chemical analysis, and risk to human health and the environment. The concentration of TCE near the center of the plume ranged from 100 to 1000 μg L^{-1}. Poplar trees inoculated with natural, TCE-degrading endophytes were planted to the contaminated soil. An endophyte is a microorganism (i.e., bacterium or fungus) living inside of the plant tissue. The poplar trees inoculated with endophytes were larger than their uninoculated controls for years 2 and 3 of the 3-year study. Additionally, the inoculated poplar trees resulted in a 95% reduction of TCE in the groundwater. Daughter products of TCE including cis-1,2-dichloroethene, trans-1,2-dichloroethene, 1,1-dichloroethene, and vinyl chloride decreased below the reporting limit. This demonstrates the benefit of using fast-growing poplar trees for effective phytoremediation of TCE.

A case study involving combined biostimulation and phytoremediation

Since approximately 1950, various gasoline, diesel, and jet fuel spills occurred across multiple sites within the former Naval Arctic Research Laboratory (NARL) in Utqiaġvik (formerly Barrow), Alaska. The average air temperature at the site is −11°C, and the soils consist of imported sand and gravel fill material, making them oligotrophic (nutrient-poor) and generally dry. Surrounding and below the fill are tundra soils. Approximately 1–1.5 m below the soil surface is permafrost, which is unconsolidated material or bedrock that has been continuously frozen for at least 2 consecutive years. In the summer, the active-layer soils thaw and then refreeze in the fall. Careful considerations of biologically limiting characteristics such as these are important when determining effective bioremediation technologies in cold regions. The petroleum contamination at the site is heterogeneous due to various spills that occurred at different times. Three main contaminated sites within the former NARL include the Bulk-Fuel Storage Facility, the Airstrip site, and the Powerhouse site.

In 1995, field mesocosms were constructed at the Bulk-Fuel Storage Facility for a feasibility study to determine the utility of using fertilizers for the biostimulation of petroleum degradation (Braddock et al., 1997). The concentration of total petroleum hydrocarbons at this site ranged from 250 to 860 mg kg^{-1} of soil. Three different levels of water-soluble fertilizer were tested. They included 100 mg nitrogen (N) and 45 mg phosphorus (P) kg^{-1} soil (low), 200 mg N and 90 mg P kg^{-1} soil (medium), or 300 mg N and 135 mg P kg^{-1} soil (high). The treated and control soils were sampled at the time of mesocosm construction and 6.5 weeks after the initial treatment. The number of gasoline-degrading bacteria, as measured through plate counts, was greatest at the lowest fertilizer rate; reduced numbers at the higher rates were attributed to salt-induced microbial osmotic stress. The low level of fertilizer also corresponded to significantly lower concentrations of measured contaminants.

In 2003, based on the results of the mesocosm study, 2,900 m^3 of contaminated soil at the Bulk-Fuel Storage Facility was treated using landfarming (McCarthy et al., 2004). Fertilizer was added to the contaminated soil in concentrations of 100 mg N and 50 mg P kg^{-1} soil 1 or 2 times depending on location and degree of degradation observed. The soil was mixed intensively over a 54-day period. After 31 days, benzene, toluene, ethylbenzene, and xylene (BTEX) were not detected in the northeastern area of the treatment site and other petroleum compounds were below target levels, indicating additional fertilization was unnecessary. The remaining areas received an additional application of fertilizer to further stimulate degradation. At the end of the study, most, but not all of the measured contaminants were below cleanup levels. Therefore, landfarming proved to be an effective bioremediation technology for weathered petroleum contaminated soils in the Arctic.

From 2016–18, contaminated soils at the Airstrip and Powerhouse sites at the former NARL were treated through a combination of in situ biostimulation (addition of nitrogen and phosphorus) and phytoremediation (Fig. 22.3; Barbato et al., 2019). For the biostimulation treatment, fertilizer in the form of 2 g N kg^{-1} soil water was added to each 25 m^2 treatment plot. For the phytoremediation treatment, Arctared fescue, annual ryegrass, and white clover were sown to each plot. Applications of the fertilizer and seeds occurred in 2016 and 2018. In addition to petroleum chemistry, also measured were gene abundances of bacteria (i.e., 16S rRNA gene) known to degrade hydrocarbons and genes involved in hydrocarbon degradation (i.e., alkane monooxygenase and cytochrome P450). The most effective bioremediation strategy differed between the two sites. At the Powerhouse site, the combination of fertilizer and plants was the most effective at lowering contaminant concentrations. At the Airstrip site, plants alone were effective for two zones, while natural attenuation was most effective for the third zone. Genes encoding for both alkane monooxygenase and cytochrome P450 were more abundant in the fertilizer and plant-treated soils. Furthermore, hydrocarbon-degrading bacteria such as *Flavobacterium*, *Arthrobacter*, *Rhodococcus*, and *Pseudomonas* were more prevalent in the phytoremediation plots. These contrasting results emphasize that, for bioremediation to be most effective, the treatment(s) selected should be customized according to the characteristics of both the site involved and the contaminants of concern.

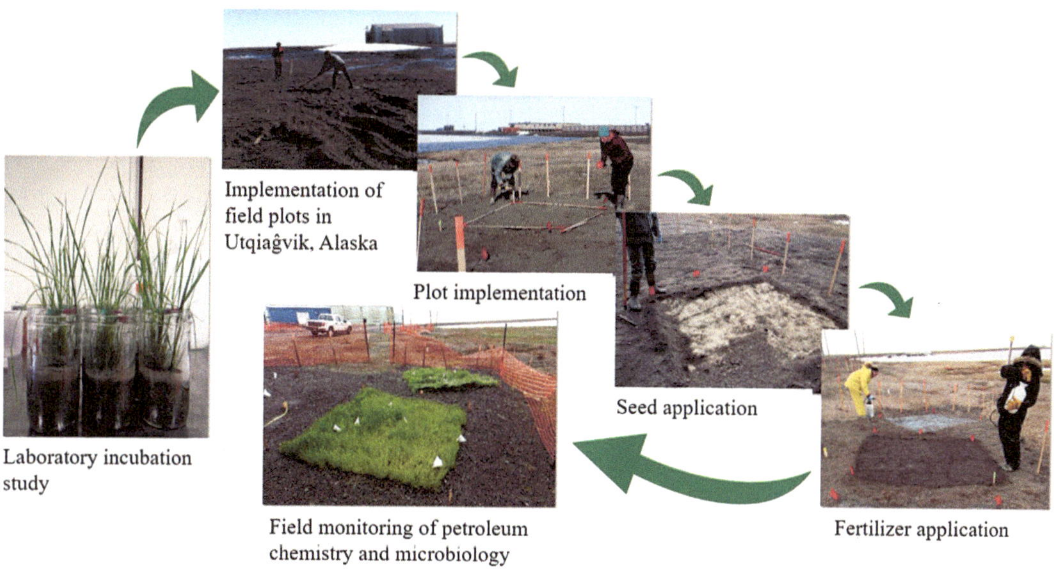

FIGURE 22.3 **Phytoremediation and biostimulation of petroleum-contaminated soils in the high Arctic.** Success from the laboratory study resulted in the implementation of the bioremediation technologies in the field. *Source: Image credit: Stacey Doherty and Robyn Barbato.*

Future of bioremediation

Bioremediation has become an integral strategy in the restoration of many contaminated environments. This is because bioremediation is a natural process that recycles the elements in contaminants rather than simply entombing them as liabilities for future generations. Furthermore, the public views bioremediation favorably, and many agencies worldwide are promoting its use for the restoration of sites damaged by environmental contaminants.

Many of the constraints that previously limited using bioremediation to treat environmental contamination have been addressed. Constraints range from scientific and technological issues to politics, regulatory concerns, and societal inertia. Research has addressed many of the scientific questions, and microbial degradation pathways are continually being identified. Key issues for implementation of bioremediation are cost relative to alternative technologies, the need for effective mixing of the system components, and suitable methods for monitoring the progress of the remediation process.

New techniques to genetically engineer microorganisms seem attractive due to the organisms' enhanced ability to break down recalcitrant chemicals. However, adding engineered organisms directly to contaminated sites requires careful consideration and adherence to policies regarding genetically modified organisms. Furthermore, introduced microorganisms may be ineffective because of competition with indigenous community and/or niche specificity. Designing the treatment strategy to offer a competitive advantage to the engineered organism is critical to its success (Janssen and Stucki, 2020). Liu et al. (2019) offer a comprehensive review of promising laboratory studies where bacteria have been successfully engineered and the associated challenges with using engineered bacteria to remediate environmental contamination.

Current areas of bioremediation research include its extension to emerging chemicals and co-contaminants, defining pathways that chemically and permanently incorporate contaminant residues into soil fractions, improving monitoring techniques, gaining a better understanding of plant and plant-microbe technologies, and applying bioremediation to contaminants broadly distributed on or near the surface of soils.

Summary

This chapter discussed bioremediation of contaminated soils and various strategies to restore these soils back to a relatively healthy state. Our discussion has covered:

- the problems that are being addressed by bioremediation, including the diversity and magnitude of soil contaminants and some of the associated risks,
- advantages of bioremediation and definitions of commonly used bioremediation terminology,
- practical considerations for selecting bioremediation,
- traditional and new monitoring approaches,
- general strategies for bioremediation,
- recent advances in measuring bioremediation efficacy,
- case studies of bioremediation, and
- brief highlights of future directions in bioremediation.

Bioremediation has become an accepted remediation technology and is continually evolving. Although many of the issues that formerly limited its use have been addressed, there are others that require further investigation. Bioremediation is a technology that is intrinsically interdisciplinary. Key roles of soil microbiology in bioremediation are: (1) understanding degradation pathways for contaminants and finding ways to promote these processes in soils, which implies that one understands both contaminant and microbial interactions with soil, and (2) using this knowledge to help design bioremediation systems and appropriate monitoring strategies.

Study questions

1. A company has asked your advice regarding the results of a chemical analysis that they just received for a site contaminated with chlorpyrifos, a popular pesticide. Is it possible to determine its bioavailability by the results from the chemical extraction technique alone?
2. You are an employee at an environmental consulting firm. Your new client recently spilled 11,000 L of a proprietary compound and contaminated 4200 cubic meters of soil. The compound has similar properties as crude oil, being toxic and lipophilic. The soil has been excavated and stockpiled on an impermeable liner, and then covered with plastic to prevent rain from infiltrating. The company has received unsolicited bids for bioventing, phytoremediation, and monitored natural remediation. The costs are all within 5% of each other. Which method or combination of methods would you choose and why?
3. What criteria must be satisfied for bioremediation to be considered a viable option for cleanup of a contaminated site?
4. What are some limitations to the successful application of bioremediation?
5. Why might bioremediation be preferred over physical and chemical technologies for remediation of contaminated sites?

References

Alahabadi, A., Ehrampoush, M.H., Miri, M., Aval, H.E., Yousefzadeh, S., Ghaffari, H.R., Ahmadi, E., Talebi, P., Fathabadi, Z.A., Babai, F., Nikoonahad, A., 2017. A comparative study on capability of different tree species in accumulating heavy metals from soil and ambient air. Chemosphere 172, 459–467.

Alexander, M., 1999. Biodegradation and bioremediation, second ed. Academic Press, Ca.

Arslan, M., Imran, A., Khan, Q.M., Afzal, M., 2017. Plant–bacteria partnerships for the remediation of persistent organic pollutants. Environ. Sci. Pollut. Res. 24, 4322–4336.

Baker, K.H., Herson, D.S., 1994. Bioremediation. McGraw-Hill, New York.

Barbato, R.A., Jarvis, S., Foley, K., Jones, R., and C.M. Reynolds. 2017. 2016 Pilot study of soils at the former Naval Arctic Research Laboratory Airstrip and Powerhouse sites in Utqia vik, Alaska. Naval Facilities Engineering Command Report.

Barbato, R.A., Doherty, S., Jones, R., Messan, K., Thurston, A., 2019. 2018 Continuation of bioremediation study of soils at the former Naval Arctic Research Laboratory airstrip and powerhouse sites in Utqia vik, Alaska. Naval Facilities Engineering Command Report.

Barron, M.G., Vivian, D.N., Heintz, R.A., Yim, U.H., 2020. Long-term ecological impacts from oil spills: comparison of Exxon Valdez, Hebei Spirit and Deepwater Horizon. Environ. Sci. Technol. 54, 6456–6467.

Boufadel, M.C., Geng, X., Short, J., 2016. Bioremediation of the Exxon Valdez oil in Prince William Sound beaches. Mar. Pollut. Bull. 113, 156–164.
Braddock, J.F., Ruth, M.L., Catterall, P.H., Walworth, J.L., McCarthy, K.A., 1997. Enhancement and inhibition of microbial activity in hydrocarbon-contaminated arctic soils: implications for nutrient-amended bioremediation. Environ. Sci. Technol. 31, 2078–2084.
Busby, R., Barbato, R.A., Jung, C., Morozova, K.A., Bednar, A.J., Bray, A.L., Milam, J.M., Smith, J.C., Indest, K.J., 2018. Photoperiod and soil munition constituent effects on phytoaccumulation and rhizosphere interactions in boreal vegetation. Water Air Soil Pollut. 229, 380.
Cunningham, S.D., Mather, M.S., 1995. Applying soil science skills in the remediation business. In: Skipper, H.D., Turco, R.F. (Eds.), Bioremediation: Science and applications, vol. 43. Soil Science Society of America Special Publication, Madison, WI, pp. 273–292.
Das, S. (Ed.), 2014. Microbial Biodegradation and Bioremediation. Elsevier, Waltham, MA.
Doty, S.L., Freeman, J.L., Cohu, C.M., Burken, J.G., Firrincieli, A., Simon, A., Khan, Z., Isebrands, J.G., Lukas, J., Blaylock, M.J., 2017. Enhanced degradation of TCE on a superfund site using endophyte-assisted poplar tree phytoremediation. Environ. Sci. Technol. 51, 10050–10058.
Errington, I., King, C.K., Wilkins, D., Spedding, T., Hose, G.C., 2018. Ecosystem effects and the management of petroleum-contaminated soils on subantarctic islands. Chemosphere 194, 200–210.
Finklea, H.C., Fontenot, M.F. Jr., 1995. Accelerated bioremediation of triazine contaminated soils: A practical case study. In: Skipper, H.D., Turco, R.F. (eds.), Bioremediation: Science and Applications, vol. 43. Soil Science Society of America Special Publication, Madison, Wi, pp. 221–236.
Fraraccio, S., Strejcek, M., Dolinova, I., Macek, T., Uhlik, O., 2017. Secondary compound hypothesis revisited: selected plant secondary metabolites promote bacterial degradation of cis-1,2-dichloroethylene (cDCE). Sci. Rep. 7, 1–11.
Gale, E.F., 1952. The Chemical Activities of Bacteria. Academic Press, New York.
Gerhardt, K.E., Huang, X.D., Glick, B.R., Greenberg, B.M., 2009. Phytoremediation and rhizoremediation of organic soil contaminants: potential and challenges. Plant Science 176, 20–30.
Guarino, C., Spada, V., Sciarrillo, R., 2017. Assessment of three approaches of bioremediation (Natural Attenuation, Landfarming and Bioagumentation–Assisted Landfarming) for a petroleum hydrocarbons contaminated soil. Chemosphere 170, 10–16.
Hinton, S.M., 1995. Bioremediation of Valdez and Prall's Island oil spill. In: Skipper, H.D., Turco, R.F. (eds.), Bioremediation: Science and Applications, vol. 43. Soil Science Society of America Special Publication, Madison, WI, pp. 211–220.
Integrated Waste Management Board, 2000. Used Oil Recycling Rate Annual Report. California.
Janssen, D.B., Stucki, G., 2020. Perspectives of genetically engineered microbes for groundwater bioremediation. Environ. Sci. Process. Impacts 22, 487–499.
Jantzen, C.E., Annunziato, K.A., Bugel, S.M., Cooper, K.R., 2016. PFOS, PFNA, and PFOA sub-lethal exposure to embryonic zebrafish have different toxicity profiles in terms of morphometrics, behavior and gene expression. Aquat. Toxicol. 175, 160–170.
Jorgensen, B.B., 1989. Biochemistry of chemoautotrophic bacteria. In: Schlegel, H.G., Bowien, B. (Eds.), Autotrophic Bacteria. Science Tech Publishers, Madison, WI, pp. 117–146.
Lee, K.Y., Strand, S.E., Doty, S.L., 2012. Phytoremediation of chlorpyrifos by *Populus* and *Salix*. Int. J. Phytoremed. 14, 48–61.
Leigh, M.B., Fletcher, J.S., Fu, X., Schmitz, F.J., 2002. Root turnover: An important source of microbial substrates in rhizosphere remediation of recalcitrant contaminants. Environ. Sci. Technol. 36, 1579–1583.
Liu, L., Bilal, M., Duan, X., Iqbal, H.M., 2019. Mitigation of environmental pollution by genetically engineered bacteria—Current challenges and future perspectives. Sci. Total Environ. 667, 444–454.
Major, D.W., McMaster, M.L., Cox, E.E., Edwards, E.A., Dworatzek, S.M., Hendrickson, E.R., Starr, M.G., Payne, J.A., Buonamici, L.W., 2002. Field demonstration of successful bioaugmentation to achieve dechlorination of tetrachloroethene to ethene. Environ. Sci. Technol. 36, 5106–5116.
Maila, M.P., Cloete, T.E., 2004. Bioremediation of petroleum hydrocarbons through landfarming: Are simplicity and cost-effectiveness the only advantages? Rev. Environ. Sci. Biotechnol. 3, 349–360.
McCarthy, K., Walker, L., Vigoren, L., Bartel, J., 2004. Remediation of spilled petroleum hydrocarbons by in situ landfarming at an arctic site. Cold Reg. Sci. Technol. 40, 31–39.
Pauget, B., Gimbert, F., Scheifler, R., Coeurdassier, M., De, A., Vaufleury, 2012. Soil parameters are key factors to predict metal bioavailability to snails based on chemical extractant data. Sci. Total Environ. 431, 413–425.

Riser-Roberts, E., 1992. Bioremediation of petroleum contaminated sites. CRC Press, Boca Raton, Fl.

Rylott, E.L., Budarina, M.V., Barker, A., Lorenz, A., Strand, S.E., Bruce, N.C., 2011. Engineering plants for the phytoremediation of RDX in the presence of the co-contaminating explosive TNT. New Phytol. 192, 405–413.

Safdari, M.S., Kariminia, H.R., Rahmati, M., Fazlollahi, F., Polasko, A., Mahendra, S., Wilding, W.V., Fletcher, T.H., 2018. Development of bioreactors for comparative study of natural attenuation, biostimulation, and bioaugmentation of petroleum-hydrocarbon contaminated soil. J. Hazard. Mater. 342, 270–278.

Scheifler, R., Schwartz, C., Echevarria, G., De Vaufleury, A., Badot, P.M., Morel, J.L., 2003. "Nonavailable" soil cadmium is bioavailable to snails: evidence from isotopic dilution experiments. Environ. Sci. Technol. 37, 81–86.

Sharma, B., Dangi, A.K., Shukla, P., 2018. Contemporary enzyme based technologies for bioremediation: a review. J. Environ. Manage. 210, 10–22.

Singer, A.C., Crowley, D.E., Thompson, I.P., 2003. Secondary plant metabolites in phytoremediation and biotransformation. Trends Biotechnol. 21, 123–130.

Skipper H.D., Turco, R.F., 1995. Bioremediation: Science and Applications, vol. 43. Soil Science Society of America Special Publication, Madison, WI.

Sullivan, K.A., 2017. Brownfields remediation: Impact on local residential property tax revenue. J. Environ. Assess. Policy Manage. 19, 1750013.

U.S. Environmental Protection Agency (U.S. EPA), 2018. BR data files for analyses: Quantity of RCRA Hazardous Waste Generated and Managed. Available from: https://cfpub.epa.gov/roe/indicator.cfm?i=54#1.

Walworth, J.M., Woodard, C.R., Braddock, J.F., Reynolds, C.M., 1997. Enhancement and inhibition of soil petroleum biodegradation through the use of fertilizer nitrogen: An approach to determining optimum levels. Soil Sediment Contam. 6, 465–480.

Xie, S., Sun, W., Luo, C., Cupples, A.M., 2011. Novel aerobic benzene degrading microorganisms identified in three soils by stable isotope probing. Biodegradation 22, 71–81.

Yergeau, E., Tremblay, J., Joly, S., Labrecque, M., Maynard, C., Pitre, F.E., St-Arnaud, M., Greer, C.W., 2018. Soil contamination alters the willow root and rhizosphere metatranscriptome and the root–rhizosphere interactome. ISME J. 12, 869–884.

CHAPTER

23

Biological control of soilborne plant pathogens and nematodes

James H. Graham[a], *Sarah L. Strauss*[b]

[a]University of Florida, Citrus Research and Education Center, Lake Alfred, FL, United States; [b]University of Florida, Southwest Florida Research and Education Center, Immokalee, FL, United States

"The research worker in his laboratory and the farmer on his fields can do nothing to the soil that does not induce some change in the activity and balance of its microbial population. There must therefore be some element of biological control, or of its reverse, in almost all disease control practices."
—*S.D. Garrett*

Soil microorganisms and plants are in a dynamic equilibrium that reflects change through space and time in the belowground environment and the biological interactions among organisms. Soil microbial interactions can be beneficial, neutral, or detrimental to the plant. If soil microorganisms cause continued irritation of the plant, an injurious alteration of a process(es) of energy utilization occurs that is recognized as plant **disease**. Affected cells and tissues of diseased plants are usually weakened or destroyed by the disease-causing agent due to alterations of normal physiological functions. The kinds of cells and tissues that become infected determine the type of physiological function that will be initially affected. For example, infection of the root can interfere with absorption of water and nutrients from soil.

The **pathogen** is the biotic agent that causes disease. The disease organism lives on or in the host plant and obtains nutrition from the host. The susceptibility of the plant will determine whether the pathogen infects and establishes a (detrimental) nutritional relationship with the host. The environment will affect host susceptibility and the activity of the pathogen (Fig. 23.1). Soil is the source of many microorganisms, including bacteria (Chapter 5), fungi (Chapter 6), nematodes and protozoa (Chapter 8), viruses (Chapter 9), and insects and mites, some of which are pathogens of plants. Many species of protozoa and microarthropods are also capable of modifying the rhizosphere microflora and acting as plant-pathogen deterrents (Curl et al., 1988). As the relationship between viruses and pathogen suppression is still relatively unknown, this chapter will focus on the first three groups of soil microorganisms as plant pathogens and biological control agents.

Principles and Applications of Soil Microbiology. http://dx.doi.org/10.1016/B978-0-12-820202-9.00023-X
Copyright © 2021 Elsevier Inc. All rights reserved.

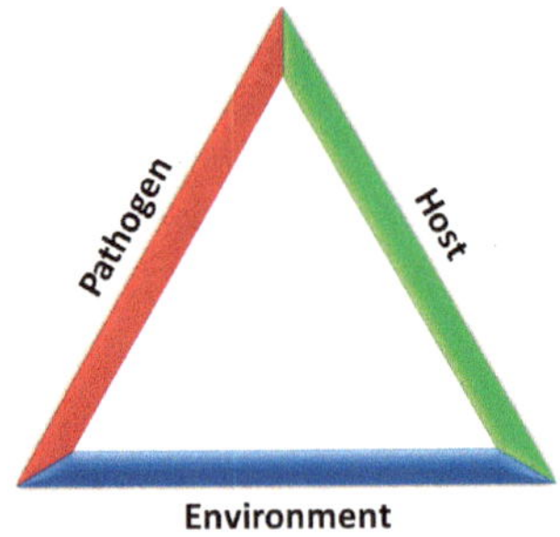

FIGURE 23.1 **The disease triangle.** Disease is related to a complex interaction of three components: pathogen, host, and environment.

Fundamental biological control concepts

Biological control of plant disease results from the reduction of inoculum of a pathogen or its disease-producing activity accomplished through one or more organisms. Biological control interactions in the belowground environment can occur in plant roots, the rhizosphere, or bulk soil. Plant roots are the site where the soilborne pathogen initially infects host tissue. The **rhizosphere** is the narrow zone of soil surrounding the plant roots that contains root exudates, sloughed roots, and large populations of soil microorganisms of various nutritional groupings (Chapter 11). Bulk soil, beyond the rhizosphere, has lower microbial numbers and activity. Biological control is usually operative in the rhizosphere but can occur in other arenas as well.

Antagonists are biological agents that interact (Chapter 10) with pathogens to reduce the inoculum or disease-producing activities through one or more of the following mechanisms:

- **Antibiosis**—the inhibition or lysis (death) of the pathogen mediated by metabolic products of the antagonist; these products include lytic agents, enzymes, volatile compounds, and other toxic substances
- **Competition**—when the antagonist reduces the pathogen's activities by restricting the supply of limited resources, such as organic and inorganic nutrients, growth factors, oxygen, or space
- **Hyperparasitism**—when the antagonist invades the pathogen by secreting enzymes such as chitinases, cellulases, glucanases, and other lytic enzymes

Plant-soil-microbial equilibrium

Generally, the soil environment changes less rapidly than the aerial environment; thus, soil microorganisms have become adapted to a relatively stable habitat. Populations of antagonists, pathogens, and host plants fluctuate within certain limits in the soil ecosystem through abiotic and biotic interactions. Slight changes in soil environmental conditions may have profound effects on populations of the pathogen and antagonist alike. Changes produced by agricultural practices tend to be varied, rapid, and ongoing, which permits little time for reestablishment of the biological equilibrium. Intensive use of fertilizers, pesticides, tillage, irrigation, and other crop management practices for selected higher-yielding cultivars has resulted in increased or decreased severity or incidences of plant disease. The replacement of a diverse plant community with a single highly vulnerable genotype (monoculture) shifts the biological equilibrium toward development of plant disease epidemics that occur less frequently in undisturbed ecosystems.

The occurrence of a plant disease epidemic indicates that one or more of the following conditions exist:

- The pathogen is highly virulent or has a high inoculum density
- The abiotic environment is more favorable for the pathogen than for the host or the antagonist
- The host plant is genetically homogeneous, highly susceptible, and continuously or extensively grown
- The antagonist(s) is absent, or in low populations, due to a lack of a favorable environment or inhibition by other microorganisms

Biological equilibrium in crop disease situations can be manipulated through the integrated use of chemical controls, such as fungicides and nematicides; cultural controls, such as sanitation or crop rotation; and biological control methods, such as plant resistance or introduction of antagonists. Usually, biological controls are subtle and operate more slowly, but can be more stable and longer lasting than chemical and cultural control measures. Biological controls should be integrated with other control measures because different methods are effective at different times and locations in the soil-plant continuum under varying conditions.

Objectives of biological control

Pathogen-suppressive soils are defined by Cook and Baker (1983) as soils in which one of the following conditions occurs:

- The pathogen does not establish or persist
- The pathogen establishes but causes little or no damage
- The pathogen causes disease, but the disease becomes progressively less severe though the pathogen persists in soil

Most soils suppress soilborne pathogens to some degree. Soils may be suppressive to pathogens in a general or a specific sense. **General suppression** of a pathogen is directly related to the total amount of microbial activity in the soil or plant at a critical time in the life cycle of the pathogen. The types of active microorganisms are probably less important than the microbial biomass that competes with the pathogen for carbon, nitrogen, and energy, or causes inhibition through more direct forms of antagonism. General suppression is often enhanced by the addition of organic matter, certain agronomic practices, or the buildup of soil fertility, all of which increase soil microbial activity. **Specific suppression** is superimposed over a background of general suppression but involves more specific effects of individual or select groups of microorganisms antagonistic to the pathogen during some stage of its life cycle. Transferability, the ability to move antagonistic properties to a non-suppressive soil, is a key characteristic of specific suppression, and the term "transferable suppression" has been used synonymously with specific suppression. Suppressive soils undoubtedly owe their activity to a combination of general and specific suppression.

Most suppressive soils maintain their activity when brought into the greenhouse or laboratory, which facilitates the assessment of their properties and mechanisms for suppression under more controlled and reproducible conditions. The strategy to determine the microbiological basis of suppressive soils is to identify the contribution of specific microbial groups. The first step is to determine whether the suppressiveness can be destroyed by pasteurization (moist heat, 60°C for 30 min), selective biocides such as novobiocin and chloropicrin, or whether harsher conditions are required such as autoclaving and gamma irradiation. Both general and specific suppression are eliminated by autoclaving and gamma irradiation of soil. General suppression is reduced but not eliminated by soil fumigation and usually survives pasteurization. Another step, which allows confirmation of the biological basis of suppression, involves transfer of suppressiveness to a conducive soil. The impact of edaphic factors on disease development in soil transfer studies is minimized when suppressive and conducive soils are diluted into a common background soil, allowing direct comparison of the introduced microbial components.

Biological control with resident antagonists

The initial objective of biological control is to maximize soil suppressiveness through the manipulation of resident antagonists. These are generally microorganisms that resided in equilibrium with a pathogen and plant in soil ecosystems before the intervention of agriculture. If the indigenous biological control antagonists are no longer present or are inadequate, introductions of antagonists (one-time, occasional, or repeated) that have the tendency to be self-maintaining may be necessary. Specific approaches to biological control of plant disease are summarized in Fig. 23.2 and are discussed further in the remainder of this chapter.

As a result of certain crop or soil management practices, some field or greenhouse soils are naturally suppressive, even though a susceptible host and soil populations of the pathogen are present. Although the exact means by which a particular instance of biological control develops is often poorly understood, pathogen-suppressive soils have been associated consistently with the following crop or soil management practices:

- Crop rotations with non-susceptible plant hosts or fallowing (no crop), which provide time for resident antagonists to displace soilborne pathogens or for pathogen propagules to die from starvation
- Soil tillage, which accelerates the displacement of certain pathogens by resident antagonists in crop residues
- Amendments of soil with composts (Chapter 24) and other organic materials, which stimulate the activities of resident antagonists or suppress the activities of pathogens
- Selective soil treatments with steam, biocides, anaerobic conditions, or solarization, which suppress or eliminate the pathogen but not resident soil saprophytes
- Use of certain cultivars of a host, which select for specific rhizosphere antagonists

In each case, suppression of the activities of the pathogen by resident antagonists occurs prior to infection of the host root (Figs. 23.3 and 23.4). The following examples of suppressive

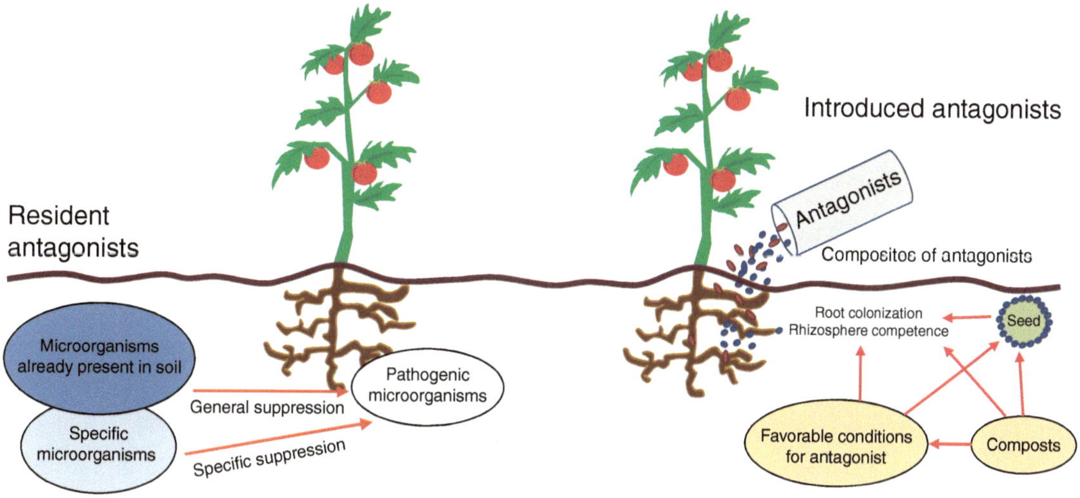

FIGURE 23.2 Biological control of soilborne plant diseases using microbial antagonists can occur with resident or introduced antagonists. Resident antagonists confer general or specific suppression. Introduced antagonists can be single or composites of antagonists that exhibit either rhizosphere competence and/or root colonization. Favorable conditions for these antagonists, such as compost, can influence competence and colonization. Seed treatments are an additional method for introducing antagonists.

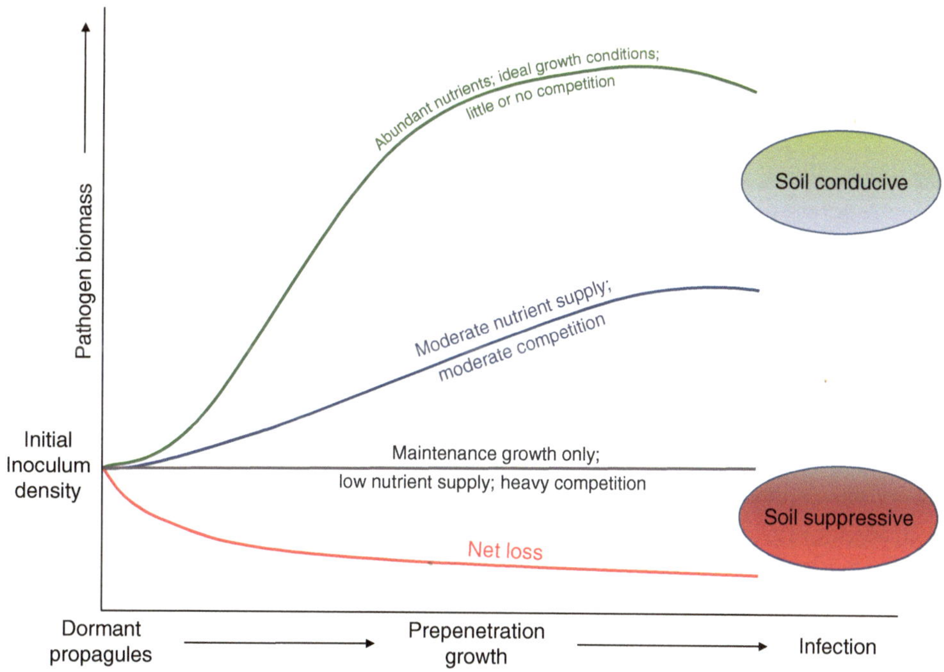

FIGURE 23.3 Possible changes in pathogen biomass near a host root. As dormant propagules go through prepenetration growth prior to infection, the pathogen biomass may increase or decrease depending on whether the soil is conducive or suppressive to disease development. *Source: Adapted from Cook and Baker (1983).*

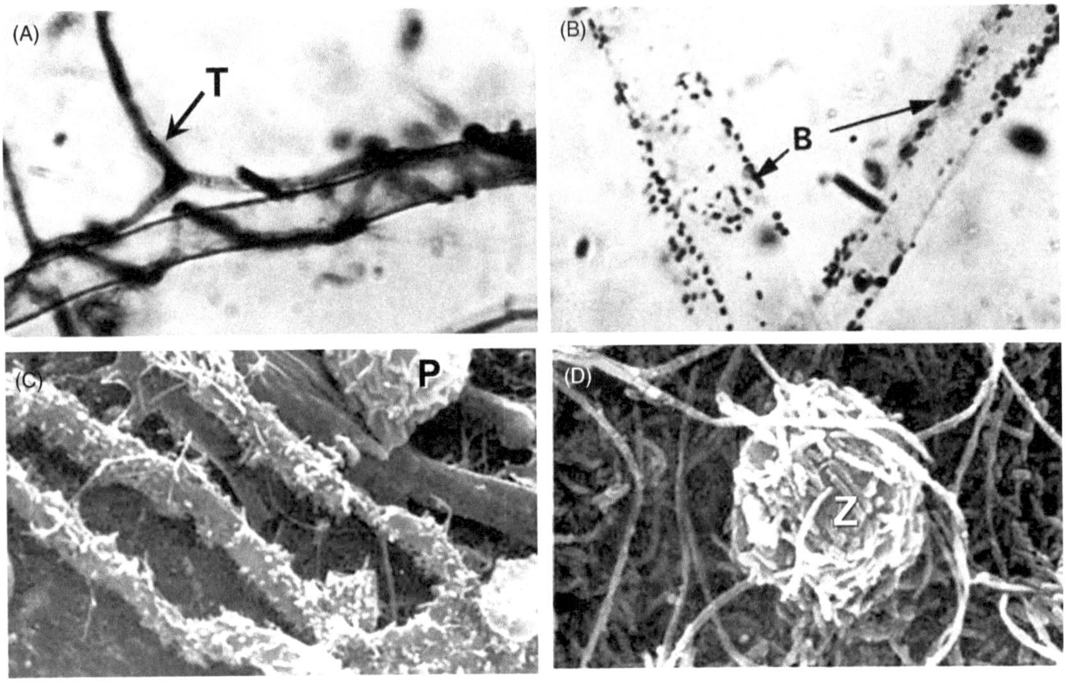

FIGURE 23.4 **Colonization of *Phytophthora cinnamomi* by soil antagonists.** (A) Hypha colonized by an antagonistic *Trichoderma* sp. (T); (B) bacteria colonizing the surface of lysed hyphae; (C) scanning electron micrograph (SEM) of hyphae incubated in a suppressive soil showing intense colonization by bacteria and presence of predatory protozoa (P); (D) SEM of *P. cinnamomi* zoospore (Z) intensely colonized by bacteria and actinomycetes in the rhizosphere. *Source: Adapted from Malajczuk (1983). Photos courtesy of N. Malajczuk.*

soils demonstrate how some of these management practices contribute to establishment of biological control through general and specific suppression of the pathogen.

General suppression: anaerobic soil disinfestation

Anaerobic soil disinfestation (ASD), also called biological soil disinfestation, is a form of selective soil treatment that alters the soil environment, resulting in general suppression of several key soilborne pathogens. While suppression can be similar to those under chemical fumigation, ASD is a biological method. To induce suppression, anaerobic conditions are created in the soil by introducing a carbon source, irrigating to field capacity, and covering the soil with an impermeable plastic tarp (Strauss and Kluepfel 2015). The combination of anaerobic conditions, increased soil temperature, changes to the overall soil microbial community composition, and production of volatile organic compounds contribute to suppression of pathogens, including *Pythium ultimum*, *Fusarium oxysporum*, *Rhizoctonia solani*, *Verticillium dahlia*, and *Ralstonia solanacearum*. Nematode pathogens have also been suppressed by ASD, including *Pratylenchus penetrans* and *Meloidogyne incognita*. The types of carbon sources, length of treatment, and soil type can influence the effectiveness of ASD pathogen suppression. Carbon sources typically include either solid (rice bran, composted broiler litter) or liquid (molasses or ethanol) organic materials. Treatment time occurs for approximately 2–10 weeks, depending on the crop and location.

General and specific suppression: take-all decline

Take-all disease of wheat, caused by the fungus *Gaeumannomyces graminis* var. *tritici*, is controlled in many areas of the world by not growing wheat or barley in the same field more often than every other year (Cook and Baker, 1983). The pathogen depletes its nutrient supply in fragments of roots and stems or is displaced by competing microorganisms in most soils during the year the host is absent. An alternative approach is to manage the disease through continuous crops of wheat or barley resulting in take-all decline. This biological control is thought to include the combined effects of: (1) competition from fungi and bacteria in root fragments, which reduces the inoculum density of the pathogen (general suppression), and (2) antagonistic microorganisms in the rhizosphere and in young root lesions, which limit infection and secondary spread of the pathogen by runner hyphae along the roots (specific suppression). The agents responsible for take-all decline are transferable by soil from field to field, are eliminated by aerated steam at 60°C or higher for 30 min, and are sensitive to the biocides chloropicrin and methyl bromide (common soil fumigants).

General suppression of the pathogen in root-fragment inoculum is partly due to nonpathogenic, saprophytic *Fusarium* spp. The fungus *Fusarium oxysporum* reduces root infection by the take-all pathogen when added with inoculum to pasteurized soil. The competitive effects of microorganisms on *G. graminis* var. *tritici* in the food base are promoted by tillage together with delayed seeding date. Tillage temporarily increases microbial activity and accelerates breakdown of fragments by exposing fresh surfaces of the residue for microbial colonization while promoting soil aeration. Some tillage is required to promote decline of the take-all pathogen in consecutive crops of wheat. Delayed seeding is beneficial because more time is allowed for displacement of the pathogen by the antagonists.

Specific suppression of the take-all pathogen by antagonistic fluorescent pseudomonads is widely implicated in take-all decline (Weller et al., 2002) because they:

- are well adapted to the rhizosphere environment
- use a diversity of organic substrates
- synthesize a variety of antifungal metabolites inhibitory to *G. graminis* var. *tritici*
- increase in number on roots with take-all lesions
- are eliminated by soil pasteurization

The most effective biological control strains of *Pseudomonas* spp. isolated from take-all decline soils produce 2,4-D acetylphloroglucinol (2,4-DAPG) (Box 23.1). This antibiotic is synthesized in the rhizosphere and is responsible for the biological control activity. Detection by specific PCR primers and hybridization probes (Chapter 4) of producers of 2,4-DAPG during wheat monoculture in Washington State has led to the hypothesis that *Pseudomonas* spp. are the major contributors to specific suppression. Suppression of the pathogen by 2,4-DAPG producers can occur during the parasitic phase of *G. graminis* var. *tritici*, when the fungus is growing on the roots, as well as during the saprophytic phase, when the fungus is in the debris and growing toward the roots of the next host.

BOX 23.1

Role of antibiotic production in specific suppression of take-all decline

Four criteria should be met to define the function of a microbial population in soil suppressiveness:

- The ability of the population to perform a suppressive function
- The presence in suppressive soils of the population at densities above the threshold required for disease control, and its absence from conducive soils
- Duplication of the function in conducive soils by addition of populations obtained from a suppressive soil with similar physical and chemical properties
- Detection of a functional activity in situ, such as antibiotic production

To investigate the role of antibiotics in specific suppression of take-all decline, PCR primers PHI2a and PHI2b were developed from sequences of *phlD*, a key gene within the 2,4-DAPG biosynthetic operon, which amplify a 745-base pair fragment from DNA of 2,4-DAPG-producing *Pseudomonas* strains (*phlD*$^+$). The *phlD*$^+$ strains were present on roots of wheat grown in take-all decline soils at populations (10^5 CFU g^{-1} root) above the threshold for take-all control. Successive cultivation of wheat demonstrated an inverse correlation between high densities of 2,4-DAPG producers and severity of take-all. The suppressiveness was lost when *phlD*$^+$ strains were eliminated by soil pasteurization. The addition of take-all decline soil to virgin soil resulted in the establishment of population densities of 2,4-DAPG producers above the threshold required for take-all control. The introduction of a 2,4-DAPG producing strain, *P. fluorescens* Q8rl-96, from a take-all decline soil into steamed or raw conducive soil at low doses controlled take-all to a similar level to that of the complementary take-all decline soil. Finally, 2,4-DAPG was detected in the rhizosphere of wheat grown in take-all decline soil at an average concentration of 19 ng g^{-1} root fresh weight, but was not detected on roots grown in conducive soil (Weller et al., 2002). These data demonstrate the important role of antibiotic-producing bacteria in disease suppression.

Specific suppression: nematode suppressive soil-decline of *Heterodera avenae*

Decline of cereal cyst nematodes, *Heterodera avenae*, on susceptible cereals grown in intensive rotation throughout Europe is one of the best-documented examples of naturally occurring specific suppression of a plant parasitic nematode. Nematode populations peak after two crops and then fall rapidly. Once natural suppressiveness is established, populations of the cereal cyst nematodes are permanently maintained at levels below the economic threshold.

Kerry and Crump (1980) demonstrated that spore populations of two principal species of nematode-parasitic fungi are much greater in soils where *H. avenae* fails to multiply. These two parasites are widely distributed in cereal-growing soils in England and northern Europe. *Nematophthora gynophila* invades cysts of nematode females as they emerge from roots and completely destroys them within a week. *Verticillium chlamydosporium* invades virgin females, which produce empty cysts; infected egg-producing females produce fewer eggs than normal and many of the eggs are parasitized. Thus, the effect of fungal parasitism is to reduce the number of females forming cysts and the number of viable eggs that cysts contain. Application of a biocide, formalin, to soil reduces spore populations of both fungi and increases nematode multiplication. The biocide effect is observed in soils where *V. chlamydosporium* and *N. gynophila* are present, but it does not occur when these fungi are absent.

Rhizobacteria and biological control

The term **rhizobacteria** has been used to describe bacteria that aggressively colonize roots (Weller, 1988). The term **plant-growth-promoting rhizobacteria (PGPR)** describes bacteria that colonize roots and cause plant growth promotion (Chapter 11) (Kloepper et al., 1980). Root-colonizing bacteria and fungi that are deleterious to plant growth but not parasitic, and therefore do not produce acute disease symptoms, are called **deleterious rhizosphere microorganisms (DRMO)** (Schippers et al., 1987). The deleterious activities include alterations in the supply of water, nutrient ions, and plant growth substances that change root functions or limit growth. The PGPR are thought to improve plant growth by colonizing the root system and preempting the establishment of DRMO on the roots. The PGPR may also promote the availability and uptake of mineral nutrients and the supply of plant growth substances. These bacteria have been shown to induce biological control through a mechanism known as **induced systemic resistance (ISR)**, which is the process of active resistance dependent on the host plant's physical or chemical barriers, activated by biotic or abiotic inducing agents (Kloepper et al., 1992). When ISR functions, application of the biological control agent to one part of the plant provides protection to a wide range of pathogens on another part of the plant. To demonstrate ISR against a root pathogen such as *Pythium aphanidermatum*, the pathogen and the inducing agent are spatially separated. For example, by splitting the root system of cucumber into two separate pots and inoculating one side with a pathogen and the other side with *Pseudomonas fluorescens* strains, it is possible to demonstrate that most strains reduce the incidence and severity of disease even though the bacterium does not come in direct contact with the pathogen (Zhou and Paulitz, 1994).

Rhizosphere and root interactions

Plants recruit bacteria into the rhizosphere with specific exudates secreted by the roots. The intrinsic ability of the bacterium to colonize root surfaces should be distinguishable from movement of propagules by mass flow of water in soil or along root surfaces. To achieve protection of subterranean plant parts, the biological control organisms must be **rhizosphere competent**, that is, capable of colonizing the root surface or rhizosphere in the presence of competitors when applied to the seed or other point source at time of planting (Chapter 11) (Ahmad and Baker, 1987). Rhizosphere-competent organisms have the ability to proliferate along the root as it develops. This ability differs from that of colonization or infection at specific points as the root encounters propagules in bulk soil. Rhizosphere-competent organisms may not only effectively protect roots against pathogens but also have other benefits, such as production of plant-growth promoting substances or enhancement of uptake of nutrients and water by plant roots. The most studied group of rhizosphere-competent organisms are rhizobacteria. Traits that contribute to rhizosphere competence include motility, attachment, growth, stress resistance, production of secondary metabolites, and survival of microorganisms (Barrett et al., 2011). In some cases, rhizosphere competence may also be specific between a strain and a plant genotype. Bacterial traits that contribute to rhizosphere competence are extracellular polysaccharides, fimbriae, flagella, chemotaxis, osmotolerance, and ability to utilize complex carbohydrates:

- **Extracellular polysaccharide (EPS)** surrounds and binds bacterial cells together to form microcolonies. The EPS protects cells against desiccation, antibacterial agents, and predators; it also aids the cells by concentrating nutrients and ions. This protection may allow an introduced bacterium to avoid displacement by indigenous microorganisms.
- **Fimbriae** are proteinaceous, filamentous appendages that function to enhance root attachment by N_2-fixing strains of species of *Klebsiella* and *Enterobacter* and the take-all suppressive strain of *P. fluorescens*.
- The importance of **flagella** for bacterial movement along roots depends on bacterial strain, plant species, or physical conditions of soil; possession of flagella does not universally promote competence. For example, non-flagellated mutants of *Pseudomonas* spp. colonize roots to the same extent as their respective wild types.
- **Chemotaxis** at a soil water potential suitable for motility may also contribute to the ability of bacteria to colonize roots. Chemotaxis would be especially important when bacteria are added with the seed in the furrow and are not initially in contact with the plant root. Chemotaxis may guide rhizobacteria to infection sites in plant roots up to several centimeters away.
- Tolerance to dry soil and low osmotic potential may aid the survival of some bacteria in the rhizosphere perhaps due in part to the rhizosphere being a drier environment as a result of plant transpiration. A relationship between osmotolerance and population size of root-colonizing strains of *Pseudomonas* species has been observed.
- Ability to use complex carbohydrates might also provide a competitive advantage. While most root exudates are readily utilized by introduced and indigenous bacteria, fewer organisms can degrade the root-tip mucilages composed of such complex carbohydrates as cellulose, hemicellulose, and pectin (Chapter 11).

Fewer rhizosphere-competent fungi have been identified for biological control. *Trichoderma* spp. are the most commonly examined. Harmon (2000) reported that mutants of the fungus *Trichoderma* that produce more cellulase than wild-type parent strains are also more rhizosphere-competent than the parent strains. Thus, rhizosphere competence of strains of *Trichoderma* spp. may be related to their ability to degrade cellulosic materials on the root surfaces. Genetic changes also have been achieved by protoplast fusion of parental strains of *Trichoderma* spp. to produce progeny with enhanced rhizosphere competence. Although all of the gene products examined are from one parental strain or the other, there is remarkable variation in morphology, growth rate, nutritional requirements, and biological control activity. Rhizosphere-competent entomopathogenic fungi have also been identified for control of the black vine weevil in strawberry production, though mechanisms for rhizosphere competence are unknown (Klingen et al., 2015).

Root colonization process

The effectiveness of introduced antagonists for biological control depends on survival, establishment, or proliferation in the infection site on the root. If a biological control agent introduced on seeds multiplies around the site of application but not along the root surfaces away from the cotyledon attachment, its suppressive effects will be limited primarily to pathogens causing seed rot and seedling diseases. In many cases, seed-applied antagonists are poorly effective in suppressing root diseases because of inadequate colonization of expanding roots. Colonization of roots by introduced organisms has been described by Weller (1988) as occurring in two phases:

- Phase I—the bacteria attach to and are transported on the elongating root tip
- Phase II—the bacteria spread locally and proliferate to the limits of the niche in competition with indigenous organisms and survive

In Phase I, introduced bacteria on seeds, seed pieces, or tubers attach to emerging roots and gain access to root exudates. As root elongation occurs, some bacteria are carried along with the root tip, whereas others are left behind as a source of inoculum on older portions of the root. Transport of bacteria continues as long as bacterial proliferation occurs at the root tip; otherwise, without multiplication, transport occurs only until the initial inoculum at the tip is diluted. Transport of bacteria by the root tip may be inefficient as tips do not always become or remain colonized. Loss of bacteria may occur by physical removal as the tip displaces soil particles, bacterial adsorption to soil particles, or because of competition from indigenous microorganisms. However, the primary limitation may be the inability of bacteria to multiply quickly enough to keep pace with the root tip, which extends rapidly through the soil (2–9 cm day^{-1}) in the zone of elongation.

In Phase II, rhizosphere-competent bacteria will multiply and survive on the root, whereas incompetent bacteria are rapidly displaced. The fate of introduced bacteria is ultimately governed by ability to compete with indigenous microorganisms. Nutrients rather than space are thought to be the limiting factor in competition during rhizosphere colonization. Bacteria tend to congregate at cell junctions where nutrients may be most abundant (Fig. 23.5). Since the carrying capacity of the rhizosphere is limited, an introduced strain must preempt indigenous bacteria if it is to become established. The distribution of bacteria along the root during Phase I, and the propagation and survival during Phase II, are profoundly affected by abiotic and biotic factors.

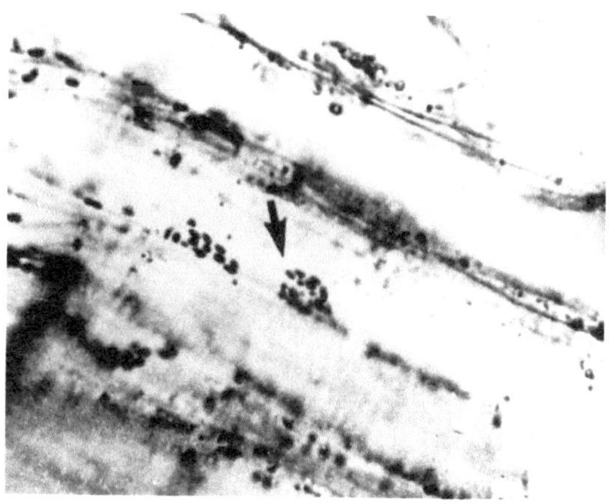

FIGURE 23.5 **Bacteria *(arrow)* proliferating at root cell junctions where exudates are most abundant.**
Source: Photo courtesy of N. Malajczuk.

Soil water potentials optimal for bacterial cell growth (−30 to −70 kPa) result in the greatest population development on roots, but establishment on roots may occur at water potentials as low as −400 kPa. Transport of bacteria along elongating roots does not require water percolation, but such water movement aids in bacterial spread. Movement through soil by water percolation is affected by bacterial characteristics, such as cell shape, size, motility, and electrostatic charge. Soil factors that influence movement of bacteria include soil type, pore-size distribution, water content, and pH. Optimum temperature for growth of bacteria in vitro may not reflect that for maximum root colonization. Optimal growth of *P. fluorescens* and *P. putida* is 25°C–30°C, but root colonization is generally greatest below 20°C. Microbial activity in soil increases as soil temperature increases; thus, better colonization at lower temperature probably reflects less competition from indigenous microflora. Plant genotype influences the quantity and composition of the rhizosphere microflora through qualitative and quantitative differences in root exudates. For example, the watermelon cultivar Crimson Sweet exhibits unusual field resistance against fusarium wilt despite years of monoculture, although it is only moderately resistant to the disease in greenhouse tests. Repeated culture of Crimson Sweet results in populations of soil microorganisms suppressive to fusarium wilt of watermelon, not only on this cultivar but also on other cultivars of watermelon when planted in the same soil (Fig. 23.6). In contrast, monoculture of other cultivars does not result in disease-suppressive populations of microorganisms. Thus, manipulation of host genotype may offer some opportunity to increase the efficiency or consistency of root colonization by introduced antagonists.

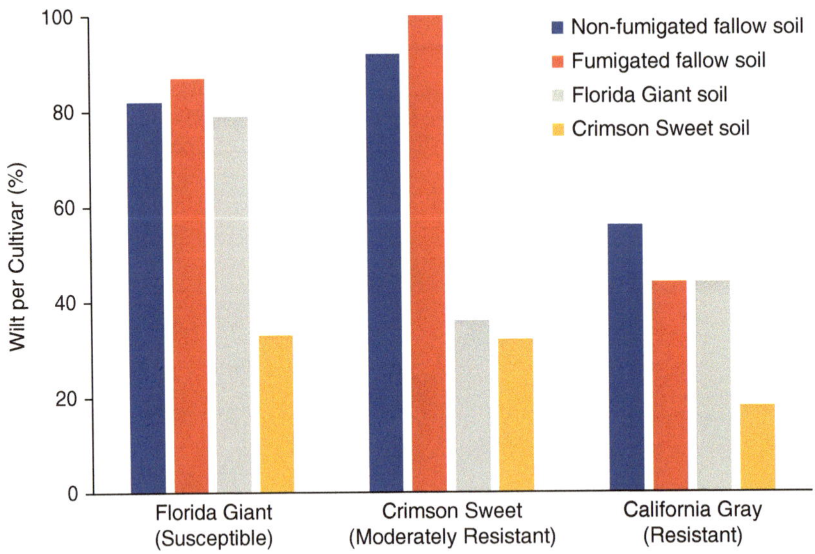

FIGURE 23.6 **Reduction in fusarium wilt when watermelon cultivars of varying susceptibility were grown in disease-suppressive soil.** Soils were from fields where the cultivars Crimson Sweet (moderately resistant) and Florida Giant (susceptible) were previously grown as monocultures. The fallow soil was not previously planted in watermelon. All four soils were infested with *Fusarium oxysporum* f. sp. *niveum* at 1.5 × 10^3 conidia g^{-1} soil. *Source: Adapted from Hopkins et al. (1987).*

Biological control with introduced antagonists

Diseases of belowground plant parts, such as seed rots, damping-off of seedlings, foot rots, root rots, and wilts, are difficult to control by chemical approaches because of the large quantities of chemicals and repeated applications required to treat the soil volume occupied by plant roots. An advantage of introduction of antagonists over chemical protectants is that microbial populations can grow from a small quantity of inoculum applied to seeds, cuttings, or as an in-furrow application, and protect belowground plant parts as they develop.

One of the most critical steps for biological control with introduced antagonists is the inoculation of the soil. Successful inoculation generally seems to require treatment of the soil that will increase the survivability of the inoculum. This includes treating the soil with steam prior to inoculation or increasing soil organic matter and resources for the added antagonists. Despite these methods, there are still only a few examples of successful commercial use of introduced organisms. However, recent advances in the technology of introducing biological control agents by encapsulation, and for improving them by selection or genetic manipulation, reflect renewed interest in the use of introduced antagonists.

Introduction of antagonists into the soil environment

The microbiological balance is altered by various environmental factors such as heat treatments, pesticides, and other chemical treatments that eliminate groups of microorganisms. Such treatment especially favors introduced strains of fungal species of *Trichoderma* or *Gliocladium* antagonistic to plant pathogens. Introduced *Trichoderma* spp. have a greater tolerance of broad-spectrum biocides than many other soil microorganisms and colonize treated soils more rapidly than other soil competitors. *Trichoderma* spp. predominate in soils after application of dilute formaldehyde, carbon disulfide, or sodium azide to soil. *Trichoderma* spp. that are resistant to a chemical when applied in combination with an amount of chemical that is sublethal to the pathogen can achieve the same effect as a lethal dose of the chemical applied alone. For example, *Trichoderma* spp. are normally sensitive to benomyl, but a benomyl-tolerant mutant of *T. harzianum* had increased colonization when benomyl was added to the soil. This mutant has a significant advantage for application with benomyl or to fields previously treated with benomyl. Strains of *T. harzianum* are also significantly more tolerant than the root pathogen *Rhizoctonia solani* to methyl bromide. Treatment of soil with sublethal doses of the fumigant can be used to kill *R. solani* with no effect on naturally occurring *Trichoderma* spp. This allows for rapid colonization of the soil with introduced *Trichoderma* spp. as well.

Trichoderma spp. can also be introduced into soils treated by solarization to control soilborne diseases. **Solarization** is accomplished by covering moistened soil in hot climates with transparent polyethylene plastic sheets, thereby trapping incoming radiation. This increases soil temperatures by 7°C–14°C, killing propagules of pathogens and weed seeds. The combination of *T. harzianum* and solarization reduces inoculum density of pathogens such as *R. solani* and *Sclerotium rolfsii*, more than either treatment alone.

Pasteuria penetrans-root-knot nematode

The bacterium *Pasteuria penetrans* has several characteristics that make it a potentially useful biological control agent against the root-knot nematode *Meloidogyne* spp., as well as other nematodes (Chen and Dickson, 1998). Bacterial populations parasitic on root-knot nematodes not only prevent reproduction but also reduce infectivity of juvenile nematodes. The spores of *P. penetrans* have morphological and biochemical features of endospores; consequently, they can tolerate environmental extremes. Spores of *P. penetrans*:

- Can be stored air-dried in soil and roots without loss of infectivity
- Have considerable tolerance of temperature extremes
- Are relatively small and therefore redistributed by percolating water

This last feature is particularly useful because spores can be applied at or near the surface of coarse-textured soils and will move down through the soil profile where nematodes are present. The obligate-parasitic nature of *P. penetrans*, however, limits its biological control potential. The soil eventually reaches a spore density that results in reduced infectivity of spore-encumbered juveniles, decreased production of infected females, and a corresponding decline in spore density of *P. penetrans*. Still, longevity of *P. penetrans* may override the need for high nematode densities in some soils.

Seed treatments with biological control agents

Biological control agents generally provide less effective control of seed and seedling root diseases, and are more variable in their performance, than are chemical pesticides. However, seed inoculation could be advantageous, as the method places the inoculum in close proximity to the developing roots and rhizosphere. There are three common methods for biological control seed treatments currently being used either by researchers or commercial companies. One method primarily used by researchers places the inoculum on the seed surface, essentially using the seed as a carrier or delivery system for the agent. After sowing, the antagonist is present at the infection site between the seed and the soil environment. Alternatively, one of the most common methods applies the biological control agents to seeds in a slurry. Though the slurry is applied uniformly, the agent may not adhere to the seed surface. Binders, such as methyl cellulose and polyvinyl acetate, are used to improve adhesion. Third, controlled hydration or dehydration treatments, known as seed priming, improve seed performance but the osmotic treatments involved in the process may adversely affect microorganisms. Solid-matrix priming with vermiculite and an osmoregulant provides a suitable medium that is nontoxic, has a high water-holding capacity, and has a favorable chemical and physical environment for biological control agents. Solid-matrix priming has been applied successfully with *T. harzianum* to provide effective and reliable biological seed treatment of a number of crops for protection against several pathogens.

Although biological control agents can provide a high level of seed protection, there has been limited success in commercializing this technology. Some reasons are that highly effective chemical pesticides are available for seed protection; these chemicals are frequently less expensive than biologicals; the shelf life of chemicals on the seeds is superior to biologicals; and the chemicals protect the seeds under a wider range of temperatures and other environmental conditions than biological control agents. However, there are opportunities to use introduced biological control agents for:

- Replacement of chemical pesticides lost to regulatory action or pest resistance and for which no adequate replacements exist

- Replacement or reduction of use of chemical pesticide in sensitive environments such as greenhouses where entry periods and worker safety are important operational issues
- Applications where biologicals accomplish tasks not possible with chemicals because of their ability to colonize the plant belowground for the life of the crop
- Organic agricultural applications where chemical pesticides are not approved for use

Biological control of crown gall

The best example of plant disease control achieved with an introduced biological control agent is crown gall. This disease is caused by the soilborne bacterium, *Agrobacterium tumefaciens*. The disease manifests itself as a disorganized and uncontrolled cell division in the host plant, usually on the roots or around the crown. A wide variety of dicotyledonous plants are susceptible to crown gall, especially stone fruits, roses, and grapes.

The majority of the bacterial genes involved for crown gall induction are plasmid-borne (Chapter 4). A portion of the tumor-inducing (Ti) plasmid, T-DNA, is transferred to the plant chromosomal DNA where it is expressed and maintained. Some T-DNA genes encode for enzymes involved in metabolism of plant growth substances that lead to production of a tumor on the plant. Other T-DNA genes code for enzymes that produce compounds called opines, which are low-molecular-weight metabolites that act as a source of nutrients for *A. tumefaciens*. The Ti plasmids are classified according to the type of opines present in the tumors they induce. Two of the Ti plasmid types are nopaline/ agrocinopine A and octopine/ agropine.

A highly successful biological control for crown gall has been developed in Australia (Kerr, 1980). The method involves inoculation of planting stock with nonpathogenic *Agrobacterium radiobacter* strain K84 that produces the bacteriocin, agrocin 84. A **bacteriocin** is a chemical that inhibits or kills closely related bacterial species. The control method involves dipping the planting material in a water suspension of K84 containing 10^7–10^8 cells mL^{-1} immediately before planting. Strain K84 has performed better than commercial chemical treatments in preventing crown gall, often providing 100% control in naturally infested soil. Strain K84 has been sold commercially in Australia since 1973 and is now used in many countries. Agrocin 84 is an adenine nucleotide (Chapter 4) that is thought to inhibit DNA synthesis. It is taken up by nopaline/agrocinopine A strains via agrocinopine permease. Part of the agrocin molecule mimics the structure of agrocinopine A, the usual substrate for agrocinopine permease. Bacterial mutant strains that do not take up agrocinopine are no longer sensitive to agrocin 84.

Production of agrocin 84 is the major, but not the only, mechanism in the control of crown gall by strain K84. Survival of the bacteria and colonization of the root surface also play important roles. The chromosomal background for the plasmid coding for agrocin 84 influences the ability of the strain to control disease. When agrocin production is transferred from strain K84 to other strains of *A. radiobacter*, the new strains do not control disease as effectively as K84. This is accounted for by the superior ability of strain K84 to colonize roots and survive in the soil and rhizosphere of the plant to be protected from crown gall. Biological control of crown gall may break down if pathogenic strains of *A. tumefaciens* acquire resistance to agrocin 84 (Box 23.2).

BOX 23.2

Minimizing the breakdown of biological control

Breakdown in biological control of crown gall by *A. tumefaciens* strain K84 appears to be related to transfer of agrocin 84 to pathogenic strains. Strains of *A. tumefaciens* have been found that produce agrocin and thereby acquire resistance to agrocin 84 because the agrocin plasmid (pAg K84) is transferred from strain K84 to the pathogenic agrobacteria. This phenomenon led to the development of a genetically engineered, transfer-deficient (tra^-) mutant of strain K84, referred to as K1026, which guards against the breakdown in biological control because part of the transfer region in the pAg K84 plasmid has been deleted. Strain K1026 has the same ability to produce agrocin 84 and crown-gall-control activity without the ability to transfer the agrocin plasmid to other agrobacteria. The use of strain K1026 in place of K84 minimizes the potential for breakdown of biological control (Jones and Kerr, 1989).

Biological control with composites of antagonists

Composites of antagonists may be selected for their capacities to inhibit a pathogen through microbial succession and therefore be used to reduce disease. This approach requires that the biological control agents be compatible. Composites of antagonists should:

- Operate at different sites, such as in the rhizosphere or in organic residues
- Effect biological control by different mechanisms, such as by competition or antibiosis
- Require different substrates for activity, such as mucigel metabolism by rhizosphere-competent fungi as compared to use of root exudate by pseudomonads
- Be compatible with the soil environment and enhanced by cultural manipulations

Composites of microorganisms have been shown to control several plant diseases. For example, a composite of fungal antagonists, selected for their ability to establish high populations in the rhizosphere of tomato, can reduce incidence of fusarium crown rot of tomato. In another example, combinations of fluorescent pseudomonads are more effective in control of take-all of wheat than each isolate alone. In both cases, the diversity of strains may better simulate the natural microflora responsible for disease suppressiveness than single species.

Biological control with composts

Biological control requires edaphic sources of organic nutrients to sustain the activities of the biological control agent. More consistent and sustained biological control of soil pathogens such as *Pythium*, *Phytophthora*, *Rhizoctonia*, and *Fusarium* has been achieved using compost-amended media, as long as variables such as consistency of parent material, salinity, C/N ratio, and other parameters of the composting process are controlled (Chapter 24). Container production and floricultural industries have used compost-amended media successfully as an alternative to biocides and soil drenches with fungicides (Hoitink and Boehm, 1999).

Similar results have been obtained in field soil, although with less consistency. In practice, composts are not consistently or naturally colonized by the diversity of biological control agents required for suppression of a broad spectrum of pathogens. Three types of specificity for biological control in compost-amended substrates have been identified:

- General suppression of *Pythium* and *Phytophthora* spp. appears to be a product of many types of biological control agents, particularly bacterial strains such as *Pseudomonas*, *Pantoea*, and *Bacillus* spp.
- Specific suppression is critical to biological control of *Rhizoctonia solani* and *Sclerotium rolfsii* through the addition of specific agents after peak heating in the composting process, but before the substantial colonization by other mesophilic microorganisms has occurred.
- Induced systemic resistance (ISR) is associated with microorganisms in composts applied to roots that induce resistance to foliar diseases. Several types of rhizosphere organisms, such as *Trichoderma* spp. and *Pantoea agglomerans*, are capable of inducing resistance, but less than 10% of the batches of composts are capable of producing ISR.

This variety of responses indicates the need for specific inoculation strategies to ensure consistency of the compost product. Organic components such as composts typically have three phases of biological control activity (Chapter 24). In Phase I, amendment with immature or non-stabilized compost not only serves as a food base for the biological control agent but also readily supports plant pathogens high in competitive or saprophytic capability and may lead to an increase in disease. In Phase II, compost use is characterized by suppression of disease because the organic component is fully colonized by soil organisms and the pathogen can no longer effectively compete for resources. In Phase III, loss of biological control activity occurs when the organic matter starts to become humified and the availability of soil organic nutrients to biological control agents becomes limiting.

Commercialization of biological control agents

Potential limitations

Because of the potential for environmental pollution, the loss of many currently registered fungicides, nematicides, and soil fumigants for control of soilborne diseases from the market is inevitable. As an example, methyl bromide, the most widely used soil fumigant, was eliminated in the US in 2005. This intensified interest in biological control agents because they have fewer negative environmental or toxicological impacts. The translation of recent research findings into products for biological control of plant pathogens has been rapid. The most successful and long-standing example is biological control for crown gall. The record for other introduced biological control treatments is far less consistent. Nevertheless, the number of private companies with programs in biological control has increased. Continued efforts in the private sector will depend on the economics of the market, future progress in research, and governmental regulation of pesticide use.

The emphasis on biological control agents in the marketplace was initially almost exclusively on singly introduced organisms, however, some commercially available composites of microorganisms have been proposed. Several reasons account for this. First, although single organisms have a relatively narrow spectrum of activity, they are more easily patented and marketed. Second, the time-limited efficacy of single applications ensures the need for repeated applications and subsequent enhanced sales. Third, the development of fermentation and formulation technology and delivery systems for individual biological control agents is similar to existing technology for mass production of microorganisms.

Overcoming limitations to commercialization

Inconsistent performance of biological control in the rhizosphere with introduced microorganisms occurs for at least three reasons:

- Varying edaphic conditions to which the introduced antagonist cannot adapt or because the strain has lost rhizosphere competence
- Antibiotic production where necessary for biological control is expressed too late for effective disease control
- Specificity of introduced antagonist for certain plant genotypes or soil environments

These limitations may be addressed through strain selection and improvement. However, research teams may become disillusioned because of the unrealistic expectations of investors, or funding may be too short-term to sustain the program until these problems are solved.

Economics is a key barrier to commercial development of biological control agents for use in agriculture. Typically, a very limited market, often represented by a single disease, cannot sustain developmental costs. The regulatory cost for registration of microbial agents with government agencies is high, especially for toxicological tests. Unfavorable economics especially reduce the prospect for commercialization of microbial agents for minor crops. This situation exists despite streamlining of the registration process specifically designed for "microbial pesticides."

Another limitation is the absence of an infrastructure to scale up and commercialize biological control agents. The production of thousands of liters of cells requires different technology than small-scale production and is beyond the capacity of most university and government researchers. This reinforces the need for teamwork between public-sector researchers and industrial scientists who have access to larger production facilities. In addition, while introduced antagonists may show promising results in suppressing pathogens in greenhouse trials, these antagonists have often not been evaluated in field tests over a full growing season (Mazzola and Freilich, 2017).

Obtaining an antagonist that will work consistently and effectively in the field may require engineering the strain with the desired traits. For example, in addition to take-all, a biological control strain of *Pseudomonas* was active against *Rhizoctonia* root rot after transformation with a gene (Chapter 4) conferring the ability to produce an additional antibiotic, phenazine-1-carboxylate. Transformed strains with the ability to control more than one disease, on more than one crop, broaden the market potential needed to justify the registration process of transgenic microorganisms.

A critical obstacle to biological control by soil amendment or seed treatment is the lack of methods to introduce microorganisms into production systems. Current storage and formulation requirements demand that most agents have a shelf life of at least one growing season when stored at room temperature. Furthermore, microbial agents are usually applied with the same equipment currently used for other seed treatments, such as foliar sprays or granular applications. A new application method will not be acceptable unless it is less expensive or fits into the current production scheme.

Biosafety and ecological concerns

Cook (1996) identified the following potential adverse effects of introduced biological control agents:

- Competitive displacement of beneficial organisms
- Allergenicity
- Toxigenicity
- Pathogenicity

These are **nontarget effects**, impacting organisms other than the intended pathogen, and may occur whether the biological control organism is native, exotic, or genetically engineered. Of these nontarget effects, all but allergenicity are also desired target effects, depending on the mechanism of biological control. Competitive displacement would become an issue if the antagonist, in addition to displacing a pathogen, displaces beneficial organisms such as mycorrhizal fungi (Chapter 12) or rhizobia (Chapter 17) that are critical to the plant's health. The spores of some fungi used as biological control agents may be allergens. Allergenicity is a concern for worker safety but is not a general public health issue and can be addressed by having workers wear appropriate masks. Toxigenicity to nontarget organisms is a concern where the mechanism of protection is a secondary metabolite that inhibits the target pathogen. This is only an issue if the toxicant is a component of the product during handling. However, in most cases the toxicant is produced after the antagonist has made sufficient growth in the habitat where it is introduced. A greater hazard is the potential for toxicity of secondary metabolites to the plant. This has been a problem with phenazine-producing strains of *Pseudomonas* spp. applied to seeds of wheat. Finally, pathogenicity to nontarget organisms is a concern in that the introduced organisms may have pathogenic effects on native insects or plants. Even greater concern exists for microorganisms with the potential to become opportunistic pathogens of humans, such as *Aspergillus ochraceus* and the clinical strains of *Burholderia cepacia*. These types of antagonists should be eliminated early in a screening process for potential biological control agents.

There has been little research on the ecological implications of introducing antagonists to a soil environment, primarily because of the wide variety of factors that can influence the soil microbial community. Strains used for biological control often are related to organisms already present in the soil microbiome, however, it's unclear if competition or dominance of an introduced strain has implications for the function of the soil microbiome.

Biological control genes for developing disease resistance

Much private investment in biological control is directed toward transformation of plants to express genes from microorganisms. In this case, the plant rather than the microorganism becomes the biological control agent. Best known is the use of the endotoxin genes from

Bacillus thuringiensis (Bt) to develop resistance to insect pests. Based on this approach, the resistance of transformed tobacco to the pathogens *Alternaria alternata* and *Botrytis cinerea* and transformed potato to the pathogens *Alternaria solani* and *Rhizoctonia solani* was developed. In both cases, the plants were modified to express the endochitinase gene from *Trichoderma harzianum* (Matteo et al., 1998). Though the possibilities for genetic transformation of plants with genes from microbial agents are exciting, limitations have been encountered. The transformation involves a single trait, even if multigenic, whereas the biological control agent rarely provides control by a single mechanism. However, the discovery of the CRISPR-Cas method for gene editing has accelerated the process of creating pathogen resistant crops (Wang et al., 2019). Multiple mechanisms are important in that they lessen the chance that the pathogen will develop resistance to the biological control agent.

Summary

Three broad categories of effort in biological control of soilborne pathogens and nematodes have been identified in this chapter. First, resident biological controls can be manipulated with cultural practices that influence the antagonist-pathogen relationship. These practices include crops with durable resistance to the pathogen, crop rotations, and other cultural practices that augment and stabilize the biological equilibrium that results in suppressivness. Second, when indigenous antagonists are inadequate, biological control agents or genes that have the tendency to be self-maintaining can be introduced one time or occasionally. Rhizosphere-competent organisms can be genetically engineered, formulated, and applied to subterranean plant parts in environments most conducive for establishment, growth, and dispersal of the introduced agent. Third, if the first two approaches are ineffective, repeated application of plant associated microorganisms can be used to reintroduce effective strains with every crop.

Although the emphasis has been on the patenting, producing, and marketing of singly introduced organisms, composites of microorganisms may be more efficacious because they simulate the response of natural microflora for disease suppressiveness. Expanded use of biological control agents will depend on the economics of their production, stricter regulation of chemical pesticides, relaxation of regulations for registration of microbial pesticides, and joint research ventures between the public and private sectors.

Study questions

1. What are the roles of plants in the dynamic equilibrium of living organisms in soil?
2. What are the methods by which microorganisms limit populations of plant pathogens?
3. How may disease of plants be reduced by biological control in situations in which populations of the pathogen are not reduced?
4. Why is it more important to integrate biological controls, as compared to chemical controls, with other plant disease management practices?
5. Which characteristics may be most important for rhizosphere competence of soilborne microorganisms?

6. What are some of the influences that plant genotype may have on the composition of rhizosphere microorganisms that are involved in biological control of plant diseases?
7. Which methods of delivery of biological control agents may be of the greatest commercial potential?
8. What are the important mechanisms in the control of crown gall by nonpathogenic strains of *Agrobacterium radiobacter*?
9. Why may the use of several biological control agents be more effective than single antagonists in the control of plant pathogens?
10. Considering the control of root-knot nematode with *Pasteuria penetrans* as a model, why may organisms with the greatest potential for biological control of a plant pathogen sometimes be the most difficult to manipulate for commercial use?

References

Ahmad, J.S., Baker, R., 1987. Rhizosphere competence of *Trichoderma harzianum*. Phytopathology 77, 182–189.

Barrett, M., Morissey, J.P., O'Gara, F., 2011. Functional genomics analysis of plant growth-promoting rhizobacterial traits involved in rhizosphere competence. Biol. Fertil. Soils. 47, 729–743.

Chen, Z.X., Dickson, D.W., 1998. Review of *Pasteuria penetrans*: Biology, ecology, and biological control potential. J. Nematol. 30, 313–340.

Cook, R.J., Baker, K.F., 1983. The Nature and Practice of Biological Control of Plant Pathogens. American Phytopathological Society Press, St. Paul, Minn.

Cook, R.J., 1996. Assuring the safe use of microbial biocontrol agents: A need for policy based on real rather than perceived risks. Can. J. Plant Pathol. 18, 439–445.

Curl, E.A., Lartey, R., Peterson, C.M., 1988. Interactions between root pathogens and soil microarthropods. Agric. Ecosys. Environ. 24, 249–261.

Harmon, G.E., 2000. Myths and dogmas of biocontrol: Changes in perceptions derived from research on *Trichoderma harzianum* T-22. Plant Dis. 84, 377–393.

Hoitink, H.A.J., Boehm, M.J., 1999. Biocontrol within the context of soil microbial communities: A substrate-dependent phenomenon. Annu. Rev. Phytopathol. 37, 427–446.

Hopkins, D.L., Larkin, R.P., Elstrom, G.W., 1987. Cultivar-specific induction of soil suppressiveness to Fusarium wilt of watermelon. Phytopathology 77, 607–611.

Jones, D.A., Kerr, A., 1989. *Agrobacterium radiobacter* strain K1026, a genetically engineered derivative of strain K84 for biological control of crown gall. Plant Dis. 73, 15–18.

Kerr, A., 1980. Biological control of crown gall through production of agrocin 84. Plant Dis. 64, 25–30.

Kerry, B.R., Crump, D.H., 1980. Two fungi parasitic on females of cyst nematodes (*Heterodera* spp.). Trans. Brit. Mycol. Soc. 74, 119–125.

Klingen, I., Westrum, K., Meyling, N.V., 2015. Effect of Norwegian entomopathogenic fungal isolates against *Otiorhynchus sulcatus* larvae at low temperatures and persistence in strawberry rhizospheres. Biol. Control 81, 1–7.

Kloepper, J.W., Schroth, M.N., Miller, T.D., 1980. Effects of rhizosphere colonization by plant growth-promoting rhizobacteria on potato plant development and yield. Phytopathology 70, 1078–1082.

Kloepper, J.W., Tuzun, S., Kuc, J.A., 1992. Proposed definitions related to induced resistance. Biocontrol Sci. Technol. 2, 349–351.

Malajczuk, N., 1983. Microbial antagonism to Phytophthora. In: Erwin, D.C., Barnicki-Garcia, S., Tsao, P.H. (Eds.), *Phytophthora*: Its Biology, Taxonomy, Ecology and Pathology. American Phytopathological Society Press, St. Paul, Minn, pp. 197–218.

Matteo, L., Woo, S.L., Fernandez, I.G., Colucci, G., Harmon, G.E., Pintor-Toro, J.A., Fillipone, E., Muccifora, S., Lawrence, C.B., Zonia, A., Sadik, T., Scala, F., 1998. Genes from mycoparasitic fungi as a source for improving plant resistance to fungal pathogens. Proc. Natl. Acad. Sci. 95, 7860–7865.

Mazzola, M., Freilich, S., 2017. Prospects for biological soilborne disease control: application of indigenous versus synthetic microbiomes. Phytopathology 107, 256–263.

Schippers, B., Bakker, A.W., Bakker, P.A.H.M., 1987. Interactions of deleterious and beneficial rhizosphere microorganisms and the effect of cropping practice. Annu. Rev. Phytopathol. 25, 339–358.
Strauss, S.L., Kluepfel, D.A., 2015. Anaerobic soil disinfestation: A chemical-independent approach to pre-plant control of plant pathogens. J. Integr. Agric. 14, 2309–2318.
Wang, T., Zhang, H., Zhu, H., 2019. CRISPR technology is revolutionizing the improvement of tomato and other fruit crops. Hortic. Res. 6, 77.
Weller, D.M., 1988. Biological control of soilborne plant pathogens in the rhizosphere with bacteria. Annu. Rev. Phytopathol. 26, 379–407.
Weller, D.M., Raaijmakers, J.M., McSpadden-Gardner, B.B., Thomashow, L.S., 2002. Microbial populations responsible for specific soil suppressiveness to plant pathogens. Annu. Rev. Phytopathol. 26, 309–348.
Zhou, T., Paulitz, T.C., 1994. Induced resistance in the biocontrol of *Pythium aphanidermatum* by *Pseudomonas* spp. on cucumber. J. Phytopathol. 142, 51–63.

Further reading

Berendsen, R.L., Pieterse, C.M.J., Bakker, P.A.H.M., 2012. The rhizosphere microbiome and plant health. Trends Plant Sci. 17, 478–486.
Cook, R.J., 1993. Making greater use of introduced microorganisms for biological control of plant pathogens. Annu. Rev. Phytopathol. 31, 53–80.
Lakshmanan, V., Selvaraj, G., Bais, H.P., 2014. Functional soil microbiome: belowground solutions to an aboveground problem. Plant Phys. 166, 689–700.
Paulitz, T.C., Belanger, R.R., 2001. Biological control in greenhouse systems. Annu. Rev. Phytopathol. 39, 103–133.
Raaijmakers, J.M., Paulitz, T.C., Steinberg, C., Alabouvette, C., Moenne-Loccoz, Y., 2009. The rhizosphere: a playground and battlefield for soilborne pathogens and beneficial microorganisms. Plant Soil 321, 341–361.
Stirling, G.R., 1991. Biological Control of Plant Parasitic Nematodes. CAB International, Wallingford, Oxon, England.

CHAPTER

24

Composting: the microbiological processing of organic wastes

David A. Zuberer[a], *Larry M. Zibilske*[b]

[a]Department of Soil and Crop Sciences, Texas A&M University, College Station, TX, United States; [b]Vice-President for Research, Texas Plant and Soil Laboratory, Edinburg, TX, United States

"What prevents us from being knee-deep in our own detritus?
Microbes, you reply. Quite so."
John Postgate. 1992. Microbes and Man. p. 96.

The word compost comes from the Middle English and Old French, compote, meaning "mixture," which is derived from Latin, *compositum*, meaning "to put together." Composting has been described as the biological decomposition and stabilization of organic substrates, under conditions that allow development of thermophilic temperatures as a result of biologically produced heat, to yield a final product that is more stable than the starting materials, free of pathogens and plant seeds, and can be beneficially applied to land.

The history of composting can be traced back at least 6000 years. Records indicate that the Sumerians managed organic urban wastes in pits for eventual applications to agricultural lands (Diaz and de Bertoldi, 2007). Historical records show that ancient Greeks and Romans were familiar with the decay processes that occurred spontaneously in stacks of plant matter or animal wastes. One of the most accurate descriptions of the composting process was given by the Knights Templar of the 13th century. Those records indicate very detailed descriptions of the process including the composition of starting mixtures, particle sizes, and dimensions and construction of windrows for best results (Diaz and de Bertoldi, 2007).

Thus, production and use of compost in ancient times indicates that this was one of the first biotechnologies. Today, heightened environmental awareness continues to drive interest in composting as a means to prevent or reduce environmental pollution and to manage and recycle organic wastes produced by society (Box 24.1). Technological innovations have modernized the process with computer controls and scientific management. While at some facilities composting has become a highly developed enterprise, nevertheless it remains a basically simple concept that the backyard gardener can master. At the center of the process are microorganisms with their remarkable capabilities for decomposing organic materials.

Principles and Applications of Soil Microbiology. http://dx.doi.org/10.1016/B978-0-12-820202-9.00024-1
Copyright © 2021 Elsevier Inc. All rights reserved.

BOX 24.1

Turning trash into treasure

Just about everyone has heard the old adage "that one man's trash is another man's treasure." Perhaps no place is that statement probably more true than in the modern business of recycling organic wastes. Today, millions of tons of organic wastes are diverted from declining landfill space to large-scale composting facilities throughout the United States (Fig. 24.2). This is especially true for yard waste trimmings, animal manures and bedding, and municipal solid waste. A survey of the states in 2017 yielded the data in the table below. One report estimates that the global compost market is expected to reach an estimated $9.2 billion by 2024 with an annual growth rate of 6.8% from 2019 to 2024.

A walk through the garden sections of any of the major "big box" stores in the United States and other related businesses reveals many compost and related products for the homeowner market. Many communities have businesses selling compost in bulk for both homeowners and landscape contractors. Some community compost facilities provide compost at low or no cost to the public. Add to this the value of the plethora of mulch products (bark mulches, etc.) being sold today, and you can truly see that materials that were once considered "trash" are now someone's "treasure."

Composting facilities in the United States (data reported, 2015–17)

Composting facility types	Total of each type	Percent of total
All composting facilities	4,713	
Yard trimmings	2,698	57.2
Yard trimming + food scraps	249	5.3
Multiple organics[a]	620	13.2
Mixed MSW	11	0.2
Biosolids	249	5.3
On-site institutional	297[b]	6.3
On-farm	354[b]	7.5
Other[c]	126	2.7
Not specified[d]	109	2.3

[a]*Taking more than yard trimmings and/or food scraps, for example, yard trimmings, food, manure, wood shavings, industrial organics, etc.*
[b]*Typically exempt from permitting or permit-by-rule thus often undercounted*
[c]*Includes sites that compost mortalities, animal by-products, community composting sites, seafood processing residuals, etc.*
[d]*Two states only provided a total number of composting facilities, and no break down by categories.*

Source: Goldstein (2017).

Composting demonstrates the powerful metabolic potential of microorganisms and vividly illustrates the vital role of environmental microorganisms in global biogeochemical cycles.

Both large-scale commercial composting industries and backyard gardeners rely on the same microbial processes, although the former may produce thousands of cubic yards of compost while the latter only a few. Composting to manage organic wastes produces several advantages:

- Compost generally poses a low risk to the environment, assuming it is free of heavy metals or hazardous organic materials.
- Compost improves soil fertility, whether in backyard gardens or reclaimed surface-mine soils.
- Composting is cost-competitive with other waste-handling technologies.

Multiple advantages of large-scale composting cited by the USEPA are listed in Table 24.1.

TABLE 24.1 Benefits of composting.

There are numerous benefits to composting. Some examples are listed below:

- Organic waste in landfills generates methane, a potent greenhouse gas. By composting wasted food and other organics, methane emissions are significantly reduced.
- Compost reduces and, in some cases, eliminates the need for chemical fertilizers.
- Compost promotes higher yields of agricultural crops.
- Compost can help aid reforestation, wetlands restoration, and habitat revitalization efforts by improving contaminated, compacted, and marginal soils.
- Compost can be used to remediate soils contaminated by hazardous waste in a cost-effective manner.
- Compost can provide cost savings over conventional soil, water, and air pollution remediation technologies, where applicable.
- Compost enhances water retention in soils.
- Compost provides some carbon sequestration.

https://www.epa.gov/sustainable-management-food/reducing-impact-wasted-food-feeding-soil-and-composting.

The goals of composting range from sanitization, that is, the elimination of pathogenic or deleterious organisms, insect larvae, intestinal parasites, and weed seeds, to reducing the bioavailable energy content of the waste, and reduction of waste volumes. Composting can even be done in a greenhouse where heat generated by the compost and the large amount of carbon dioxide produced are beneficial to plant production in cold climates. Composting also helps to bioremediate wastes containing polycyclic aromatic hydrocarbons (Kästner and Mahro, 1996), pesticides (Kästner and Miltner, 2016), and petroleum-contaminated soils (Wu et al., 2020) (Chapter 22). Composting has been accepted for several decades as a "Process to Significantly Reduce Pathogens (PSRP)" in the treatment of wastewater sludges (US EPA, 1993).

The inputs and outputs (or products) of composting systems are shown in Fig. 24.1 and two examples of large-scale composting operations are shown in Fig. 24.2. The microbes can be viewed as the central driving force or "engine" that substantially transforms the inputs (substrates, water, and oxygen) into products (mainly compost, new microbial cells, carbon dioxide, and heat). An understanding of the microbial activities occurring in composting is necessary to comprehend both the beneficial aspects of composting, and the limitations of the process.

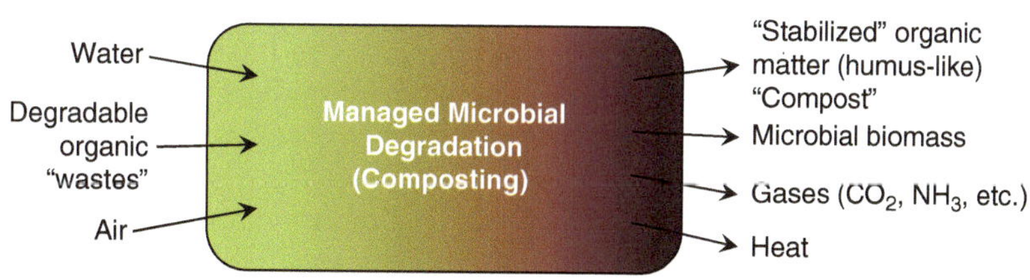

FIGURE 24.1 **Compost inputs and outputs.** Fresh compostable substrates are converted through microbial action into microbial biomass, stabilized humus-like organic matter (compost) and several gases, mainly CO_2.

FIGURE 24.2 **Two types of large-scale composting processes.** (A) A windrow pile being turned by a mechanical turner. Note the steam coming from the hot compost as the pile is inverted. (B) The Beltsville aerated static pile employs a perforated pipe system which can draw or push air through the pile to keep it aerobic and to some extent control the temperature. Odors are removed by running the exhaust gases through a pile of screened compost to filter out odor-causing gases. This method was developed by scientists at the USDA-ARS Research Center in Beltsville, Maryland (Willson et al., 1980). *Source: Part A: Photo courtesy of Engel and Gray, Inc., Santa Maria, CA.*

Compost microbiology

Changes in microbial populations during composting

Most organic substrates carry an indigenous population of microbes from the environment, and it is therefore generally not necessary to add any type of inoculum. Representatives of the major microbial groups—bacteria (including actinobacteria also called actinomycetes) and fungi—are normally present when the composting process begins. The meso- and macrofauna (Chapter 8) are also important at the beginning stage of composting. Sow bugs, millipedes, centipedes, beetles, mites, springtails, and earthworms break down coarse organic materials into smaller pieces. This *comminution* (grinding) of the coarse fragments is mainly a preparative step that accelerates the rate of decomposition by greatly increasing their surface area-to-volume ratios, improving access of the microbial populations to the substrates. Many protozoa are also active during the early phase of composting, processing smaller bits of particulate and colloidal organic materials and preying upon microbial populations. There is a degree of complementation in activities carried out by each of the microbial groups, and the overlapping of decomposition activities promotes rapid destruction of the more complex polymers in the substrates.

Measurements of microbial community physiology have shown that enormous changes in population distribution occur during composting. Figs. 24.3 and 24.4 and Table 24.2 show

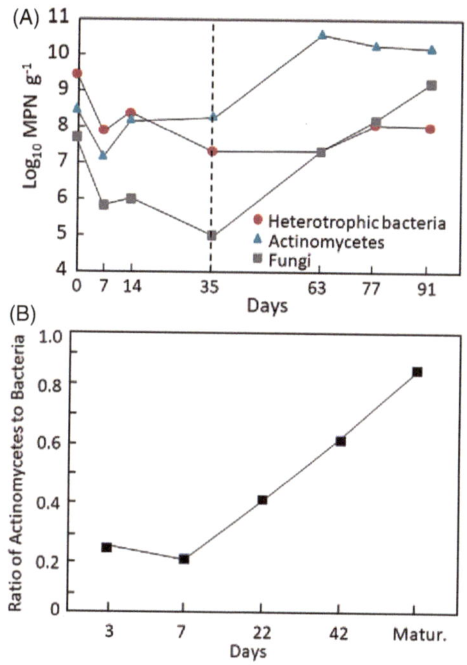

FIGURE 24.3 Microbial population changes during the composting of animal manures. (A) Numbers of bacteria, actinomycetes, and fungi during composting of poultry litter with yard waste trimmings. *Dashed vertical line indicates approximate end of the thermophilic stage.* (B) Ratios of actinomycetes to bacteria based on actinomycete and bacteria 16S RNA gene copy numbers during the composting of swine manure with wood chips and sawdust. Day 3—mesophilic stage, days 7, 22, 42—early-, mid-, and late-thermophilic stage and day 53—during maturation. *Source: Part A: Data from Tiquia (2002); Part B: Data from Wang et al. (2014).*

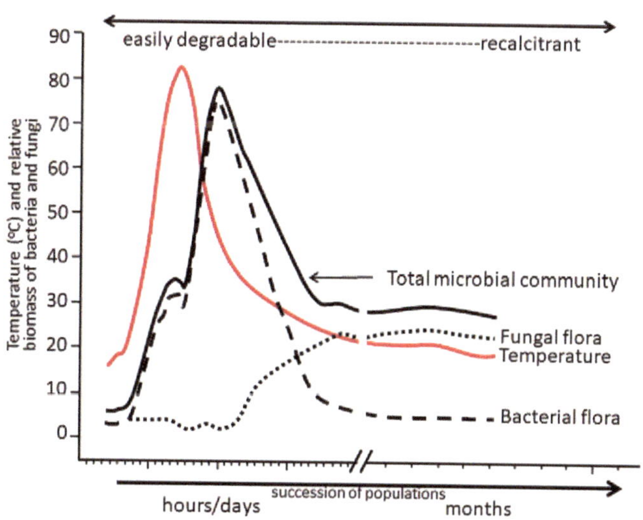

FIGURE 24.4 Relative changes in the biomasses of the bacterial, fungal, and total microbial populations during the temperature transitions in a compost pile. *Source: Redrawn from Insam and de Bertoldi (2007), with permission.*

TABLE 24.2 Genera of bacteria and fungi reported to occur in composts.

Bacteria	Fungi	
Achromobacter	*Absidia*	*Nectria*
Acidovorax	*Acremonium*	*Oedocephalum*
Acinetobacter	*Actinomucor*	*Oidiodendron*
Alcaligenes	*Agaricus*	*Paecilomyces*
Amphibacillus	*Alternaria*	*Penicillium*
Arthrobacter	*Arthrinium*	*Peziza*
Bacillus	*Arthrobotrys*	*Phanerochaete*
Brevibacillus	***Aspergillus***	*Phoma*
Brevibacterium	*Aureobasidium*	*Preussia*
Cellulomonas	*Beauvaria,*	*Pseudeurotium*
Clostridium	*Scolecobasidium*	*Pseudogymnoascus*
Corynebacterium	*Botryotrichum*	*Rhizomucor*
Cytophaga	*Botrytis*	*Rhizopus*
Enterobacter	*Cephalosporium*	*Scopulariopsis*
Flavobacterium	*Chaetomidium*	*Sepedonium*
Klebsiella	***Chaetomium***	*Sordaria*
Micrococcus	*Chrysosporium*	*Sporotrichum*
Micromonospora	*Cladosporium*	*Stachybotrys*
Nocardia	*Clitopilus*	*Stibella*
Paenibacillus	*Coniothyrium*	*Talaromyces*
Pseudomonas	*Coprinus*	***Thermoascus***
Pseudonocardia	*Corynascus*	***Thermomyces***
Rhodococcus	*Doratomyces*	*Thielavia*
Saccharomonospora	*Emericella*	*Trichocladium*
Serratia	*Epicoccum*	*Trichoderma*
Streptomyces	*Fusarium*	*Trichophaea*
	Geotrichum	*Trichothecium*
	Gilmaniella	*Trichurus*
	Gliocladium	*Ulocladium*
	Humicola	*Verticillium*
	Malbranchea	*Volutella*
	Mortierella	*Zygorhynchys*
	Mucor	
	Myceliophthora	

Boldface, fungal genera containing known thermophilic species.
Source: Data from Ryckeboer et al. (2003). See also Insam and de Bertoldi (2007).

some of the major microbial participants as composting progresses from the *mesophilic stage* (approximately 20°C–40°C) into the *thermophilic stage* (above 40°C and up to about 80°C) and lastly through a gradual cooling period, the *stabilization* or *curing stage* (Fig. 24.5B). This final stage is sometimes broken into a cooling stage followed by a "maturation" stage.

Bacteria are the most numerous microbial group in compost and are responsible for much of the decomposition. Actinobacteria are common and often impart a dull sheen and an earthy odor to the compost. Their numbers have been shown to increase markedly (Fig. 24.3A) during the maturation phase, a phenomenon consistent with their description as "late colonizers" using the more recalcitrant substrates and microbial cell remains left after the thermophiles have run their course. In a recent study (Fig. 24.3B), Wang et al. (2014) determined that the ratio of actinomycetes to bacteria in a swine-manure compost increased from 0.18 to 0.25 in the early stages to 0.86 as the compost matured. The increase in fungal numbers (MPN technique) reported by Tiquia (2002) is consistent with the increases in fungal biomass reported by Insam and de Bertoldi (2007) (Fig. 24.4). The actinobacteria, along with the fungi, decompose much of the cellulose, hemicelluloses, and lignin-containing polymers in organic materials especially in the later phases of composting. The fungi also produce filamentous growth of cottony threads that permeates the compost matrix, giving the outer layer a grayish-white appearance. The complementary activities of each microbial group increase the arsenal of tools available for the degradation of a vast array of complex substrates under constantly changing environmental conditions.

The relative biomasses of bacteria, fungi, and the total microbial population in a compost pile are shown in Fig. 24.4. Note the marked increase in fungal biomass during the later stages

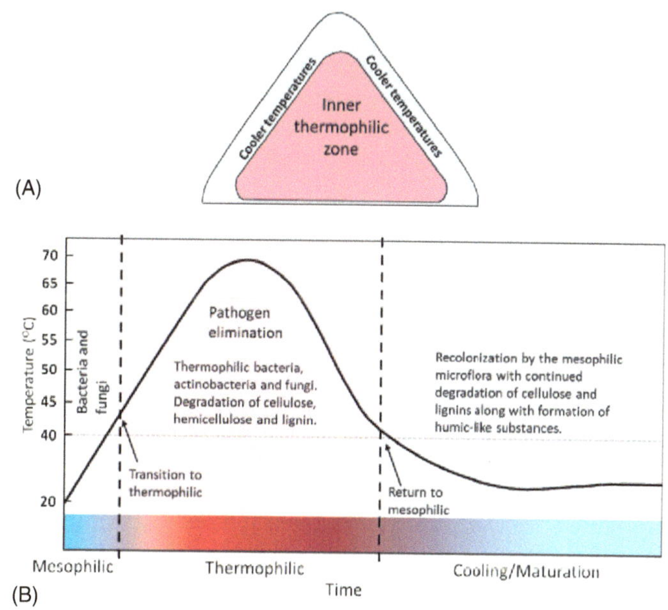

FIGURE 24.5 Temperature zones (A) within a compost pile or windrow and an examples of a 3-phase (B) composting temperature curve.

of composting consistent with the increased fungal numbers shown by Tiquia (2002). Conditions of moisture, aeration, temperature, and substrate quality constantly change during the composting process and with those changes come shifts in microbial communities adapted to the changing environment. The increased fungal biomass late in the composting process suggests they too, along with the actinobacteria, are involved in degrading the more recalcitrant compounds (e.g., lignin and its derivatives) in the initial substrates as well as remnants of microbial cells and their products.

Enzymatic activities during composting

Microbes in the compost cannot directly metabolize the insoluble particles of organic matter. Rather, they must produce hydrolytic extracellular enzymes (Chapter 13) to depolymerize the larger compounds to molecules that are water-soluble. At this point, microbes transport the substrates across the plasma membrane to complete the degradation process.

The activities of several enzymes have been monitored during composting (Wei et al., 2012; Tiquia, 2002). For example, cellulase activity decreases the amount of cellulose in the substrates by about 25% in about 3 weeks. Lipase, protease, cellulase, asparaginase, arylsulfatase, phosphomonoesterase, and amylase activities rise and fall during the successive stages of composting. Activities of all these enzymes decrease sharply during the thermophilic stage, probably due to heat inactivation. Denaturation of enzymes is often correlated with the death of the microbe. This suggests that the apparent recovery of microbial numbers and enzymatic activity in the pile interior after the thermophilic stage is due to reintroduction, during turning, of organisms that survived on the outer, cooler parts of the pile (Fig. 24.5A).

So, we see that composting is at its very essence a microbial process, and its rate can be controlled by any factor that affects the microbes involved in the process. Lack of suitable substrates, a moisture content or temperature in the compost outside the optimum range, and problems with oxygen diffusion into the pile are the most common rate-limiting factors in composting. Some of these factors are discussed in the following section.

Physical and chemical factors affecting the composting process

Aeration (O_2 availability)

As decomposition of organic wastes is more rapid under aerobic conditions, it is essential that compost systems maintain an adequate supply of O_2 to the decomposer communities during the composting process. This is especially true during the early stages of the process when readily decomposable substrates (sugars, proteins, fats) are being rapidly consumed by aerobic microbes with concomitant high rates of O_2 consumption. A limited O_2 supply leads to the development of anaerobic conditions within the pile, and decomposition is markedly slowed as microbial populations switch from respiratory to fermentative metabolisms. This is generally accompanied by the production of foul-smelling compounds such as organic acids and reduced sulfur compounds such as mercaptans. To overcome problems associated with poor aeration, bulking agents are often used to increase the porosity of the compost pile. Static systems like windrows are turned periodically to reintroduce oxygen as well as to mix materials from the cooler outside layer of the pile into the interior so as to expose them to the thermophilic decomposers. Still other composting systems use various methods to move air through the pile to maintain adequate aeration for rapid decomposition (Fig. 24.2B).

Moisture

Moisture is essential for the composting process. Decomposition is carried out in an essentially aquatic environment, that is, the water films on the surfaces of substrate particles. Soluble components from the solid substrates and wastes from microbial metabolism diffuse through the films of moisture on the compost solids. Optimum moisture content for composting is generally between 40% and 60% by weight. Too much moisture interferes with oxygen accessibility and slows composting. Too little moisture hinders diffusion of the soluble molecules, likewise, slowing the rate of composting. See Chapter 2 for a review of the relationships between particle size, oxygen diffusion, and moisture holding capacity.

Temperature

Properly constructed compost piles will generally heat spontaneously within a few days due to trapped heat generated by the metabolic processes of the mesophilic decomposer microbes. Their activities pave the way for the thermophilic community to become active (Fig. 24.5B). During decomposition of organic substrates, microbes release a large quantity of energy of which about 40%–50% is captured by the microbes to produce energy and cells for growth; the remaining energy is lost as heat in the compost pile. This large amount of "lost" heat causes an increase of temperature in the pile on the order of 70°C–90°C. Indeed, these high temperatures can inhibit microbial growth, slowing the biodegradation of organic matter. The high temperatures usually last a few days after which the compost pile returns to cooler temperatures (30°C–45°C) and further decomposition of the wastes continues. The interior zones of the pile are well insulated against temperature changes, although the exterior layers may remain much cooler especially when ambient temperatures are low during the cooler months. This is another reason why frequent turning is necessary, as it mixes those materials into the interior where they can be decomposed. As time proceeds and substrates become exhausted, the pile cools down and recolonization by mesophilic communities (bacteria, actinobacteria, fungi) continues the decomposition and "stabilization" of the organic materials into and throughout the curing phase of the composting process. The benefits of the "thermophilic stage" are that it allows rapid decomposition of wastes and sanitizes the waste materials by killing pathogenic microbes, insect larvae, and weed seeds.

C/N ratio

Another important factor in composting is the ratio of carbon to nitrogen in the starting material. The optimal starting C/N ratio for most types of wastes is 25–30. Pile materials with an excess of available carbon relative to nitrogen content (i.e., "wide" C/N ratios) can lead to immobilization of available N as the microbes attack the readily available C source. This "tie up" of available N slows the decomposition process. In contrast, materials with low C/N ratios (nitrogen rich) are frequently highly degradable and can lead to rapid growth and subsequent O_2 depletion within the pile.

Stages in the composting process

The composting process is generally described as occurring in three stages (Fig. 24.5B). The three-stage process includes the mesophilic and thermophilic phases followed by a prolonged "cooling/maturation" phase. Fig. 24.2 illustrates two types of large-scale compost operations. In the *turned windrow method*, described as a static pile method (Fig. 24.2A), compostable substrates are piled into long (up to 100 m) roughly triangular- or trapezoid-shaped rows 2–3 m tall. The windrows are periodically turned with a mechanical windrow turner (Fig. 24.2A) to mix the outer layers of material into the center of the pile where they are degraded. Turning often reinitiates a second thermophilic heating event as the newly buried outer materials are decomposed releasing heat. The windrow system is described as a batch system since only one batch of feedstocks comprises the pile from start to finish during the composting process.

In other static batch systems, more elaborate means of aerating compost piles have been developed. For example, the Beltsville process (developed by the USDA-ARS in Beltsville, Maryland, United States) provides aeration to compost piles via a series of pipes underlying the pile (Fig. 24.2B). A blower mechanism pulls air through the pile and into pipes near the bottom of the compost pile that run parallel to the length of the windrow. Air may also be pumped from the pipes upward through the pile. This approach obviates the need for turning the pile to achieve aeration. These kinds of windrows are called *aerated static piles*. Forced aeration, as it allows the compost process to proceed with minimal or no turning, permits piles to be larger. Direction of airflow is sometimes reversed in order to minimize odors emanating from the pile. Air drawn through the pile when odor levels are high can be filtered through activated charcoal (or biochar) or finished compost to absorb odor-causing compounds. Aerated static piles are popular because they provide a higher level of control of the process, minimize the need to turn the pile, and encourage rapid rates of decomposition. For example, complete composting of municipal biosolids can occur within 21 days.

Mesophilic phase

The mesophilic phase is characterized by the explosive growth of mesophilic bacteria and fungi. Mesophiles are defined as organisms that grow at moderate temperatures between 20°C and 45°C with an optimum growth temperature in the range of 30°C–39°C (Fig. 24.5).

Metabolism of the simpler compounds by mesophiles causes physical and chemical changes in the decomposition environment in the pile. These eventually select for those microbes better able to cope with the new substrates and with the increasing heat generated by the mesophiles. Numbers of bacteria have been shown to increase by 2–3 orders of magnitude during the mesophilic phase. Actinobacteria apparently are not as active in the preparative processes carried out by other mesophilic bacteria, perhaps due to their inability to compete, but their numbers rise during the thermophilic stage. Likewise, fungal counts remain stable (on the order of 10^5 and 10^6 g^{-1}) in the initial and final stages of the process, both of which are much cooler than the thermophilic stage. However, fungal numbers generally decline during the thermophilic phase with the exception of a few thermophilic species (e.g., species belonging to the genera *Aspergillus, Chaetomium, Humicola, Mucor, Penicillium,* and *Thermomyces*). Thermophilic fungi generally have temperature maxima around 60°C thus they are not active at the higher temperatures where thermophilic bacteria dominate the composting process.

Mesophilic bacteria and fungi present at the start of the composting process begin to give way to thermophiles when the temperature reaches about 40°C (Fig. 24.5).

Thermophilic phase

Compost piles effectively restrict the free dissipation of heat generated during the decomposition of organic substrates, resulting in an increase of pile temperature and a transition to the thermophilic community. Their metabolism continues to raise internal temperatures of the compost pile from 70°C to 80°C and above. Decomposition rates are most rapid during the earlier thermophilic stage but decrease as temperatures rise above 60°C. The thermophilic stage is driven by a mixed population of thermophilic bacteria and actinobacteria and the more heat-tolerant fungi. It is during this stage that much of the breakdown of cellulose, hemicellulose, proteins, and fats occurs. Thermophilic organisms have unique properties that allow growth at elevated temperatures (Box 24.2). Recent studies have shown that thermophilic methane-producing archaeans are present during the composting of cattle manure. And, although not thermophiles, ammonia-oxidizing archaeans have been repeatedly found during composting. An exhaustive listing of bacterial and fungal genera that have been reported in composts can be found in an excellent paper by Ryckeboer et al. (2003) and the chapter by Insam and de Bertoldi (2007).

Thermophilic temperatures in the pile are necessary to ensure rapid stabilization and **pasteurization** of the compost, eliminating harmful organisms. United States governmental regulations for composting Class B biosolids require that the internal temperature of the compost pile be 40°C or higher and remains at 40 degrees Celsius or higher for 5 days; during that period, the temperature must exceed 55°C for 4 hours (US EPA, 1993). Temperature standards have also been set by the USDA National Organic Program (https://www.ams.usda.gov/rules-regulations/organic/handbook/5021). For an in-vessel or static aerated pile, compost should maintain a temperature of between 55°C (131°F) and 77°C (170°F) for a minimum of 3 days. For a windrow system, this temperature must be maintained for 15 days and the pile turned a minimum of 5 times within that time period. They further state that the active

BOX 24.2

When the "heat is on"!

Thermophiles have several mechanisms that impart heat resistance and allow them to grow at temperatures high enough to cook other microbes. Their membranes contain saturated fatty acids that preserve cell integrity at high temperatures. Their proteins can maintain flexibility at high temperatures thereby avoiding inactivation due to denaturation at elevated temperatures. Even their DNA has a special reverse gyrase that allows coiling of the molecule into a more heat resistant form. *Pyrodictium* is a member of the Crenarchaeota that has an optimum growth temperature of 105°C and can maintain growth at 115°C. This archaean grows in hot springs like those at Yellowstone National Park in the United States.

composting process is finished when the pile returns to ambient temperature levels. Detailed temperature records are required to be maintained by compost facility operators.

Enteric pathogens, such as *Escherichia coli* and *Salmonella* spp., normally survive less than 1 hour once the compost enters the thermophilic stage. Other noxious organisms, such as weed seeds and intestinal parasite eggs, may be more durable but are often eliminated or greatly reduced during composting.

Curing phase (compost maturation)

As the thermophilic stage nears completion, reaction rates slow and further chemical and physical changes in the compost occur more slowly. The composting process is not yet complete at this stage, as the compost probably contains intermediates of decomposition that may have bad odors, attract pests, and adversely affect crop seed germination and root growth and function. The compost may even constitute a fire hazard due to release of trapped, combustible gases.

Anaerobic microsites always occur in composting systems and contain many facultative anaerobes. More than half of the decomposition in these sites is strictly anaerobic, resulting in fermentation products that accumulate during composting. These products are often reduced compounds of nitrogen and sulfur, for example, mercaptans, volatile amines, fatty acids, and hydrogen sulfide that produce a bad odor in uncured compost. In addition, some of the compounds in uncured composts can interfere with seed germination or other plant activities. Acetic acid and phenolic compounds are often thought to be responsible for inhibition of seed germination, but others are probably involved as well. Some composts can spontaneously combust. Rapid substrate decomposition in the center of the pile depletes oxygen and can result in the production of methane and other combustible gases that tend to diffuse outward. Heat generated in the active process at the center of the pile may allow the reaction to reach the activation energy needed to ignite combustible gases diffusing from the inside of the pile to the outer regions, where oxygen is present to support the ignition. Allowing time following the thermophilic phase for mesophiles to complete the destruction of those compounds can eliminate these potentially deleterious metabolic intermediates. This so-called **curing phase** is necessary to ensure the elimination of those substances that interfere with plant growth or give compost a bad odor.

It is during the long curing stage that the slow degradation of remaining lignins and other bio-resistant compounds occurs with the concomitant formation of resistant "humus-like" compounds and microbial remnants similar to those comprising soil organic matter (SOM; Chapter 13), and which are characteristic of mature composts. Interestingly, Danon et al. (2008), using molecular techniques, documented continual changes in the microbial community of a biosolids based compost during a prolonged (336-day) curing/maturation period. Their analyses were done after the point at which the compost producer had declared the compost finished and ready for use. They found that "the microbial community continued to change long after the compost organic matter and other biochemical properties had stabilized."

pH and C/N ratio

Other changes occur in the compost matrix as decomposition proceeds (Fig. 24.6). During the mesophilic stage, pH (Fig. 24.6A) initially declines as a result of the release of organic acids when rapid growth causes declining O_2 concentrations and sugars are fermented. The decline

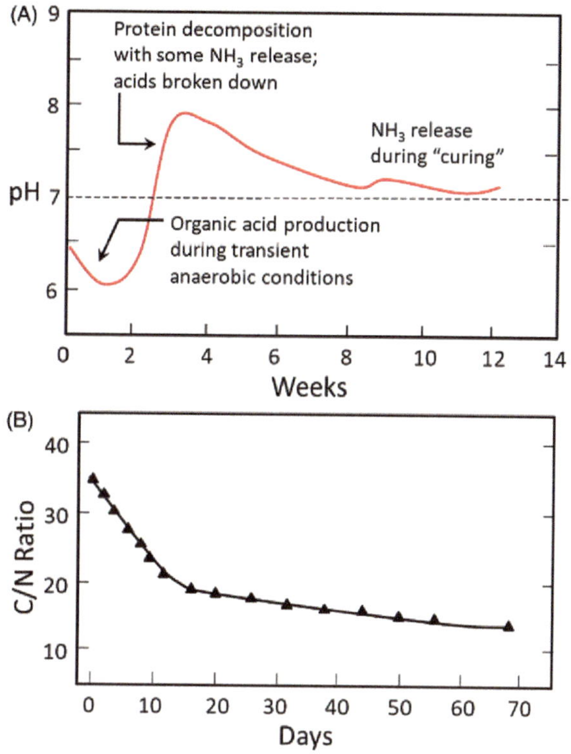

FIGURE 24.6 Changes in pH (A) and C/N ratio (B) during composting.

is transient however and, as the heating phase begins and the populations shift, organic acids are consumed and decomposition of proteins and polysaccharides commences raising the pH. During this time, there may be a release of ammonia due to mineralization of the N-containing organics. Eventually the pH declines to near neutral values as the pile cools and undergoes curing. Characteristic of the composting process is the observed decline in the C/N ratio (Fig. 24.6B) as decomposition proceeds; a result of the export of carbon as CO_2 from the pile materials, whereas the nitrogen is tightly cycled through the compost microbes. Eventually the C/N ratio reaches an equilibrium value near that of the remaining microbial biomass.

Raw materials for composting

Compostable substrates

Most organic wastes can be successfully composted (Table 24.3). For example, paper and food wastes often comprise 50% of municipal solid waste (MSW). Both are well suited to composting. Biosolids (sewage sludges), animal manures, and agricultural crop and food processing wastes are also excellent candidates for composting. A portion of these wastes, however, is still placed in landfills. In the past, yard wastes (primarily leaves, garden crop residues, grass clippings, and shredded brush) were also placed in landfills. Most states have enacted legislation

TABLE 24.3 Some feedstocks and bulking agents commonly used in composting.

Compostable materials	Bulking agents
Green yard wastes	Woody yard wastes
Lawn clippings	Wood chips/tree trimmings/bark
Garden trimmings	Leaves
Fruit and vegetable scraps	Newspaper and cardboard
Crop residues	Sawdust and shavings
Tea bags, coffee grounds and filters	Straw
Municipal solid waste	Domestic paper goods (napkins, etc.)
Treated sewage sludge (biosolids)	Shredded tires (biosolids composting)
Manures (cattle, horses, swine, poultry, rabbit)	

Compiled from multiple sources. See, for example, Dickson et al. (1991) and Rynk (1992).

limiting or prohibiting yard-waste disposal at landfills. This has led to a resurgence of interest in community-wide composting facilities and a correspondingly greater number of these facilities. Many organic waste products of farms, industries, municipalities, and private residences can be effectively composted, eliminating or greatly reducing a large portion of the bulk delivered to landfills. The amount of yard wastes composted has increased dramatically over the past three decades with the vast majority now being composted instead of landfilled (Fig. 24.7, Box 24.1).

In practice, most natural and many human-made organic materials are potential substrates for microbial processing. Table 24.3 lists some of the materials commonly used as feedstocks

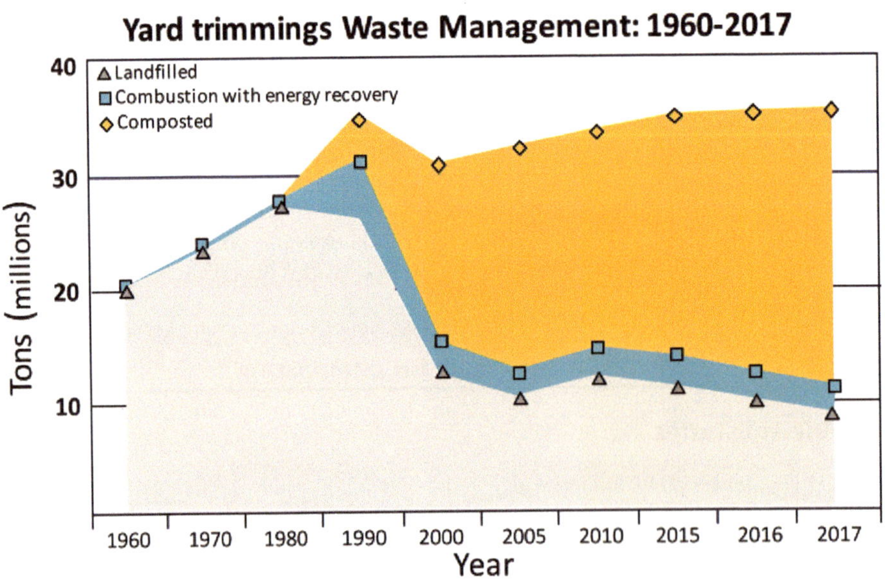

FIGURE 24.7 **Yard waste trimming management from 1960-2017**. *Source: From https://www.epa.gov/facts-and-figures-about-materials-waste-and-recycling/yard-trimmings-material-specific-data.*

TABLE 24.4 Concentrations of selected nutrients in representative compostable substrates.

Substrate	N	P	K	Ca	Mg	C/N
			(% of dry weight)			
Chicken manure	4.51	0.85	0.75	1.85	0.38	7
Cattle manure	1.51	0.52	0.59	1.00	0.27	18
Grass clippings	1.25	1.10	2.00	0.02	0.15	27
Alfalfa	2.45	0.25	1.79	1.40	3.90	15
Corn Stover	0.95	0.13	1.17	0.40	0.10	42
Oat Straw	0.60	0.09	1.25	0.60	0.15	90
Rice Hulls	0.60	0.13	0.67	0.37	0.12	85
Fish processing Wastes	9.00	7.00	0.83	1.45	0.14	4
Pine sawdust	0.14	0.02	0.10	0.08	0.03	225
Lime-stabilized Biosolids	3.65	1.16	0.43	3.60	0.40	14
Starch processing Wastes	0.11	0.03	0.06	0.02	0.03	312
Mixed green weeds	2.30	0.34	1.33	0.09	0.05	21
Mixed papermill Sludge	0.88	0.09	0.07	6.94	0.30	61

Various sources.

for composting, and Table 24.4 lists the nutrient contents of selected compostable substrates. Notice the wide range in C/N ratios, a property that can have substantial effect on the decomposition of the substrates. Those materials with very high C/N ratios are often employed as bulking agents due to their inherently slow degradation rates. Substantial quantities of compostable wastes are generated in the food processing industry. Compost piles constructed with too much easily degradable materials can become anaerobic due rapid decomposition and failure of O_2 resupply to meet the microbial demands of the aerobic decomposers. This is especially true for piles lacking adequate porosity for effective gas exchange.

Bulking agents

If the materials being composted do not contain sufficiently large particles (e.g., biosolids slurry) to achieve acceptable porosity for good aeration then another material must be added. A common solution is to combine a coarse, relatively insoluble material, that is, a **bulking agent**, along with the more soluble substrates. The mixture of particle sizes improves aeration and can be effective in establishing and maintaining appropriate moisture content for composting. Common bulking agents are tree bark, sawdust, wood chips (Fig. 24.8A), and straw. Shredded rubber tires (Fig. 24.8B) have been used as a bulking agent in the composting of biosolids (sewage sludge); these do not degrade during biosolids composting and therefore are usually removed by sieving and reused when new piles are constructed. Sawdust is particularly useful for adjusting the moisture content of very wet substrates such as biosolids. Biochar is a fairly recent addition to the list of bulking agents for composting. Biochar (Fig. 24.8C) is the product that remains after various organic materials (e.g., crop residues, wood chips, tree bark, straw) have been subjected to pyrolysis (thermal decomposition) in the

(A)
(B)
(C)

FIGURE 24.8 **Bulking agents used in composting.** (A) Yard waste trimmings. (B) Wood chips with shredded tire pieces used for bulking biosolids during composting. (C) Biochar. *Source: USDA Forest Service photo by Deborah Page-Dumroese. https://www.usda.gov/media/blog/2019/07/24/after-fire-wood-waste-put-work, public domain.*

absence of oxygen (see Chapter 14). It is essentially charcoal and it is finding usage in composting applications (Sanchez-Monederoa et al., 2018). Among these are its use as a bulking agent for composting, use as an agent for odor control, and as a soil amendment with beneficial properties for soil quality and potential for carbon sequestration (Chapter 14).

Alternatives to "hot composting"

Up to this point, we have been discussing the so-called "hot composting" processes and their characteristic sequences of events. Large-scale operations rely heavily on these methods but there are other composting processes that can convert organic wastes into useable, high-value end products. One such large-scale process is that of vermicomposting.

Vermicomposting

This is a special kind of composting that emphasizes the use of earthworms to hasten the preparation step associated with mesophilic microbes. Burrowing worms play an important part in the cycling of nutrients in soil and can significantly affect composting. Earthworms are attracted to the high concentrations of organic matter in composting environments. The worms create tunnels in the compost through which air exchanges easily, and their movements continuously mix substrates in the pile, improving the contact between substrates and microbes. Increased aeration and mixing reduce composting time, sometimes by two-thirds, and can eliminate the need for turning the pile. These activities benefit the physical environment of the compost microbes.

Worms also have a biochemical effect on composting. They ingest organic debris, partly digesting it and reducing the particle size of the substrates. Both activities enhance the microbes' abilities to decompose the organic material in the worms' excreta. In addition, worms secrete a thick mucus containing polysaccharides, proteins, and other compounds along their paths (the "drillosphere") through the compost. Bacteria and fungi use these readily available substrates as additional nutrient sources. The product of vermicomposting is called **vermicasts**, or simply, **casts**. This is a variant of the term **castings**, which describes the granular debris composed of soil aggregates and plant material held together with polysaccharides that surround holes made by the common earthworm, *Lumbricus*, in soil. The latter are also called **middens**. The most commonly used earthworm in vermicomposting is *Eisinia fetida* (commonly called "red wigglers"). Other earthworms are undoubtedly involved, but *Eisinia fetida* is the worm most often sold for "inoculating" compost piles.

Because vermicomposting proceeds at lower temperatures, some concern about the sanitization of biosolids containing human pathogens exists, but after 30–50 days, human pathogens are apparently destroyed by the process. Numerous reports indicate that pathogens are reduced to levels to meet safety standards during vermicomposting. Various mechanisms for the pathogen reduction are proposed including microbial inhibition due to intestinal enzymes, production of coelomic fluids with antibacterial properties, and stimulation of indigenous microbes leading to competition and antagonism. The subject of pathogen reduction is discussed in comprehensive reviews by Edwards and Subler (2010) and Swati and Hait (2017).

Cold composting

Cold composting, as its name implies, does not involve a thermophilic phase (Illmer, 2002; Rishell, 2013). Rather, materials to be composted are simply added to a pile or a constructed bin of some type and allowed to decompose slowly. Many backyard composters use cold composting to convert leaves, grass clippings, and vegetable scraps into compost suitable for use in gardens and lawns. The upside of cold composting is that it requires little effort on the part of the user, the pile does not have to be turned frequently, if at all, and moisture from rainfall is usually sufficient to maintain microbial activity. The downside of cold composting is that it can take a long time (many months or longer) to degrade and stabilize the materials and it does not destroy weed seeds and insects that may be added with the raw materials. Also, one must take care to not add materials to the pile that will attract rodents and other animal pests.

Compost inoculants or "starters"

Since composting is a microbiological process, it has often been proposed that exogenous microbes (inoculants or "starters") might be added to speed up the process of decomposition. To this end, various preparations have been available commercially for many years.

While there is some research suggesting that application of specific microbial preparations can shorten the duration of the composting process, the preponderance of the scientific literature suggests that little, if any, benefit can be achieved through inoculation with the various microbial preparations available. The prevailing thought is that, if a starter is desired, adding some finished compost or soil to the fresh materials to be composted provides a good source of microbes. Moreover, in nearly all cases, the substrates making up the compost pile are already colonized by the decomposer microbes.

Compost testing and use

Compost testing

The subject of compost maturity and stability is a complex issue (Wichuk and McCartney, 2010). Many different parameters have been proposed as means to identify the point at which a compost is mature and stable some of which are listed in Table 24.5. In fact, the terms maturity and stability are sometimes used interchangeably. Stability typically refers to that stage at which the easily degradable materials are gone and only a dark, slowly degradable material with an earthy smell remains. In other words, the organic materials have been semi-stabilized with respect to further microbial degradation. Maturity, on the other hand, refers to the degree of decomposition of phytotoxic substances within the compost and its suitability for plant growth.

Compost *stability* has historically been tested by determining oxygen uptake or carbon dioxide evolution. The more mature the compost, the less oxygen is taken up or carbon dioxide evolved as the readily decomposable materials have been exhausted. Another simple test of compost stability is latent heat production. A sample of the compost is placed in an insulated container and the temperature is monitored over time. Raw compost self-heats to

TABLE 24.5 Tests for the evaluation of compost maturity/stability.

Compost parameters	
Physical:	**Biological:**
Pile temperature	Respiration
Color and odor	O_2 uptake
Organic matter reduction	CO_2 evolution
Spectrophotometry	Self-heating test
Chemical:	Enzyme activity
C/N ratio	Phytotoxicity—plant bioassay
pH	Germination tests
NH_4^+ and NO_3^-	Plant growth
Cation exchange capacity	
Humification parameters	
Dissolved organic C	

Wichuk and McCartney (2010)

65°C or 70°C in 2 or 3 days, whereas mature compost normally heats to less than 30°C. Care must be taken that the materials being tested are not too dry as that can negatively influence microbial activity giving rise to low levels of respiration and hence a false reading of greater stability than what might actually exist.

Biologically based tests of compost *maturity* often include tests for the presence of phytotoxins or seed germination inhibitors. Some phytotoxicity in composts may be related to salt injury to the plants while some may be related to the low-molecular-weight fraction (e.g., organic acids) of the compost. Interestingly, the high-molecular-weight fraction may stimulate plant growth. Because composts can fertilize the soil and improve soil physical conditions, testing often includes an estimation of the mineralization rate. This is important for farmers because they must be able to accurately predict the amount of a nutrient that will be supplied by the decomposing compost in the field. With regard to testing maturity and stability of compost, there is general agreement that no one particular parameter is suitable and that more than one parameter should be assessed, yet there is incomplete agreement among researchers and practitioners as to which parameters are the most suitable.

Compost use

Many composts are good sources of nutrients for growing crops, containing useful amounts of nitrogen, phosphorus, and sulfur but they are best looked at as low-nutrient, slow-release organic fertilizers (Table 24.6). They differ from inorganic fertilizers in that compost requires further microbial decomposition (mineralization) to release most of those nutrients. The mineralization of the organic compounds results in "metering" of the nutrients to the plants. However, reliance solely on organic nitrogen sources often fails to meet the demands of actively growing crops as their nitrogen contents are generally on the order of 1%–2% and only a fraction of that becomes available during the first year of application to soil. Supplemental fertilizer often must be added to meet immediate crop demand. The mineralization rate tests mentioned above allow a producer to better estimate the supply of nitrogen that can be expected from the compost application. As soil organic matter increases with repeated use of composts, the need for supplemental fertilizers often becomes less severe. On average, 10% of the compost is mineralized during the first year after soil application but mineralization of food-waste compost can release up to 25% of its nitrogen.

TABLE 24.6 Selected properties and elemental contents of some finished composts.

	Compost Type			
Parameter	**Biosolids**	**Yard wastes**	**Food wastes**	**MSW**
pH	6–6.5	5.8–7.2	—	—
Moisture (%)	20–35	34–61	40	45
Bulk density (kg m^{-3})	772	304–400	352	352–481
Total N (%)	2.21	0.7	0.95	—
P (%)	0.97	0–0.03	0.13–0.46	—
K (%)	0.27	0.02–0.36	0.58–0.89	—

*For comparison with starting materials see Table 24.4.
Data adapted from Diaz et al. (1993) and Satriana (1974).

In subsequent years, the annual decomposition rates fall to between 1% and 3%, resembling the decomposition rates of native soil organic matter. In addition, composts provide the same benefits that are generally ascribed to organic matter. These include increasing soil aggregation and stabilization, soil aeration, water infiltration, water holding capacity, and cation exchange capacity.

It should be pointed out that some nutrients, particularly, phosphorus can be problematic in the use of composts for land application (Box 19.8). For example, composted dairy manure can contain high levels of phosphorus. The high phosphorus content derives from the ration fed to the dairy cows. Phosphorus levels remain high during the composting process and, in fact, increase relative to the starting concentration due to volume reduction of the organic fractions. Historically, compost application rates were based on levels needed to achieve a given rate of nitrogen addition without much regard to the amounts of phosphorus that were being applied. Use of high rates of such composts can lead to runoff and perhaps even some leaching of phosphorus from the application site with the potential for negative consequences in nearby receiving waters such as streams and lakes.

Composts have been used in land reclamation efforts and other applications such as facilitating the establishment of vegetation on newly graded roadside embankments. While they serve a distinct positive role in such applications, here too problems can arise. As mentioned earlier, compost recommendations are frequently based on estimated nitrogen needs of the plants in question. An example of this is the heavy application of composts during the establishment of vegetation on newly established slopes along roads and highways. Such high rates of compost applications can lead to excessive P in the compost "blanket" and underlying soil. The soluble P leached from such sites can find its way into ditches and streams, where it can lead to excessive growth of algae and other nuisance vegetation.

Compost can also be used, with varying degrees of success, for biocontrol of some plant diseases (Litterick and Wood, 2009; Fuchs, 2002). Diseases caused by *Pythium* have been partly controlled with compost application. Antagonistic microbes such as *Trichoderma, Pseudomonas fluorescens*, and *Acinetobacter* have been inoculated into finished compost, which was found to be suppressive against the pathogens, *Fusarium* and *Rhizoctonia*. Studies have also shown that soil amended with non-inoculated compost can increase microbial populations antagonistic to pathogenic fungi in the rhizosphere of tomato (De Brito Alvarez et al., 1995). The suppression is often associated with the production of siderophores (Chapter 19) in the rhizosphere, suggesting a role for compost stimulating natural antagonists in the rhizosphere. Some stimulation of microbial populations and dehydrogenase and protease activities also occurs when compost or compost extract is added to soil.

The extent of compost use has grown to include fertilization of golf courses, athletic and recreational turf areas, home lawns and gardens, landfill caps, roadside grade stabilization, and agricultural fields. Composts have become marketable products for industries and municipalities, further reducing compost operation costs.

Other organic soil amendments

Compost extracts (a.k.a. "compost teas")

Over the past few decades, interest in using compost extracts (teas) in various agricultural settings has increased along with the increased interest in organic and sustainable agriculture.

This increased interest is driven in part by the needs of organic producers to control plant pests without using synthetic pesticides and other chemicals. However, interests in using compost teas also extend to their use as substitutes or supplements in fertilization programs.

The term "compost extract" refers to solutions prepared by steeping compost in water, usually in 1:5 or 1:10 ratio, and then filtering the compost from the solution to obtain the extracted nutrients and microbes. Extracts differ from "teas" in that there is no prolonged "fermentation" period involved in their making. Compost tea is the term generally used to refer to the filtered product of compost fermented in water with or without additives such as molasses, rock dust, selected microbes, or other organics. It is important to distinguish between aerated compost tea (ACT) and non-aerated compost tea (NCT). Aerated compost tea is prepared by immersing compost in water while maintaining active aeration, generally through sparging with air or allowing the circulated solution to percolate through compost suspended in mesh bags. In contrast, NCT is not aerated and generally some degree of anaerobiosis occurs during the fermentation. Regardless of the preparation method, "compost teas" are purported to contain beneficial microorganisms, nutrients, and organic compounds that have positive effects on plant growth when applied as foliar sprays or as a seed and soil drench. However, such preparations do not supply substantial amounts of plant macro- or micro-nutrients when applied at low volumes (Quilty and Cattle, 2011). Perhaps the situation is best summed up by the conclusion of Edmeades (2002) that "these products do not contain sufficient concentrations of plant nutrients, organic matter, or plant growth substances (pGSs) to elicit increases in plant growth when applied as recommended" (2–20 L ha^{-1}).

While there are reports of positive effects on plant growth and soil properties resulting from the use of compost teas, the scientific literature is inconsistent. One of the major drawbacks of compost teas is the lack of reproducibility from one batch to the next. Differences in the starting materials, conditions under which the solutions are prepared (e.g., aerobic vs. anaerobic), selected additives, and other factors combine to make the production of consistently reproducible products rather difficult. In short, glowing recommendations for the use of such dilute solutions should be viewed with skepticism.

However, the literature regarding control of foliar pathogenic fungi is a bit more robust (Scheuerell and Mahaffee, 2002). Many researchers and practitioners report effective control of foliar diseases through applications of compost teas. Still, as with experiments conducted with soils, the effects of compost teas as biocontrol preparations have shown inconsistent results. Control of foliar pathogens is thought to derive from competition between the pathogen and antagonistic microbes or their products (e.g., antibiotics) applied with the sprayed or drenched preparations but the mechanisms remain obscure. While there have been some successes in this area, researchers are quick to point out that much remains to be learned about the best ways to produce consistent effective preparations, their modes of action in controlling pathogens, and best methods for economic delivery in greenhouse and field applications. St. Martin et al. (2020) provide a comprehensive discussion of the research aimed at better understanding the microbiology and effect of compost and compost tea using -omics approaches (genomics, metagenomics, metaproteomics, metaprotegenomics, metatranscriptomics, and metabolomics).

Commercial humates (humic substances)

Because of the multifaceted benefits ascribed to SOM (Chapters 13, 14), there has been a long-standing effort to produce commercial products reminiscent of "humus" for application as soil amendments with the hope of improving soil quality. Non-soil-derived materials like lignite coal (a low-grade soft brown coal) and Leonardite (an oxidation product of lignite) have been used as large-scale source materials for extraction of commercial humates or added directly to soil as granular preparations. Like soil humic substances, the humic substances from lignite are extracted with strong alkali solutions. However, the composition and properties of humic acids derived from lignite are quite different than those of humic substances extracted from soil. "Coal humates are essentially free of such biologically important compounds as proteins and polysaccharides" found in soil humic substances (Stevenson and Cole, 1999). And, as pointed by Stevenson and Cole (1999), "mined humates are to a large extent inert owing to their combination with mineral matter." Nevertheless, some of these extracts have been marketed commercially for many years, and many are still being sold today. The efficacy of these products has been a subject of contention for as long as they have been on the market (Box 24.3).

As with any product purchased to accomplish some form of enrichment of the soil ecosystem, consumers would be well advised to check the label for actual contents of the "active ingredient(s)" whether they be microbes, humates, or plant nutrients (e.g., N, P, K). From this

BOX 24.3

Commercial humates

Commercially available humates (coal humates) are oxidized lignite material that is associated with lignite; a soft brown coal. The oxidized lignite material, often referred to as Leonardite, is not suitable for fuel because of its low heating potential, but it does contain from 30% to 60% "humic acid" by weight (Stevenson and Cole, 1999). The mined lignite material is crushed and frequently amended with inorganic fertilizer. The lignite can also be treated to produce ammonium humate fertilizer. Both products are generally expensive compared to the comparable amount of nutrients in inorganic fertilizer. Often the marketing claims include that the humates increase plant growth, nutrient uptake, buffering capacity, cation exchange capacity, microbial activity, and organic matter levels. Additional claims can be that the humates decrease soil bulk density and enhance soil aggregation.

If you applied a humate material containing 60% humic matter at 500 kg ha^{-1} to a soil containing 2.5% organic matter, the humate application would increase the organic matter level from 56,000 kg (a hectare slice = 2,240,000 kg: $0.025 \times 2{,}240{,}000$ kg = 56,000 kg) to 56,300 kg or 2.51% organic matter, an increase of 0.01%. This would be an expensive method of increasing soil organic matter levels.

The nitrogen content of oxidized lignite without fertilizer fortification is generally less than 1.5%. The addition of 500 kg humate ha^{-1} would add approximately 7.5 kg nitrogen ha^{-1} and, if the nitrogen is mineralized at an annual rate of 5%, the mineralized nitrogen applied would be about 0.4 kg ha^{-1}, or less than the nitrogen contained in precipitation.

In normally productive soils in humid climates, it does not appear that application of commercially available humates at recommended rates would have any significant beneficial effects based on C or N added.

BOX 24.4

Calculating actual application rates of "active ingredients" for non-traditional agricultural and horticultural products

Many non-traditional soil amendment products on the market instruct the user to apply the product at a rate of 1 gallon per acre. Some even specify 1 gallon per 10 acres. Let's analyze the 1 gallon-per-acre rate to determine just how much product (active ingredient, AI) is actually applied.

The "one-gallon-per-acre dilution":

One gallon is equal to approx. 3800 mL (3785.4, milliliters or cm^3)

One milliliter (mL) = 20 drops:

So... 20 drops × 3800 mL = 76,000 drops gal^{-1}.

One acre (ac) = 43,560 ft^2. Thus, one-gallon ac^{-1} = 76,000 drops / 43, 560 ft^2 = 1.75 drops ft^{-2} (0.088 mL or 0.0029 oz.).

But, that's just the volume of liquid product that would be applied. Frequently, these types of products contain just a few percent of "active ingredient(s)." Thus, a product containing 10% "natural plant extracts" would deliver just 0.175 drops of AI ft^{-2}.

Consider the application rate of a "humic acid" (HA) product containing 12% humic acids (HA) and applied to lawns at a rate of 16 oz. 1000 ft^{-2} (an actual "off the shelf" application rate):

16 oz. = 473 mL (9463 drops) spread over 1000 ft^2 = 0.47 mL (9.46 drops) ft^{-2}

At 12% humic acid content: 0.47 mL × 0.12 = 0.056 mL HA ft^{-2}

9.46 drops × 0.12 = 1.12 drops HA ft^{-2}

If the surface 6 inches (15 cm) of soil contains 21 kg soil ft^{-2} (based on an acre-furrow-slice of 2,000,000 lbs.) and the average dissolved organic carbon content is around 300 mg C kg^{-1} then that one square foot contains 6,300 mg C (6.3 g) of dissolved organic C. Thus, one can calculate that that 1.12 drops of HA preparation delivers an infinitesimally small contribution of humate C to the existing soil C pool.

information, calculate how much of the active ingredient will actually reach the soil when applied at the recommended rate specified on the product label and consider that in the context of what a reasonably fertile soil already contains. Such calculations can be quite informative (Box 24.4). Remember, a soil having just 2% organic matter contains 10,524 kg (23,200 pounds) organic C per hectare (organic matter on average contains 58% C) along with literally thousands of kilograms of microbial biomass.

Summary

Composting is a microbial process in which organic matter progresses through stages of decomposition and stabilization in a relatively short time including the early mesophilic, intermediate thermophilic, and final curing phases (cooling and maturation). It is characterized by an early period of self-heating in which heat generated by rapid microbial metabolism is produced faster than it can dissipated from the site of decomposition. Temperature increases and substrate changes elicit a succession of microorganisms that reflect the change from mesophilic to thermophilic stages and the changing nature of the organic substrates themselves. During the curing phase, mesophilic conditions return and further decomposition stabilizes the compost. Finally, after a "maturation" stage the compost is relatively stable and becomes a beneficial product with multiple uses in landscapes, agricultural settings, and land reclamation projects. Applied to land, finished compost can function as a slow-release fertilizer as it is further acted upon by soil microbes, and it serves as a soil-improving amendment through modest increases in SOM and soil physical properties.

Study questions

1. What is the purpose of composting?
2. How does composting differ from decomposition of organic substrates in soil?
3. What is the role of oxygen in composting?
4. Why is so much heat produced during composting?
5. What chemical and physical characteristics of organic substrates promote rapid composting?
6. What successional changes do microbes undergo during composting?
7. What is the "stabilization" or "curing" phase of composting?
8. What are the advantages and disadvantages of using compost as a fertilizer?

References

Danon, M., Franke-Whittle, I.H., Insam, H., Chen, Y., Hadar, Y., 2008. Molecular analysis of bacterial community succession during prolonged compost curing. FEMS Microbiol. Ecol. 65 (1), 133–144, doi:10.1111/j.1574-6941.2008.00506.x.

De Brito Alvarez, M.A., Gagné, S., Antoun, H., 1995. Effect of compost on rhizosphere microflora of the tomato and on the incidence of plant growth-promoting rhizobacteria. Appl. Environ. Microbiol. 61 (1), 194–199.

Diaz, L.F., de Bertoldi, M., 2007. History of composting, pp.7–24. Diaz, L.F., de Bertoldi, M., Bidlingmaier, W., Stentiford, E. (Eds.), Compost Science and Technology, Vol. 8, Elsevier, p. 364.

Diaz, L.F., Savage, G.M., Eggerth, L.L., Golueke, C.G., 1993. Composting and Recycling Municipal Solid Waste. Lewis Publishers, Boca Raton, FL, p. 296.

Dickson, N., Richard, T., Kozlowski, R., 1991. Composting to Reduce the Waste Stream—A Guide to Small Scale Food and Yard Waste Composting. NRAES-43—Northeast Regional Agricultural Engineering Service.

Edmeades, D.C., 2002. The effects of liquid fertilisers derived from natural products on crop, pasture, and animal production: a review. Aust. J. Agric. Res. 53, 965–976.

Edwards, C., Subler, S., 2010. Human pathogen reduction during vermicomposting. In: Edwards, C.A., Arancon, N.Q., Sherman, R.L. (Eds.), Vermiculture Technology: Earthworms, Organic Wastes, and Environmental Management. CRC Press, Taylor & Francis Group, Boca Raton.

Fuchs, J.G., 2002. Practical use of quality compost for plant health and vitality improvement. In: Insam, H., Riddech, N., Klammer, S. (Eds.), Microbiology of Composting. Springer, Berlin, Heidelberg, Available from: https://doi.org/10.1007/978-3-662-08724-4_36.

Goldstein, N., 2017. State of Organics Recycling in the U.S. BioCycle.

Illmer, P., 2002. Backyard composting: general considerations and a case study. In: Insam, H., Riddech, N., Klammer, S. (Eds.), Microbiology of Composting. Heidelberg, Springer, Berlin.

Insam, H., de Bertoldi, M., 2007. Microbiology of the composting process. In: Diaz, L.F., de Bertoldi, M., Bidlingmaier, W., Stentiford, E. (Eds.), Compost Science and Technology,, Vol. 8, Elsevier, p. 364.

Kästner, M., Mahro, B., 1996. Microbial degradation of polycyclic aromatic hydrocarbons in soils affected by the organic matrix of compost. Appl. Microbiol. Biotechnol. 44, 668–675, https://doi.org/10.1007/BF0017250.

Kästner, M., Miltner, A., 2016. Application of compost for effective bioremediation of organic contaminants and pollutants in soil. Appl. Microbiol. Biotechnol. 100, 3433–3449, 10.1007/s00253-016-7378-y.

Litterick, A., Wood, M., 2009. The use of composts and compost extracts in plant disease control. In: Walters, D. (Ed.), Disease Control in Crops: Biological and Environmentally Friendly Approaches. Blackwell Publishing Ltd, pp. 93–121.

Quilty, J.R., Cattle, S.R., 2011. Use and understanding of organic amendments in Australian agriculture: a review. Soil Res. 49, 1–26.

Rishell, E., 2013. Backyard Composting. Virginia Cooperative Extension. VT/0113/w/HORT-49P.

Ryckeboer, J.J., Mergaert, K., Vaes, S., Klammer, D., De Clercq, J., Coosemans, H.I., Swings, J., 2003. A survey of bacteria and fungi occurring during composting and self-heating processes. Ann. Microbiol. 53, 349–410.

Rynk, R., 1992. On-Farm Composting Handbook. Northeast Regional Agricultural Engineering Service, Cooperative Extension, Ithaca, NY, NRAES-054.

Sanchez-Monederoa, M.A., Cayuelaa, M.L., Roiga, A., Jindob, K., Mondinic, C., Boland, N., 2018. Role of biochar as an additive in organic waste compost. Bioresource Technol. 247, 1155–1164.

Scheuerell, S., Mahaffee, W., 2002. Compost tea: principles and prospects for plant disease control. Compost Sci. Utiliz. 10 (4), 313–338, doi: 10.1080/1065657X.2002.10702095.

Satriana, M.J., 1974. Large Scale Composting. Noyes Data Corp, Park Ridge NJ, p. 269.

St. Martin, C.C.G., Rouse-Miller, J., Barry, G.T., Vilpigue, P., 2020. Compost and compost tea microbiology: the "-omics" era. In: Meghvansi, M., Varma, A. (Eds.), Biology of Composts. Soil Biology,, vol. 58, Springer, Cham.

Stevenson, F.J., Cole, M.A., 1999. Cycles of Soils: Carbon, Nitrogen, Phosphorus, Sulfur, Micronutrients, second ed. John Wiley and Sons, New York, p. 448.

Swati, A., Hait, S., 2017. A comprehensive review of the fate of pathogens during vermicomposting of organic wastes. J. Environ. Qual. 47, 16–29.

Tiquia, S.M., 2002. Evolution of extracellular enzyme activities during manure composting. J. Applied Microbiol. 92, 764–775.

U.S. EPA., 1993. Yard Trimmings: Material-Specific Data. Available from: https://www.epa.gov/facts-and-figures-about-materials-waste-and-recycling/yard-trimmings-material-specific-data.

Wang, C., Guo, X., Deng, H., Dong, D., Tu, Q., Wu, W., 2014. New insights into the structure and dynamics of actinomycetal community during manure composting. Appl. Microbiol. Bio-technol. 98, 3327–3337.

Wei, H., Tucker, M.P., Baker, J.O., Harris, M., Luo, Y., Xu, Q., Himmel, M.E., Ding, S.Y., 2012. Tracking dynamics of plant biomass composting by changes in substrate structure, microbial community, and enzyme activity. Biotechnol. Biofuels. 5 (1), 20, https://doi.org/10.1186/1754-6834-5-20.

Wichuk, K.M., McCartney, D., 2010. Compost stability and maturity evaluation—a literature review. Can. J. Civ. Eng. 37, 1505–1523.

Willson, G.B., Parr, J.F., Epstein, E., Marsch, P.B., Chaney, R.L., Colacicco, D., Burge, W.D., Sikora, L.J., Tester, C.F., Hornick, S., 1980. Manual for Composting Sewage Sludge by the Beltsville Aerated-Pile Method. EPA600-8-80-022. Washington, DC.

Wu, M., Guo, X., Wu, J., Chen, K., 2020. Effect of compost amendment and bioaugmentation on PAH degradation and microbial community shifting in petroleum-contaminated soil. Chemosphere 256, 126998.

Further reading

Brown, S., Angle, J.S., Jacobs, L. (Eds.), 1997. Beneficial Co-Utilization of Agricultural, Municipal and Industrial By-Products, Kluwer Academic Publishers, p. 430.

Epstein, E., 1997. The Science of Composting. Technomic Publishing Co, Lancaster, PA, p. 487.

Insam, H., Riddech, N., Klammer, S. (Eds.), 2002. The Microbiology of Composting, Springer-Verlag Berlin Heidelberg, New York, p. 632.

CHAPTER

25

Soil microbial influences on "One Health"

Ian L. Pepper[a], *John P. Brooks*[b]

[a]Water and Energy Sustainable Technology Center (WEST), University of Arizona, Tucson, AZ, United States; [b]USDA-ARS, Genetics and Sustainable Agriculture Unit, Mississippi State, MS, United States

"The earth will open and bring forth salvation"— ***inscription on Selman Waksman's headstone.***

The previous chapters detailed the incredible abundance, activity, and diversity of microorganisms in soil, some of the major biogeochemical cycles these organisms control, and how we can manage these organisms for beneficial processes such as bioremediation of organic pollutants and biological control of plant pathogens. However, as the COVID-19 (SARS CoV-2) pandemic of 2020 painfully reminds us, some microorganisms also cause human disease. Although most human pathogens are not directly associated with soil, it is important to recognize the potential for soil to serve as a buffer against, a temporary reservoir for, and/or even a source of pathogenic organisms. As we complete our discussion of the principles and applications of soil microbiology, this chapter briefly describes examples of how the soil microbiome can impact soil, plant, animal, and human health—often for the better but sometimes for the worse.

What is "One Health?"

The original "**One Health**" concept was based on the fact that human health is inextricably linked to the health of animals and the environment. The practice of one health has been around for centuries, having been postulated by Greek philosophers. In fact, the Greek word, hygia, describes good hygiene as a condition of well-balanced harmony with the body and the universe. Furthermore, work by Hippocrates defined the equilibrium between the environment and human health. This is particularly true with respect to infectious agents, which have significantly increased in recent years, including the spread and emergence of epizootics, zoonoses, and epidemics (Box 25.1).

Principles and Applications of Soil Microbiology. http://dx.doi.org/10.1016/B978-0-12-820202-9.00025-3
Copyright © 2021 Elsevier Inc. All rights reserved.

BOX 25.1

Definitions of infectious agents and outbreaks

Infectious agent/outbreak	Definition
Epizootic	An epidemic outbreak of disease in an animal population, with the potential for extension to humans
Zoonotic (zoonosis)	A disease which can be transmitted to humans from animals
Epidemic	Rapid spread of human infectious disease to a large number of people in a given population within a short period of time, usually 2 weeks or less

TABLE 25.1 Examples of disease transmitted to humans from animals.

Disease	Infectious agent	Route of transmission/host
West Nile fever	West Nile virus	Mosquitoes
Avian influenza	Bird Flu virus	Birds
COVID-19	SARS CoV-2	Suspected animal
Ebola virus disease	Ebola virus	Monkeys, Gorillas
Tapeworm	*Taenia saginata* or *T. solium*	Livestock
Plague	*Yersinia pestis*	Fleas
Salmonellosis	*Salmonella* spp.	Livestock, many animals
Malaria	*Plasmodium* spp.	Mosquito

More than half of the recognized diseases in humans are due to pathogens capable of infecting or colonizing multiple hosts. As such, we as humans, have an interdependence on the health of animals and on the conditions of their rearing environment and ultimately the greater environment. Examples of specific disease resulting from animals or insects are shown in Table 25.1.

The globalization of health risks has led to an expanded vision of "One Health" to include the human-animal-ecosystem interface (Destoumieux-Garćon et al., 2018). Van Bruggen et al. (2019) argued that the health conditions of all organisms in an ecosystem are interconnected through the cycling of subsets of microbial communities from the environment, in particular soil, to plants, animals, and humans, and back into the environment. Increasingly then, soils are being recognized as a significant part of the environment and critical to human, plant, and animal health. Thus, if the environment is thought to be a reflection of soils and plants, then conceptually "one health" must include the soil-plant-animal-human ecosystem (Fig. 25.1). This in turn has resulted in a formal definition of "one health" (Box 25.2). Overall, the inclusion of soil within the "one health" concept is revolutionizing the importance of soil, through the growing realization that soil does much more than just allow for plant growth.

When discussing one-health, one of the most common applications is with antibiotic resistance and zoonotic pathogens (e.g., pathogens capable of colonizing multiple hosts such as humans and other animals). Overall, humans are closely related to animals at the genetic level. We share 95% genetic similarity with pigs and 99% with apes, thus sharing of pathogens is a

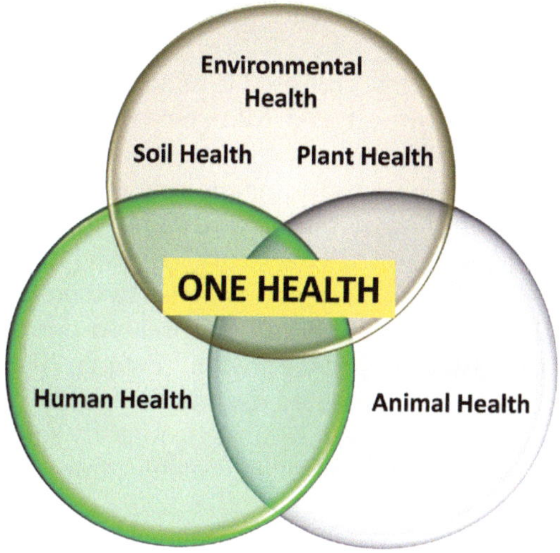

FIGURE 25.1 **Conceptual model of "One Health."**

> **BOX 25.2**
>
> **Definition of One Health**
>
> One Health is a collaborative, multisectional, and transdisciplinary approach working at local, regional, national and global levels—to achieve optimal **health** and well-being outcomes recognizing the interconnections between people, animals, plants, and their shared environment.
>
> *One Health Commission, a globally focused 501 C3 organization, chartered in Washington, DC.*

distinct possibility that one health aims to address. For instance, the misuse of antibiotics at the farm level (e.g., pig production) potentially can select for antibiotic resistant pathogens, such as *Salmonella*, that can also infect humans. Thus, this is a one-health issue, as efforts by world agencies to reduce or optimize antibiotic use in animal production can have wide-reaching effects on the prevalence of antibiotic-resistant pathogens entering the environment and food chain. The complexity of these problems should not be lost, as there are numerous interactions occurring and webs of links between the environment-animal-human ecosystem, that is, "One Health."

Influence of microbes on soil health

Evolution of soil health

Soils were classically defined as the product of the five soil forming factors by the Russian geologist Dokuchaev in the 19th century and endorsed by Hans Jenny in the 20th century.

They have also colloquially been defined as the "skin of the earth" and the "most complicated biomaterial on the planet."

The concept of **soil health,** sometimes referred to as soil quality, has been widely discussed since the 1970s when the term was introduced. More recently, Doran (2002) defined soil health as "the capacity of a living soil to function, within natural or managed ecosystems boundaries, to sustain plant and animal productivity, maintain or enhance water and air quality, and promote plant and animal health"—a definition eerily similar to the "One Health" concept. A different concept is to think of soil as a "living entity" and soil health as analogous to human health. Both must be in a state of well-being with respect to their physical, chemical, and biological characteristics. Likewise, neither should be diseased nor compromised, and ideally, both should function sustainably at optimal potential (Pepper, 2013).

Enhanced interest in soil health manifested itself in the formation of the United States Soil Health Institute in 2018, which spearheaded efforts to quantify soil health. The Institute identified 19 physical, chemical, and biological soil parameters termed as the Tier One Soil Health indicators (Table 25.2).

However, there has also been a recognition that microbial indicators are more sensitive to changes in soil use and management, and that biological indicators are underutilized, but show great potential (Bünemann et al., 2018). In other words, soil microbes and their activity are vital to soil health.

In fact, a close examination of Table 25.2 shows that all of the "biological" Tier One Soil Health indicators are governed by soil microbes. Crop yield is directly dependent on soil microbes through the plant root/rhizosphere interaction and indirectly through soil microbial transformations including carbon and nitrogen mineralization. Soil microbes also control organic matter degradation and hence total soil organic carbon. Soil microbes also influence many of the "physical" Tier One indicators by enhancing soil structure and water stable aggregations, which in turn affects available water holding capacity, bulk density and infiltration rate. Thus, it is not surprising that soil microbes are critical to soil health.

Role of the soil microbiome in soil health

Soils contain vast amounts of soil microorganisms capable of utilizing most or all available nutrients present within the soil. However, total soil microbial biomass is typically 2%–3%

TABLE 25.2 Tier one chemical, physical, and biological soil health indicators.[a]

Chemical	Physical	Biological
Base saturation	Available water holding capacity	Crop yield
Cation exchange capacity	Bulk density	Nitrogen mineralization
Electrical conductivity	Infiltration rate	Organic carbon
Nitrogen	Erosion rating	Carbon mineralization
Micronutrients	Penetration resistance	
pH	Texture	
Phosphorus	Water stable aggregation	
Potassium		

[a]*Soil Health Institute (2018).*

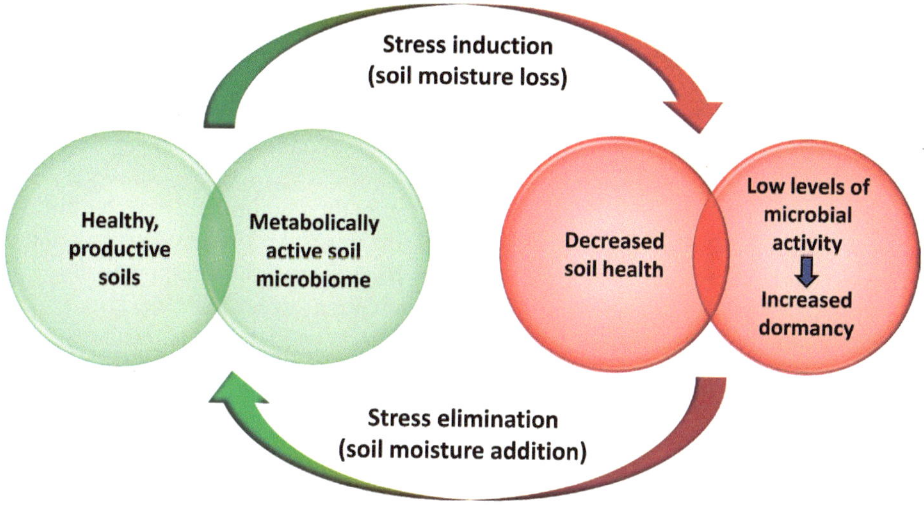

FIGURE 25.2 **Conceptual model of the influence of soil moisture stress on microbial activity.** *Source: Adapted from Ikner et al., 2021.*

of the total soil organic matter content, and is rarely greater than 4.5% (Anderson, 2003). Additionally, active soil microorganisms normally comprise only a small fraction of the total microbial biomass with typical values being 0.1%–2% and rarely exceeding 5% in soils (Blagodatskaya and Kuzyakov, 2013). Thus, in a pristine soil perhaps only 1% (active microbes) of 2% (total soil organic matter) is metabolically active. However, potentially active microbes including dormant microbes can comprise anywhere from 10% to 60% of the total microbial biomass (Blagodatskaya and Kuzyakov, 2013).

Inactive or dormant microbes can result from lack of nutrients and/or soil moisture stress. These two factors can be co-dependent with decreasing soil moisture limiting solute diffusion as moisture films become thinner and disconnected (Chapter 2). In contrast, increasing soil moisture increases liquid diffusion rates, providing microbes with nutrients. A conceptual model of this is illustrated in Fig. 25.2 including the reversible nature of stress induction and stress elimination. Clearly then, soil health can be dramatically affected by soil microbial communities (Ikner et al., 2021).

Quantification of soil health via soil microbial activity

Traditional estimates of soil microbial counts and activity

Since the formation of the Soil Health Institute in 2018, there has been increased interest in utilizing the living component of soils as an indicator of soil health, namely soil microorganisms and soil microbial activity. The gold standards for evaluating microbial counts and activity have been heterotrophic plate counts (HPCs) and dehydrogenase activity. However, HPCs do not give significant information on soil health for a variety of reasons (Box 25.3). First, HPC counts obtained via dilution and plating result in a low sensitivity of detection of the total soil microbial population since many bacteria are viable but non-culturable with

BOX 25.3

Reasons why heterotrophic plate counts (HPCs) are not a good indicator of soil health

- Only 1%–10% of total bacterial community is typically recovered by dilution and plating cultural assays
- Dilution plating is an impressive methodology but prone to significant error
- Soil bacterial communities can be primed by the addition of water in as little as 5 min; thus, adding soil to water during the dilution procedure can resuscitate bacterial populations, potentially inflating "in situ" HPC levels
- Selects for members of the "rare biosphere"

common methods. This in turn has given rise to the concept that culture-dependent assays capture "the rare biosphere" and that cultured bacteria are actually atypical members of the soil microbial community (Shade et al., 2012). Likewise, dehydrogenase activity values as reflected by triphenyl formazan (TPF) measurements do not appear to be sensitive enough to give useful information on soil health (Ikner et al., 2021).

Estimates of soil microbial activity via the adenosine monophosphate (AMP) stress Index

A quantitative assay for soil microbial activity based on the AMP Index was recently proposed by Ikner et al., 2021. The AMP Index is the quantitative ratio of soil microbial community AMP to ATP. As discussed in Chapter 3, ATP is the "molecular unit of currency" for many intracellular energy processes within living cells. As energy is used, ATP is converted to AMP. Thus, if the AMP Index is high within soil communities, the overall metabolic activity of soil microbial communities is low. Conversely, if the AMP Index is low within the same soil community, this represents a condition where ATP predominates and the metabolic state of activity is high.

As discussed in Box 3.3, ATP levels in soil have been used as an indicator of overall microbial biomass and activity. More recently, Keasler et al. (2013) assayed both ATP and AMP, using a LuminUltra test kit, to evaluate the AMP Indices of microbial communities within wastewater. Based on their data, they proposed that the AMP Index could be used to quantify the level of stress on the microbial community where:

AMP Index	Interpretation
< 1	No stress: active microbes
1 to < 10	Increasing stress conditions
>10	Extreme stress conditions, dormancy predominates

Ikner et al., 2021 modified the LuminUltra assay for use as a biological indicator of soil health. To determine whether the AMP Index could indicate the metabolic status of soil microbial communities, samples of five diverse soils were collected in the Tucson, Arizona (USA) area in the middle of summer. At the time of collection, there had been no rainfall for several weeks, and a high temperature record of 18 consecutive days above 105°F (40.5°C) had been broken. Hence, these non-irrigated soils contained microbial communities that were both heat stressed and moisture stressed.

AMP Indices were assayed for the heat stressed and moisture stressed soil microbial communities within the five different soils. These stress conditions were relieved by the addition of moisture to the soils, and the AMP Indices again assayed (Fig. 25.3). Additional experiments evaluated the effects of wetting/drying cycles on the AMP Indices. Soil moisture was shown to regulate soil microbial activity and dormancy. Moisture stressed soil microbial communities were shown to respond metabolically to the addition of water within 5 min. Overall, the AMP Indices appeared to be good indicators of soil microbial metabolic activity and soil health.

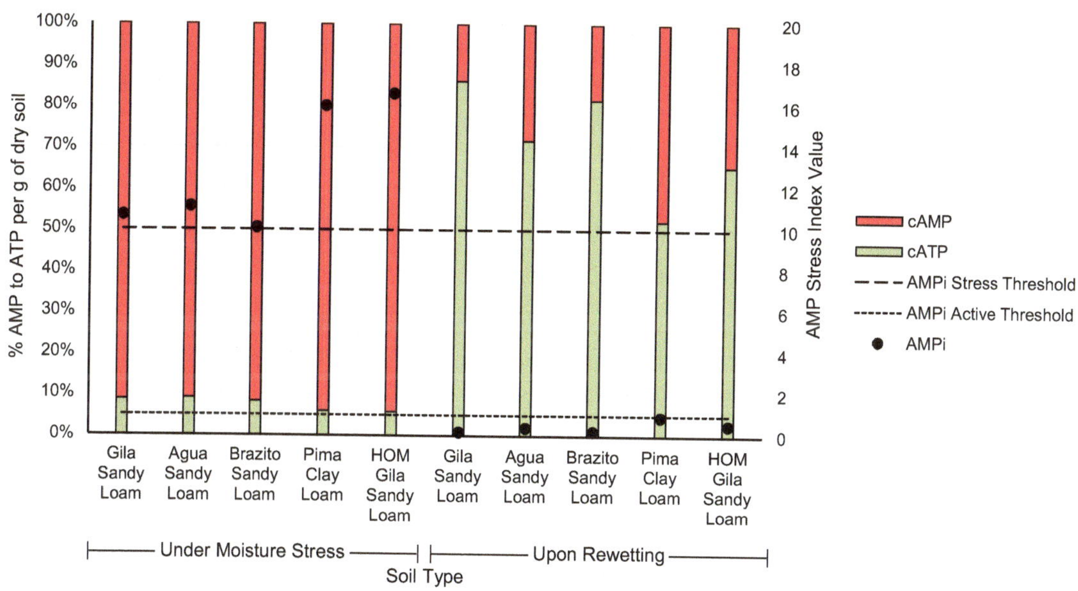

FIGURE 25.3 **Relative AMP, ATP, and AMPi (AMP Index) values under soil moisture stress and 24 h after rewetting.** *HOM*, high organic matter.

Influence of soil microbes on plant health

Soil-root-plant continuum

Plant roots are anchored in soil, and the soil in close proximity to the root system is known as the rhizosphere (Chapter 11). Due to root exudates, which function as microbial substrates, soil microbial numbers and activity in the rhizosphere are enhanced relative to bulk soil not under the influence of plant roots. This rhizosphere soil microbiome is in direct contact with the root microbiome, and both microbiomes directly affect the plant host. The root microbiome consists of bacterial and fungal endophytes that live within the plant root. Thus, a soil-root-plant continuum influences plant health. Ultimately, the plant host is influential in selecting the members of these microbiomes (Berendsen et al., 2012). Specifically, the root endophytic community is comprised of members selected by the host plant from within the rhizosphere soil microbiome (Bulgarelli et al., 2012). For example, under the influence of drought conditions and soil moisture stress, the soil microbial community shifts toward enhanced numbers of low-moisture-tolerant species. Concurrent changes in plant metabolism due to low moisture conditions results in altered root exudation patterns, which affects the selection of members for the root microbiome. Specifically, an increase in Gram-positive bacteria relative to Gram-negative bacteria occurs, since Gram-positive bacteria have thicker peptidoglycan cell wall layers and can be endospore formers, making them more tolerant of soil moisture stress. Drought conditions also result in increased numbers of slow-growing Gram-positive oligotrophic bacteria, which survive more readily in low-nutrient environments. Low soil moisture environments result in lower nutrient availability due to discontinuous moisture films, which limit substrate diffusion. Gram-negative copiotrophic bacteria prefer high-nutrient, high-soil-moisture conditions, and the levels of these bacteria decline rapidly in dry soils (Kurm et al., 2017). As the soil microbiome changes, this is reflected in the root microbiome and the status of plant health. However, how are these changes in the soil-root-plant manifested? The answer is communication between soil and root bacteria and between bacteria and the plant host.

Microbial communication: bacteria-bacteria

The soil environments that contain microbes are complex and subject to rapid change. In order to be successful, bacteria must be able to sense and rapidly respond to these changes through the expression of specific genes and metabolic pathways. Additionally, it is now well documented that bacteria produce signals that allow communication between cells (Pierson et al., 2015). To do this, cells communicate by emitting specific chemical signals into the environment, which is inhabited by many different microbes. If the concentration of the signal reaches a level that other cells can perceive, known as the threshold concentration, then the gene expression of those organisms becomes modified.

The cell-cell communication is important for coordinating gene expression within a single population of bacteria (**intraspecies signaling**); between bacterial populations (**interspecies signaling**); and between bacteria and other organisms (**interkingdom signaling**). **Quorum sensing** is the regulation of gene expression in response to levels of diffusible signal molecules. The concentration of the signals usually correlates with population density. In other words, a sufficient number of cells or a quorum must be present in order for gene expression to occur. These quorum-sensing bacteria produce and release chemical signal molecules or **autoinducers** that control gene expression of the whole bacterial population. In soils, Gram-negative bacteria utilize N-acyl homoserine lactones (AHL) as chemical signals (Scott et al., 2006). In contrast, Gram-positive bacteria utilize γ-butyrolactones or small peptides as their primary communication molecules (Pierson et al., 2015).

Microbial communication: bacteria-plant host

Microbes preceded plants as soil inhabitants, thereby allowing mechanisms of communication to evolve over time. The most well studied communications between plants and bacteria include the *Rhizobium*-legume symbiosis (Chapter 17) and the pathogenic association between *Agrobacterium tumefaciens* and plant hosts (Chapter 23). In both cases, extensive interkingdom signaling results in bacterial invasion of plant tissues, the former being beneficial, and the latter detrimental. This communication is believed to be widespread and bidirectional (Pierson et al., 2015). In contrast, most plant-associated bacteria colonize only the plant root surface. However, here, bacterial-produced AHL signals can alter the expression of plant defense genes, known as **induced systemic resistance (ISR)** in plants. These effects have been shown to beneficially increase resistance of tomatoes to pathogens, or detrimentally reduce resistance to pathogens. AHL signals can also modify plant development and plant protein patterns.

Overall, these studies show that soil microbes are inextricably associated with plant roots, and that there is a profound influence of soil microbes on plant health.

Influence of soil microbes on animal health

The environment plays a major role in human and animal health, whether the influence comes from exposure to pollution, pathogens, or beneficial abiotic or biotic interactions. While the negative influence of environmentally transferred pathogens on animal health is known, the benefits of environmental microbes have not been studied as thoroughly. Movement of pathogenic microorganisms or parasites from soil to animals occurs, such as helminth ova parasitism in livestock, humans, and poultry. Zibaei et al. (2017) suggested that the presence of naturally infected "free-range" broiler chickens could serve as zoonotic hosts of *Toxocara* spp. (a parasitic roundworm), which cause toxocariasis infections in humans. Additionally, there is increasing evidence suggesting that antimicrobial-resistant bacteria originating in livestock and human wastes can contaminate the environment, potentially affecting human health (Pepper et al., 2018; Scott et al., 2019). However, the beneficial effects of the soil microbiome on animal health is not as obvious or well studied.

Animal diseases caused by microbes in soil

There are numerous animal diseases caused by pathogens found in the soil. Bacteria, parasites, animal viruses, and fungi can be found in soil, causing a number of animal diseases costing millions of dollars to animal agriculture. Many of these pathogens transiently inhabit soil awaiting their next host, while others are soil inhabitants (e.g., geophilic).

There are some animal bacterial diseases (e.g., *Bacillus anthracis* (anthrax) and *Coxiella burnetii* (Q fever)) which can cause problems in either free-roaming (e.g., organic) or industrialized animal rearing operations (e.g., concentrated animal feeding operations). Few of these pathogens are vectored to the host directly from soil; most exposures occur through indirect exposure such as through contaminated forage. One example of direct exposure is through **geophagia** (i.e., direct consumption of soil). For instance, poultry operations are often built on unused farmland repurposed for poultry rearing. Birds, whether industrialized or wild, often feed on soil or litter. This increases the possibility for exposure to pathogens such as *Clostridium perfringens* (gas gangrene), a common spore-forming soil bacterial member, particularly in soil where fecal matter has been deposited. *C. perfringens* causes necrotizing enteritis in poultry, most often in industrialized poultry operations, which allows for rapid spread throughout the population; it is typically treated with antibiotics, but mortality rates can reach 1% per day for days to weeks (Van Immerseel et al., 2004). Thus, geophagic activity can expose birds to soil microorganisms, both beneficial and pathogenic.

Ruminants such as cattle and non-ruminants such as horses share unique interactions with the soil since they forage on grass for days at a time. As such, these animals share microbiota between themselves and the environment and vice versa. For instance, Cross et al. (2019) reviewed common zoonotic diseases associated with animals and identified anthrax, brucellosis, tularemia, and Q fever as bacterial diseases with natural reservoirs. *Bacillus anthracis* and *Coxiella burnetii* cause anthrax and Q fever, respectively, in both humans and animals. In animals, exposure to *B. anthracis* occurs during foraging where a large number of spores can be inhaled or ingested, typically causing death shortly after symptoms are evident. *C. burnetii* causes a mild disease in ruminants often leading to cattle still-births. More recently, equine botulism (*C. botulinum*) was hypothesized to be soil-borne. A study by Gerber et al. (2006) preliminarily indicated that *Clostridium sordellii* and *C. bifermentans* could produce lethal toxins capable of causing pasture myodystrophy in horses.

There are numerous animal parasitic diseases originating from contaminated soil. For instance, Zibaei et al. (2017) described broiler chickens infected with the nematode *Toxocara* spp. and the potential for broilers to act as paratenic hosts, which is an organism that acts as an intermediate, often asymptomatic, carrier but is not a host within a parasitic life cycle. Many "free-range" birds are susceptible to infection with nematodes as these birds feed on earthworms, which are also known to be paratenic hosts. Berenji et al. (2015) measured soils in Northern Iran and determined approximately 10% of public park soils were contaminated with *Toxocara* spp. While the risk is most likely lower for commercial chickens, their microbiome and health are still highly influenced by their surrounding environment as they still maintain their geophagic activities through consumption of litter.

Otranto and Deplazes, (2019) recently reviewed the potential nematodes of wild carnivores and their potential risk of zoonotic transmission to humans. They identified several soil-transmitted nematodes, such as *Toxocara* spp., *Baylisascaris procyonis*, *Strongyloides stercoralis*, *Ancylostoma* spp., and *Capillaria aerophila* to name a few. Still other genera of parasitic worms such as *Trichinella*, *Ascaris* (roundworm), and *Taenia* (tapeworm) are more well known, given their impact among humans throughout the world and leading to the long-held belief about avoiding consumption of undercooked pork and beef. In particular, *Taenia solium* (pigs) and *T. saginata* (cattle) are two well-known causes of taeniasis or tapeworm and are known to utilize the soil as a transmission vector.

Fungal and prion diseases are not as well-known as their bacterial and viral brethren. Fungal diseases are common among plants, but animals such as turtles, frogs, and bats are also susceptible to fungal pathogens. Many of these fungi are saprophytic and are commonly found in the soil, but for the most part, they are merely opportunistic pathogens, rarely affecting healthy animals. Avian respiratory aspergillosis can be a problem in poultry production. As with most fungal pathogens, animals are constantly exposed to aerosolized fungal spores, but only under circumstances where the balance between animal and fungal agent is imbalanced, infections are expected (Seyedmousavi et al., 2018). In addition, prions, "pathogenic" proteins, can cause fatal neurological disease in animals and humans. Many prion diseases (e.g., scrapie) are not identified until after years of "infection" and proliferation throughout the central nervous system. Saunders et al. (2012) recently described the transmission of prions via the environment, with a focus on the effect of soil type. They reviewed the literature and identified instances of scrapie recurrence following herd reintroduction, indicating that the environment served as a reservoir. Given the simplicity of prions, they can survive for years on soil particles, with particular affinity for clay-based soils, albeit with limited transport as a result.

Influence of soil microbes on the animal and human gut microbiome

There is substantial evidence to suggest that the environmental microbiome plays a direct role on our health; for example, the low prevalence of asthma patients in populations highly exposed to the environmental microbiome (von Mutius, 2016). It is also well known that animals interact with their immediate environment, whether it is via consumption of dirt (e.g., geophagia) or indirectly via consumption of bedding material, (e.g., poultry litter) or forage. Humans do not typically consume soil, but accidental consumption can occur. In some parts of the world, women who are pregnant will often consume soil in an attempt to ingest nutrients, most often associated with nutrient deficiency (Bisi-Johnson et al., 2010). Additionally, children often consume soil and can be exposed to pathogenic helminths as a result (Bisi-Johnson et al., 2010). Inadvertent ingestion of soil particles occurs when working outdoors or living in rural communities. This ingestion of soil or particles from our surrounding environment undoubtedly exposes animals and humans to a plethora of environmental microorganisms; the question is, "How much influence do these intentionally and unintentionally consumed soil particles influence our gut microbiome?"

The positive influence, if any, on the gut microbiome of intentional or unintentional soil consumption is not understood. In a previous section we discussed the influence of soil microbes on plant health, which indirectly affects the human gut microbiome by improving the quality of plant macro- and micro-nutrient uptake. However, Blum et al. (2019) reviewed the literature and demonstrated that over time, we as a civilization have lost contact with environmental microorganisms through application of the common principles of hygienic societies. Loss of soil biodiversity, due to use of agrochemicals and mono-crop management systems, may have exacerbated the negative health impacts resulting in human lifestyle diseases associated with the intestinal microbiome (Box 25.4).

So, you may ask, "...how do we change or acquire microorganisms, when we have our own microbiota?" For instance, Candela et al. (2012) reviewed the literature suggesting that the intestinal microbiota, while relatively stable, responds to changes in the environment, particularly changes in diet. These changes are mainly associated with changes to the quality of food, as in the case of changes from carnivore to vegetarian diets. However, Hehemann et al. (2010) demonstrated that environmental bacteria could transfer carbohydrate degrading enzymes to the gut microbiota via seaweed consumption. Thus, the same process may also be possible with soil as the vector. However, it is likely that ingestion of soil microbes is due to indirect consumption, such as soil particles attached to the roots of consumed forage or crops

BOX 25.4

The hygiene hypothesis

The "hygiene hypothesis" was first proposed by researchers studying the increase in allergic and autoimmune diseases. The increase in these diseases was described as a "post industrial revolution epidemic"; thus, Strachan (1989) suggested a possible explanation for the trend. In his study, he eliminated viral infections as a precursor to allergic diseases but considered that viral infections early in life may limit allergic diseases. These exposures were termed "unhygienic contact" with older siblings or acquired prenatally.

Basically, this was the birth of the "hygiene hypothesis," whereby exposures to an unhygienic environment could lead to a more protective immune system as well as less hyper-immune diseases such as allergies later in life.

or, in the case of humans, ingestion of soil particles attached to unwashed root vegetables. In fact, this is considered as a risk pathway for ingestion of soil or fecal borne pathogens such as *Salmonella* on crops (Brooks et al., 2012). In that study, it was assumed that up to 2% of soil particles remain attached to root crops and, under specific circumstances, potentially attached to low lying freshly consumed crops such as lettuce.

Hannula et al. (2019) recently demonstrated that foliar-feeding insects could acquire soil-based microbiomes. The study showed that caterpillars allowed to feed on intact plants had a more complex microbiome, similar to soil microbiomes, than those of caterpillars fed detached leaves from the same plants. The authors inferred that the changes in the soil microbiome were translated to the leaves of intact plants and that caterpillars directly in contact with soil were often found with beneficial members of the soil microbiota within their microbiome. In another study, the mouse gut microbiota was shown to be influenced equally by their diet as well as the soil microbiome of bedding within their cages (Zhou et al., 2018). Additionally, hygiene level (e.g., increased ingestion of soil particles) was suspected of influencing the early gut microbiota. The use of a healthy, foreign gut microbiota inoculated onto bedding material was shown to positively influence the gut microbiota of early developmental mice, particularly compared to both sterile and probiotic-inoculated bedding material. The key to most of these findings suggests that influencing the gut microbiota must occur at an early stage of life. It appears that the human gut microbiota can also be influenced by the environment; however, the influence on the gut microbiota can be challenging to separate from that of other factors such as diet (de Filippo et al., 2010).

Influence of soil microbes on human health

In the previous section we demonstrated that it was possible for the external environment to (positively?) influence the gut microbiota of animals and possibly humans. However, we are well aware of the negative influence of soil microbial pathogens on human health

(Pepper, 2013). The soil environment can serve as a conduit for transfer of a number of environmental or transient pathogens (e.g., those passing through the environment), such as *Pseudomonas aeruginosa* and *Salmonella* spp., respectively.

Human disease caused by soil microbes

The practice of "geophagia" is common amongst children and pregnant or lactating women, often associated with nutrient deficiencies, and is recognized as an eating disorder. Geophagia is not relegated to one part of the world, sex, culture, or religion; in fact, it is part of the cultural norm for some. As McLoughlin (1987) described in the southern US, some pregnant women traditionally ate clay believing that some properties of the material would prevent vomiting and promote healthy children. However, this deliberate consumption of soil introduces the human gut microflora to many soil microbes otherwise not typically encountered. For instance, the consumption of soil can lead to exposures of *Ascaris lumbricoides* and other parasitic worms as well as bacteria. The soil environment also harbors a number of indigenous pathogens such as *Clostridium perfringens* and *C. botulinum*, which are well-known orally-transmitted pathogens (Bisi-Johnson et al., 2010). We can also be exposed to indigenous pathogenic fungi without consuming soil but rather by merely breathing soil particles (e.g., dust) laced with *Coccidoides immitis*, *Cryptococcus neoformans*, or *Histoplasma capsulatum* spores, which cause "Valley Fever," pneumonia-like respiratory symptoms, and histoplasmosis, respectively. Frequently, these pathogens require specific environmental conditions to be of concern and often cause mild respiratory symptoms, but they can nevertheless be cause for concern in certain sub-populations.

Pathogenic microorganisms can also be introduced into the soil via agronomic management, such as land application of biosolids or manure, from wastewater treatment plants or concentrated animal feeding operations, respectively. These pathogens include some of the "celebrity" bacterial and viral pathogens highlighted on the evening news during outbreaks associated with fresh produce or similar products. *Salmonella* spp., *Escherichia coli* O157:H7, *Listeria monocytogenes*, norovirus, hepatitis A virus, and other enteroviruses highlight the most infamous. Brooks et al. (2012) evaluated the risks and the pathways of exposure associated with many of these fecal/oral pathogens and their human exposures via direct soil-ingestion, inhalation, and indirectly via consumption of fresh food crops. Many of these pathogens are quickly inactivated (e.g., no longer viable) when exposed to ambient environmental conditions, often inactivating within days to weeks after introduction to the soil. Generally, risk associated with exposure to these transient pathogens via contaminated soil is low unless specific conditions are met, such as high levels of contamination and exposure almost immediately following introduction of the pathogen to the soil.

In many cases these pathogens may be introduced to the soil by wild animals, such as feral pigs, fowl, and vermin, but, as previously stated, the risks of exposure and subsequent infection are generally low. Similarly, domestic and wild animals and other higher-order organisms are responsible for deposition of parasites. Many of these parasites have complex life cycles involving multiple hosts, one of which ultimately deposits the cyst or environmentally stable form into the soil. Ingestion of soil or ingestion of undercooked meat from grazing cattle or sheep are ways that these parasites gain entry to humans. Parasitic worms infect approximately 2.5 billion people worldwide according to the US Centers for Disease Control.

In some cases, parasitic worms are often correlated with geophagia (Bisi-Johnson et al., 2010). Other parasites which are commonly found in fecal-contaminated soil are *Cryptosporidium parvum* and *Giardia lamblia*. Their environmentally stable forms, oocysts and cysts, respectively, can often be recovered from soils contaminated with the feces of many wild as well as domestic animals.

Likewise, exposure to geophilic pathogens, such as *Clostridium tetani*, can occur by breaking the skin or deeper tissue with a contaminated sharp object. *Bacillus anthracis* is another bacterium commonly found in soil capable of causing human anthrax, but it typically requires a large exposure to anthrax spores, either via inhalation or cutaneous exposure, as can happen when veterinarians are dealing with colonized animals. Still other soil bacteria are opportunistic pathogens such as species of *Acinetobacter*, *Burkholderia*, *Ochrobactrum*, and *Stenotrophomonas*, causing diseases from necrotizing fasciitis to lung infections but only under specific circumstances such as deep-tissue wounds or immunocompromised patients. *Legionella pneumophila* is also commonly found in soil and is a reason that potted plants are not allowed in the rooms of immunosuppressed patients in most hospitals. In most cases, we are constantly exposed to these pathogens and they rarely result in disease. Fungal pathogens, which feed on the keratin of skin (e.g., dermatophytes) can also gain access to the body via soil and many are commonly found in the soil. The most common fungi found in soil, *Aspergillus* spp., can be pathogenic, but are rarely symptomatic or known, particularly given how often we are exposed to *Aspergillus* spp.

Antibiotics and antibiotic resistance

Simultaneously a recent and past concern is the emergence of **antibiotic-resistant bacteria**. While this is not a recent emergence (indeed, it has been around since the first bacteria fought over the same environmental niche or energy substrate), it is more recently a growing concern given high profile **nosocomial** (hospital acquired) infections that are no longer treatable with common antibiotics. Since the first antibiotic was discovered by Alexander Fleming, the battle between antibiotic usage and staying one step ahead of fast-replicating, evolving organisms has been one that was almost always doomed for failure. Bacteria are constantly evolving and developing mechanisms to cope with their environment, whether that environment is rhizosphere soil or a deep tissue wound. Even with proper use of antibiotics, emergence of resistance frequently occurred. That being said, we most likely contributed to a more rapid emergence by abusing and misusing antibiotics, whether that was at the clinical or farm level, which coupled with increasing "urban sprawl" and a demand for high quality animal-based protein, created a situation where interactions between humans and the environmental reservoirs of antibiotic resistance were more common place.

Numerous review articles have covered the emergence of antimicrobial (e.g., antibiotic) resistance (AMR) and the roles that humans, animals, and environment contribute to the issue (Scott et al., 2019). For instance, the complexities involved in AMR span the entirety of biological life. Bacteria have evolved, maintained, and shared defense mechanisms against antibiotics, particularly the natural ones, over the entirety of 3.5 billion years, with no end in sight. The very fact that many, if not all, of these genes originated in the environment confounds humanity's ability to "solve" the issue. There are no clear routes of exposure when it comes AMR, because the **antimicrobial-resistant bacteria** (ARB) may have emerged because of large scale concentrated animal feeding operations, human clinical exposure (e.g., nosocomial), or indirect selection due to an unknown anthropogenic chemical capable of selecting for resistance or merely selecting for another unrelated genetic trait located on the same chromosome or plasmid. The mere vastness of ARB in the upper few centimeters of soil is overwhelming to comprehend (~10^{15} ARB per acre furrow slice) (Pepper et al., 2018). It is well documented that the land application of manure or wastewater treatment biosolids can contribute to this amount, but the contribution is often orders of magnitude less than what was already there. The question in this case is "do we only need to know about the pathogenic ARB?," thus ignoring the rest of the population. The short answer is no. While the tendency to focus on just the pathogenic bacteria is understandable, non-pathogenic ARB are typically far more numerous and serve as a reservoir of resistance; they share many genetic features with pathogenic bacteria, increasing the potential for transfer of resistance genes to related pathogens. However, the pathogenic ARB are the ones that cause the most acute problems and should be where often limited resources are prioritized for maximum effect. One issue with any form of antibiotic treatment is the idea that you treat the entire "body," and this can impact many members of the microbiome, whether the targeted pathogen or commensal bacteria. The same goes for any sort of antibiotic treatment in animal or crop agriculture.

Despite the prevalence of ARB, it is important to recognize that just because an environment has ARB, it doesn't mean that there is a direct route of exposure. In fact, reaching you is a "marathon race" for most bacteria as there are many intervening environmental, societal, and safety factors before any environmental organism can reach the public.

Other natural products derived from soil microbes

In addition to natural antibiotics, soil microorganisms produce other natural bioactive products. There are a number of review articles highlighting the identification of **bioactive compounds** from endophytic bacteria, actinomycetes, and fungi. Most of these compounds range in activity from antibiotic to anticancer compounds. Some have other medicinal qualities such as free-radical scavenging compounds. Some of these compounds are phyto-bioactive, meaning they are related to plant growth-promoting metabolites and thus indirectly promote human health as a result (Singh et al., 2017). For instance, *Bacillus* spp. and actinobacteria have been shown to produce aromatic compounds often considered as plant-growth-promoting (Ek-Ramos et al., 2019). However, in this section, we will focus on the microorganisms capable of producing bioactive compounds directly affecting human health.

As discussed in Chapters 1 and 5, soil actinomycetes (e.g., *Streptomyces* spp.) have been the source of many antibiotics used to control human, animal, and plant infections; but many other compounds with potential human health benefits are also produced by soil microorganisms. For instance, Kumar et al. (2017) demonstrated the isolation and genetic potential associated with myxobacteria isolated from Indian soil. In this case, the authors were interested in isolated myxobacteria capable of producing antibacterial compounds against *Bacillus cereus* and *Pseudomonas syringae*, as well as anticancer compounds capable of inhibiting pancreas and prostate cancer cell lines. While the potential for producing bioactive compounds against cancer does not usher in a cure, it does provide pharmaceutical researchers potential targets for other drugs, and the fact that these are biologically produced means that manufacture can be scaled up to production using engineered microorganisms, often within "cell factories" such as *E. coli*. Many genera described as actinomycetes are known producers of bioactive compounds, including *Streptomyces*, *Polymorphospora*, *Micromonospora*, and *Microbispora* to name a few (Masand et al., 2015). For instance, *Streptomyces* sp. YIM66017 has been reported as being able to produce flavensomycinoic acid, which is capable of activity against breast cancer lines (Masand et al., 2015). Many of the compounds serve dual purposes, being cytotoxic against cancer cell lines as well as antifungal, as is the case with 3-acetonylidene-7-prenylindolin-2-one, isolated from *Streptomyces* (Masand et al., 2015).

Fungal endophytic agents are also capable of producing bioactive compounds. For example, Deshmukh et al. (2018) reviewed the literature for mangrove-associated fungi as a source of novel metabolites with biological activity. Mangroves are salt-tolerant coastal, often intertidal, forest ecosystems; thus, they represent a unique ecosystem from which promising bacteria and fungi can be found. This distinctive and harsh environment can create competitive niche environments where production of novel bioactive compounds can flourish. For example, species of *Pestalotiopsis* are known as producers of new and varied metabolites with anti-microbial, -fungal, -tumor, and -oxidant activities. Often the identification of these compounds must be done under unique in vitro conditions, whereby the production of the compound is only considered under biologically competitive conditions involving co-culture of the fungus of interest and an antagonistic bacterium. Compounds such as phomazine, epicorazine, and exserohilone have all been isolated from mangrove endophytic fungi, such as *Phoma* sp. (Deshmukh et al., 2018).

Role of soil microbes in "Metro Nature" and subsequent human health benefits

"**Metro nature**" refers to the creation and integration of green spaces within urbanized metropolitan areas with the intention to improve urban livability (Wolf and Robbins, 2015). This includes the creation of parks, urban forests, greenbelts, exercise paths, or riparian corridors that are privately or publicly owned. Since much of nature originates from a healthy soil, which in turn relies on an active soil microbial community, it follows that soil microbes contribute significantly to metro nature. Potential benefits of exposure to nature are presented in Table 25.3. A quintessential example of metro nature is the iconic Central Park in New York,

TABLE 25.3 Health benefits from contact with nature.

Health/well-being benefits
Reduced stress
Better sleep
Improved mental health
Reduced depression
Reduced anxiety
Greater happiness, well-being
Reduced aggression
Lower blood pressure
Reduced congestive heart failure
Reduced obesity
Reduced diabetes
Reduced mortality

Adapted from Frumkin et al. (2017).

FIGURE 25.4 **An example of "metro nature," the iconic Central Park in New York.** *Source: Reproduced from Environmental and Pollution Science, 2019, third ed., Elsevier, Inc.*

which is critically important to millions of community residents for recreational and other activities (Fig. 25.4).

Wolf and Robbins, (2015) reviewed the literature to describe the confluence of metro nature and human health. Recent research has described the loss of gut microbiome biodiversity (Blum et al., 2019), and one possible cause for this drop is the fact that, as human populations become more urbanized, we are not exposed to environmental microorganisms at the same magnitude that we once were under more rural or agrarian conditions. In other words, the "hygiene hypothesis" suggests that exposure to more microorganisms at an early stage of life leads to less immunity-based diseases, such as asthma (Box 25.4). In theory, placing "green" environments within increasingly populated urbanized areas would increase the population's exposure to environmental microorganisms, which according to the "hygiene hypothesis" would result in improved human health. Even locations typically considered more concrete or "gray" than "green," such as subway systems, increase our exposures to environmental bacteria. But "green" and "gray" exposures could increase contact with environmental pathogens and antimicrobial resistance as previously stated in the sections above. Certainly, these exposures increase our contact with human bacteria and viruses. In theory, any environmental pathogen potentially could be part of "metro nature" exposures; however, exposure to wild animals, manure of domesticated livestock and poultry, or more exotic natural environments would be limited within an urbanized area. That being said, the fact that we are increasingly moving toward a highly technological, hygienic, and urbanized existence, "green" environments may be the only conduit toward boosting exposures to environmental microbes, thus giving our immune system a much needed "booster."

Summary and concepts for the future

The concept of "one health" clearly links environmental health to human health through interactions with soil, plants, and animals. A review of this chapter emphasizes the importance of soil microbes in these complex interactions and how soil microbes can impact all aspects of our everyday lives, including: what infects us; what heals us; what we eat, drink, and breathe; and even the composition of our gut microbiome. Given the quantity and speed of new technological advances, it remains to be seen whether or not we can manipulate soil microbial communities to improve environmental and human health. Possibilities include: genetic manipulation of plant chromosomes; "designer" rhizospheres with specific root border cells; and engineered microbial inoculants with desirable traits for rhizobial and mycorrhizal root/microbial interactions. Finally, note that development of new soil-based probiotics could improve human health by directly introducing soil microorganisms into the gut.

Study questions

1. Discuss the connections between soil health and one health. In what ways are the two concepts similar and different?
2. List and discuss three ways that soil microorganisms impact each of plant, animal, and human health.
3. Discuss how ATP-based assays can be used to measure microbial activity and soil health. What are some advantages and disadvantages of these assays?
4. What is a geophilic pathogen? List two examples.
5. What is geophagia? Is this practice beneficial or harmful?
6. Discuss the concept of metro nature and how this relates to the hygiene hypothesis and modern society.

References

Anderson, T.H., 2003. Microbial eco-physiological indicators to assess soil quality. Agric. Ecosyst. Environ. 98, 285–293.

Berendsen, R.L., Pieterse, C.M.J., Bakker, P.A.H.M., 2012. The rhizosphere microbiome and plant health. Trends Plant Sci. 17, 478–486.

Berenji, F., Movahedi Rudy, A.G., Fata, A., Tavassoli, M., Mousavi Bazaz, M., Salehi, G., Sangani, 2015. Soil contamination with *Toxocara* spp. Eggs in public parks of Mashhad and Khaf, north east of Iran. Iran. J. Parasitol. 10, 286–289.

Bisi-Johnson, M.A., Obi, C.L., Ekosse, G.E., 2010. Microbiological and health related perspectives of geophagia: An overview. Afr. J. Biotechnol. 9, 5784–5791.

Blagodatskaya, E., Kuzyakov, Y., 2013. Active microorganisms in soil. Critical review of estimation criteria and approaches. Soil Biol. Biochem. 67, 192–211.

Blum, W.E.H., Zechmeister-Boltenstern, S., Keiblinger, K.M., 2019. Does soil contribute to the human gut microbiome? Microorganisms 7, 287.

Brooks, J.P., McLaughlin, M.R., Gerba, C.P., Pepper, I.L., 2012. Land application of manure and class B biosolids: An occupational and public quantitative microbial risk assessment. J. Environ. Qual. 41, 2009–2023.

Bulgarelli, D., Rott, M., Schaeppi, K., ver Loren van Themaat, E., Ahmadinejad, N., Assenza, F., Rauf, P., Huettel, B., Reinhardt, R., Schmelzer, E., Peplies, J., Gloeckner, F.O., Amann, R., Eickhorst, T., Schulze-Lefert, P., 2012. Revealing structure and assembly cues for *Arabidopsis* root inhabiting bacterial microbiota. Nature 488, 91–95.

Bünemann, E.K., Bongiorno, G., Bai, Z., Creamer, R.E., De Deyn, G., de Goede, R., Fleskens, L., Geissen, V., Kuyper, T.W., Mäder, P., Pulleman, M., Sukkel, W., van Groenigen, J.W., Brussard, L., 2018. Soil quality—A critical review. Soil Biol. Biochem. 120, 105–125.

Candela, M., Biagi, E., Maccaferri, S., Turroni, S., Brigidi, P., 2012. Intestinal microbiota is a plastic factor responding to environmental changes. Trends Microbiol. 20, 385–391.

Cross, A.R., Baldwin, V.M., Roy, S., Essex-Lopresti, A.E., Prior, J.L., Harmer, N.J., 2019. Zoonoses under our noses. Microbes Infect. 21, 10–19.

De Filippo, C., Cavalieri, D., Di Paola, M., Ramazzotti, M., Poullet, J.B., Massart, S., Collini, S., Pieraccini, G., Lionetti, P., 2010. Impact of diet in shaping gut microbiota revealed by a comparative study in children from Europe and rural Africa. Proc. Natl. Acad. Sci. USA 107, 14691–14696.

Deshmukh, S.K., Gupta, M.K., Prakash, V., Sudhakara, M., Reddy, 2018. Mangrove-associated fungi: A novel source of potential anticancer compounds. J. Fungi 4, 101.

Destoumieux-Garćon, P., Mavingui, P., Boetsch, G., Boissier, J., Darriet, F., Duboz, P., Fritsch, C., Giraudoux, P., Le-Roux, F., Morand, S., Paillard, C., Pontier, D., Sueur, C., Voituron, Y., 2018. The One Health concept: 10 years old and a long road ahead. Front. Vet. Sci. 5, 14.

Doran, J.W., 2002. Soil health and global sustainability: translating science into practice. Agric. Ecosyst. Environ. 88, 119–127.

Ek-Ramos, M.J., Gomez-Flores, R., Orozco-Flores, A.A., Rodríguez-Padilla, C., González-Ochoa, G., Tamez-Guerra, P., 2019. Bioactive products from plant-endophytic Gram-positive bacteria. Front. Microbiol. 10, 463.

Frumkin, H., Bratman, G.N., Breslow, S.J., Cochran, B., Kahn, Jr., P.H., Lawler, J.J., Levin, P.S., Tandon, P.S., Varanasi, U., Wolf, K.L., Wood, S.A., 2017. Nature contact and human health: a research agenda. Environ. Health Perspect. 125, 075001.

Gerber, V., Straub, R., Frey, J., 2006. Equine botulism and acute pasture myodystrophy: New soil-borne emerging diseases in Switzerland? Schweiz. Arch. Tierheilkd. 148, 553–559.

Hannula, S.E., Zhu, F., Heinen, R., Bezemer, T.M., 2019. Foliar-feeding insects acquire microbiomes from the soil rather than the host plant. Nat. Commun. 10, 1254.

Hehemann, J.H., Correc, G., Barbeyron, T., Helbert, W., Czjzek, M., Michel, G., 2010. Transfer of carbohydrate-active enzymes from marine bacteria to Japanese gut microbiota. Nature 464, 908–912.

Ikner, L., Foster, A., Stark, E., Pepper, I.L., 2021. The AMP Stress Indicator: A new biological indicator of soil health. Soil Biol. Biochem., In review.

Keasler, V., Bennett, B., Keller, C., Whalen, P., Cairns, J., De, R.M., Paula, 2013. Expanding the microbial monitoring toolkit: evaluation of traditional and molecular monitoring methods. Int. Biodeterior. Biodegradation 81, 51–56.

Kumar, S., Yadav, A.K., Chambel, P., Kaur, R., 2017. Molecular and functional characterization of myxobacteria isolated from soil in India. 3 Biotech 7, 112.

Kurm, V., van der Putten, W.H., De Boer, W., Naus-Wiezer, S., Gera Hol, W.H., 2017. Low abundant soil bacteria can be metabolically versatile and fast growing. Ecology 98, 555–564.

Masand, M., Jose, P.A., Menghani, E., Jebakumar, S.R.D., 2015. Continuing hunt for endophytic actinomycetes as a source of novel biologically active metabolites. World J. Microbiol. Biotechnol. 31, 1863–1875.

McLoughlin, I.J., 1987. The picas. Brit. J. Hosp. Med. 37, 286–290.

Otranto, D., Deplazes, P., 2019. Zoonotic nematodes of wild carnivores. Int. J. Parasitol. Parasites Wildl. 9, 370–383.

Pepper, I.L., 2013. The soil health-human health nexus. Crit. Rev. Environ. Sci. Technol. 43, 2617–2652.

Pepper, I.L., Brooks, J.P., Gerba, C.P., 2018. Antibiotic resistant bacteria in municipal wastes: Is there reason for concern? Environ. Sci. Technol. 52, 3949–3959.

Pierson, III, L.S., Maier, R.M., Pepper, I.L., 2015. Microbial communication: bacteria/bacteria and bacteria/host. In: Pepper, I.L., Gerba, C.P., Gentry, T.J. (Eds.), Environmental Microbiology, 3rd Edition. Elsevier Inc.

Saunders, S.E., Bartz, J.C., Bartelt-Hunt, S.L., 2012. Soil-mediated prion transmission: Is local soil-type a key determinant of prion disease incidence? Chemosphere 87, 661–667.

Scott, H.M., Acuff, G., Bergeron, G., Bourassa, M.W., Simjee, S., Singer, R.S., 2019. Antimicrobial resistance in a One Health context: exploring complexities, seeking solutions, and communicating risks. Ann. N. Y. Acad. Sci. 1441, 3–7.

Scott, R.A., Weil, J., Le, P.T., Williams, P., Fray, R.G., von Bodman, S.B., Savka, M.A., 2006. Long- and short-chain plant-produced bacterial N-acyl homoserine lactones become components of phyllosphere, rhizosphere and soil. Mol. Plant-Microbe. Interact. 19, 227–239.

Seyedmousavi, S., Bosco, S., de Hoog, S., Ebel, F., Elad, D., Gomes, R.R., Jacobsen, I.D., Jensen, H., Martel, A., Mignon, B., Pasmans, F., Piecková, E., Rodrigues, A.M., Singh, K., Vicente, V.A., Wibbelt, G., Wiederhold, N.P., Guillot, J., 2018. Fungal infections in animals: A patchwork of different situations. Med. Mycol. 56, S165–S187.

Shade, A., Hogan, C.S., Klimowiez, A.K., Linske, M., McManus, P.S., Handelsman, J., 2012. Culturing captures members of the soil rare biosphere. Environ. Microbiol. 14, 2247–2252.

Singh, M., Kumar, A., Singh, R., Pandey, K.D., 2017. Endophytic bacteria: a new source of bioactive compounds. 3 Biotech 7, 315.

Strachan, D.P., 1989. Hay fever, hygiene, and household size. Brit. Med. J. 299, 1259–1260.

Van Bruggen, A.H.C., Goss, E.M., Havelaar, A., van Diepeningen, A.D., Finckh, M.R., Glenn Morris, Jr., J., 2019. One Health—Cycling of diverse microbial communities as a connecting force for soil, plant, animal, human and ecosystem health. Sci. Total Environ. 664, 927–937.

Van Immerseel, F., De Buck, J., Pasmans, F., Huyghebaert, G., Haesebrouck, F., Ducatelle, R., 2004. *Clostridium perfringens* in poultry: an emerging threat for animal and public health. Avian Pathol. 33, 537–549.

Von Mutius, E., 2016. The microbial environment and its influence on asthma prevention in early life. J. Allergy Clin. Immunol. 137, 680–689.

Wolf, K.L., Robbins, A.S.T., 2015. Metro nature, environmental health, and economic value. Environ. Health Perspect. 123, 390–398.

Zhou, D., Bai, Z., Zhang, H., Li, N., Bai, Z., Cheng, F., Jiang, H., Mao, C., Sun, X., Lu, Z., 2018. Soil is a key factor influencing gut microbiota and its effect is comparable to that exerted by diet for mice. F1000Research 7, 1588.

Zibaei, M., Sadjjadi, S.M., Maraghi, S., 2017. The occurrence of *Toxocara* species in naturally infected broiler chickens revealed by molecular approaches. J. Helminthol. 91, 633–636.

SI Units and Conversions*

Base and supplementary SI (Système International) units.

Quantity	Unit	Symbol
Amount of substance	mole	mol
Electric current	ampere	A
Length	meter	m
Mass	kilogram	kg
Thermodynamic temperature	kelvin	K
Time	second	s

Preferred (P) and acceptable (A) units for several quantities.

Quantity	Application	Unit	Symbol
Concentration	Known molar mass (liquid, and solid material)	mole per cubic meter (P)	$mol\ m^{-3}$
		mole per kilogram (P)	$mol\ kg^{-1}$
		mole per liter (A)	$mol\ L^{-1}$
		gram per liter (A)	$g\ L^{-1}$
	Unknown molar mass (liquid and solid material)	gram per kilogram (P)	$g\ kg^{-1}$
		gram per cubic meter (P)	$g\ m^{-3}$
		gram per liter (A)	$g\ L^{-1}$
	Known ionic charge	mole charge per cubic meter (P)	$mol_c\ m^{-3}$
		mole charge per liter (A)	$mol_c\ L^{-1}$
	Gas	mole per cubic meter (P)	$mol\ m^{-3}$
		gram per cubic meter (A)	$g\ m^{-3}$
		gram per liter (A)	$g\ L^{-1}$
		liter per liter (A)	$L\ L^{-1}$
		microliter per liter (A)	$\mu L\ L^{-1}$
		mole per liter (A)	$mol\ L^{-1}$
		mole fraction (A)	$mol\ mol^{-1}$
Exchange parameters	Exchange capacity	mole charge of saturating ion per kilogram (P)	$mol_c\ kg^{-1}$
		centimole charge of saturating ion per kilogram (A)	$cmol_c\ kg^{-1}$
Light	Irradiance	watt per square meter	$W\ m^{-2}$
	Photon flux density (400–700 nm)	micromole per square meter per second	$\mu mol\ m^{-2}\ s^{-1}$

*Appendix tables adapted with permission from American Society of Agronomy, Crop Science Society of America, Soil Science Society of America. 2021. Chapter 7. Units and Measures. In Publications Handbook and Style Manual. P. 7-01 – 7-12. ASA-CSSA-SSSA, Madison, WI, USA.

Quantity	Application	Unit	Symbol
Water potential	Driving force for flow	joule per kilogram (P)	$J\ kg^{-1}$
		kilopascal (A)	kPa
		meter of water in a gravitational field (A)	M

Derived SI units with special names.

Quantity	Name	Symbol	Expression in terms of other units	Expression in terms of SI base units
Activity (of a radionuclide)	becquerel	Bq		s^{-1}
Celsius temperature	degree Celsius	°C		K
Conductance	siemens	S	A/V	$m^{-2}\ kg^{-1}\ s^{3}\ A^{2}$
Electric potential, potential difference, electromotive force	volt	V	W/A	$m^{2}\ kg\ s^{-3}\ A^{-1}$
Electrical resistance	ohm	Ω	V/A	$m^{2}\ kg\ s^{-3}\ A^{-2}$
Energy, work, quantity of heat	joule	J	Nm	$m^{2}\ kg\ s^{-2}$
Force	newton	N		$m\ kg\ s^{-2}$
Frequency	hertz	Hz		s^{-1}
Power, radiant flux	watt	W	J/s	$m^{2}\ kg\ s^{-3}$
Pressure, stress	pascal	Pa	N/m^2	$m^{-1}\ kg\ s^{-2}$
Quantity of electricity, electric charge	coulomb	C		s A

Prefixes and log scales for SI units.

Prefix	Symbol	Numerical	Log_{10}	Exponential	
exa	E	1,000,000,000,000,000,000	18	10^{18}	1E18
peta	P	1,000,000,000,000,000	15	10^{15}	1E15
tera	T	1,000,000,000,000	12	10^{12}	1E12
giga	G	1,000,000,000	9	10^{9}	1E9
mega	M	1,000,000	6	10^{6}	1E6
kilo	k	1,000	3	10^{3}	1E3
hecto	h	100	2	10^{2}	1E2
deca	da	10	1	10^{1}	1E1
no prefix		1	0	10^{0}	1E0
deci	d	0.1	−1	10^{-1}	1E-1
centi	c	0.01	−2	10^{-2}	1E-2
milli	m	0.001	−3	10^{-3}	1E-3
micro	μ	0.000001	−6	10^{-6}	1E-6
nano	n	0.000000001	−9	10^{-9}	1E-9
pico	p	0.000000000001	−12	10^{-12}	1E-12
femto	f	0.000000000000001	−15	10^{-15}	1E-15
atto	a	0.000000000000000001	−18	10^{-18}	1E-18

Conversion factors for SI and non-SI units.

To convert column 1 into column 2, multiply by	Column 1 (SI unit)	Column 2 (non-SI unit)	To convert column 2 into column 1, multiply by
Length			
0.621	kilometer, km (10^3 m)	mile, mi	1.609
1.094	meter, m	yard, yd	0.914
3.28	meter, m	foot, ft	0.304
1.0	micrometer, μm (10^{-6} m)	micron, μ	1.0
3.94×10^{-2}	millimeter, mm (10^{-3} m)	inch, in	25.4
10	nanometer, nm (10^{-9} m)	Angstrom, Å	0.1
Area			
2.47	hectare, ha	acre	0.405
247	square kilometer, km^2 (10^3 m)^2	acre	4.05×10^{-3}
0.386	square kilometer, km^2 (10^3 m)^2	square mile, mi^2	2.590
2.47×10^{-4}	square meter, m^2	acre	4.05×10^3
10.76	square meter, m^2	square foot, ft^2	9.29×10^{-2}
1.55×10^{-3}	square millimeter, mm^2 (10^{-3} m)^2	square inch, in^2	645
Volume			
9.73×10^{-3}	cubic meter, m^3	acre-inch	102.8
35.3	cubic meter, m^3	cubic foot, ft^3	2.83×10^{-2}
6.10×10^4	cubic meter, m^3	cubic inch, in^3	1.64×10^{-5}
2.84×10^{-2}	liter, L (10^{-3} m^3)	bushel, bu	35.24
1.057	liter, L (10^{-3} m^3)	quart (liquid), qt	0.946
3.53×10^{-2}	liter, L (10^{-3} m^3)	cubic foot, ft^3	28.3
0.265	liter, L (10^{-3} m^3)	gallon	3.78
33.78	liter, L (10^{-3} m^3)	ounce (fluid), oz	2.96×10^{-2}
2.11	liter, L (10^{-3} m^3)	pint (fluid) pt	0.473
Mass			
2.20×10^{-3}	gram, g (10^{-3} kg)	pound, lb	454
3.52×10^{-2}	gram, g (10^{-3} kg)	ounce (avdp), oz	28.4
2.205	kilogram, kg	pound, lb	0.454
0.01	kilogram, kg	quintal (metric), q	100
1.10×10^{-3}	kilogram, kg	ton (2000 lb), ton	907
1.102	megagram, Mg (tonne)	ton (US), ton	0.907
1.102	tonne, t	ton (US), ton	0.907
Yield and rate			
0.893	kilogram per hectare, kg ha^{-1}	pound per acre, lb $acre^{-1}$	1.12
7.77×10^{-2}	kilogram per cubic meter, kg m^{-3}	pound per bushel, bu^{-1}	12.87
1.49×10^{-2}	kilogram per hectare, kg ha^{-1}	bushel per acre, 60 lb	67.19
1.59×10^{-2}	kilogram per hectare, kg ha^{-1}	bushel per acre, 56 lb	62.71

To convert column 1 into column 2, multiply by	Column 1 (SI unit)	Column 2 (non-SI unit)	To convert column 2 into column 1, multiply by
1.86×10^{-2}	kilogram per hectare, kg ha^{-1}	bushel per acre, 48 lb	53.75
0.107	liter per hectare, L ha^{-1}	gallon per acre	9.35
893	tonnes per hectare, t ha^{-1}	pound per acre, lb acre^{-1}	1.12×10^{-3}
893	megagram per hectare, Mg ha^{-1}	pound per acre, lb acre^{-1}	1.12×10^{-3}
0.446	megagram per hectare, Mg ha^{-1}	ton (2000 lb) per acre, ton acre^{-1}	2.24
2.24	meter per second, m s^{-1}	mile per hour	0.447
Specific surface			
10	square meter per kilogram	square centimeter per gram	0.1
1,000	square meter per kilogram	square millimeter per gram	0.001
Pressure			
9.90	megapascal, MPa (10^6 Pa)	atmosphere	0.101
10	megapascal, MPa (10^6 Pa)	bar	0.1
1.00	megagram per cubic meter, Mg m^{-3}	gram per cubic centimeter, g cm^{-3}	1.00
2.09×10^{-2}	pascal, Pa	pound per square foot, lb ft^{-2}	47.9
1.45×10^{-4}	pascal, Pa	pound per square inch, lb in^{-2}	6.90×10^3
Temperature			
1.00 (K - 273)	Kelvin, K	Celsius, °C	1.00 (°C + 273)
(9/5 °C) + 32	Celsius, °C	Fahrenheit, °F	5/9 (°F - 32)
Energy, work, quantity of heat			
9.52×10^{-4}	joule, J	British thermal unit, Btu	1.05×10^3
0.239	joule, J	calorie, cal	4.19
10^7	joule, J	erg	10^{-7}
0.735	joule, J	foot-pound	1.36
2.387×10^{-5}	joule per square meter, J m^{-2}	calorie per square centimeter (langley)	4.19×10^4
10^5	newton, N	dyne	10^{-5}
1.43×10^{-3}	watt per square meter, W m^{-2}	calorie per square centimeter minute (irradiance), cal cm^{-2} min^{-1}	698
Transpiration and photosynthesis			
3.60×10^{-2}	milligram per square meter second, mg m^{-2} s^{-1}	gram per square decimeter hour, g dm^{-2} h^{-1}	27.8

To convert column 1 into column 2, multiply by	Column 1 (SI unit)	Column 2 (non-SI unit)	To convert column 2 into column 1, multiply by
5.56×10^{-3}	milligram (H_2O) per square meter second, $mg\ m^{-2}\ s^{-1}$	micromole (H_2O) per square centimeter second, $\mu mol\ cm^{-2}\ s^{-1}$	180
10^{-4}	milligram per square meter second, $mg\ m^{-2}\ s^{-1}$	milligram per square centimeter second, $mg\ cm^{-2}\ s^{-1}$	10^4
35.97	milligram per square meter second, $mg\ m^{-2}\ s^{-1}$	milligram per square decimeter hour, $mg\ dm^{-2}\ h^{-1}$	2.78×10^{-2}
Electrical conductivity, electricity, and magnetism			
10	siemen per meter, $S\ m^{-1}$	millimho per centimeter, $mmho\ cm^{-1}$	0.1
10^4	tesla, T	gauss, G	10^{-4}
Water measurement			
9.73×10^{-3}	cubic meter, m^3	acre-inches, acre-in	102.8
9.81×10^{-3}	cubic meter per hour, $m^3\ h^{-1}$	cubic feet per second, $ft^3\ s^{-1}$	101.9
4.40	cubic meter per hour, $m^3\ h^{-1}$	US gallons per minute, $gal\ min^{-1}$	0.227
8.11	hectare-meters, ha-m	acre-feet, acre-ft	0.123
97.28	hectare-meters, ha-m	acre-inches, acre-in	1.03×10^{-2}
8.1×10^{-2}	hectare-centimeters, ha-cm	acre-feet, acre-ft	12.33
Concentrations			
1	centimole per kilogram, $cmol\ kg^{-1}$ (ion exchange capacity)	milliequivalents per 100 grams, $meq\ 100\ g^{-1}$	1
0.1	gram per kilogram, $g\ kg^{-1}$	percent, %	10
1	milligram per kilogram, $mg\ kg^{-1}$	parts per million, ppm	1
Radioactivity			
2.7×10^{-11}	becquerel, Bq	curie, Ci	3.7×10^{10}
2.7×10^{-2}	becquerel per kilogram, $Bq\ kg^{-1}$	picocurie per gram, $pCi\ g^{-1}$	37
100	gray, Gy (absorbed dose)	rad, rd	0.01
100	sievert, Sv (equivalent dose)	rem (roentgen equivalent man)	0.01
Plant nutrient conversion			
	Elemental	*Oxide*	
2.29	P	P_2O_5	0.437
1.20	K	K_2O	0.830
1.39	Ca	CaO	0.715
1.66	Mg	MgO	0.602

Index

Note: Page numbers followed by "f" indicate figures, "t" indicate tables, and "b" indicate boxes.

B

H

I

K

L

M

N

O

P